7th International Forum on Industrial Design 2019

IOP Conference Series: Materials Science and Engineering
Volume 573

Luoyang, China
17 – 19 May 2019

ISBN: 978-1-5108-9454-9
ISSN: 1757-8981

Printed from e-media with permission by:

Curran Associates, Inc.
57 Morehouse Lane
Red Hook, NY 12571

Some format issues inherent in the e-media version may also appear in this print version.

Institute of Physics
Dirac House, Temple Back
Bristol BS1 6BE UK

Phone: 44 1 17 929 7481
Fax: 44 1 17 920 0979

techtracking@iop.org

Additional copies of this publication are available from:

Curran Associates, Inc.
57 Morehouse Lane
Red Hook, NY 12571 USA
Phone: 845-758-0400
Fax: 845-758-2633
Email: curran@proceedings.com
Web: www.proceedings.com

Table of contents

Volume 573

2019 7th International Forum on Industrial Design

17–19 May 2019, Luoyang, China

Accepted papers received: 21 June 2019
Published online: 2 August 2019

Preface

Papers

IOP Conf. Series: Materials Science and Engineering **573** (2019) 011001 doi:10.1088/1757-899X/573/1/011001

Preface

Industrial Design is important in the development of global industrialization. In order to boost the development of Industrial Design, the 2019 7th International Forum on Industrial Design was held in Jinkaiyue Hotel(Luoyang Sports Center),China during May 17-19, 2019 to gather the scientists, scholars, engineers and students from the Universities all around the world and the industry together to present ongoing research activities, and hence to foster research relations between the Universities and the industry. IFID2019 is hosted by Luoyang Government, organized by Luoyang Science and Technology Bureau, Luoyang Productivity Promoting Center, Luoyang Creative Industry Design Center and Guangdong Academy of International Academic Exchange, co-organized by Northwestern Polytechnical University, Lanzhou University of Technology, Henan University of Science and Technology,GOTH DESIGN Co.LTD, Low Carbon Design Society of Hong Kong, Xihua University, North Minzu University, Zhengzhou University of Aeronautics, Zhengzhou University of Light Industry, Nanchang University, Xi'an University of Science and Technology, Xi'an Jiaotong University, Hebei Agricultural University, Tropical Agricultural Machinery Research Institute of Chinese Academy of Tropical Agricultural Sciences,School of Design and Art of Shenyang Jianzhu University, School of Design Art of Dalian University of Science and Technology, College of Mechanical Engineering of Jilin Engineering Normal University, Bengbu University and School of Information Science and Engineering of Dalian Polytechnic University. The forum is held every year to make it an ideal platform for people to share views and experiences in Industrial Design and related areas. In this edition, IFID2019 is providing a venue of high quality state-of-the-art research papers in the following areas of interest:

- Discussion on the theory of design innovation
- Research on innovative design methods
- Personalized, stylized, and characteristic re-creation of industrial design
- Human-machine system research

More than 200 submitted papers were received, which were reviewed by appropriately qualified experts in the field selected by the Conference Committee. After that tight selection procedure, only 115 papers were accepted for oral presentation and publication in this proceeding.

On behalf of the organizing committee and scientific committee of the IFID 2019, we would like to express our gratitude to all the participants, presenters, technical reviewers, and all collaborative partners for their contributions to the conference. We would like also thank the contributing authors for submitting papers and for being very patient to revise them multiple times based on the recommendations of the assigned reviewers.

As well, we would like to thank the technical committee members, who get the papers from the contributors being reviewed by suitable experts. We have received valuable feedbacks from all the participants, which will help us in organizing such conference in a better way in the coming future. Additionally, we place on record our sincere gratitude to the chairperson of each and every conference session, for decently conducting the oral presentation of the technical papers and coordinating the discussions among the participants. The discussions and deliberations during the sessions are sure to be valued by the participants.

IOP Conf. Series: Materials Science and Engineering **573** (2019) 011001 doi:10.1088/1757-899X/573/1/011001

Finally, we are deeply indebted to IOP Publishing for excellently bringing out this conference proceedings through the *IOP Conference Series: Materials Science and Engineering*.

The Committee of IFID2019

IOP Conf. Series: Materials Science and Engineering **573** (2019) 011001 doi:10.1088/1757-899X/573/1/011001

Committees

Conference Chairman:
Prof.Changde Lu, Industrial Alliance for innovation and design of the Silk Road, China

Conference Vice-chairman:
Prof.Jianning Su, Lanzhou University of Technology, China
A.Prof.Xuesheng Pei, Henan University of Science and Technology, China

Organizing Committee:
President.JEONG SUKJUNE, GOTH DESIGN Co.LTD, Korea
Commander.Guanxian Zhi, Low Carbon Design Society of Hongkong, China
Prof.Kaining Meng, Xihau University, China
Prof.Chunyang Mo, North Minzu University, China
Prof.Zhongyan Hu, Zhengzhou University of Aeronautics, China
Prof.Jianjun Ren, Zhengzhou University of Light Industry, China
Researcher.Xuan Wang, Nanchang University, China
Prof.Dengkai Chen, Northwestern Polytechnical University, China
A.Prof.Qing Feng, Xi`an University of Science and Technology, China
A.Prof.Hongwei Li, Xi'an Jiaotong University, China
A.Prof.Xiyin Gao, Hebei Agricultural University, China
Prof.Linghao Zhang, China Industrial Design Association/ School of Design of Jiangnan University, China
Anil R. Kumar,Associate Professor,San Jose State University, USA
Ari Widyanti, Ph.D. , Bandung Institute of Technology, Indonesia
Chih-Yung Tsai , Ph.D. , University of Taipei, China
Jaehyun Park, Ph.D. , Incheon National University (INU), Korea
A.Prof.Rui Gou, Southwest Jiaotong University, China
Prof.Que Weimin, Peking University, China
Prof.Tongzheng Song, National Taiwan University of Science and Technology, China
Dr.Kaida Xiao, School of Design, University of Leeds, UK
Prof.Rungtai Lin, National Taiwan University of Arts, China
Prof.Bernhard E. Bürdek, University of Art and Design Offenbach am Main, Germany
A.Prof.Mehdi Toloo, Technical University of Ostrava, Czech Republic/ Editor in Computers&Industrial Engineering at ELSEVIER
Prof.Bin He,Shanghai University, China
Ph.D.Shih-Wen Hsiao,National Cheng Kung University, China
A.Prof.Lifang Yang,Harbin Institute of Technology, China
Dr.Chanprapha Phuangsuwan,Rajamangala University of Technology Thanyaburi, Thailand
Dr. Fadhlan Hafizhelmi Kamaru Zaman, Universiti Teknologi MARA, Malaysia
Prof. Shuping Xiong, Korea Advanced Institute of Science and Technology, South Korea
A.Prof.Wei Li,vice-president of School of Mechanical and Electrical Engineering,Southwest Forestry University, China
A.Prof.Weiwei Wang,vice-president of School of Art & Design, Shaanxi University of Science and Technology, China

IOP Conf. Series: Materials Science and Engineering **573** (2019) 011001 doi:10.1088/1757-899X/573/1/011001

IOP Conf. Series: Materials Science and Engineering 573 (2019) 011002 doi:10.1088/1757-899X/573/1/011002

Peer review statement

All papers published in this volume of *IOP Conference Series: Materials Science and Engineering* have been peer reviewed through processes administered by the proceedings Editors. Reviews were conducted by expert referees to the professional and scientific standards expected of a proceedings journal published by IOP Publishing.

IOP Conf. Series: Materials Science and Engineering **573** (2019) 012001 doi:10.1088/1757-899X/573/1/012001

Enlightenment of Design Education in Purdue University: Research and Practice on Assessment Method in Project-Based Learning of Industrial Design Discipline

Zhijuan Zhu, Huai Cao, Xue Chen and Ren Long*

Department of Industrial Design, School of Mechanical Science and Engineering, Huazhong University of Science and Technology, Wuhan, Hubei, 430074, China

*Corresponding author's e-mail: longren@hust.edu.cn

Abstract. Due to the characteristics of industrial design, the project-based curriculum learning method has been adopted by more and more universities. However, there is a lack of relevant research on the evaluation of course results in project-based learning. In practice, the single evaluation method for the final operation is usually adopted, often with the subjective nature of the teacher, which has many disadvantages. Learning from the evaluation method of Purdue University's industrial design specialty, through the project-based learning practice of the Department of Industrial Design in Huazhong University of Science and Technology, a combination of periodic assignments evaluated by teacher and final assignments evaluated by experts was used to obtain a more objective score. This method has achieved good teaching results. This research is expected to promote the improvement of the evaluation method of the design discipline curriculum.

1. Introduction

Project-based learning (PBL) is a form of contextual learning[1]. PBL is gradually adopted by more and more majors in higher education institutions with its vivid teaching method, scientific and reasonable teaching process, and teaching objectives of positively facing the market[2]. In the professional course teaching of industrial design, PBL has become a common used teaching method. In the professional curriculum, practical topics were frequently introduced in the teaching procedure to guide students to understand the design concept, experience the design process, and latter achieve the purpose of learning.

PBL is a model for organizing learning around projects[3]. The method let students acquire knowledge in the process of solving problems through challenging and complex topics[4]. In China, some scholars gave specific teaching paths and methods starting from the background of social needs to improve students' practical ability[5-6]. They also proposed that PBL is a learning method that emphasizes student-centred learning, interdisciplinary, collaborative learning, independent inquiry, and practice[7]. However, there is a lack of relevant research on the assessment of course performance in PBL currently.

In the actual course teaching of colleges and universities in China, due to the differences in the characteristics of specific courses, from the theoretical teaching courses to the design courses, the performance assessment method are different. The scores of most professional courses in other majors are based on the score of the examination, and the final score is obtained in combination with the usual scores. While in the industrial design discipline, PBL is utilized in professional courses, and many professors evaluate students' performance based on the design work or design report of the final class operation as the sole score. This method may cause two problems: First, from the student's point of view, the final design work is the only work needs to be provided, it is easy to cause the phenomenon of being lazy and rushing to work hard before submitting assignments, so the learning effect is poor. Secondly, because the preferences and standards of different people are different, the scores given by the teacher will have different degrees of personal subjectivity and cannot give objective results.

2. Method

In the industrial design courses of many American universities, PBL methods are also adopted. By observing the teaching procedure of many design courses, as a visiting scholar at Purdue University, the assessment methods were generalized. The evaluation mode is diversified (see Table 1). In assessment of product design courses such as "Problems in Industrial Design", presentation is of importance, and professors including visiting scholars are invited to participate in the evaluation of the results of the presentation. Some teachers also use the method of exhibition of course works, such as "Information Visualization" to evaluate students' performance. At the end of the term, students are required to exhibit their design works, and the effect of the exhibition will be the most important factor to evaluate their performance.

Table 1. Assessment methods of 5 courses in Purdue University.

	Courses	Assessment items	Ratios	Remarks
1	**Interactions Design Studies**	Mobile phone interface design	20%	Independent assignments
		Personal website design	30%	
		Interactive information visualization design	20%	
		Laboratory cooperation project	30%	PBL mode, Group assignments
2	**Information Visualization Design**	Course participation	10%	PBL mode, Including both independent assignments and group assignments
		Classwork	20%	
		Design project results exhibition	70%	
3	**Human Factors in Engineering**	Test 1	25%	Three exams, with each 25% weight
		Test 2	25%	
		Final examination	25%	
		Assignments	25%	
4	**Design Cognition and Learning**	Weekly class discussion	20%	Focus on class discussion and after-school thinking
		Weekly after class feedback	10%	
		Summary	20%	
		Course project research	30%	
		Final course feedback	20%	
5	**Problems in Industrial Design**	presentation	100%	PBL mode Several professors attend the presentation to evaluate

According to the above analysis, PBL mode was utilized in Huazhong University of Science and Technology. Participators are junior students with the major of industrial design. During the procedure of the course, both independent assignments and group assignments are required. In addition, a new assessment method with a combination of periodic assignments evaluated by teacher and final assignments evaluated by experts is used to obtain a more objective score for students.

The final grade of the course is calculated as follows:

$$F= (S_1R_1+ S_2R_2+\ldots\ldots+SnRn) +SR \qquad (1)$$

F is the final score of students, S_1, S_2, ...S_n are scores of periodic assignments, R_1, R_2, ...R_n are ratios of assignments, which are decided by the teacher according to the difficulty of the assignment. S is the score of final assignment, R is the ratio.$R_1+R_2+......+R_n+R=100$. (According to the characteristic of the major, in practice, n is less than 3.)

3. Practice teaching process

3.1. Participant

In this paper, 104 students of Huazhong University of Science and Technology attend the course *Product Design II*. They are junior students at the time.

3.2. Introduction of the course

Product Design II is a professional advanced course based on the course of *Product DesignI*, and is an important subject core course of industrial design. Through the design theory guidance and design topics in the course, students learn to understand and master the basic theory of product design, design procedures, design methods and main performance methods of design. The proposition design exercises of this course are mainly based on the design of mechanical equipment such as digital equipment. Through the comprehensive design training of the course, students can understand and familiarize with the methods of electromechanical product design, understand related materials and processing techniques, and further improve performance techniques. The course also provides an in-depth explanation of the integrated design of product design methods, enabling students to master the methods of coordinating and solving several problems, such as products and markets, individual and overall products, product details and overall, product serialization, etc. in actual design, and improving students' practical design practices. The objective of the course is to improve students' ability of using design methods, and to cultivate their ability of observing, analysing, solving problems, in addition, comprehensive modelling and design capabilities are also focused.

3.3. Course arrangement and grade assessment

During the procedure of the course, there are 56 lessons in total. According to the syllabus requirements, there are 3 assignments, including 2 periodic assignments and 1 final assignment. The 2 periodic assignments are the basis of the final assignment. Each assignment accounts for a certain percentage of the course grade. In the follow-up course, PBL methods are adopted. The teacher specifies several topics, for students to choose, and set the deadline of the final closing assignment. Then students set up their design team (no more than 5 students in each group), and develop their own schedules according to the deadline. Later they combine the knowledge acquired in the course with the previous courses, and complete the final complete design report for presentation within the specified time.

In the final assessment of students' grades, the periodic assignments evaluation mode and the final assignment evaluation mode are combined. According to the formula (1) described above, the final scores of the course is calculated as follows:

$$F=S_1*15\%+S_2*15\%+S*70\% \qquad (2)$$

F is the final score of the student, S_1 is the first assignment result, S_2 is the second assignment result, and S is the course completion result. Among them, S_1 and S_2 embody the staged results of the course. The class teacher scores the knowledge points according to the actual operation. S is the result of the graduation work and adopts the expert evaluation mode. In this course, the proportion of each grade is set by the instructor according to the amount of work and difficulty. In the course evaluation method of the course, since S_1 and S_2 respectively examine the material process structure and product brand recognition (PI) concept, it is easier for the teacher to evaluate the mastery of the theoretical knowledge points and give objective scores. S is a comprehensive large-scale work performance, thus the single class teacher is subjective in evaluation, and the proportion of the score is also high, therefore the expert evaluation is adopted. The model can obtain more reasonable and objective scores for students. (See Table 2)

Table 2. Course composition and assessment method.

Course composition	Periodic assignments assessment		Final assignment assessment
Assignments	Assessment 1	Assessment 2	Assessment 3
Ratios	15%	15%	70%
Rating person	Teachers of the course		Experts (Including teacher of the course, professors, post graduate students, industrial designers, mechanical engineers, etc.)
Objective of assignments	Basic theory of the course		PBL method
Examination point	Mastery of specific knowledge points		Comprehensive design capability

In the final course, the final assignment was presented by students and evaluated by experts. Experts were consist of 10 members, including 2 teachers of the course, 3 professors from design major, 3 graduate students, 1 mechanical engineer, and 1 professional industrial designer. The teachers and graduate students have more experience in design research, and can evaluate the students' assignment from the perspective of research; the invited mechanical engineer is the chief engineer of the designated project and has a good understanding of the product, especially experienced in the structural process; the invited professional industrial designer is also the designer of the project, who not only understand the project but also had many practical design project experience before. A comprehensive evaluation by such a professional team can evaluate students' works from different perspectives, can lead to more objective scores. In addition, through the experts' comments after the presentation, students can better understand their own advantages and disadvantages, so as to obtain the views of experts from different background, then they can have a clearer direction in the future learning. Therefore, this evaluation procedure is also a valuable additional learning opportunity for students.

Among the scores of the course completion assignments, 10 members of experts responded from four aspects according to the course objectives, including the content (30%), quality (40%), attraction (20%) and integrity (10%) of the student's presentation, and then the scores of each group were rated. The scoring projects are designed according to the course content requirements and convenience of scoring.

After the presentation, the scores obtained by each group are input into the EXCEL software, and the average score of the expert group is obtained, thereby the score S of students in the formula (2) obtained. Combine the scores of S1 and S2 given by the teacher, and use formula (2), the final score of students generated.

4. Results and discussion of teaching practice

4.1. Result of PBL

From the teacher's point of view, PBL method requires the instructor to have overall control during the course, thus the higher demands on the class teacher are required.

During the process of course learning, the use of PBL methods, combined with practical topics, allow design results to be practically adopted, and mobilize the initiative of students greatly from passive learning to active learning. Students are grouped by themselves, and their design progress is arranged according to the strengths and weaknesses of their respective members. Cooperation and division of labor are carried out within the group. The teacher's guidance is more targeted and effective. Judging from the quality of the final design reports submitted by the four classes in two years, the level of submitted work is significantly better than in previous years.

From the perspective of the completion report submitted by students, all students expressed that they learned a lot through the PBL method course. The following is an excerpt from the course completion of some students:

"From receiving this topic to previous research, analysis, summary, positioning, sketching, solution determination, modelling to the final report, this is a very complete design process for me. The relevant analysis and knowledge we have learned are also applied throughout the design, and I gained a deeper understanding of the theoretical knowledge learned in the classroom through practice."

"In the process of doing my homework, I encountered many difficult problems. From the unclear idea in the beginning to the in-depth design through the guidance of the teacher, our team got the final design through five design stages. We experienced the happiness of design. This process allows us to learn a lot."

"In the process of project design, we have learned a lot from the previous research, the various analysis in the medium term, to the later modelling and rendering. After the design experience, we have made the design process clearer. The final design of the program is the results of many revisions and discussions, and we benefit from the course a lot."

4.2. The effect of evaluation method

From the point of view of the class teacher, it is easy to give a reasonable score for the evaluation of the stage work because the knowledge of the work is clear. In the comprehensive design of PBL, the experts give scores according to the scoring items and various scores, which makes the individual's performance subjectively small and has good operability.

From the student's point of view, the entire course of the learning process has a phased goal, and the student's mental state can remain active because of the staged homework scoring. After learning that the final assignment will be presented and the final results will be judged by experts, students feel the pressure and get a lot of motivation, therefore the enthusiasm for completing the coursework is well mobilized. The combination evaluation model can achieve more fair and appropriate results, and also play a very good role in mobilizing students' enthusiasm for learning.

5. Conclusion

In the course of industrial design, the use of PBL method allows students to learn actively. The evaluation method of course performance can be improved from a single class operation evaluation method to a combination of stage work and final class work evaluation. In the evaluation of the course work, the mode of expert evaluation is more effective than the class teacher, so that it can get objective and fair results and better mobilize students' initiative. The teaching and evaluation method lead to good effect, and it has good operability to spread.

Acknowledgments

We would like to thank our colleagues and other friends who attended the course presentation as experts and help us to evaluate students' performance. We would also thank Circodesign Company in China for providing the project for the course. This work is supported by the Teaching Research Project of Hubei Province (No. 2016075) and the Teaching Research Project of Huazhong University of Science and Technology (No. 2018056).

References

[1] K. Koedinger, A Corbett. (2006) The Cambridge Handbook of the Learning Sciences. Ed. R K Sawyer. New York: Cambridge University Press.

[2] Sun, Q. Q., Sun, H. C. (2015) Exploring the realization process of project-based teaching in industrial design——Taking the collection of rainwater and semi-automatic cleaning bins as an example. Journal of Inner Mongolia Normal University(Educational Science), 28: 152-154 (in Chinese)

[3] Jones, B. F., Rasmussen, C. M., Moffitt, M. C. (1997) Real-life problem solving: A collaborative approach to interdisciplinary learning. Washington, DC: American Psychological Association.

[4] Thomas, J. W., Mergendoller, J. R. (2000) Managing project-based learning: Principles from the field. Paper presented at the Annual Meeting of the American Educational Research Association, New Orleans.

[5] Du, H. B., Hu, H. Q. (2014) Exploration of a new model of industrial design project-based teaching. MEIYUAN, 1： 9-12 (in Chinese)

[6] Lv, W. L., Huang, Y., Liu, F. (2011) Primary discussion of project teaching methodology reform in environment design. ZHUANGSHI , 2: 139-140(in Chinese)

[7] Long, R., Cao H., Li, J. (2014) The Research and Practice of Product Design in a Guiding Robot. In: 2014 the 3rd International Conference on Industrial Design and Mechanics Power (ICIDMP2014).

IOP Conf. Series: Materials Science and Engineering **573** (2019) 012002 doi:10.1088/1757-899X/573/1/012002

Research on Fuzzy Hierarchy Optimization Model of Product Family Parameters Based on Flexible Design of Clothing

Zhe Li [1], Huan Tian [2 *]

[1] Chinese academy of lifestyle design, Beijing institute of fashion technology, Beijing 100029, China

[2] Department of Electronics and Information Engineering, Lanzhou Vocational Technical College, Lanzhou,730070, China

[*]Corresponding author's e-mail: 120839127@qq.com

Abstract. In this paper, by means of summarizing the current situation of product family design, methods of product family design based on the common platform is discussed. The concept of the fuzzy hierarchical optimization model which is the main model type is advanced. Further, this method educes product family design methodology under multipurpose constraint environment, which is used to achieve response to customer requirements. Moreover, this method is al so used to integrally considering the comparability and diversity so that the application can lend itself to attain the flexibility of product family design and to support the dynamic customer requirements.

1. Introduction

The term "mass customization" first appeared in the 1970 book of the prophet Alvin Toffler's book The Impact of the Future. In 1993, Joseph Pine II gave a complete description of the concept of mass customization [2]. But it has only been paid attention to by people in recent years. Mass customization was originally applied to the mechanical manufacturing industry and was gradually extended to other industries. The successful implementation of Levi Strauss & Co.'s mass-customized brand “Personal Pair” jeans proved that the production method is equally applicable to the apparel manufacturing industry. Garment mass customization is an advanced garment enterprise production and management model that integrates garment enterprises, customers, suppliers and the environment, and makes full use of the latest research results of human beings in various technical fields [3]. One of its core ideas is to require apparel companies to produce personalized clothing at times and costs similar to mass production. First of all, it is a philosophy that guides garment enterprises to participate in market competition. It requires garment enterprises to consider the relationship with customers from the perspective of long-term interests. "Customer satisfaction" is one of the highest pursuit goals, attracting and "permanently" retaining customers. Secondly, it is a way for garment enterprises to organize and manage production. It requires enterprises to provide customers with any variety and quantity of clothing by speed, cost and quality similar to mass production through rational allocation of resources and optimization of operation process. Third, it is a guideline for designers to design customized products, requiring designers to fully consider the individualized variants of the clothing in the design process [4].

With the increasingly fierce market competition, the global market has shifted from the seller's market to the buyer's market. Consumer demand tends to be diversified, flexible and personalized. The target market of products is becoming more and more fine. The focus of competition among enterprises is not only how to satisfy the maximum satisfaction. Consumer demand is also focused on how to shorten product development cycles, improve product quality and reduce product cost. In order to improve the competitiveness of products, manufacturers are looking for effective strategies to reduce product costs and development cycles. Mass Customization is generated in this context. It is a production model that based on the individualization of each customer and provides customized products and services with low cost and high efficiency in mass production [1].

By analysing the existing user requirements and combining the predicted requirements, the product family design extracts the deformation parameters according to the customized method to be used, and designs a family of products to form a deformable product model, which provides a theoretical basis for the rapid design of customized products. The essence of product family design is to identify and exploit the commonalities and similarities in design and manufacturing. This similarity includes the similarities between user needs and market competition to construct a product family architecture. A reasonable product family architecture not only facilitates the communication between the marketing department and the user, determines the configuration relationship of each module, but also reflects the optimal configuration of the manufacturing process, thereby reducing the cost loss caused by the variation. In this process, because of the necessity to analyze the user's needs, and consider the relationship between different products in the design, it must reflect the constraints of the manufacturing process. Therefore, how to obtain a public core platform becomes the key to the development of product family architecture.

2. PRODUCT FAMILY DESIGN METHOD BASED ON MAPPING TECHNOLOGY

An important idea of mass customization design is to realize the transformation from reactive to pre-determined design ideas. To this end, it is necessary to meet the requirements by planning and designing a family of products. The product range includes product lines over time and product lines at a certain point in time [6]. In the process of designing a product family, it is necessary to respond to the dynamic changes in user requirements so that the designed product family is highly adaptable to changing needs. Another feature of the product family design is that it is designed under a number of target constraints.

The dynamic changes in user demand mainly include changes over time and changes at a certain point in time, which will be passed downstream and affect the composition and structure of the product. The product family design to achieve user needs and the mapping between user requirements, product parameters and product functions need to be carried out in two steps. The first step is to achieve a mapping between user requirements and product parameters, and the second step is to implement product parameters. And mapping of product features.

The relationship between user requirements and product parameters is generally vague at the beginning of the design. The mapping between the two is the core and foundation of mass customization. The final planned product family is to plan the design as a whole based on all the above changes. Therefore, the result of mapping user requirements and product parameters is to form a product parameter for the entire product family rather than for a single product. set. A reasonable analysis of the collection, through the allocation of indicators to divide the changes of each product member in different periods and the same period, can achieve the planning of the entire product family.

The change in demand causes a change in the value of the corresponding product parameter, and the degree of this change is different for different user requirements, and the change of the product parameter brings about the change of the specific component selection. The two levels of association mapping realize the impact of changing user requirements on the final parameters of a functional model of the product family, and finally obtain the product family component module relative to the series of products (series products at different time points and over time) The overall requirements of the product family can be designed on top of this.

3. MATHEMATICAL PLANNING METHODS FOR PRODUCT DESIGN

3.1. Solution definition

A typical traditional optimization method is to find the optimal value to express the performance index. A performance indicator is given below:

$$p = f(d)$$

Get the minimum value by finding d * and get:

$$f(d^*) = inf\{f(d)\,|d \in Z\}$$

This result is the minimum solution

$f(d^*) = \min(f(d))$ of the design problem analyzed and formulated. Similarly, in order to meet performance metrics with target values, you can define a minimum, through the change of goals:

$$p = |f(d) - \tau|$$

This method represents all the optimization problem solving formulas.

3.2. Basic concepts of optimization

The template is used to format your paper and style the text. All margins, column widths, line spaces, and text fonts are prescribed; please do not alter them. You may note peculiarities. For example, the head margin in this template measures proportionately more than is customary. This measurement and others are deliberate, using specifications that anticipate your paper as one part of the entire proceedings, and not as an independent document. Please do not revise any of the current designations.

Constraint: A constraint is a performance evaluation metric and a target value pair used to constrain the final solution to exhibit additional characteristics beyond the set Z. Local constraints: Calling such a constraint an inequality constraint is represented by g_j. For the convenience of calculation, the upper limit of these expressions is usually converted to 0.

$$g_j(x) \leq 0 \qquad (1)$$

Equality constraint: This type of constraint condition defines that some design variables are related at the same time, and h_j is used to represent equality constraints. Similarly, the equality constraint h_j is deformed to obtain:

$$hj(x) = 0 \qquad (2)$$

3.3. Objective function

Beyond the constraint condition, the required performance metric is to identify an objective function that is a function that is managed to minimize. Definition: The objective function refers to determining which structure is optimized for the other performance index f by minimizing f in the Z zone.

3.4. Standard form

When the product performance index is converted into g_j and h_j constraints and a single index is selected as the objective function, the expression of the optimization problem is formed [7]:

$$\begin{aligned} &\text{Minimize} && f(\vec{d}) \\ &\text{Obey} && \vec{g}(\vec{d}) \leq 0 \\ & && \vec{h}(\vec{d}) = 0 \qquad (3) \end{aligned}$$

Here$(\vec{d}) \in Z = Rn$, all performance indicators are real values. Each point $\vec{s} \in Z$ that satisfies the constraint of (3) is a feasible point, and all such points constitute a feasible space S. The expression of this optimization problem forms the standard form of nonlinear programming problems.

4. OPTIMIZATION MODEL OF PRODUCT FAMILY PARAMETER DESIGN

According to the idea of product family design, the design variables are divided into platform variables and differentiated variables. Here, the platform variable is assumed to be x, and the differentiated variable of each personalized product is y^i. For all products, the value of the platform variable x is the same, and the non-platform variable y^i has different values depending on the product, that is to say, all products have their own differentiated variable y^i, and the product's direct differential variable is none. Mutually affected, but both are affected by the platform variable x.

Product design plays an important role as a link between the market and production. On the one hand, the products designed must meet the needs of the market. On the other hand, the products produced must be adapted to the actual production capacity of the company and can be lower. Cost to produce. Therefore, the optimized design of the product family should realize the integration of market, design and manufacturing, and achieve a good connection between the three ones [8]. In this paper, a two-tier planning with fuzzy coefficients (objective function and constraint condition expression with fuzzy coefficients) is applied to study the optimization model of product family parameter design. The specific model is as follows:

$$\min F(x, y(x))$$
$$s.t. H(x) \le 0$$
$$y(x) = (y^1(x), y^2(x), \dots, y^N(x))$$

$$y^i(x) = \arg\min f_i(x, y^i)$$
$$s.t.\ g^i(x, y^i) \le 0$$
$$i = 1, 2, \dots, N \qquad (4)$$

The upper objective function is F (x, y), the decision variable is the platform variable x, the lower optimization target is fi (x, y^i), and the decision variable is the differential variable y^i.

In the fuzzy coefficient planning, the following two points are pointed out:

- For any fixed membership degree, the cut off of an arbitrary fuzzy number is a closed interval (when λ=0, the cut set of λ with $\tilde{e}$ is defined as the support set of the fuzzy number), that is $\tilde{e}(\lambda) = [e^L(\lambda), e^U(\lambda)]$,the real number is $\tilde{e}(\lambda) = [e^L(\lambda), e^U(\lambda)]$.
- A cut set for any fuzzy vector $\tilde{e} = (\tilde{e_1}, \tilde{e_2}, \dots, \tilde{e_n})$, the cut set of λ with e˜ , which is $\tilde{e_i}(\lambda) = [e_i^L(\lambda), e_i^U(\lambda)]$. If you take a value in each $[e_i^L(\lambda), e_i^U(\lambda)]$, you can get a vector $e(\lambda) = (e_1(\lambda), e_2(\lambda), \dots, e_n(\lambda))$. Because of $e_i(\lambda) \in [e_i^L(\lambda), e_i^U(\lambda)]$, get the expression $e(\lambda) \in [e^L(\lambda), e^U(\lambda)]$.
- It is not difficult to see from the model (4) that the upper objective function is a cost function related to the manufacturing capability of the enterprise, and is mapped to the platform function, and its optimization direction is minimized.

Consider the manufacturer's production function as $Q = f(x_1, x_2, \dots, x_n)$, where Q is the yield of the combination of inputs $(x_1, x_2, \dots, x_N), x_i(i = 1, 2, \dots, N)$ is the input amount of the i-th element. If the price of the i-th element is p_i, the manufacturer's production cost is $TC = p_1x_1 + p_2x_2 + \dots + p_Nx_N = \sum_{i=1}^{N} p_ix_i$. For a given productivity Q, the manufacturer's cost minimization of production behavior can be expressed as: $\sum_{i=1}^{N} p_ix_i$

$$\min\{TC\} = \min \sum_{i=1}^{N} pi\,x\,i$$
$$s.t. f(x_1, x_2, \dots, x_N) = Q \qquad (5)$$

- The Lagrange function of the optimization problem (1) is:

$$L = \sum_{i=1}^{N} pi\,x\,i - \lambda\,(f\,(x1, x2, \dots, x\,N) - Q)$$

where Lagrange multiplier is λ. The first-order condition for minimizing vendor costs is:

$$\lambda \frac{\partial f}{\partial x_i} = p_i \ (i = 1, 2, \dots, N).$$

From formula (2) can get:

$$TC = p_1x_1 + p_2x_2 + \cdots + p_Nx_N = \sum_{i=1}^{N} p_ix_i = \lambda \sum_{i=1}^{N} \frac{\partial f}{\partial x_i} x_i$$

Known by the definition of the cost function:

$$TC = C(Q) = \lambda \sum_{i=1}^{N} \frac{\partial f}{\partial x_i} x_i$$

In the formula, C(Q) is the cost function, and the Lagrange multiplier is the shadow price of Q, that is, the increase per unit of production leads to an increase in the total cost. Therefore, the marginal cost for the optimal decision of the manufacturer is:

$$\lambda = \frac{\mathrm{dTC}}{\mathrm{dQ}} = C'(Q)$$

For enterprises, even if the optimized design products have good performance, it is not feasible for the products produced by the company to be not profitable. Therefore, it is used as the upper-level decision-making problem of the two-tier planning and as a guide for the lower-level decision-making problems. The lower-level decision-making problem is the optimization problem of multiple slaves. The essence of the problem is the optimal design of the actual product, which embodies the optimization design of multiple products according to market demand. So, the objective function is f_i (x, y^i). Because the objective function would be the optimization of multiple product parameters of the i-th product, so it is mapped to the differentiation function, that is, the lower-level optimization problem in the formula (6) model:

$$\begin{aligned} & y^i\,(\mathrm{x}) = \arg\min y_i\,(\mathrm{x}, y_i) \\ & \mathrm{s.t.}\, g_i\,(\mathrm{x}, y_i) \le 0 \\ & \mathrm{i} = 1, 2, \ldots, \mathrm{N} \end{aligned} \qquad (6)$$

5. EMPIRICAL RESEARCH

People have different requirements for clothing in different seasons. The frequency of people updating clothes is also increasing. They are required to wear different clothes at home, work, social, leisure, travel and other occasions, resulting in diversified market demand for clothing. Everyone chooses their favorite clothing according to their own personality, so that different seasons, different occasions and even each clothing have different needs due to different personality differences. Therefore, the clothing demand market is multivariate. There are no metrics for the individualized factors of clothing, and they cannot be quantitatively described and have ambiguity. Consumers' individualized demand for clothing is also vague. People have basic goals in the minds of the various clothing needs, but the boundaries of the goals are not clear, and it is usually unclear what exactly they want [5].

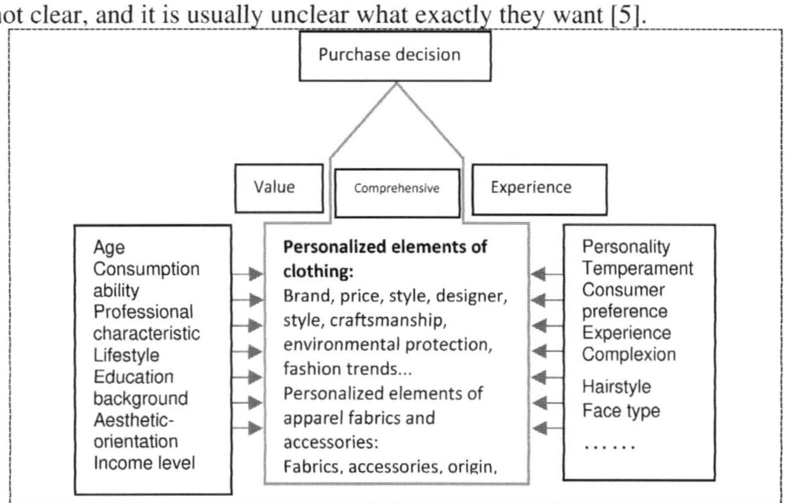

Figure 1. Customer Personalization Needs Schematic Diagram of Individualized clothing Design Decisions

The final decision of the choice can be expressed by function (7):

$$Y_s = \sum_{\substack{i=1 \\ j=m=1}}^{\substack{i=n \\ j=m=k}} \left(\xi_i^{ind} x_j^{com} + \xi_{im}^{ind} x_{jm}^{com}\right) + x_{ec}$$
$$i = 1,2,\ldots,n; j = m = 1,2,\ldots,k \qquad (7)$$
$$\xi_i^{ind} \geq 0;\ x_j^{com} \geq 0;\ \xi_{im}^{ind} \geq 0;\ x_{jm}^{com} \geq 0;\ x_{ec} \geq 0$$

In function (7), i is the number of individualized demand items for the customer, j is the number of individualized design items for the garment, m is the number of fuzzy items designed for the individualized clothing, and Ys is the final choice for the customer to purchase their own clothing and external factors.

For the individualized factorization of the customer, ξ_i^{ind} is the choice bias of the individualized design choice of the garment, x_j^{com} is the individualized design of the garment, ξ_{im}^{ind} is the choice bias of the different factors of the customer's individual demand for the fuzzy factor of the garment, x_{jm}^{com} is the fuzzy factor in the design of the clothing personality, x_{ec} is various impact of external factors on the final decision of the customer. Because of the different ξ_i^{ind} *and* ξ_{im}^{ind} between different customers, the Y_s is also different. Under different moods and objective environments, the choice bias of each factor in the individual's personality changes the choice of clothing factors, and the Y_s may be uncertain due to the existence of uncertain variables.

Equation (7) reflects the ambiguity of the customer's selected target. The designer's suggestion is to increase the value of x_{ec} *and* Y_s in the neutralization. The result will reduce the ambiguity ratio and improve the decision-making speed of the customer. Converting a model with a fuzzy target into a lower-level objective function is the satisfaction function of the decision maker. For the enterprise, only the specific expression of the cost function is determined in the model, where the price coefficient and the disturbance variable meet the fuzzy number of a certain distribution. It can be operated to optimize the product family design and obtain satisfactory product parameter values. It can be imagined that the designer brand will be welcomed in the era of mass customization of clothing.

6. CONCLUSION

Differentiated variables y^i optimized for each product after determining the platform variable x Optimized design issues that meet the design requirements (based on market demand) reflect the integration of market (consumer demand) and design (personalized design capabilities).The upper-level decision-making problem reflects the manufacturing capabilities of the enterprise, while the lower-level decision-making issues reflect the market and design requirements, so that the two-tier decision-making problem realizes the integration of market, design and manufacturing.

Acknowledgments

(1) Funded by Beijing Apparel Industry Digitalization Engineering Technology and Research Center

(2) Project name: Building a "design capital" to promote cultural and technological integration (Scientific research); Affiliation type: Beijing Municipal Science and Technology Commission Beijing Key Research and Development Program; Project title: Development and application of flexible design and intelligent manufacturing service platform for apparel; Project number:Z171100005017004

References

[1] Geng Guoning, Gu Jianxin, Li Renwang. Mass customization production [J]. Computer Aided Design and Manufacturing, 2000(1), pp. 50-51.

[2] Pine B Joseph. Mass Customization - the new frontier of corporate competition. Beijing: China Renmin University Press, 2000.

[3] Li Renwang et al. Preliminary study on mass customization and its implementation methods. China Mechanical Engineering, 2001 (4), pp. 405-408.

[4] Wu Dichong. Several studies on production models. Journal of Zhejiang Institute of Engineering, 2002(1), pp. 38-41.
[5] Wu Dichong, Gu Jianxin. Analysis of mass customization characteristics. Journal of Textiles, 2005 (6), pp. 151-152.
[6] Wang Aimin, Meng Mingchen, Huang Jingyuan. Research on demand-driven product family design method through dynamic QFD [J]. China Mechanical Engineering, 2003 (4), pp. 667-670.
[7] Robertson D, Ulrich K. Planning for product platform [J] Sloan Management Review, 1998, 39 (3), pp. 19-31.
[8] Kevin N. Otto, Kristin L. Wood. Product design [M]. Beijing: Electronic Industry Press, 2005.

The Application of Fractal Art in Ceramic Product Design

Xingrui Lin[1], WenmingLiu [2*]

[1]Student Management and Service Center, Liaoning University of International Business and Economics, Dalian, Liaoning,116052, China

[2] School of Design and Art, Shenyang Jianzhu University, Shenyang, Liaoning,110168, China

[*]Corresponding author's e-mail: 83373734@qq.com

Abstract,Starting from the meaning of fractal geometry and fractal art,this paper discusses the artistry of fractal,and draws a conclusion through the introduction and analysis of four kinds of classical fractal patterns,that is,fractal art can be called real art,It can be found that fractal art has great artistic charm and practical value in ceramic product design by applying fractal art pattern to the field of traditional ceramic daily necessities and handicraft design,However,as a designer,the art pattern of fractal processing cannot be directly attached to the surface of ceramic products,Further researches and innovation are needed to make fractal art more suitable for ceramic products and enhance the artistic value of ceramic products.

1.Fractal Geometry and Fractal Art

The fractal geometry has been applied for nearly 40 years since the mathematician Mandelbrot discovered the fractal theory in the 1970s,Its research fields include medicine,biology,soil mechanics,physics,graphics,seismology,music,computer and so on,It has been widely used in textile industry,wallpaper industry,product packaging,product design and so on.

The basic idea of fractal geometry is that objective things have self-similar hierarchical structure and are composed of individuals with similar characteristics,Although fractal geometry belongs to the category of mathematics,it can find its aesthetic existence form in nature,which makes fractal geometry have practical value in the field of art.

Fractal art belongs to the concerned fractal in fractal geometry,that is,the self-similarity description of graphics in different dimensions (self-similarity is the similarity of local and global shapes in graphics),the detail structure of which can be enlarged infinitely,Fractal art can be classified into fractal art by using the same basic unit and calculating formula,even if it is not a professional fractal.

Fractal art only appeared after fractal geometry was produced for a period of time,Although fractal geometry can produce inexhaustible and diverse shapes,it was not regarded as a work of art in the early day,until people gradually discovered the artistic value of fractal geometry.

By transforming the equation into visible fractal patterns,fractal art realizes the organic combination of modern science and technology and aesthetic art,and achieves the fusion of mathematics and art through the transformation and combination of orderly and disordered patterns,Different from the asymmetric beauty of Euclidean geometric form,aesthetic works with artistic value are formed.

Fractal art is a branch of visual communication art,Nowadays,fractal artists have penetrated into the fields of product design,art design,graphic design and so on,Fractal art as the visual expression of electronic image,the artist quickly accepted its artistic value,and created a series of aesthetic products through mathematics,Although fractal art is realized by computer computing,it still needs a certain

IOP Conf. Series: Materials Science and Engineering **573** (2019) 012003 doi:10.1088/1757-899X/573/1/012003

artistic foundation to design a good fractal work.

2.The artistic value of Classification Art

Is fractal art real art? Some have affirmed, others have been tit-for-tat,and there has been a long debate on this issue.

In fact,the conclusion is very clear,If fractal art patterns give people a feeling of beauty,then it can be called art,In that case,of course,we do not have to explore how it came about,When we can think of others,we are impressed by the amazing workmanship of nature,which has produced countless wonders of beauty,and we have been deeply impressed by these beautiful sights,but we cannot deny that they are valuable aesthetics because they come from nature and are not artificially crafted.

The patterns obtained by fractal technology are not necessarily works of art in the direct sense,We need to process and deal with them,People need to affirm the existence of fractal art,even though the results derived from a few fractals have no artistic value,but more derivative results are artistic,Fractal is a new concept emerging in modern times,Whether you appreciate it from the perspective of science or aesthetics,fractal art will give many people a feeling of beauty.

Now let's look at some classic fractal models: Mandelbrot, Julia, Newton, Nova

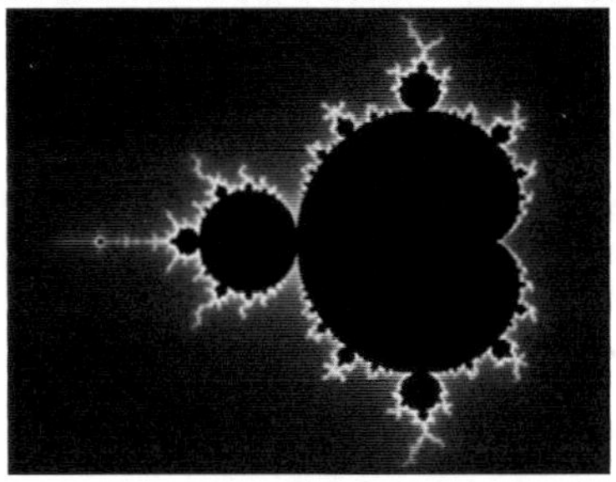

Figure 1,The fractal graph of Mandelbrot

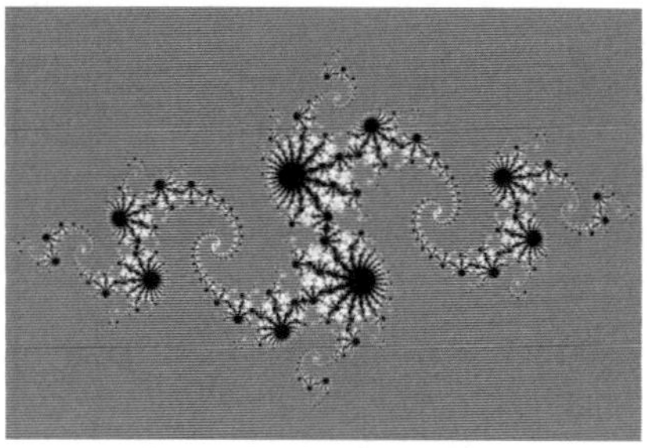

Figure 2,The fractal graph of Julia

Figure 3,The fractal graph of Newton

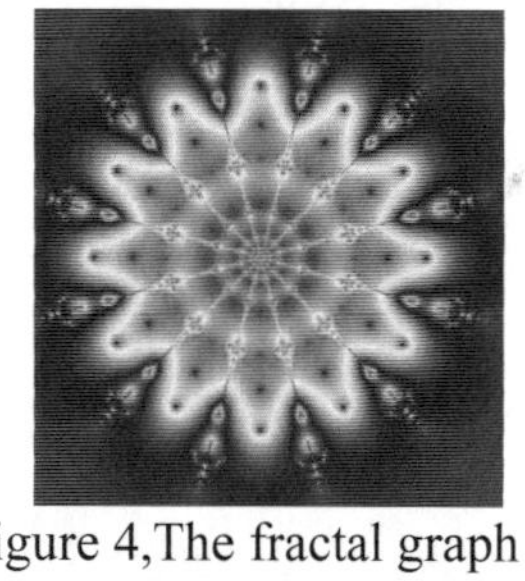

Figure 4,The fractal graph of Nova

Figures 1 to 4 have a traditional aesthetic sense of Euclidean geometry,reflecting the criteria of traditional aesthetics,such as balance,harmony,symmetry,etc,Apart from that,these pictures also contain many new art forms that are superior to traditional aesthetics,For example,its equilibrium is dynamic equilibrium,which is the equilibrium of all parts of the image in the process of change,its harmony is mathematical harmony,and the most special one is the symmetry in fractal art,which is neither horizontal nor vertical symmetry,In essence,the symmetry of fractal art is the symmetry of the small part and more part of the picture,or the symmetry of the part and the whole.

From the above fractal figure,fractal art as a new art category can be called real art.

3.Application of Fractal Art pattern in Ceramic Handicrafts

The patterns processed by fractal art can give people a feeling of beauty and convey a sense of beauty different from traditional art,These fractal works of art are mainly derived from the formula based on computer,Then the aesthetic view is used to process the art in the later stage,which produces amazing works of art.

After appreciating the effect of fractal art,the application value of fractal art in ceramic products is analyzed,The art of typing is not simply to attach the pattern of fractal processing to the object,but to merge the modern science and technology with the traditional technology into one,and to form a new form of visual impact.

3.1 The fractal patterns on teapots

The decorative patterns on the traditional teapot are usually plants and poems,We tried to apply the fractal patterns in figure 1 and figure 2 to the teapot design,as shown in figure 5 and figure 6:

Figure 5,The fractal teapot of Mandelbrot

Figure 6,The fractal teapot of Julia

The application of the above two fractal patterns in teapot gives people a different feeling from the traditional flower and bird landscape patterns and makes people have a harmonious aesthetic feeling,Although these decorative patterns are not familiar to us,they give people a kind of endless changeable dynamic feeling,make people have fresh visual experience and make people's imagination continue to extend,To appreciate the same fractal works at intervals will give people a different feeling and thus produce ever-changing visual effects,The pattern in figure 5 is similar to the traditional pattern,but it feels more profound than the traditional pattern.

3.2 The fractal patterns on bowls

The environment and frequency of use of bowls are different from those of teapots,We cannot use complex patterns similar to Rococo,For the use of the bowl of fractal patterns,we should avoid too complex,People usually prefer to favor traditional artistic effects,so the fractal images used in bowls should be concise without losing a certain depth,From this point of view,a reasonable fractal pattern can be obtained,as follows:

Figure 7,The fractal bowl of Julia

Figure 8,The fractal bowl of Newton

By enlarging and shrinking the same fractal pattern and applying it to different places,we can give people different psychological feelings,Compared with the traditional pattern design,fractal pattern has a novel design idea,and it is more easily accepted by people through the innovative treatment of the pattern,Figure 7,figure 8 contains the traditional feeling,while producing dynamic rhythmic effects.

3.3 The fractal patterns on a vase

Vase is also one of the common ceramic ornaments in life,By applying fractal patterns to the external design of the vase,the results are as follows:

Figure 9, The fractal flower vase of Mandelbrot

Figure 10, The fractal flower vase of Newton

Figure 11, The fractal flower vase of Nova

Figure 12, The fractal flower vase of Julia

The fractal patterns in figures 9 to 11 give a different feel,Although vases are not as popular as bowls and teapots,they are common,Vases have a certain attention to patterns,traditional vase decorative patterns are usually flowers,scenery and figures mainly,The design requirements of these patterns are high,the occupation area is large,the design method is complex,and the design difficulty is high,Let's look at the effect of applying fractal effects to the above vase,The fractal patterns in figure 11 and figure 12 are relatively close to the traditional flower pattern decoration of a vase,Compared with the traditional decorative pattern,fractal flowers show a dynamic effect that natural flowers do not have,and are similar to the vase itself,The integration of the vase made the vase more dynamic,Figures 9 and 10 are different from traditional vases and make people feel more modern and modern,The fractal patterns on these vases are different,but there is a relative unity in the changes,that is,the motility of the patterns.

4.The conclusion

As a new subject,the artistic value and importance of fractal geometry have been paid more and more attention,and its application has been extended to various fields of product design,We can see the application value of fractal in the design of ceramic products such as teapots,bowls,vases and so on from the previous analysis,Fractal art will be widely used in the design of ceramic products.

As the early application of fractal art,fractal art plays an important role in the design of ceramic decorative drawing,However,the fractal design pattern used in ceramic decoration pattern is limited to the application of fractal pattern directly in the overall drawing of ceramic art,which makes fractal art design become "fractal art mapping" to some extent,Although this direct application confirms to some extent the feasibility of fractal art in ceramic design,as a designer,it is not recommended that the art pattern of fractal processing be directly attached to ceramic products,For fractal patterns and ceramic products Combine mode and style to carry out necessary research,study and innovation,With the continuous development and progress of fractal art,in view of the application of fractal art,we should innovate constantly on the basis of adapting to products,so that fractal art can be integrated into product design and realize the real unity of science and technology and art,So that fractal art in the future design field can be more and more far,more and more widely used,and in the field of product design to create a higher artistic value.

IOP Conf. Series: Materials Science and Engineering **573** (2019) 012003 doi:10.1088/1757-899X/573/1/012003

References:

[1] Yao Hui,Gao Wei,Ren Zhongming,(2018) Sustainable Packaging Design in regular Fractal Geometry,Green package,09: 73-79.

[2] Mao Yalong,He Jingtang,Guo Weihong,Title (2014) the Lingnan Architectural School and its creation from the Fractal Perspective,South Architecture,01: 88-93.

[3] Cheng Yi-fang,(2015) the application of fractal art pattern in decorative design,Building Engineering Technology and Design,12: 321-321.

[4] Li Zaibo,Gao Ying-ying,(2016) Product Decoration Design and Application based on Fractal Image,Design,069: 120-121.

[5] Luo Yun,(2017) Feasibility of Chinese traditional pattern used in modern fashion design,Dyeing and finishing Technology,05: 8-10

Research and Design of Intellectual Products for People with Visual Impairments

Tong Huo[1], Songming Liu[2], Shuangjing Fan[2] and Yi Liu[1*]

[1] Fashion Accessory Art and Engineering College, Beijing Institute of Fashion Technology, Beijing, 100029, China

[2] School of Information Engineering, Beijing Institute of Fashion Technology, Beijing, 100029, China

[*] gxyly@bift.edu.cn

Abstract. The paper mainly focuses on designing intellectual products for people with visual impairments. By analyzing physiological features, psychological characteristics and pattern of behavior of people with visual impairments and comparatively analyzing the status of current intellectual products for people with visual impairments on the market, the research designed the products based on LEGO EV3 model and proposed three functionally intellectual products design for people with three different visual impairments symptoms: automatic user tracking guide stick for the blind, traffic signals identifying guide stick for color blindness people and dim light alarm for night blindness people. The proposed products will help people with visual impairments go out freely and independently to enjoy a convenient life in the future.

1. Product demand of people with visual impairments

1.1. Physiological features of people with visual impairments

Vision, hearing, taste, touch and smell are five sensory systems from which people gain information from the outside world. These sensory systems are physiological basis of human cognition. Among them, vision is the most important sensory system no matter for human or diurnal animals. It can help us to get accurate information over a wide range and remote distance. As for people with normal vision, 90 percent of external information is obtained by the visual system. The visual system of both humans and animals generate senses by the stimulation of light, and eyes are the organ that humans use to receive and feel light.

People with visual impairments refers to people who, due to innate or acquired causes, have partial or total obstacles in the structure or function of the visual organ and after treatment they are still unable or even hard to see objects. Researches show that, due to visual obstacles, people with visual impairments rely more on touch, hearing and other sensory to make up for the missing visual system function[2]. With the help of other four sensory system, people with visual impairments can get new information from outside world which will help them to update existing memory or experience and form their own cognition of the object world. Therefore, the particularity of physiological features of people with visual impairments are the fundamental basis which guide the LEGO EV3 model design in this thesis.

1.2. Psychological characteristics of people with visual impairments

The definition of visual impairments is very broad which includes blurred vision, high hyperopia or myopia, color blindness, color weakness, night blindness, etc. In daily life, people with different symptom of visual impairments will encounter different difficulties. Because of the visual sensory deficiency, most people with visual impairments may have vague, fancy or uncertain cognition of the world, their psychological characteristics have startling similarity [2], which can be concluded into five aspects:

a. Loneliness: Because of the partial or total obstacles of vision, people with visual impairment would not like to go out. In a long period, they rely on limited methods like internet to obtain information of the world which makes them feel loneliness.

b. Dependence: People with visual impairment usually lack of abilities to take care of themselves, and their limited life experiences make them can't bare too much stress. Therefore, when they want to complete something independently, it is easy for them to generate an avoidance mentality and a dependence on guardians.

c. Inferiority: Most people with visual impairment over care the deficiency of their physiological functions which make them doubt their own abilities and feel inferiority.

d. Lack of security: Visual system deficiency will cause many troubles for people to gain information on time, which will make them feel afraid of the environment and people around them.

e. Sensitive self-esteem: According to the survey, people with visual impairments are more sensitive than others. Because of their vulnerable position in society, they hope to get the understanding and attention of the society, but at the same time they do not want to be looked down upon by others.

2. Current status of products for people with visual impairments

There are some products for people with visual impairments on the market at present. Most of them use braille or fixed form of irregular tactility to convey message, which design is relatively lacking of intellectualization, but mainly depending on sensory system. These products such as braille household appliances, braille reading and other products are designed for users who are familiar with braille, however, for those users with visual impairments but who are not familiar with braille, it undoubtedly will become a new difficulty. For the blinds, when they rely solely on listening to locate an item, they may not be able to avoid obstacles on the road. However, if there are intelligent products which can help the blinds to predict the existence of obstacles in advance and bypass it, the blinds will be able to take care of themselves. For people with visual impairments who cannot distinguish colors properly, there are some color comparison cards that are specifically designed for their vision which can practice their ability to distinguish color. However, if intellectual products can automatically recognize color and convey message to the users, it will be more convenient for color blindness people no matter they are at home or going out. In addition, the lack of perception of different intensity of illumination is also a major problem for some of people with visual impairments. As the weather getting dark, their vision gradually become blurred. If something can remind them to turn on the lights before this, they can avoid some danger which caused by not seeing clearly. The research is to design more intelligent and convenient products to help people with visual impairments living a convenience life.

3. Research of product design

3.1. Guide stick for the blind

Some people with visual impairment will choose guide stick as their own "eyes". If it is out of the reach, searching the stick will become a problem. To solve the problem, we make the stick find its owner by using intelligence LEGO products. If there is an unreachable distance between the user and the guide stick, by pressing the remote controller, the guide stick will lock the remote controller by infrared ray and keep forward until facing the user. During the process of achieving the function that the guide stick can find its owner automatically. We choose the infrared sensor and the collision sensor. In the process of EV3 program, the design chooses the loop module, steering module and comparison module.

The infrared sensor of LEGO is used to lock the owner's position.

The stick is located the position by receiving the infrared ray produced by remote control, then it adjusts the angle of guide stick until the distance between the stick and its owner is less than the setting value. It will stop in front of the owner. During the process of going forward, the design avoids the obstacle by using the collision sensor. While the returned value is true, it will turn right 90 degrees and keep going towards the owner to achieve the function of obstacle avoidance. (Figure 1)

3.2. Guide stick for color blindness people

Among the people with visual impairment, the color blindness symptom takes a large part. The research should take it into consideration, especially when they are going out, it always arises the security problem that they can't tell the traffic lights. During the process of achieving the function to help the color blindness tell the traffic lights, color sensor and the infrared sensor are chosen. In the process of EV3 program, the loop module, comparison module and steering module are chosen.The color sensor has color mode, reflected illumination mode and environmental illumination mode. The infrared sensor has close range mode, control mode, near beacon mode and guided beacon mode. The design will detect the recognized color by using the color mode .

During the process of going forward, the product avoids the obstacle by using the infrared sensor. While setting at the close range mode, it can measure the distance between the sensor and the obstacles. The measurement uses 0%-100% instead of centimeter or feet. If the measurement is lower than the setting value, the procedure to avoid obstacles will be triggered. (Figure 2)

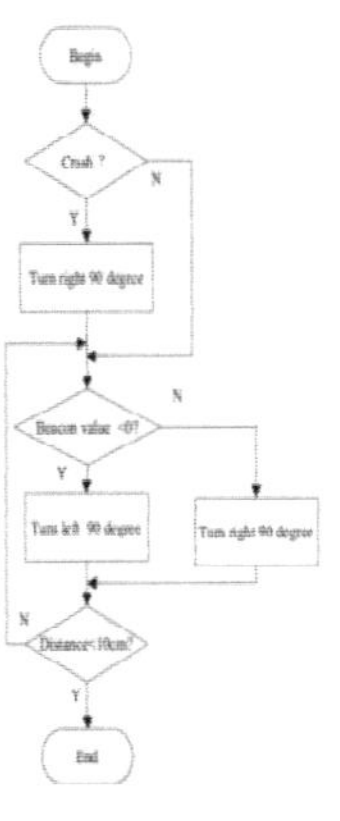

Figure 1. Flow chart of guide stick for the blind.

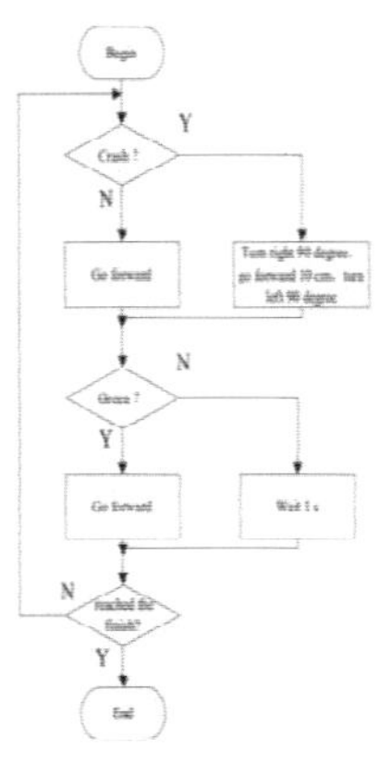

Figure 2. Flow chart of guide stick for color blindness people.

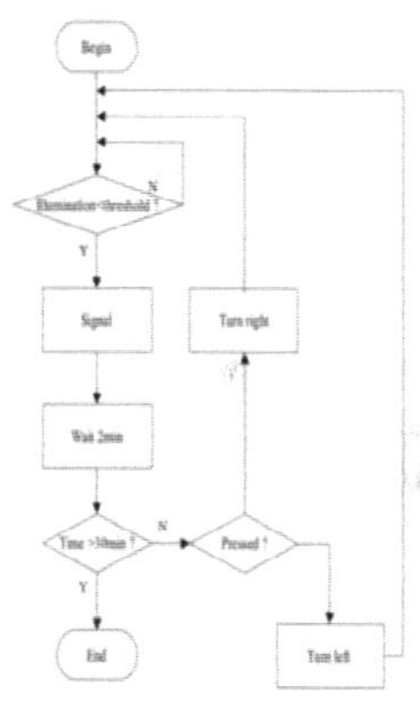

Figure 3. Flow chart of annunciator for night blindness people.

3.3. Annunciator for night blindness people

Night blindness is a kind of the visual impairment. Its symptom appears under different intensity of lights or the symptoms of things cannot be seen clearly. Due to lack of vitamin A and lesions of the body, they may get short-term night blindness or acquired night blindness. Facing with the night blindness people, the recommended signal will be sent before the environmental illumination lower than the setting value to avoid the problem of inconvenient movement of night blind patients under the dark light condition. The color sensor and the collision sensor are chosen, and set the color sensor into

IOP Conf. Series: Materials Science and Engineering **573** (2019) 012004 doi:10.1088/1757-899X/573/1/012004

environmental illumination mode to sense the outside illumination, at the same time, avoiding the obstacle in low illumination environment. In the process of EV3 program, the loop module, comparison module and steering module are chosen. Under testing environment, the road are set that illumination from strong to weak with some unknown obstacle. Before testing, the range of thresholds at which the color sensor senses illumination are measured, and make it the variable under control. If the illumination higher than the setting value, the sound will play. After the sounds plays, it will stop for a second and keep going. If the collision sensor is touched during the journey, it will be turned to the left to avoid obstacles. If the illumination remains the same value, it will repeat continuously, and the ring is set to ring for no more than 30 minutes, and the program will terminate automatically .(Figure 3)

4. Product prototype based on LEGO EV3

4.1. Automatic user tracking guide stick for the blind

The prototype for this direction design is shown in Figure 4. The modules used in this design are: infrared sensor, EV3 programming module, collision sensor, motor module, infrared tele-controller.

Set the infrared sensor of EV3 in the beacon working mode, and the button of the infrared tele-controller keeps pressing (continuous signal is sent). Using the "guided beacon" mode to achieve the function of advancing toward the target, the infrared sensor can receive the signal from the infrared tele-controller and identify the position of the beacon and then proceed to the position of the beacon. The guide stick advance until the adjacent beacon value of the infrared sensor is less than 10%, and it will stop when less than 10%. The collision sensor is used to avoid obstacles. The guide stick will turn right 90 degrees when the return value of the collision sensor is true, then continue to identify the position of beacon. (Figure 5)

Figure 4. Design product structure of automatic user tracking guide stick for the blind

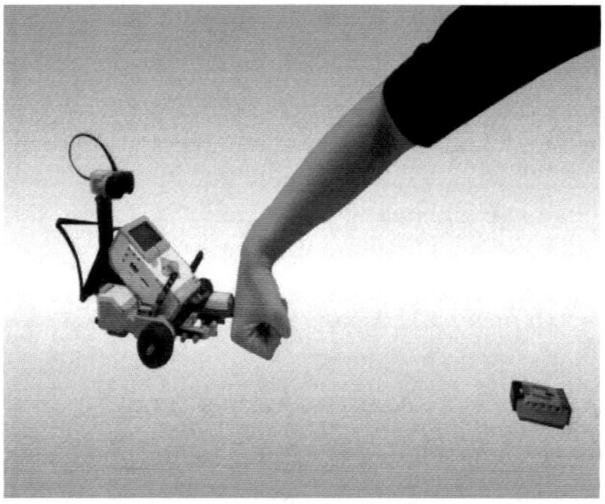

Figure 5. the function of obstacle avoidance.

4.2. Traffic signals identifying guide stick for color blindness people

The prototype for this direction design is shown in Figure 6. The modules used in this design are infrared sensor, motor module, EV3 programming module, and color sensor.

Setting the EV3 color sensor to color mode and the infrared sensor to close distance mode, then the color sensor will recognize the color and the guide stick continue to advance when the color turns to green, else the guide stick will stop. The robot will turn right 90 degrees in advance about 10 cm, and then turn left 90 degrees. When the measurement result of the infrared sensor is less than 50%. (Figure 7, Figure 8)

Figure 6. Design product structure of Traffic signals identifying guide stick for color blindness people.

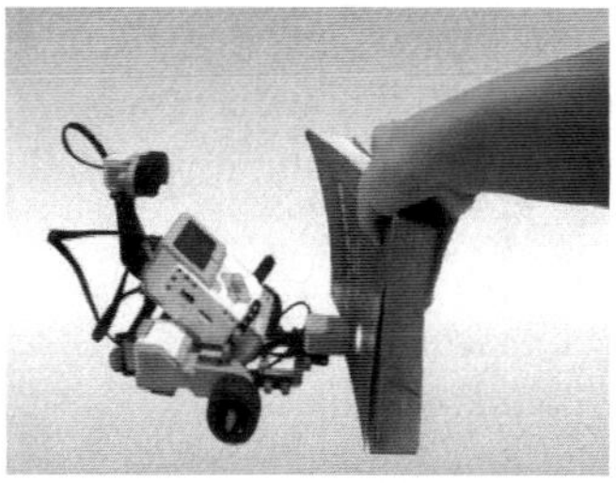

Figure 7. Go forward when encounter green.

Figure 8. Stop when encounter red.

4.3. Dim light alarm for night blindness people

The prototype for this direction design is shown in Figure 9. The modules used in this design are: color sensor, and two EV3 programming module, collision sensor, and motor module.

Setting the color sensor of EV3 in ambient light mode, the module repeats ringing to prompt when the light intensity under the threshold, then the waiting block pauses the program, the ringing is repeated and the actuator is pressed.

Figure 9. Design product structure of Dim light alarm for night blindness people.

5. Conclusion

People with visual impairments are vulnerable groups in society. The pain points of such groups have caused public's concern and need to be solved urgently. The pertinence and humanization of intelligent product design for people with visual impairments is the fundamental condition for helping them to back to society independently, and help them to recover self-esteem and self-confidence. Based on the current situation of intelligent product design for people with visual impairments, the research found that the design function of existing products is vague during the research process, and it is not advisable to make one product serve all visually impaired people. It can be concluded that only by paying close attention to the different pain points of people with different symptoms and visual impairments, and designing for different pain points, the smart products can serve such groups better. The LEGO EV3 model used in this design can help designers realize the expected functions and synthesize the product prototypes for function testing quickly. In addition, the modular programming form also reduces the burden on the design team. As a matter of fact, the Sensors based on the LEGO EV3 model still have some limitations. So they are just designed to provide some design ideas for intelligent product design innovation for visually impaired people. There still has a long way to go for in-depth design research and perfect product application.

Acknowledgments
This research was financially supported by Beijing Institute of Fashion Technology College of Special Plan Young Top-notch Talent Project(BIFTBJ201803) ，Beijing Institute of Fashion Technology College Student Research Training Program (2018) ，Beijing Institute of Fashion Technology Teaching Reform and Innovation Key Program (ZDJG-1709) ，Textile Vision China National Textile And Apparel Council Higher Education Teaching Reform and Innovation Program (2017-08) ，and Open Project of Digital and Interactive Media Key Laboratory (KF2017-04).

References
[1] Wei,J.Y.,Research on the Design of Visual Barrier Auxiliary Products in Public Space [J]. Popular Literature, 2017 (08): 106.
[2] Huang,L.Y.,Research on the Innovative Design of Daily Necessities for Visually Impaired Persons[J]. Packaging Engineering, 2018, 39(14): 114-117.
[3] Liu,L.L.,Research on the design of visual impairment aids in public space [J]. Popular Literature, 2017 (08): 106.
[4] Zhu,Y.B.,"Senses and Acts"——Analysis and Research on Public Facilities Products for Visually Impaired Persons[J].Art Science and Technology,2016,29(03):324.
[5] Wen, Y.Q, Dong, J.X, A preliminary study on product design methods for visually impaired people [J]. Beauty and the Times (I), 2011 (01): 76-78.
[6] Lu,C.C.,Li G A. Research on the design of emotional products and packaging for visually impaired people [J]. Shanghai Packaging, 2017 (06): 27-29.
[7] Li,S.M.,Design of barrier-free environment for visually impaired people[J].Shanxi Architecture, 2010, 36(11): 14-15.
[8] Huang,Y.,Wan,X., Research on the Design of Metro Guide System Based on the Needs of Visually Impaired Persons——Taking the Design of Chengdu Metro Guide System as an Example[J]. Industrial Design, 2017(08): 105-106.
[9] Chen,X.E.,Design and Research of Visually Impaired Action Navigation Service [A]. Industrial Design Research (3rd Series) [C]: Industrial Design Industry Research Center, 2015:12.
[10] Ding,H.Y.,Yang,J.,Research on the Synthetic Design of Travel Wear Equipment Based on the Characteristics of Visually Impaired People [J]. Journal of Shandong Institute of Arts and Crafts, 2018(01): 64-68.

Exploration of Practice Teaching Mode of Product Design Based on Three-dimensional Printing Technology

Wei Shu[1*], Fuliang Sun[2]

[1] Art school, Huangshan University, HuangShan, AnHui, 245041, China

[2]Art school, Huangshan University, HuangShan, AnHui, 245041, China

[*]Corresponding author's e-mail: shuwei@hsu.edu.cn

Abstract. As a highly interdisciplinary and practice major, product design is expected to achieve the cultivation of innovative talents with artistic accomplishment, engineering knowledge, business awareness and design skills within limited learning time. Technology and art are regarded as the core of its cultivation. In the traditional industrial processing system, mass industrial production is taken by product design practice as the premise. Nevertheless, the maturity of rapid prototyping system represented by three-dimensional printing technology and its combination with whole-process digitalization and intelligent production bring about a great reduction in individualized customization and flexible production costs, which leads to new development opportunities for product design practice. This paper is devoted to the discussion of the practice teaching mode of product design in the context of three-dimensional printing technology.

1. Introduction

The third industrial revolution, represented by three-dimensional printing technology and marked by the application of digitization, artificial intelligence manufacturing and new materials, has arrived, while a brand new industrial era will be created by three-dimensional printers, the Economist magazine said in April 2012.

On May 8th, 2015, the State Council promulgated the first ten-year action guiding principle of Made-in-China 2025, which put forward the requirements of insisting on "innovation-driven, quality-first, green development, structural optimization and talent-oriented"[1].Have been mentioned repeatedly six times in this document, three-dimensional printing technology runs through the key sections of background introduction, improvement of national manufacturing innovation capacity, in-depth integration of informatization and industrialization, and breakthrough development in key areas. Integrated into the main line of promoting intelligent manufacturing, it has become a crucial link of the strategic plan [2].At the same time, the document also highlights the significance of talent demand, emphasizes to improve the talent cultivation system from R&D, transformation, production to management of Made-in-China and focuses on the cultivation of "three types of high-end talents in shortage": High-level professional and technical talents and innovative talents, outstanding entrepreneurs and high-level management talents and those are proficient in both manufacturing technology and information technology.

2. Characteristics and Advantages of Three-dimensional Printing Technology

Three-dimensional Printing (3D Printing) technology is an object construction technology based on digital model files with bondable materials such as powder metal or plastic by printing layer by layer [3]. As a rapid prototyping technology of relative maturity at present, it forms three-dimensional solid model through material stacking by virtue of mainly the method of additive manufacturing, with the American patent on the layered manufacturing method to form topographic maps proposed by J. E. Blanther in 1892 as its technical prototype [4]. After that, Charles W. Hull invented the world's first 3D printer in 1983 and founded 3D Systems in 1986, which developed the current general STL file format and launched the SL-based 3D industrial printer in 1988. Subsequently, three-dimensional printing and molding technologies such as SLA (Stereo lithography Appearance), SLS (Selected Laser Sintering), FDM (Fused Deposition Modeling), 3DP (3D Printing) and PUG (vacuum casting) came into being.Compared with traditional manufacturing methods, the revolutionary change of production and processing concept brought about by three-dimensional printing technology, the major technical features of which are manifested as: Rapid free prototyping due to the absence of molds in manufacturing equipment; Full digitalization and high flexibility of the production process; Capable of manufacturing almost infinitely complex geometric structures; Current technology can be applied to the manufacture of materials of a number of varieties; Capable of achieving the arbitrary composite manufacturing of multi materials (e.g. functionally gradient structures).These features have brought three-dimensional printing technology with the following advantages in actual production:

Firstly, apart from shortening the processing and manufacturing cycle, three-dimensional printing technology can also bring a significant reduction in production costs. In particular, it enables human beings to achieve freedom in the field of processing by breaking through the limitations of traditional processing and manufacturing methods on processing of complex shapes.

Secondly, three-dimensional printing technology has been widely applied in special fields including medical, aerospace, jewelry manufacturing and so forth. For example, in the medical field, the Ninth People's Hospital affiliated to Shanghai Jiao Tong University manufactured an artificial pelvis with three-dimensional printing technology and successfully transplanted it into the patient's body to promote its rehabilitation.

Thirdly, with high plasticity, the finished products produced by three-dimensional printing technology can be applied to both two-dimensional and three-dimensional. Layer-by-layer printing will be adopted in processing, while the processing accuracy of the printer can be as accurate as 0.01 mm per layer.

Finally, three-dimensional printing technology demonstrates great advantages in individualized customization. For traditional industrial manufacturing, which achieves product cost reduction through mass production, individualized production is difficult to realize. However, the possibility of low-cost individualized customization has been brought about by three-dimensional printing technology.

3. Comparison of Mainstream Three-dimensional Printing Equipment

Three-dimensional printers based on different technologies including "SLA, SLS, SLM, FDM, DLP, MJM..." with have assumed increasingly maturity in recent years, while industrial equipment with large print size, high accuracy, wide materials compatibility and desktop equipment with smaller print size, acceptable accuracy and single print material are constantly emerging. However, desktop printing equipment is widely adopted and accepted by users in teaching activities that are limited by the cost of equipment and consumables. Printers of this type are mostly based on fused deposition molding (FDM) and photosensitive resin selective curing technology, the latter of which is divided into two categories: stereo lithography appearance (SLA) and digital light processing (DLP). We are now carrying out a comparison of desktop three-dimensional printers based on FDM, SLA and DLP technologies,as shown in table 1.

Table 1. Comparison of SLA, FDM and DLP Equipment

Name	Stereo Lithography Appearance	Fused Deposition Molding	Digital Light Processing
Abbreviation	SLA	FDM	DLP
Consumable	Liquid photosensitive resin	Wire such as ABS, PLA, etc.	Liquid photosensitive resin
Method	Laser beam (point curing)	Heating devices (line extrusion)	Digital projection light source (surface curing)
Example		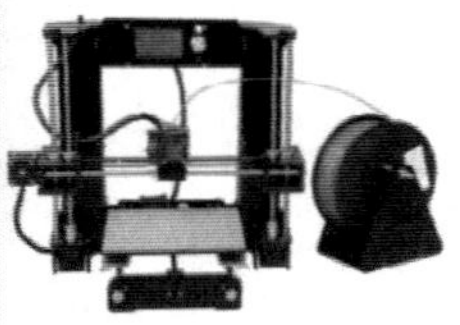	

The size of FDM three-dimensional printer with flexible structure can be either large or small. Thanks to the low price of equipment and consumables, it has gained considerably high popularity. Nevertheless, it assumes poor stability and slow printing speed in the production of large molding size. The major molding way of extruding consumables through the extrusion head brings obvious step effect. In addition, the processing accuracy and fluidity are both affected by the diameter of the nozzle and are mutually restricted.

The SLA and DLP three-dimensional printers which adopt light molding method are better in accuracy. Among them, DLP emits scattered sector light, while SLA emits laser of and approximately straight line, which determines its performance in accuracy is better than DLP.

SLA three-dimensional printer also completes printing through the motion of optical axis. Despite that it can print large size in theory, the current technology fails to support the manufacture of large SLA three-dimensional printers. It mainly cures the material through the laser points in forming, which can achieve higher processing accuracy.

The digital light source of projector is adopted by DLP three-dimensional printers, which is different from SLA three-dimensional printer which uses laser and cures materials with scattering sector light. The accuracy will be difficult to be ensured if the size is enlarged.

In General, in terms of molding size: FDM > SLA ≈ DLP; in terms of molding accuracy: SLA > DLP > FDM; in terms of equipment price: SLA > DLP > FDM; in terms of price of consumables: SLA ≈ DLP > FDM.

In accordance with the equipment and its application introduced at various additive manufacturing exhibitions in recent years, the relevant equipment of Fused Deposition Molding (FDM) has achieved the expected goal of being usable, use-friendly and cost-effective. In addition to meeting the application requirements of rapid prototyping in product design field in terms of accuracy, it also presents satisfactory printing time and stability, which makes it sufficient to customize small batch and non-durable design samples. Compared with other types, the price of FDM three-dimensional printer equipment is greatly reduced yet the high stability of consumables is maintained. Due to the lack of technical barriers, the post-processing and surface treatment processes of processed products are becoming increasingly mature, while the processing time can be shortened by simultaneous printing of multiple devices when necessary. As the most simple and direct way to meet the demanding fields including three-dimensional printing services, assistant education and teaching and so forth, the relatively improved supporting resolutions in technology and equipment are provided by FDM three-dimensional printers.

4. Application in the Field of Education

With its characteristics and advantages, three-dimensional printing technology has been widely applied in fields of rapid development of new products, individualized manufacturing, etc. People take advantage of the technology to produce extremely complex structure that fails to be coped with by traditional technology and make design models and samples to achieve the optimization of design and the improvement of product function. Providing new teaching methods for the education industry at the same time, three-dimensional printing technology can be applied to most disciplines as a general technology, including various types of education such as formal learning, informal learning and training. It is especially suitable for the related fields of design and engineering or for the areas where there is a demand for rapid prototyping [5],as shown in table 2. Except for the convenience of assistant teaching for various disciplines, three-dimensional printing technology is also conductive to stimulate and promote students' innovative design. More importantly, the mastery of three-dimensional printing technology will exert a far-reaching impact on their technical literacy and future career development [6].

Table 2. Application of Three-dimensional Printing in Teaching of Various Disciplines

Subject	Majors and fields	Application method
Historiography	Fields including archaeology, cultural relics, museums, etc.	Help in the restoration of historical relics, the reproduction of fragile articles, the production of display models, etc.
Natural Science	Fields including Mathematics, Physics, Chemistry, etc.	Production of three-dimensional geometric blocks, molecular models, etc.
Engineering	Fields including architecture, aerospace, electronics, machinery, etc.	Production of scale model, rapid prototyping and replacement parts as well as small batch trial production, etc.
Agronomy	Fields including forestry, zoology, aquatic products, etc.	Production of animal and plant models, garden sand tables, etc.
Medicine	Fields including basic medicine and clinical medicine, etc.	Production of models of molecules, viruses, organs, etc.; Processing of artificial joints and organs, etc.
Management	Fields including real estate development and management, land resource management, etc.	Production of various sand table models for properties, terrain, city, etc.
Art	Fields including animation, sculpture, design, etc.	Production of 3D models of works, design prototypes, etc.

5. Three-dimensional Printing and Product Design Education

Practice teaching in product design teaching takes teachers putting forward practice projects and leading students to carry out design practice training focusing on the project as its major pattern, while low-cost three-dimensional printing technology promotes the popularization of rapid prototype design schemes in design practice. Before three-dimensional printers being introduced into practice teaching, CNC equipment was mainly adopted in rapid prototyping of product design. However, for all that such equipment is capable of realizing the production and processing of design samples, it is expensive and the cost of its consumables is relatively high as well. In addition, competent professionals are required to solve the related technical problems arising in the transformation from CAD to CAM of design schemes, while the operators of control equipment in processing are also expected to undergo special training before dealing with complex technical parameters. Hence, in spite a number of colleges and universities have invested a lot in their product design teaching practice in the early stage to complete the relevant supporting system, only a few colleges with sufficient commercial project support and favorable communication with production enterprises are capable of completing the whole teaching practice process. What's more, resulting from the restrictions of engineering literacy of product design students, they tend to withdraw from these complex practice processes, while the high cost also leads to rapid prototyping practice eventually become a link of internship, in which most of the students' design schemes for practice teaching stay merely at design drawings.

Having been devoted to the research of three-dimensional printing technology since 1992, the Laser Rapid Prototyping Center of Tsinghua University is the earliest institution committing to the field in China, accompanied by the subsequently gradual carrying out of relevant researches in other areas of China. Purchasing related equipment in 2010and starting the exploration of the application of related technology in the teaching practice of product design, relevant practice of the university has been promoted by virtue of the relatively mature and cost-effective FDM technology.

The introduction of rapid prototyping technology supported by three-dimensional printing enriches the link of practice in product design teaching, while it is especially worth noting that it enables the design scheme to break through the boundaries of the drawings and allows students to carry out a series of practices such as geometric model, rapid prototyping and sample trial production under the guidance of teachers. Based on this, we make the following amendments to the design practice process in teaching,as shown in Figure 1:

Figure 1. Design Practice Process before and after the Introduction of Rapid Prototyping in Teaching

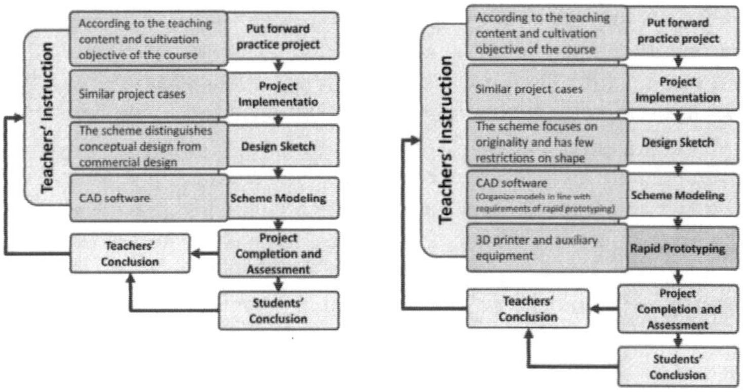

Firstly, the practice of rapid prototyping is added. A number of advantages of FDM technology enable students to complete simple design prototype in limited teaching time, which focuses on the overall pattern and basic structure without the demand for surface treatment. It is mainly applied to the research of structure, the optimization of schemes, the revision of design and the assistance of evaluation of design scheme upon the completion of practice projects. Digital design is adopted in the whole process of practical design and guidance of teachers and students, merely requiring that the digital and analog demand of rapid prototyping is taken into account in the process of CAD.

Secondly, the innovation of product form will be more concerned by the design schemes. The convenience of three-dimensional printing in complex modeling achieves a significant reduction in the energy investment in engineering drawing and the favorable workability of product modeling in design practice. The controversial patterns or nodes can be presented in the form of physical objects by designers resorting to three-dimensional printing technology, so that the design schemes can be compared through all-round perception and finally enables to select the design scheme which are difficult to be realized in drawings.

Furthermore, the practice of rapid prototyping is conducive to the cultivation of students' favorable engineering habits. Compared with traditional processing, despite that rapid prototyping is subject to fewer engineering constraints, plenty of engineering problems still have to be taken into account in the process of converting the design from paper to physical objects, such as the specific scale of design schemes, the human-machine relationship of design components, the coordination methods between different components, the operation mode of simple mechanisms, etc. Merely focusing on the visual proportionality of effect diagrams in the past, the introduction of rapid prototyping will require students to carry out the construction of design schemes as well as the study of product nodes and detailed structures in line with the actual size of products since the initial stage of design, which indicates that the design schemes will go beyond the appearance [7].

Finally, rapid prototyping achieves the promotion of design evaluation accuracy, design guidance and the materialization of design schemes. In the design evaluation, the comparison between object and drawing as well as the combination of effect diagrams and rapid prototyping demonstrate the state of nodes and mechanisms, so that the accuracy of design expression can be improved by comprehensively exhibiting the design schemes. By right of their convenience in storage, exhibition and research, objects of rapid prototyping is capable of facilitating the communication between students and assumes a considerable guiding and reference significance for the subsequently design practice of the similar type.

6. Conclusion

By virtue of the incomparable advantages of traditional manufacturing methods, accompanied by the rapid development of related equipment of three-dimensional printing technology are new opportunities brought to product design education. In addition to materializing students' design practice, the technology can also enable designers genuinely percept their creativity and conduct the experience and evaluation in person of their design. Considering the inevitable shortcomings and limitations of the current three-dimensional printing technology itself, its subordinate status in product design education has to be clarified. That is to say, three-dimensional printing technology is merely a path to realize the rapid prototyping technology and an approach providing assistance for designers to carry out design optimization and scheme evaluation. In addition, students should be reminded that three-dimensional printing, as the representative of additive manufacturing, should only be regarded as a processing method, while its processing components are not likely to actually replace the mass products due to the constraints of material, surface technology and accuracy. Therefore, it is inappropriate to ignore other processing methods because of it. At the same time, various materials and engineering knowledge should be absorbed in order to realize the continuous improvement of knowledge system. Besides, we are also reminded of the significance of focusing on cutting-edge technology and mastering the pulse of the times by the change of practice teaching modes of product design brought about by three-dimensional printing technology. As an emerging technology going back to dreariness after the uproar, three-dimensional printing assumes a number of advantages worthy to be affirmed. Despite that researches on it have never stopped in the relevant fields so far, there is still a long way to go before it achieves the genuine counter-balance with traditional manufacturing. The impetuous and speculative attitude as well as blindly exaggeration and promotion of some capital in dealing with it will only lead to the opposite effect.

Acknowledgment

Key Project of Education Research of AnHui,Research on the Influence of 3D Printing Technology on Contemporary Design Education，SN：2015jyxm336.

References

[1] Si,J.N.(2015)Made in China 2025 is the Guiding Principle of Action. China Industry News, (A01)
[2] Ren,W.Y.(2017)Desktop 3D Printer Optimization Design based on FDM Molding process. Hefei University of Technology,HeFei
[3] Zhang,J.(2016) Research on Teaching Design of Cultivating Creative Thinking through 3D Printing. Shenyang Normal University,ShenYang
[4] Huang,W.d.(2014)Additive Manufacturing and Innovative Design.In: High-level Forum on International Development Trends of New Materials in 2014. XiAn.23-27
[5] Li,Q.,Wang,Q.(2013)3D Printing: An Emerging Learning Technology. Journal of Distance Education,08:29-35
[6] Xiong,J.Q.(2017)Research on the Application of 3D Printing in Education.Journal of Nanchang Normal University,09:21-25
[7] Shu,W.,Sun,F.L.(2018)Research on the Construction of "Internet Plus" Innovation and Entrepreneurship Education Platform based on Studios of Product Design.Journal of Jiamusi Vocational College,10:385-386

The Application Research of Mortise and tenon structure in Cultural and Creative Products

Wenming Liu [1], Jiajun Leng [1*]

[1] School of Design and Art, Shenyang Jianzhu University, Shenyang, LiaoNing, ,110168, China

*Corresponding author's e-mail: 3601925429@qq.com

Abstract. This paper takes the form of Mortise and tenon structure as the entry point, analyzes the structural characteristics, style and use advantages, studies the application of Mortise and tenon structure in Cultural and Creative products design, discusses the development of the following products in current and Development issues such as shape, function, and practicality of future products. By combining the Mortise and tenon structure in the design of the product design, the application of Mortise and tenon structure in the product design should follow the rules and principles of the design in order to realize the application and research combination of the Mortise and tenon structure. It provides reference for the design practitioners in the application of Mortise and tenon structure and product design.

1. Preface

Since the development of product design, the Mortise and tenon structure is the essence and foundation of classical architecture and furniture in China. The unique connection structure method makes China's architecture and furniture have unique structural styles and characteristics in the world. Mr Liang Sicheng once said: "No matter which ancient city building, or the cornerstone of the foundation, there is history in virtually telling and even singing about the changes of time. "Architecture is the history and life of a nation, and structure is the foundation of architecture. The structure of architecture and furniture created by the experience of life and the crystallization of wisdom, which is closely related to people's lives, is the most intuitive and true way to understand this nation. The research and mastery of Mortise and tenon structure is of great significance in the research of product design that seeks the characteristics of Chinese culture.[1]

2. Mortise and tenon structure

The Mortise and tenon structure refers to a combination of bumps and joints used on two wood members. This is the main structural form of ancient Chinese architecture, furniture and other wooden equipment. Due to the high requirements for the material processing accuracy of the Mortise and tenon structure, it has not been commonly used in other fields except construction and furniture.[2] Nowadays, with the continuous updating of processing equipment and the continuous advancement of process technology, The components and products of the Mortise and tenon structure are more and more well known and applied.

Mortise and tenon structures can be classified into three types:

2.1. Surfaces of objects joint:

The Surfaces of objects joint is mainly through the two or more surfaces joint of different objects, or two edges of the joint, or the surface and the edge of the joint. Such as 'rabbet tenon' (pic.1), 'dovetail tenon' (pic.2).

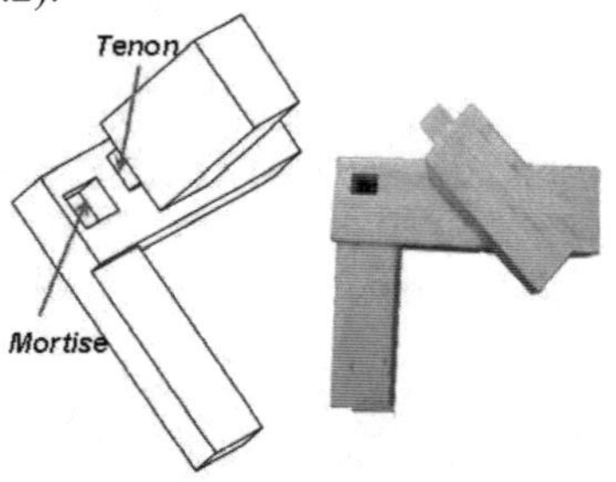

pic.1 rabbet tenon

pic.2 dovetail tenon

2.2. Point joint

Point structure joint is mainly used for t - joint, corner joint, cross joint and extension joint of straight and arc wood. Such as "transparent tenon" (pic.3), etc.

2.3. Component combination

There is also a kind of construction method that combines three components and links them together. This method is more complex and special except for using some of the mortise and tenon joint structures mentioned above. Such as the common "horse sales" (pic.4), etc.

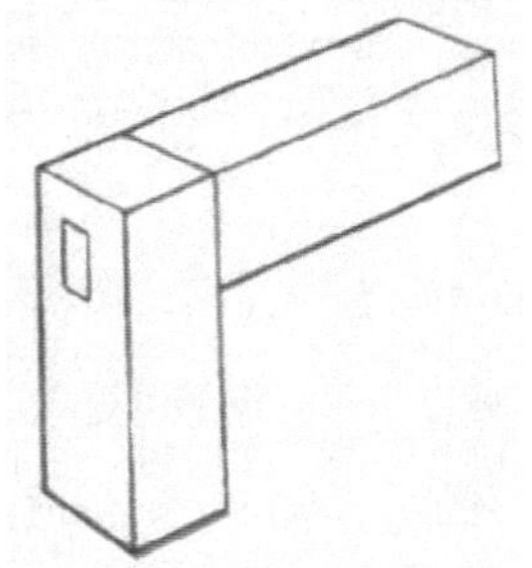

pic.3 transparent tenon

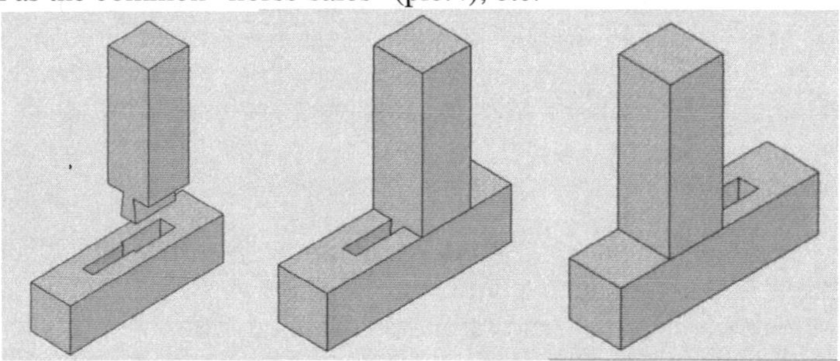

pic.4 horse sales

Through the above research, it can be found that the mortise and tenon structure has strong pertinence in the joint mode, and different applications have the corresponding joint mode.[3] Therefore, under the corresponding conditions of use, the product presents a unique shape and style. The application of mortise and tenon structure in product design provides a variety of joint modes. Increased the flexibility and pertinence of mortise and tenon structure products greatly.

3.The advantages of tenon and tenon structure using in product design

By studying and considering the characteristics of mortise and tenon structure, the mortise and tenon structure has the following advantages in product design:

3.1. Modular design

The biggest characteristic of mortise and tenon joint structure is not to need to use other affiliated nails or glue, but by using the principle of mortise and tenon joint natural joint way objects can be strong together, each fittings has independent function, geometric connection with a consistent interface and consistent input and output interface unit, in a series of products can be reused and swaps, related modules of permutation and combination to form the final product. Through the combined configuration of modules, products with different requirements can be created to meet customers' customized requirements. The reuse of similarity can simplify the procurement, logistics, manufacturing and service resources in the whole product life cycle, facilitate the update of worn parts in use, and extend the service life of products. On the one hand, modular design can shorten product development and manufacturing cycle, increase product series, improve product quality, and quickly respond to market changes; On the other hand, it can reduce or eliminate the adverse impact on the environment, and facilitate reuse, upgrading, maintenance and disassembly, recycling and disposal of discarded products.

3.2. Structural modeling innovation

The components of mortise and tenon are various in types and shapes. The rational application of components has a strong combinability. Different components can be combined flexibly. Many fittings have similarity in connection elements, it also makes the similar products on the premise of meet the use function on the modelling with the possibility of a variety of style, can highlight the fittings in product structure makes the product on the modelling is novel and chic, out of the modelling of products in the people's cognitive habits can overlay the features of the product at the same time the potential function of mining products, even in today's increasingly improve the processing technology, structure of mortise and tenon joint fittings through the improvement of the processing is simple, fast and keep the traditional structure inheritance and conforms to the contemporary people's aesthetic pursuit, Breaking through the limitations of previous modeling, it also has the function expansion and transformation to adapt to people's needs for new life products and personal preferences for personalized improvement design.

3.3. Green design

The structure of bismuth is mainly wood, and the science of structure can be revealed by the characteristics of wood itself. With the continuous maintenance of materials maintenance and construction technology, material processing technology is becoming more and more perfect. Many old wood materials appear in people's lives. The processed concrete components can be used as secondary processing of the structural parts of the product, which also fits the trend and principle of sustainable design. In addition, in modern product design, whether in terms of product form or functional structure, it should be consistent with the selected materials, and use the characteristics of natural products to follow the natural laws. Fully consider regional environmental factors, save energy, reduce pollution, and fully integrate natural materials with design ideas, show products that are integrated and harmonious, and show environmentally friendly green design concepts.

4.Application analysis of mortise and tenon structure combined with cultural and creative product design

Cultural and creative industry is an emerging industry with creativity as the core under the background of economic globalization, which emphasizes the development and marketing of intellectual property rights through technology, creativity and industrialization by individuals or teams depending on main culture or cultural factors.[4] Advanced cultural product design has entered an era that emphasizes more on personality, culture and function. People begin to consciously seek for cultural products that can reflect their cultural identity, the evolution and development of folk customs and national culture. Master graphic design master hara kenzai once advocated "defamiliarization of known things is a kind of creation". Abandon the existing product solidification thinking, start from the product essential function, and then shape its image. Restore the product to its original attractive appearance in the "material" dimension. In fact, digging into the roots of different perspectives, such as nationality, regionalism and culture, is an in-depth study of a nation and a civilization.

Nowadays, there are a lot of cultural and creative products combine their functions with tenon and mortise structures skillfully in design. Combining mortise and tenon structure, the product design is more durable and novel than similar products in terms of modeling and installation. Works as the "design culture" in a card case design, for example (pic.5), the product is very clever combination of mortise and tenon joint structure and a business card case, refining the function of the card case, in the design adopted the sycee tenon mortise and tenon joint structure (dovetail) way to split, will be split into two responsible for serving the card's box. Use the weight of the tenon to replace the function of

IOP Conf. Series: Materials Science and Engineering **573** (2019) 012006 doi:10.1088/1757-899X/573/1/012006

the clamp of the business card to prevent the business card from slipping. The whole convenient installation and disassembly, suitable for the use of all age groups, dovetail tenon trapezoid design and interface interlocking, in the transverse can make the business card structure will not produce displacement resulting in product structure broken, component fall off and other problems. The two components that hold business CARDS are designed to be portable and easy to handle. Product parts easy to update, extend the service life of the product, with the series of product parts can also carry out replacement, combination, simple design is also convenient for future product function upgrade, exquisite and practical.

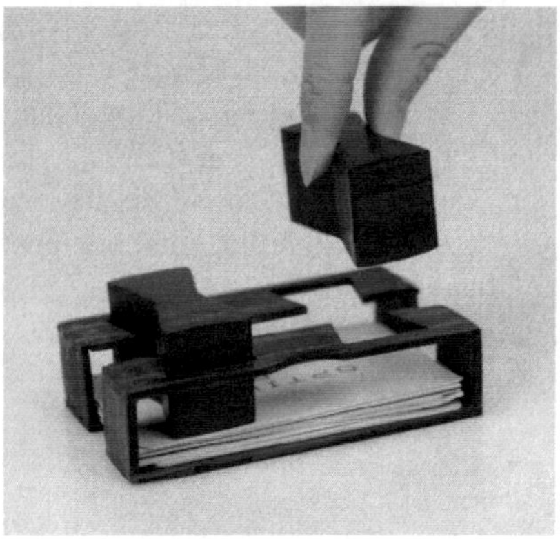

pic.5 card case design

"NUDE" (pic.6), which has won the red point award from Germany, the golden point award from Taiwan and the red cotton award from China successively in 2014, adopts the component concept of luban lock (yin and yang lock). Lu ban lock is a traditional civil and architectural fixed binding device in ancient China. It is completely supported by the connection of its own structure without nails and ropes. Design adopted three: the NUDE coatrack is short of a total of six after cutting cutting sticks, each on a long wooden stick has two different directions corresponding to the notches of short stick diameter above to connect fully, can make the short stick, for notches below, as a short stick support, products with a short stick as connectors, the connection between each other can have the support of the entire product, oblique slot can increase the force area of sticks, and at the same time change short stick force direction makes the short stick won't because of the heavy compression fracture, six stick after combination mutually share the stress make products more durable, At the same time, the concise components make the product development cycle, parts update, reuse and replacement more convenient, which is conducive to the extension of the service life of the product.

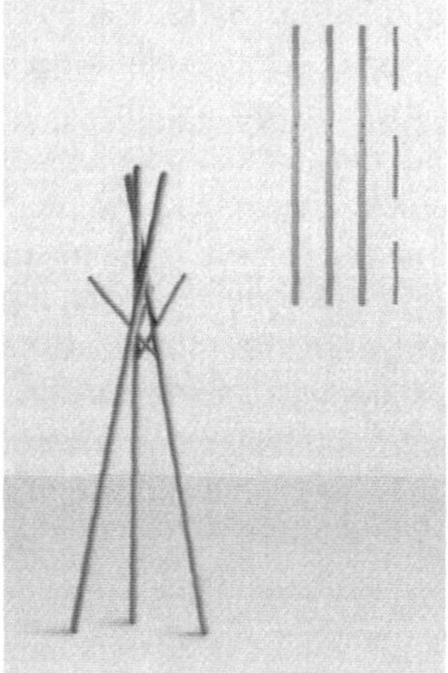

pic.6 NUDE coatrack

IOP Conf. Series: Materials Science and Engineering **573** (2019) 012006 doi:10.1088/1757-899X/573/1/012006

According to the analysis of the above product examples, the application of mortise and tenon structure in product design should consider the following aspects:

First of all, to design a detailed analysis of the features of the product from the problems of existing products, and the corresponding to the solution of the question, at the same time to conduct the thorough research to the functional aspects of mortise and tenon joint structure, with existing mortise and tenon joint structure fitting, a configuration to solve actual problem as the breakthrough point, develop fitting the actual effect to solve the problems need to be solved by products.

Secondly, the design direction of mortise-and-tenon structure products should pursue diversification of products, rather than limit the inherent style and style of mortise-and-tenon structure. Give full play to the advantages of mortise and tenon structure modeling, appropriate to the original structure of the transformation to adapt to the actual needs of the product, the final product should be convenient, simple, practical as the design criteria.

Thirdly, in the design, more consideration should be given to the changes in materials and structures, so as to break through the previous commonly used materials and seek for changes in the texture of products to enrich the changes in the products themselves.[5]

5.Research conclusions

Through the above product case summary, mortise and tenon joint structure in the application of the product, we should first consider the mortise and tenon joint structure in product design which should be aimed at the application of mortise and tenon joint structure mining its internal design ideology, the mode of thinking to solve the problem of the mortise and tenon joint structure to the real application of mortise and tenon joint structure to solve the problems which encountered in the design and creative points. Secondly, in the direction of product design should not blindly pursue "Chinese feeling". The primary significance of product design and production is that it should have functions and can solve the problems brought to users in daily life. Only when it can be truly accepted and used can it subtly make people accept and think about its internal cultural symbols. Therefore, how to make people accept is the essence of product production and design, we should not emphasize too much "Chinese elements", "Chinese style", but should pay more attention to the value of the product itself. Finally, in terms of design, more consideration should be given to changes in material and structure application. Mortise and tenon structure that is in line with product functions should be reasonably selected and the mortise and tenon structure should be redesigned in response to changes in functions. Only by breaking through the conventional thinking mode, can mortise and tenon structure be better used to solve corresponding problems.[6]

To sum up, we can find that the application of mortise and tenon structure in the design of cultural and creative products is not simply the superposition and solution of design, modeling, materials and other problems, but a problem that can be raised to a higher level of design thinking and design philosophy. The application and research of mortise and tenon structure in product design still need to be studied and discussed in the future, which is not only a design process, but also the result of the "re-shaping" of culture.

References

[1] Hara,K.(2006) Designing Design.Shandong people's publishing house,Shandong.
[2] Wang,S.(2003) Ming Style Furniture Treasure.Cultural relics press,Beijing.
[3] Wang,S.(2010) Research on Ming-style Furniture.SDX joint publishing company,Beijing.
[4] Su,Y.(2010) Product form design;.Shanghai Scientific & Technical Publishers,Shanghai.
[5] Lefteri,C.(2017) Materials for Design.Publishing House of Electronics Industry,Beijing.
[6] Brunner,R.(2012) Do You Matter?:How Great Design Will Make People Love Your Company.China Renmin University Press,Beijing.

IOP Conf. Series: Materials Science and Engineering **573** (2019) 012007 doi:10.1088/1757-899X/573/1/012007

The Research on Culture Resources and the Development of Cultural and Creative Industries of Quanzhou "Maritime Silk Road"

Wang Xing-he

College of Cultural Communication, Liming Vocational University, Quanzhou, Fujian, 362000, China

*Corresponding author's e-mail: qzwangxinghe@163.com

Abstract. The regional culture should be the core of the research and development of cultural and creative industries. The resourceful Quanzhou "Maritime Silk Road" culture constitutes a comparative advantage in the development of cultural and creative industries. The proposal of the strategies for the development of Quanzhou "Maritime Silk Road" cultural and creative industries on this basis aims to create industrial value while promoting and protecting Quanzhou "Maritime Silk Road" culture.

1. Introduction

Quanzhou, Fujian Province is the only starting point of Maritime Silk Road recognized by UNESCO and the Pioneer Zone of the 21st Century Maritime Silk Road in the Belt and Road Initiative. With its long history of overseas trade and its own cultureby taking in the essence of different cultures, Quanzhou formed a unique national culture featuring richness, diversity and inclusiveness, and presentation . At the same time, its local government and people pay greater attention to the Culture of Maritime Silk Road in Quanzhou by holding various events such as the 2013 Minnan International Cultural Festival, being elected as a Cultural Capital in East Asia in 2014, Building 21st Century Maritime Silk Road Pioneer Zone and the 14th Asian Art Festival in 2015, 2016 CCTV Spring Festival Gala in Quanzhou, and Maritime Silk Road International Symposium, etc..All these fully proves that the charm of Quanzhou Culture of Maritime Silk Road has gradually become recognized and accepted by Chinese people with cultural consciousness and confidence.

Research on the cultural resources of culture of Maritime Silk in Quanzhou has seen much progress. For example, from the perspective of culture inheritance and promotion, it is proposed that Quanzhou, with its distinct feature of culture of Maritime Silk Road, should strengthen the exchange and academic research in this respect so as to promote the cultural brand [1]; from the perspective of cultural exchange, it is proposed to explore the origin between the Maritime Silk Road and southern opera of Quanzhou[2]. However, study concerning the application of culture of Maritime Silk, such as cultural and creative industry is relatively few. In Quanzhou, a place with distinct cultural resources of Maritime Silk Road, the key to develop and innovate the cultural and creative industry of the culture of Maritime Silk Road is to dig deep its culture and propose appropriate development strategies.

2. Comparative Advantages of Quanzhou Culture of Maritime Silk Road

Quanzhou, as an important starting point of the Maritime Silk Road in ancient China, has many related precious cultural relics, which help to create its unique culture of Maritime Silk Road [3]. Double

IOP Conf. Series: Materials Science and Engineering **573** (2019) 012007 doi:10.1088/1757-899X/573/1/012007

Towers in Kaiyuan Temple and arcades show people the rich Quanzhou Culture of Maritime Silk Road with its distinctive characteristics of the times; religious relics such as Laojun Rock, Qingjing Temple, and Mani Temple fully display the diversity and inclusiveness of Quanzhou culture; Liyuan Opera, Gaojia Opera, and Puppet performances show the refinement and potential of Quanzhou culture.

2.1. Cultural Richness

Quanzhou, boasting the essence of Minyue Culture, Central Plain Culture, Marine Culture and profound culture of its own, is the starting point of the Maritime Silk Road, the first historical and cultural city recognized by the State Council, and the only city in China that has three types of UN intangible heritage [4]. At present, Quanzhou is home to 802 cultural sites under government protection at all levels, including 31 at the national level, 96 at the provincial level, and 601 intangible cultural heritages at all levels. The level and quantity of cultural sites in Quanzhou are among the top 3 in China. Therefore, there is a saying: To see under-ground, go to Xi'an; to see above-ground, go to Quanzhou. Qingjing Temple, one of the four ancient temples of Chinese Islam, and Anping Bridge, the longest stone bridge and masterpiece of ancient bridge in China are among one of the first national key cultural relics under protection. Kaiyuan Temple, the largest Buddhist temple in Fujian Province and the highest stone tower with wood-like structure in China, and Luoyang Bridge, which is known as the First Bridge, are among the second batch of national key cultural relics under protection. In addition, the ruins of Dehua Qudou Palace Kilnand and the ruins of Nankeng Kiln, which stand in the third batch of national key cultural relics under protection. Rich culture relics of Maritime Silk Road provide a unique resource advantage and broad space for Quanzhou to develop its cultural industries. Table 1 is the list of national key cultural relics under protection in Quanzhou.

Table 1.List of National Key Cultural Relics under Protection in Quanzhou.

Number	Batch number	Name
1	The first batch of national key cultural relics under protection	Qingjing Temple (North Song Dynasty), Anping Bridge (South Song Dynasty)
2	The second batch of national key cultural relics under protection	Kaiyuan Temple (Tang Dynasty), Zheng Chenggong Mausoleum (Ming Dynasty)
3	The third batch of national key cultural relics under protection	Luoyang Bridge (North Song Dynasty), Tianhou Palace (South Song Dynasty), Chongwu City Wall (Ming Dynasty), Laojun Rock Statue (Song Dynasty), Jiurishan Cliff Stone Carving (Song Dynasty), Ruins of Dehua Qudou Palace Kiln Site (Song Dynasty to Ming Dynasty), Islamic Holy Tomb (Tang Dynasty), Mani Temple
4	The fourth batch of national key cultural relics under protection	Manichean Stone Carving (Yuan Dynasty)
5	The fifth batch of national key cultural relics under protection	Quanzhou Government Confucious' Temple (Song Dynasty to Qing Dynasty), Ancient Residential Complex of Cai Family (Qing Dynasty)
6	The sixth batch of national key cultural relics under protection	Cizao Kiln Site (Song Dynasty to Yuan Dynasty), Ancestral Hall of Ding Family in Chendai(Ming Dynasty), ShilangHouse, Cihe Tomb (Qing Dynasty), Dejimen Site (Song Dynasty to Ming Dynasty), Zhenwu Temple, Liusheng Tower, Jiangkou Wharf, Ancient Architecture in Quanzhou Port (Song Dynasty

	to Yuan Dynasty), Anxi Confucius Temple (Qing Dynasty), Dunes Site of Anshan, Jinjiang City (Shang Dynasty to Zhou Dynasty), Nan'an Wuta Rock Tower (Song Dynasty), Huian Qingshan Palace (Ming Dynasty to Qing Dynasty), Longshan Temple in Anhai, Jinjiang (Ming Dynasty to Qing Dynasty), Qingshuiyan Temple in Anxi County (Qing Dynasty), Yang's Residence in Tingdian, Licheng District (Qing Dynasty), Lin's Residence in Nan'an (Qing Dynasty), Zhongxian Mansion in Nan'an (Qing Dynasty),Li GuangdiHouse andShrine in Anxi County (Qing Dynasty), Stone Buddha Statue of Xizi Temple in Jinjiang City (Song Dynasty), Stone Buddha Statue and Cliff Stone Carving of Nantian Temple in Jinjiang City (Song Dynasty and Ming Dynasty)

2.2. Cultural Diversity and Inclusiveness

Quanzhou Culture of Maritime Silk Road inherits the Wuyue Culture of the Central Plain, is the advanced regional culture created by the Han people who moved to southern Fujian and the aborigines, and takes the essence of Nanyang culture, Arab culture and Western culture. Thus Maritime Silk Road Culture becomes more thriving and flourishing being an important part of Chinese traditional culture. With distinct local characteristics and rich cultural connotations, it is a wonderful work of Chinese culture with great diversity [5]. The maritime trade during the Song and Yuan Dynasties brought Quanzhou prosperity of "People from different places in the same market" and "Businessmen from different countries when sea rising" in business [6]. Trade development also brought cultural diversity and inclusiveness to culture with the religion as the typical representative. Quanzhou is known as a religious resort with the harmonious gathering of Buddhism, Taoism, Islam, Catholicism, and Christianity, leaving quantities of precious religious relics, making it a spectacle in the history of both Chinese and Western culture. Therefore, Quanzhou is also known as the World Religious Museum and has become the first World Multicultural Exhibition Center designated by UNESCO. Typical representatives include Laojun Rock, Kaiyuan Temple, Qingjing Temple, Mani Hermitage and so on. Distinctive culture of various religions reflects the inclusiveness of culture, offering the idea, form, style and color of cultural and creative products a kind of unique potential, providing rich materials and thinking modes for the development of cultural products.

2.3. Cultural Presentation

Quanzhou Culture of Maritime Silk Road has much to present and perform. Nowadays, Quanzhou has four world-class intangible cultural heritages and 31 national ones. Among them, two are on the List of Representatives of Intangible Cultural Heritage of Humanity, one on the List of Intangible Cultural Heritage in Urgent Need of Protection, and one on the List of Outstanding Practice of Intangible Cultural Heritage of UNESCO. Therefore, Quanzhou has become the only city in China that has three types of intangible heritages recognized by the UN. Table 2 is the catalogue of world-class intangible cultural heritage acquired by Quanzhou. Table 3 shows the catalogue of national intangible cultural heritage acquired by Quanzhou.

Table 2. Catalogue of World-Class Intangible Cultural Heritage in Quanzhou.

Number	Type	Name
1	Representativesof Intangible Cultural Heritage of Humanity	Southern Opera of Quanzhou, Skills of Building Chinese Traditional Wood Structure (Minnan Folk House)
2	List of Intangible Cultural Heritage in Urgent Need of Protection	Skills of Building Watertight Compartment in Ships

IOP Conf. Series: Materials Science and Engineering 573 (2019) 012007 doi:10.1088/1757-899X/573/1/012007

3	List of Outstanding Practice of Intangible Cultural Heritage	Training Program for Fujian Puppet Show Inheritance

Table 3. Catalogue of National Intangible Cultural Heritage in Quanzhou.

Number	Type	Name
1	Folk music	Southern Opera of Quanzhou, North Tube of Quanzhou
2	Folk dance	Chest-beating dance in Quanzhou
3	Traditional drama	Liyuan Opera, Gaojia Opera, Puppet Show (Quanzhou Pickup Puppet Show), Puppet Show (Jinjiang Puppet Show), Dacheng Opera, Gaojia Opera (Ke Style)
4	Folk art	Hui'an Stone Carving, Lantern (Quanzhou Lantern), Yongchun Paper Weaving, Bamboo Weaving (Anxi Bamboo and Rattan Weaving) (Extension)
5	Traditional craftsmanship	Dehua porcelain firing technique, Paper-cutting (Quanzhou <Li Yubao> Paper Cutting), Puppet Head Carving (Jiangjia Puppet Head Carving), Skills of building Watertight compartment in Ships, Oolong Tea Making (Tie Guanyin Making skills), Skill of Building Minnan Traditional Houses Building (Quanzhou residential buildings), Skill of Building Traditional Residence in Minnan (Hui'an Traditional Architecture), Skill of Building Traditional Residences in Minnan (Traditional Residences in Nan'an)
6	Customs	Hui'an Women's Dress, Lantern Festival (Quanzhou Lantern Festival Custom), Lantern Festival (Lantern Customs in Dongshi, Mintai), Lantern Show (Baba Latern in Yingdu, Nan'an), Xunpu Women's Customs, Dragon Boat Festival (Lotus Gathering in Anhai), Dragon Boat Festival (River Crossing Customs in Mintai, Shishi), Folk Beliefs (Qingshui Master)
7	Traditional medicine	Health preserving of traditional Chinese Medicine (Lingyuanwanying Tea)
8	Folk literature	Story of Chen San and Wu Niang
9	Acrobatics and Athletics	Wuzu boxing, Quanzhou Lion Dancing

Quanzhou, renowned for its operas and dramas, has formed a long-standing mechanism for the exchange between ancient operas such as Southern Opera of Quanzhou, Liyuan Opera, Gaojia Opera and Puppet Show with the operas along the Silk Road. Quanzhou is also renowned as The Hometown of Drama, The City of Puppets and The Capital of Southern Opera. For example, Southern Opera, the world-class and national intangible cultural heritage, is closely related to the Maritime Silk Road. It kept the Quanzhou dialect in the ancient Chinese language and was influenced by Persian culture, making itself the crystallization of traditional music exchange between China and foreign countries [7]. It is the cultural and religious exchange between the belief brought by merchants from all over the world who have stepped on the land of Quanzhou for the past thousands of years and the cultures of Quanzhou and Central Plain that have made Quanzhou what it is today. Nowadays, Southern Opera, carrying the love for hometown and families of people overseas, has been sung on the Maritime Silk

Road for thousands of years. Rich cultural connotations, diversified forms of performance, and outstanding display methods are conducive to the development of cultural and creative products.

3. Research on the Development of the Cultural and Creative Industry Based on the Comparative Advantages of Quanzhou Culture of Maritime Silk Road

3.1. Strengthening Policy Guidance

State Council's "Several Opinions on Promoting the Integration and Development of Cultural Creation, Design Services, and Relevant Industries" (Issued by the State Council [2014] No. 10) clearly stated that cultural and creative industry is of great important in promoting economic upgrading, improving living quality, and enhancing the soft power of culture. For example, the value of cultural industry in Quanzhou in 2017 increased by 26.9 billion yuan, a year-on-year increase of 20%, 5 percentage points higher than the GDP growth rate of the same period last year. It contributed 5.4% to GDP, and the proportion continued to increase, making itself the pillar of the national economy of Quanzhou. In recent years, the Quanzhou Municipal Party Committee and the Municipal Government have set up a leading group and a related office for cultural reform and development with Mayor as the leader. It has also issued a series of important policies supporting cultural industry, which has escorted the industrial development in Quanzhou. Strengthening policy guidance with emphasis on supporting projects, implementing the working mechanism of planning, signing, operating, producing, and increasing capital so as to fully promote the transformation of the cultural and creative industry are beneficial to the inheritance and development of Quanzhou Culture of Maritime Silk Road, to rich cultural resources, to strengthen cultural self-confidence, and to be the sustainable development of the cultural and creative industry.

3.2. Promoting Platform Building

It is necessary to build a public service platform for cultural resources and the cultural and creative industry with platform building as the carrier. It is feasible to build a municipal-level public service platform featuring the combination of Research and Development, public innovation, Internet plus, training, and communication, entrust qualified universities to set up courses related to the development of cultural resources and the cultural and creative industry in order to cultivate professional personnel of cultural resources and cultural innovation as well as talents in theoretical research. Meantime, it is necessary to support qualified vocational colleges to strengthen cultural and creative majors so as to cultivate design talents with good cultural and artistic tastes [8], and encourage universities and enterprises to jointly set up cultural research centers and laboratories to improve the R&D of products, promote application, strengthen research innovation and Industry-Academy-Research cooperation so as to facilitate industrial upgrading.

3.3. Promoting Industrial Integration

It is necessary to accelerate the integration of the culture of Maritime Silk cultural resources and the cultural and creative industry by grasping a cross-border integration platform. Enterprises shall be guided to follow a new design mode of "Culture of Maritime Silk Road plus cultural and creative industry", develop products related to this culture, and create a new cultural format. At the same time, it is important to integrate culture with the tourism industry, highlight the hand-made cultural and creative products and tourism featuring the culture of Maritime Silk Road, organize competitions of tourism product design with the Tourism Bureau, and actively develop a new sightseeing tour route so as to create an industrial tour integrating production, display, experience, leisure, sightseeing and shopping.

3.4. Focusing on R&D Innovation

It is necessary to strengthen the mass entrepreneurship and innovation and promote the inheritance, innovation, and development of the Culture of Maritime Silk Road. First, it is important to increase the introduction of new technologies. Advanced processing techniques at home and abroad shall be introduced and new technologies, new materials, new processes, and new equipment shall be actively applied in design, production and marketing of cultural and creative products so as to improve productivity efficiency. Second, it is necessary to guide independent R&D, set up quality testing centers meeting the standard of the industry, and in the meantime, encourage R&D institutions and design companies to develop and innovate together so as to become the R&D backbone of the cultural and creative industry of Maritime Silk Road. Third, it is necessary to strengthen Industry-Academy-Research cooperation, explore the development mode of the "Culture of Maritime Silk Road plus modern technology", and take advantage of universities and scientific institutions to apply new materials and new technologies in the fields of ceramics, carvings and incense products so as to accelerate the transformation of products.

3.5. Increasing Market Expansion and Publicity
It is necessary to strengthen the strategic thinking of globalization, emphasizing both domestic and foreign markets so as to increase market share. First, it is important to go global by taking measures such as organizing overseas exhibitions for enterprises concerning the Culture of Maritime Silk Road, guiding enterprises to make new breakthroughs in the orders of the Canton Fair and Xiamen Cultural Fair, and encouraging companies to set up off-line stores and marketing outlets in large-and medium-sized cities in China to increase domestic market share. Second, it is important to develop new marketing channels. By using the e-commerce platform, expanding new consumption format, and taking advantage of online trading platform and global platform, the mode of "buy global, sell global" can be promoted and market share can be expanded. At the same time, it is necessary to build and promote the brand of Quanzhou Culture of Maritime Silk Road in an all-round way by increasing investment and producing promotional videos for Quanzhou entrusting professional institutions to promote the Visibility of Cities and Enterprises.

4. Conclusion
Quanzhou Culture of Maritime Silk Road boasts rich cultural resources and comparative advantages for the development of the cultural and creative industry. The key to innovate and develop the cultural and creative industry of Quanzhou Culture of Maritime Silk Road is to dig deep and integrate its cultural resources and propose appropriate development strategies. As the cultural and creative industry develops, we should also promote Quanzhou Culture of Maritime Silk Road and enhance the city's reputation.

Acknowledgments
[Fund Project] Quanzhou City College Youth Leadership (Professional) Leader Training Plan; 2018 Intelligent Manufacturing Research Center for Applied Technology Collaborative Innovation Center of Higher Vocational Colleges in Fujian Province

References
[1] Lin,L.Z. (2016) Analysis of construction Quanzhou "Maritime Silk Road" culture. Culture Journal，6：46-47.
[2] Wang,Y.H. (2015) The Origin of the Foreign Musical Factors in Quanzhou Nanyin's "Four Famous Genealogy" and Its Incoming Path. Music Research，5：5-8.
[3] Yu,Z. (2010) Pondering on the Haisi Culture Theme Park in Quanzhou. Decoration， 210：100-101.
[4] Wang,X.H. (2015) On the Creative Tourism Products Design Based on Quanzhou Regional Culture. Journal of Liming Vocational University，2：87.
[5] Baidu Encyclopedia. (2015) Minnan Culture.https: //baike.baidu.com/item.
[6] Guang,M. (2015) Maritime Silk Road and China's five major ports. China Green Pictorial，7：

46-47.
[7] Wan,W. Z.(2012) Singing Method of Nanyinin Fujia. Journal of Quanzhou Normal University，30：23-24.
[8] State Council Bulletin. (2017) Notice of the General Office of the State Council on Forwarding the Chinese Traditional Crafts Revitalization Plan of the Ministry of Culture and Other Departments. http://www.gov.cn/zhengce/content/2017-03/24/content _5180388.htm-.

IOP Conf. Series: Materials Science and Engineering **573** (2019) 012008 doi:10.1088/1757-899X/573/1/012008

The power analysis technique in determining sample size for military equipment test and evaluation

Yun Bo[1], Zhiyuan Du[1] and Xuejun Liao[2]

[1] Graduate School, Space Engineering University, Huairou, Beijing, 101416, China

[2] Space Engineering University, Huairou, Beijing, 101416, China

Corresponding author's e-mail: yunmails@163.com

Abstract. Test and evaluation is a statutory behaviour to verify that certain design of military equipment has met established operational requirements. The determination of its sample size directly affects the scientific rigor and impartiality of test identification. This paper summarizes the basic theory of power analysis, discusses the algorithm of performing power analysis, and some suggestions for parameter setting of power analysis in different applications are given.

1. Introduction

In statistics, power refers to the probability of being able to reject the null hypothesis correctly under the significance of α, usually in the form of (1-β), when the alternative hypothesis is true [1]. Where α and β are the probabilities of the errors of type I & II. Simply put, α, β and (1-β) are the probabilities of "wrong judgment", "missed judgment" and "right judgment" respectively.

The so-called power analysis is a statistical technique to study the relationship between α, β, sample size N, and effect size ES [2]. The effect size ES refers to the actual difference between the true value of the tested index and the comparison value.

Before the power analysis, though impossible to get the true value, ES does exist, and can be estimated by historical or simulation data, so it can be regarded as fixed. Therefore, efficacy analysis generally has three applications [3]: one is to specify α and β before the test to get the required N; the other is to specify α and N before the test to get the actual value of β; the third is to determine N before the test and give the relative importance of α and β, which is generally expressed in the form of ratio $q=\beta/\alpha$, and then determine the appropriate value of α and β.

Because the power analysis provides a theoretical basis for determining the sample size, it effectively enhances the scientific rigor and impartiality of test & evaluation. In addition, test and evaluation, as an acceptance behaviour of weapon equipment design results, is bound to contain a large number of hypothesis testing contents for indicators. Therefore, the application of research efficacy analysis in test and evaluation is of great theoretical and practical significance. In this paper, the single-tailed t-test [4] on the right side is taken as an example to discuss the algorithm implementation of power analysis, study the relationship between various parameters of power analysis, and then give relevant suggestions in practice.

2. Algorithm implementation

Take the single-tailed t test on the right side as an example, when the null hypothesis is true, the distribution curve is the t-distribution [5] probability density curve on the left side of FIG. 1 (when the statistic t equals to the comparison value); otherwise, it is the non-central t-distribution [6] probability density curve on the right side. If α and N are determined, a critical value t_c can be determined by using the t distribution in figure 1, and then the integral of the probability density function of the non-central t distribution less than t_c can be used to calculate β.

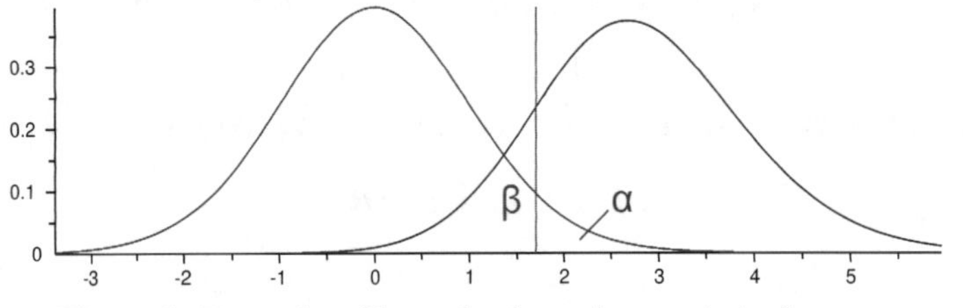

Figure 1. Example of hypothesis testing statistical power

The algorithm implementation of these calculations is shown in the three flow charts in FIG. 2-4.

Note that the meaning of each symbol in the flowcharts is: *μ0*-mean of null hypothesis, *μ1*-mean of alternative hypothesis, *σ*-standard deviation of sample, *N*-sample size, *q*-the ratio of *β*&*α*, tol-error allowed by the program, step-the step size of *α*, pdf-probability density function; and the meaning of "change sign" in the flowchart is that the tested value changes from negative to positive or from positive to negative。

3. Power analysis parameter setting study

3.1. Sample size N and statistical power (1-β)

Using Cohen d [7] as the measure of *ES*, i.e. $ES = {}^{\mu_1 - \mu_0}/_{\sigma}$. When *ES* and α remain unchanged, the statistical power (1-β) is consistent with the increasing trend of sample size N by observing the calculation results in table 1(ES=0.5, α=0.05). For example, reducing the requirement for (1-β) from 0.9855 to 0.8483 can reduce the sample size from 60 to 30.

This is mainly because the smaller the N is, the smaller the non-central parameter [6] is, that is, the closer the distance between the t-distribution and the non-central t-distribution is, and the smaller the N is, the larger the sample standard deviation is, that is, the wider the probability density curve is, which results in the larger overlap coverage of the two distribution curves. If α remains constant, it inevitably leads to an increase in β, i.e., a reduction in statistical efficacy (1-β).

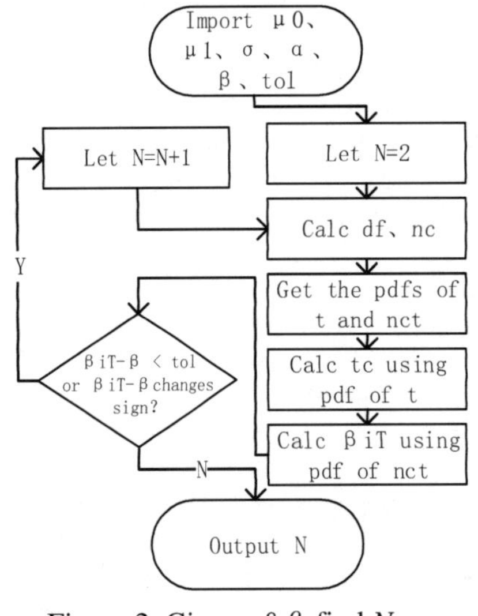

Figure 2. Given α&β, find N

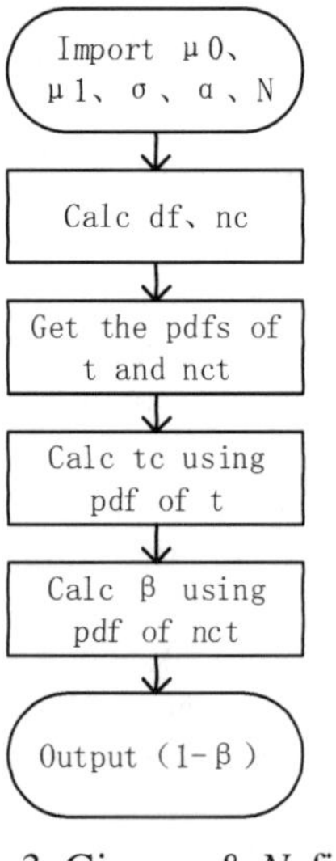

Figure 3. Given α & N, find β

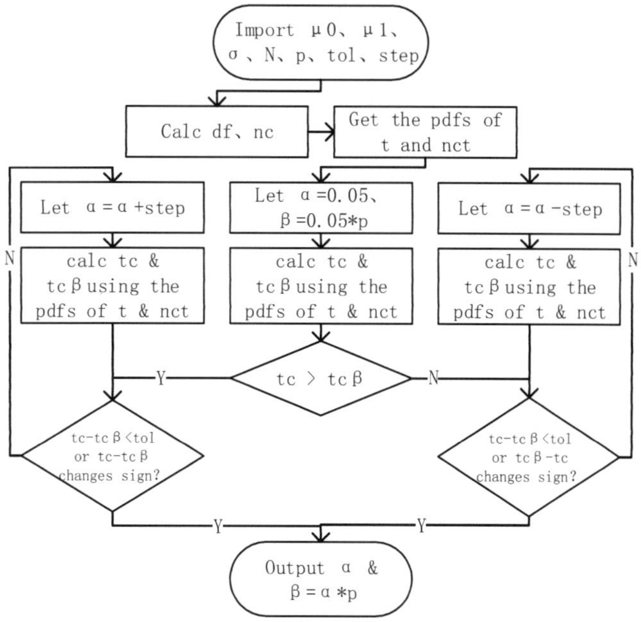

Figure 4. Given N & q, findα & β

However, statistical power (1-β) cannot be reduced indefinitely, as shown in table 1. When N is 15 and 10, (1-β) is only 0.5781 and 0.4273. Take military equipment test & evaluation as an example, if a certain performance of a certain equipment does meet the requirements of the index, but under such a poor power, the probability of "passing" the correct evaluation is close to 0.5, which is almost the same as the decision based on coin toss.

According to statistical practice, 0.8 is generally set as (1-β). By observing table 1, it can be seen that a value between 15 and 30 is appropriate for the sample size N at this time (the accurate value is N=27).

Table 1. The variation trend of (1-β) with N

1 - β	N	Critical t
0.9996	100	1.6604
0.9855	60	1.6711
0.8483	30	1.6991
0.5781	15	1.7613
0.4273	10	1.8331

3.2. Sample size N and significance level α

When *ES* is fixed (*ES*=0.35 in the table), it can be seen from the observation in FIG. 5 that if you want to ensure statistical power as high as possible (in this case, (1-β) =0.8), you can loose the significance level α, that is, increase its value, so as to reduce the sample size *N*. For example, raising the requirement for α from 0.05 to 0.1 would reduce the sample size from 52 to 38.

The reason for this is the same as for 2.2, except that the (1-β) values are fixed and the values of alpha are calculated.

However, α could not also be loosen indefinitely. As shown in figure 5, when N<38, α would be greater than 0.1, which means that if the design under test does not achieve the desired purpose, in 10 tests, there could probably be more than one "passed". In the field of procurement of military equipment, this risk is often unbearable, because they have a bearing on the success or failure of front-line operations and the life and death of soldiers.

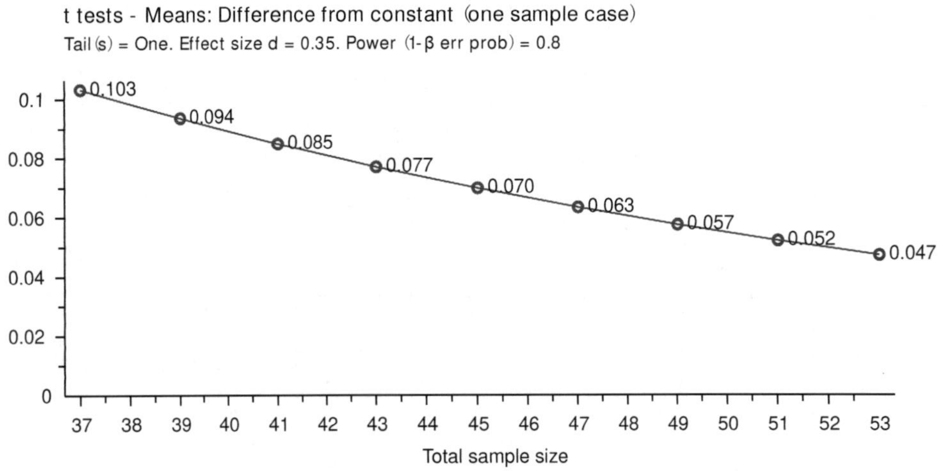

Figure 5. Trend analysis of significance level α with sample size N

3.3. Sample size N and effect size ES

As can be seen from the calculation results in FIG. 6, when there is no reason to relax the significance level α, and statistical power (1-β) is not expected to be sacrificed, but there is evidence that the effect size *ES* is larger than previously thought, then, the demand of sample size *N* can also be reduced. For example, when *ES* changes from 0.5 (medium effect size) to 0.8 (large effect size), the sample size requirement decreases from 27 (rounded up from 26.14) to 12 (rounded up from 11.14).

The statistical principle of this trend is that a larger *ES* will cause a greater deviation between the t-distribution and the non-central t-distribution, which will result in less overlap between the two distributions, and thus a smaller *N* can meet the requirements of alpha and (1-β). Taking the graph of FIG. 1 as example, when *N* is smaller, the corresponding curves of the t-distribution and the non-central t-distribution will be closer and become wider, but the larger *ES* will push them further, thus reducing the influence of the reduction of *N*.

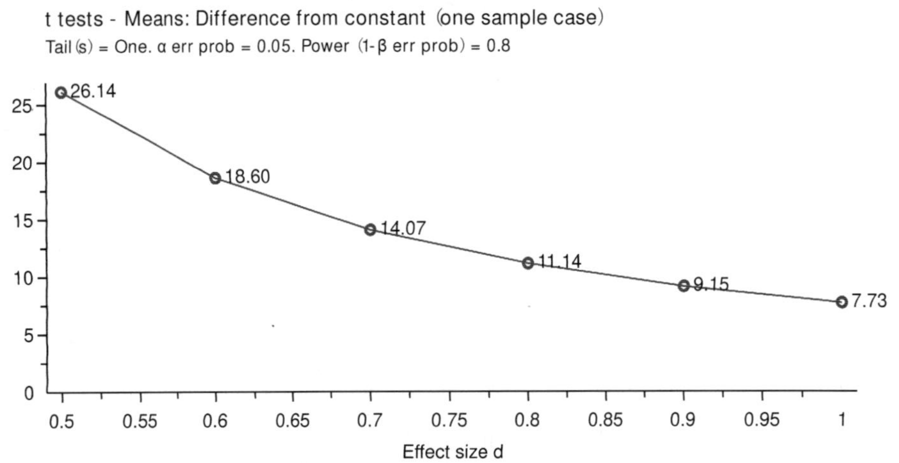

Figure 6. The variation trend of sample size *N* with statistical effect *ES*

4. Suggestions

In order to improve the comprehensive benefits of test and evaluation, it is usually expected to reduce the sample size *N* of the tested equipment as much as possible. However, based on the above analysis, when reducing the sample size N in test, the preconditions shown in table 2 should be considered and appropriate adjustment measures should be taken.

Table 2. Adjusting sample size based on power analysis

The focus of the test	Approach	Application scenarios
To avoid "wrong judgment"	Fix α, increase β	In order to strictly control the quality of the projects with mature technology or wide application in the front line of the army, you would rather miss the design which has met the demand.
To guarantee the "right judgment"	Fix β, increase α	For major scientific and technological innovation projects or projects with forward guidance, if the design really meets the needs, and the risk of wrong judgment is tolerable, try to select it.
Both	Fix both	For projects with relatively mature technology or extensive front-line military applications, under the premise of strict quality control, there is evidence to prove that the degree of performance better than the index requirements is greater than previously expected, so the sample size requirements can be reduced without sacrificing the probability of correct selection.

5. Conclusion

Taking t test as an example, this paper expounds the basic principle, calculation method and basic application of statistical power analysis in the test of military equipment design, the basic way of adjusting test sample size by using power analysis and the precondition that should be paid attention to are given, which can better serve the related work of equipment design and acceptance。

When analyzing the power of F test, chi-square test and other hypothesis test methods, the basic principle is consistent with the content of this paper, and the solution method, application premise and applicable scenarios are basically similar, therefore, in practical work, the content elaborated in this paper can be adjusted appropriately according to the type of hypothesis test to meet the actual needs.

References

[1] Gravetter, F.J., Wallnau, L.B. (2014). Essentials of Statistics for the Behavioural Sciences, 8e. Cengage Learning Asia Pte. Ltd, Singapore.
[2] Wikipedia (2018). Power (statistics). https://en.wikipedia.org/wiki/Power_(statistics).
[3] Mayr, S., Erdfelder, E., Buchner, A., Faul, F. (2007) A short tutorial of GPower. Tutorials in Quantitative Methods for Psychology, Vol.3(2), p.51-59.
[4] Wikipedia (2019). Student's t-test. https://en.wikipedia.org/wiki/Student%27s_t-test.
[5] Wikipedia (2019). t-distribution. https://en.wikipedia.org/wiki/Student%27s_t-distribution.
[6] Wikipedia (2019). Noncentral t-distribution. https://en.wikipedia.org/wiki/Noncentral_t-distribution.
[7] Cohen, J. (1988). Statistical Power Analysis for the Behavioural Sciences (2nd ed.). Lawrence Erlbaum Associates, New York.

Applied Research on "Fallibility" Chinese Character Aesthetics in Package Design

Dong Yan[1*], Dong Yang[2]

[1]Dalian University of Science and Technology, Dalian, Liaoning, 116052, China

[2]School of Fine Arts, Shenyang University, Shenyang, Liaoning, 110000, China

[*]Corresponding author's e-mail: 274261074@qq.com

Abstract. Chinese character is a symbol with the function of information communication, its design thought and method are of widespread significance to package design. The thesis has introduced the "Fallibility" ideas into Chinese character design to analyze the relation between "Fallibility" of Chinese characters and the identification of Chinese characters, and how to shape "Fallibility" features. Visual communication needs to break people's common visual habits to manifest most abundant and profound content in simplified and symbolized form, besides, by extracting highly precise graphic language and easily understandable arrangement order, it can deliver expected connotation. The thesis studies how to use "Fallibility" idea for reference to form directive design method, so as to make Chinese character more flexible and exploit more free space for theoretical research and practice of package design.

1. Introduction

"Fallibility" in Chinese character recognition makes people broadly recognize Chinese characters. As a symbol to bear information, characters are required to be accurate, standard and clear in information communication, but in visual design art, character design shall realize maximally visual differentiation on the basis of being recognized. Modern Chinese character art design strives for design method with open visual appeal, as shown in figure 1.

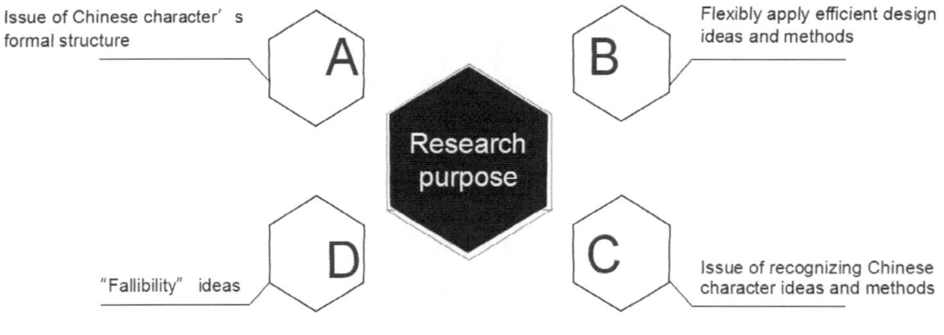

Figure 1. Chinese character design research relation

2. Current status of research

2.1. Current status of domestic research

During the long-term evolution, Chinese characters manifest configurative features like one character in several typefaces and one character in several fonts. On the one hand, Chinese character font presents its diversity and increases visual variation in application, so as to offer various font basis for further art creation. On the other hand, configurative features of different typefaces also reflect predecessors' open and inclusive attitude toward recognition and application. It's believed that Chinese character "Fallibility" can play to the fullest. In Chinese character information structure, by maintaining key and stable part as the feature and adopting the rest parts for transformation, characters won't lose its ideographical meaning no matter how diversely font transforms in application. Therefore, both functionality and artistry can be possessed. Regionally divided, there are mainland China and HongKong, Macao and Taiwan.

2.2. Current status of foreign research
In foreign countries, character design research initiated early, attracting great emphasis and owning diversified methods. In spite of China- the birthplace of Chinese character, countries and districts which have been impacted by Chinese character are from Southeast Asian neighboring China. In history, districts which had set Chinese character as communication language and cultural carrier are mainly Japan, North Korea, Vietnam and so on. From the aspect of visual communication design, as for foreign designers, Chinese character is a kind of graphic language. In other words, for the discrepancy in cultural background and cognitive concept, they directly design and apply Chinese character to produce distinctive visual effect.

2.2.1. Korean character design. Early in 1985, Ahn Sang-soo systematically researched phonetic Korean character from 600 years ago, found the beauty of Korean character and created a Korean typeface named with his family name Hangul. Subsequently, he designed typefaces lee, myrrh and mano, and then provided abundant and practical resources for brand creativity in Korean graphic design, as shown in figure 2.

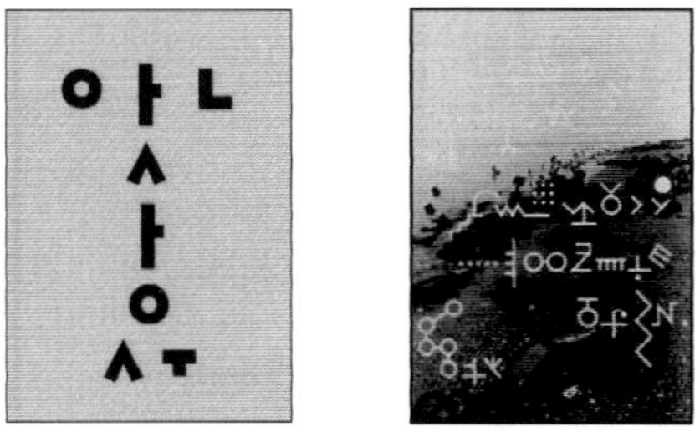

Figure 2. Poster design Ahn Sang-soo (Korea)

2.2.2. Japanese character design. Japan may probably be the country which has adopted the most Chinese characters. Chinese character design in Japan is mainly precise and forms its own series of aesthetic concepts. Shiraki Akira, a Japanese design, has skillfully applied Chinese characters as graphs into graphic design. When designing, he has also introduced "Fallibility" principle into typeface. For design requirements, some parts of character have been exaggerated, the conclusive graphic characters can not only be readable and recognizable, but also be more beautiful in form, as shown in figure 3.

Figure 3. Character design Flower, Shiraki Akira (Japan)

3. Origin of "Fallibility" philosophical idea

K·R·Popper, a famous western philosopher of science, proposed scientific knowledge rule in his Conjectures and Refutations published in 1962: a given problem situation- tentative theory- error elimination- more and more interesting problems, as shown in figure 4.

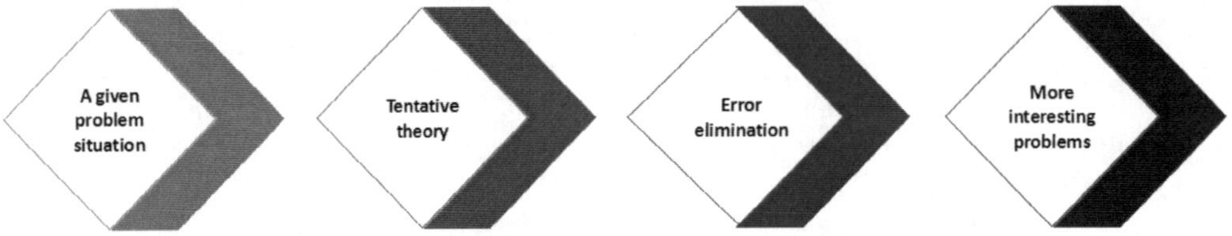

Figure 4. Scientific knowledge rule in "Fallibility" Philosophy

Popper's "Fallibility" idea is of important theoretical value and practical significance, and of vital guiding significance to scientific development. "Fallibility" principle highlights subjective initiative in scientific research process, stresses on connotation, finds out a key point for scientific development, and proposes to hold a scientific attitude of positively learning from error.

3.1. Take a new look at Chinese character design from the aspect of Chinese character font "Fallibility"

"Fallibility" philosophical principle offers a brand new perspective for recognition. It has not only re-positioned "false" opinion and value from ideas, but also provided more beneficial reference to methods. As a scientific discovery, trial-and-error method is rational to a certain extent, in addition, it reflects actual status of scientific research and development, and then leads to methods of discovery and trial. In conclusion, this thinking mode is indeed worthy of reference and study.

3.2. Chinese character development and evolution in the view of "Fallibility"

Diversification of Chinese character font makes it more flexible in design and abundant in visual variation. Different typefaces have different constitutive features, which can present predecessors' inclusiveness and openness toward Chinese character font cognition and application. Chinese character font "Fallibility" can be appropriately played in this thinking mode. In the process of Chinese character development, important parts of Chinese character structure are maintained as features, and other parts are applied to be varied, even being richly varied, the original meaning of Chinese character can still be kept, so the vitality of character can be fully presented from both function and art.

3.3. Relation between "Fallibility" and the developmental evolution of various calligraphies

The evolution from inscriptions on bones or tortoise shells of the Shang Dynasty to inscriptions on ancient bronze objects, small seal style, an ancient style of calligraphy current in the Han Dynasty and regular script is actually the process of Chinese character steps to simplification and abstraction. For instance, after being evolved, an ancient style of calligraphy current in the Han Dynasty has completely separated from graphic feature, broken away from illiberality of Chinese character font combination, besides, Chinese character has transformed from graph to character, from visualization to abstraction. Chinese character font can be divided into parts, and part can be disassembled into strokes. Basic strokes have become the most perfect fundamental elements in character font spatial combination, based on various arrangements and combinations, they can constitute more than 200 parts and then form Chinese characters with different structural relations. Aesthetic elements of Chinese characters lay the foundation for calligraphy. Calligraphy's aesthetic image, theory and artistic spirit are inseparable from false or true, yin or yang and symmetrical arrangement and aesthetic pursuit of Chinese character modeling.The developmental evolution of Chinese character makes archaeologists and litterateurs agree that inscriptions on bones or tortoise shells of the Shang Dynasty is the earliest and most systematic Chinese calligraphy. Restricted by writing instruments at that time, most characters were carved by knife, strokes were simple, neat, slim, averagely distributed, mostly squared and folded, and unvarnished, so subsequent writing brush and knife used for seal cutting have been imposed with a far-reaching impact.

4. Chinese character "Fallibility" idea in package design

In modern Chinese package design, graphic character design adopts basic strokes for design through methods like addition, variation, subtraction， connection and so on, highlights symbolic meaning and commodity feature and attribute of Chinese character, emphasizes on form and decoration, conforms to Chinese character form, corresponds to basic character strokes, extends character imagination and pictograph, so as to make characters easily known, readable, abundant in connotation, and possess various techniques. Although character structure has been arranged, processed, changed and recreated, it's obvious that recognition of Chinese character can not be decreased in package design, for connotative font "Fallibility" and Gestalt effect mutually affect each other. As shown in figure 5.

竹叶青 中国绿茶
ZHUYEQINGTEA

Figure 5. Zhuyeqing Tea Logo by Chen Youjian

If Chinese character elements are separated and recombined into packaged design, the same visual feeling will be generated. In package design of Seiko watch, the designer has directly presented product feature in package, as this type of watch is exquisitely designed, four Chinese character Dong, Nan, Xi, Bei (representing directions like East, South, West and North) are organically combined with numbers like 3, 6, 9, 12 (representing time scale) on plate, it has broken people's existing visual habits and thinking mode, and then bestowed watches with abundant imagination and connotation. As shown in figure 6.

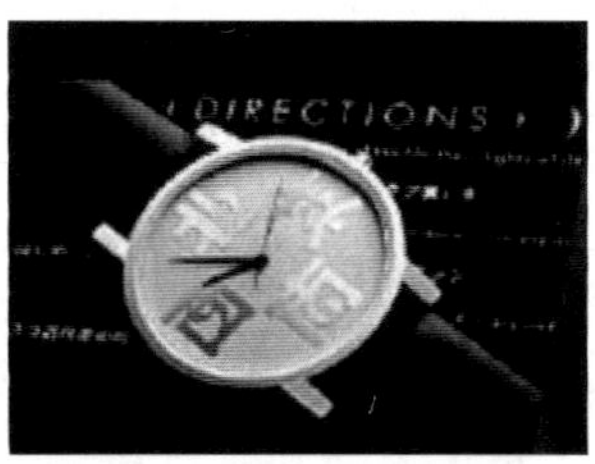

Figure 6. Packaging design of Seiko watch

As Chinese character strokes are well-arranged and varied in both static and dynamic statuses, character layout will be skillfully changed to better conform to the topic, and satisfy with consumers' psychological demands, so a relaxed, humorous and casual visual manifestation will bring about considerate care and understanding to people. In terms of real package design, a kind of typeface won't be directly applied. When designing package for Chinese cultural products, historical brands and local specialty, a certain kind of calligraphy will be directly applied, however, while creating and innovating calligraphy in modern product packaged design, it will generally integrate times feature and product feature for variation, so as to form an enjoyment for consumers to easily recognize and accept, and innovative fonts that conform to product selling. As shown in figure 7.

IOP Conf. Series: Materials Science and Engineering **573** (2019) 012009 doi:10.1088/1757-899X/573/1/012009

Figure 7. Font design in package

In cognitive psychology, illusory effect means that as two or more simulations rapidly present, their feature or parts will be mistakenly recombined23. As shown in figure 8, the package design of Food is the Paramount Necessity of the People(民以食为天) dinnerware adopts illusory effect of Chinese character recognition, by dismantling and recombining "民"（people）and "食"(food), a brand new deconstructed Chinese character is presented. "民"can be seen in "食", and the concept of Food is the Paramount Necessity of the People(民以食为天) is integrated into package design to present diversified psychological imagination and visual enjoyment.

Figure 8. Packaging design (Food is the Paramount Necessity of the People)

In semiotics, entropy is used to present information capacity contained in unit symbol, its unit is bit. Chinese scholars have concluded in data investigation that entropy of a Chinese character is 9.65 bit, far more than those of Latin letter and Slavonic letter in Indo-European language24. Character information capacity has provided more expression methods to designers, endowed Characters more brand information, and burdened greater brand mission, as shown in table 1.

Table 1. Language entropy

Language	French	English	Japanese	Chinese
bit	3.96	4.03	4.81	9.65

Character, part and stroke are highly flexible, and can offer an abundance of possibilities in design. In designing label, logo, poster, book, package and electronic media, as for Chinese characters to be applied, both part and entirety, periphery and connotation can be varied according to actual requirements, recognizable parts can be maintained, and the rest parts can be varied.

5. Conclusion

Cultural in modern package design, design with Chinese character as elements can deliver the most direct and accurate information. Through exploring Chinese character structural form, character design can better combine form, information communication and graphic creation, merge character font "Fallibility" with artistic associations, so as to make information communication more accurate, rapid, appropriate and unique, endow character with incomparable flexibility in design, and expand a diversified space for character's formal creative design. At present, there have appeared some new calligraphy systems which can't be strictly classified into any category, for they are not written according to a certain type, actually, they are created by combining with design environment, principle or written by computer programs or generated by comprehensively applying several methods. A great amount of information delivered by Chines characters can offer designers more design methods, so

Chinese character definitely deserves to be the best element in Chinese package design. The existence of Chinese character recognition "Fallibility" makes character "Vague" in recognition. By means like deformation, recombination of two different graphs, reversion, simulation, decomposition and reconstruction, addition, graph combination and artistic conception, design with Chinese characters as elements is processed in an artistic way. Chinese character font "Fallibility" is well manifested in character design.

References

[1] Chen Yuanchuan, Chinese Character Design, Beijing: China Architecture & Building Press, 2013: 111-121

[2] Li Congqin, Chinese Character and Chinese Design, Jinan: Rong Bao Zhai Press, 2015: 204-220

[3] Kan Tai-Keung, Applied Visual Communication Design, Shanghai Literature & Art Publishing House, 2012: 49-61

[4] Sun Dejin, Research on Chinese Character Teaching for Foreigners, Beijing: The Commercial Press, 2006: 99-106

[5] Chen Zhong, Cognitive Linguistic Research, Jinan: Shandong Education Press, 2011: 118-130

Research on the Innovative Design of Museum Cultural and Creative Products in Liaoning

Dong Yan[1*]

[1]Dalian University of Science and Technology, Dalian, Liaoning, 116052, China

[*]Corresponding author's e-mail: 274261074@qq.com

Abstract. The thesis is aimed at profoundly dissecting issues existing in Liaoning museum cultural and creative brands like weak brand awareness, superficial and homogeneous museum element design, ornament-orientation, weak practicability, inferior product marketing, weak digital products and sales platform, so as to classify and analyze main causes in details. Meanwhile, it proposes to construct museum cultural and creative products with Liaoning regional and cultural features, so as to establish a visual identification system for cultural and creative products, excavate classical symbols for museum collection, recreate cultural elements of museum cultural heritage, emphasize marketing method and develop digital cultural and creative products and online sales platform. So, it offers targeted solutions and suggestions of practical reference value and outstanding application value, gradually, operation philosophy with distinguished features and development ideas of cultural and creative products will be formed.

1. Introduction

Based on cultural resources collected in museums, the development of various cultural and creative products provides an important method for Chinese culture to creatively transform and develop in an innovative way, and then make China Dream and socialist core values win great support from people, besides, it is a vital channel to push forward Chinese culture to the whole world and enhance national cultural soft power, and also an important way to enrich people's spiritual and cultural life and then satisfy diversified consumption demands. Meanwhile, being the necessary requirement of cultural institutions to strengthen their serving capacity, improve serving level and enrich serving content, it is of vital significance to make excellent traditional culture adapt to modern culture and coordinate with modern society, so as to bring forth the new through the old and educate people through culture.

2. Current status of research on museum cultural and creative product design

2.1. Current status of domestic and foreign researches

There are over 3400 museums in China. In May, 2013, Cultural and Creative Products Council of Chinese Museums Association was founded in Beijing. In 2015, the release of Regulations on Museums has indicated the direction for the development of museum cultural and creative products, and provided new power for museum cultural and creative work. In May, 2016, Ministry of Culture, National Development and Reform Commission, Ministry of Finance and National Cultural Heritage Administration jointly issued Several Suggestions on Promoting Cultural Heritage Institutions to Develop Cultural and Creative Products which has created an upsurge of cultural and creative product design among all museums in China. As for museum cultural and creative product development,

IOP Conf. Series: Materials Science and Engineering **573** (2019) 012010 doi:10.1088/1757-899X/573/1/012010

Taipei Palace Museum started relatively early and developed comparatively mature. It's roughly estimated that Taipei Palace Museum has released over 2400 kinds of cultural and creative products, the sales revenue in 2013 approached NT900 million dollars, which could directly compete with its ticket revenue of NT1 billion dollars and has thus become an important income of Taipei Palace Museum. The best seller of cultural and creative products was a tape label with a phrase "I Know" (commonly said by ancient Chinese emperors), subsequently, Jadeite Cabbage- a hanging ornament for mobiles ranked the second, it has been an outstanding feature for Taiwan cultural and creative industry to develop culture into business. In recent years, Beijing Palace Museum has enlarged and strengthened its research and development of museum cultural and creative products, so as to create a series of products with Palace Museum cultural connotation and distinguished contemporary features which are widely favored by people of all ages. Beijing Palace Museum has reached 900 million yuan in the revenue of cultural and creative products in 2014. In the end of 2016, there were more than 8700 kinds of cultural and creative products in Beijing Palace Museum, 8 types of online apps were averagely downloaded by over 1 million times, physical stores have attained sales a highest daily revenue of over 100 thousand yuan, and achieved a total revenue of over 1 billion yuan. What a popular scene.

2.2. Current status of research in Liaoning

Over the past few years, Liaoning museums have also spared great efforts in cultural and creative products, but the effect is yet to be promoted. In Liaoning, there are many famous museums like Shenyang Palace Museum, Lushun Museum, China Industrial Museum, Marshal Zhang Mansion Museum, 9.18 Historical Museum and so on, all of them are abundant in cultural deposits and collect a diversity of resources. In 2014, Shenyang Palace Museum has set up department of cultural industry, in the next year, by adhering to the concept of vivifying cultural heritages, Shenyang Palace Museum have created diversified cultural and creative products for different groups of consumers, released over 100 pieces of cultural and creative products, like Queen's Arrival Series, Qing Dynasty Baqi Wine Series and so on, and established sales sites in museum, but these products have neither attracted much attention nor well sold, no online platform is available at the moment. In July, 2017, Liaoning Provincial Museum hosted an exhibition called Art & Life: Liaoning Provincial Museum Cultural Products Exhibition which consists of three parts like Jade and Dragon Brilliance, Ancient and Contemporary Glamour, Glorious Treasure, more than 90 pieces or sets of independently researched and developed cultural and creative products were exhibited, but many exhibits were not able to be purchased in Liaoning Provincial Museum cultural shops. Cultural and creative products in Liaoning museums are classified as follows, in figure 1.

Figure 1. Classification of cultural and creative products in Liaoning Museums

3. The importance of cultural and creative products on regional cultural and economic development

3.1. Research method of museum cultural and creative products based on design semiotics
Based on design semiotics theory, the thesis theoretically and practically explores on how to innovatively develop cultural and creative products in Liaoning museums. Besides, by combining with theories related to design, semiotics, psychology, aesthetics and being rooted in design semiotics, it studies how to integrate cultural elements of cultural heritages with modern product forms and then forms emotional experience required by consumers' life aesthetics, proposes to construct a design semiotics interpretation model for cultural heritage elements, including semantics, syntax, pragmatics and context, and preliminarily constructs a cultural and creative products design system called innovative design of museum cultural heritage and cultural element recreation.

3.2. Advantage in spreading cultural essence in Liaoning museums
Original and unique design method is adopted to translate and present the most representative museum elements and related culture, so as to attract public to be fond of museum culture. Innovative and scientific means shall be adequately used and combined with industrial development, therefore, cultural resources can be integrated with modern production and life to spread culture, develop industry and increase benefits. As a result, cultural value and practical value can be organically unified to reach the goal of delivering culture essence in Liaoning museums.

3.3. Advantage in promoting popularity and good reputation of Liaoning regional culture
To promote the popularity and good reputation of a region, diversified forms of cultural art are indispensable to appear on various and carriers, like literature, film, music and so on. Raise the Red Lantern sets off the unique tourism brand of Shanxi Merchants' Courtyards. Cultural and creative products play a great role in strengthening popularity and influence of sightseeing places. For instance, a tape label with a phrase "I Know" (commonly said by ancient Chinese emperors) released by Taipei Palace Museum was all the rage in mainland and Taiwan, and impacted the development trend of cultural and creative products in mainland tourism. As a result, a successful museum cultural and creative product can similarly integrate into life to construct a carrier of review and memory for audiences. Furthermore, owing to promotions and sales on various platforms, the popularity and well reputation of Liaoning regional culture can be definitely enhanced.

3.4. Advantage in optimizing museum service experience and satisfying diversified consumption demands.
Motivated by new museum thoughts and cultural and creative industry, museums have turned from emphasis on article management to focus on serving for people. Museums own an abundance of precious collections, through deriving applications, offering creative design and providing service and experience of cultural and creative products, they can not only increase economic value, but also provide audiences quality services. By deeply excavating value connotation and cultural elements of cultural resources, and widely applying various carriers and forms, the research has developed cultural and creative products which integrate artistry and practicability and adapt to modern life requirements, so as to satisfy with diversified consumption needs.

4. Research method of museum cultural and creative product design
The thesis profoundly investigates the development status of main museum cultural and creative products in Liaoning, comprehensively master amount, scale, price, targeted group, product design form (visual identification system and collection symbol application), sales approach and digital research and development of museum cultural and creative products, analyzes favored factors and adverse factors that affect cultural and creative product development in Liaoning and opportunities

and challenges confronted by it, emphasizes on analyzing existing problems, and dissects related causes. Thought of museum cultural and creative product design is illustrated in figure 2.

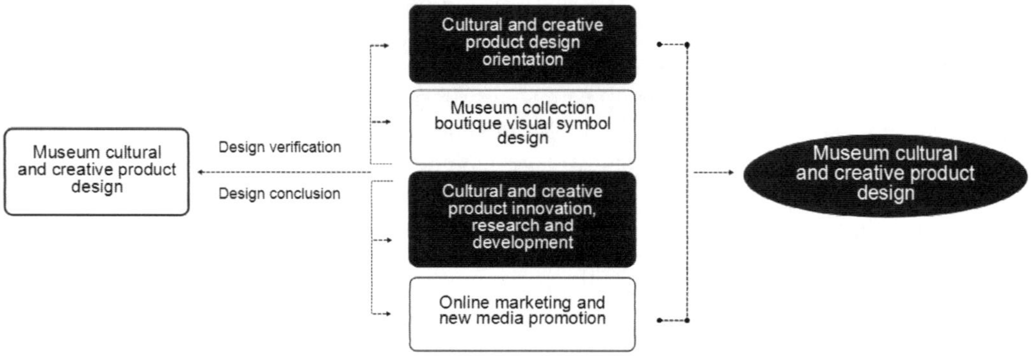

Figure 2. Museum cultural and creative product design

4.1. Construct brand visual identification system for cultural and creative products

At preliminary stage of developing museum cultural and creative product design, it's necessary to formulate design specification brochure and unified package design specification, this is the brand image of a museum. Brand system can integrate museum cultural element resources- scattered and isolated factors, besides, with the systematic support of brand communication, cultural heritage can thus be linked and integrated with cultural and creative industry.

4.2. Excavate cultural symbols of classic museum collections for innovative design

In the field of museum, Louvre will be naturally connected with Mona Lisa, Van Gogh Museum will be linked with Sunflowers. Therefore, representative cultural symbols of museum collections lay the foundation for all design development. By extracting, classifying, reconstructing and recreating cultural heritage elements, cultural connotation can be integrated into cultural and creative products, equiping them with typical museum cultural genes, connotation and prospect, so as to promote their features and competitiveness.

4.3. Cultural and creative products can be more practical through recreating cultural elements of museum cultural heritage

By studying how museum cultural and creative products develop based on cultural heritage elements, elements like ornamentation, appearance and texture are extracted from cultural heritage, after designing and integrating symbols, new cultural and creative products will be made. At present, cultural and creative design of Liaoning museums stress more on history, knowledge and appreciation, but neglect enjoyment and practicability. On this occasion, the development of cultural and creative products makes traditional art and culture truly approach the public.

4.4. Cultural heritage interpretation and innovative design research based on design semiotics

Symbol is a new means of communication, a method to simplify things and objects, and a combination of signifier and signified. The research applies design semiotics to arrange, interpret and discover intrinsic value in cultural heritages, adopts modern design means to apply presentation and image of cultural heritage elements into designing cultural and creative products, so as to solve problems in cultural integration and harmony of function and significance during the process of cultural and creative design process. Through proposing the idea of cultural and creative product innovative design based on design semiotics, the research interpret and discover cultural heritage elements from four dimensions, including semantics, syntax, pragmatics and context.

5. Conclusion

Cultural innovation is rooted in culture, but its freshness lies in innovation. Although Liaoning museums universally pay attention to cultural and creative field and also carry out some practices, the actual consequence still reflect severity and entanglement in various problems which need to be solved. Based on theory and practice, the research exactly points out that Liaoning museum cultural and creative industry shall present its regional features, manifest museums' feature and character, inherit culture, create high taste and serve for life. Meanwhile, relying on related theories like design, semiotics, marketing and aesthetics, together with design practices, the research is based upon regional museum collection resources, discovers cultural connotation in collections, develops diversified, practical, well-designed, digitally applied and featured cultural and creative products. As a consequence, after being creatively designed, lofty museum culture can truly approach the public, more young people will be attracted to focus on traditional culture, and then work hard to create a path for efficiently spreading and inheriting traditional Chinese culture. From the aspect of design and semiotics, the research puts forward the idea that construct brand visual identification system and excavate classic symbols of museum collections. On the basis of brand theory and life aesthetics, it lays emphasis on practicability and enjoyment of cultural and creative products. From the view of new media and marketing, it proposes to release cultural and creative products in museum featured exhibitions, construct online marketing and digital cultural and creative products and so on.

References

[1] Pei Huining, Ma Yu, System Design Method of Souvenirs, Design, 2012
[2] Lou Ying, The Development Strategies of Tourism Souvenirs in Xinjiang, Xinjiang University, 2011
[3] Bai Jing, Research on Development and Marketing Model of Museum Souvenir Products, Nanjing University of the Arts, 2013
[4] Huang Meixian, The Concept and Practice of Cultural Creative Industries Development in Museums in Taiwan, Southeast Culture, 2011
[5] Xiong Ziying, Analytical Research on Designing Museum Cultural Souvenir Products, China Academy of Art, 2015

IOP Conf. Series: Materials Science and Engineering **573** (2019) 012011 doi:10.1088/1757-899X/573/1/012011

Research on national cultural creative products design from the perspective of internationalization

Wenming Liu [1] and Xiaoxiao Bao [1, *]

[1]School of design and art, Shenyang jianzhu university, Shenyang, Liaoning, 110168, China

[*]Corresponding author's e-mail Email: 624831665@qq.com

Abstract. Through the analysis of the development and status quo of cultural creative product design and the analysis of the characteristics of national cultural concepts, the author tries to propose the key points of the national cultural creative product design method from the perspective of internationalization. Using the literature research method, it sorts out the development and current situation of the design concept of cultural and creative products; it also carries out functional classification of cultural and creative products; and it discriminates the concept of national culture; Finally, through the summary of the existing research directions and contents of the combination of national culture and creative product design, it is proposed that from the perspective of internationalization, designers can only accurately convey the image of national culture to users through creative products if they have a deep understanding of all aspects of national culture. Through the grasp of three aspects of emotion, inheritance and interactivity, it can help designers to effectively carry out innovative design of national cultural products from the perspective of internationalization.

1. Introduction

Since entering the 21st century, the increasingly mature modern information technology has profoundly affected the global economic cooperation and cultural communication. The geographically restricted communication methods have been broken, and gradually formed a more extensive "global village." In this context, global trade goods began to become homogenized, and thus began to lose the uniqueness of regional characteristics in terms of recognition. In the field of industrial products, the global production and sales of goods have led to the blurring of the geographical characteristics of the brand, and sometimes it is difficult to identify where it comes from.

2. The development of cultural and creative product design concepts

The design activities of creative products also belong to a cultural creation activity. In the "Bauhaus Declaration" published in 1919, "Industrial design is for people, not for products", which reveals the cultural essence of industrial design. With the advent of the era of knowledge economy, the interaction between culture and commerce, culture and economy is getting closer and closer, the power of culture is becoming more and more prominent, the development of added value of product culture is accelerated, and the proportion of humanization in product design is improved. The unification of product value, aesthetic value and cultural added value has become more and more urgent. Gradually, the idea of product design began to change from the market-oriented to the cultural-oriented connotation, and a culture-oriented design innovation boom was launched worldwide. In 1997, the British government first proposed the development of creative culture industry. British modern culture reveals some contradictory temperament, and design works often contain mixed features of tradition and rebellion. The Japanese government has been actively developing the cultural and creative industries and implementing the strategy of "culture-based" to become the most developed country in the Asian cultural and creative industries [1]. Japanese creative design is usually based on clear functions and simple and convenient operation. In order to effectively meet the growing domestic cultural market demand and promote the development of related industries, the State Council issued a document in 2014: "Promoting the development of new and high-end service industries such as cultural creativity and design services......It is an inherent requirement for the development of an innovative economy, the promotion of economic restructuring and the transformation of development modes, and the acceleration of the transition from 'Made in China' to 'Created in China' [2]." But the overall level of cultural and creative product design in China is still uneven. Designers float on the surface when refining cultural elements, usually just symbolic combination of elements and formal decoration, lacking in-depth exploration of the integration of product functions and cultural factors. Designers float on the surface when refining cultural elements, usually just symbolic combination of elements and formal decoration, lacking in-depth exploration of the integration of product functions and cultural factors. In the aspects of new material selection, mining material semantics, visual texture re-engineering, and forward-looking creativity in the application of popular color science, it shows a state of laissez-faire. Therefore, how to express the profound national cultural connotation through design, enrich the modern design language and enrich people's spiritual world will be the problems we need to face together.

3. Characteristics of cultural and creative products

The design products have practical significance on the one hand because of human use, and on the other hand, they have social significance because of different consumption choices of people. Therefore, in addition to the convenient and comfortable functional attributes, designers must simultaneously consider individual and social, local and global product forms and symbolic meanings in the context of globalization, this requires designers to fully consider the usage situation of the product. According to different usage scenarios, cultural and creative products can be generally divided into the following four categories: holiday gifts and social gifts for social activities; event activities with publicity functions, art exhibitions, derivatives around film and television animation; tourist souvenirs with human characteristics ; exquisite daily necessities and technology daily necessities that improve the quality of life and taste. These products are manufactured and processed in a way that is not much different from everyday products, so we should recognize that the basic attributes of cultural and creative products are products with practical functions. It is what people need in their daily lives and can be industrialized. Second, it has an artistic attribute. Without the value of goods, the artistic value of Wenchuang products cannot be discussed. [3]

4. The Connotation Concept and Characteristics of National Culture

Culture is the sum of material wealth and spiritual wealth created by human beings in the process of social and historical development. It records and inherits the historical civilization of mankind. It is a complete, concrete and historical unity with commonality and individuality [4]. The interpretation of "culture" in "Ci Hai" is that "culture is the sum of the social forms that occur and develop on the basis of certain production methods of material data in history [5]." National culture is the core factor for distinguishing between many ethnic groups. If a nation loses its own unique culture, then a nation will disappear. This disappearance is not necessarily the disappearance of the physiological individual as a member of the nation, but the disappearance of the ethnic group [6]. Therefore, the first nature of national culture is unity, as the crystallization of wisdom formed by the accumulation of time by members of the entire nation, the core content of national culture is bound to be followed by all ethnic members. No matter how far you are in your hometown, you will always show a consistent national identity in your heart; The second is uniqueness, the members of the nation are the creators of national culture. Many ethnic groups live in one place but still maintain distinct national characteristics, precisely because of their uniqueness; The third is binding, national culture profoundly affects the way of thinking and life tradition of national members, has an irreplaceable role in its psychological level, and forms a unified psychological identity, so it has a unified binding force for ethnic members. Even within a thousand miles, everyone accepts the same traditional norms and thus exhibits consistent cultural identity. This kind of constraint only affects members of the nation and does not bind other members of the nation.National culture can be divided into three levels from shallow to deep: Symbol level including national totem symbols, physical symbols,national language and text symbols [7]; Spiritual level including the aesthetic tastes, ways of thinking, and psychological state of people determined by values and ideology; Institutional level including customary systems and rules and regulations. It concludes the national culture as a dynamic system with rich levels and dimensions and continuous development and improvement [8]. The three national cultural levels of symbol, spirit and system are relatively independent and interrelated. They interact and interact with each other and shape the overall image of the nation. The shaping of this image is a kind of cultural capital, which can make the nation unique, form a brand and face the world.

5. The Influence of National Culture on Product Innovation Design

In the new century, modern art design has been more closely linked with culture. The development of art design has become an important cultural issue and social issue. Any art form is based on the ideology, emotions, and aesthetic laws of the nation. For the repetitive and simple return of national culture, we should enter the consciousness of globalization to inherit and use creativity and transcendence.

Design is an integral part of social culture. Different cultures have different spiritual customs and cultural psychological structures or cultural psychological logics, reflecting different values and aesthetic concepts. They play an important role in the design of industrial products, construction, clothing and environmental construction. The influence of national culture on product innovation design is mainly as follows: First, it affects the way designers and the public think: designers are people with a strong local cultural experience, and culture is different because of its unique particularity and situationality. Similarly, users or consumers are also local. Design has never been a purely personal act that can be arbitrarily, from the explicit traditions of materials, tools, language, expression techniques, to the implicit tradition of cultural thinking and aesthetic understanding of design understanding, affecting every designer and consumer. Second, it affects the design principle: the adoption of global culture or local national culture in product or brand design is the primary consideration depending on the group and object. A basic question is, Under what conditions is national culture more effective? Third, it affects the formal system of design: in an increasingly global context, how to develop modern art design activities. In the era of building a new art design culture with national and national cultural characteristics, designers are paying more and more attention to the understanding, application, inheritance and development of the traditional elements of national culture in the practice of art design. The distinctive features and linguistic meanings of many national cultural symbols are increasingly integrated into the concepts and works of art design. Fourth, it affects the evaluation criteria of design: the shape of the product is the time condensation of space, the static expression of dynamic narrative, which conveys the eternal information in an instant form, which is the starting point of product narrative [9]. The purchase and use of national cultural products is actually a process of communication and dialogue with the national culture represented by the products, and it is also a process for users to decode national culture. Only by deeply understanding the natural humanities, historical economy, politics and other aspects of national culture can designers effectively convey the image of national culture to users through innovative products.

6. The purpose of cultural creative product design from the perspective of internationalization

The role of culture in the development of globalization has become more and more prominent. The cultural economy with the production and consumption of cultural concept products as the mainstay of the industry is becoming more and more important. But in fact, companies lack clear and solid theoretical guidance on how to effectively use cultural elements to shape products and brand image, which leads to frequent encounters with Waterloo in practice, and even lead to consumer resentment and resistance. Blindly emphasizing locality, emphasizing national characteristics will lead to difficulties in the interpretation of symbolic systems in the process of product circulation, misunderstanding, misunderstanding and even inability to interpret, not only can not accurately convey national culture, and even lead to misunderstanding or rejection of the symbol system Rejection of its own. Therefore, the design of cultural and creative products from the perspective of internationalization should fully consider the following key points.

(1) Emotional: The essence of creative design is to create and express new meaning and meaning through products. The meaning of the product can only be reflected when the consumer and the product interact and interact with each other. This kind of deep psychological activities and ways of thinking are the connotations that cultural factors give to products. It is also the unique emotional value that the national culture gives to the product. Each user will form a desired goal for a product through a combination of his knowledge, experience, memory and imagination. This goal is not a simple function or form, but a collection of many elements. These elements can be divided into three levels in nature: the outer layer is mainly including material aspects such as color, shape, texture and other factors; behavior layer is corresponding to the operation, function and other factors in the ritual customs; psychological layer is the highest level of consciousness, spirit and affection. Therefore, when designing regional cultural products, designers must thoroughly understand the spirit of national culture, and deeply understand the user's psychological needs, through the combination of modeling, color, function, material, structure and other design elements to meet the user's emotions. Linking cultural characteristics with the user's emotional experience, enabling them to have a strong inner resonance in the process of use and cognition, and communicate emotionally with the product to understand the cultural and spiritual content of the product.

(2) Inherited: The core of the design of cultural products should be accurate and effective design concepts. It's not how good a single product or a few products are, but a systematic design that can be systematically thought through. Whether it is the pre-planning of product categories, the design of mid-term design elements or the setting of post-use situations, it should be a key link in this system. Each link needs to fully consolidate and express the design concept of this product. They are indispensable. For example, in the early stage of product planning, it is necessary to consider the functional categories, size, packaging color and other elements of the product; in the middle of the design, it should be multi-dimensional thinking from the aspects of appearance, scale, and process materials. At the same time, we must also consider the relationship between the user's use situation and the way of use after the product is sold, and guide the user through reasonable reminders. In the traditional Chinese garden design, the designers unify the architecture, the landscape and the interior through a combination of virtual and real, encirclement and sharing, and interspersed with the borrowed scenes, forming a perfect effect of the landscape inside and outside the garden, and the grass and wood complement each other. This is the best example of our systematic product innovation design.

(3) Interactivity: Product innovation design becomes a co-creation model between designers and users in the context of national culture. Under this model, product design is no longer only the way the designer outputs and the user accepts, instead, the designer has transformed into a bridge between national culture and users and has become a “third party”. As an audience of regional national culture, users are no longer individual consumers in the traditional sense, but become a group, which makes the designer's object no longer a personal needs experience into a diverse group of needs. In the process of innovative design of national cultural products, designers need to pay attention to the interests and needs of most people in the national cultural audience, emphasizing the relationship between the group and the regional national culture, and at the same time, the designers need to integrate into the group and hold the culture. The identity and perspective of the individual is co-authored with the audience to propose a more complete, integrated and sustainable design approach.

As a kind of international communication medium based on non-verbal symbol system, designers should have a strong sense of international communication when designing products, and realize the naturalization, rationalization and universalization of national culture based on the specific consumer groups in the international market. This makes the product a medium for self-narrative. However, if the designer puts aside the diversity of culture, it is a misunderstanding and misunderstanding of the national culture to design products with the generalized and homogenous "characteristics" concept. True internationalization is a multicultural style that can be widely accepted by people of different cultural backgrounds. National cultures and characteristics that are overemphasized are often only welcomed as short-term appreciation and hunting. It is possible to truly highlight the national characteristics and build a national image by tapping the "image" of a more universal national culture and absorbing more colorful product forms and vocabulary expressions. If the designer grasps the design concept of national cultural creative products from the above three aspects, it can better consider the user's functional requirements and emotional expectations for cultural and creative products based on decoration and design.

7. Conclusions

Through the analysis of the status quo and characteristics of cultural and creative products, it is recognized that the future design development trend will inevitably be culturally oriented. In order to better study the combination of national culture and design, firstly, the in-depth analysis and comparison of the concept and characteristics of the national culture is carried out, and the influence of national culture on the four aspects of creative product design is summarized. Finally, a brief summary of the elements of creative product design using ethnic culture was made. After the author has conducted more extensive thinking and communication from communication, semiotics, and psychology, it puts forward three aspects of national culture for the purpose of creative product design, hoping to provide design ideas and references for the design of national cultural creative products from the perspective of internationalization.

References

[1] Otmazgin N K. Contesting soft power: Japanese popular culture in East and Southeast Asia [J]. International Relations of the Asia-Pacific, 2008, 8(1):73-101.
[2] The State Council promotes the development of cultural creativity and design service industry [J]. Printing Journal, 2014 (04): 73.
[3] Zhang Yan. Research on regional cultural creation products based on situation construction [D]. Zhejiang Gongshang University, 2018.
[4] Liu Yuan. Research on Several Issues of Advanced Culture [M]. Tianjin: Tianjin People's Publishing House, 2005: 5.
[5] Yuan Jinqiu. From Culture to Ideology: Ideology of Culture[J]. Academic Forum, 2012, 35(09): 153-158.
[6] Ba Yuxi. A Theoretical Analysis of the Differences between National Culture and Regional Culture[J]. Journal of China Three Gorges University (Humanities and Social Sciences), 2014, 36(02): 95-98.
[7] Cao Mengyi. Analysis of the use of traditional national cultural symbols in art design [J]. Caizhi, 2013 (35): 280.
[8] Wei Haixiang. The Construction of National Culture under the Background of Paper Globalization[J]. Academic Research, 2008(02): 114-117.
[9] Tian Longdu. Chinese Design and "Made in China" in the Perspective of International Communication [J]. Packaging Engineering, 2018, 39(20): 43-47.

IOP Conf. Series: Materials Science and Engineering **573** (2019) 012012 doi:10.1088/1757-899X/573/1/012012

Study on Banpo Painted Pottery Patterns in Fabric Pattern Design

Wei Shu[1*], Fuliang Sun[2]

[1] Art school, Huangshan University, Huangshan, AnHui, 245041, China

[2]Art school, Huangshan University, HuangShan, AnHui, 245041, China

[*]Corresponding author's e-mail: shuwei@hsu.edu.cn

Abstract With rich varieties demonstrate ingenious patterns, the decorative patterns of Banpo painted pottery is regarded as full demonstration of the modeling conception and artistic creativity of people at that time. In today's fabric pattern design, traditional Chinese cultural elements are constantly being integrated to achieve people's pursuit of national, artistic and personality of fabrics. Based on the interpretation of patterns in Banpo painted pottery and from the aesthetic perspective of modern design, this paper explores its innovation in modern fabric pattern design by extracting and applying the classical elements of patterns in Banpo painted pottery.

1. Introduction

It is the painted pottery that was originally adopted as ritual vessel in the primitive ceremonies and made ritual vessel of great significance in the ceremonies [1]. In view of the fact that painted pottery is taken as daily necessities by people and its production is to meet the demands of the production, life and agricultural development of human beings, it is bound to the embodiment of people's production and life experience as well as cultural accumulation. In primitive times, people survived by relying on heaven and there was a close relationship between painted pottery and heaven, which endowed the two major functions of painted pottery: Offering sacrifices to heaven and serving as food utensils.

Among the Banpo painted pottery with a history of about 7000 years, the decorative patterns of Banpo painted pottery are considered to be the most distinctive, which is a crucial component of the traditional culture and visual art of ancient human beings. The decorative methods and surface texture of Banpo painted pottery include colored painting, stamping, scratching, stacking and so forth, of which the paintings with color pens and brushes are undoubtedly the dominant, that is, to draw black colors on brown-red (or orange-yellow) pottery fetuses [2]. It can be said that the patterns of Banpo painted pottery are the ancestor's whole sustenance for natural life, social life and spiritual life. Based on the analysis of the classical patterns and artistic charm of Banpo painted pottery, the connection between patterns and modern aesthetics is taken advantage of in this paper to enrich the design language of modern fabric patterns.

2. Classical Patterns in Banpo Painted Pottery

Banpo people observed and understood the ever-changing and colorful world with innocent and childish eyes as well as painted it on potteries with colored drawings [3]. Being concise and with profound implications, the patterns of Banpo painted pottery are adequate reproductions of the spiritual world of human beings thousands of years ago.

2.1. Fish Patterns(Figure 1)

As the decorative patterns with variable forms and compositions in Banpo painted pottery pattern, the fish patterns are mostly in the corresponding form of triangle and arranged counterclockwise, which are either figurative, abstract, semi-abstract, geometric, alone or composite, and are all characterized by vividness, conciseness as well as highly likeness in both forms and spirits.Formatting author names.

The side image of the fish swimming in water is selected by the fish patterns. The head, body and fins of realistic fish pattern are complete while the head is the most life-like. The approximate triangular head, the open mouth, the inside of the mouth and even the teeth are clearly painted. On the contrary, the Abstract fish patterns assume isosceles triangle heads and rhombic rectangle bodies, etc. [4]. Instead of focusing on the in-depth depiction of images, the fish patterns gradually got rid of the realistic style of describing the natural forms of fishes and evolved into the freehand geometric patterns in the later stage, which assume the symbolic meaning of holding things to speech will.

Figure 1. Fish Patterns

2.2. Human Face Patterns (Figure 2)

As the symbolic pattern of Banpo painted pottery, the human faces in human face patterns are mostly round, with eyes open or closed, the upper eyebrows and the lower philtrum are black while the middle part being white. The artistic effect of decoration is enhanced by such contrast of black-white-black colors. In the late Banpo style, it was frequently painted on gourd-shaped bottles or other thin neck bottles. Despite that the face patterns at that time had formed the concise geometric human faces after a long period of exaggeration and deformation, the facial features were still recognizable and emitted a strong artistic flavor.

Figure 2. Human Face Patterns

2.3. Human Face Fish Patterns (Figure 3)

The human face fish pattern is especially prominent in the patterns of Banpo painted pottery, where the human faces are painted as a circle, while above the eyes was painted with a black triangle of "high crown" (perhaps a hairstyle). The ears of the characters were painted with a pair of up-warping hooks and decorated with two opposite fish patterns, while the mouth was composed of two intersecting blank slants, which were also decorated with a pair of opposite fish patterns on both sides [5]. In addition, the front face pattern was also painted on the side of the fish heads, while such strange and mysterious image symbolized "comparing people to fish". Perhaps the human face fish pattern depicted a prehistoric wizard who was carrying a fish in his mouth and praying for a good harvest in both fishing and hunting.

Figure 3. Human Face Fish Patterns

2.4. Geometric Patterns (Figure 4)

Geometric patterns with distinct formal features are both independent and interrelated, both unique in form and profound in meaning. The possibilities of geometric patterns in Banpo painted pottery include: The first is point, of extension of which can produce long point patterns, short line patterns, long line patterns, fine line patterns, thick line patterns, wide band patterns, etc.; The undulation of points can produce curves such as ripple patterns, wave patterns, etc.; The rotation of points can produce S-shaped patterns, spiral patterns, etc. The second is that several major geometric patterns may have evolved from animal patterns. The representative geometric patterns can be divided into two categories, in which the spiral patterns are evolved from bird patterns, while the wavy curve patterns and vertical patterns are evolved from frog patterns [6]. Banpo geometric pattern is an imagery of the natural law of the solar and lunar cycle, such as the wide band pattern as the earliest form of painted pottery patterns. The implication of wide band pattern is extended by carving on it, through which the wide band pattern is divided from one to two and then from two to four, which embodies the worship of nature. Certain formation law of natural forms is followed by the emergence of geometric pattern decoration, while the pure decoration process from complex to simple and then from simple to complex is experienced [7].

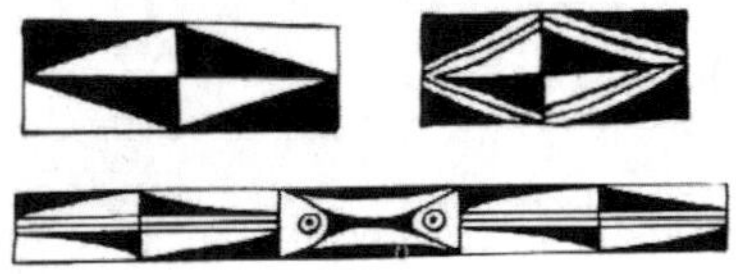

Figure 4. Geometric Patterns

3. Artistic Charm of Banpo Painted Pottery Patterns

Except for being the products of meeting the demand of cultural records and communication, the patterns of painted pottery also demonstrate high aesthetic value and artistic characteristics, which enhance the aesthetic feeling of pottery decoration and promote the development of painted pottery. The artistic forms manifested by Banpo painted pottery patterns not only necessarily contains the unique logic and thought of the ancestors, but also taken on the ability of comprehending and expressing pure aesthetic feeling [8].

3.1. Right Alignment

The two sides continual composition has been widely adopted in the plain patterns on Banpo painted pottery, which shows a clear direction and a fixed trend. The fish patterns almost all present the right-facing phenomena, with the big mouths and big heads to the right and scissors tail to the left; A right-rotating form is adopted by swirling patterns i.e. clockwise rotation; The raised sharp angle of Xiyin patterns also points to the right; The leaf patterns are mainly inclined about 40 degrees to the upper right, while the heads in bird patterns are inclined to the right while the wings and tails to the left. It can be concluded from the above that "right alignment" is adopted in the arrangement of patterns in Banpo painted pottery as the dominant direction, which is considered to be both of amusement and embodiment of the form of patterns full of interest and charm.

3.2. Realism

Realism is to achieve a more realistic and objective reflection of the prototypes through the similarity between pattern images and natural forms, which is also the easiest choice of artistic ways to manifest the real life of society [9]. Fish patterns are depicted concretely in Banpo painted pottery basins with fish patterns, in which the fish eyes, teeth, whiskers, fins and tails are all depicted completely; The proportion of fish is close to the real with few exaggeration. Thin straight lines and triangles are adopted to form the shape of fish while the head-up view of the side perspective is taken, which is a preliminary summary of the natural form of fish [10]. The realism of fish patterns is regarded as the embodiment of the ancestors' feelings for living environment and their expectation of the harvest of fishing and hunting.

3.3. Shadow-writing
Shadow-writing refers to extract the image features and dynamic characteristics of objects with "shadows" while the details of the objects are neglected. The image of the fast-moving deer in Banpo painted pottery was sketched out in a few simple strokes without any details at all. The deer's expression in running is caught, while the necessary features in real life, such as antlers, tails, legs and rounded body, are merely outlined in general. Such generalized and concise artistic processing achieves the proper transmission of the dynamic charm of deer. As a vivid and interesting pattern that transcends the details of the image and reaches the verve directly, the portrayal of deer pattern is the manifestation of both the crisp and sharp painting techniques of the ancients and the highly generalized beauty of the objects.

3.4. Deformation
In accordance with the textual research, the deformation of the patterns of painted pottery is a gradual process from realistic animals to abstraction and symbolization: From realistic fish pattern to geometric pattern; from bird pattern to whirlpool pattern; from frog pattern to wave pattern. There are a number of changes in Banpo fish patterns. Some fins in the fish pattern are simplified into right-angled symmetrical triangles, some mirror fish patterns are abstracted into composite up-down-left-right symmetrical triangles, while some black and white fish patterns evolve into black and white triangles. According to the verification from books, firstly, the deformation of fish patterns reveals the ancient people's formal thinking to a great extent. Formal laws including symmetry, continuity, intersection, change are displayed to meet the demands of visual aesthetics; Secondly, the deformation of fish patterns implies witchcraft thinking. The mysterious witchcraft has been accompanied by human beings since its birth, while the deformation of fish patterns embodies the wizard's subjective thinking of objective things. In any case, deformation is to add new colors to life and thereby is derived from the beauty of chaos.

The decorative patterns of Banpo painted pottery also assume unique charm, including empathy, such as the swimming fish and the running deer depicted on the painted pottery, which places the anticipation of catching more fish and deer on the pottery; Decoration, such as the painted pottery patterns carved on the outer and inner walls of pottery with novel patterns as well as fluent and clean brushwork, which presents obvious decorative characteristics; Symbolization, such as human face fish pattern and human face geometric pattern, which carried a fish in the mouth and the triangle is like a hair bun or hat, which might be the representation of the totem or national emblem of Banpo clans; All these are to express the desire for a wonderful life by transforming the perception of nature into patterns and describing them on articles for daily use.

4. Current Situation of Fabric Pattern Design
In the current fabric market, similar and uniform pattern can be found everywhere, which is mainly due to the popularity of the Internet and the development of computer-aided design software. Digital printing technology has been adopted by an increasingly number of designers, while major websites have also been taken as inspiration for fabric pattern design. The numerous Internet image materials are copied, pasted, imitated and tailored by designers, while these simple and non-creative design elements lead to the vapidity of fabric pattern. At the same time, the design of fabric pattern is also tending to internationalization. Lack of excavation of traditional culture, fashion patterns on fashion weeks are frequently imitated by designers, which result in the single form of design works and the lack of visual elements with appeal and national cultural characteristics.

5. Design of Fabric Patterns of Banpo Painted Pottery Patterns
The formation of a new design concept is frequently inspired by other things, while the expression techniques can be referred, imitated and learned. The similarities between the pattern design of Banpo painted pottery and fabric pattern design include the form of composition, such as realism, deformation, symbolism, etc.,; And the content of pattern, such as patterns of animals, characters,

plants and geometries, which all could be the sources of artistic creation for designers. Achieve the transmission of unique traditional artistic value and aesthetic view of the times by looking for the "basic form" of fabric pattern design in Banpo painted pottery and applying it to fabric pattern design in four manners, namely, color harmony, approximate repetition, scattered point continuity and Skeleton Filling, by virtue of modern computer aided design software.

5.1. Color Composition

As an essential part of pattern composition, color is interrelated with shape and pattern and play its role independently. Color is one of the crucial elements in fabric pattern design and its design determines the success of fabric. Remarkable color design attracts more consumers by virtue of its expressive power, while excellent color contrast design can achieve full transmission of the visual effect of the image. Pop art is regarded as the embodiment of the significance of color design and the exploration of the mystery of colors. As shown in Figure 5, with the basic shape and rectangle as the background, the four fish patterns present a clear primary and secondary, dense and balanced composition with the background, which forms a sharp contrast by adding bright and prominent colors. In terms of artistic form, the manifestation of rich color collocation increasingly highlights the main pattern, such as adjacent color, contrast color, similar color, similar color and so forth. It is not limited to a single way of expression that has made the mysterious fish patterns emit a ardent sense of color and thus enhance the visual impact; In terms of structure forms, four different fish are arranged as the basic unit into two groups in different positions and present regular repetition and continuity, which produces rhythmic beauty. Reasonable application of color design as well as the fluctuation, gradual change, staggering and contrast techniques are all capable of achieving effective improvement of the artistic appeal of fabric patterns.

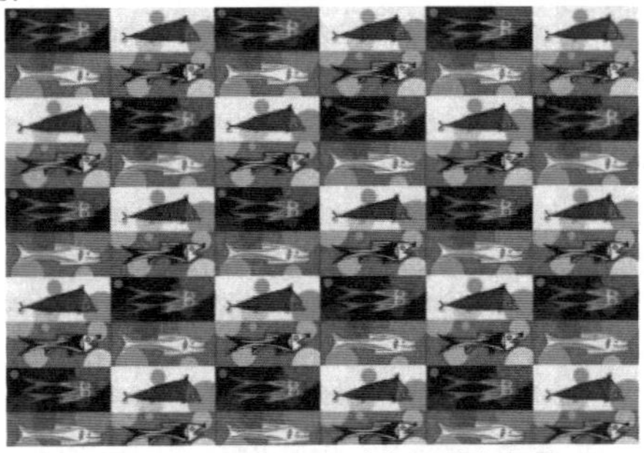

Figure 5. Color Composition: Fish Patterns

5.2. Approximation Repetition

Approximation refers to the similarity and incompleteness of images, which adds interest by repeating a variety of graphics. Being similar in shape, factors such as the color, size and direction can be designed in approximation. A perfect approximation is expected to be the continuous adjustment of the visual elements as a whole; Repetition refers to the arrangement of the same basic form, which is for the purpose of unity and thereby being orderly yet inevitably monotonous. Nevertheless, the human face patterns painted by the ancients on Banpo painted pottery are repetitions of the similar human face patterns. As can be seen in Figure 6, taking the human face patterns in Banpo painted pottery as the original form and stripping it from the traditional context and putting it into modern aesthetics to redesign, in which the ancient design method is adopted and repeated by following the basic shape. Changes and innovations in reshaping, adjusting and color are carried out in the extraction of face patterns to form quite similar graphics for rearrangement and repeated combination, thus giving new vitality and strong cartoon interest to traditional patterns. Approximation repetition of fabric pattern design demonstrates a unified and varied form, which improves the lively and clear effect as well as exhibits the integration of traditional and modern aesthetics, a new idea of fabric pattern design is thereby created.

Figure 6. Approximation Repetition: Human Face Patterns

5.3. Scattered point continuity

Scattered point continuity is a distributed continuous arrangement of dots in accordance with a certain distance and direction in the frame designed with one or more patterns as a basic unit. Without direct relationship with the major pattern, the auxiliary patterns are arranged around the major one in a form of irregular diffusion. As shown in Figure 7, three typical human face fish patterns are taken as the major pattern, while certain shapes of human face fish patterns are filled with graphics and matched with colors in the process of design, which breaks the stereotype of traditional human face fish patterns and forms a color pattern with modern aesthetic taste. Around it, the patterns of triangular, deer and ripple are scattered freely, which are distributed in the form of sparseness, density, non-uniformity and irregularity. The disorderly pattern combination is extended and expanded as a basic unit through the way of continuous formation to the four directions, thus avoiding monotony and rigidity. Another way of traditional pattern in fabric pattern design is created by the distinct major pattern and scattered auxiliary patterns.

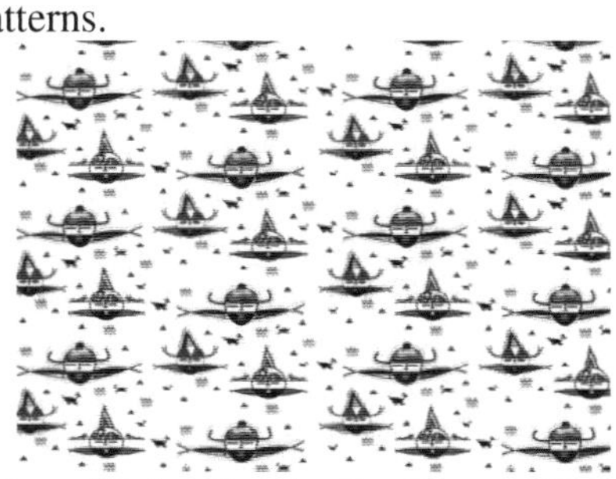

Figure 7. Scattered Point Continuity: Human Face Fish Patterns

5.4. Skeleton Filling

As the design of a set of identical or similar basic forms, the existence of skeleton helps to achieve the internal unify of the design. Skeleton plays a jurisdictional role in the layout of images in design and determines the relationship between images in design beforehand. In the design, the skeleton firstly fixes the position of the basic shape, and secondly divides the picture into several skeleton units of the same size and shape, while the filling plays a coordinating and changing role by enriching the content of the skeleton. As shown in Figure 8, the shape of Banpo painted pottery is taken as the basic form of the skeleton, which assumes the same size, uniform direction and strong sense of order in the arrangement. Rigorous skeleton frames are arranged repeatedly in a diagonal way, which forms regular skeleton while repetition is for unification; After that, the geometric patterns of Banpo painted pottery are refined, in which patterns such as diagonal triangular, corrugated, grid, diamond, etc. are filled into the skeleton by means of two sides and four sides continual The complex and novel picture effect is presented through organization and change. The filled patterns promote the formation of new images through the decoration of the skeleton, which provides greater possibilities for the inheritance of artistic charm of Banpo painted pottery.

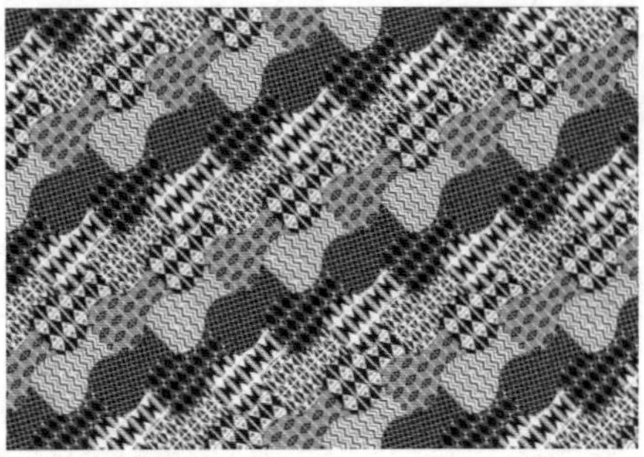

Figure 8. Skeleton Filling: Apparatus Type, Geometric Patterns

6. Conclusion

An increasingly number works integrating both traditional culture and fashion can be created as well as a diversified development path of fabric pattern design can be sought only by deeply rooting pattern design in the soil of traditional art and culture as well as achieving the integration of the essence of traditional culture and art with modern design. In conclusion, apart from inspiring the designer's creation, Banpo painted pottery pattern in fabric pattern design can also be refined the shapes, structures, application modes of traditional patterns, relying on which the redesign can be carried out with modern technology and design methods, which will endow Banpo painted pottery pattern a new visual perception and promote the formation of national and regional Chinese style of textiles through its application in fabric pattern design and integration with fashion trends, thus transferring the charm of traditional Chinese culture to the whole world.

Fund

Key Project of Excellent Youth Talent Support Program of the Education Department of AnHui, Shu Wei, SN：gxyqZD2016298

Research project of Huangshan University (Huizhou culture), Study on the Design of Visual Image System of Ancient Village Tourism Brand—Taking Huizhou Ancient Village as an Example, SN：2014xhwh008

References

[1] Zhang, F. (2003) Chinese Art: Course and Spirit. China Renmin University Press, Beijing.
[2] Weng, J.Q. (2006) Form and Implication: Eight Lectures on Chinese Traditional Decorative Art. Peking University Press, Beijing.
[3] Chen, J. (2009) Chinese Ceramics. Baihuazhou Literature & Art Press, Nanchang.
[4] Song, Y.C. (2015) History of Chinese Fine Arts. Tsinghua University Press, Beijing.
[5] Weng, J.Q. (2006) Form and Implication: Eight Lectures on Chinese Traditional Decorative Art. Peking University Press, Beijing.
[6] Jiang, S.Q. (2007) Brief History of Painted Pottery Art. Shanghai People's Fine Arts Press,Shanghai.
[7] Zhang, K. (2015)Design Reconstruction of Decorative Patterns of Xi'an Banpo Painted Pottery. Modern Decoration, 3: 140-141.
[8] Kong, L.Q. (2003) History of Chinese Ceramic Painting Art. Southeast University Press, Nanjing.
[9] Kong, L.Q. (2003) History of Chinese Ceramic Painting Art. Southeast University Press, Nanjing.
[10] Zhang, P.C. (1990) Atlas of Chinese Painted Pottery. Cultural Relics Press, Beijing.

The Design of Face Check-in Scheme on Conference Industry

Jiankang Zhang[1*], Yuxin Zhou[2] and Xuying Zhang[3]

[1,2,3] School of International Economics & Tourism Management, Zhejiang International Studies University, Hangzhou, Zhejiang310012, China

[*]Corresponding author's e-mail: 88018456@163.com

Abstract. The application of face check-in scheme has a positive impact on the conference industry,which can improve the service quality of the exhibition industry,strengthen the MICE activities site management,and boost the deeper development of the Intelligence exhibition.The API of face check-in scheme provides five interface,which named the conference operation interface (it is used to create and modify meetings),the conference information inquiry interface,the user registration interface,the face verification interface,and check-in status feedback interface.It can be effectively realized the conference forms publishing and managing,and fill in the registration form online,free check-in APP,unified management of check-in data and other functions.

1. Introduction

With the advent of the information era,the exhibition industry develops rapidly.Among them,Among them,the development trend of the conference industry which belongs to the exhibition industry is great.In a broad sense,the conference industry carries out conferences and activities by means of scale,centralization and modernization.It grows with the development of the market economy.In recent years,the rapid development of relevant information technology has had a tremendous impact on the transformation and upgrading of the industry.Currently,the evolution of biometric technology is strong.Face recognition technology has matured.The face check-in scheme is gradually being applied to quite a few activities,such as meeting sign-in,annual meeting sign-up,salon sign-in,training sign-in,event sign-in,class sign-in,conference sign-in,contest sign-in,etc.It achieves high efficiency of staff management and forms detailed statistics.In this context,it is of great practical significance to strengthen the face sign-in exhibition scheme.

2.Basic concept of face conference check-in scheme

The face conference check-in scheme uses zero-cost development throughout the entire process.Provide with the fulfill creation of the conference,conference registrations,on-site face sign-in,data information platform,the scheme can be applied to the organization and management of offline meetings,to improve the management efficiency and user experience of meeting activities.In the event registration process,developers combine information technology such as AI and big data,and release a meeting.Invitation,registration of participants' personal information,sign-in at the meeting,background data information statistics service,and in accordance with to the scenario requirements of the meeting activities,provide a variety of personalized face venue check-in solution,to name only a few,large screen face recognition,gate face recognition and many other check-in modes,to make the conference activities have a better experience.The application of face check-in scheme has a positive impact on the conference industry.First of all,the scheme can improve the service quality of the convention and the exhibition industry.The customer is the core of the service industry,the third pillar industry of the convention and the exhibition industry.The service quality of the traditional convention and exhibition industry in the exhibition activities is relatively deficient.

Combined with the face sign-in scheme,users can better experience the conference activities,to accelerate and improve the progress and efficiency of the entire process of conference activities,so as to improve the service quality of the entire conference,and to improve the service quality of the conference and exhibition industry.There is one more point,the application of face sign-in scheme can strengthen the on-site management of exhibition and conference activities.In the conference and exhibition activities,face sign-in scheme can strengthen management ability.Through its powerful database,the program can analyze the personal information data of each user and effectively figure out the emergencies of on-site management and even the unexpected events of exhibitions.Face check-in scheme is widely used in the conference and exhibition industry,which optimizes and upgrades relevant modules such as conference and exhibition services and management.And it has a greater power to adapt to the trend of wisdom for the exhibition and the conference industry.The continuous development of the Internet + era,and the convention industry gradually developed from traditional exhibition to information,technology and exhibition intelligent,it has greatly promoted the sustainable development of the exhibition industry and its transformation and upgrading.

3.The association function of large-scale network face conference check-in scheme technical interface

3.1 Conference operating interface
The conference operation interface is mainly used for creating meetings and updating conference content.relationship.Calling the conference operation interface can set the content of the conference,such as the conference title,the conference theme and so on,and can effectively manage the prompt copy and background pictures of users who successfully otherwise failed to brush their faces on the free sign-in APP.

3.2 Conference information query interface
The conference information query interface is mainly used to obtain the detailed information of a meeting.The information acquired is relevant to the meeting created when the developer calls the conference interface.It is convenient to grasp the specific situation of the meeting and to deal with the unexpected situation in time.

3.3 Face registration interface
The registration interface is mainly used to receive the pictures of face information uploaded by developers. (Developers can develop their own H5 page for user registrations,insert the face image of acquisition control,and then conveniently call the user registration information collected by the face registration interface.)

3.4 Face verification interface
Face Verification Interface is mainly used to verify the authenticity of the user information and to brush face and check in.Customers need to develop host service to forward APP requests.For Example : the host address format is schema://host[: port#]/path/.../

3.5 Check-in state feedback interface specification
The check-in state feedback interface is mainly used to feedback the check-in status of conference users.The check-in state feedback interface is mainly used to feedback the check-in status of conference users.Calling this interface can obtain the status information of the customer check-in and all the information that the user fills in when registering,understand the general situation of the participants,and play a good role in the management of conference activities.Of course,for the use of

check-in state feedback interface,the organizer can do personalized action such as check-in wall,lottery,etc.to meet the needs of different scenarios of the conference,increase the interaction between participants and meetings,and enhance the user experience.

3.6 Check-in state feedback interface specification
Users who use Face Check-in Active Push Service need to provide some standard data receiving address(the data receiving address specified when the host creates the meeting)

4.Functional of face check-in scheme

4.1 Conference form publishing and management
In the scheme,the organizer can customize the content of the form required for the registration of the event according to the conference activity(for example,the conference title,meeting time,the meeting venue and other conference related content).Then it will be published online and extended to the majority of users.This facilitates the organizers to understand and master the participants' information data in time,and to collect and organize the participants' information data efficiently and conveniently.

4.2 Fill in the registration form online
Participants can complete the entry of the information registration,register information and acquire two-dimensional code by filling in the registration form online.This enables real-time dialogue between the host and the user.Participants register online anytime,anywhere,to close the distance between the participants and the sponsors,adding interactive elements,laying the foundation for the later sponsors to provide a better service for attendance.

4.3 Free sign-in to APP
Free check-in to APP supports a variety of modes of check-in,such as face-brushing check-in,two-dimensional code,etc.Only the iPad (ipad4. mini2) device is needed to complete the check-in operation.The scheme provides various forms of matching check-in choices,and plays a convenient role for participants to verify their identity information,and the quality of service of the organizers has also been improved.

4.4 Conference operating interface
Backstage management forms and check-in data,with a comprehensive improvement of the efficiency of activity registration and check-in data analysis,can better respond to the emergencies of exhibition activities,the first time to respond.It can also predict the future trend of exhibition activities and promote the sustainable development of conference activities.

5.Successful cases of face conference check-in scheme

5.1 Conference operating interface
In 2018,the 14th china exhibition economic cooperation CEFCO conference was successfully held in Qingdao,China.The participants of conference came from 18 overseas countries and regions.In all,about 1000 participants attended the conference.The conference is different from the usual,and the committee of the conference has adopted face check-in scheme to solve the problem of sign-in.This system is jointly designed by Baidu artificial intelligence team and Kunlun,the official registration service provider of the conference,for BBS.

Kunlun has been cooperating with baidu since December 2017,and joined the baidu AI ecological partner program.This plan is applied to the latest generation of "e zhantong" registration system developed by kunlun.And the conference was customized on-site attendees registration,a new generation of supporting hardware self-service face recognition and the certification machine and face recognition sign-in system for registering personal information. This system has been well received by the audience and exhibitors.Not only greatly reduced the waiting time of the audience,but also avoided the bad phenomenon of in the traditional exhibition.Not only greatly reduced the waiting time of the audience,but also avoided the bad phenomenon of in the traditional exhibition.And lay a good foundation for the sustainable and in-depth development of the conference.

5.2 Face check-in scheme for industry conferences

The 2018 China IT leaders summit successfully held in Shenzhen,China.Face recognition is the most important part of the summit's security system,including many emerging black technologies.In this summit,all staff,participants and even relevant media reporters need to apply by email in advance,getting the invitation code from the summit organizing committee,Submit personal information and photos according to the prompts on the page,and then you can enter the meeting site to attend the meeting after you check in by scanning your face ID.Enter the meeting site to attend the meeting after you check in by scanning your face ID.In addition,participants also need to verify their personal identity face information before entering the venue.

In particular,this summit adopted the intelligent conference face recognition system.After the participants arrive at the scene,the system's camera can automatically capture the participants' profile picture,Conduct information comparison,after successful comparison.In the identification window,the word "successful" and the information of participants will pop up automatically,and the relevant certificates of participants will be printed out.Face check-in scheme realizes the unity of people,time and space,which greatly improves the efficiency of the meeting and ensures the safety of summit participants.

5.3 Academic conference application of face check-in scheme

The symposium of Zhejiang association of traditional Chinese medicine was successfully held in Hangzhou,China,this academic seminar invited the first group of famous Chinese doctors from Zhejiang province to give special lectures.There were about 500 people signing in at the main venue,and different channels were set up according to the identities of the guests.Besides,for the Chinese food and dinner check-in of this academic seminar,participants are also required to quickly identify their identities by scanning their face ID.The reception information of VIP guest scholars will be pushed in a real time,and the one-to-one reception service will be provided.

According to the identities of different guests,conference halls were divided into several different channels,for example,2 representatives,following the channel of education center,etc.After arriving at the conference,scholars attending this academic seminar need to scan the face for recognition according to their personal identity information and then sign in.The average time is 5 seconds per person.After the successful identification of participants,the seminar site can print label paper to complete the personal registration.Face sign-in scheme is a more accurate way to identify attendees through face recognition.It provides the guests and scholars with more humanized and more detailed services,and greatly improves the service quality and management ability of the academic seminar.

6.Conclusion

The rapid development of face check-in technology has had a huge impact on the conference industry,which provides attendees with a better experience,improves the service quality of the whole conference,and improves the flexibility and safety of the conference.The rapid development of face check-in technology has had a huge impact on the conference industry,which provides attendees with a better experience,improves the service quality of the whole conference,and improves the flexibility and safety of the conference.How to prevent face detection face detection algorithm attack,how to deal with the uncertainty of face image acquisition process,face plastic model uncertainty.How to prevent face detection face detection algorithm attack,how to deal with the uncertainty of the process of face acquisition,face plastic model uncertainty and so on the need to be considered.With the development of an emerging field,along with opportunities and challenges,relevant government departments should strengthen the management of this technology field,unify the standards of this technology field,and determine the entry threshold.Relevant policies and regulations should be issued to regulate the technical field,promote the transparency,rationality and legalization of the technical field,and better serve the development of the conference industry.In accordance with the general trend of intelligent exhibition,I believe that face recognition technology will greatly contribute to the sustainable development of the conference industry.

Acknowledgment

This paper is a periodical research result of Zhejiang University student science and technology innovation activity plan project research on the creation of application scene and service module design of intelligent exhibition and convention.

References

[1] Cong, H.B.(2013) Zhou Liandong.Research on the Development Model and Promotion Mechanism of Smart Exhibition Industry in Ningbo J.Beijing Economy,03: 38-39.

[2] Yan, S.H.(2016) Problems and Countermeasures of Exhibition Industry Development Under the Support of Network Technology J.Economic Development Research,02: 220.

[3] Yang, Y.C.(2017) Application Research of Smart Exhibition Technology in Industry Development Trend J.Science and Technology Innovation Guide,07: 166-168.

[4] Zhang, J.Y.(2018) Research on the Application of Smart Exhibition Technology in Exhibition Activities -- a Case Study of Hangzhou International Animation Festival J.Times Finance,03: 223-224.

[5] Hao, R.(2015) Research on the Development of Exhibition Economy Under the Internet Technology J.China Business & Trade,2015,01: 140-141.

[6] Yang, Y.C.(2017) Development Status and Strategies of Smart Technology in Exhibition Industry J.China Science and Technology Information,16: 116-118.

[7] Fan, J.Y.(2017) Research on the Influence of VR Technology on the Development of Exhibition Industry and Countermeasures J.Modernization of Education,51: 305-306.

[8] Han, L.Y.(2017) The Application of VR Technology and Big Data in Exhibition Industry J.Economic Vision, 22: 174-175.

A Research on Electric Car Styling Design and Low Aerodynamic Drag

Yanlong Li[1], Hui Zhu[2*]

[1] School of Automotive Studies, Tongji University, China

[2] Shanghai Automotive Wind Tunnel Center, Tongji University, China

Email: mail_dragon@tongji.edu.cn; hui.zhu@sawtc.com

Tel: 86-13818845461

Abstract: With the development of the market and the automotive industry, the styling and aerodynamic performance of a car take a more and more important role. This paper takes a research on electric car styling design which focuses on low aerodynamic drag. The design starts with an ideal low-drag shape, and then develops the shape into a car styling. With the method of Computational Fluid Dynamics, and after several times of iterations between design refinement and aerodynamic optimization, the Coefficient of Drag (C_D) finally resulted in 0.19, which is quite low for electric cars for saving energy and further the running range.

1. Introduction

With the development of the automotive industry, electric car is going to take a leading role in the future. The purpose of car styling is not only to get a good look, but also to express other engineering features, especially the aerodynamic drag. According to the fluid dynamic Equation (1), lower Coefficient of Drag (C_D) means lower drag force, hence less energy consumption, less environment pollution, and longer running range as well.

$$F_D = \frac{1}{2}\rho v^2 C_D A \quad (1)$$

where,

F_D is the air drag force,

ρ is the density of the air

v is the speed of the car relative to the air

A is the frontal area of the car

C_D is the coefficient of drag, a dimensionless number

Thanks to the technology, an electric car could have a much flexible architecture, this is very helpful to get a much free shape, which might also very helpful to lower the drag. Since the C_D is proportional to the aerodynamic drag force (drag for shot), how to make the C_D lower through design will be the key point of the research.

The design of Low-drag car has a long history. As early as 1921, the Rumpler Tropfenwagen, also called 'tear drop car', gets a C_D of 0.28, which is much lower than the average level of 0.6 at that time[1].

Quite a few companies launches low-drag cars thereafter, for example the Tatra 77 Type in 1934 from Czechoslovakia was the first mass production car, has a C_D of 0.2455[1][3]; the Audi 100 C3 in 1982 has a C_D of only 0.30, much lower than the average C_D of that time, the aerodynamic body gives the Audi 100 higher top speed than other cars of similar engine size.

For electric cars, the GM EV1 in 1996 has a C_D of 0.19[4]. In recent years, the Volkswagen's XL1 has a low C_D of only 0.186[5]. Although the C_D of these cars are quite low, but their body size are very small and have only 2 seats, which is not practical enough for local market.

In recent China, electric cars boom up very quickly. However, many of the designs are directly from traditional engined cars. Even cars from the so called "New Manufacture Force" are also quite similar to the traditional car types, especially similar to SUV cars (Figure 1), which usually has a large, tall, boxy body and relatively higher C_D than a common passenger car.

So how to make an electric car practicable, and also make it a low-drag will be a challenge.

Figure 1. The NIO ES8 electric SUV.

2. Material and Methods

2.1 Research of Ideal Basic Shape

Accord to the aerodynamic studies, 'Half-car-body' and Jaray Type are typical low-drag forms[4]. So the design starts with half-car body P1 (the "P1" stands for the 1st stage of the Proposal, shown as Figure 2). The size is set as 4380×1900×1500 (L×W×H, mm), which is close to a common C-segment car. With Computational Fluid Dynamics (CFD for short), the P1 comes out with a quite low C_D of 0.06.

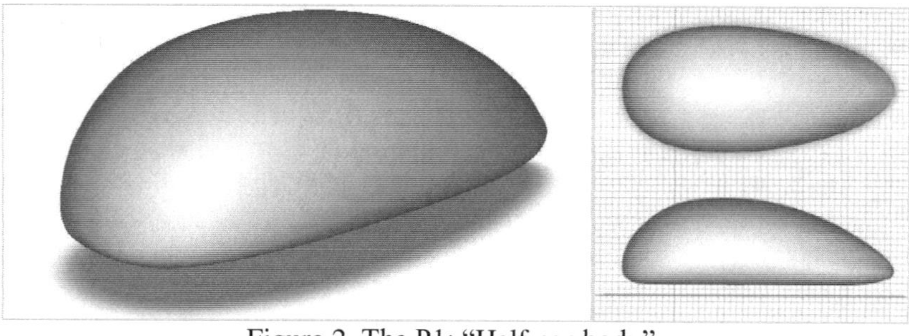

Figure 2. The P1: "Half-car-body".

In order to get aware of how much do the wheels influence the aerodynamic drag, a 'half-car-body' 3D model with wheels is built (the P2, shown as Figure 3), and keep the same ground clearance with P1. The C_D of P2 increases sharply to 0.135.

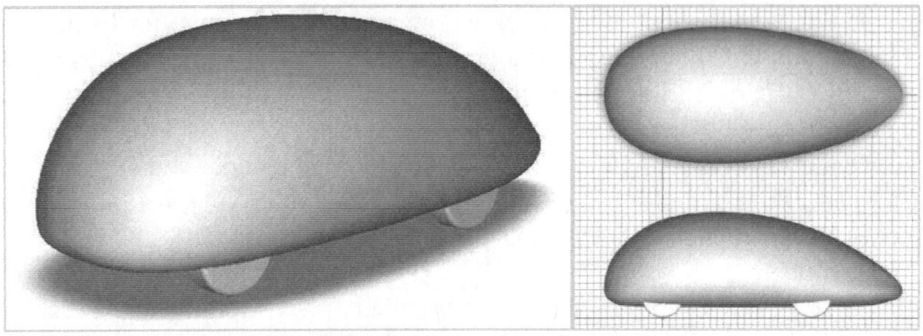

Figure 3. The P2: "Half-car-body" with 4 wheels.

2.2 *Early Concept*

Since the air flow is greatly affected by the wheels, which have a tight relationship between the body, it is important to lower the C_D by optimizing the shape that around the wheels.

Based on the ideal shape, the early sketches mainly concern on how to get a form from 'half-car-body' in order to get a base of good aerodynamics. The side surfaces are quite smooth to avoid bumps. In order to reduce the flow interference of the wheels, the rear wheels are covered, while the front wheels keep open for steering function (the P3, Figure 4), and the size is set as 4670×1700×1447 (L×W×H, mm).

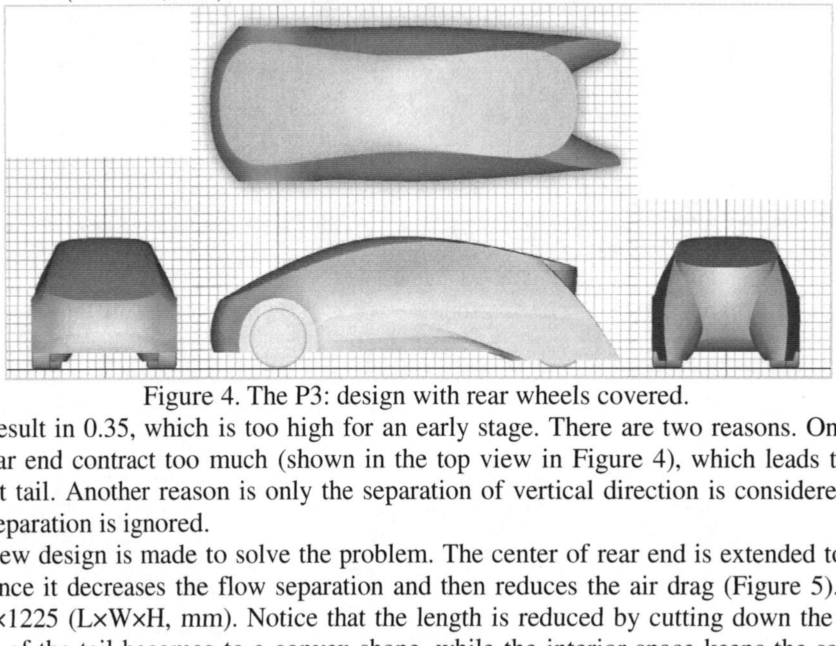

Figure 4. The P3: design with rear wheels covered.

The C_D result in 0.35, which is too high for an early stage. There are two reasons. One is that the center of rear end contract too much (shown in the top view in Figure 4), which leads to premature separation at tail. Another reason is only the separation of vertical direction is considered, while the horizontal separation is ignored.

Then a new design is made to solve the problem. The center of rear end is extended to fill out the concave, hence it decreases the flow separation and then reduces the air drag (Figure 5). The size is 4100×1700×1225 (L×W×H, mm). Notice that the length is reduced by cutting down the rear fender, the concave of the tail becomes to a convex shape, while the interior space keeps the same. The C_D comes out as 0.25, which is a very quick drop from the P3.

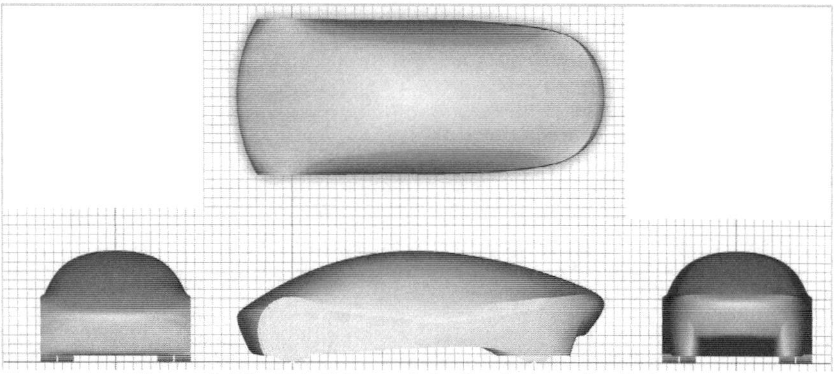

Figure 5. The P4: 3-D model of redesigned tail shape.

Since the front overhang of P4 design is a very short, the front end is quite flat from top view (Figure 5). The airflow concentrates on the front end and result in a large high-pressure zone (Figure 6), which leads to a relatively higher drag.

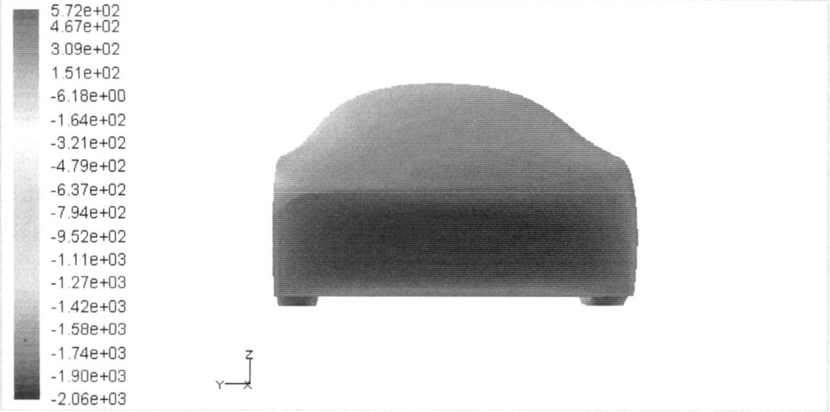

Figure 6. The P4: large high-pressure zone of the front end.

In order to solve this problem, the front overhang is extended forward for a rounder front end to reduce the high-pressure zone. The size is set as 4205×1700×1225 (L×W×H, mm), a little bit longer than P4 by adding length to the front overhang (Figure 7 B). Comparing P5 (Figure 7 B) with P4 (Figure 7 A) , the P5 has a much smoother transition from the front end to the side body, which also result in much smoother airflow, and leads to an obvious reduction of airflow separation and drag (Figure 8). The C_D comes out to be 0.21.

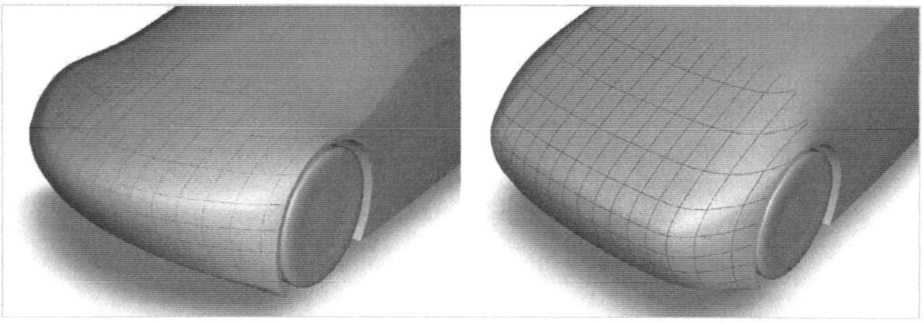

A B

Figure 7. Front ends comparison of P4 (left) and P5 (right).

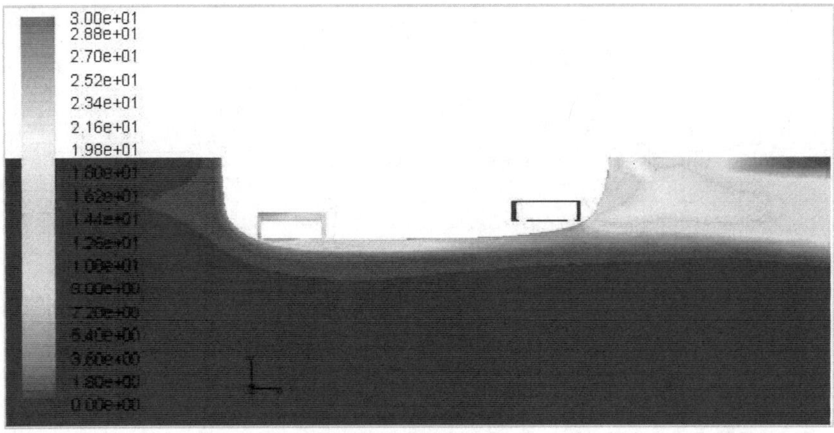

Figure 8. Turbulence intensity distribution of airflow at the height of 500mm for P5

2.3 Design refinements

With the aerodynamic performance improves obviously, the totally shape, however, gradually breaks away from the original design during the iteration. P6 (Figure 9) shows the new sketch where the cabin, roof, pillars and other details are re-designed, and Kamm-tail is also used to lower the drag.

Keeping the main features and aerodynamic achievements of P5, the design refinements are: the body side is designed as flush as possible in order to remain the low drag feature; some subtle feature lines are added on the side to avoid being boring; the shoulder line goes across the front wheel arch and runs into the front end to give a sleek look. The size is 4425×1800×1270 (L×W×H, mm), and the interior is a 2+2 layout, with 95% manikin in the front and 50% at rear, this makes the design more practical (Figure 10).

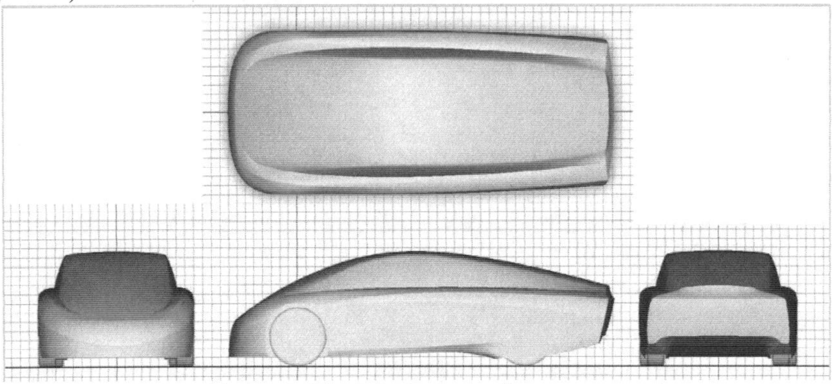

Figure 9. The P6: design refinements.

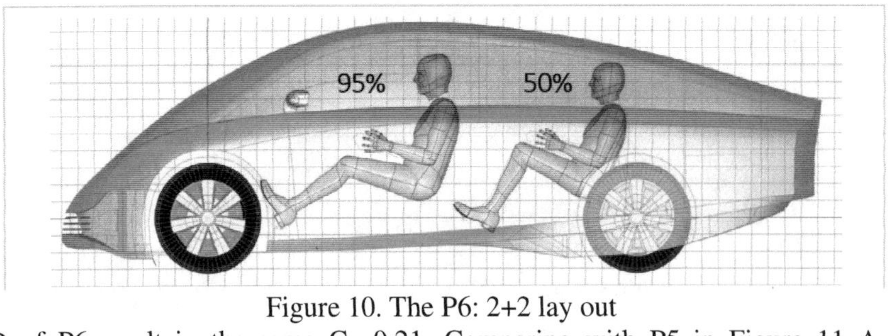

Figure 10. The P6: 2+2 lay out

The CFD of P6 result in the same C_D 0.21. Comparing with P5 in Figure 11 A, the airflow separation on tail of P6 (Figure 11 B) decreases on both intensity and area. Analysis shows that there

IOP Conf. Series: Materials Science and Engineering **573** (2019) 012014 doi:10.1088/1757-899X/573/1/012014

are two reasons for this: firstly, the 'Kamm-back' design of P6 helps to decrease the separation intensity at tail; secondly, the chin of P6 is risen about 30mm from P5, which allows more airflow to run from the under clearance of the car, and then reduces the separation zone at tail.

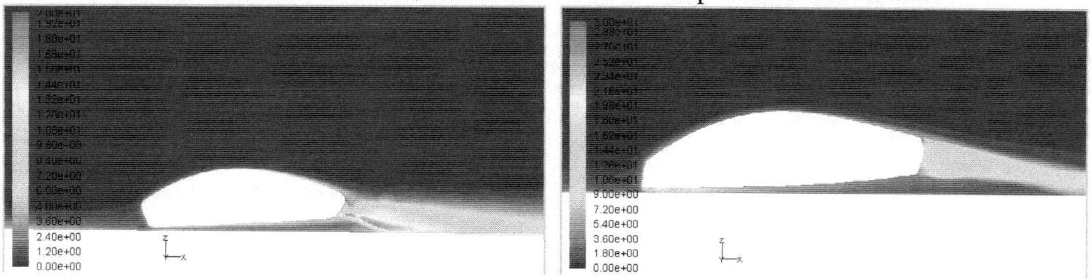

A B

Figure 11. The turbulence intensity distribution of airflow on Y0 of P5 (left) and P6 (right).

According to the research of P6, the front chin rises again and reaches to a new height of 200mm, which is the new design P7 (Figure 12. A). Figure 12. B shows that the separation at tail is reduced greatly, this leads to an apparent decrease of C_D, which comes out to be 0.19.

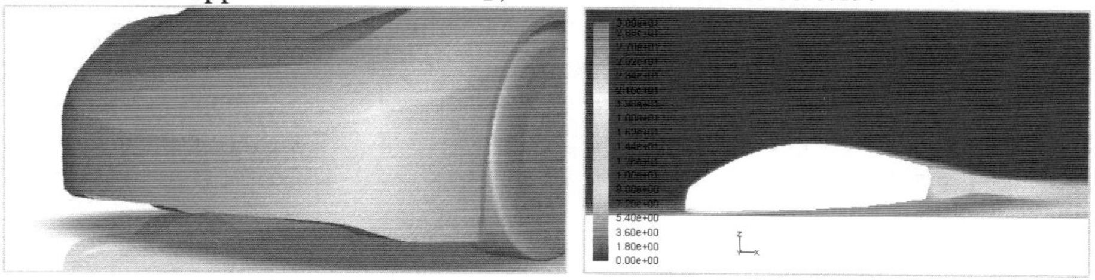

A B

Figure 12. The P7: the chin rises to 200mm from the ground

Based on the P7, the front styling is refined again according to the original design, the main refinement is adding feature lines at the front corner to make the shape more dynamic and less boring (Figure 13). Since the changes of surfaces near the feature lines are so subtle that they rarely affect the airflow. Thus the C_D remains 0.19 by CFD.

Figure 13. The P8: final design

3. Results

1) The wheels play an important role of the whole C_D. The C_D of 'half-car-body' is only 0.06, while the C_D of 'half-car-body' with wheels increase sharply to 0.135.

2) The shape of tail has a great effect on flow separation at the rear end, a concave tail concept with C_D of 0.35, while a redesigned tail decreases it to 0.25.

3) A rounder nose of car will give a better transition from the front end to side body, which makes the air flow much smoother, and leads to a lower C_D of 0.21.

4) The height of chin also has obvious effects on the flow separation at the rear end. Higher chin allows more air runing through the ground clearance, and result in lower C_D to 0.19.

5) Subtle styling design refinements have very few effect on air drag, which keeps the C_D as 0.19.

4. Discussion

1) Covering the rear wheel leads to a lower drag. If the front wheel could also be covered, the C_D should decrease either. In order to meet the steering function, the front track could be reduced to give the front wheel a turning space under the cover.

2) The shape of tail has a great effect on flow separation at the rear end. A convex shape is much better than a concave one. However, a too convex shaped tail maybe unpractical because it will waste too much of length. Hence the cut-tail design is a better choice for both low drag and efficiency of tail length.

3) The height of chin also affects the flow separation at the rear end. Usually a lower chin could reduce the air flow that runs through the underbody and result in smaller friction and irregular flow. In this case because the underbody is set to flat to fit the battery of an electric car, the result is different.

5. Conclusions

1) For the better commercialization of electric cars in the future, styling design will be an essential element of low aerodynamic drag, which will be helpful for energy saving, and also be helpful to extend the driving range as well.

2) Iteration between design refinement and CFD optimization is an effective way to research low-drag styling. Usually, starts with a low-drag shape and then develops it into a low-drag car is easier than starts with a car and then tries to make it low drag.

3) Although styling has a tight relationship with aerodynamics, subtle styling design refinements, however, have very few effect on air drag. This is quite useful when designs a car from early low-drag concept, and then adds detail features to a final styling without too much increase of C_D.

Acknowledgements

This research is financially supported by the National Key Fundamental Research and Development Program (973 Program, Grant No. 2011CB711203), The Energy Consumption Law and Optimization Design of Aerodynamics and thermodynamic System of Distributed Drive Electric Vehicle.

Reference

[1] Hucho, W. (1998) Aerodynamics of Road Vehicles. Sae Intl, Michigan.
[2] Lyons P. (2003)10 Best Ahead-of-Their-Time Machines. Car and Driver, Vol. 49(1):73.
[3] Gabor J. (2008) Optimization and computational fluid dynamics. Springer, Vol. 2008(1):196.
[4] Hu XJ. (2014) Automotive Aerodynamics. China Communications Press, Beijing.
[5] Volkswagen to debut XL1 PHEV prototype at Qatar Motor Show (2011). http://www.greencarcongress.com/2011/01/xl1-20110125.html#tp.

The Application Research of Fuzzy Mathematics in Design Quality Evaluation of Industrial Product

Wenming Liu [1*], **Jiajun Leng** [1]

[1] School of Design and Art, Shenyang Jianzhu University, Shenyang, LiaoNing, ,110168, China

*Corresponding author's e-mail: 83373734@qq.com

Abstract. The applying of Fuzzy Mathematics theory of evaluation and decision method is applied to single or comprehensive evaluation of design quality of industrial products by using scientific mathematical model and the formula model which can be applied directly is given by examples. It solves the problem that there is no scientific, accurate and comprehensive evaluation on the fuzzy concept of industrial product design quality. Introducing one scientific evaluation method which can evaluate design quality of industrial products by using Fuzzy Mathematics for relevant people who are engaged in judging design quality. It also provides relatively direct and specific reference basis and direction for industrial design practitioners to improve design quality.

1. Preface

With the rising steadily of the industrial design in China, industrial capability in China has also increased all these years, new industrial design products have also emerged, but the quality of some of these new products are only common level. How to make a scientific and comprehensive evaluation of the quality of industrial products design, what kind of industrial products are good products? This issue has been raised.According to the commonly used evaluation method, the experts of the relevant professional disciplines score the product design quality, and take the average value after removing the highest and lowest scores for all the scores.The evaluation of the design quality of industrial products by this method can only obtain a specific value.But the evaluation results are not objective and comprehensive.If we use the evaluation method in fuzzy mathematics theory, the product design quality can be evaluated more scientifically, which can make the evaluation reasonable, more scientific and comprehensive.

The industrial product design quality evaluation elements usually have the following aspects:

① Functional elements:

Function refers to the utility or function of the product, and the ability to be accepted. The product needs to have a certain function to be able to produce and sell. Therefore, the product is essentially the carrier of the function, and the realization of the function is the ultimate goal of the product design.

②Structural elements:

If the product function is the external connection between the system and the environment, then the structure is the connection of the elements within the product system.The structure is the bearer of the product function, structure determines the realization of the product function. The structure is both the bearer of the function and the product form.

③Human factors:

Human factors are related to people which can be extended to human factors engineering. It covers a wide range of factors including ergonomics, psychological, sociological elements and aesthetic elements.

④ Formal elements:

Form generally refers to the form of things under certain conditions. Form is the external manifestation of the internal structure of a product in design terms. There are many factors related to morphology, the formal aesthetic law lays the foundation for the evaluation factor selection.

⑤color elements:

Color is one of the most sensitive form factors that can infect people's aesthetic emotions, color can also cause people's aesthetic pleasure. Colors, like the form, have similar language functions and can convey semantic meaning. In the color design, the association is often generated by color, on the other hand, color and form are regarded as symbols together, and the color symbol suggestion function is utilized to convey the intention.

⑥ Environmental elements:

Environmental factors must be considered in product design because products causing environmental problems at every stage of the life cycle.

2. Single evaluation and comprehensive evaluation.

A certain factor or part of an object or category is evaluated according to certain criteria, called a single evaluation. Obtaining an overall evaluation of an object or categories of objects from a multitude of individual evaluations is called a comprehensive evaluation. The purpose of comprehensive evaluation is usually to hope that several objects can be sorted according to a certain meaning, pick the best or the worst objects from these objects. called the decision process. Comprehensive evaluation is also a type of decision-making process.

3. Evaluation method:

3.1. the traditional method:

3.1.1. the total score method:

Set 'n' factors for one object, we evaluate a sore 'S_i' for every factor, find sum.

$$s = \sum_{i=1}^{n} s_i$$

'S' is the evaluate standard of the object. For example, the results of the pentathletes, the college entrance examination qualifying line, etc., are all adopted

3.1.2. weighted average method:

Set 'n' factors for one object, we evaluate a sore 'S_i' for every factor,Given the different emphasis on each factor, factors can be assigned different weights $a_i = (i = 1,2,\ldots,n)$, It represents the percentage of the 'i' factor in the judgement. Available sum

$$s = \sum_{i=1}^{n} a_i \bullet s_i \qquad (\sum_{i=1}^{n} a_i = 1)$$

As the standard for judgement, such as in job determination, grading usually use this method.

3.2. Fuzzy method:

With the traditional method of evaluation, the result is always single, that is, the evaluation result is expressed with a numerical value, but sometimes that is not comprehensive and accurate.In particular, the evaluation of industrial product design quality is a factors always be influenced, and the factors involved in the object to be evaluated are often fuzzy. The traditional evaluation method is obviously not scientific, comprehensive or accurate enough. Therefore, we can adopt the method of fuzzy comprehensive evaluation for the design quality of industrial products.

To scientifically, correctly and comprehensively evaluate a specific object, it is necessary to give appropriate comments on several aspects of the object (called elements), and then make a comprehensive evaluation. For example, to evaluate the design quality of an industrial product, the product's "functional elements", "structural elements", "human factors", "morphological elements", "color elements", "environmental elements" and so on should be evaluated first, and then integrated.

It is always denoted $U=\{u_1,u_2,\cdots,u_m\}$ as the evaluation factor set. Like evaluate the industrial product, we can use $U=\{u_1,u_2,\cdots,u_m\}$, $u_1,u_2,\cdots,u_m$ respectively express functional elements, structural elements and human factors, etc.

'V' represent the comment set. 'V' can be a finite set, for example V=(A,B,C,D,E), or V=(level 1, level 2,…,level n), we usually can call that V=(v_1, v_2,…,v_n), 'V' could also be the interval on the real axis, for example V=(0, 1), or V=(0, 100) etc. For the case where V is a finite set, we can also take v_1, v_2, …v_n quantitative. As a subset of [0,1] or [0,100](for example, if a product is scored on a percentage scale, then V = {1,2...), 100}).

Set the fuzzy vector that has been given

$$A=(a_1,a_2,\cdots,a_n)$$

And fuzzy matrix

$$R=\begin{pmatrix} r_{11} & r_{12} & \cdots & r_{1m} \\ r_{21} & r_{22} & \cdots & r_{2m} \\ \cdots & \cdots & \cdots & \cdots \\ r_{n1} & r_{n2} & \cdots & r_{nm} \end{pmatrix}_{n\times m}$$

The fuzzy synthesis operation is defined as

$$AOR=B=(b_1,b_2,\cdots,b_m)$$

Among them

$$b_j=(a_1*r_{1j})*(a_2*r_{2j})*\cdots*(a_n*a_{nj}) \qquad (j=1,2,\cdots,m)$$

Operator $M(*,*)$ is a symbol of operation. Two algorithms are introduced here.

(1)"Dominant factor type"

According to the law of Max - min, the box "Sunday afternoon" instead of ", "use" $\vee$ "instead of " ".

$$b_j=\mathop{\vee}_{i=1}^{n}(a_j\wedge r_{ij}) \qquad (j=1,2,\cdots,m) \tag{1}$$

Or

$$b_j=\max\{\min(a_1,r_{ij}),\cdots,\min(a_n,r_{nj}) \tag{2}$$

(2)"Weighted average"

By the ordinary matrix multiplication symbol, that is, " ' instead of' , '+', ' instead of ' ".

$$b_j=\sum_{i=1}^{n}a_i r_{ij} \qquad (j=1,2,\cdots,m) \tag{3}$$

For example: The product appraisal department identifies an industrial product. From the perspective of design, we must consider the three elements of form, color and human factors. Take the domain 'U '.

$$U=\{u_1(\text{form}),\ u_2(\text{color}),\ u_3(\text{human factors})\}$$

The evaluation is divided into four levels, that is, take the comment set as 'V '

$$V = \{v_1(\text{very good}),\ v_2(\text{better}),\ v_3(\text{average}),\ v_4(\text{bad})\}$$

When it comes to form evaluation, 50% of people think it is "very good", 40% think it is "better", 10% think it is "average" and no one thinks it is "bad"

$$(0.5,\quad 0.4,\quad 0.1,\quad 0)$$

Similarly, the evaluation of color is

$$(0.4,\quad 0.3,\quad 0.2,\quad 0.1)$$

The evaluation of human factors is

$$(0,\quad 0.1,\quad 0.3,\quad 0.6)$$

In this way, an evaluation fuzzy matrix is obtained

$$R = \begin{pmatrix} 0.5 & 0.4 & 0.1 & 0 \\ 0.4 & 0.3 & 0.2 & 0.1 \\ 0 & 0.1 & 0.3 & 0.6 \end{pmatrix}$$

The following is the weight of the three factors in the evaluation of the product, which can be obtained by two methods:

(1) The simplest method, directly by the experts to agree;

(2) through the statistical experts to determine the score

The experts participating in the evaluation believe that the main requirements for the product are: shape elements and human factors, color elements are ranked in the least important position, so the weight vector can be

$$A = (0.5,0.2,0.3)$$

The following two algorithms are used to calculate the comprehensive evaluation of the product.

The principal factor prominence algorithm is first used, which can be obtained from the previous formula (1) or (2)

$$B = AOR = (0.5 \quad 0.2 \quad 0.3)\begin{pmatrix} 0.5 & 0.4 & 0.1 & 0 \\ 0.4 & 0.3 & 0.2 & 0.1 \\ 0 & 0.1 & 0.3 & 0.6 \end{pmatrix} = ((0.5 \wedge 0.5) \vee (0.2 \wedge 0.4) \vee (0.3 \wedge 0),$$

$$(0.5 \wedge 0.4) \vee (0.2 \wedge 0.3) \vee (0.3 \wedge 0.1), (0.5 \wedge 0.1) \vee (0.2 \wedge 0.2) \vee (0.3 \wedge 0.3),$$

$$(0.5 \wedge 0) \vee (0.2 \wedge 0.1) \vee (0.3 \wedge 0.6)) = (0.5,0.4,0.3,0.3)$$

But because of

$$0.5+0.4+0.3+0.3=1.5$$

Than normalize it, Divide each item by 1.5 times, and the final comprehensive evaluation result is

$$(0.33(\text{very good}),0.27(\text{better}),0.20(\text{average}),0.20(\text{bad}))$$

Using weighted average algorithm, which can be obtained from the previous formula (3)

$$B = AOR = (0.5 \quad 0.2 \quad 0.3)\begin{pmatrix} 0.5 & 0.4 & 0.1 & 0 \\ 0.4 & 0.3 & 0.2 & 0.1 \\ 0 & 0.1 & 0.3 & 0.6 \end{pmatrix} = ((0.5 \times 0.5) + (0.2 \times 0.4) + (0.3 \times 0),$$

$$(0.5 \times 0.4) + (0.2 \times 0.3) + (0.3 \times 0.1), (0.5 \times 0.1) + (0.2 \times 0.2) + (0.3 \times 0.3),$$

$$(0.5 \times 0) + (0.2 \times 0.1) + (0.3 \times 0.6)) = (0.33,0.29,0.18,0.20)$$

Because of

$$0.33+0.29+0.18+0.20=1.0$$

Has been normalization, so the final comprehensive evaluation result is

$$(0.33(\text{very good}),0.29(\text{better}),0.18(\text{average}),0.20(\text{bad}))$$

According to the above two algorithms, when morphological elements, color elements and human factors are taken into consideration for this industrial product at the same time, there is a large proportion of the comprehensive evaluation results, which are respectively 60% and 62%.

As can be seen from the above example, the comprehensive evaluation should be carried out in the following steps:

(1)Select the factor set 'U' and comment set 'V'. The principle is to be comprehensive while addressing the principal contradiction. In this way, we can better simulate people's thinking and avoid some unnecessary troubles.

(2)Single factor evaluation vector is determined through survey statistics. The number of respondents should be enough and representative, and then the evaluation fuzzy matrix 'R' can be obtained from each single factor evaluation vector.

(3)Determine the weight vector 'A'. One method is by authoritative experts and representative people according to the importance of the factors to determine the degree.

(4)Select the appropriate algorithm. Two algorithms are commonly used: weighted average and principal factor prominence.

In general, the results of the two algorithms are much the same. Note the characteristics of these two algorithms:

Weighted average algorithm is often used in the case of many factor sets, it can avoid information loss;

The principal factor prominence algorithm is used to avoid the situation that the data in the fuzzy matrix is very different

One of the "naughty" data interference.

In addition, a large number of statistical calculations show that the number of factor sets cannot be greatly different from the number of comment sets when the dominant factor type is used; otherwise, the weight value will be too small and the single-factor evaluation data will be "overwhelmed".

4.Research conclusions

Fuzzy mathematics is a relatively scientific, accurate and comprehensive method to evaluate the design quality of industrial products. The fuzzy mathematics method can make the expression of the evaluation result more concrete and comprehensive, and make the evaluation of the fuzzy concept of product design quality clear. This method is simple and easy to use, and can directly use the mathematical formula template enumerated in this paper to set different evaluation factor sets and comment sets for different products, and then determine the single factor vector and weight vector, and select one of the weighted average or dominant prominence algorithm for calculation according to the need. In this way, a relatively scientific, accurate and comprehensive evaluation conclusion can be drawn. The evaluation conclusion can reflect the design quality of a product in the specific factors on which the performance of good or poor. This not only provides a scientific evaluation method for professionals engaged in product design, but also provides relatively direct and specific reference basis and direction for product designers to improve design and improve design quality.

References

[1] Wu,X.(1993) Product System Design.China Light Industry Press,Beijing.
[2] Peng,Z.(2003) Fuzzy sets and applications.Wuhan University Press,Wuhan.
[3] Wang,F.(1988)Fuzzy mathematics and engineering science.Harbin Institute of Shipbuilding Engineering Press,Harbin.
[4] He,Z.(1983)The application of fuzzy.Tianjin Science and Technology Press,Tianjin.
[5] Chen,S.(2004)Cyclical iteration model and unified theory of fuzzy cluster, recognition and optimum decision.Journal of Dalian University of Technology,Dalian.
[6] Duan,B.(2004)Interval number-based multiple fuzzy optimization model.Journal of Hohai University,Nanjing.
[7] Zhang,X.(2003)Advance in fuzzy comprehensive evaluation methods theory and application.Journal of Shandong university of architecture and engineering,Shandong.

Research on Innovative Application of Modular Design in University Student Apartment Furniture

Gu Yu, Chen Dai, Suwan Huang, Liqing Gan and Weixia Gao*

[1] School of Art and Textile and Clothing Engineering, Changshu Institute of Technology, Changshu, 215500, China

*Corresponding author's e-mail: 1046978972@qq.com

Abstract. Based on the analysis of the application status of modular furniture in college students'apartment furniture, and taking the excellent design examples as the reference plane, this paper explores the innovative application forms of modular furniture from the perspective of function, shape, color and space, with a view to creating a private space and one belonging to them for the current college students in the fast-paced college life.

1. Introduction

Modular furniture can be disassembled and combined at will, and matched freely according to users'needs, thus forming a product with new visual effects and functions. The design of apartment dormitory furniture is not only related to the comfort of students'accommodation, reflecting the humanistic concern of the school for students, but also an important part of college students' apartments, which participates in the construction of university brand to a certain extent. With the emergence of modular design, there are more possibilities for the furniture design of college students'apartments[1].

2.Modular design and furniture of college student apartments

2.1.Modular design

In product design, design modularization means that the product can be divided into several modules, each module has its own independent function and has the same input and output port, so that users can combine each module perfectly according to their own needs to form a new form, which will have new visual effects and functions. Modular products have the characteristics of relative independence, interchangeability and versatility. Firstly, each module is relatively independent and has its own function and form, which can exist independently. Secondly, each unit's products have a consistent access point, which is conducive to the interchange between modules. Finally, the modular generalization between horizontal and vertical is conducive to the realization of the generalization of cross-series products.

2.2.University student apartment furniture

Apartment is an important place for college students to live and study on campus. It is also the basic position for them to perceive social reality and form interpersonal relationships. College students'apartment furniture is different from traditional household furniture. The use environment is dormitory, the living area is tense, the living environment is single, the space is relatively narrow, and the main object of service is students. The types of dormitory furniture are less than traditional household furniture, mainly including tables, chairs, beds, cabinets and so on. With the development of modern society, students'demand for dormitory furniture is diversified. College students'apartment furniture should not only meet the functional needs of students for dormitory design, but also become a participant and shaper of dormitory culture.Therefore, its design is no longer a simple solution to the functional and morphological problems, but should deal with the multi-symbiotic relationship among desks, chairs, beds, cabinets and other apartment furniture, so as to realize the innovative application of modularization in university student apartment furniture[2].

3. A comparative analysis of the current situation of furniture design of university student apartments at home and abroad

3.1.Domestic situation

Domestic accommodation system mainly arranges for undergraduates to enter the dormitory when they enter school. Postgraduates and doctoral students can apply for dormitory. In 2001, China's Ministry of Education issued "Several Opinions on the Standards for the Construction of Apartments for College Students", which stipulated that there were four undergraduates with an average floor area of $8m^2$ per student, two postgraduates with an average floor area of $12m^2$ per student, and one doctoral student with an average floor area of $24m^2$ per student. From the founding of the People's Republic of China to now, there are obvious changes in the facilities, shape and number of student dormitories. In terms of building volume, there are low-rise, multi-storey and high-rise buildings; in terms of the plane composition of dormitory buildings, there are corridors, short corridors and units; in terms of the spatial layout of residential units, there are hotel-style and family-style; in terms of space function, the bedroom is the prototype, combined with living room, toilet, kitchen, balcony and other spaces, and is no longer limited to the "421" standard of the ministry of education. Various spaces are combined with each other, and the combination modes are changed to form various unit nests.

The survey shows that although many colleges and universities are committed to improving construction facilities and living conditions of students, they are still insufficient. In fact, many schools are not up to the standard. It will take a long time for the situation to change. Per capita living space is small, and many dormitories are in disorder. Some colleges and universities have made some alterations, and still use them. The evaluation index of safety facilities and accommodation area in dormitory buildings is low. Of course, this is also related to the large number of students enrolled in colleges and universities at present, and the construction of student apartments no longer depends on state funding. Nowadays, the overall level of social economy and science and technology has greatly developed, and the psychological characteristics, quality of life and ways of college students have also changed and improved significantly. The results of big data analysis and statistics show that college students'demand for dormitory furniture is no longer a single functional level, but pay more attention to sleep, storage, leisure, information-based learning and other multi-functional furniture. However, at present, many college students'apartment furniture in China are more traditional, mainly strong, durable, single layout, can not meet the needs of students at different stages, different personalities, lack of innovative design features as shown in Figure 1.

Figure 1. Apartment furniture of domestic colleges and universities

3.2.Status quo abroad

Foreign research on university students'furniture is earlier than the domestic, which is related to the

IOP Conf. Series: Materials Science and Engineering **573** (2019) 012016 doi:10.1088/1757-899X/573/1/012016

development psychology research earlier than domestic research. Its research results are relatively rich, and its application in the field of practice is more extensive, which is not overnight to catch up with. Compared with the domestic situation, the big difference is that many foreign universities do not require students to live in school apartments. Most international students choose to rent a house, not a dormitory, that is, to share a villa or apartment with others, basically one-room.

The environment in the dormitory area is pleasant and beautiful, which makes people feel relaxed and happy, friendly and comfortable, easy to communicate freely, active life, and conducive to academic discussions. The accommodation conditions are better, but the quantity is limited, and the price is more than several times higher than that in China. Overseas dormitories are roughly classified as dormitory collective type, hotel type, apartment type and residential type. At present, we are discussing the type of collective dormitory. According to the survey, it is similar to the domestic design of space and furniture. It is different from the most common four-person room in China. The most common type of room in foreign countries is double room and single room. As far as equipment arrangement is concerned, students in developed countries come from complex sources and age gap is relatively large. Schools will try to arrange students with similar culture, diet and living habits together.

The application of modular design in foreign countries is obviously more than that in domestic countries. Its design is mainly simple and generous modular furniture, which can be used independently and randomly, so that dormitory space does not need to be restricted to the division of functional areas.

4.Case Study on furniture design of modular college student apartments

4.1.Scholarship in university of Bangkok

Through the study of Fabrica, an Italian exchange research institute in Treviso, the concept of co-living space is a "space scholarship" offered by the Thai government to seven economically disadvantaged students, which not only provides sleeping and learning functions, but also includes space settings for communication and interpersonal relationships. We can clearly see that the application of modularization and intellectualization in the field of student dormitories is completely feasible. Figure 2 shows the overall residential effect.

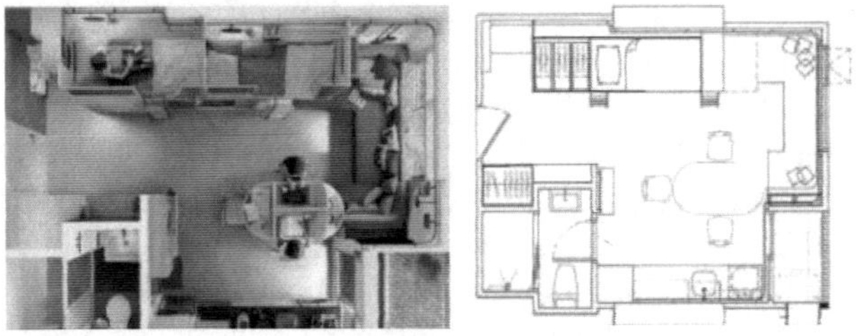

Figure 2. Space Scholarship Dormitory

Its modular design features in the girls'dormitory are as follows:

- Unlike previous upper deck designs, each student's sleeping area has its own stairway entrance. The designer surrounded the ceiling with an iron frame for students to air their clothes.
- In the design of personal privacy, the designer has also made great efforts. Every student's sleeping area has added bedpanels and curtains, which open when needed and close when sleeping. A light shield is also attached to the outside of the lamp. Turning on the light at night can reduce the brightness of the light and control the direction of the light, so as not to disturb the roommate with poor sleep quality, as shown in Figure 3.

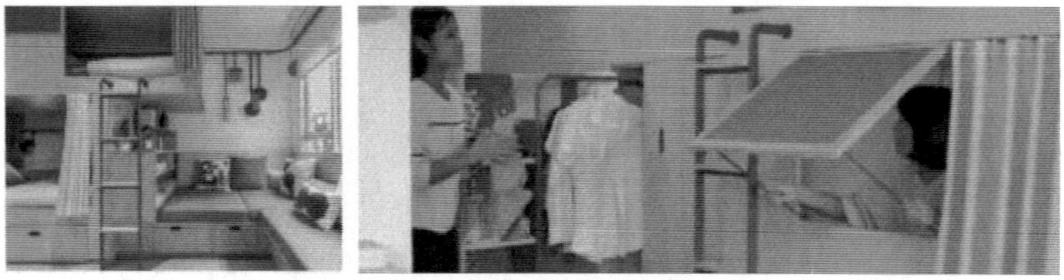

Figure 3. Personal privacy space processing

- About acceptance: using the space of edges and corners to do acceptance. The shelves drawn from the upper and lower layers can hang clothes or put some small items, as shown in Figure 4.

Figure 4. Inclusion space processing

- Independent learning space: dining table and desk design. Without interrupting each other, it creates an atmosphere of small libraries as shown in Figure 5.

Figure 5. Multifunctional dining table

The spatial planning of the male dormitory is slightly different. Four boys are divided into two separate bedrooms, each with its own desk.

- Humanized design: considering the height of boys, designers will open the distance in the design of beds.
- Desk design: the ceiling is also surrounded by a circle of iron frame, can hang some lightweight furniture, when used can also be moved according to location. The lamp can also be moved as needed.
- Colour collocation: the overall tone is mainly log color, simple and textural. In order to reflect the youth and vitality of university life, the designer also added bright yellow, orange, blue and other colors to make the dormitory more lively and bright as shown in Figure 6.

Figure 6. Furniture design of boys'dormitory

4.2.Tietgenkollegiet in university of Copenhagen

Tietgenkollegiet is a student dormitory near the university of Copenhagen South Campus. The student dormitory has won many architectural awards and has been hailed by some media as "the coolest student dormitory in the world". Tietgenkollegiet's design was inspired by Tulou in Fujian, China, so

Chinese students in Denmark kindly call this dormitory "Geda Tulou". The whole building has 360 single or double dormitories, 30 kitchens and public areas. These kitchens are like boxes hanging inside the circle building. They are very technological. Tietgenkollegiet is not only unique in appearance, but also innovative in interior decoration. Designers Aggebo and Henriksen have boldly applied their Nordic modular design concepts to the interior decoration of student dormitories, as shown in Figure 7.

Figure 7. Furniture design of Tietgenkollegiet dormitory

5.Design strategy of university apartment furniture based on modularization concept

Colleges and universities are an important base for talent cultivation in China. With the reform and development of higher education and the continuous progress of society, the material and spiritual functions of college students'dormitory construction are extending towards diversification and depth. The influence of university environment on talent cultivation is both imperceptible and far-reaching. From the above description, the current modular university apartment furniture design has a lot of room for development. Not only can it solve the current problem of overcrowded dormitories in major universities, but also bring many conveniences to students'lives, improve the quality of life of students, and also make parents more comfortable. The advantage of modular design of dormitory furniture for college students is obvious. It not only greatly increases the space utilization rate of dormitory for college students, but also has more flexibility. By disassembling modules, dormitory furniture can be varied, and dormitory life for college students can be more practical and flexible. The modular structure realizes many functions, and embodies the concept of humanization and humanistic care[3].

5.1.Diversified furniture design and attaching importance to use habits and taking into account private need.

Furniture design is related to college students'physiological and psychological needs and ways of use. The attitudes, interests, emotions and needs of students in their daily activities will change with the use of different situations. Therefore, the facilities provided for their activities in apartments should be adjusted accordingly, with a variety of characteristics. From the perspective of space division, the limited space in the apartment, so we should break the traditional standardized living form to meet the diversified needs. The modular design is applied to the design of apartment furniture in colleges and universities. Based on the function analysis, modular system is established by decomposition layer by layer, which makes it possible to unify standardization and individualization. Different products can be assembled by selecting modules. We should be able to change from privacy to semi-openness and then to openness at any time, realize flexible functional adjustment, and satisfy students'various demands of study, life and communication[3].

5.2.Expanding residential space of apartments to realize individualization

Humanized design is the highest goal pursued by modern design and the inevitable product of the development of social civilization. At present, the standardized allocation of apartments makes it difficult for students to embody their personality and freedom, which is bound to be unable to adapt to the future trend of use. The age stage of college students also determines that the accommodation environment should be younger and personalized. The effective way to solve the problem of individualization is to introduce the mechanism of "elastic design" into apartment furniture design, that is, under the condition of the original furniture volume, furniture can be properly shifted, folded, pulled out and retracted. Flexible design enables furniture to have a variety of functions and forms, which can be obtained in relatively small spaces.

5.3. Intelligent home equipment

With the socialization of university construction and the advent of the information age, education is becoming more and more modern. Intelligent education environment also needs modern accommodation environment to adapt to it, and information-based learning media has also changed the way students live and live. Students will use computers to study, complete homework, entertainment and access to information has become a common phenomenon in modern colleges and universities, electronic information technology and traditional apartment furniture combination of new products have emerged, and has a strong development potential. For example, the lamp not only maintains the original use function, but also charges the electronic equipment as a mobile power supply, but also charges the mobile phone wirelessly, bringing endless fun and new experience for college students'life and learning[3].

5.4.Reasonable use of decorative soft environment to create cultural atmosphere

Furniture design not only requires innovation in function, but also in material, color and soft decoration, so as to make apartment furniture more individualized and innovative. The surveys show that the overall design of apartments is almost the same at present. Furniture modelling is very similar and the color is single. Take the color as an example, the existing apartment furniture hardly takes into account the problem of color matching. On the whole, it generally adopts single colors, such as log color or milky white, which is general and monotonous, and the environment seems rigid and indifferent. In the design, we should reasonably control the decorative elements such as color and texture. We can provide several different color modules for the users, for example, the local color of furniture used by boys and girls can be different, and different functions can also be combined with color to reflect. The introduction of local culture can be considered, such as incorporating the typical water culture into the design, helping students integrate into the city, so as to better study and life.

6.Summary

As far as modular design of apartment furniture in colleges and universities is concerned, further research can be done on the diversity of connection modes of "No Tools" concept and the evaluation system in the future to improve the flexibility between modules, achieve the goal of easy and convenient disassembly and reorganization of furniture modules, and better realize the modular use of furniture. Scientific and reasonable evaluation system is conducive to the further optimization of modular design of apartment furniture, so as to provide more comfortable living space for contemporary college students.

Acknowledgments

This research was financially supported by Research on Furniture Design of College Student Apartments Based on Modularization (Grant NO.XJDC2018171), Research Project of Changshu Institute of Technology (Grant NO.KYZ2017103Z), Jiangsu Provincial Philosophy and Social Science Fund(Grant NO.2018SJA1373).

References

[1] Zhao Jinghao. (2018) Design and research on furniture combination change mode of student apartments in colleges and universities. Hebei University of Science and Technology, Shijiazhuang.

[2] Ou beno. (2013) Application and research of multi-functional furniture in small household. Changchun University of Technology, Changchun.

[3] Yuli. (2014) Research on interior space and furniture design of university student apartments . Art education research, 20: 87-87.
[4] Wang Wenning, Zhang Yuanqun. (2010) On humanization of student apartment furniture design. Packaging Engineering, 12: 22-24.

IOP Conf. Series: Materials Science and Engineering **573** (2019) 012017 doi:10.1088/1757-899X/573/1/012017

Study on Industrial Heritage Regeneration Design for Creative Industry Park in Nanchang City

Guo Yang

Construction Engineering Management Department, JiangXi Institute Of Economic Administrators, Nanchang, Jiangxi Province, 330008, China

*Corresponding author's e-mail: 798796716@qq.com

Abstract: Along with the development of the urban economy and the society, some of the industries in the city cannot meet the economic demand in the late 21st century due to the technological progress, the economic transformation, the employment change and the social and political transformation. Therefore, some of the city's space has lost its original function and meaning. Many factories and office buildings that are no longer suitable for the development of urban center need to be moved out of the center of the city. These urban spaces have the industrial memory of the city and are the historical heritage of industrialized cities. From the point of view of the layout of creative industry, this paper conducts planning study on the space of cultural and creative parks oriented to the reconstruction of industrial heritage, and explores the development mode of the cultural and creative industrial park, which is suitable for the cultural characteristics of Nanchang city and has a forward-looking vision to provide reference for the construction of cultural and creative park, the protection of industrial heritage, the adjustment of industrial structure and the improvement of urban competitiveness of Nanchang.

1. Introduction

In recent years, since Nanchang government attaches great importance to the development of cultural industry, cultural industry has gradually become an important force to promote the adjustment of economic structure and change the way of development. To lay out and build a cultural industry cluster, to strengthen and do excellent cultural brand, to promote the integration of cultural industry and other industries, creative cultural industry has become a new supporting point and new highlight of supply-side reform and upgrading in Nanchang.

The adjustment of industrial structure and the upgrading of supply-side reform in Nanchang make more and more industrial land in cities lose their original use value. The combination of creative industries with these industrial buildings and sites provides a direction for preserving the industrial culture of the city and transforming it into an industrial heritage to regenerate new cultural values and use values. The preservation of industrial heritage and the development of cultural and creative industries are a new opportunity and a intense collision between history and new ideas with strong impact force.

2. Challenges to the Design of Industrial Heritage Creative Industry Park

Nanchang, as the provincial capital of Jiangxi Province, has a complete industrial system. With the development of the times, some industrial zones in the city center have gradually withdrawn from the stage of history and are faced with the problems of developing industrial waste land regeneration and utilization. According to the investigation, there are five industrial heritage parks in Nanchang, which are responsible for recording the memory of industrial civilization, improving the quality of the city and optimizing the living environment. Their basic status is as follows (Table 1).

Table 1. Industrial Heritage Regeneration Design Projects in Nanchang City.

Project name	Region	Industrial history	Current Status
699 Cultural Creativity Park	Shanghai Road, Qingshan Lake District	The original site is Jiangxi Hua'an Knitting General Factory. In 2009, because of the difficult operation of the enterprise, conducting the enterprise reform and creative transformation.	Creative Park, a complex sets all kinds of creative design, cultural and educational training, film and television and music production, artists studio, exhibition and performing arts, commercial office and other supporting facilities in one.
Camphor forest	66 Sheshan road, Donghu district	The original site is Jiangxi Chemical Fiber Factory.Conducting creative transformation in 2011.	A commercial complex that sets green ecotourism, creative and cultural tourism, industrial heritage tourism as one. At the same time, it is also a AAA-class tourist attraction.
Nanchang Diesel Engine Plant Community	Dinggong Road West Lake District	The original site is Nanchang Diesel Engine Factory, which was reformed and moved away from the current site in 2002.	Make use of the site space, shape the public activity square, implant the facilities that correspond to the history of Nanchang Diesel Engine Factory, build the industrial cultural district.
AVIC Hammer Park	Donger Road, Xinxiqiao, Qingyunpu District	The original site is Nanchang Hongdu Machinery Factory, the birthplace of the first aircraft made by the people's Republic of China.	Relying on Hongdu industrial civilization, AVIC Hammer Parkopened landmarks of fashion culture zone, creative culture zone, science and technology culture zone and aviation culture zone.
791 Art District	113 Sanjing road, Donghu district	The original site is Jiangxi Pharmaceutical Factory, which was later creatively transformed into 791.	Creative block, mainly with the art museum and private studio as the main body.

The modern industrial sites in Nanchang City have obvious characteristics of machine production, and have pioneering significance in the long river of history. Among them, there are many outstanding sites in history with profound cultural background and mass foundation. However, the current reconstruction design still faces great challenges. The lack of regional culture, the rigidity of space integration and the insufficient capacity of ecological restoration of the deign still hinder the regeneration and utilization of industrial heritage.

2.1The lack of use of regional culture

The so-called industrial heritage must bear the history of struggle of a city. In the process of activating the history and memory of these cities, the accurate expression and regeneration design of regional culture should be the priority among priorities. Reasonable cultural expression shall be based on the correct interpretation of the culture, designers shall start from the cultural sites that the projects rely on. No matter what new technology and new structure is used to break the past human-living wisdom, even if the visual impact is put first, the aesthetic fatigue will still appear under the textual research of time. The existence of industrial heritage as a culture is a respect for the city and a record of history. Once separated from the blood of the city and run in a opposite direction, it may make people happy at the very beginning, but it splits the undertaking vein of history and lacks a far-reaching intention.

At present, the reconstruction of the Industrial Heritage Creative Industry Park in Nanchang City has generally adjusted the site of the original site in order to better serve the business of the park. Although the symbol characteristics of the park are relatively uniform and obvious, they are general relatively scattered, which cannot preserve the industrial civilization memory very well.

2.2Reasonable Use of Space

The site reconstructed with industrial heritage often increases the design difficulty for designers. Since the original industrial site is not custom-made for public services, the space functions need to be converted. In this process, ratio scale problem, spatial function problem, and efficiency problem arise accordingly. At present, the function of creative industry park with the theme of industrial heritage in Nanchang is generally biased towards commercialization so that the atmosphere of R & D and creation is not strong enough, the exhibition area is missing or the utilization rate is not high and so on.

This article takes the park exterior space matching, the interior space design and the architectural intention of the transformation as the investigation object. In view of the effect of the satisfaction degree of the building space transformation, the supporting facilities and the overall functions of the park, this article made corresponding questionnaires to the visitors to the 5 creative parks in Nanchang City. 204 valid questionnaires were collected,including:48 from 699 park, 42 from Camphor Woods, 33 from 791 block, 49 from China Aviation City, and 32 from Nanchang Diesel Engine Plant Community.

According to the investigation and evaluation, the comprehensive evaluation grade of the Camphor Forest,699 Park and AVIC Hammer Parkis relatively good. The whole function of the park is relatively perfect, the industrial building reconstruction effect is good, and the surrounding supporting facilities are complete. However, the public leisure plazas in 699 Park and Camphor Woods are often used for commercial activities, resulting in the lack of public space in the park. The satisfaction degree comprehensive evaluation of 798 Block and Nanchang Diesel Engine Plant Community is relatively low, and the interviewees generally think that there is no perfect functional space and the function is too single. Although the location is in downtown, and the surrounding supporting is better, the space in the block is narrow, and the activity place is insufficient.

2.3 Landscape Renewal and Reuse of Brown Field

In the process of development and utilization of waste industrial land, there is often an objective or intentional environment pollution. This kind of contaminated land is referred to as the "brown field". Because of the contaminated characteristics of “brown field”, the development of this kind of land is more complicated than that of other sites. The restoration and development of the "brown field" can bring more economic and social opportunities through the improvement of the original land ecology, and can also promote the land use value around the “brown field”. "Brown field" has provided new service facilities and activities for the city through the regeneration design, uses renewed ecological restoration methods to change the pollution of the former abandoned industrial land itself and its adverse impact on the surrounding environment, which can bring the potential of the site itself and its convenient transportation and space advantages into full play, thus attracting more commercial and economic investment, leading to the prosperity of the surrounding economy. Earlier industrial land in Nanchang lacked ecological concept, destroyed the natural environment, lacked green space system, and the rigid public infrastructure destroyed the water cycle of the area, which was not conducive to ecological balance and self-restoration.

3. Strategies of Industrial Heritage Regeneration Design in Creative Industry Park

3.1Continuation of industrial historical and cultural values and creation of cultural atmosphere in creative industries

The industrial heritage can be used as a carrier to record the development of human industrial history as well as to show the cultural landscape of human historical heritage, which is a witness of modern urban development and human industrial civilization. The architectural materials used in these industrial facilities with distinctive characteristics of the times, the enclosed space and the production atmosphere are as memorable as the epic psalms. Therefore, in the process of park planning and reuse, it is necessary to adhere to the principle of protecting history and culture, pay attention to the organic continuation of site history and culture, fully respect and retain as much as possible the original environment, planning layout, the historical and cultural status of architectural space of the site, as well as its industrial historical scenes and segments. For the whole city, this is also the protection of the continuity of urban economic, social and historical cultural development. In addition, while paying attention to the historical and cultural value of the original industrial heritage, it is also necessary to cultivate a good creative cultural atmosphere for the creative industry which infuses new life into the industrial heritage in the planning. The impact of historical culture and creative culture is the best means of sustaining and strengthening the cultural atmosphere.

Therefore, in the process of planning and reuse of the park, the two cultures should be brought together in a symbiotic relationship. While the industrial historical culture are protected and continued, a favorable cultural atmosphere should be created and collided for the creative industry at the same time.

3.2Flexible Treatment of Space Utilization

The planning of creative industry has its particularity. First of all, the distribution of different types of construction space, public space (including exhibition and performance, leisure and entertainment public space) in the park and the distance between the different areas and the intimate distance, the preferences and needs of creative work and so on, affect the creative workers' choice of office location, thereby affecting the distribution of functions in the campus. At the same time, the uncertainty of the choice will make different types of creative enterprises mixed in a region, so that the division of land can not be distinguished clearly. In addition, most creative workers prefer the mode of integrating living space with office space, and better closely connected with the public space of leisure and entertainment, and the whole park is like an organic whole, connected with each other. Rigid functional design will shackle creative flash, so the use of space needs to be rich and flexible.

3.3Decentralization of Design Rights and Macro-control of Planning

In order to increase the richness of space and the diversity of forms in the park, attract more creative workers to enter, and create a more fashionable and fashion-forward creative space and atmosphere with its unique tide sense, in the spatial planning of the park, creative workers should be given the greatest degree of freedom and shall decentralize past of the design rights. For example, the reconstruction and design of the interior space of the building can be decided by the creative workers themselves. For the facade of the building, a certain review mechanism shall be required to ensure the unity of the diversity and coordination of the whole park.

3.4Ecological Restoration and Regeneration

Compared with the other heavy industry cities, the industrialization of Nanchang city is relatively weak, the physical and chemical "brown" are not very prominent, but the damage of the industrial facilities to the land is not allowed to be small. In the opportunity of reconstruction, culture and ecology should be combined, and sponge city construction should be combined to transform it from the decaying organization of the city to the "city lung" of renewing the natural ecology of the city. In recycled design, designers try to make use of the existing buildings and facilities on the site and give them new use functions. By doing this, it can save materials while reducing the energy consumed during the production, processing and transportation of materials and reduce the damage to the original ecological environment and embody the ecological design method.

4. Conclusion

In the background of the supply-side reform and the development of the cultural industry, the migration of industrial land is the inresistable trend. A great deal of industrial waste land appears in people's view. For the sentimentally attachment to the city memory, the protection of the city development history and the restoration of the city ecology, these industrial land can become the industrial heritage, and create new vitality. The reconstruction of the city is the development, the memory preservation of the city can also have the development. This development is the process of realizing the land value and the cultural value, incorporating the missing function in the urban change in its original regional characteristics, and producing the value-added effect. More and more industrial waste land can be transformed into creative cultural industrial park through careful planning to continue the historical and cultural values, and restore the urban ecology.

References

[1] WAIBEL, W.（2012 ）The Beijing 798 Art Zone:A maturing creative cluster?. M,ZIELKE, P. Pacific News .

[2] Justin O'Connor,Xin Gu.（2010）Developing a Creative Cluster in a Postindustrial City: CIDS and Manchester. The Information Society

[3] Hans Mommaas. (2004) Cultural Clusters and the Post-industrial City: Towards the Remapping of Urban Cultural Policy.Urban Studies .

[4] Frederick Steiner. (2011) Landscape ecological urbanism: Origins and trajectories . Landscape and Urban Planning .

[5] Ren X.（2011）Building globalization:transnational architecture production in urban China.

IOP Conf. Series: Materials Science and Engineering **573** (2019) 012018 doi:10.1088/1757-899X/573/1/012018

Exploring product collaborative design under big data architecture with user experience scenarios

Chi Ningjun [1], Wu Zhiling [2]

[1] College of Art, Xi'an University of Science and Technology, Shaanxi, 71000, China

[2] College of Art, Xi'an University of Science and Technology, Shaanxi, 71000, China

[1]406646868@qq.com

[2]1443159748@qq.com

Abstract: In order to establish a user-oriented product design model, it is necessary to analyze existing user collaboration methods. By analyzing the existing user collaboration methods, the existing user collaboration methods are summarized. A sales store based on user experience scenarios and simulated design experiences is proposed. Through the tangible products and intangible experience, build a positive communication platform for users and designers. With face recognition, multi-channel camera, sensor technology, help users to participate in the design. Deeply explore the potential consumer needs of users under the big data architecture. It will help designers to better exert their own value in design research, and at the same time better assume responsibility in design research, so that design and manufacturing and consumer demand can reach full consistency and avoid waste of resources.

1. Introduction

In a new era of rapidly changing consumer demand, consumer subjects, consumption patterns, consumption structures, and consumption concepts have undergone “subversive” changes. Today, while product design has been designed to meet the needs of the user experience, it is still difficult to achieve a personalized, customized design for a specific user or user group. Consumers as the most varied elements, different consumer groups, design entry points will be different, designers only through the existing research methods to collect design data designed products, which cannot meet the consumer aesthetics and consumption of consumption pursue. Therefore, how to effectively realize the communication between consumers and designers has become a hot topic in the field of design.

2. The development of user collaborative design in the design field

The concept of “Design Participation” was first proposed at the Design Research Society conference in 1971. The original intention was to bring users to the design field. The focus of the research was mainly on improving the aesthetics of users, with the aim of more users understand the designer's design ideas [1]. Churchman proposed the concept of user collaborative design in the 1970s, but it did not mention the specific implementation method and implementation steps [2]. In recent years, the British Design Association has defined the user collaborative design in more detail: building a design that engages more people. Therefore, it emphasizes the participation of users [4]. The current research on user collaborative design is mainly based on three aspects: business management, general design theory and engineering design decision. In the field of product design, the way users participate in designing product collaborative design has not been effectively developed. Today, some brands have already deployed experience stores, such as MUJI, IKEA, Xiaomi House and so on. Although these brands built a user experience product scene, attracted a large amount of customer traffic, completed a certain amount of product sales, but did not play a role in the auxiliary product design and product upgrade. If you can build an experience scene with face recognition, multi-channel camera, sensor technology. Covering different consumer groups for comprehensive design, obtaining a large amount of feedback information, and transmitting this information to designers to help designers better complete product design, which will greatly improve design efficiency and bring greater economic benefits.

3. The way users design in the field of collaborative design

The ultimate goal of product design is to serve consumers. Consumer satisfaction is the market orientation and the design orientation [5]. Currently, an important part of designer designing a product is the research on the target user in advance, which can also be called "simple user collaborative participation design" in a certain extent. The existing research methods mainly include user interviews, questionnaires, scenario surveys, eye movement experiments, usability tests, A/B tests, and focus groups [6]. User interviews, questionnaires and scenario surveys are the three most commonly used survey methods. User interviews are the most direct and effective way to obtain real ideas and needs of users through long-term communication with users [7]. However, in the interview, it takes a long time for the user to communicate. Questionnaire survey is the most common survey method. Its advantage lies in the wide coverage of the survey and the ability to obtain more people's feedback for data statistics and analysis. However, the questionnaire will largely influence the user's answer [8]. Although the semi-closed and open-ended questions in the questionnaire can make the subjects more thinking, but the users usually complete the questions in a short period of time, lacking a lot of rational thinking and analysis, which will make the answer superficial. Scenario survey can help designers get some details that are ignored in the design process, but the construction of the scene has its own limitations. The scene experience can't cover a large number of users, and the collected information is still not perfect.

The research methods above, time, environment, population and other factors are all controllable factors. The participants have a certain degree of psychological pressure. The form of users participating in the research process is passive. Under such conditions, it is not fully feedback. The true feelings of the user. The research methods and the imperfect research content make the whole product design industry relatively weak in terms of innovation and insight, and are facing bottlenecks of continuous innovation.

4. User-designed product design new method

4.1. Experiential sales store design concept and development process

The main goal of design is to use the product as a bridge of communication between the designer and the user, and to exert the user's participation advantage through the construction of the communication platform [11]. In the new product design method of user collaboration, build a product experiential sales store, build a new social experience store platform centered on the user, and focus on the scene design of the experience store. Unlike existing traditional convenience stores, experiential sales stores record users as separate individuals. The positioning of the electronic number plate is issued to a typical user on the premise of giving the reward. Users who have received the electronic number plate will be the researched user. The whole research and development process mainly involve layout concept, competitive product research, market analysis, user research, computer virtual scene construction, shop location selection, related visual design, product selection, product procurement, product display, media promotion, testing stage, and official operation Stages.

4.2. Experiential sales store design

Simulate the design of an experience-based sales store based on user experience scenarios, and using big data and the internet, to upgrade the consumer experience, promote the reconstruction of people-goods-field, and explore how to better achieve user collaborative design. The layout of the experiential sales store breaks the previous category layout principle and changes to the lifestyle scene mode. The store is divided into several experience zones for different groups of people, including entertainment zone, health zone, kitchen zone and children zone, as shown in Figure 1. The store features face recognition, RFID tag recognition, multi-channel camera and sensor technology, target detection and tracking system technology. Through the scene design of product and leisure entertainment, the user research population covers a wider range, and the crowd category is more comprehensive. The product is used as a medium, and the store is used as a platform to complete the relationship between the designer and the wider user group. "Interactive communication", which is a good solution to the difficulty of finding the subject and the limitations of the subject in the user survey.

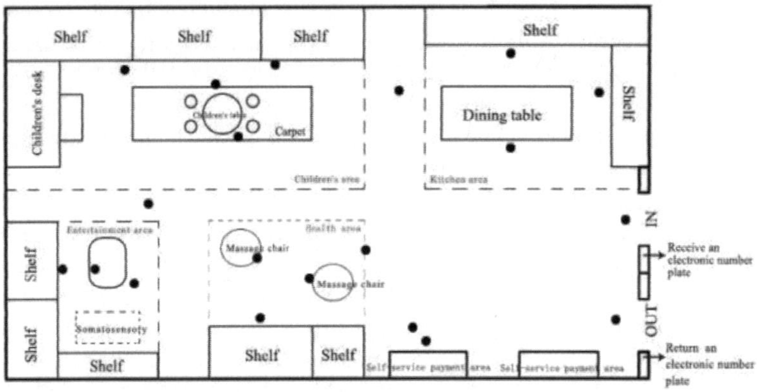

Figure 1. Experiential sales store layout design

The user experience process of the experiential sales store is: attracting consumers by rewards, participating in user collaboration, and participating consumers receiving the positioning electronic number plates when entering the store. The local user group in the store's inbound and outbound flow is issued with a positioning electronic number card (200 number plates per day). Return the electronic number plate when consumers leave the store.

4.3. Individual sample tracking analysis

Taking human flow analysis as an example, individual sample tracking analysis is performed. A qualitative user portrait is created by the persona method to derive a series of design insights. First, the user files are divided into two categories according to gender: male and female; according to age group: juvenile (7-17 years old), youth (under 45 years old), middle-aged (45-59 years old), old age (over 60 years old). Then select the typical consumer of conducts individual sample tracking analysis in four time periods (Morning: 8:00-10:00/10:00-12:00, Afternoon: 14:00-16:00/16:00-18:00) on Sunday (In the rest day consumers flow rate is large, the test data is more accurate), and the qualitative user portrait is shown in Figure 2. Then, the in-store camera, motion sensor, direction sensor, distance sensor and other devices are activated to perform individual sample tracking analysis. A stream of photos is taken every 5 seconds, and computer graphics are processed to form a density thermal map. The photo of the flow of people and the heat map of the flow density are shown in Figure 3. From the four time periods on Sunday, each unit took 20 minutes, and each representative time group selected six representative heat flux maps for analysis. The heat flux diagram of the flow density is shown in Figure 4, Figure 5, Figure 8, and Figure 9, and the peak pattern is formed by superimposing 6 layers of human thermal density layers, as shown in Figure 6, Figure 7, Figure 10, Figure 11.

IOP Conf. Series: Materials Science and Engineering **573** (2019) 012018 doi:10.1088/1757-899X/573/1/012018

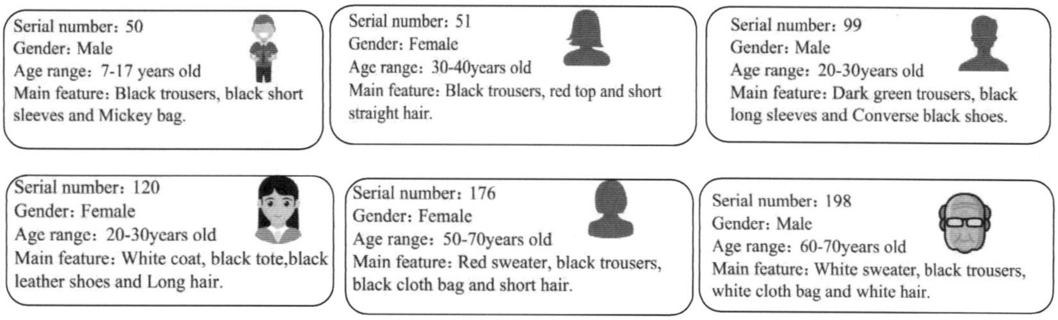

Figure 2. Qualitative User Portrait

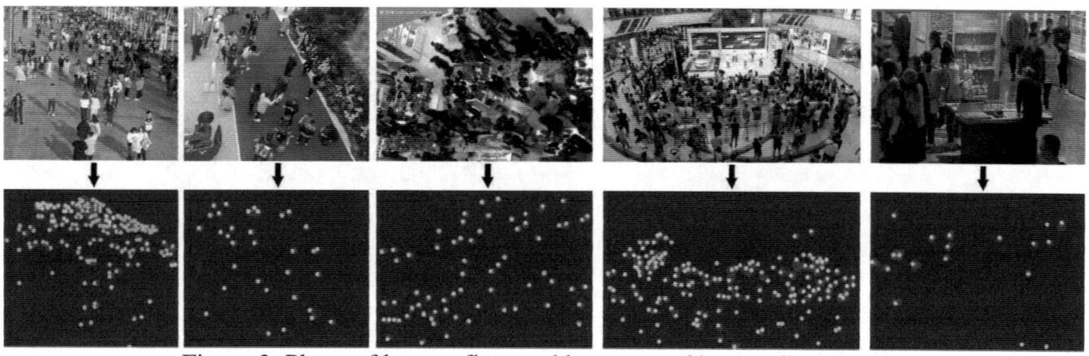

Figure 3. Photo of human flow and heat map of human flow density

In the following flow density thermal map, the number 50 consumer is marked with a red square; the number 51 consumer is marked with a pink triangle; the number 99 consumer is marked with a blue rectangle; the number 120 consumer is marked with a sky blue semicircle The consumer numbered 176 is marked with a yellow heart; the consumer numbered 198 is marked with an orange petal shape.

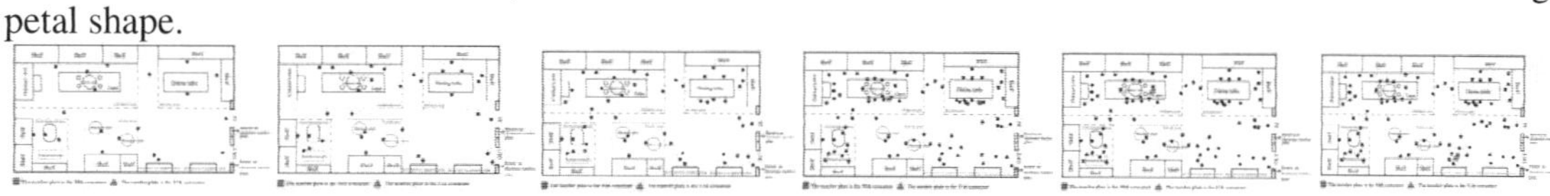

Figure 4. Morning: 8:00-10:00 People's flow density thermal map

Figure 5. Morning: 10:00-12:00 People's flow density thermal map

By selecting six typical human flow density thermal maps in the morning: 8:00-10:00, the peak density can be formed, as shown in Figure 6; in the morning: 6:00:00-12:00, 6 typical human flow density thermal maps, A peak density is formed, see Figure 7.

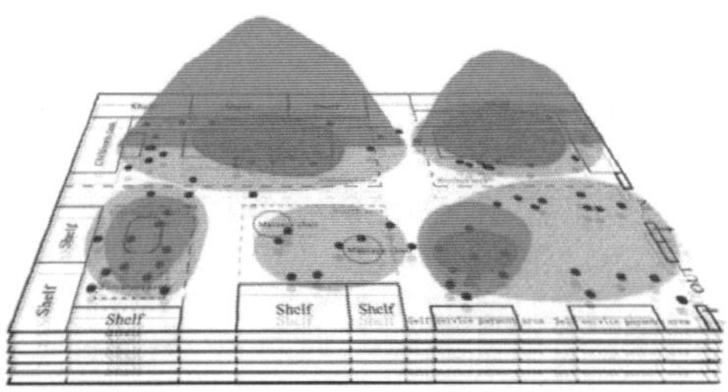

Figure 6. AM: 8:00-10:00 density peak map

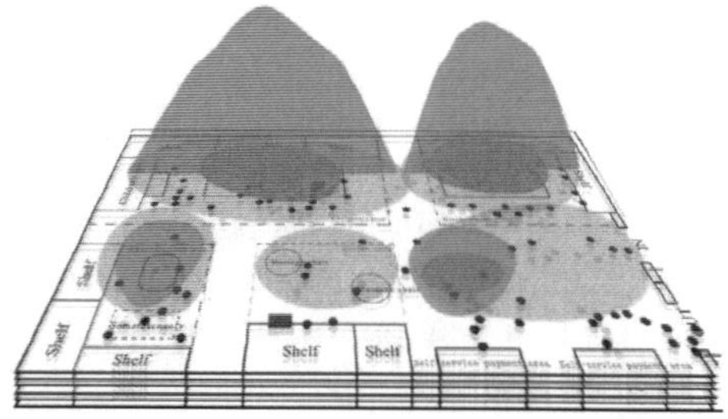

Figure 7. AM: 10:00-12:00 density peak map

Figure 8. Afternoon: 14:00-16:00 People's flow density thermal map

Figure 9. Afternoon: 16:00-18:00 People's flow density thermal map

By selecting the typical thermal flow pattern of 6 typical abortions in the afternoon: 14:00-16:00, the peak density can be formed, as shown in Figure 10; in the afternoon: 6 typical human flow density heat maps from 16:00-18:00, a peak density can be formed, see Figure 11.

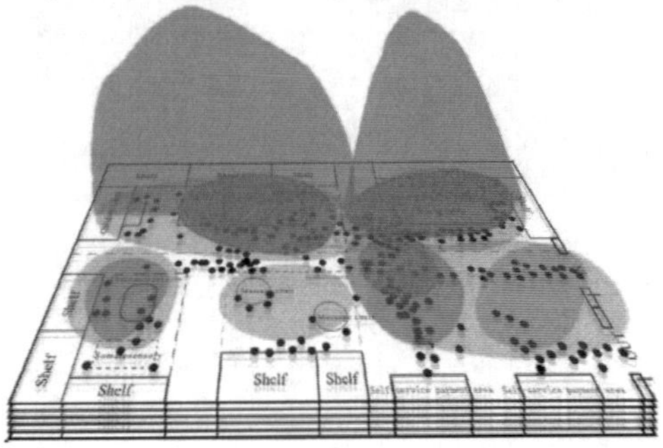

Figure 10. PM: 14:00-16:00 density peak map

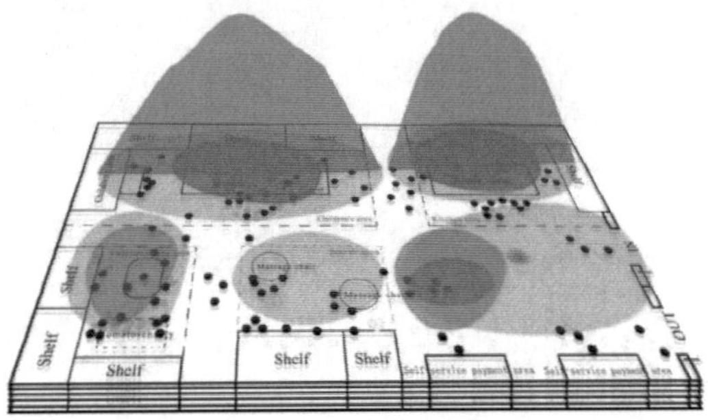

Figure 11. PM: 16:00-18:00 density peak map

Put the four sets of peak maps together for comparative analysis, as shown in Figure 12.

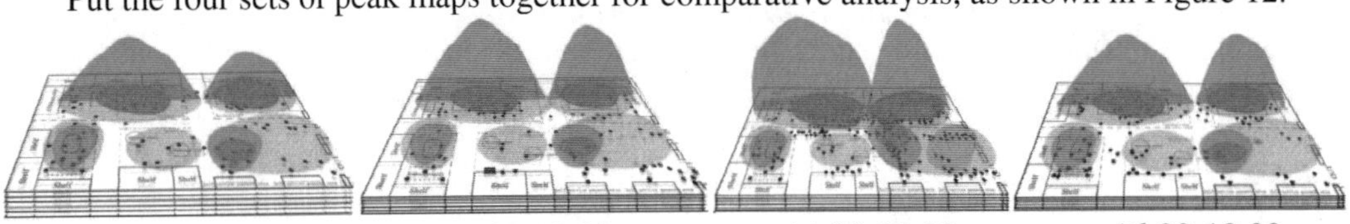

8:00-10:00 10:00-12:00 14:00-16:00 16:00-18:00

Figure 12. Peak graph comparison analysis

Through the collection of people's current density in 4 time periods, the analysis shows that: (1) In the day of Sunday, 8:00-10:00, the customer in the store is the least, and the afternoon is 14:00-16:00. The consumer is the most. (2) Through the peak flow chart of the four periods of time, it can be analyzed that the population density of the children's area and the kitchen area is the most concentrated, and it can also be defined as the area most favored by consumers. (3) Analysis of the preferences of people of different age groups by the length of stay of 6 typical users in the area. Numbers 50 and 51 have the longest stay in the children's area. The number 99 consumer has the longest stay in the health zone. The number 120 consumer also stays in the health zone for the longest time and finally completes the payment. The number 176 consumer has the longest stay in the kitchen area.

4.4. Other sample tracking techniques

In the subsequent information collection, the direction sensor is used to record the movement trajectory of the consumers in each area, and the data is analyzed at which position the flow rate of people is the most, and the reason is analyzed to provide information for the change of the subsequent scene style. By recording the length of time that the consumer experiences the product through the distance sensor, the consumer's interest in the product can be obtained for different age groups, and the design direction is determined for the designer. The gravity sensor records the number of times the consumer experiences the product, analyzes the reason why purchase, and the reason why not purchase. Through face recognition, the expression changes in the consumer experience product are recorded, and then the reasons why the consumer likes the product and the reasons for dislike are analyzed. Through face recognition, multi-channel camera, sensor technology, target detection and tracking system information collection, generate massive data information, transmission background processing, which help designers to explore user spending habits and consumption direction, combined with user needs to adjust and optimize the product.

5. Conclusion

This paper innovates a new approach to product design for user collaboration. Today, after the 80s, 90s, and 00s, as the core consumer of the market, they pursue psychological satisfaction and emotional needs, and pursue quality, diversification, individuation, and convenience. In order to meet the needs of the main consumer groups, build a product experiential sales store, through the way of product scene partitioning, with face recognition, multi-channel camera, sensor technology, target detection and tracking system, and the use of big data and Internet to help consumption Collaborate in the design phase of product design. This method can help designers to explore the potential needs of consumer entities and propose targeted product design solutions. Moreover, this method of user collaboration is not limited in the field of product design, but can also be applied to other design fields.

References

[1] Yu Conggang,Zhao Jianghong,Tow Product Design Patterns Based Data driven.Packaging Engineering.Vol.34.4.Feb.2016
[2] Li Weichao ,Jia Yiwei ,Zhao Haixia.Research on the Mining of Users Needsbased on Big Data Analysis.Innovation Science and Technology.Vol.201， No.11.Nov， 2016.
[3] Shao Xiuli, Yu Qiqi, Yang Yajing, Ren Zhijuan.A collaborative valve design system based on Web3D.Computer Age.Vol.7.2018.
[4] LIU Lanlan.Virtual reality based collaborative product appearance design under human computer interaction.Modern Electronics Technique.Vol. 41 No. 7.Apr. 2018
[5] ZHANG Bo, QI Xiaoli ,WANG Cong.Analysis and Research of Process Design Synergy in Product Design Process.Henan Science and Technology.Vol. 6.2018.
[6] YIN Xuemei，ZHOU Junhua，ZHU Yaoqin.Data modeling and data-driven method in collaborative design of complex products.Journal of Computer Applications.Vol.38.No.10.2018.
[7] LUO Shijian,WENG Jianguang.Patterns of Tacit Knowledge Based on Ethnography in Product Design.Chinese journal of mechanical engineering.vol.44.No.4.April,2008.
[8] Li Yuemei,Sun Shouqian,Pan Yunhe.Multi-Agent Conflict Resolution in Cooperative Design.JOURNAL OF COMPUTER-AIDED DESIGN& COMPUTER GRAPHICS.Vol.14,No.2.Feb., 2002.
[9] KA RA NG E LEN N E, T . HO AN G N . S imulatio n base d de sign for large complex computer based system [A] . 1994, Proceedings of Systems Engineering of Compute r - Based Systems [C] . 1994. 116- 123.
[10] Shan hongbo,yan haifeng,zhou shanjin,yu haiyan,wang shuguang,cheng qixing.Key Technologies and Realization of Multi user Collaborative Design System in Product Conceptual Design.Journal of Donghua university(Natural Science).Vol.41.No.4.Aug.2015
[11] MCLELLAN H.Experience Design，Cyber Psychology &Behavior[J].Design Studies，2000，3（1）：64—68.
[12] LU Jian-zhou. Research on the Method of Product Experience Design[J].Packaging Engineering，2014，35（6）：86—89.
[13] Yang Qiuqiong, Cao Jian, Zhang Youliang. Research and realization of constraint management system for cooperative roduct development[J]. Mechanical Science and Technology, 2000, 19(4): 638～ 640
[14] Lynn S G. Designing services that deliver[J]. Harvard Business Review, 1984:39-133.

IOP Conf. Series: Materials Science and Engineering **573** (2019) 012019 doi:10.1088/1757-899X/573/1/012019

A Study of Formal Model of Interactive Behaviour in the Whole System for Implicit Action and Context

Bo Li*, Yanmin Xue

Department of Art and Design, Xi'an University of Technology, Xi'an, Shaanxi, 710054, China

*Corresponding author's e-mail:ekinshow@sina.com

Abstract. Formal techniques have been used to analyze a variety of properties of interactive behaviour in a number of systems. To have a better understanding of the implicit of the interactive behaviour, it is necessary to find a general method to describe the task. In this study, we proposed a formal model to illustrate the implicit task. We validated the effect of this model by an experiment in the simulated coach bus station by receiving transition information by the holder held device. This paper analyses the whole system, the rooms, the public displays, the mobile devices and the timeliness of information in the system. The model made the interactive process clearer and provided a basic framework to study the related issues of orientation in the interactive system in the future.

1. Introduction

Formal approaches bring rigor and automation to usability engineering. Models capture key features of the interactive behaviour of the design and are then subjected to systematic analysis using verification techniques such as model checking (Clarke, Grumberg and Peled, 1999). The appropriate selection of models and verification techniques is critical to the value of the analysis. Otherwise models can bias the analysis and verification techniques can take the focus away from important safety or business-critical features of the designed interactive system. Two categories of automated verification techniques can be identified. An important challenge if these techniques are to be used effectively and efficiently is that there be generic models of interactive behaviour that can be taken off the shelf by the engineer and instantiated to the particular requirements of the system under design.

The research focuses on formal methods such as statechart diagram as a different sort of model for interactive systems. Instead of modelling the user like cognitive models, these models consider the system and the context for the system. They can then be reasoned about to ensure that they function as they should. This is particularly important in safety critical system, like coach bus control or medical monitors, where the cost of failure is never acceptable.

Although formal techniques are widely used in the analysis of interactive systems there are reasons why an appropriate set of tools, suitably designed to be usable by system engineers, could be of value in the portfolio of techniques used to assess interactive systems. This paper describes the role of formal techniques in modelling and analyzing interactive systems, discusses unfulfilled opportunities and speculates about the removal of barriers to their use. It also presents the opportunities that a clear expression of the problem and systematic analysis techniques may afford.

The mode is based on concurrently executing automata that can be synchronized by using labels on the state transitions: a transition can occur in two parallel automata if complementary labels occur on arcs in the two automata and both arcs are enabled at the same stage in the execution of the system.

The approach used Uppaal, which additionally focused on continuous time (Behrmann, David and Larsen, 2004). Models based on these techniques are good for describing behaviour of actions in the interactive system in terms of the actions to which they will lead. Models can be used to analyze the mode characteristics of an interactive device (Gow, Thimbleby and Cairns, 2006). There are also textual approaches to modelling labeled transition systems such as there. For example, SPIN (Holzamann, 2003) captured the model using a process algebra called Promela. Other approaches, for example Alloy (Jackson, 2006) and Modal Action Logic (Campos and Harrison, 2001), captured relations between states described textually in terms of typed attributes of the states.

2. Verification in development

The value of formal (mathematically based) models of interactive behaviour of systems is in providing precise descriptions of features of the interactive behaviour of the design early in the process. The features focus on user action, whether explicit or implicit, and the observable effects that action has in the context. This makes it possible to analyze characteristics of the behaviour of the system before expensive commitments have been made to implementation. In our research, a model approach is taken that share a number of features in common. It is diagrammatic approaches based on finite state model which capture properties of the system in term of states and transitions between states.

As briefly summarized, analysis explores behaviours of a device. An important feature of the behaviours of interest is that they correspond to envisaged uses of the device. The mobility of the designed hand-held device may have consequences for interaction. The analysis is designed to relate to a number of concerns including, for example, whether these behaviours lead to confusing mode shifts or whether there are situations in typical sequences where it is impossible to recover. As discussed in Campos (1999), the process of establishing usability requirements, modelling systems and checking the models against requirements can be thought of as a cycle consisting of four steps, as represented in Figure 1. These steps may involve discussion between software or system developers on the one hand and human factors specialists on the other. The cycle is as follows:

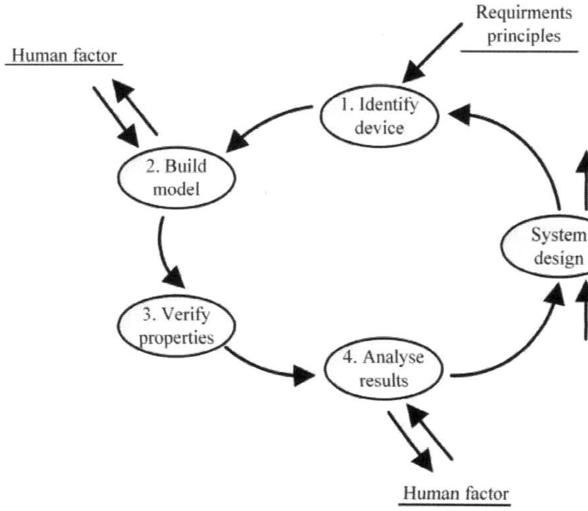

Figure 1. The circle of verification in development.

(1) Identifying usability requirements: Based on usability guidelines or heuristics or specific concerns derived from a requirements elicitation. They are used to assist the development of a model to be used for the analysis. They involve device descriptions or whole system descriptions – a device might be a mobile phone, the system is the whole system in which the mobile phone is embedded. These requirements are expressed as formal properties.

(2) Building model and formulating properties: Once the initial requirements have been expressed, a model of the interactive device or system and a set of properties are formulated. This step benefits from the participation of someone with specific user or domain expertise. Models capture

assumptions about the user's view of the device or how users are embedded along with devices inside the system. Properties capture constrains that if broken represent potential failures in the use of device or system. This stage could be achieved by re-using an old model or instantiating a generic model.

（3） Verifying properties: The next stage is to verify whether the properties hold of the device under the constraints imposed by the environment—for example, a property requiring the visibility of actions but only over the paths that relate to expected tasks. Constraints are either expressed in the properties themselves or through additional models capturing relevant characteristics of the environment in which the device is to be considered. The properties could be examples of general property patterns.

（4） Analyzing the results: Finally the consequences of the results are analyzed. The technique used in our research is model checking. The method of analysis involves checking a property. If the answer is positive then, under the assumptions captured by the model, the usability property is true of the system described by the model. If the answer is negative then the reasons why are investigated. A rationale explaining why the property does not hold is provided by the reasoning tool, in this case a trace representing the confluence of conditions in which the property fails. A negative result might indicate that the device model is incomplete. It may therefore be necessary to determine whether all relevant aspects of the device have been considered. Alternatively it might point to a situation where the assumptions made about the user must be refined. For example, it may be necessary to eliminate specific courses of action from the expected user behaviour. Finally, it might be a genuine usability problem.

3. Modeling implicit action and context in the whole system

We provide a model of the system that can be used to check properties for the coach bus station. The model focused on usability issues associated with interaction with a mobile device within a system. Usability issues may also be associated with the characteristics of the whole system, for example the way in which information is presented to users in different locations within a built environment. The issues that are of concern are about whether implicit actions will have desired effects—for example, will a relevant message be sent to a user within a certain time after arrival at a given location.

Many factors affect users of built environments. These include the texture and physical characteristics of the environment and where information displays are situated. They are also affected by whether relevant information is received by them in a timely way. Consider the example of a coach bus station. In each space within the station there is a public display of messages about transit information to passengers that occupy the space at any one time. Each passenger carries a mobile phone and receives messages on this phone that are specifically relevant to their information and location. Action is implicit, in the sense that when a passenger walks into a space the location sensor notes the fact and as a result information may be distributed to the passenger's mobile device and possibly to the public display if it is the first time someone for a designated coach bus arrives in the space. The design that is modelled adopts a simple set of techniques for deploying information users sufficient to provide a basis for the discussion. Many schemes are feasible for combining public displays with private information, see for example Gilroy et al. (2006). The scheme used here is illustrative of a range that could equally be addressed by the techniques described.

In this case the implicit action occurs when a user moves into a space. The effect of this movement is that a sensor in the space recognizes the presence of the user's mobile device and triggers the system to provide information that is relevant.

3.1 User requirements for the system

The user concern is translated into the following properties of the model:

P1 When the passenger moves into a location then the transit status information is presented to the passenger's hand-held device within some maximum delay time.

P2 Information on public display should reflect the current state of the system within a maximum delay time.

IOP Conf. Series: Materials Science and Engineering **573** (2019) 012019 doi:10.1088/1757-899X/573/1/012019

P3 When the passenger enters a new location, the sensor detects the passenger's presence and the next message received concerns transit information and updates the passenger's hand-held device with information relevant to the passenger's position and stage in the embarkation process.

P4 When the passenger moves into a new location then if the passenger is the first from the coach bus to enter, public displays in the location are updated to include this transit information.

P5 When the last passenger on a particular transit in the location leaves it then the public display is updated to remove the transit information.

3.2 Characteristics of the coach bus station model

Since properties P1 and P2 are timing properties, a model that incorporates time is adopted for this system. The model is split into a number of processes describing:

Sensor: the activity within a room, including the mechanism for sensing the arrival and departure of passengers. This process updates the room-based display to show transit information for those passengers that are in the room.

Passenger: the passenger that receives specific messages relating to transit and location in the coach station. The passenger moves from room to room.

Dispatcher: the centralized dispatcher that sends messages regularly.

The dispatcher is critical to the timing characteristics of the design and is the basis for exploring the system to ensure that timing guarantees are satisfied. It is assumed that the dispatcher is human machine system. The dispatching of messages involves human intervention at some level and this is the primary source of delay. This exploration is carried out by adjusting the rate and the order of distribution of messages, as well as passenger arrival volumes. The illustration in Figure 2 distributes messages in strict order with a delay between broadcasts specified by the constant workload. The model makes a simple set of assumptions about messages and the way they are tagged for recognition. Tags are associated with coach bus number and location. In practice, of course, the station system would be more elaborate than this. Many of the potential complexities of the model are removed by focusing on the specific requirements with the proviso that checking the properties does not bias the analysis exclusively to a certain class of properties, thereby providing a model that can be used as a structure for implementation.

Two types of process receive information from the dispatcher. The sensor process (Figure 3) combines the behaviour of the public display with the room sensor. Each room or space in the station contains a sensor (how the sensor is implemented is not the concern of this level of modelling). The generic sensor is instantiated for each space: entry hall, stand 1, stand 2, main hall, gate and so on. The other type of process is the passenger process （Figure 4） that describes the passenger and the relevant behaviour of the passenger's mobile device.

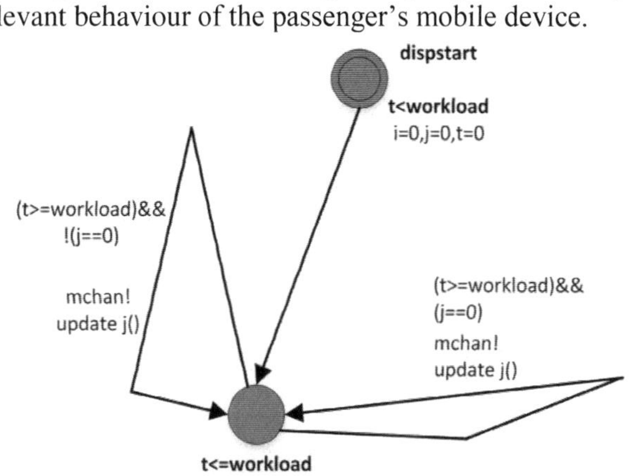

Figure 2. The dispatcher process

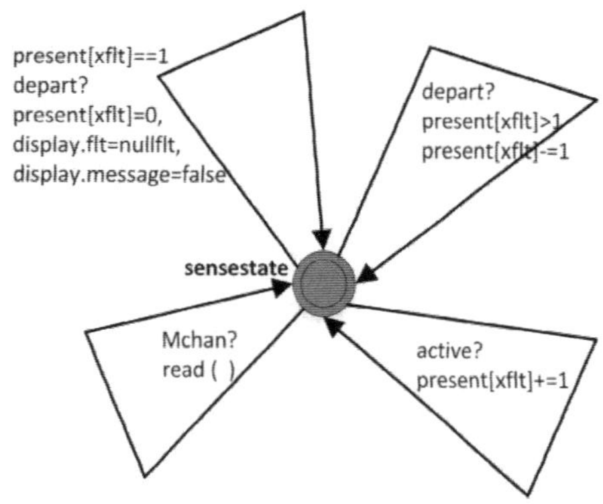

Figure 3. The sensor process

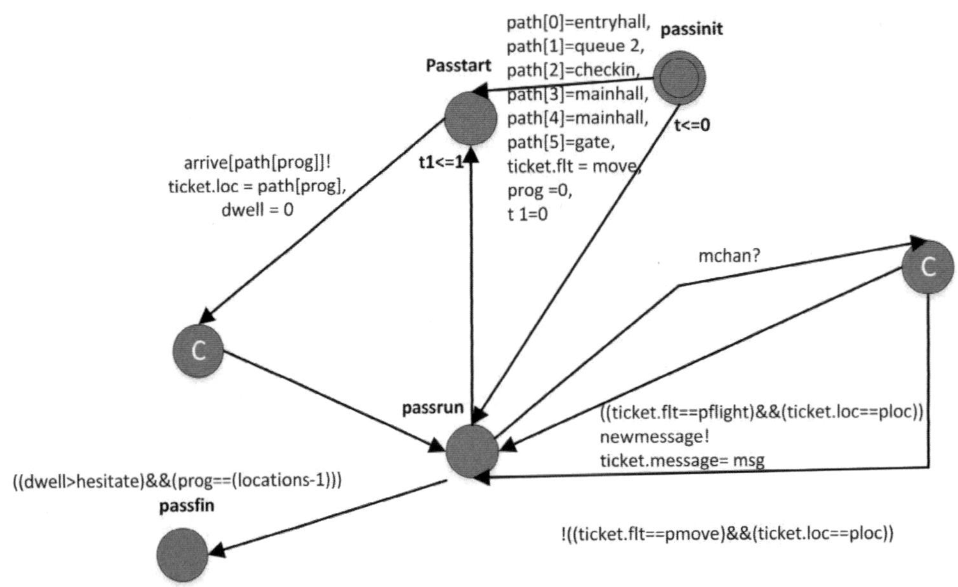

Figure 4. The passenger process

The sensor process (Figure 3) describes the key interaction features of:

- The public display located in the room
- The sensor updating its knowledge of whether passengers are present in the space—this assumes an interaction between the sensor and the passenger device. The sensor communicates by means of three channels.
- It receives messages that have been distributed to it from the dispatcher by means of the channel mchan.
- It receives requests from the passenger' hand-held devices (via *arrive*) where they arrive in the room that relates to the sensor.
- It receives requests from the passenger's hand-held devices (via *depart*) when they leave the sensor's room.

When the sensor receives a message from the dispatcher, the function read ( ) checks the tags on the message and if the location tag coincides with the location of the sensor then the display is updated. Of course, a realistic implementation of this system would update a transit information array for display each time a relevant message is received. The array updating mechanism is not of interest to interaction analysis.

When the sensor receives a message from the arrive channel this signals the entry of a passenger. The array present [ ] keeps a count of the number of passengers present for a particular transit and is incremented with the arriving passenger's coach bus number. When the sensor receives a message from the depart channel then the array is decremented using the departing passenger's coach bus number. If the result of this is that there are no passengers for a particular coach bus left in the room then the transit information is removed from the display.

As a consequence, when the last passenger moves out of the space this information is cleared from the display. When the passenger is newly arrived in the space then the array present is incremented and so next time a message arrives about this coach bus the information will be displayed for the first time.

The passenger process (Figure 4) describes the activity of the passenger and the key features of their mobile phone. This activity has a number of characteristics:

- The passenger is given a specific path to follow. This is defined in the array path, one would anticipate more than one passenger defined associated with different paths.
- The process notifies the room sensor that it has arrived. The passenger ticket is updated to

IOP Conf. Series: Materials Science and Engineering **573** (2019) 012019 doi:10.1088/1757-899X/573/1/012019

point to the current location.

- The passenger moves to a state where it receives messages from the dispatcher via mchan. If the received message is tagged with the passenger's current location and the passenger's coach bus number then the mobile phone display is updated.

4. Discussion

P1 requires that when the passenger moves into a new location in the coach station then move status information is presented to the passenger's hand-held device within a maximum delay time (this is defined by the variable *maxdelay*). It must be updated within a period of delay after a passenger enters the location (regardless of coach bus number) the relevant message will be received by the passenger within the time delay. Two transitions are of interest in the passenger process (Figure 4). The first occurs as the passenger moves into the new location and the second occurs when the passenger receives a message from the dispatcher that matches the coach bus number and location of the passenger. P1 is checked by introducing an observer process (Figure5) and adding a communication (*newroom*) in the passenger process (Figure 4) to signal arrival in the new location and similarly a communication （*newmessage*） to signal receiving a relevant message. If the message does not arrive while the passenger is in a location (this time is determined in the model by the variable dwell) then the observer will deadlock. The deadlock occurs because the observer is waiting to receive notification that a message has been received but instead receives a message from the passenger that it has entered another room. When appropriate diagnostics are switched on, deadlock generates a trace that can then be further analysed to work out why the system does not satisfy the properties.

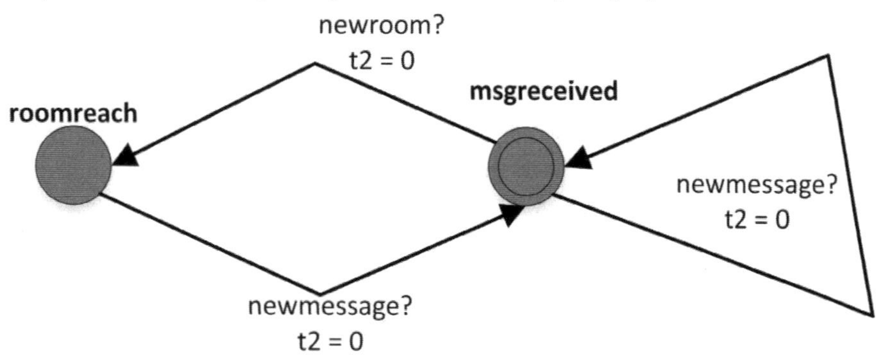

Figure 5. The observer process

In practice the generalized deadlock property is very computer intensive and on a no-frills specification PC the Uppaal system ran out of memory after three hours execution. The specific properties relevant to P1 and P2 relate to checking that fresh messages arrive within the maxdelay interval. In the case the properties was checked experimentally for different values of maxdelay. When the property failed, and the passenger did not receive the message in time, the diagnostic trace was investigated to check which assumption caused this failure to occur thereby allowing the designer to explore alternative strategies for distributing messages so that passengers would receive timely updates.

5. Conclusions

The advantage of these simple models is that they can be explored using model checking. These approaches explore paths exhaustively to check whether the model satisfies the property. The model therefore is described in one of the languages above. The model checker converts it into an appropriate form for analysis. The analyst provides properties, often expressed in a temporal logic, to analyze the model. The model checker then attempts to find a counter-example to the property, or an instance where the property is true, and provides the analyst with a sequence of states that describes the counter-example or true instance. Model checking has the advantage that it is a decidable algorithmic approach that is therefore relatively easy to use. The disadvantage of the technique is that there are a number of properties of interactive systems that cannot be adequately analyzed using these approaches, for example representation properties.

The problems with using formal techniques relate to how well the techniques scale, how easy it is to use the methods and whether the methods bias the analysis so that key issues may be ignored. It is not contended that this type of approach is the answer to all of the usability engineering problems. However, if usability engineering is to be a true engineering discipline, it does need rigour and formality. Formal methods of software engineering have a role to play during interactive system design and analysis. They enable a type of reasoning that is more thorough and repetitive, without precluding the use of other more traditional approaches to usability analysis. How to best integrate the two worlds is a challenging and interesting research area. No single current technique has the expressive power to capture, and enable the analysis of, all interesting facets of an interactive system.

Instead, different techniques should be applied as needed.

Acknowledgement
Financial support provided by the Humanities and Social Sciences Fund of the Ministry of Education, China (17JAZH100) & the Special Project on Humanities and art of Xi'an University of Technology (106-451017004).

References
[1] Clarke, E.M., Grumberg, O. and Peled, D.A. Model Checking [M]. Cambridge, MA: MIT Press, 1999
[2] Behrmann, G., David, A. and Larsen, K.G. 'A tutorial on Uppaal'. Formal Methods for the Design of Real-time Systems[M]. Springer Lecture Notes in Computer Science, Berlin: Springer, 185: 00-36, 2004
[3] Gow, J., Thimbleby, H. and Cairns, P. Automatic critiques of interface modes [M]. Springer Lecture Notes in Computer Science, Berlin: Springer, 3941:201-12, 2006
[4] Holzmann, G.J. The SPIN Model Checker [M]. Primer and Reference Manual. Harlow, UK: Addison Wesley, 2003
[5] Jackson, D. Software Abstractions: Logic, Language and Analysis [M]. Cambridge, MA: MIT Press, 2006
[6] Campos, J.C. and Harrison, M.D. Model checking interactor specifications[M]. Automated Software Engineering, 8:275-310, 2001
[7] Campos, J.C. Automated Deduction and Usability Reasoning [D]. PhD thesis, Department of Computer Science, University of York, UK, 1999
[8] Gilroy, S.W., Olivier, P.L., Cao, H., Jackson, G. D., Kray, C. and Lin, D. Cross-board: Crossmodal access of dense public display[C]. International Workshop on Multimodal and Pervasive Service (MAPS06), Lyon, France, 2006
[9] NijunLi , Tongchi Zhou , Lin Zhou , Zhenyang Wu . Action recognition by Huffman coding and implicit action model [C]. 2015 IEEE International Conference on Computational Intelligence and Virtual Environments for Measurement Systems and Applications (CIVEMSA), 2015
[10] Y. P. Huang, G. Z. Wang, T. S. Chang, T. H. Chen. Three-dimensional virtual touch display system for multi-user applications[J]. Display Technol., vol. 9, no. 11, pp. 921-928, 2013
[11] Bonnechère, B.: Serious Games in Physical Rehabilitation: From Theory to Practice [M]. Springer, Brussels , 2017
[12] Garner, T.: Applications of virtual reality[M]. In: Echoes of Other Worlds: Sound in Virtual Reality, pp. 299–361, 2018

IOP Conf. Series: Materials Science and Engineering **573** (2019) 012020 doi:10.1088/1757-899X/573/1/012020

Exploring museum service experience using gamification-based personas with distinct motivations

Shunli Liu[1] and Muhammad Zaffwan Idris[2*]

[1]Faculty of Art, Computing & Creative Industry, UniversitiPendidikan Sultan Idris, TanjongMalim, Perak, 35900, Malaysia

[2]Faculty of Art, Computing & Creative Industry, UniversitiPendidikan Sultan Idris, TanjongMalim, Perak, 35900, Malaysia

[*]Corresponding author's e-mail: zaffwan@fskik.upsi.edu.my

Abstract. In the context of museums facing challenges from the cultural and leisure market, a desirable experience that can be tailor-made for different type of visitors can serve as an indicator to measure brand equity. To explore the museum service experiences, the Bartle Test of Gamer Psychology was introduced for identifying personas representing different visitors' motivations to visit the museum. With the personas being recruited, this study will employ the adapted rapid ethnography as the research approach to gain in-depth and detailed insights into the perceptions from the user's perspective.

1. Introduction

Although the museum is considered the top of the cultural pyramid, the challenges from the cultural & leisure market to the museum cannot be underestimated. In order to cope with this challenge, the branding strategy in the era of symbolic consumption can help visitors to speed up their selection and enhance the museum's brand equity. As different individuals may hold different perceptions since a brand is individuals' gut feeling about a product or service, the Bartle Test of Gamer Psychology based on players' type theory will be introduced to identify personas to explore museum service experiences of different user categories of motivation.

2. Museum and branding

Cultural strength can show comprehensive national strength. Notwithstanding, in the aspect of attracting new audiences, the museum faces challenges which come from the competition both within non-profits industry and the culture and leisure market industry (include the leisure FPOs [Farmer Producer Organizations]). Despite its increasing competition, museum branding is still a neglected area of research [1-3]. To a large extent, non-profit organizations, including museums, hesitate to adopt branding techniques due to seeming too commercial, resource commitment, and other risks [4].

Despite this reality that non-profit organizations are reluctant to adopt branding strategy, what needs to be acknowledged is social economy have shifted from mass production to mass customization. This is causing us to originate our choices more from symbolic characteristics [5]. For museum, brands help visitors to speed up choice-making, decrease perceived risks, and progressively help the museum survive in competitions.

Regarding the brand and branding, Neumeier defined a brand as individuals' gut feeling about a product, service, or company, and each person creates his or her version of it. A company can be stated to have a brand when enough individuals arrive at the same feeling [5].Through the definition, it shows that different individuals may hold different perceptions for the same thing because the brand is a personal perception. Although the companies or organizations cannot control the personal perception version, a company is able to affect those personal impressions in the way that the company wants individuals to see its brand. This process of brand building is branding [5-6].

3. Service experience design

Having defined what is meant by brand, move on now to discuss services in the brand. According to Murphy [7], the concept of branded things has been expanded successfully to include services, and non-tangible factors have become progressively to distinguish services from one another. Service undoubtedly needs to be designed. Specifically, Service Design is the design of how a service deliver an experience to their customers over time [8]to improve customers' service experiences.

According to the definition of the service design described above, it can be concluded that what service design (SD) provides is user experience (UX). Many studies concluded that experiences are a feeling of extreme emotional excitement[9-10], that is, something extraordinary or memorable that differs from everyday life. In detail, SD extremely focus on the experience of individuals in their own context during all stages of the design process[11]. Therefore, SD can serve as an approach to explore UX for museum. Further, Camarero, et al[3]concluded that repeating a satisfying experience in future dates can even serve as a measure of brand equity for cultural organizations. For the museum, it is not only the educational institution. To attract visitors in the leisure market, other experiential values of museum brand identity ought to be considered to integrate into the brand communication by employing SD approach.

4. Gamification-based visitors' motivations

Until now, this study has focused on SD and museum UX and concluded that service experience design emphasize the feeling of individual rather than the so-called general public. Based on this conclusion, the following will discuss the personal motivations of museum users. A study provided new insights into another potential problem: "How can potential visitors be motivated to visit the museum continuously?" [12]. In recent years, there has been an increasing amount of literature on Gamification, and authors believed that Gamification is a way to improving user motivation, engagement, and experience [12-14].

Specifically, Gamification is the use of elements of game design in non-game contexts [15]. In this definition, the term 'non-game contexts' refers that the purpose of using game elements is to help accomplish something non-gaming. Further, Bartle [16] drew on an extensive range of sources to identify four types of players: achievers (A), socializers (S), explorers (E), and killers (K) (we re-name killers as attackers). In other words, the four player types make up four categories of 'fun'. In brief, achievers want to book results, socializers want to collaborate, explorers want to understand, and attackers want to win. According to the analysis of axes, Bartle found that achievers and explorers were more interested in the game world (environment) than in other players, while attackers and socializers were more inclined to focus on players. When turn to the other axis, attackers and achievers were actively interested in acting, while socializers and explorers prefer to interact in an environment of Gamification. These four types of players identified by Bartle will help researchers use the visitors' motivation to explore museum UX.

IOP Conf. Series: Materials Science and Engineering **573** (2019) 012020 doi:10.1088/1757-899X/573/1/012020

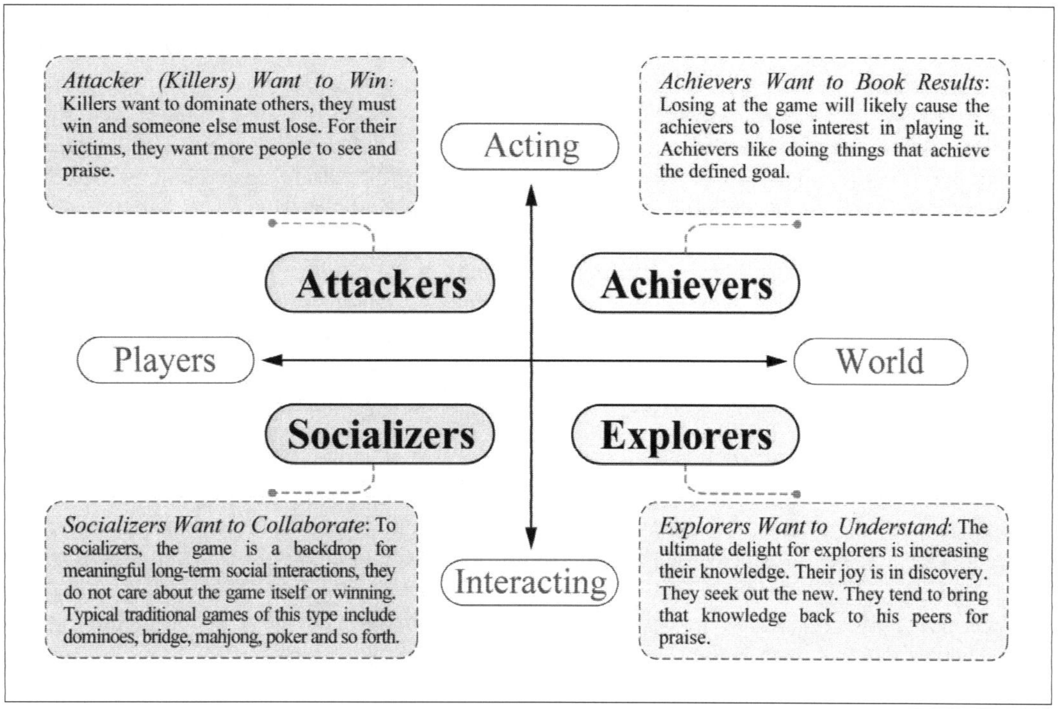

Figure 1. Types of player in game (Adapted from Bartle, 2004).

5. Introducing the Bartle Test of Gamer Psychology

Based on the above taxonomy established by Bartle, Andreasen and Downey developed a questionnaire (GamerDNA test) for the Bartle Test of Gamer Psychology to identify gamers' playing style preference that predominates in a group of players. According to Konert, Göbel, and Steinmetz [17], this test already collected data of more than 200,000 recipients. However, this version of the Bartle test is now offline. Thus, based on the same underlying data/questions as the GamerDNA test, Dr Matthew Barr who works at the University of Glasgow has implemented his own version.

We corresponded with Barr and he gave the consent for us to use his implementation of the online test in our research. Furthermore, he emailed us the questions pool and some principles for building the test. Drawing on an extensive range of sources, especially the mail conversation with Barr, we know that there are 39 questions in the pool, in detail the number of groups for each combination are: S/A (7), S/E (6), S/K (7), E/A (6), E/K (7) and K/A (6). Accordingly, the script takes 30 questions at random but checks that there are an equal number of questions for each player type combination. In this way, one of the choices relates to one specific playing style preferences. Thus the questionnaire works by asking questions for each combination of two different Bartle playing style preferences (e.g., socializer vs. achiever) and adding the numbers to get your own score. For instance, if one participant chooses the first option in the following *socialiser vs. achiever* question, he or she adds 1 to his or her socializer (S) score; conversely, if the second option is selected, it will add 1 to achiever (A) score.

Are you more comfortable, as a player on a MUD:

A. Talking with friends in a tavern? (+S)

B. Out hunting orcs by yourself for experience? (+A)

Consequently, the result of the Bartle Test is known as the 'Bartle Quotient'. Specifically, the 'Bartle Quotient' is computed according to the individual answers that grant a 200% entirety in four categories of preferences, without any separate style achieving above 100%. For example, a user get a result with "100% socializer, 50% explorer, 30% achiever, and 20% killer", which indicates the player's motivation for playing is collaborating or interacting with others more than alternative style of interests. This result may be abbreviated as SEAK.

6. Identifying personas to explore the museum service experience

As a basic SD tool, personas usually represent a group of people with shared interests, common behaviour patterns, or demographic and geographical similarities[18]. In order to recruit personas to explore museum service experience, the Bartle Test questionnaire will be distributed to our students. Actually, most respondents have a dominant characteristic, which determines their overall preference. In the end, participants can be judged which player type he or she predominantly belongs to according to which category achieve the highest score among the four. In addition, another essential criterion is the difference between the highest score and the second highest score is extremely large.

With the four player types being identified, this study will initially involve four personas (one achiever, one socializer, one explorer, and one attacker) as respondents in order to get in-depth and detailed data. When referring to research approach, the ethnography will be used in this museum UX investigation. In fact, many design disciplines which emphasize users' experience in their own context at whole stages has employed the ethnography as a research approach [11]. In detail, the ethnography is an effective approach to gain in-depth and detailed insights into the perceptions from the user's perspective, which provides a holistic way to understanding UX with services. However, researchers find it difficult to adopt conventional ethnography due to the major issue of severe time consumption. This is because in most cases, it is not possible to spend months or even weeks in the museum to gather data. Thus, researchers have sought adaptations of traditional ethnography [19]. It is hoped that these adapted strategies will enhance the practicality of field research and come up with richer field experience in a shorter timeframe. Under the adapted rapid ethnography, observation methods will be employed by taking advantage of the Service Safari and Shadowing techniques.

7. Conclusion

To conclude, this work has employed Gamification perspective to identify and create personas to help the museum better understand its different types of users. The ultimate goal of this research is to enhance the museum branding through SD experience. In order to explore the service experience in museum, this article mainly emphasizes individuals' motivations rather than the so-called general public. There are 34 provincial-level administrative regions in China, and each province has at least one or several provincial-level museums. Additionally, there are nearly 5,000 museums of various levels in China. Therefore, the widespread application of research results will be worthy of attention.

References

[1] Belenioti, Z. C., Tsourvakas, G., & Vassiliadis, C. A. (2017). A Report on Museum Branding Literature. Strategic Innovative Marketing, (September), 229–234. https://doi.org/10.1007/978-3-319-56288-9

[2] Ober-Heilig, N., Bekmeier-Feuerhahn, S., & Sikkenga, J. (2014). Enhancing Museum Brands with Experiential Design to Attract Low-involvement Visitors. Arts Marketing: An International Journal, 4(1/2), 67–86. https://doi.org/10.1108/AM-01-2014-0006

[3] Camarero, C., Garrido, M. J., & Vicente, E. (2010). Components of Art Exhibition Brand Equity for Internal and External Visitors. Tourism Management, 31(4), 495–504. https://doi.org/10.1016/j.tourman.2009.05.011

[4] Ritchie, R. J. B., Swami, S., & Weinberg, C. B. (1999). A Brand New World for Nonprofits. International Journal of Nonprofit and Voluntary Sector Marketing, 4(1), 26–42.

[5] Neumeier, M. (2005). The Brand Gap: Revised Edition (2nd ed.). Peachpit Press.

[6] Holland, D. (2006). Branding for Nonprofits: Developing Identity with Integrity. New York: Allworth Press.

[7] Murphy, J. (1988). Branding. Marketing Intelligence & Planning, 6(4), 4–8.

[8] Sandoval, S., & Sortland, O. (2013). Building A Service Design Explanation. Retrieved from http://odasortland.com/service-design-explanation/

[9] Ober-Heilig, N., Bekmeier-Feuerhahn, S., & Sikkenga, J. (2014). Enhancing Museum Brands with Experiential Design to Attract Low-involvement Visitors. Arts Marketing: An International Journal, 4(1/2), 67–86. https://doi.org/10.1108/AM-01-2014-0006
[10] Zhou, H. Y. (2014). Zhou Hongyi's Own Statement: My Internet Methodology. Beijing: China CITIC Press.
[11] Segelström, F., Raijmakers, B., & Holmlid, S. (2009). Thinking and Doing Ethnography in Service Design. IASDR: Rigor and Relevance in Design, (January 2009), 1–10.
[12] Döpker, A., Brockmann, T., & Stieglitz, S. (2013). Use Cases for Gamification in Virtual Museums. Informatik 2013, Informatik Angepasst an Mensch, Organisation Und Umwelt: Tagung Vom 16. - 20. September 2013 in Koblenz; 43rd Meeting of the German Informatics Society, (September 2013), 2308–2321.
[13] Zichermann, G., & Cunningham, C. (2011). Gamification by Design: Implementing Game Mechanics in Web and Mobile Apps. US: O'Reilly Media, Inc.
[14] Hamari, J., & Koivisto, J. (2015). Why Do People Use Gamification Services? International Journal of Information Management, 35(4), 419–431. https://doi.org/10.1016/j.ijinfomgt.2015.04.006
[15] Deterding, S., Khaled, R., Nacke, L., & Dixon, D. (2011). Gamification: Toward a Definition. CHI 2011 Gamification Workshop Proceedings, (January 2011), 12–15.
[16] Bartle, R. A. (2004). Designing Virtual Worlds. US: New Riders.
[17] Konert, J., Göbel, S., & Steinmetz, R. (2013). Modeling the Player, Learner and Personality: Independency of the Models of Bartle, Kolb and NEO-FFI (Big5) and the Implications for Game Based Learning. In Proceedings of the 7th European Conference on Game Based Learning (ECGBL) (pp. 329–335).
[18] Stickdorn, M., Hormess, M. E., Lawrence, A., & Schneider, J. (2018). This Is Service Design Doing: Applying Service Design Thinking in the Real World. O'Reilly Media, Incorporated.
[19] Millen, D. R., Drive, S., & Bank, R. (2000). Rapid Ethnography: Time Deepening Strategies for HCI Field Research. In Proceedings of the 3rd conference on Designing interactive systems: processes, practices, methods, and techniques (pp. 280–286).

IOP Conf. Series: Materials Science and Engineering **573** (2019) 012021 doi:10.1088/1757-899X/573/1/012021

A Concept Innovation on Urban Personal Mobility

Yanlong Li

School of Automotive Studies, Tongji University, China

Email: mail_dragon@tongji.edu.cn

Tel: 86-21-69589204

Abstract: With the urbanization around the world, the automotive industry tries a lot on the challenge of transportation, especially on electric personal mobility. The paper poses a totally new concept that a car is small enough to be contained in a common elevator, which not only makes it possible to get into an apartment house and successfully solve the problems of charging and parking without further request of infrastructure, but also able to extend the range greatly by riding a metro, gives greater freedom for personal mobility. As a combination of a light scooter and a 4-wheeled car, and a house appliance, the concept is hi-tech, high efficiency, comfortable, safe, fashionable, and also has a striking urban styling. During the research, a method called OKARF is developed, which stands for Objective Knockdown and References Filtering. The implementation is to knockdown the objective into tasks, list out the references related to the tasks, filter out the references to generate the concept, which will be the new solution for the objective.

1. Introduction

The urbanization meets a great challenge of transportation around the world. In Asia, the number of urban residents increased from 234 million to 1 billion from 1950 to 1990, and is projected to reach 3.4 billion by 2025[1].

Despite heavy congestion, there is still a desire for personal mobility in urban because no other means of transportation has so far offered the same mix of freedom, comfort, utility and security as the automobile [2]. Cars are well suited for conveying multiple passengers over long distances at high speeds, but inefficient for providing personal mobility within cities [3]. Moreover, congestion is not the only problem, parking, energy and pollution are also serious issues that urge the car companies to try for compact, electric, personal transportation (Table 1).

The 2003 Toyota PM is a one-seat car for personal mobility, also features a wearing concept rather than a driving one [4]. The 2009 GM EN-V (Electric Networked-Vehicle) is a 2-seat urban electric concept with Segway technology features significant autonomous operation [5] [6] [7], and maintains the core principle of personal mobility – freedom [8]. The 2012 Renault Twizy is a completely innovative two-seater design with 4 wheels. On a typical urban journey, it enables a time saving of 25% - including parking [9]. The 2013 Toyota i-Road Concept offers a comfortable, enclosed environment for two occupants seated in tandem, provides a safe, intuitive and enjoyable, helmet-free driving experience [10].

All the designs are interesting and meaningful for research of the future transportation. However, the limits of the electric vehicles are still troublesome. According to a survey in Beijing China, over 50% of the interviewees claim that the biggest worry about electric vehicles lies in insufficient range and inconvenient charge [11]. So according the specifications of the designs (Table 1), range, charge and parking are still big problems to be solved, especially for big cities.

Table 1. Specifications of the four personal transportations.

	Toyota PM Concept	**GM EN-V Concept Jiao**	**Renault Twizy**	**Toyota i-Road Concept**

Length (mm)	1750-2650	1500	2320	2350
Width (mm)	1470	1425	1190	870
Height (mm)	1220-1860	1640	1460	1445
Range (km)	-	40	100	50
Weight (kg)	-	450-500	450	300
Seats	1	2	2	2

2. Material and Methods

2.1 The concept

As the battery technology and charging infrastructures are difficult to be solved in short time, the paper address a new solution to solve these problems. Since the objective is to search for a better transportation, so the idea of designing a more advanced traditional car must be abandoned. However, pick inspirations from current things may still work. For example, a scooter is light and compact, a car is comfortable and safe, charging is never an issue for a household appliance, etc. So how to make all these happen on a single thing becomes the new possible way (Figure 1).

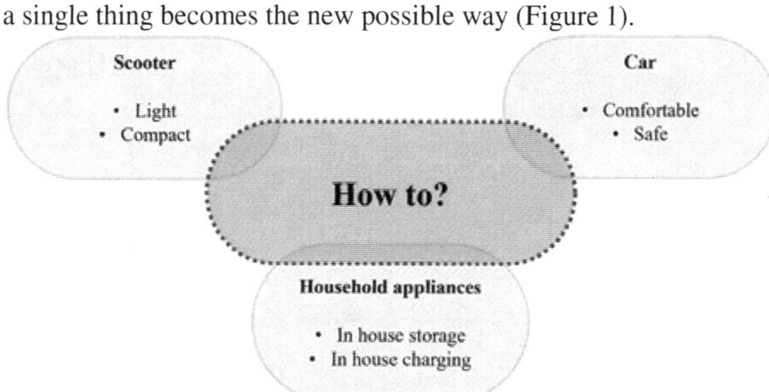

Figure 1. Searching for the concept.

There must be plenty of answers, but the easiest one may be this: a household appliance that is big enough to contain a rider inside, and also small enough to be contained in an elevator then go out from the apartment to move around.

The specifications of common elevators are show in Table 2, and the following design target is set according to it. Here the dimensions of the concept is set as 1100×1400×2200 mm (width×depth×height), makes the idea suited for most of the situations.

Table 2. Design targets according to the common elevators.

	Apartment Elevators	**Public Elevators**	**The design target** (Retraced / Stretched)
Width (mm)	1100	1100	Less than 900
Depth / Length (mm)	1400	1400	Less than 1200 / 2000
Height (mm)	2200	2200 / 2300	Less than 2000 / 1600
Load capacity (kg)	400-600	630-1250	100

Weight (kg)	-	-	Less than 200
Top speed (km/h)	-	-	5 / 50
Range (km)	-	-	100 + Metro's travel

This concept should not be confused with a traditional wheel chair, which is usually unclosed and low-speed for physical disabilities, while the new concept will focus on daily mobility for normal persons including the elders. The basic image could be a seat in a wheeled capsule (Figure 2. A). Since the capsule mainly moves around as the speed of a scooter, the wheel-base need to be big enough to keep stable when running, this makes the capsule has a foldable architecture (Figure 2. B), which also makes the center of gravity lowering down greatly for better stability.

However, because of the limit capacity of battery, the range is still a problem. Thanks greatly to the idea of being contained in an elevator, the car could also moves into a metro station elevator, and then carried by the metro for a much longer distance (Figure 3).

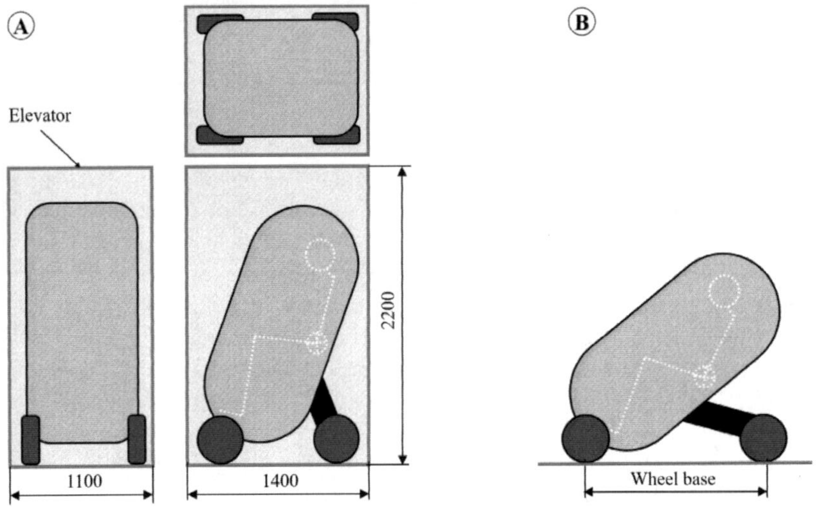

Figure 2. The basic concept and dimensional limits (mm).

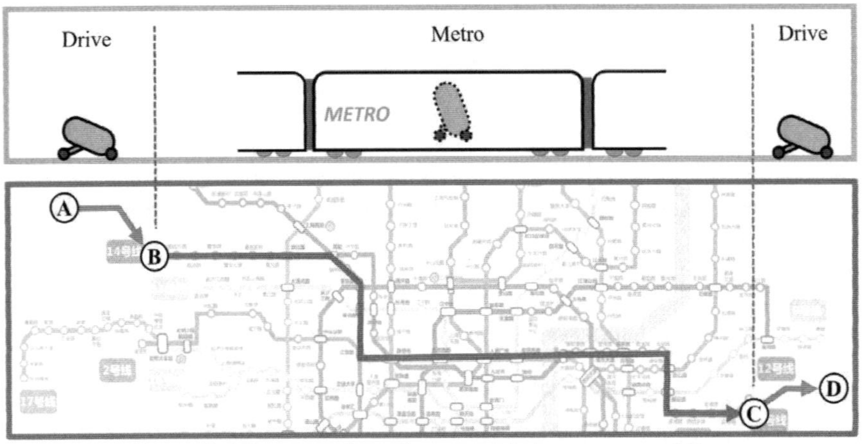

Figure 3. The extension of range: driving + metro riding.

2.2 *The design*

In order to represent the totally new concept for future consumers, some key elements are concerned to express the sense of clean, electric, fashion, light, hi-tech and colourful life. The design image is shown in Figure 4.

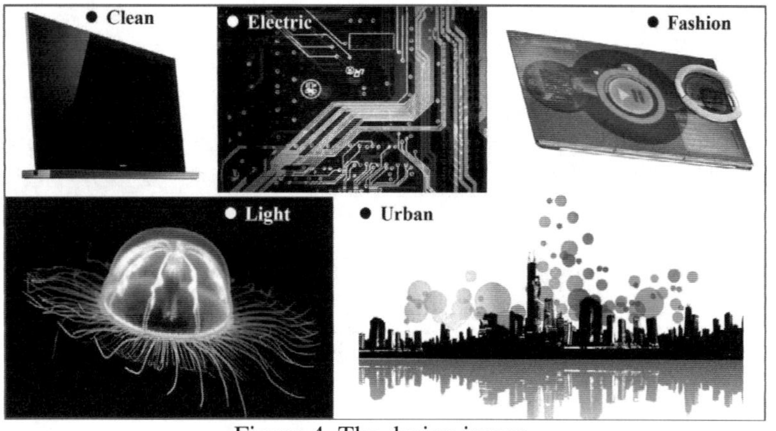

Figure 4. The design image.

The design features are shown in Figure 5. The wraped glass A gives a panorama vision and bright cabin; the front lamps B, rear lamps O, and the turning flash E are all of LED, which gives an elaborate look; the front hatch C can rotate up around the pivot N, and the pivot H implies the function of zooming the wheelbase; the lines M shown in red color represent the sense of electric, gives a strong hi-tech feeling; the flush body D could be painted in different colors to make the car stylish and fashionable; the front wheel F is smaller for better package of feet space and steering room, while the rear wheel I is bigger for better driving performance; part of the side body is mesh G and is benefit of ventilation; the battery J stores the energy for range of 100 km and top speed of 50/km; the carbon-fiber K gives a light and strong structure for the whole car; the camera L is used as a rear view mirror and is also helpful during reversing.

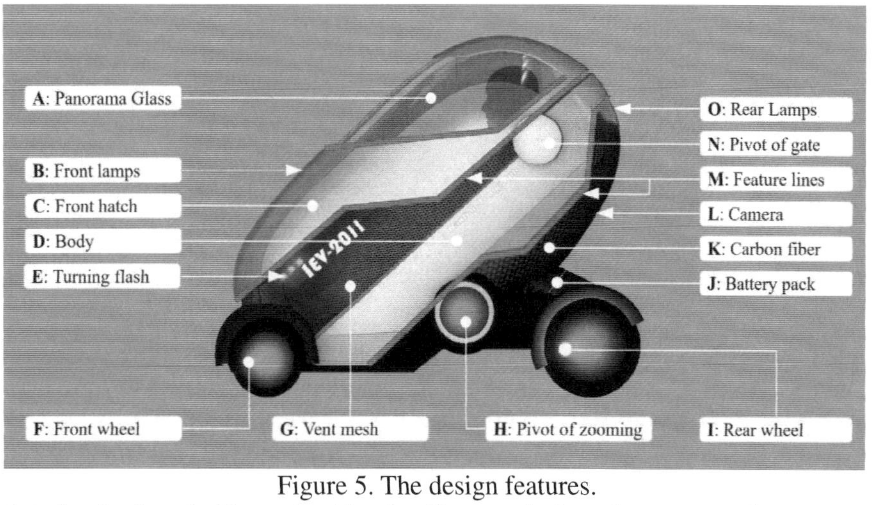

Figure 5. The design features.

The final design is show in Figure 6 with miscellaneous body colors presented.

IOP Conf. Series: Materials Science and Engineering **573** (2019) 012021 doi:10.1088/1757-899X/573/1/012021

Figure 6. The final design.

2.3 The functions

There are two basic modes of the car, one is fold-up and the other one is stretched.

According to the concept, the car could be contained into an elevator, hence the length of the fold-up mode and the width are set 100mm smaller on each side of the elevator, shown with A and B in Figure 7, and B also shows the opened helmet for the passenger could operate the elevator from inside of the car. C is the stretched mode and D is shows the front hatch opened for the passenger's ingress and egress.

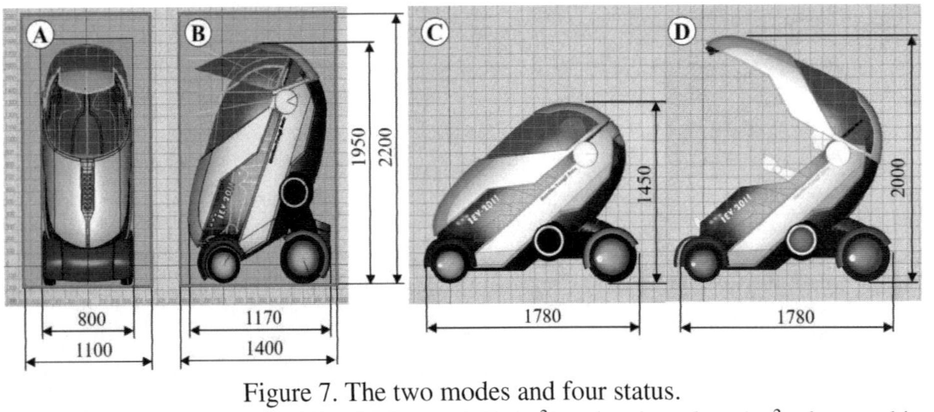

Figure 7. The two modes and four status.

Notice that the footprint is only 800×1170mm (0.936m^2), takes less than 1m^2 when parking, this is very space-saving comparing with a common compact car whose foot print is over 8 m^2.

When at the fold-up mode, the car can climb through the slope (Figure 8. A) at low speed of 5 km/h (similar to walking speed), reverses into the elevator (Figure 8. B) with the assistance of rear view camera and intelligence technology, and then moves into the room for parking and charging with standard household power. (Figure 8. C).

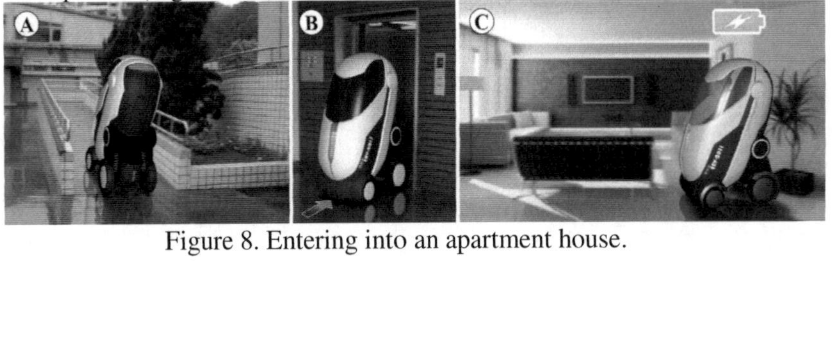

Figure 8. Entering into an apartment house.

Also when folds up, the car could get onto the metro platform (Figure 9. A) through the elevator and ride the metro for a much longer trip (Figure 9. B).

Figure 9. Riding the metro.

3. Results

1) The concept is a combination of the advantages of a scooter, a car and a house appliance, and comes out as a hi-tech, high efficiency, comfortable, safe, fashionable personal mobility solution, and also a striking urban styling.

2) One of the highlights of the concept is that the size is compact enough for being contained in a common elevator, which effectively solves the problems of parking and charging with standard household power while without the further request of infrastructure. The foot print is only 1/8 of a common compact car.

3) The concept of being contained in an elevator also result in a very effective solution for expanding the range of 100km by riding a metro when necessary, while despite the bottleneck of current battery technology. This gives greater freedom for personal mobility, and also is an innovative combination of personal mobility and public transportation.

4. Discussion

Comparing with a common car, the concept is much lighter, greener, and more environmental friendly, so it is possible for the government to set a preferential policy to promote, just like the bus priority lane for public transportation. This will greatly encourage the whole green mobility trend.

This concept can also be used in small cities and towns, where the daily activity region is not so big and the driving range is enough regardless of no metro lines there, saving a lot of road and parking resources.

Beside the private mode, the concept can also be used for car-sharing, which also could be much high-efficient than traditional cars.

With the development of network and intelligence technology, the concept can run in formation, communicate with the other road users for better active safety, and linked to the infrastructures as well. In a word, it can be a new highlight of the traffic ecosystem.

5. Conclusions

1) A good conceptual idea could speak louder than technology. Usually the capacity of battery is the key factor for the long range of an electric car, however, it is difficult to achieve a breakthrough on battery technology in short time. Furthermore, the charging points and other facilities are also limited greatly by the traditional infrastructures. In this case, a car that is small enough of being accessible into an elevator is the key idea, which avoids the challenges of battery technology and limits of infrastructure, and makes a small electric car possible to run a long range, easy to park, and easy to charge as well.

2) For a typical objective, there are always different ways to deal with. Here a method called *Objective Knockdown and References Filtering* (OKARF for short) is presented. The route map is shown below, also shown in Figure 10.

Get start: determine the objective. In this case the objective is urban personal transportation.

Proceed: Knockdown the objective into tasks as A, B, C, etc. Here the tasks are range, charging, parking, riding experience, etc.

Essential: List out the References. Here the references of range are car, scooters, etc., and references for charging are charging pile, lots, etc. There could be a lot of references for more choices later.

Key step: Filter out the references and generate the concept, hence the concept is the new solution for the objective. In this case, a concept of mobile capsule is the solution for the personal transportation.

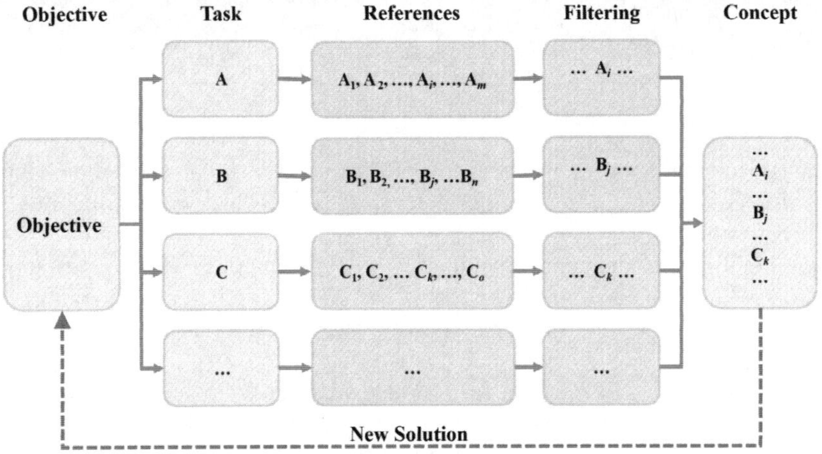

Figure 10. The OKARF method.

Acknowledgements

This research is financially supported by the National Key Fundamental Research and Development Program (973 Program, Grant No. 2011CB711203).

Reference

[1] Spencer N and Butler D (2010) Cities: The century of the city. Nature 467:900–901.

[2] Borroni-Bird, C. E. (2012). Personal Urban Mobility for the Twenty-First Century. Energy, Transport, & the Environment.

[3] Mitchell, W. J., Burns, L. D., & Borronibird, C. (2010). Reinventing the automobile: personal urban mobility for the 21st century. MIT Press.

[4] http://www.allcarindex.com/auto-car-model/Japan-Toyota-PM/.

[5] Jim Motavalli (2010-03-24). "G.M. EN-V: Sharpening the Focus of Future Urban Mobility". New York Times. Retrieved 2010-10-05.

[6] GM Unveils EN-V Concept: A Vision for Future Urban Mobility gm.com, 2010-03-25

[7] Jim Motavalli (2010-03-24). "G.M. EN-V: Sharpening the Focus of Future Urban Mobility". New York Times. Retrieved 2010-10-05.

[8] https://www.netcarshow.com/gm/2010-en-v_concept/.

[9] https://www.netcarshow.com/renault/2012-twizy/.

[10] https://www.netcarshow.com/toyota/2013-i-road_concept/.

[11] Sun, S., Zhang, L. (2015). A Design Research on Electric Personal Mobility in Urban. Industrial Design (6), 74-78.3011.

Emotional experience evaluation method of interaction task based on EEG technology

Zhiang Chen, Li Lin*

School of Mechanical Engineering, Guizhou University, Guiyang, Guizhou, 550025, China

Corresponding author's e-mail: *linlisongbai@163.com

Abstract: In past studies, the emotional experience of interaction task was quantified mainly by the subjective evaluation. This method is susceptible to subjective factor, which can cause the deviation of result. To solve this problem, an emotional experience evaluation method of interaction task based on objective Electroencephalograph(EEG) technology was proposed. Supported by the theory of cognitive neuroscience and EEG, the same family of products with different interaction processes were selected as a sample of experiment, and interactive tasks were set. Then the emotional experience evaluation experiment of interaction task was conducted to obtain objective EEG physiological information. Based on the Power Spectral Density analysis(PSD) and the event-related synchronization/desynchronization theory, characteristic values were calculated. Next, Correlation between the EEG characteristic value and the emotional experience of interactive tasks was judged by Pearson Correlation Coefficient. Finally, the partial least squares regression(PLS) was selected to construct the calculation model. Smartphones were used as the interaction case, and the model's performance was verified. The results showed that the proposed calculation model could achieve the similar evaluation effect as the subjective evaluation, and the deviation was smaller.

1. Introduction

In the research of human-computer interaction, the emotional experience of product interaction system is one of the priorities. Researchers in neuroscience, psychology, and cognitive science had shown that human emotional responses could cause electrophysiological changes in the brain[1]. Therefore, to get user's Electroencephalograph(EEG) signal is an effective way to study the emotional experience of the product interaction system.

In the related research on emotional experience of interaction systems, Yang et al.[2] used the category hierarchy method to decompose the design project of the interface in the interaction task. And a mapping model between the affective image and design project was established to effectively guide the design of the interaction task interface. Lin[3] studied the emotional state of the car driving operation. Then an emotion detection method based on driver characteristics and vehicle running state was proposed. These studies mainly used subjective evaluation to quantify the emotional experience of the interaction system. The research based on this method has the characteristics of simple operation and high efficiency, but it is easy to be affected by subjective factor and lead to deviation of evaluation results[4].

In the EEG study, the EEG produced by the left and right frontal lobe is directly related to the emotional state[5]. The high energy response of the left frontal lobe represents positive feedback, while the high energy response of the right frontal lobe represents negative feedback[6]. The Power Spectral

Density(PSD) of five common EEG of α, β, θ, δ, and γ has been used to evaluate emotional states[7]. The PSD of the α wave is related to the inactive state of the brain and the positive valence state of the emotion. And the smaller the PSD, the more active is the brain[8].

Therefore, this study uses EEG measurement technology as a key means to quantify the emotional experience of interaction tasks, and constructs the calculation model based on EEG. This method solved the problem of subjective decision-making by designers when conceiving product prototypes and evaluating interaction design schemes, and provided objective user knowledge and scientific basis for designing product interaction prototypes that met the user's emotional experience. We took smartphones as examples to verify the effect of the interaction task emotional experience calculation model.

2. Construction of an emotional experience evaluation method of interaction task based on EEG technology

Common interaction tasks are selected based on the characteristics of the product interaction system. Products with the same function and different interaction processes were chose as samples in emotional experience evaluation experiment. The set of experimental subjects is set as $B = \{b_1, b_2, b_3, \cdots, b_m\}$, where m is the number of subjects. And the set of product samples is $C=\{c_1, c_2, c_3, \cdots, c_n\}$, where n represents the number of product samples. Subjective evaluation of interaction tasks usually includes six aspects: pleasure, autonomy, fluency, responsiveness, immersion, and overall functionality[9]. This study chose pleasure and immersion as the evaluation object of interaction emotional experience.

The experiment is to obtain the EEG data of the subject during the interaction task. The experimental steps are as follows: First, Subjects need to sit for 60 seconds and record EEG data in relaxed state. Then, the task guidance appears on the screen, and next the interaction task is performed. After completing all the interaction tasks of one product sample, the emotional experience of this product were evaluated by Subjects. Then subject experiments with another product sample after a little rest. The experimental flow of one product sample is shown in Figure 1.

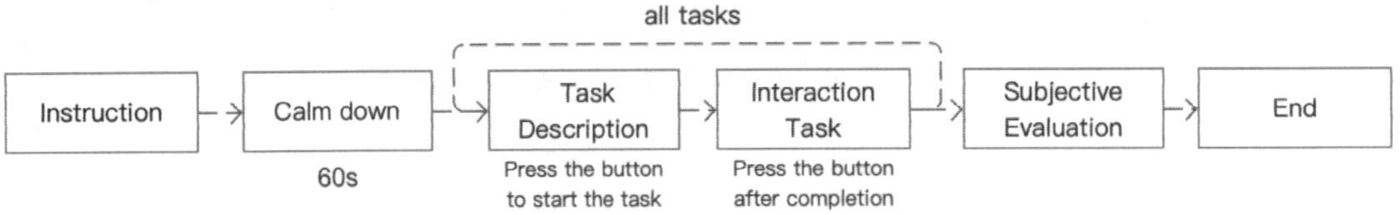

Figure 1. Emotional cognitive experiment flow of interaction tasks based on EEG.

If the emotional experience score of the m-th subject(B_m) on the n-th product sample(C_n) is D, the average score of all subjects on the n-th product sample can be expressed as Equation (1), where $m = 1, 2, 3, \cdots, M$.

$$y_m(C_n, D) = \frac{1}{M}\sum_{m=1}^{M} y(B_m, C_n, D) \tag{1}$$

The EEG equipment that the subject is wearing will record the EEG data during the interaction task. If the EEG data of the m-th subject(B_m) on the n-th product sample(C_n) is E_p, and p is the number of EEG indexes. The average value of each EEG index of the n-th product sample used by all of the subjects can be expressed as Equation (2), where $m = 1, 2, 3, \cdots, M$.

$$y_m(C_n, E_p) = \frac{1}{M}\sum_{m=1}^{M} y(B_m, C_n, E_p) \tag{2}$$

This study will obtain the PSD of α waves (electrodes AF3, AF4, F3, F4, F7 and F8) in the frontal lobe of the subjects as physiological indicators of emotional response. There may be differences in α waves generated between subjects. In order to actually respond to the degree of α activation, the Equation (3) is used to calculate the characteristic value of EEG indexes, which is based on the theory of Event-related EEG synchronization and desynchronization[10].

$$F = \frac{A - R}{R} \tag{3}$$

Where F represents the characteristic value of the α wave, and A and R are the PSD of the α wave in the task phase and the rest phase, respectively.

When the screened EEG index(E_p^{θ}) is converted into formula $\theta(E_p)$, the screened EEG index expression is as follows in Equation (4):

$$E_p^{\theta} = \theta[\frac{1}{M}\sum_{m=1}^{M} y(B_m, C_n, E_p)] \tag{4}$$

To predict the emotional experience score of the interaction task in product sample through the EEG indexes, an emotional experience calculation model of interaction tasks based on EEG is established. The emotional experience score is taken as the dependent variable y, and the EEG indexes are the independent variables x. Then the theoretical model of experience evaluation can be described as equation (5):

$$Q = f[E_p^{\theta}, y_m(C_n, D)] \tag{4}$$

where Q is the quantitative value of the interaction task emotional experience. f is the data-mining technology of the emotional experience calculation model. And E_p^{θ} and $y_m(C_n,D)$ represent the characteristic value of EEG indexes and the emotional experience score, respectively

3. Method Verification: Smartphone Case Study

Because of the interaction tasks are simple and interaction processes are diverse, the smartphone is used as the experimental sample.

3.1. The emotional experience evaluation experiment of interaction tasks based on EEG

3.1.1. Objective

Get the EEG data of the subjects during the interaction task of the smartphone.

3.1.2. Subjects

Fifteen industrial design students participated in the experiment. The subjects' ages ranged from 20–26 years old and averaged 24.323 years. All of the subjects were confirmed to be right-handed, to avoid individual differences in changes in the left and right brain hemispheres[11].

3.1.3. Experimental Samples

Four common smartphones with operating systems of iOS, MIUI, EMUI, and Smartisam OS were used as experimental samples. For the built-in program in the smartphone, three tasks were selected as the experiment task, the common function of to find the destination in the map, to download APP in the appstore, and to input the specified text in the notes were selected.

3.1.4. Experimental Equipment

Smartphone: To avoid the impact of product appearance, smartphones with the same screen size, appearance, colour and other factors were used as the experimental sample. After focus group discussion, iphone 7 Plus, Xiaomi Note 3, Huawei P10 Plus and Jianguo Pro four smartphones were selected. Each smartphone's body was black, the screen size was 5.5 inches, and the appearance of the four smartphone is shown in Figure 2.

Figure 2. The appearance of four smartphones.

EEG equipment: The EEG hardware used in the experiment was the Emotiv EPOC+ wireless head-mounted EEG device with 14 channels. Sample rate is 124 Hz per second. The electrode positions were AF3, F7, F3, FC5, T7, P7, O1, O2, P8, T8, FC6, F4, F8, and AF4 (see Figure 3). In addition, CMS and DRL were used as reference electrodes.

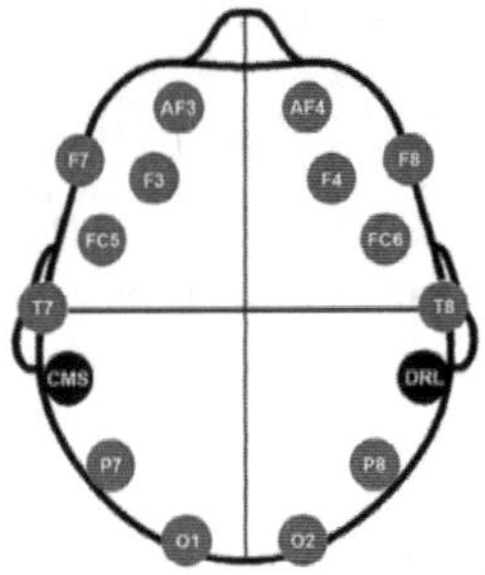

Figure 3. The electrode positions.

Experimental platform: Two computers were required in this experiment, task computer with E-prime program and EEG computer connected to Emotiv EPOC+. The task computer was connected to the EEG computer through a serial cable and sends the Mark during the experiment. The Mark will mark the time point in the EEG data as an experimental event. The experimental platform is shown in Figure 4.

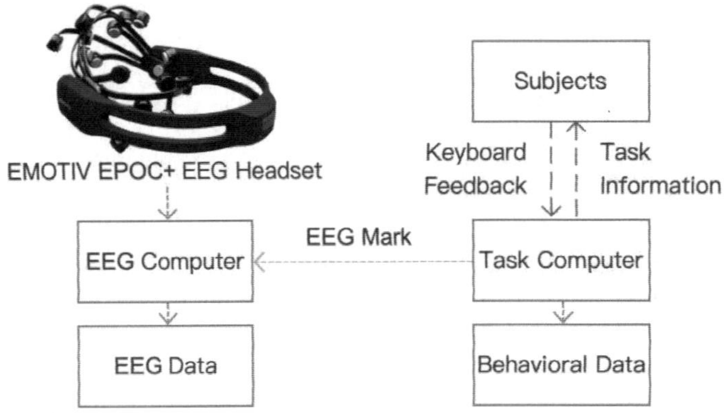

Figure 4. Emotional cognitive experiment platform of interaction tasks based on EEG.

3.1.5. Experimental Process

The specific experimental process is as follows: First, The experimental instruction was displayed; Next, the subjects were asked to calm their mind and record the EEG for 60 seconds of rest. Then, jump to the description page of the first experiment task. Hit the space bar to start the task; After completing the task, press the space bar to go to the next task description page. 3 tasks of one smartphone will be arranged as a group. After one group was completed, the subjects needed to score 1~5 points on the overall emotional experience of this smartphone through the keyboard.

The duration of each task for each phone will last 2 to 3 minutes. Together with the preparation time, it took about 35 minutes for each subject to complete the experiment. The experimental flow is shown in Figure 1.

3.2. Data processing

3.2.1. Subjective evaluation

Table 1 shows the subjective evaluation results of the interaction task emotional experience. The evaluation contents were pleasure and immersion, with a score of 5 points.

Table 1. The results of subjective evaluation.

Content	Smartphone	N	Mean(M)	Standard deviation(SD)	Standard error
pleasure	1	12	4.13	0.34	0.21
	2	12	3.86	0.43	0.23
	3	12	3.23	0.68	0.35
	4	12	3.77	1.02	0.19
immersion	1	12	3.84	0.69	0.19
	2	12	3.57	0.43	0.26
	3	12	2.84	0.85	0.34
	4	12	3.68	0.46	0.37

3.2.2. EEG data

First, EEGLAB 14.1.2b was used to preprocess EEG data. The main steps included channel position importing, event value reading, reference electrode setting, band-pass filtering, EEG previewing, artifact removal, EEG segmentation, baseline correction, and noise components removal. Next, the PSD values of α from the electrodes AF3, AF4, F3, F4, F7 and F8 were extracted, and the result is shown in Table 3. It can be found that the relative α wave high energy of each electrode is at a higher level during the rest phase. It is generally reduced during the task phase, and the inactive state appears, indicating that a positive emotional state has occurred.

Table 2. The PSD values of α from the electrodes AF3, AF4, F3, F4, F7 and F8(M±SD).

		AF4	AF3	F4	F3	F8	F7
Rest phase		0.29±0.18	0.28±0.19	0.27±0.17	0.29+0.18	0.28±0.16	0.29±0.15
Task phase	Smartphone1	0.15±0.08	0.14±0.09	0.14±0.05	0.14±0.04	0.14±0.08	0.13±0.07
	Smartphone2	0.13±0.07	0.15±0.04	0.13±0.05	0.14±0.06	0.13±0.06	0.14±0.04
	Smartphone3	0.14±0.06	0.11±0.07	0.12±0.07	0.11±0.08	0.12±0.07	0.12±0.04
	Smartphone4	0.12±0.08	0.13±0.06	0.11±0.06	0.13±0.09	0.13±0.07	0.13±0.02

The characteristic values of EEG indexes of the α wave were calculated by the Equation (3) in order to effectively reflect the activation degree of the brain. After calculation, the EEG characteristic values as shown in Table 3 were obtained. These characteristic values would be used as EEG indexes to establish an Emotional experience calculation model of interaction tasks.

Table 3. The EEG characteristic values of the interaction tasks.

	AF4	AF3	F4	F3	F8	F7
Smartphone1	0.483	0.500	0.481	0.517	0.500	0.552
Smartphone2	0.552	0.464	0.519	0.516	0.536	0.517
Smartphone3	0.517	0.607	0.556	0.621	0.571	0.586
Smartphone4	0.586	0.535	0.593	0.552	0.536	0.552

3.3. Construction of emotional experience evaluation model based on smartphone interactive task

3.3.1. Determination of independent variable and dependent variable

Between 6 sets of α characteristic values and 2 sets of interaction task emotional experience mean values, The correlation analysis was used to gain pearson correlation coefficient(see Table 4). The absolute value of the correlation coefficient between the emotional experience mean values and the α characteristic value approaches 1, indicating that there is a linear relationship.

Independent variable: α characteristic values of the electrodes AF4, AF3, F4, F3, F8 and F7 in the frontal area of brain.

Dependent variable: The mean value of the emotional experience of the interaction task: pleasant and immersive.

Table 4. Pearson correlation analysis.

	AF4	AF3	F4	F3	F8	F7	Pleasure	Immersion
AF4	1	-0.086	0.798	-0.005	0.321	-0.322	-0.149	0.077
AF3	-0.086	1	0.524	0.969	0.711	0.955	-0.849	-0.803
F4	0.798	0.524	1	0.551	0.640	0.314	-0.587	-0.356
F3	-0.005	0.969	0.551	1	0.858	0.866	-0.953	-0.917
F8	0.321	0.711	0.640	0.858	1	0.486	-0.971	-0.919
F7	-0.322	0.955	0.314	0.866	0.486	1	-0.675	-0.666
Pleasure	-0.149	-0.849	-0.587	-0.953	-0.971	-0.675	1	0.965
Immersion	0.077	-0.803	-0.356	-0.917	-0.919	-0.666	0.965	1

3.3.2. Construction of multiple linear regression model

In table 4, the absolute value of the correlation coefficient between each α characteristic value approaches 1, indicating that there is a correlation between data. Therefore, the partial least squares regression(PLS) was used to establish the mathematical model, which could solve the adverse effects from the multiple correlation of variables[12]. The mean value of pleasure and immersion was the dependent variable Y. The α characteristic value was used as independent variables X. And the regression coefficients were obtained by PLS.

Two multiple linear regression models were obtained as shown in Equations (6) and (7).

Pleasure:

$$Y_1 = 0.231x_{\alpha AF4} - 0.742x_{\alpha AF3} + 0.362x_{\alpha F4} - 2.640x_{\alpha F3} - 8.500x_{\alpha F8} + 0.658x_{\alpha F7} + 9.467 \quad (6)$$

Immersion:

$$Y_2 = 1.735x_{\alpha AF4} - 0.599x_{\alpha AF3} + 2.108x_{\alpha F4} - 3.454x_{\alpha F3} - 12.027x_{\alpha F8} + 1.768x_{\alpha F7} + 9.111 \quad (7)$$

3.4. Model verification

To verify the evaluation effect of the model, reschedule the interaction task experiment of 4 smartphones. Subjective evaluation values and α characteristic values from each subjects were obtained. Bring each characteristic values into the Equations (6) and (7) to get the model calculation value. The mean(M) and standard deviation(SD) of each kind of evaluation value from 10 subjects were calculated.

The paired sample t test was performed between the subjective evaluation mean values and the model evaluation mean values. According to the significance level of t test results, the critical t value was 2.776. The mean of each evaluation value and the paired sample t-test result are shown in Table 6. The range of each evaluation value are 1 to 5 points. According to the table 6, the absolute value of t value is less than 2.776, indicating that there is no significant difference between the model calculation value and the subjective evaluation value. The practicability of the model is proved, and it has the effect similar to the subjective evaluation value.

IOP Conf. Series: Materials Science and Engineering **573** (2019) 012022 doi:10.1088/1757-899X/573/1/012022

The SD of each evaluation value are shown in Table 6. SD reflects the dispersion degree of the data. The SD of the subjective evaluation values in the table is generally larger than the SD of the model evaluation values. It is indicated that the deviation of the model evaluation value is smaller.

Table 5. The mean of each evaluation value and t values of paired samples t test t values.

	Evaluation Value	Sample1	Sample2	Sample3	Sample4	t value	Critical Value
Immersion	subjective	3.68	4.31	3.65	3.75	0.123	2.776
	model	3.74	4.23	3.72	3.68		
Pleasure	subjective	4.21	3.91	3.86	3.78	-0.420	
	model	4.17	3.89	3.92	3.82		

Table 6. The standard deviation of each evaluation value.

	Evaluation Value	Sample1	Sample2	Sample3	Sample4
Immersion	subjective	0.510	0.311	0.602	0.842
	model	0.159	0.185	0.269	0.245
Pleasure	subjective	0.311	0.481	0.301	0.334
	model	0.159	0.215	0.196	0.371

4. Conclusion

This study proposes an emotional experience evaluation model of interaction task based on brain electrophysiological indexes. The research shows that the evaluation model based on physiological signal can achieve the similar evaluation effect with the psychological cognition-based evaluation method, and the deviation of the evaluation results is smaller. This method circumvents the uncertainty that may be caused by subjective evaluations, and provides product decision makers or designers objective user data support when selecting product interaction schemes for specific emotional attributes.

This study discusses the impact of EEG indexes on the emotional experience of product interaction tasks. However, the human body also has other physiological indexes such as eye movement, skin electricity, pulse, etc. Whether these indexes have an impact on the emotional experience of product interaction tasks needs further study. In addition, research on the triggering factors of each index will also help to directly relate physiological changes and design changes, and more effectively guide the design of product interaction schemes.

Acknowledgments

Funding: This project was supported by National Natural Science Foundation of China (grant numbers 51865003 and 51465007) and Guizhou Provincial Science and Technology Plan Project (grant numbers [2018] 5781).

References

[1] Yang, M.Q., Lin, L., & Milekic, S. (2018) Affective image classification based on user eye movement and EEG experience information. Interacting with Computers, 30(5): 417-432.
[2] Yang, M.Q., Zhao R., Xi H.L., & LIN, L. (2018). Interface design based on C-MKE model. Journal of Graphics, 39(4): 635-641.
[3] Lin, Q.W. (2014) Reseaarch on driving behavior based on emotional computing. Guangdong Uniersity of Technology.
[4] Gao, X.Q., Wang, Y.Y., Ge, L.Z., (2005) The combination of eye movement technique and EEG technique--a new approach to cognitive research. Chinese Journal of Ergonomics, (1): 36-37+44.
[5] Davidson, R.J. (1995) Cerebral asymmetry, emotion, and affective style. MIT Press, Cambridge.
[6] Hakamata, A., Ren, F., & Tsuchiya, S. (2008) Human emotion model based on discourse sentence for expression generation of conversation agent. In 2008 International Conference on Natural

Language Processing and Knowledge Engineering (pp. 1–8).
[7] Kim, M.K., Kim, M., Oh, E., & Kim, S.P. (2013) A Review on the Computational Methods for Emotional State Estimation from the Human EEG [Research article].
[8] Balconi, M., & Mazza, G. (2009) Brain oscillations and BIS/BAS (behavioral inhibition/activation system) effects on processing masked emotional cues.: ERS/ERD and coherence measures of alpha band. International Journal of Psychophysiology, 74(2), 158–165.
[9] Wang, C.C., & Hsu, M.C. (2014) An exploratory study using inexpensive electroencephalography (EEG) to understand flow experience in computer-based instruction. Information & Management, 51(7), 912–923.
[10] Pfurtscheller, G., & Lopes da Silva, F.H. (1999) Event-related EEG/MEG synchronization and desynchronization: basic principles. Clinical Neurophysiology, 110(11), 1842–1857.
[11] Provins, K.A., & Cunliffe, P. (1972) The Relationship Between E.E.G. Activity and Handedness. Cortex, 8(2), 136–146.
[12] Geladi, P., & Kowalski, B.R. (1986) Partial least-squares regression: a tutorial. Analytica Chimica Acta, 185, 1–17.

IOP Conf. Series: Materials Science and Engineering **573** (2019) 012023 doi:10.1088/1757-899X/573/1/012023

Discussion on the Aesthetic Concept of Chinese Traditional Culture in Modern Furniture Design

Shuang Wu [1], Wenming Liu [2*]

[1] Arts Institute, Ningxia Institute of Science and Technology, Shizuishan, Ningxia,753000, China

[2] School of Design and Art, Shenyang Jianzhu University, Shenyang, Liaoning, 110168, China

*Corresponding author's e-mail: 83373734@qq.com

Abstract. Chinese traditional culture is the source of cultural self-confidence. This article starts from the Chinese furniture design in the new era, through the collation and analysis of Chinese traditional culture, discusses its inherent power, expounds the inevitable demand of modern furniture design; From the perspective of the new expression of the traditional culture aesthetics, it provides new ideas for the design of Chinese furniture in the context of the new era.

1. Introduction

China's industrial design concept started in the 1980s. Because it was imported from abroad, it was greatly influenced by western design. Therefore, China's industrial design has developed limited, and the products lack independent innovation.

Chinese culture has a long history and presents unique cultural characteristics. Furniture is also an important material carrier. The traditional culture shows the excellent oriental aesthetic. Contemporary furniture design is more like a comprehensive competition of national culture. Combining the chinese culture with new application of modern science and technology and applying it reasonably in furniture design is an important subject for modern industrial design. The organic combination of traditional aesthetics and technology can not only improve the aesthetic value of the product itself, but also show the cultural heritage of Chinese characteristics and enhance the soft power of national culture.

2. The basic performance of Chinese traditional aesthetic concept

The aesthetic concept of Chinese traditional culture is a unique way of thinking, is the essence of the long-term practice life, and the display of the basic characteristics of the nation.

Under the overall guidance of the traditional cultural aesthetic concept, the Chinese nation's creation activities show a unique national spirit. Among them, the holistic view, the harmonious view and the dialectical view are the typical representatives.

2.1. The holistic view of traditional culture

The holistic view of Chinese traditional culture can also be called a systematic view, which is an important aspect of traditional culture. Chinese traditional philosophy is based on integrity and intuition. It is also the backbone of traditional cultural aesthetics. This view holds that human beings and the entire human society are an organic whole and inseparable from the world. This is also the Basic theory of holistic view [1].

2.1.1. Heaven and man. The traditional Chinese philosophical classic "I-Ching" first proposed the concept of holistic view, and built a complete world with gossip. "Easy to have Tai Chi, is to live two instruments, two instruments and four images, four elephants and gossip" is the earliest holistic view, laying the foundation for the traditional holistic thinking.

Figure 1 is the Taiji Bagua map. The ancients started from the two abstract symbols of Yin (▬ ▬) Yang (▬▬), combined with the long-term observation of nature, unified the integration of orientation, five elements, yin and yang,and incorporated into the natural world. Everything is everything, this is the holistic view of thinking, which has strong stability and inclusiveness, affects and restricts the overall behavioral norms and thinking methods of the Chinese nation, and builds a world view of "Heaven and Man"[2]. The overall holistic and systematic view of the world in Zhouyi makes the Chinese nation better at adopting a multi-angle and all-round thinking mode rather than a partial and individual approach to thinking and solving problems.

Figure 1. Taiji Bagua map

2.1.2. Existence between to be or not to be. The traditional Chinese aesthetic advocates use a holistic and comprehensive perspective to analyze and solve problems, and emphasizes the interdependence of the two sides of the contradiction, so it also has a strong dialectic. This means that the reason why "real" can work is precisely because of the "virtual". We must not be able to grasp one aspect of anything, ignore the "virtual" where it is, and consider it comprehensively. Therefore, "virtual" and "real" are interactions and constant transformations. In the traditional architectural space design, the designer uses a large number of entities to construct a virtual space, and its typical representative is the use of the screen. The main function of the screen is to divide the space. The space after the division is the ideal space for the user to use, so that a combination of real and virtual meets the psychological needs of the user. Everything in heaven and earth is an independent individual composed of "virtual", "real", "having" and "none". The combination of different individuals constitutes a relative whole. The world is built on such a foundation and is constantly developing. This kind of "with or without life" also enriches our thinking system.

2.2. The concept of harmony in traditional culture

In addition to the overall characteristics of Chinese traditional thinking, there are also characteristics of harmony. It is formed on the basis of holistic thinking, and it also plays an important role in promoting the aesthetic development of Chinese traditional culture. The ancient Chinese believed that the world was connected and pursued a harmonious and stable life. "Harmony" means the relationship between proper and stable, and pays attention to the internal relationship between the links. Especially in the field of furniture, its style, materials and other aspects pay attention to the pursuit of harmony and unity, but also create a specific cultural atmosphere.

2.2.1. Harmony of shape. Furniture design is an important part of the design field. In the long-term development, Chinese traditional furniture has formed a unique style and formed an independent

Chinese furniture system, which embodies traditional Chinese aesthetics and philosophical thinking. An important part.

In the design of the backrest chair, the backrest and the chair plate mostly appear in a straight line shape. Such a straight line shape is regular, straight and full of rhythm, which makes the overall shape clean and stable, and reflects the "just" of the furniture; On the shape of the back of the chair, the two ends of the brain extend outwards, and are slightly upturned, curved in shape, with rounded ends at both ends, forming a soft curve overall, and the back of the chair is curved from the side, the overall curvature Naturally, it conforms to the curve of the person's back to satisfy the comfort of sitting, and shows the "softness" of the furniture. It can be said that such a design will be a perfect combination of straightforward and succinct and smooth, and its beauty of "rigid and soft" shows the unique artistic aesthetic style of Chinese furniture, and is also the embodiment of the aesthetic harmony of Chinese traditional culture.

2.2.2. Harmony of material selection. The Chinese pay attention to the harmonious and unified relationship between man and nature, and it is a deep emotional communication. In the application of Chinese furniture materials, this concept is fully demonstrated. In particular, Ming style furniture retains the texture and natural color of the original materials and retains its natural characteristics. This is the ancient people's respect for nature and performance of thought [3].

At the same time, the ancient Chinese people also pay attention to the harmonious application of "adapting to local conditions" in the selection of furniture, and establish a harmonious relationship between man and nature.

Wood has always been the material of choice for Chinese-style furniture. It can be seen that "wood" represents vitality and motivation in Chinese traditional culture, so the ancients also gave more emotion and understanding to "wood"; on the other hand, ancient Chinese forests cover a wide area and wood is obtained. It is relatively easy and easy to process [4]. Therefore, the ancients chose wood as the main material for building and furniture manufacturing, and can carefully design according to the texture and texture of the wood, which retains the natural attributes of the material and shows the artistic height. The aesthetic value of these reflects the harmony of traditional cultural aesthetics.

3. The necessity of contemporary design and application of traditional culture aesthetics

3.1. The important position of industrial design in the new era

In 1980, the International Federation of Industrial Design Associations defined the following for industrial design: "In terms of mass-produced products, the materials, structure, form, color, surface finish and decoration of the products are given by training, technical knowledge, experience and visual experience. With new quality and specifications, and solving problems in publicity and display、market development ,called industrial design [5].

In the contemporary era, industrial design is not only the expression of technical means, but also a necessary means to create a corporate brand effect, playing an important role in modern society. It can be said that industrial design can not only transform science and technology into market economy products, but also feed market information into the enterprise, thereby promoting enterprise development and enhancing enterprise vitality. It is a powerful driving force for social and economic development and a national soft power.

3.2. The necessary integration of industrial design and traditional cultural aesthetics

Chinese traditional philosophy advocates practical reason. The main feature of this mode of thinking is to pay attention to practicality. It is precisely this kind of promotion of pragmatism that stimulates the creativity of the ancients and promotes the positive development of the creation activities.

As we enter the 21st century, China has become the world's second largest economy. China's development is based on traditional culture. National culture is the cornerstone of our continuous advancement. At the same time, the industrial design of modern society is not a purely artistic creation activity, nor a single scientific and technological activity. It is a cultural re-creation of products. Therefore, the essence of industrial design is the cultural design of the product. In the design competition environment, we cannot ignore the necessary connection between products and culture, and recognize the important role of traditional culture to satisfy people's spiritual pursuits.

In this context, it is required that contemporary industrial design should be a combination of cultural aesthetics and art. As a cultural carrier, it is also not allowed to have cultural aesthetics and satisfy people's higher demand for cultural spirit. This is also the culture and The inevitable requirements of science and technology for contemporary industrial design.

3.3. The innovative consciousness of traditional aesthetic culture concept in modern furniture design
Furniture is artistic carrier of traditional cultural aesthetics. It can make a comprehensive presentation of human social civilization. Now, people's demand more aesthetic level, and the satisfaction of mental emotions can bring psychological satisfaction to users. This requires the designer to grasp the pulse of the times, understand the concept of the times, experience the interest of life, and make the furniture and the user have a broader spiritual resonance.

Figure 2 is a companion table. The main body of the table is cut from a piece of wood and divided into two parts. The texture of the wood is completely preserved. This design also shows the natural charm of the product. At the same time, the two parts of the table are treated in the form of yin and yang. This kind of design, combined with the natural sensibility brought by the product materials, makes a reasonable display of the oriental meaning. Figure 3 is breeze chair. The whole work retains the overall shape of the Ming furniture chair. The simple linear strip is used to express the beauty of the armchair. The seat is made of western paper rope, and the leather is used later. The comprehensive application of different materials highlights the beauty of the work, and its design essence is rooted in the traditional Chinese aesthetic concept, inheriting the culture, expressing the beauty of Chinese philosophy in the works, and conveying humanistic care.

Figure 2. Companion table

Figure 3. Breeze chair

4. Create featured furniture design

4.1. Strengthen the integrity of traditional aesthetics
To build industrial design products with Chinese characteristics, we must grasp the whole in design, strengthen the overall view of Chinese traditional culture, and establish an overall thinking concept in line with contemporary society.

Chinese traditional culture grasps the relationship between various elements as a whole and forms an organically related whole. This concept is still of great significance to contemporary industrial design. In terms of styling, structure, and crafts, we must consider the overall concept so that each specific part of the product becomes an integral part of the whole. We must inherit the traditional holistic view of thinking and strengthen the exploration of product analysis and design methods. Continuously improve the overall concept of thinking, so that the products have the symbolism of the traditional cultural spirit, highlighting the industrial design products with Chinese characteristics.

IOP Conf. Series: Materials Science and Engineering **573** (2019) 012023 doi:10.1088/1757-899X/573/1/012023

4.2. Pay attention to the harmony of traditional aesthetics

The Chinese culture believed that the world is connected and pursued a harmonious and stable relationship. This requires that in the design activities, we should pay attention to the practicality of the products in the modeling, materials, etc., from the perspective of people, pay attention to people's life requirements. This concept also requires us to adapt to the natural ecological environment as much as possible in the selection of product materials, reduce material consumption and waste, make rational use of existing resources, pay attention to the application of recycled materials, and promote the natural and simple aesthetic concept.

Modern industrial design should learn from the excellent traditional culture, trace the harmonious beauty under the traditional cultural aesthetic, and explore its rich spiritual connotation. Fully understand the harmonious view of traditional culture, refine its essence, re-create, rationally apply cultural points, show the aesthetic heritage of traditional culture, and embody the spirit of the times.

5. Conclusion

In the context of the new era, China's industrial design needs to inherit and promote its excellent traditional culture, extract its national characteristics, and form a complete and independent design concept, in order to occupy a place in the world competition.

Furniture is a model representative of traditional Chinese design, and it is the product of the ancients who followed the laws of nature. Designers should pay attention to the overall grasp, emphasize the integration and unity of each part, and use a comprehensive perspective to recognize thing and emphasize the overall concept of overall unity and contrast between reality and reality; At the same time, in the choice of styling and materials, we must also conform to the traditional cultural aesthetic concept, conform to the laws of nature. It is necessary to comprehensively consider the problems of furniture modeling, material selection, aesthetics, economic benefits, etc., and look at the design of contemporary furniture from the perspective of dialectical development.

At the same time, contemporary furniture design should pay more attention to its internal aesthetic and spiritual values, it should not only focus on the surface of modelling,but also conform to the modern design concept; Only by grasping the pulse of the times can we create a new era of industrial design with Chinese characteristics and give it a new vitality, let the traditional concept of Chinese traditional culture show a new look in the contemporary era, and demonstrate the cultural confidence of Chinese characteristics.

References

[1] Liu,G.Z.(2013) Thinking on the structure mechanism of industrial design industry in China. Design, 163: 51–59.
[2] Wang,Z.F. (2006) The beauty of great ease. Peking University Press, Beijing.
[3] Li,Z.H. (2002) Chinese Aesthetics. Guangxi Normal University press,Nanning.
[4] Zong,B.H.(1987) Aesthetics and artistic conception. Chinese People's Publishing House,Beijing
[5] Wang,S.Z(2002) History of world industrial design. China Youth Publishing House,Beijing

IOP Conf. Series: Materials Science and Engineering **573** (2019) 012024 doi:10.1088/1757-899X/573/1/012024

Application and Research of Plant Landscape Elements in Catering space

Zhiyu Feng[1*], Chen Wang[2] and Zixing Jiang[3]

[1]Design School, Shanghai Jiao Tong University, Shanghai, 200240, China

[2]Tongji Architecture Group., Ltd, Shanghai, 200092, China

[3]Tencent Music Entertainment Group, Shenzhen, Guangdong, 518057, China

[*]Corresponding author's e-mail: 544091292@qq.com

Abstract. In recent years, with the improvement of the quality of life, people enjoy the food culture and put forward higher requirements for environmental culture, which makes the catering culture develop rapidly. More and more restaurants use space design techniques to "soften" the design of the space, such as different lighting, art decoration and the proportion of color, which make the space attract customers' eyes. At the same time, a variety of plants are increasingly being integrated into the design of the catering space. They often play the role of softening the transition, changing the line of sight, and illuminating the space. However, these plants are often treated as individual fillers in space and are not pre-designed as a design element, systematic and planned, and often do not achieve long-term desired results.

1. Introduction

From the past, people simply went to restaurants to eat, but now restaurants and diners gradually integrate and interact. Many restaurants are developing in a multi-functional and diversified direction. The competition in the catering industry is becoming increasingly fierce, and the selectivity of customers is getting bigger. It is precisely because of this that the restaurant has made great innovations in space design and interactive experience development. At the same time, as the concept of green environmental protection has become meaningful, green restaurants have gradually become popular.

Therefore, in the design of catering space, the use of plant landscapes is increasingly extensive and is often used as an embellishment or decoration or in conformity with good intentions and color heads. For example, bamboo stands for sturdy, high-spirited and vigorous, and is usually used in Chinese restaurants. However, in terms of plant landscapes' effects and long-term performance, these restaurants only use the plant landscape as a single decorative element to assist the design of the entire space and have not been systematically and planned designed. The plant landscape itself is not treated as an element and expressed in the language of design, thus making the effect of the catering space not achieve the expected effect of the design and the long-term goal.

The plant landscape is incorporated into the interior design as an element of restaurant design to reflect the new concept and new methods of the new era catering space design, and to increase the diversified expression of the catering space design. Usage of its natural attributes can explore advocating green environmental protection concept, pursuing a more natural and harmonious environment, enhancing the taste of space and creating a unique atmosphere, so that people can relax

IOP Conf. Series: Materials Science and Engineering **573** (2019) 012024 doi:10.1088/1757-899X/573/1/012024

their emotions, reduce fatigue and leave a beautiful and deep impression while enjoying the catering space.

2. Overview of plant landscape elements

In order to better integrate the plant landscape and the catering space design, the plant landscape is regarded as an element to assist the design of the catering space, making it a part of the space design. First of all, the plant landscape is an image represented by individual plants or communities, and is a scene that people can actually feel through the senses. Elements are a factor that is especially important in design languages. It can be seen that plant landscape elements are one of the crucial elements of landscape plant landscaping or garden plant configuration, and a critical design factor in design language. It is closely related to the ecological environment and green environment that human beings depend on. Plant landscape elements are mainly used to create green landscapes that put people in a vibrant natural environment.

Secondly, in the catering space, the plant landscape elements are design elements that can change the spatial layout, enhance the visual effect, and optimize the ecological environment in the space design, which can effectively divide the function of the catering space and enhance the visual attraction, as well as improving the catering environment of the restaurant.

3. The function of plant landscape elements

Plant landscape elements have three important functions: ecological function, spatial function and aesthetic function.

3.1. Ecological function

The plant landscape elements can make the catering space in a better environment, giving the diners an ideal catering space. Because most plants have strong adsorption capacity, the environmental quality is effectively regulated. Restaurant design can use this function to adjust the indoor micro-environment, because the restaurant is a space that requires high environmental quality, such plants can effectively reduce dust pollution and gas pollution, purify the air, manufacture oxygen, absorb radioactive materials, adjust temperature and air humidity to improve the microclimate.

3.2. Spatial function

3.2.1. Different body shapes, forming different spatial effects

In the restaurant design, the smaller plant landscape elements can usually be used as decoration or embellishment to enhance the atmosphere of the catering space and reduce visual fatigue. The larger plant landscape elements can be used as a space partition to block the line of sight or become a visual focus(figure1).

Fig 1: Small plants for decoration, large plants for partitions

Fig 2: Plant landscape elements form a semi-enclosed space

3.2.2. Spatial definition function

Plant landscape elements act as a flexible spatial transition element, using thin and tall trees. It can achieve the effect that the wall or other substantial space media can't achieve, and avoid the space to produce a relatively rigid barrier to achieve the definition of space. In addition, the plant landscape elements also have the function of functional partitioning, so that the space has the primary and secondary points, distinguishing the key parts of the space and the corner parts, and its visual effects and actual functions are often more natural than the use of walls or partitions(figure2).

3.2.3. Spatial enclosure function

Plant landscape elements can also replace the traditional space enclosing method, and the space is separated from the outside by clever and flexible, forming an area and making the space more vivid. For example, plant landscape elements can be used to form semi-enclosed spaces. The use of a higher plant combination on one side of the open space constitutes a one-sided closure, resulting in a more intimate spatial interface, which also becomes the line of sight for people. At the same time, the plant landscape elements can also be well-enclosed spaces, have strong concealment, and are relatively isolated from the surrounding environment. Moreover, the line of sight and streamlines are also strictly restricted.

3.2.4. Spatial guidance function

Plant landscape elements can be arrayed and rhythmic in the transportation route, which connect different function and play a guiding role in catering space. For example, the usage of low shrubs and ground cover can define the space, which guide people to effectively. Moreover, such plants are able to connect the two spaces and produce good visual effects.

3.3. Aesthetic function

3.3.1. Color space

Due to the excessive use of space and the distribution of functions, it is easy to neglect the overall color tone of the space. However, if the plant landscape elements are formed into a color space, the problem of unbalanced and boring color of the catering space can be greatly reduced. Through the design of plant landscapes and display methods, the color space formed by plant landscape elements in the catering space has played a strong role in beautifying the environment. If the color of the plant landscape elements is considered together with the main color of the space itself, the visual attraction of the design of the catering space can be increased by the fusion and contrast of the two colors. It means that the color of the plant landscape elements can make the visual effect of the entire catering space shines.

3.3.2. Softening space

In the design of the catering space, most of the geometry of the straight line structure, such as the turning point, the starting part, the ending part, and the visual center point of the space, these key parts will inevitably appear abrupt. In this case, the plant landscape plays a transitional role, changing the rigid connection of the space, not only has the effect of eye-catching drainage, but also can shift people's sight. Different textures can soften the catering space effectively, so that primary and secondary points in the space give the overall comfortable visual effect.

3.3.3. Growth changes bring different views of the landscape

The shape, color, foliage and its growth cycle of plants have high appreciation value. Different growth stages can produce different visual effects, and the visual center of the composition plays a very good role in beautifying the visual environment. In addition, the difference in color and texture of any kind of plant forms a small range of features, which can improve the identification of the landscape, thereby creating an artistic conception.

4. The relationship between plant landscape elements and catering space

4.1. Types and forms of plant landscape elements
The plant landscape has been greatly improved and introduced, and there has been great innovation and development in the modeling. In recent years, there have been many novel, chic and pleasing kinds of form, such as display type, hanging type, wall hanging type, planting type. To be specific, many restaurants choose plants that are mainly based on spider orchids, and they are suspended. The French restaurant, Joe Lubson, is decorated with planting landscapes in the restaurant, giving the whole catering space an active and relaxed atmosphere. The plant landscape bears the role of beautifying space and active atmosphere in various spaces, which can better highlight the design effect and has the finishing touch.

4.2. Plant landscape elements and restaurant style expression
As one of the decorative elements, the plant landscape elements, in the catering space, must be unified with the overall style, or complemented by the combination of plant landscape and catering style, in order to set off the restaurant style and achieve harmony and perfection, therefore the space can please diner.

5. Spatial design principles of plant landscape elements
The design principle of a catering space is reflected in the aesthetic principles applied in the design of this space, and integrates many aspects of professional knowledge, which often reflects the design connotation of this catering space. The plant landscape elements will also reflect its spatial design principles in the catering space through its own aesthetic principles.

5.1. Equilibrium
Equilibrium is a sense of balance in the aesthetics principle, and this balance harmonizes the layout on both sides of the axis to achieve a visual stability[1]. In the catering space, the display of plant landscape elements also has a virtual or real axis. When the plants are distributed on both sides of the axis, the space is in a state of balance and stability, and the space at this time is more visually harmonious.

5.2. Concise
The simplicity of the display of plant landscape elements does not mean that the visual effect is simple and monotonous. In the catering space, it is not the more landscape elements are used, the better, but the control and suitability necessary in the design. It is more suitable for the elements of space utilization, and sometimes it can achieve good results in the catering space even if only one landscape element is used.

5.3. Contrast
The contrasting technique is applied more in plant display. It can be divided into color contrast, shape contrast, volume contrast and texture contrast [2]. The effect of contrast is reflected by the strength of contrast, which is essentially against each other. To highlight each other's personality, but also serve the overall effect when applied.

5.4. Rhythm
Rhythm is a regular or irregular intermittent reproduction, and the spatial distance in which it is repeated is the rhythm. In the design of plant landscape elements, certain landscape elements reappear in the same or similar shape, texture or color, or enhance the visual rhythm in the catering space by placing similar elements together.

5.5. Unification

Any display technique of a plant landscape element has its center, which is the theme. The various means used by the plant landscape elements in the display of the catering space, from the selection of plant species to the use of landscape elements, are unified in the theme. However, there is no change in unity. There should be a master in the space. The host and the guest are distinct. In the configuration of plant landscape elements, there is a need for color change and unity of plant species, so that it appears in the design of the catering space.

5.6. Primary and secondary and diversity

Plant landscape elements should emphasize the primary and secondary relationship in the design of the catering space, in order to highlight the key points. This can be achieved by distinguishing between the primary visual plant and the auxiliary plant in the plant landscape element, or by the primary visual space and the secondary visual space formed in the space. This is also a common method which combines the primary and secondary organic to create an orderly and distinctly landscaped effect.

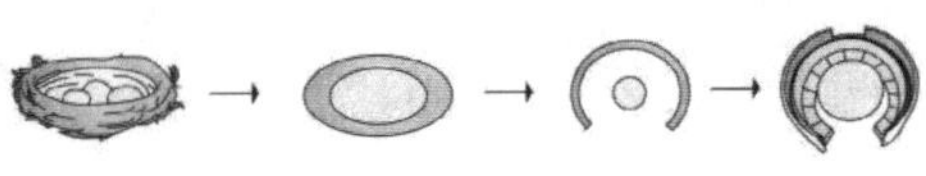

Fig 3: The deck is evolved from the bird's nest

Fig 4: The trunk shape of the deck evolved from the tree

The work is a forest-themed western restaurant. The restaurant mainly uses plant landscape elements as the main design element. The first floor is a common catering area, and the second floor is a card seat that evolved from a combination of trees and nests(figure3&figure4).

Since the deck is extended upwards on the basis of a plane, this part has become the visual center in the space first. Secondly, the auxiliary of the plant landscape elements is added to the deck to make it become main visual space, weakening catering area downstairs(figure5). After entering the restaurant, diners' visual attention is directly concentrated on the second floor deck(figure6). Therefore, it not only highlights the key points, but also plays a role in beautifying the space environment.

5.7. Sequence

In the catering space, the spatial sequence can be divided into four phases: start phase, transition phase, climax phase, and end phase. In the beginning, an attractive design focus is created, in order to allow the diners entering the restaurant to pass through the plant landscape display, and to move according to the desired visual effect. The transition can be achieved by a gradient in the form, texture or color of the plant landscape elements. The climax stage is the main body of the sequence principle. The design of the landscape elements in the catering space needs to attract enough attention of the diners, or stimulate emotions and even generate satisfaction. Finally, the climax is about to return to calm, and landscape elements can be used in a small amount to achieve a feeling of endless taste for the diners.

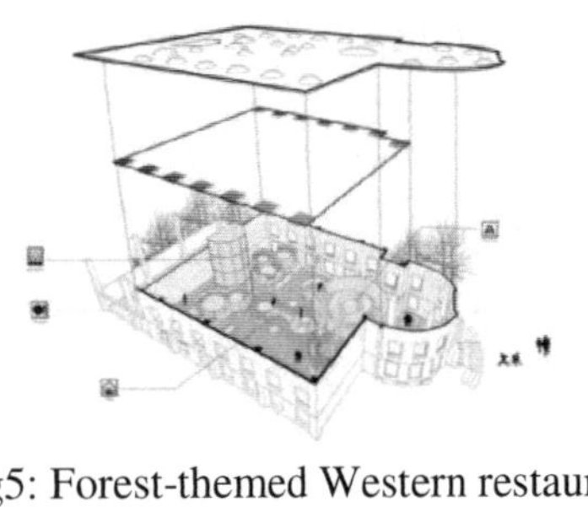

FIg5: Forest-themed Western restaurant

Fig 6: Combine the deck and the trunk

6. The use of plant landscape elements

6.1. Display of plant landscape elements
With the pursuit of new and different food and beverage space, its space is no longer limited to the conventional scale law, and various different proportions of space appear, which promotes the diversified expression of the catering space style. The use of plant landscape elements has also broken through the limitations. It has broken through the previous single use method, and has incorporated the landscape elements into the design language and become part of the space design.

Plant landscape elements generally have the following display methods.

6.1.1. Regular formula:
Use the arrangement of rules to arrange the plant landscape elements neatly. In the complex space of structural lines, the regular display way can make the space look less messy, but can weaken the heavy space structure to give people a heavy sense.

6.1.2. Natural style:
In the layout, the application of frame, clip, and borrowing is not limited to the framework of indoor space. The layout mode gives people the feeling of returning to nature.

6.1.3. Mixed type:
Several display methods are combined to design and arrange plant landscape elements through different display methods, and a lively and flexible atmosphere can be created in the regular space of the line.

6.2. The Artistic Conception and Aesthetic Thought in the Catering Space
The design of the catering space mainly combines the various elements of the times, historical context, environmental factors and other creative concepts. On the basis of reasonable spatial planning and functional zoning[3], uses specific scene techniques to reflect different spatial concepts and structures. Individuality and emotion expression meets people's spiritual needs. The integration of plant landscape elements is a fusion of food and environment, conveying different spatial experiences to the diners[4].

7. Conclusion
With the rise of the catering culture, restaurant design has become increasingly popular in people's lives, and people are becoming demanding on the restaurant environment. As a new design element, plant landscape elements, together with other design elements, are added to the design of the catering space, and also bear the responsibility of innovative catering space design. Therefore, plant landscape elements need to use systematic, comprehensive and planned design language to coordinate the overall space. Through reasonable allocation of plant landscape elements, people pursue a more natural and harmonious environment, enhance space taste and create uniqueness. The atmosphere allows people to relax their emotions, reduce fatigue and leave a beautiful and deep impression while enjoying the catering space.

References
[1] Xuwei Zhang& Chenqin Yang. (2012). Discussion on design principles and development trends of indoor greening in restaurants.[D]
[2] Feng Wang. (2015). Discussion on Indoor space greening. [D]
[3] Aiyin Ren. (2008). Indoor greening design research. Shanxi Agricultural Science,, 36(8), 59-61.
[4] Tschudy, R. H., et al. "Disruption of the terrestrial plant ecosystem at the Cretaceous-Tertiary boundary, Western Interior." Science 225.4666 (1984): 1030-1032.

IOP Conf. Series: Materials Science and Engineering **573** (2019) 012025 doi:10.1088/1757-899X/573/1/012025

Optimal Design of Product Culture Image Modeling Based on FNN Model and PSO Algorithms

Li Lin*, Yanzhe Zhao

School of Mechanical Engineering, Guizhou University, Guiyang, China 550025

*Corresponding author's e-mail: linlisongbai@163.com

Abstract: In order to solve the problem of strong subjectivity of cultural image modeling design in today's market and single optimization result caused by product cultural image transformation design method, the evaluation method of product cultural image modeling based on FNN model and the optimization design method of product cultural image modeling based on PSO algorithm were both proposed. Firstly, the representative samples were determined, and the cultural images were sorted out and classified to quantify the shape parameters of the samples; then, the membership function of the product shape parameters was established to fuzzify them, and the mapping relationship between the "product shape parameters-cultural image fitness" was obtained by using three-layer BP neural network, and the evaluation method of the product cultural image modeling was established; By updating the speed and position of representative sample population, adjusting its modeling parameters based on PSO algorithm, the optimization design mechanism of product cultural image modeling was established. Finally, taking the design optimization design of Miao embroidery products in Songtao area of Guizhou Province as an example, the results showed that the optimization design method could well adapt to the optimization design of cultural image modelling and had injected new vitality into cultural creative design.

1. Introduction

Cultural products can not only meet the basic needs of users for functions, but also bring users cultural emotional enjoyment, making products more popular with consumers. Today's wave of cultural design makes the purpose of design not only to give products meaningful functions, but to convey a cultural spirit and concept to users. The purpose of this paper is to find an optimal design method that can make the product modeling elements and cultural images one by one correspond to each other, so as to improve the accuracy of cultural images in product expression.

Cultural image is a cultural symbol with relatively fixed meanings and profound associations[1]. Cultural products, with their unique modeling elements and rich cultural connotations, make people feel the profound connotation behind products and trigger corresponding images. Document [2] applied Chinese traditional architectural elements to furniture design. Document [3-4] respectively explored the design strategy oriented by cultural image recognition and the application of semiotic theory to cultural image design in Jiangnan region. However, the above-mentioned literature lacks effective integration with computer technology in the study of the relationship between cultural image and product form, which is not conducive to the development of computer-aided image design.

At present, many scholars at home and abroad have carried out relevant research on how to apply evolutionary algorithm to product image modeling design. For example, document [5] put forward the optimization design method of product image modelling based on genetic algorithm, used multi-dimensional scale method and morphological analysis method to analyze product modelling features, and then obtained the quantitative relationship between modelling features and perceptual images by a quantitative method, and finally carried out genetic optimization to obtain the design scheme. In literature [6], the mapping relationship between high-speed train design module library and sensory image set was obtained by using text mining method and BP neural network, which provided effective assistance and support for high-speed train modeling design. Literature [7] established and improved the neural network model of product design evaluation. The feasibility and validity of the model were validated by taking the design of garden tool mower as an example. The mapping relationship between design parameters and product perceptual image was obtained by using support vector machine in document [8]. In document [9], the improved particle swarm optimization algorithm was used to optimize the radial basis function artificial neural network, and a prediction model was established. The color emotion of each user was predicted by the model. The above scholars have achieved good research results in seeking the mapping relationship between perceptual image and the elements of morphological design.

However, the above research focuses on the matching between the implicit perceptual needs[10] of users and the elements of product shape, and lacks the design method of mapping cultural images into product shape. Therefore, with the deepening of research and the help of computer technology, effective improvement measures are provided to solve the problems of strong subjectivity in cultural image modelling design and single optimization result caused by image transformation design method. Based on this, this study uses the fuzzy neural network to objectively evaluate the perceptual part of the cultural image which is difficult to say clearly, and to establish the mapping relationship between the cultural image and the shape. In order to provide reference and support for the further programmability and intellectualization of creative product design, a computer-aided optimization design method for cultural image products is proposed, which combines the excellent searching ability of particle swarm optimization algorithm to optimize and innovate product multi-modeling schemes.

The full-text framework is shown below.

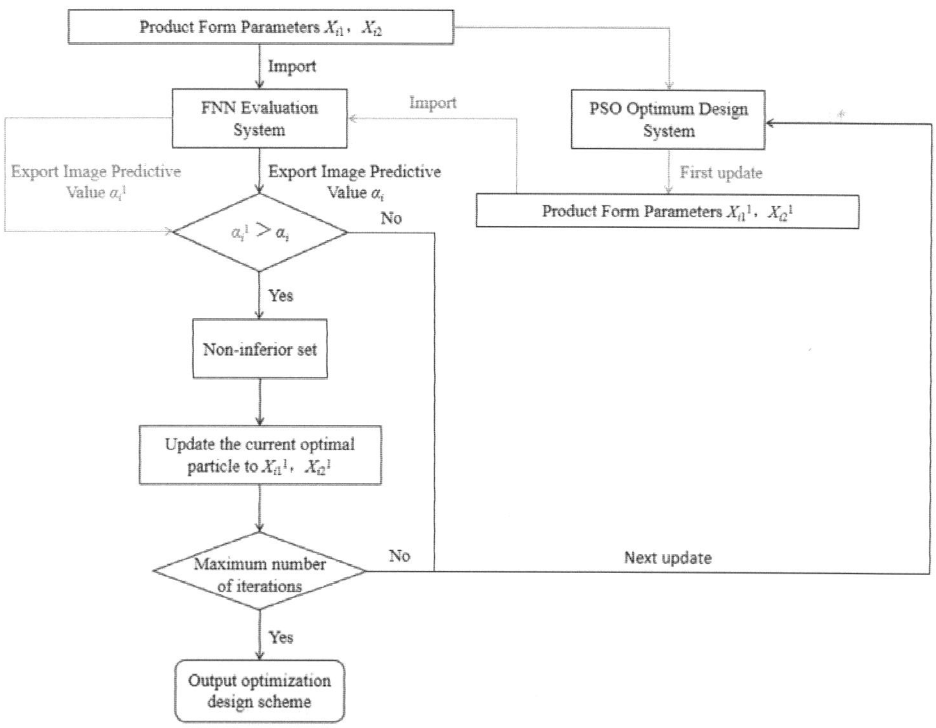

Figure 1. Optimum design process of cultural image modelling

2. Initialization of Product Culture Image Modeling Design

Initialization of cultural image modeling of cultural products includes collecting information of cultural image perceived by users and parameterization of product shape. Because the product color factor can interfere with the user's perception of the product, the color factor is not taken into account in the analysis of product modeling parameters. The research process of this part can be divided into three stages (Fig. 2).

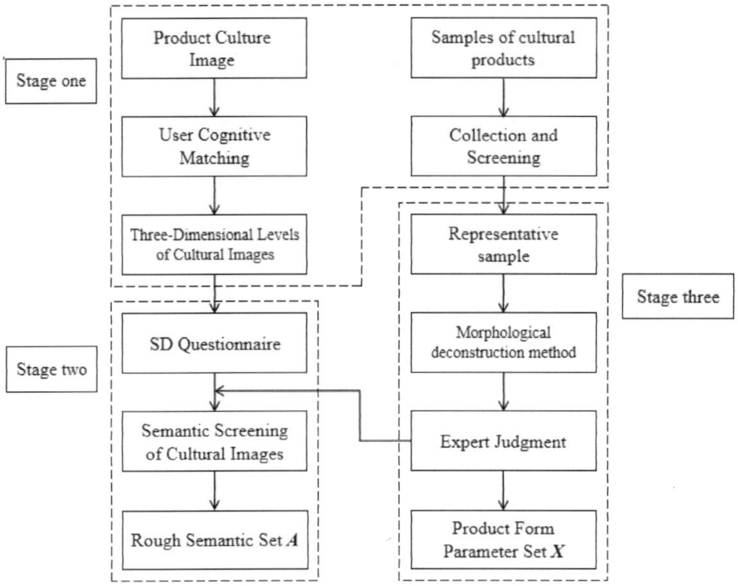

Figure 2. Initialization of Product Culture Image Modeling Design

2.1. *Collection of Product Samples and Analysis of Cultural Images*

Collect product samples in cultural theme websites, newspapers, books, magazines, and related literature, then conduct preliminary screening. Through the analysis of three levels of consumer cognitive[11] matching model of regional cultural image, it is concluded that the implicit value layer of cultural image contains the most cultural semantic information and has the most cultural identification. Therefore, exploring the spiritual connotations contained in cultural products and translating the semantics and vocabulary are the preconditions for the next step of cultural image semantics selection.

2.2. *Selection of the Meaning of Cultural Images*

For the semantic selection of cultural images in a certain region, through field research, expert interviews, reading literature and other methods to collect images and image vocabulary describing cultural products, combined with Semantic Differential Method[12] (SD), Rough Semantic Set **A** which is more in line with the local cultural image is selected from a large number of vocabulary.

$$A=\{A_1,A_2,\cdots,A_n\}$$

In formula, $A_i = \{a_{i1}, a_{i2}\}$, $i = 1, 2, \ldots, n$, a_{i1} is the image of things, a_{i2} is the perceptual image. Therefore, Rough Semantic Set **A** contains two subsets, namely, object image set **F** and perceptual image set **I**.

$$F=\{a_{11},a_{21},\cdots,a_{n1}\},\quad I=\{a_{12},a_{22},\cdots,a_{n2}\}$$

2.3. *Establishment of Product Form Parameter Set*

The preliminary screened samples of cultural products in the first stage are classified as representative sample set **N**. x is the number of representative samples, The sample set is:

$$N=\{N_1,N_2,\cdots,N_x\}$$

According to the content of "Morphology" in the outline of art education in China, the formal genes of patterns can be divided into force image and perceptual force field[13], which can be used as the basis of product morphological deconstruction. Quantitative indicators of force image and field are proposed by design experts (Fig. 3, Fig. 4). They are summed up as force image coefficients l and field density coefficients ρ. Thus, the dimension of shape parameters of each sample is 2. The sample dimension is expressed as:

$$X_i = \{X_{i1}, X_{i2}\}$$

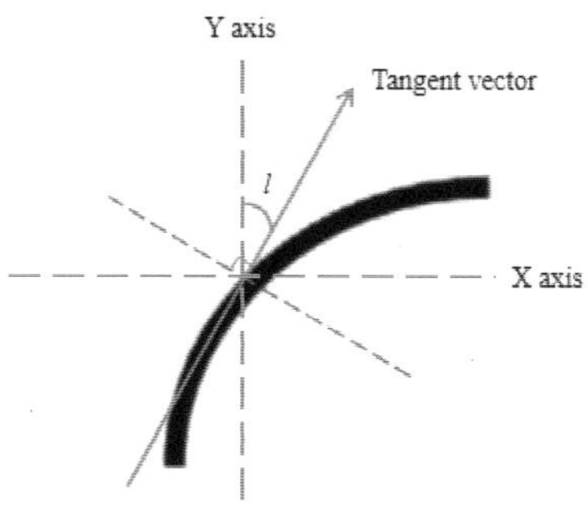

Figure 3. The force image coefficient l of the structural line (unit: degree)

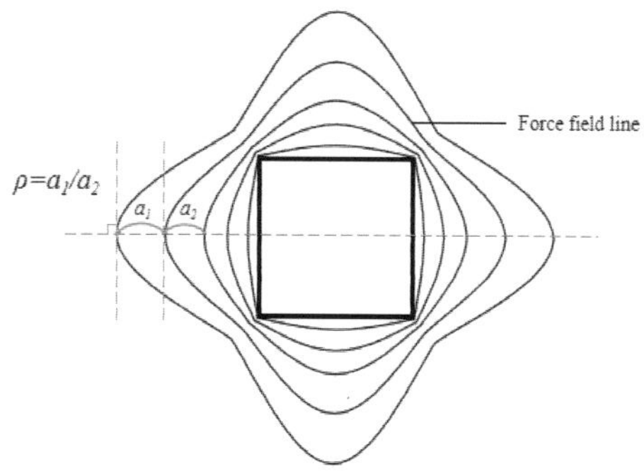

Figure 4. The field density factor ρ of the main contour (unit: 1)

The morphological parameter set $\boldsymbol{X}$ of representative sample set $\boldsymbol{N}$ is:

$$X = \begin{bmatrix} X_{11} & X_{12} \\ X_{i1} & X_{i2} \\ \cdots & \cdots \\ X_{x1} & X_{x2} \end{bmatrix}$$

In the matrix, $i = 1, 2, ..., x$.

3. Cultural Image Evaluation of Product Modeling Based on FNN Model

A cultural image evaluation system is constructed by using simple fuzzy reasoning theory and three-layer forward BP neural network, which provides criteria for judging the quality of product cultural image modeling and terminating operation. In view of the fact that the force image coefficient l and field density coefficient ρ involved in this study can not determine their magnitude and degree, the two coefficients are fuzzified. The evaluation architecture is as follows: the first layer is the data preprocessing layer, which takes the shape parameters of the product as input variables and imports them into the fuzzifying neurons; the second layer is the middle layer, which contains many hidden nodes and plays a role of connecting transition[14]; the third layer is the output layer, which outputs the actual evaluation value and cultural image fitness value of the sample rough linguistic meaning set $\boldsymbol{A}$, respectively. This value indicates whether the sample fits the local cultural image. The structure of the network is shown in Fig. 5.

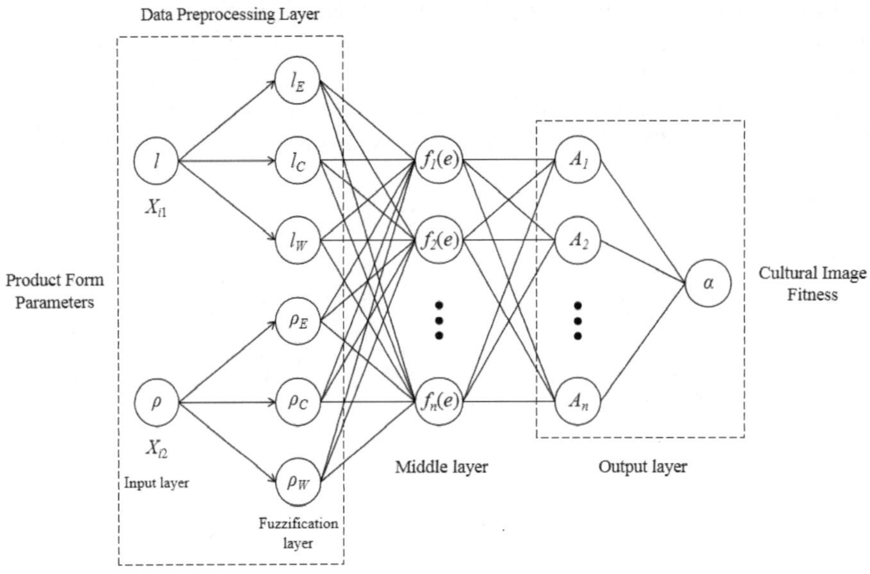

Figure 5. Cultural Image Evaluation Network Structure

3.1. Data preprocessing

The input variables ($X_{i1} = l$, $X_{i2} = \rho$) are fuzzified by membership function. Each fuzzy input variable is divided into three fuzzy subspaces, which are expressed by extreme (E), general (C) and weak (W) linguistic variables respectively, and the membership function is a triangular distribution function.

3.2. Intermediate and Output Layer Processing

By consulting literature [15], the empirical formulas for determining the number of nodes in the middle layer are as follows:

$$n_l = \sqrt{n+m} + a$$

In the formula, n_l is the number of nodes in the middle layer, n is the number of nodes in the input layer, m is the number of nodes in the output layer, and a is a constant between 1 and 10[15]. After the initial value of the number of nodes in the middle layer is determined by empirical formula, the optimal number of nodes in the middle layer is determined by numerical experiments, that is, increasing the number of nodes gradually, and training with the same sample, from which the number of nodes corresponding to the minimum error is determined[16].

3.3. Derivation process and evaluation process between layers

The input product shape parameters are transformed into the cultural image evaluation value of the product through a certain form. In this study, i represents ordinal number of products, $X_{i1} = l$, $X_{i2} = \rho$.

3.3.1. Derivation from Fuzzy Layer to Intermediate Layer. Input value of j node in middle layer $I_j = w_{(\mu1)j}\mu_1 + w_{(\mu2)j}\mu_2 + w_{(\mu3)j}\mu_3 + w_{(\mu4)j}\mu_4 + w_{(\mu5)j}\mu_5 + w_{(\mu6)j}\mu_6 + \theta_j$, the output value of middle layer $y_j = f_j(I_j)$, $j = 1, 2, \ldots, n$. As shown in Fig. 6.

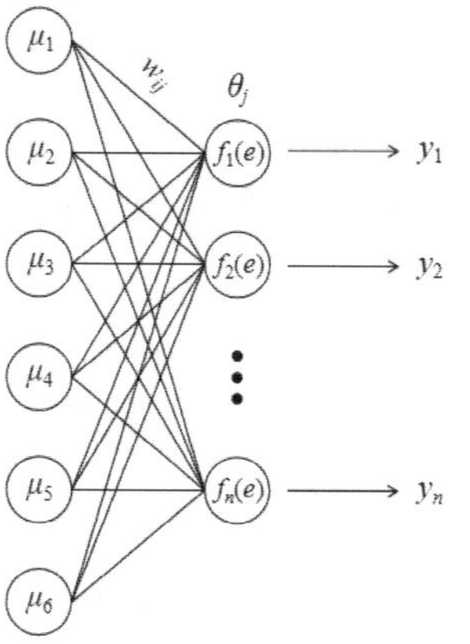

Figure 6. Derivative diagrams from the fuzzification layer to the middle layer

3.3.2. Derivation from Intermediate Layer to Output Layer. $f_z(A_z) = f_z(w_{(y1)z}y_1 + w_{(y2)z}y_2 + \ldots + w_{(yn)z}y_n + \theta_z)$, $z = 1, 2, \ldots, n'$. All the output results are regarded as the actual evaluation values of the rough semantic set $\boldsymbol{A}$ of cultural images. After error analysis with the expected output, the error is conversely transmitted. The weights and thresholds are updated to make the actual evaluation values close to the expected output, as shown in Fig. 7.

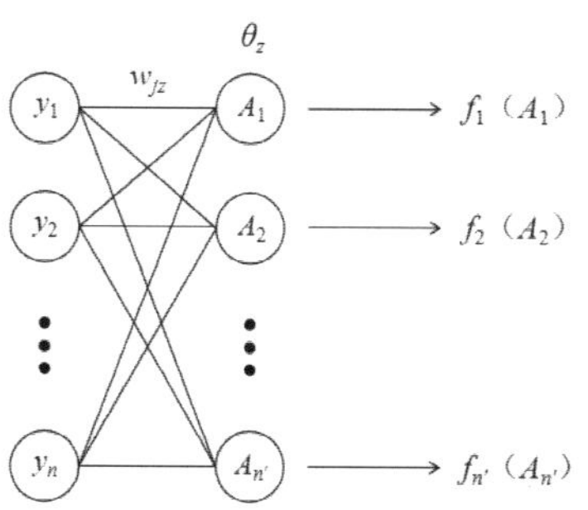

Figure 7. Derivative diagrams from the middle layer to the output layer

3.3.3. One-way Derivation in Output Layer. The first layer of the output layer participates in the reverse transmission of errors, while the second layer does not. There is only forward transmission of numerical values between the two independent layers. After updating the weights and thresholds of the network, the adjusted actual evaluation value z_1, z_2, …, $z_{n'}$ is input into the α node. The weights $w_{z\alpha}$ of each evaluation value are obtained according to the statistical data of the questionnaire. The threshold θ_α makes the fitness value change within a certain range, The range of values is 0~1. The cultural image fitness $\alpha_i = w_{1\alpha}z_1 + w_{2\alpha}z_2 + \ldots + w_{n'\alpha}z_{n'} + \theta_\alpha$, then normalize it.

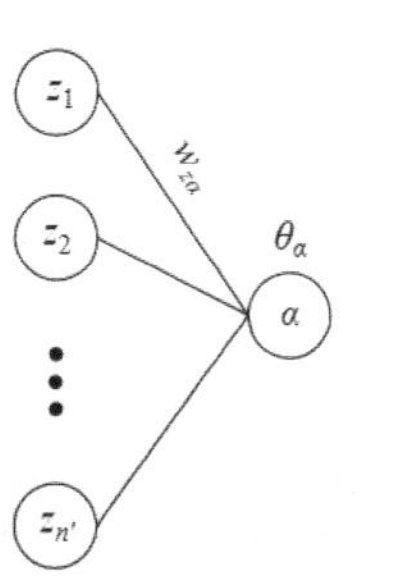

Figure 8. Unidirectional derivation diagram in output layer

3.3.4. Ending Conditions of Network Adjustment. If the error between the actual evaluation value and the expected output is within the set range, the update of the termination weights and thresholds will result in good network performance. After inputting the shape parameters of the product, the actual evaluation value and the cultural image fitness α are output, $\alpha \geq 0.5$ means it consists with local cultural images, the closer to 1, the higher the fit; $\alpha < 0.5$ means it deviates from local cultural images, the closer to 0, the higher the deviation.

4. Optimum Design of Product Culture Image Modeling Based on PSO Algorithms

Design activities are not only the process of integrating several elements to achieve a specific goal or satisfy a certain demand, but also the process of searching for the best solution through a certain design procedure or method. Because each designer has different understanding of cultural image, he often relies on personal experience and intuitive judgment to design the image of cultural products.

To solve this problem, an optimal design of product cultural image modeling based on PSO algorithm is proposed, The flow chart is shown in Fig. 9.

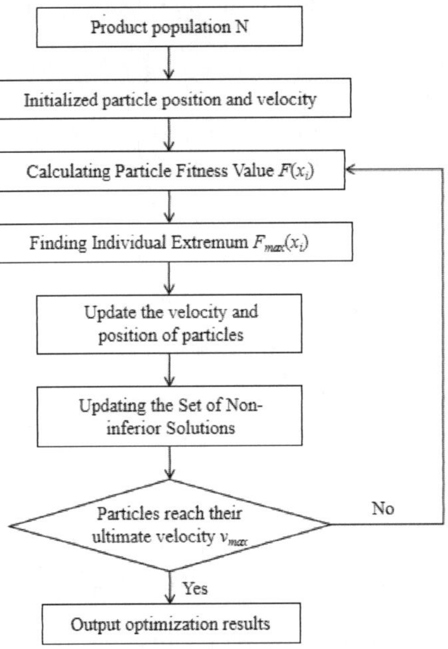

Figure 9. Optimum Design Flow Based on PSO Algorithms

4.1. Particle population parameterization

Consider each cultural product sample as a particle unit in the algorithm x_i, The representative sample set $\boldsymbol{N}$ is the initial population of particles. If the two morphological parameters X_{i1} and X_{i2} of the product are regarded as the two search dimensions of the algorithm, the particle searches for the individual optimal position x_{ib} with a certain speed v_i in the two-dimensional target search space.

$$x_i = (X_{i1},\ X_{i2}),\quad v_i = (V_{i1},\ V_{i2})$$

When the fitness value $F(x_i)$ of particle x_i reaches its maximum, individual optimal solution:

$$F_{max}(x_i) = F(x_{ib})$$

4.2. Determination of fitness value

The fitness value of the algorithm is the basis for judging the position of particles. The goal of optimum design of cultural image modelling is to get a more suitable product form for local cultural image. Therefore, the fitness value of particles is cultural image fitness α.

$$F(x_i) = \alpha_i$$

4.3. Particle position and velocity updating

This study focuses on the optimal design of product cases, not on the optimization of solution groups, so the "social part" of particles is zero. That is, the acceleration factor $c_2 = 0$. The updating formula of particle position and velocity is as follows:

$$v_i(t+1) = \omega \times v_i(t) + c_1 \times rand_1 \times (x_{ib} - x_i)$$

$$x_i(t+1) = x_i(t) + v_i(t+1)$$

4.4. Optimum Design Flow Based on PSO Algorithms

The initial position of particle x_i is recorded as $x_i(0)$, and the initial velocity $v_i(0) = 0$. Input the shape parameters X_{i1}, X_{i2} to calculate the image predictive value α in the product modeling culture image evaluation system based on FNN model, Taking α as the historical optimal solution $F(x_{ib})$ of particle x_i. Update the particle's velocity and position every time according to the above formula, import its morphological parameters into the cultural image evaluation system to calculate the fitness value $F_t(x_i)$ of generation t new particle $x_i(t)$, compare it with individual extreme value $F(x_{ib})$. If $F_t(x_i) \geq F(x_{ib})$, Then it is regarded as a non-inferior solution and the individual extreme value is updated to $F_t(x_i)$, all non-inferior solutions constitute a set of non-inferior solutions. If the particle has reached the limit velocity v_{max}, the updating of the non-inferior solution is stopped.

5. Example analysis

5.1. Parameterization of product form and Semantic Analysis of Cultural Images

Based on the analysis of the correlation between the selected 30 images of seedling embroidery samples in Songtao area and the local culture, 12 representative samples were simplified (Fig. 10). The cultural image semantics is summed up as the sentence pattern of "object image - perceptual image". Rough semantics set A (Table 1) is selected through SD questionnaire. After quantitative description of the structure line and main contour line of the pattern sample, the sample morphological parameters (Table 2) are obtained.

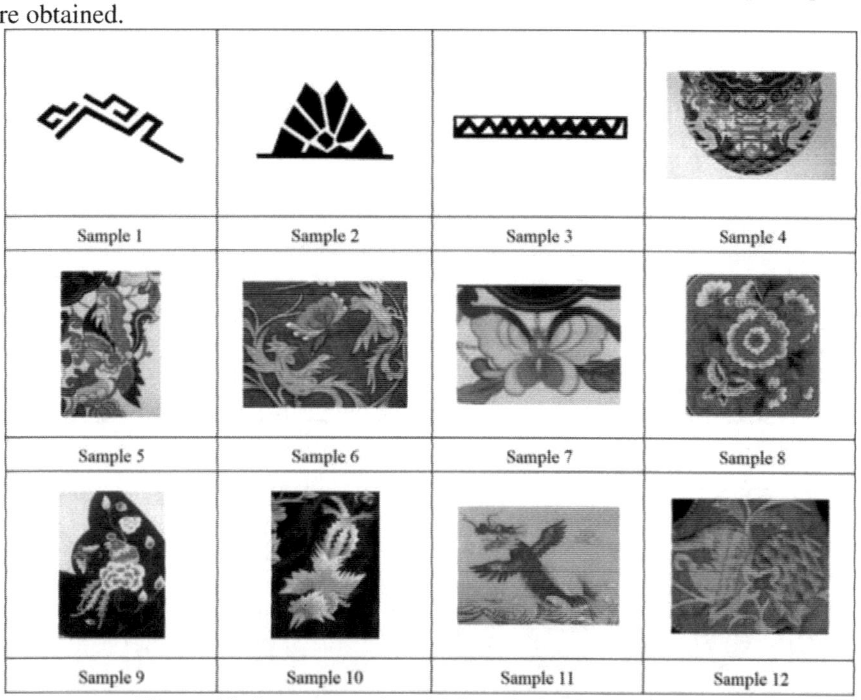

Figure 10. Representational Miao Embroidery Samples

Table 1. Rough Semantic Set A (F-I)

Imagery semantics 1	Flying horse——Animal worship
Imagery semantics 2	Lofty ridges and towering mountains ——Fear of Nature
Imagery semantics 3	Harmony between Luan and Phoenix ——Husband and wife and aesthetic feeling
Imagery semantics 4	Ichthyosaurus diving——Indomitable feeling
Imagery semantics 5	Birds fall on Peony——Need To Breed
Imagery semantics 6	Butterfly Dance Blossoms——A sense of happiness in life

Table 2. Sample morphological parameters

Sample serial number	Sam-ple 1	Sam-ple 2	Sam-ple 3	Sam-ple 4	Sam-ple 5	Sam-ple 6	Sam-ple 7	Sam-ple 8	Sam-ple 9	Sam-ple 10	Sam-ple 11	Sam-ple 12
force image coefficients l	62.1	13.4	90.0	56.2	34.8	78.5	22.6	45.7	86.3	30.8	52.9	17.2
field density coefficients ρ	3.5	5.8	1.2	6.7	8.1	4.3	2.7	7.9	6.5	3.2	4.0	6.1

There are six groups of vocabulary in Semantic Set $\boldsymbol{A}$. The evaluation values of each sample are T_1, T_2, T_3, T_4, T_5 and T_6. Morphological parameters of each sample X_{i1}, X_{i2} (62.1, 3.5), (13.4, 5.8), (90.0, 1.2), (56.2, 6.7), (34.8, 8.1), (78.5, 4.3), (22.6, 2.7), (45.7, 7.9), (86.3, 6.5), (30.8, 3.2), (52.9, 4.0), (17.2, 6.1) constitute the morphological parameter set $\boldsymbol{X}$, sample number i = 1, 2, ..., 12.

5.2. *Establishment of FNN Evaluation System*

The membership function of force image coefficient l is:

$$\mu_W = \begin{cases} 1 & x \le 0 \\ \dfrac{64.15 - x}{64.15} & 0 < x < 64.15 \\ 0 & x \ge 64.15 \end{cases}$$

$$\mu_C = \begin{cases} 0 & x \le 0 \\ \dfrac{x}{64.15} & 0 < x < 64.15 \\ \dfrac{128.3 - x}{64.15} & 64.15 \le x < 128.3 \\ 0 & x \ge 128.3 \end{cases}$$

$$\mu_E = \begin{cases} 0 & x \le 64.15 \\ \dfrac{x - 64.15}{64.15} & 64.15 < x < 128.3 \\ 1 & x \ge 128.3 \end{cases}$$

According to the membership function, the size of image coefficient l can be judged to guide the follow-up optimization design. Similarly, the membership function of field density coefficient ρ can be obtained. The number of nodes in the input layer $n = 6$, the number of nodes in the output layer $m = 6$, $a = 1$, and the number of nodes in the middle layer $n_l = 4$. Setting initial weight $w_{ij} = 0.2$, $w_{jz} = 0.5$, according to the percentage of the sample of each semantic vocabulary in the questionnaire, conclude $w_{1\alpha} = 0.08$, $w_{2\alpha} = 0.06$, $w_{3\alpha} = 0.29$, $w_{4\alpha} = 0.17$, $w_{5\alpha} = 0.22$, $w_{6\alpha} = 0.18$. Set initial threshold $\theta_j = 0.6$, $\theta_z = 0.3$, the activation function $f(x)$ of each node chooses S-type function and inputs the morphological parameters X_{i1} and X_{i2} of each sample into the network for training. After 4286 training sessions, the error between the adjusted actual evaluation values of z_1, z_2, z_3, z_4, z_5, z_6 and the expected output of T_1, T_2, T_3, T_4, T_5 and T_6 is close to 0.1. The training process is shown in Figure 11 and the output of the network is shown in Table 3.

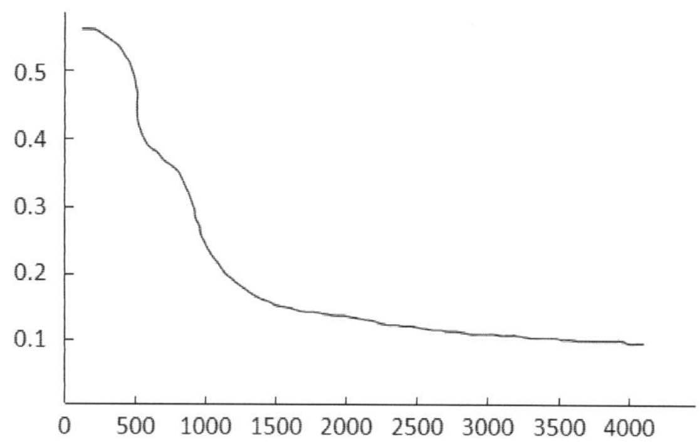

Figure 11. Convergence process of network

Table 3. Expected Output and FNN Output

Number	Sample 1	Sample 2	Sample 3	Sample 4	Sample 5	Sample 6	Sample 7	Sample 8	Sample 9	Sample 10	Sample 11	Sample 12
T_1	0.78	0.86	0.68	0.45	0.37	0.30	0.90	0.29	0.35	0.88	0.85	0.46
z_1	0.81	0.80	0.59	0.52	0.28	0.36	0.83	0.20	0.41	0.79	0.76	0.55
T_2	0.22	0.35	0.10	0.60	0.58	0.54	0.21	0.51	0.50	0.54	0.57	0.48
z_2	0.18	0.46	0.15	0.53	0.62	0.44	0.28	0.60	0.57	0.61	0.52	0.40
T_3	0.66	0.27	0.20	0.75	0.81	0.43	0.26	0.52	0.90	0.56	0.73	0.44
z_3	0.62	0.21	0.25	0.82	0.73	0.38	0.22	0.60	0.85	0.65	0.77	0.53
T_4	0.80	0.66	0.90	0.34	0.61	0.59	0.42	0.20	0.49	0.72	0.82	0.42
z_4	0.86	0.57	0.82	0.41	0.68	0.53	0.49	0.25	0.56	0.66	0.90	0.34
T_5	0.33	0.84	0.87	0.90	0.10	0.12	0.89	0.51	0.79	0.38	0.23	0.45
z_5	0.41	0.78	0.82	0.80	0.05	0.16	0.79	0.60	0.73	0.34	0.27	0.53
T_6	0.11	0.15	0.10	0.19	0.43	0.46	0.90	0.78	0.12	0.17	0.28	0.13
z_6	0.18	0.07	0.12	0.23	0.51	0.55	0.83	0.72	0.20	0.26	0.35	0.09
α	0.89	0.76	0.41	0.52	0.80	0.68	0.87	0.63	0.79	0.85	0.96	0.46

5.3. Optimum Design Results of PSO Algorithms

Let inertia weight $\omega = 0.8$, acceleration factor $c_1 = 0.5$ and $rand_1 = 1$, then the velocity and position update formula of particles is as follows:

$$v_i(t+1) = 0.8v_i(t) + 0.5(x_{ib} - x_i)$$

$$x_i(t+1) = x_i(t) + v_i(t+1)$$

Gradually update the speed and position of individuals and populations. If the limit speed $|V_{max}| = 10$ is reached, the operation is terminated and the final modeling scheme of each pattern sample is output. Because of the limited space, the final modeling scheme is shown in Figure 12, taking Sample 1 as an example.

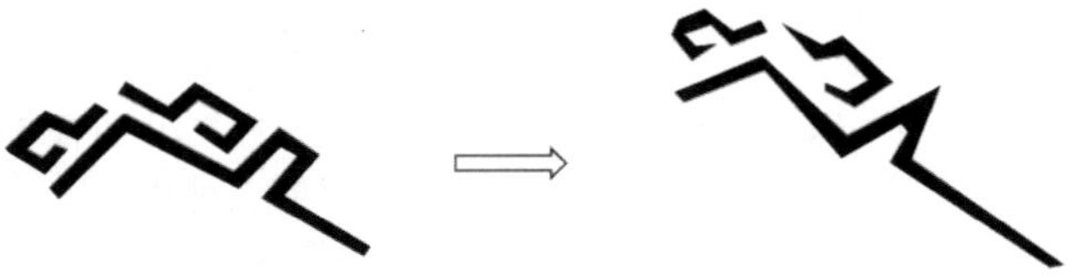

Figure 12. Sample 1 Final Modeling Scheme

5.4. Verification of optimization design results

Using SD method, 23 Miaoxue experts and 56 design teachers were assessed with questionnaires. The results were normalized and compared with the original samples to verify whether the image of the optimized scheme retained the original cultural image. Take Sample 1 as an example, It's evaluation value $T_1 = 0.89$, $T_2 = 0.12$, $T_3 = 0.46$, $T_4 = 0.82$, $T_5 = 0.25$, $T_6 = 0.07$, Compared with the evaluation value of the original sample (see Table 3), the new pattern "Image Semantic 1" and "Image Semantic 4" have high evaluation value, which shows that the new pattern has the same cultural image as the original pattern.

6. Conclusion

(1) In the process of identifying and evaluating cultural images, many concepts that are difficult to explain Daoming are transformed into numerical values for quantitative analysis by using the powerful reasoning ability of fuzzy logic of fuzzy neural network and the unique parallel and distributed information processing ability of artificial neural network[17], so as to determine the attributes of cultural images and establish a cultural image evaluation system with good accuracy.

(2) The particle swarm optimization is used to search for the best particle in the solution space[18], which can make the initial sample achieve the local optimum without deviating from the specific cultural image, and turn the disadvantage of PSO into advantage, so as to simulate the divergent thinking for the innovation of cultural image modeling design.

Acknowledgements

Foundation Project: National Natural Science Foundation Project(51865003，51465007); Science and Technology Project of Guizhou Province(Qian Kehe Platform Talents[2018]5781).

About the author

1.Li Lin(1973-), female, born in Nanchong, Sichuan Province, professor, doctor of Southeast University, supervisor of master's degree. Research directions: product design, perceptual engineering, traditional cultural creative design. E-mail: linlisongbai@163.com;
2.Yanzhe Zhao(1994-), male, from Jinan, Shandong Province, graduate student. Research direction: product design, computer aided design. E-mail: zyz2414508137@163.com.

References

[1] Liu J. (2007) Type：Behavior，Image and Cultural Connotation [J]. Central China Architecture，25(1)：64-65.
[2] Xing H.T，Qiao L. (2017) Application of Chinese Traditional Architectural Elements in

IOP Conf. Series: Materials Science and Engineering **573** (2019) 012025 doi:10.1088/1757-899X/573/1/012025

Furniture Design [J]. Art Evaluation, 13: 145-147.
[3] Su D. (2014) Research on design strategy oriented by cultural image recognition--Taking pot design as an example [D]. Zhejiang University of Technology.
[4] Jiang Y.Z. (2014) Product design research reflecting regional cultural image [D]. Nanjing University of Aeronautics and Astronautics.
[5] Xu J, Sun S.Q, Zhang K.J. (2007) Optimal Design of Product Image Modeling Based on Genetic Algorithms [J]. Journal of Mechanical Engineering, 43(4): 53-58.
[6] Pu J.Z, Li Y.L, Liu Z.X. (2017) Image Modeling Design of High-speed Train Based on Text Mining and Neural Network [J]. Mechanical Design, 09: 105-109.
[7] Ni M.N, Sun Z.H, Wang Z, et al. (2016) Application of BP neural network model for perceptual image evaluation of product modeling [J]. Journal of Donghua University (Natural Science Edition), 42 (4).
[8] Su J.N, Zhao H.J, Wang R.H, et al. (2015) Product image modeling optimization design based on support vector machine and particle swarm optimization [J]. Mechanical design, 1.
[9] Li M.S, Xu Q.Y, Gao D.M, et al. (2017) Colour Decision-Making System Integrating Mixed Intelligence Method and Multi-user Images [J]. Journal of Computer Aided Design and Graphics, 29 (11): 2091-2099.
[10] Luo S.J, Zhu S.S, Sun S.Q, et al. (2004) Research on product conceptual design technology based on integrated knowledge [J]. Journal of Computer Aided Design and Graphics, 16(3): 261-266.
[11] Zhu S.S. (2009) Research on Regional Cultural Images in Modern Product Design [J]. Packaging Engineering, 5: 209-210.
[12] OSGOOD, GEORGE J. (1957) The measurement of meaning[M]. Urbana: University of Illinois Press.
[13] Xin H.Q. (1999) Morphological Composition [M]. China Academy of Fine Arts Press. Hangzhou.
[14] Su J.N, Fan Y.F, Zhang S.T, et al. (2011) Product modeling design based on Kansei Engineering and Neural Network [J]. Journal of Lanzhou University of Technology, 37 (4): 47-50.
[15] Shen H.Y, Wang Z.X, Gao C.Y, et al. (2008) Determination of the number of hidden layer units in BP neural network [J]. Journal of Tianjin University of Technology, 24 (5): 13-15.
[16] Xu L.P, Shang D, Chen X.Y. (2014) Application of Fuzzy Neural Network in CT Diagnosis of Lung Cancer [J]. Journal of Zhengzhou University (Medical Edition), 2: 191-194.
[17] Wu J.D. (2012) A New Evaluation Approach to Mechanical Kinematic Concept Using Fuzzy Neural Network[J]. Advanced Materials Research, 421:666-669.
[18] Lee K.Y, Park J.B. (2006) Application of Particle Swarm Optimization to Economic Dispatch Problem: Advantages and Disadvantages[C]//IEEE Pes Power Systems Conference and Exposition. IEEE, 188-192.

IOP Conf. Series: Materials Science and Engineering **573** (2019) 012026 doi:10.1088/1757-899X/573/1/012026

Research on Ideological and Political Teaching of Product Design Course for Industrial Design Major

Shuying Guo

College of Mechanical Engineering, Jilin Normal University of Engineering Technology, Changchun City, Jilin Province, 130052, China

Corresponding author's e-mail: 29012610a@qq.com

Abstract. The main purpose of curriculum ideological and political teaching is to form synergy effect of curriculum system. With the main channel of classroom teaching, ideological and political theory courses and other courses are taught together. Based on the related literature research and research, this paper analyzes the nature and characteristics of industrial design. The course combines the relevant contents of the ideological and political theory course and product design research, so as to strengthen the value orientation and correct guidance of students in the teaching of professional courses.

1. Research overview

General Secretary Xi Jinping emphasized at the National Conference on Ideological and Political Work in Colleges and Universities that we should make good use of the main channel of classroom teaching and all kinds of courses should go with the ideological and political theory courses in the same direction to form a synergistic effect. College education is an important way of Ideological and political education for contemporary young people. It shoulders the important task of training excellent builders of socialist cause. Therefore, we must adhere to the correct political direction. The main purpose of curriculum ideological and political education is to promote the transformation of University Ideological and political education curriculum to curriculum ideological and political education, emphasizing that the important professional courses of various majors in universities should be included in the curriculum content that can guide students to establish correct values and world outlook.

Although great achievements have been made in the research of curriculum ideological and political affairs in major domestic universities, there are still some problems. The erroneous tendencies are manifested in two aspects: first, Intentionally add ideological and political elements to the curriculum. The combination of curriculum content is too rigid and lacks rational cohesion. The second, Some people set up projects in the name of Ideological and political affairs to carry out unreasonable curriculum reform. In fact, not all courses are suitable for embedding ideological and political content and carrying out curriculum reform. Through the analysis of production background, cultural background and professional spirit of industrial design specialty, this paper studies the combination of professional course content and ideological and political course content. Finally, the ideological and political system of product design course is formed. Drawing lessons from the concepts and principles of recessive education, we inadvertently achieved the function of education.

IOP Conf. Series: Materials Science and Engineering **573** (2019) 012026 doi:10.1088/1757-899X/573/1/012026

2. Research on Ideological and Political System of Product Design Course

The core program of Ideological and political education in product design course is to connect with the cultural background of teaching in the process of imparting professional knowledge. This background can be either an excellent traditional culture or a history and current achievement of industry and professional development. The content of Ideological and political education is mainly embodied in the background of construction, of course, also related to the content of the course itself.

2.1. Ideological and Political Research of Socialism with Chinese Characteristics in Product Design Course under the Background of New Era

In the report of the Nineteenth National Congress of the CPC, Xi Jinping put forward a scientific judgment on the new historical orientation of China's development. As socialism with Chinese characteristics enters a new era, the main contradiction of our society has been transformed from the contradiction between people's growing material and cultural needs and backward social production to the contradiction between people's growing needs for a better life and unbalanced and inadequate development. As for the people's growing need for a better life, there are two aspects: one is to put forward higher requirements for the quality of material and cultural life, and the other is to put forward higher requirements from the perspective of the overall development of human beings and the overall progress of society.

The people's need for a better life includes the yearning for good products. A good product design can improve the quality of life of users and meet their needs for certain functions of the product from a micro point of view. From a macro perspective, it is conducive to the development of the product industry and contributes to the progress of society. From the scientific judgment of the transformation of contradictions, we can see that the main obstacle hindering social progress and development is not backward social production, but unbalanced and insufficient development. The productivity level of our country has improved significantly on the whole, which is no longer the situation of backward social production. Faced with the new social needs, we still have many problems in product design, development and product innovation. The problem of unbalanced and inadequate supply of products is evident in the context of the new era, which has become the main factor restricting the satisfaction of people's growing needs for a better life.

In product design, we must pay attention to the insufficient imbalance and solve it. The imbalance is mainly manifested in the imbalance of regional development, the imbalance of supply structure and the imbalance of group development ability. The insufficient development mainly refers to the insufficient innovation ability, the low development ability and level, and the quality and efficiency need to be improved. Therefore, in the stage of market research and conceptual design of new product development, it is necessary to clarify the technical means necessary for the realization of new product, whether it meets the basic needs of consumer groups, and whether it solves some social contradictions. The goal of this stage is to strive for better quality, higher efficiency and more sustainable development.

2.2. Analysis of Ideological and Political Modules in Product Design Course

Product design course is an important professional core course for industrial design specialty. The course content can be divided into three modules: product design process, product system design and product design case analysis. Through the previous research and analysis, the first step in product design process, that is, Market Research and defining the design concept stage, can be combined with the theory of socialism with Chinese characteristics and the theory of the initial stage of socialism to analyze how to solve people's growing demand for a better life and imbalance in the field of material life by means of product design. There are three aspects of value orientation of Ideological and political module in this part of curriculum: demand orientation, technology orientation and innovation orientation.

2.3. Application of Marxist Theory in Ideological and Political Education of Product Design Course

For the study of Marxism, we should first establish the philosophical belief that Marxist philosophy is a scientific world outlook and methodology. The philosophical thoughts of Chinese leaders of several generations on state leadership and construction are the theoretical basis of the new development of Marxist philosophy in China and socialism with Chinese characteristics. Product design course is an important professional core course for industrial design specialty. In the course of course teaching, it is the first task to inculcate students' correct concepts and methods.

In the process of teaching professional courses, we should educate students on the basic principles of Marxism. For example, in the process of explaining the systematic view of products, students can grasp the basic viewpoints, standpoints and methods of Marxist philosophy by combining the material theory of the world. Combining the knowledge content of Marxist world outlook and methodology in practical teaching, we can help students set up correct outlook on life and values. In the teaching of product system composition, students are trained to analyze and solve problems with the viewpoints and methods of Marxist philosophy. In the teaching process of product function, structure, form, color, human factor, material and environment, students are guided to analyze and solve problems with the theoretical knowledge of dialectical materialism, substance, practice, connection and development, and knowledge in Marxist philosophy, so as to achieve a certain value orientation.

3. Research on Product Design Talents Training Based on Chinese Traditional Culture

3.1. A Study on the Course Ideology and Politics of Chinese Traditional Culture

Chinese traditional culture is the fundamental creativity of the achievements of Chinese civilization. It is also the overall inheritance of morality, various cultural ideas and spiritual concepts in the national history. The cornerstone of Chinese traditional culture is the basic thought of Taoism and Confucianism. Taoism and Confucianism, such as "Taoism follows nature", "Benevolence, righteousness, courtesy, wisdom and faith", play a guiding role in product design. Chinese traditional culture includes Chinese traditional festivals, as well as the characteristic cultures of different regions and nationalities. The ideological and political content system of product design course can not be established without the support of Chinese traditional culture. In our efforts to move from "Made in China" to "Design in China", we will undoubtedly experience difficulties and obstacles. Therefore, in the course of curriculum teaching, it is particularly important to cultivate students'love for specialty and their determination to contribute to the development and revitalization of the industry, which involves the cultivation of family and country feelings.

The feeling of family and country is one of the excellent traditional cultures of the Chinese nation. Its basic connotation is the main body's recognition of the community and the ideas and concepts that promote its development. The external manifestation of the feeling of family and country is to attach importance to family, mind the world, national spirit and patriotism. For the personnel training of product design, it is important to build professional pride, professional loyalty, and realize the leap from loving profession, loving profession to loving country. Therefore, in the process of education, we should not only connect closely with traditional culture, but also innovate and surpass traditional culture. The feeling of the country is epochal. In contemporary society, this excellent cultural tradition transcending nationality and ideology plays a great role in social construction, national development and national unity.

3.2. Ideological and Political Talents Training Objectives of Product Design Course

The feeling of family and country studied in this paper refers to a kind of thought and idea that individuals, under the influence of Chinese traditional culture, have a high degree of recognition of the value community and promote it to develop in a positive, positive and benign direction. The individual refers to the students majoring in industrial design in the course of teaching, and the value community refers to the whole industry of industrial design in China. The fundamental purpose of Ideological and political education in the course is to promote students' understanding and love of industrial design, and then to form a sense of professional pride. Only by recognizing and being loyal to a major, can they participate in the work to the maximum extent possible. The industry contributes and realizes its own value. In order to stimulate students to study hard, study hard and innovate constantly, we should sort out the well-known persons and events at home and abroad in the work positions related to product system design in the course of development of this major, study their essence, nature and contribution to the country and nation, and implant them into the content system of Ideological and political course. In addition, we choose well-known alumni to start businesses and serve the country, so as to educate the students of this major in the value orientation of their national feelings.

4. Research conclusion

According to the analysis of this paper, the task of product design course for industrial design specialty is refined, and the task analysis is carried out with the corresponding ideological and political theory. Then the fundamental purpose of solving the task is achieved through the realization of the goal and the guidance of the value. The specific research results are shown in the table1.

Table 1 Ideological and Political System of Product System Design Course

Ideological and Political System of Product System Design Course				
Task Proposal: Content Basis of Product Design Course	Task Analysis: Application of Philosophical Method and Implantation of Ideological and Political Contents		Task Solution: Teaching Objectives of Course Ideological and Political Education	
			Curriculum objectives	Value orientation
Market Research and Analysis\ Definition of Design Concept	Socialism with Chinese Characteristics has entered a new era	People's growing need for a better life	Product direction	Demand orientation
		Unbalanced development	product positioning	Technology oriented
		Inadequate development	Product concept	Innovation oriented
Product System Design	Marxist Philosophy	Universal Connection and Eternal Development of the Material World	Function and structure	Material orientation
		The Dependence of the Spiritual World on the Material World and the Active Role of Consciousness	Form and color	Aesthetic orientation
		Examination and Evaluation of Practical Result	Human factors and dimensions	Humanized Orientation
		Dialectical Materialism, Motion Properties of Material and Material	Material and Technology	
		Theory of Sustainable Development	environmental factor	
Case Analysis of Product Design	Chinese Traditional Culture	Taoism and Confucianism	Product form	Cultivation of cultural connotations
		National culture	Humanization design	Cultivation of Humanistic Care
	Family feelings	Famous Product Designer		Professional Identity
		Outstanding Alumni		Professional loyalty
		Industry Events		Industry pride

IOP Conf. Series: Materials Science and Engineering **573** (2019) 012026 doi:10.1088/1757-899X/573/1/012026

References

[1] Mei Weihui, Meng Ying. Innovation and entrepreneurship education in Chinese universities: role orientation and action strategies of government, universities and society [J].Higher education research. This reference has one entrie but it is not numbered (it uses the 'Reference (no number)' style.
[2] Cheng Xi, Zhang Pingping. Course Construction of Innovation and Entrepreneurship Education in Colleges and Universities Based on Core Quality Concept [J]. This reference has one entrie but it is not numbered (it uses the 'Reference (no number)' style.
[3] Sukhomlinski. On Moral Education and Comprehensive Development [J]. Foreign Educational Materials, 1980. This reference has one entrie but it is not numbered (it uses the 'Reference (no number)' style.
[4] Wu Ni. Constructing the Moral Education Mechanism of the Interaction of Ideological and Political Course and Philosophical and Social Science Course. This reference has one entrie but it is not numbered (it uses the 'Reference (no number)' style.

Research and practice of express transportation service based on ecological thinking

Di Liu[1*], Kaining Meng[2] and Fuxi Ouyang[3]

[1]School of Art and Design, Xihua University, Chengdu, Sichuan, 610039, China

[2]School of Art and Design, Xihua University, Chengdu, Sichuan, 610039, China

[3]School of Art and Design, Xihua University, Chengdu, Sichuan, 610039, China

[*]Corresponding author's e-mail: 541066550@qq.com

Abstract. This paper attempts to apply the ecological thinking mode to the service of express transportation industry, and puts forward the corresponding solutions to many pollution problems existing in the current express transportation industry.This paper analyzes and summarizes the existing problems of express transportation industry, through the methods of field investigation, data collection and data analysis, probes into the present situation and pollution problems of express transportation cost at present, and looks for the opportunities and related strategies in which ecological thinking mode can be involved. Improve the transportation quality, control the logistics cost, and use the ecological thinking mode to establish the ecological express transportation service process of saving resources, protecting the environment and promoting the sustainable development of economy and society.

1. Introduction

In recent years, air pollution has become more and more seriously. "Haze" weather has been playing out in many places across the country, and PM2.5 values in many cities have been continuously setting new records. People are particularly sensitive and worried about the environment, especially the air pollution. However, among many environmental pollution factors, logistics industry plays an important "contributor" role. Most logistics enterprises in China still follow the traditional operation and management mode, taking their own economic profit as the guidance, lacking a certain sense of environmental protection,existing some non-environmental protection phenomena in the operation process such as waste of resources,environmental pollution, which is particularly prominent in the transport link. In order to alleviate the environmental pollution pressure caused by express transportation, meet people's environmental protection needs, and save the occupation of environmental resources by express transportation, this paper uses the ecological thinking mode to control and optimize the express delivery service, improve the utilization efficiency of logistics energy, which can not only reduce the cost of express delivery, but also reduce the environmental pollution caused by express delivery activities, and promote the ecological and sustainable development of express delivery.

2. The present situation and problems of express transportation industry in China

Developed countries have implemented the sustainable development strategy earlier, with higher environmental standards and stronger environmental awareness of consumers. In the field of express delivery, they pay more attention to "ecological" and have been well implemented and applied. However, China started relatively late in this respect and lacks the inherent power to build ecological transport.

Transportation, as one of the most important logistics functional elements, is also the biggest environmental pollution source. The pollution caused by express delivery mainly exists in three aspects: the pollution caused by the running of cars, the pollution caused by the goods themselves in the process of freight transport, and the noise pollution. The existing express delivery system is not a healthy ecosystem, but more oriented to economic interests. According to the state post bureau (Figure 1), in 2018, the number of "double 11" parcels will exceed 1.87 billion, an increase of 25% over the same period last year. Behind these huge volume of express data, there is a large number of express delivery, the waste of resources and environmental pollution caused in the transport is immeasurable.

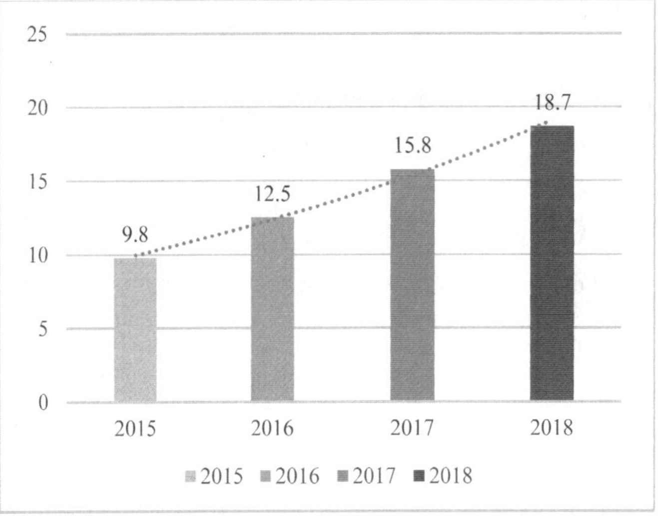

(Source: home post office website, census and statistics department)

Figure 1: Statistics of express delivery volume during the singles' day (11.11-11.16) (unit: 100 million pieces).

2.1. High cost of express delivery

The main reasons for the high cost of express transportation are improper cooperation of transportation mode, backward organization and means of transportation, and unreasonable distribution of transportation means. First of all, Chinese enterprises' logistics system function is not strong, their coordination between the transport and warehousing functions are lack of coordination, and various modes of transport do not cooperate properly. Moreover, China's phenomenon of express roundabout transport mode is serious, the phenomenon of unsmooth transport information flow caused by empty vehicles and vehicles overload transport is serious. Finally, the freight structure of China's logistics industry is unreasonable, and the goods that should be transported by rail for medium and long distances are borne by road transportation, thus driving up logistics costs.

2.2. Enterprise independent transport, lack of cooperation

China's express delivery enterprises are mostly independent operation, lacking of cooperative operation enterprises. Many express delivery enterprises have some scattered transportation business, which can't fit a whole car. They can't wait because of the time requirement of transportation, even if other enterprises have similar transportation tasks, they can only send vehicles without a full load due to lack of information or lack of cooperation, resulting in the waste of resources and the increase of transportation costs. At the same time, most urban distribution enterprises operate independently, which makes it difficult to form a win-win community.

2.3. Less sufficient attention to reverse logistics

Reverse logistics refers to the process of reverse flow from the downstream to the upstream along the supply chain starting from the end client in order to regain the use value of returned goods, recycled or abandoned products. Reverse logistics is a special link in the supply chain. It is more difficult for people to control and plan reverse logistics than forward logistics. Therefore, reverse logistics in China is limited to the recycling and utilization of waste materials, household garbage classification and other primary behavior, as the enterprise's own behavior of reverse logistics activities is not common.

3. Ecological thinking and express delivery service

In the face of many environmental pollution problems in the transportation of express delivery industry, a change in thinking mode must take place: that is, to jump out of the current prevailing economic thinking mode and transform into ecological thinking mode. Ecological thinking is to regard the ecological environment of human survival and development as a whole with orderly operation, and regard human beings as the constituent part of the organic connection of nature. Generally speaking, the mode of ecological thinking mainly lies in thinking about people's economic and social activities due to their ecological survival relationship with nature. However, this kind of thinking should be realized on the premise of thinking about the social survival relationship between people and solving the ecological problems between man and nature through the solution of social ecological problems. Therefore, the mode of ecological thinking is not only based on the relationship between human beings and nature, but also the social and ecological relationship between human beings.

Express transportation service is an industry that seriously pollutes the environment with waste gas and noise, and at the same time consumes a lot of fuel and energy (Table 1). China's modern logistics enterprises pay attention to the economic value they create, without considering the consequences of non-sustainable development on the environment and resources. Our daily living environment has been threatened by the express delivery activities, so we must use ecological thinking to improve the problems existing in the modern express delivery system, and build an ecological express delivery system from the relationship between human and nature and the ecological survival of people.

Table 1: Main pollution in express transportation.

	Direct consequences	Indirect consequences
Pollution from cars themselves	Automobile exhaust, solid particulate matter, inhalable dust, dust, etc	Carbon monoxide, Sulfur dioxide, photochemical smog and acid rain
Pollution in the transport of goods itself	Chemicals, harmful substances, cement, sand, etc	

4. Ecological approach of express transportation service in China based on ecological thinking

The main reason for the above environmental pollution problems lies in a series of unreasonable problems in the express delivery service process. Based on the mode of ecological thinking, starting from the social survival relationship between people and the ecological survival relationship between people and nature, we should think about people's economic and social activities, optimize the express delivery process, and make it more ecological in the transport service process (Figure 2).

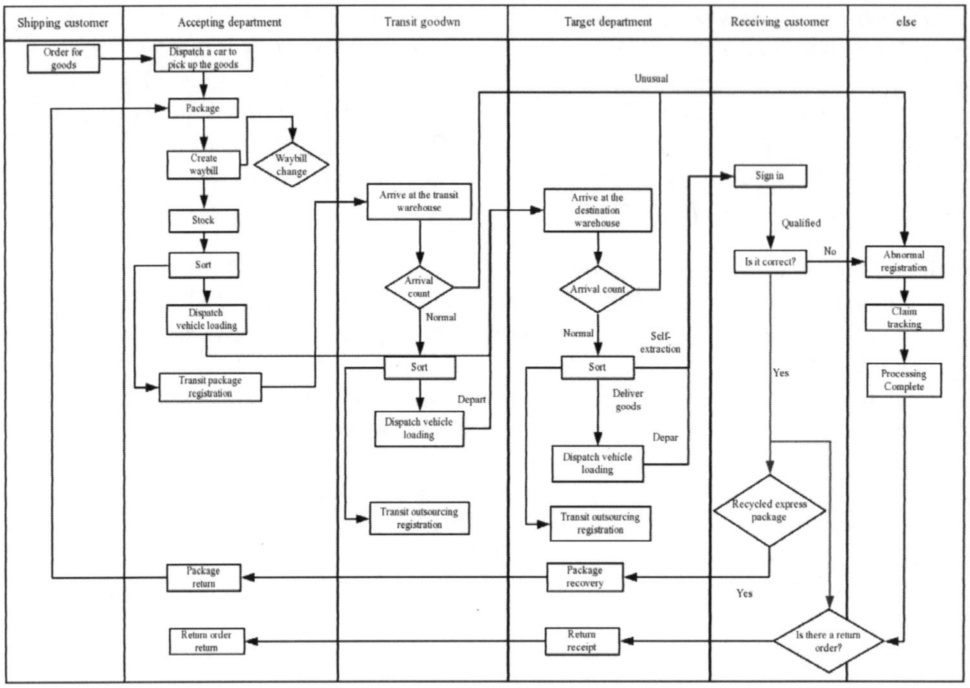

Figure 2: Optimization of express delivery service flow based on ecological thinking.

4.1. Development of new energy and new energy transportation

In a comprehensive comparison of the advantages and disadvantages of railway, highway and air transportation (Table 2), highway transportation is the first choice for many enterprises due to its advantages of punctuality, low cost, safety and flexibility. Therefore, in the express logistics system, multi-modal transportation is adopted to realize the cooperative transportation of multiple means of transportation. Most of the energy used for transportation in China is coal and oil whose exhausting gas also does a lot of damage to the environment. The existing new energy has solar energy, wind energy, and so on. The existing new energy has solar energy, wind energy, and so on, now the country is also fully research and development of new energy vehicles, for this reason, if the express transport industry is to achieve the goal of energy saving and emission reduction, they needs to introduce a large number of environmentally friendly new energy vehicles, using new energy vehicles in the short-distance transport .

Table 2: comparison of advantages and disadvantages of express delivery mode.

	Advantages	Disadvantages
Air freight	Speed is fast, Lower breakage, Good safety	Small capacity, High cost, Unable to transport heavy and explosive items
The railway	Large cargo capacity, Low energy consumption, The cost is low	Large construction investment, Time consuming, Not good for short haul, Flexibility is poor
Highway	Suitable for short distance transportation, flexible and easy loading	Long distance transportation is expensive, Easy to pollute the environment and consume much energy

4.2. Enterprises reasonable layout of the transport network, and active joint distribution

Enterprises should formulate reasonable transportation routes, that is, through the optimization of transportation routes and networks, reducing roundabout transportation and repeated transportation, so as to reduce transportation costs and save resources. For many distribution tasks is for customers less demand, less than the vehicle transport, usually in the case of not full load of transport, so that the vehicle utilization rate cannot be fully used. In order to avoid the occurrence of this kind of situation, the information of multiple express enterprises should be shared to implement joint scheduling. In this scheduling mode, the scattered transport tasks of different enterprises are "carpool", and then the appropriate driving route is selected according to the location of each demand point, and the goods are sent to each demand point in turn to realize joint distribution, which can maximize the efficiency of resource utilization and thus maximize the economic benefits.

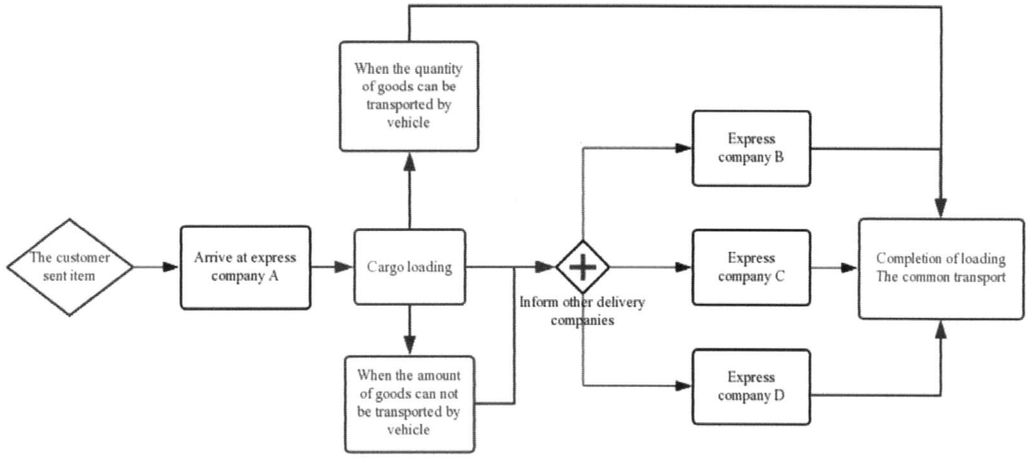

Figure 3: Flow chart of cooperative carpooling transportation of express enterprises based on ecological thinking.

4.3. Express package design improvement

In countries with sound environmental protection laws and regulations, the transportation and recycling of packaging materials have become an important responsibility and obligation of enterprises. Redesigning packaging for transport-friendly purposes has multiple benefits for virtually all businesses in the supply chain. Innovative packaging design can help reduce the number of vehicle use, so as to save physical transport costs, to achieve environmental protection, to achieve the ecological logistics transport.

4.4. Large-scale implementation of reverse transport

Reverse logistics transportation plays an important role in ecological transportation, so enterprises should pay more attention to it. Reverse logistics is divided into recycling and waste logistics. For the repaired defective products, returns and other renewable resources, express delivery enterprises can cooperate with other stakeholders in the supply chain, following the recycling and recycling principles of circular economy, and establishing the recycling and recycling system, so as to constitute an ecological recycling and reusable logistics system. For wastes, actual needs are selected for classification, processing, packaging and storage, and assigned to special sites for disposal. Through the recycling and reuse of wastes, material resources can be recycled, thus saving resources, greatly reducing the cost of enterprises, reducing environmental pollution, adhering to the concept of "ecological thinking".

5. Conclusion

The essence of express delivery is service, and the process of express delivery service needs to be paid attention to improve service quality. This paper has carried on the optimization design to the existing express delivery service flow, using the ecological thinking of unreasonable part in transportation service process improvement.Then the development of new energy transportation has been proposed , laying out of transportation network reasonably, carrying out a joint distribution, improving the express package design and implementing the retrograde transport and process optimization approaches, so as

IOP Conf. Series: Materials Science and Engineering **573** (2019) 012027 doi:10.1088/1757-899X/573/1/012027

to alleviate the express transportation pollution to the environment, to use of limited resources to maximize utility. The optimization of express transportation service process is to curb the harm to the environment in the process of transportation and form an ecological express transportation cycle system.

References

[1] Zhang,Y.X.,(2018) Logistics and transportation cost control under green economy. J. Logistics engineering and management.,40(07):52-54.
[2] Peng,M.Y.(2014)General situation and development of domestic reverse logistics. J. China logistics and procurement., 20:74-75.
[3] Miao,C.(2014) On ecological thinking mode and its construction. J. Academic Exploration., 06:11-15.
[4] Feng,Y.(2015) Research on ecological logistics system in China. J. The Chinese market., 37:49-50.
[5] Liu, J.F. (2017) Study on development countermeasures of e-commerce logistics distribution under the background of big data. J. Business economics., 02:98-99.
[6] Wong. X.G. (2017) Rethinking the characteristics and innovative development of China's logistics industry.J. China's current economy., 31(03):8-15+2.
[7] Ma, S.S. (2017) We will promote green transport. J. China logistics and procurement., 17:38-39.

Research on Eco-design of Express Packaging Based on System Thinking

Fuxi Ouyang[1*], Kaining Meng[2] and Di Liu[3]

[1] School of Art and Design, Xihua University, Chengdu, Sichuan, 610039, China

[2] School of Art and Design, Xihua University, Chengdu, Sichuan, 610039, China

[3] School of Art and Design, Xihua University, Chengdu, Sichuan, 610039, China

[*]Corresponding author's e-mail: 939079941@qq.com

Abstract. The development of online shopping has boosted express delivery market, and it has caused heavy burdens to environment. The most serious problem is the waste of resources and environmental pollution because of excessive packaging waste. This paper uses the system thinking and ecological design method to take the internal and external systems of the express packaging as a whole research object . Through dialectical analysis and high integration， the elements of the internal and external systems are coordinated with each other, and then the optimization of express packaging is realized. This paper achieves the real ecological express packaging design by studying the ecological express packaging system and considering various factors inside and outside the express packaging system.

1. Introduction

Nowadays, the development of online shopping has boosted the development of the express delivery market, but also caused heavy environmental burdens. The most serious problem is the waste of resources and environmental pollution because of excessive packaging waste. [1] Through the research of express packaging, this paper explores the problems of express packaging with systematic thinking and seeks solutions to realize the ecologicalization of express packaging, fully embodying the strategy of mutual coordination and sustainable development of environmental protection and economic development. Most of existing researches are focusing on individual fields of express logistics and transportation such as boxes packaging and transportation.Therefore,they are not multi-angled analyses. The trend of ecological packaging is irreversible,and complicated as the ecological packaging is, the use and recycling are essential. Only by considering both internal and external factors of the express packaging system can we achieve a true ecological cycle.

2. Research background

The development of online shopping has boosted the development of the express delivery market. According to the statistics of the State Post Bureau, the volume of China's express delivery business is increasing at an alarming rate from 2013 to 2017, from 9.19 billion to 40 billion. In just five years, this figure has actually quadrupled. The prosperity of the courier industry has also led to a heavy environmental burden while promoting economic development rapidly. One of the most critical problems is the waste of resources. As the consumption of products increases, the scale of logistics packaging for commodity packaging boxes is also growing, and due to a large number of disposable packages and excessive packaging, it will inevitably consume too much natural resources.

Table 1. Traditional design versus system design

Traditional product design	Eco-product design based on system thinking
Take the product as an individual part, considering aspects like profit, function, aesthetics,environmental conditions and efficiency, corporate image and overall quality separately.	Consider the relationship between products and the environment in the whole life cycle, and decompose and combine different factors of design to realize the real eco-express packaging design.

3. Internal and external factors of product design based on system thinking

In the product design based on system thinking, the designer is required to consider not only the requirements of the basic elements such as the shape and function of the single object, but also effects on environment based on system thinking. Product eco-design refers to the integration of environmental factors into product design, taking into account the environmental impact of the entire life cycle of the product during the design phase, thereby helping the decision-making direction of the design, and then minimizing the negative impact on environment by improving the design idea.

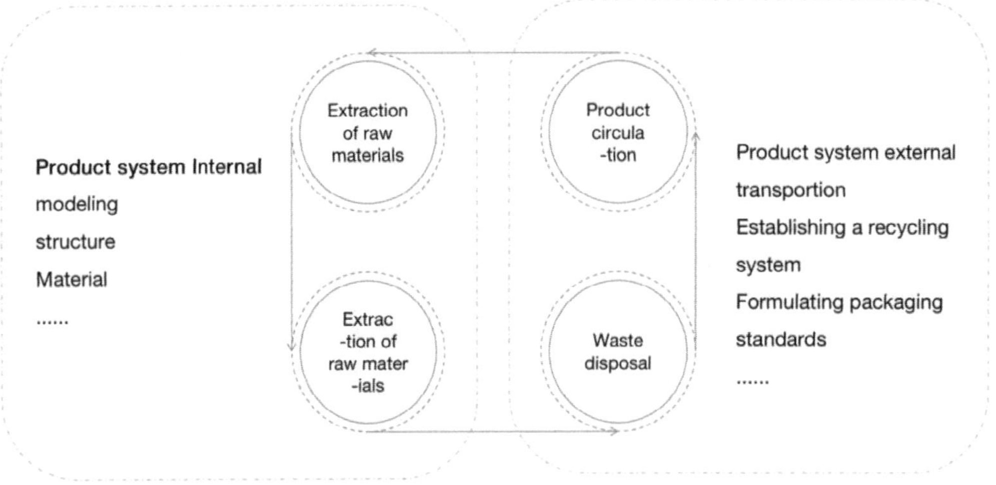

Figure 1. system flowchart.

3.1 Product internal system

In the product life cycle, the extraction of raw materials to the manufacture of products is the formation process of the product, thus forms the internal system of the product. The product internal system consists of relatively stable elements and structures and has relatively independent functions. Elements constitute the internal system of the product. The structure is the method and order in which several elements are related to each other and interact. The purpose of connecting the product elements through the organic structure is the product function. The realization of the product function is the combination of internal product system and the external environment. [5] By the process of combination, the element and structure of the product transforms into a deep structure to achieve product functions.

3.2 Product External System

In the product life cycle, from product circulation to waste disposal, energy regeneration and reuse are the realization of product function, forming an external system of products. There are various factors affecting the external system of the product, such as the market environment, the state of the consumer (including age, gender, consumption concept, cultural taste, custom, etc.) and national policies and

IOP Conf. Series: Materials Science and Engineering **573** (2019) 012028 doi:10.1088/1757-899X/573/1/012028

regulations, etc., which may functionally affect the product. Moreover, the realization of function usually means the interaction between products and people with different lifestyles. Consumers don`t understand and use the same product in the same way in different situations, which makes the function of the product complicated and diverse.

4. The ecological express package internal system based on system thinking

4.1 Ecological packaging

Ecological packaging should be a packaging that is harmless to the natural environment and human health, which can be reused and can promote the sustainable development of the national economy. [6] That is to say, the whole process of packaging products including raw material selection, product manufacturing, use, recycling and disposal should meet the requirements of ecological environmental protection. For instance, saving resources and energy, reducing volume, even avoiding waste generation, recycling, being able to incinerate or degrade.

4.1.1 Packaging modularization. The modularization of the package determines the basic dimensions of the package.The smallest packaging unit is a carton. A number of cartons are placed on the pallet, and then several pallets are finally placed in containers or vans for long distance transport. In order to reduce the logistics cost, it is a very simple and intuitive way to maximize the product quantity in the packaging. If the modular packaging design is adopted, the packaging efficiency will improve, thereby the logistics cost will be reduced .

4.1.2 Large-scale and containerized packaging. It is conducive to the logistics system in the process of loading, unloading, relocation, storage, transportation, etc., making the operation in these links more effective, and it is good for reducing unit packaging, saving packaging materials and packaging costs, and protecting the cargo. Those advantages can be achieved by various methods such as containers,container bags,trays and others. [8]

*4.1.3 Packaging reuse and waste disposal.*Using general-purpose packaging, and no special arrangement for returning to use; using reusable packaging, which can be used many times,such as beverage and beer bottles, etc.; realizing material-step-utilization,which means the material can be used for other purposes after packaging process; recycling the packaging material and then convert it to produce other objects.

4.1.4 Exploiting new packaging materials and packaging equipment. It is a new trend that making the packaging multi-function to realize various functions of packaging with less material. [10]

4.2 Ecological material

On the premise that the packaged product meets its design function, the less the material and energy consumption in the packaging manufacturing and using process, the better the impact on the environment. Environmental impacts include: effects on air, water consumption and pollution to water, and impacts on soil, heat and energy consumption in manufacturing and use, possibilities of waste reuse and the service life of packaging material, etc. [11]

Table 2. Ecological packaging and ecological materials

Ecological packaging	**Ecological material**
Packaging modularization.	On the premise that the packaged product meets its design function, the less the material and energy consumption in the process of packaging manufacturing and use, the better the impact on the environment.
Large-scale and containerized packaging	
Packaging reuse and waste disposal	
Exploiting new packaging materials and packaging equipment.	

5. Eco-Express Packaging External System Based on System Thinking

5.1 Ecological transport
Ecological logistics aims to reduce pollution and reduce resource consumption. It uses advanced logistics technology to plan and implement logistics activities such as transportation, storage, packaging, and loading unloading, circulation processing. [12] While mitigating the harm of logistics to the environment in the logistics process, the purification of the logistics environment is realized, and the logistics resources are fully utilized.

5.2 Establishing a recycling system
Reusing packaging material and establishing a recycling system. It is recommended that government takes effective measures to speed up the classification and recycling of express packaging. [13] Government can cooperate with large-scale e-commerce companies such as Suning Tesco, Jingdong, and some express delivery enterprises to implement the express package recycling program, and establish pilot packaging callback and recycling center at cargo picking up store or logistical distribution center for packaging material`s classified recycling. [13]

5.3 Formulating express packaging standards
The government and express industry associations should become the makers of express packaging standardization. In terms of packaging size, mainly corrugated boxes and packaging bags, the government and associations should formulate a set of rules that meet the needs of the express delivery industry based on the actual situation of express packaging.

6. Conclusions
The eco-design of courier packaging based on system thinking proposed in this paper no longer treats courier packaging parts as a separate link, but from a perspective of eco-friendly it takes product system as a goal to achieve an overall optimal system through scientific decomposition, modular design and combination optimization.

This paper discusses the implementation of ecological design based on system thinking through practical cases, and verifies the feasibility of the method. Through analysis, it is proved that the technology and art, function and form, economy and efficiency ,macro and micro all can be balanced , by applying the ecological and systematic design method and considering the relationship between product and environment in the whole life cycle and also by decomposing and combining different design factors. Therefore, the products involved answer the requirements of sustainable development better than common ones.

References
[1] Wang, X. (2017) Countermeasure Research on Recycling Utilization of Express Packaging Resources. J. Electronic Commerce., 01: 1-2+66.
[2] Yang, F. (2017) The Art of Writing a Scientific Article. J. Chinese Market., 23:36-38.
[3] Gao，S. (2017) Research Status and Progress of Green Packaging Materials in China. Inner Mongolia Science and Technology and Economy., 17:3+6.
[4] Li, X. (2018) Analysis on the Development and Countermeasures of Green Logistics in China. J. Market Research., 02:6-8.
[5] Zhu, S. (2018) Countermeasures and Trends of the Development of Green Logistics Packaging Industry in China. J. Logistics Technology and Application., 12:166-169.
[6] He, Y. (2018) Problems and Countermeasures of Express Package under Low Carbon Economy. J. Green Packaging., 03:84.

IOP Conf. Series: Materials Science and Engineering **573** (2019) 012029 doi:10.1088/1757-899X/573/1/012029

Study on the Ecological Education of Product Design

Kunpeng Li[1*], Kaining Meng[2], Yuli Zhong[3]

[1]School of Art and Design, Xihua University, Chengdu, Sichuan, 610039, China

[2]School of Art and Design, Xihua University, Chengdu, Sichuan, 610039, China

[3]School of Art and Design, Xihua University, Chengdu, Sichuan, 610039, China

[*]Corresponding author's e-mail: 524977881@qq.com

Abstract: At present, although the design of our country has made progress to a certain extent, in the process of its gradual development, there is still a waste of resources and the pseudo-ecological design emerges one after another. Therefore, this paper mainly analyses the problems existing in design education, and puts forward a new concept of design education based on the view of ecological education, and makes a useful supplement to deepen the concept of ecological design. Let students grasp the relationship between design and humanity, design and ecology, design and society, design and ethics comprehensively. On this basis, try hard to explore the teaching model which is beneficial to the development of product design education in our country, promote the development of product design education in our country to a greater extent, and cultivate innovative talents of ecological consciousness.

1.Introduction

The world uses nearly 260 million tons of plastics whole year, of which 170 million tons are disposable, and only about 14% of the plastic packages are recycled and most of the waste plastics are discarded or landfill. In a study published in the journal Royal Society Open Science, an open access journal founded by the Royal Society on Feb. 27, 2018, researchers found plastic fibers in six deep-sea trenches and nine sites along the Pacific coast. At least 72% of the digestive system contained plastic fibers and particles, and 84% of the lamellae found micro-plastic fibers. The startling data also suggest that there may be no more marine ecosystems in the world that are not affected by plastic pollution. The ocean contains at least 5.25 trillion pieces of plastic, weighing 268940 tons, and these plastic debris returns to humans in various forms. Now, scientists have found plastic particles in people's feces. If we only concerned with the immediate interests, it will hurt human beings finally. Only when the concept of ecological is deep-rooted in people's mind, the ecological environment can be effectively improved.

Therefore, design education should not only pay attention to the combination of theory and practice, but also train responsible designers, carry out ecological education, put more emphasis on the relationship between design and humanities, design and ecology, design and ethics, design and society. Only in this way can we really promote the progress of design in our country.

2.Current situation of domestic product design education

Theoretical knowledge education is the place where design education needs to be perfected. Design practice should be based on abundant theories. Theory plays a guiding role in practice. Only by mastering the theoretical knowledge firmly can student carry on the design practice better. Let students carry on the design practice because of the curriculum project blindly, will only let they carry on the imitation continuously, but cannot really grasp the design innovation consciousness. Students can only design the product, and do not take the social value and economic value of the product into account. Such a design is irresponsible. It is the lack of theoretical guidance of ecological education that leads to the emergence of more and more pseudo-ecological products in the market.

Responsible design was first proposed by the American design theorist *Victor Papanek*. One of his most important books "*Design for the real world*" offers a new perspective on the purposiveness of design, namely, design should seriously consider the use of the earth's limited resources, and design should serve to protect the limited resources of the earth we live on.

3.About ecological education and ecological design

Ecological education is mainly to realize the need of sustainable development and to create ecological civilization society, and to integrate the relevant ideas, principles and methods of ecology into the process of modern education. Carrying out ecological education is the fundamental way to improve the public's ecological consciousness and shape ecological civilization. It can provide spiritual resources for solving the contemporary ecological crisis and realizing sustainable development. As shown in table 1, the performer, educational content and expected results of ecological education are shown.

Table.1 The framework of ecological education

	Performer	Educational Contents	Expected Results
Ecological Education	Government Enterprise School Home Publicity Department Mass Organizations	Ecological Theory Ecological Knowledge Ecological Technology Ecological Culture Ecological Health Ecological Security Ecological Value Ecological Philosophy Ecological Ethics Ecological Identification Ecological Aesthetics Ecological Civilization	Ecological view of nature Ecological world view Ecological ethics Ecological values Sustainable development Ecological civilization view

In the field of product design, the most intuitive embodiment of ecological education is ecological design. Ecological design requires priority to be given to ecosystem attributes throughout the product life cycle. In other words, it is a harmonious and unified whole system of human-environment-product, rather than a narrow focus on people's own interests and needs. It makes design follow nature and serve ecology, and emphasizes the harmony and unity among the whole system composed of people, nature, environment and products.

The new industrial revolution has pushed the world into a new and complex design era, and the world today will undoubtedly make greater progress than in the past. However, we cannot fail to see that the world today is also full of more contradictions and crises than before, such as the increase in population, the energy crisis, the serious destruction of the natural environment and ecological balance and so on, all of which threaten human beings all the time.

It is imperative to promote ecological design in China. Ecological design is not only a technical consideration, but also a conceptual change. And the application of ecological design in product design requires designers to give up the excessive emphasis on the appearance of products and focus on the innovation in the real sense.

4.Discussion on the New idea of the Development of Ecological Education in Product Design

This paper uses the concept of ecological education to discuss the educational development of product design, as shown in figure 1, mainly studies and analyzes the relationship between design and humanity, design and ecology, design and society, society and ethics, and on this basis, constructs the framework of product design ecological education.

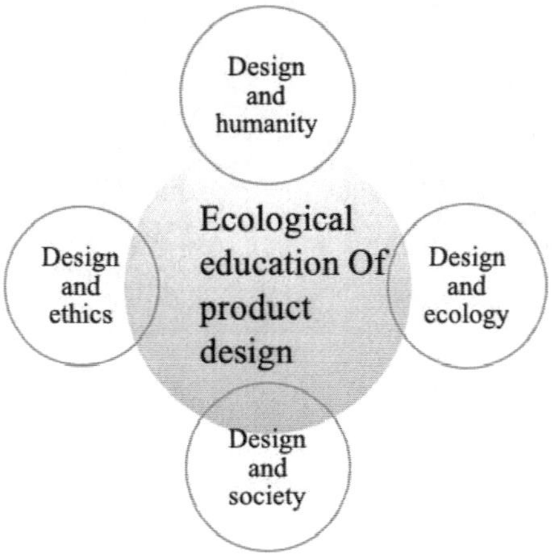

Figure.1 The framework of product design ecological education

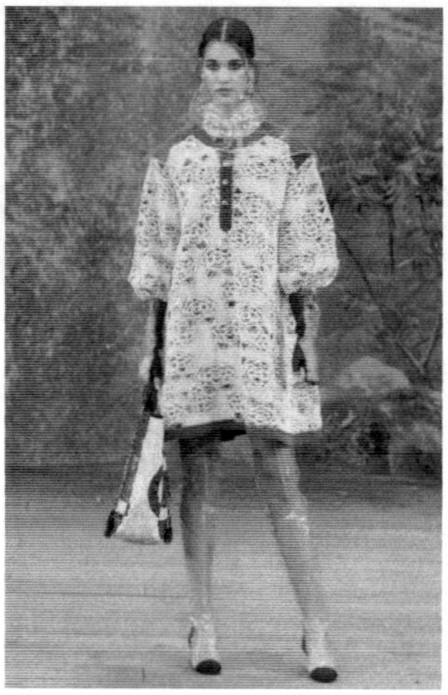

Figure.2 Chanel womenswear

4.1. Design and humanity

As far as product design is concerned, the humanistic view of design is carried out around each link of people, product, environment and society. The cultural background, environment and concept of people and society constitute the humanistic view of product design. And in terms of design education, designers trained by universities and colleges should pay more attention to social hot issues and take safeguarding the long-term interests of the society as the fundamental starting point of design activities, instead of only competing for immediate interests.

Nowadays, the society tends to be pluralistic, and the main contradiction of our country has been transformed into the contradiction between the increasing needs of the people for a better life and the unbalanced and inadequate development. And diversified design is also needed to meet the increasing spiritual needs of the people. In the field of clothing, as shown in figure 2, Chanel initiated the revolution of women's clothing liberation and prevailed in the world, liberating the clothing styles that bound women's bodies and enabling women to obtain psychological and physiological freedom. Good design not only makes consumers satisfied from the function, but also makes consumers enjoy the meaning of design and maximize humanistic care.

4.2. Design and ecology

The progress of science and technology and the development of human civilization are accompanied by a series of problems such as the exhaustion of resources and environmental pollution. And the harm brought by the destruction of the environment is gradually approaching the daily life of human beings. In the past, product design led to a great deal of waste of natural resources in the process of pursuing commercial interests. Therefore, design should be used as means to solve ecological problems and guide people to pay attention to ecological problems.

Ecological design is not only a new strategy to protect the environment, but also the need of developing circular economy. As shown in figure 3, the plastic can be used as road paving tiles by crushing it and then processing it into different shapes, reducing plastic pollution as well as recycling.

Figure.3 Floor tiles made of plastic

4.3. Design and society

Social design models take into account factors related to all human beings, including physiological, cognitive, psychological, social and cultural factors. It can be said that everyone is a designer, because design is the most basic activity of human beings. In order to survive, human beings will always make various instruments to solve problems in life. Reviewing the development history of design, the design forms in different stages are based on and serve the society.

Siemens hearing aid is a typical example, convenient and easy to carry, according to the physiological structure of the ear. Soft materials are selected and the color is close to the skin. Moreover, after receiving the external sound source, it can be transmitted to the ear more clearly.

4.4. Design and ethics

The ethics of design develops and progresses from the coordination between human and nature and the social environment. On the ethical level of design, there are three following aspects: first, it refers to the relationship between design and natural. For example, "green design", "ecological design" and "sustainable design" are advocated, with the emphasis on the status of natural environment. Second, it refers to the ethical boundary in the design. Third, the relationship between design ethics and the establishment of social order. Design ethics raises concerns about special groups of people who are physically disabled. Now design ethics are increasingly reflected in product design, such as the braille library. Good design takes the ethical value as the guidance, carries on the moral ethics norm to the design activity personnel all the time, creates good social atmosphere.

5.Conclusion

Design should be sustainable and design education should foster responsible designers. And attention should be paid to the cultivation of design theory: social view of design, social ecological view, humanistic view of design, ethical view of design. Designers need to study how to make the products more in line with the sustainable development of human beings and the needs of human survival. It is necessary not only to minimize the environmental load and resource consumption of products throughout their life cycle, but also to meet the requirements of improving the quality of life. Therefore, it is necessary to put the thought of ecological education into the product design education, and develop innovative design talents with ecological thinking.

References

[1] Papanek, V. (2013) Design for the Real Word. M. CITIC, China
[2] He, R.K. (2000) History of Industrial Design. M. Beijing Institute of Technology press, China
[3] Tao, H.Y. (2013) A study on the modernity of Chinese design from the perspective of sociology. D. Chinese National Academy of Arts, China
[4] Wang, Z., He, W. C., Wang, T. M.,(2006) Ecological design -- design for the sustainable development of society .J., Journal of Beijing University of Aeronautics and Astronautics, 2006(03):28-31+36.
[5] Tian, Q. H.,(2018) Industrial design innovation and industrial design education development .J.,

Estate and Science Tribune, 17(08):140-141.
[6] Tian, H.(2018) To interpret the existence and development of design art from the Angle of sociology .J., Art education, 2016(08):281-282
[7] Zhi, Y.(2018) The new value of design in the context of consumer society .J., Art and Design, 2(Z1):26-28.
[8] Xiong, X.F., Zhao, Y.Y.,(2018) Analysis of public service facilities based on design ethics .J., Packaging Engineering, 39(06):240-244.

IOP Conf. Series: Materials Science and Engineering **573** (2019) 012030 doi:10.1088/1757-899X/573/1/012030

Design of Persuasive APP Based on IOS System-- Taking Intelligent Mask as an Example

Li Wei[1], MaoWangCheng[1], Zhang Meng[1], Bo Meng Zhe[1]

[1] School of Mechanical and Transportation, Southwest Forestry University, Kunming, Yunnan, 650224, China

[*] Corresponding author's e-mail: 772074913@qq.com

Abstract. Based on the research of existing air monitoring and the research of APP products, the author is based on experience design and emotional design, from the perspective of human-computer interaction design, to explore better communication between users and intelligent products at emotional level. Through the analysis of the emotional design of products, we completed a set of persuasive APP design based on IOS flat system. Based on DOAS's working principle, we can get real-time data from environment and connect with mobile's APP, it is convenient for users to choose their own travel, to solve a series of health problems caused by environmental problems and people are concerned about their own health .

1.Introduction

With the rapid development of China's economy, it has greatly improved the living standards and family income of the general public, and it has also brought about the problem of environmental degradation in various places. According to the survey data of air quality problems in China last year, China's environmental problems are extremely serious. Although the government has taken various measures, it has no immediate effect on environmental control. At present, there are many healthy APP products and wearable devices that are becoming more and more perfect, but there are not many smart wearable devices that are specifically targeted at environmental issues. For the continuity of user experience, masks are the most direct solution and equipment to solve the health problems caused by environmental pollution. Studying the availability of smart masks and designing relevant apps is to achieve good human-computer interaction. The important way to implement the design concept of "central" and "human-machine-system" is to pay attention to environmental issues as an important manifestation and concrete realization of concerns about their own health problems. Based on the experience design and emotional design theory, this paper creates environmental data monitoring and user data based on the analysis of product emotional design and the environmental monitoring algorithm based on DOAS working principle. The interface of the main functions such as dynamic sharing interaction, regional real-time data statistics, and user communication in the same area at the same time, feedback data to the user through the extraction of official data and GPS positioning, and can record environmental data when connected with the terminal. The author has completed the design of data visualization as a functional requirement, which provides reliable data protection for user travel.

2.Environmental monitoring data and design investigation based on IOS platform

In this study, wearable smart device was discussed that adjusts the electrostatic protection size through the data monitoring of GPS positioning and mobile phone applications. It is proposed that an environmental quality assessment of the external environment can be performed indoors by obtaining official data to determine whether the wearable device is required to be out. In an outdoor environment ,the data can be analysed and detected by the optical quality sensor, the environment can be evaluated, and the protection settings can be adjusted by the app to achieve self-protection.

2.1. Requirement analysis of environmental monitoring algorithm

As an environmental monitoring system, it is necessary to real-time monitor the environment changed, and the monitoring results of data and images directly fed back to users. Whether users are pollen-allergy or block pm 2.5, it could be able to measure the environmental data that they need to make corresponding adjustments of the electrostatic protection of the mask according to the environment. Therefore, there is a need for an accurate algorithm for real-time data collection and analysis monitoring that provides real-time feedback to the user on behalf of the user's air quality data in their environment.

The environmental monitoring system consists of three parts: data collection, background data processing, and mobile application. Through data acquisition of sensors, data simulation conversion is transmitted to the background for data analysis and processing. The system uses the database for screening analysis, when the harmful substances in the air and the concentration exceed the standard, the system sends a warning notice to the mobile phone, and the data is displayed through the mobile application interface. Display various pollutants and pm2.5 concentration curves in the APP, and show them to users, users can intuitively and quickly understand the changes in airborne pollutants and pm2.5.

2.2. Applied research on persuasive design

A good lifestyle requires the individual, the public, and the society to work together in many ways. The correct behavior requires a guided process and persistence. This APP product is a self-health persuasion platform based on environmental testing, providing users with real-time environmental information and reminder services. The platform allows users to have a precise understanding of the pollution situation in their environment and prompts users to make corresponding feedback interactions. In persuasive design, the app can persuade people to pay attention to their environment and their health, and the continued use of this app will protect the health of themselves and their families.

In the process of user use, the interaction between the application and the user is mainly reflected in: guide the user to promote environmental awareness and protect their own health through persuasion, and let users generate intent to use the product. Provide a social platform for users through the mobile application platform when the user's behavior is persuaded to change, collect data and share data feedback. The persistence of these user behaviors is achieved by different persuasion methods. In the process of user interaction with products and services, Persuasive design uses different motivations, trigger factor to attract users' abilities, motivations, and trigger points. At the stage where user behavior is persuaded to change, through the attention to the details of the user's behavior, use different methods of persuasion to achieve the goal. It will eventually become one of the long-term lifestyles that users must own.

2.3. Design survey and result analysis based on smart wearable devices

In the design process of smart wear matching APP, the number of applications and installation rates of smart wearable connection apps is increasing. Questionnaire survey found that people are highly accepting smart wearable devices, but did not pay attention to the user experience in the connection between smart hardware and APP, and insensitive to the guiding interactions in UI design. Therefore, in the APP design, the following aspects should be considered:

- At present, the smart mask on the market has a single function, cannot be connected with the APP, cannot carry out the user experience, and has no detection function. In the face of no reference data, only official data can be used, and users cannot take corresponding measures;
- Data visualization should ensure that data analysis is clearly visible and the data displayed in the app is intuitive and easy to read;
- Enhance communication and interaction between users and improve user viscosity;
- Set automatic defense on function in APP to reduce manual operation by users.

3.Analysis of APP development process

Mask is software based on real-time air quality detection, it is combined with a hardware smart mask to protect users from air pollution sources and PM2.5. When the Mask is linked to the mobile app, it can send the complete access data of the environment the user is in. The app provides a real-time air quality report and corresponding regional air quality map. Get data in real time from the user's environment and turn it into actionable advice, such as: when the PM2.5 concentration exceeds the normal value; the application will give operational advice and defense by adjusting static electricity protection. From the concept to the product entity, after the hardware preparation is completed, before the development of APP program, it is necessary to carry out a well-organized product analysis. During the development process, the rationality of the functions and interactions is judged by the product requirements Sections, subsections and subsubsections.

The use of sections to divide the text of the paper is optional and left as a decision for the author. Where the author wishes to divide the paper into sections the formatting shown in table 2 should be used.

3.1. Demand analysis

Mask is an environmental monitoring and protection software, it has a clear guidance from the user's needs. Through the functional analysis of product requirements, divide the application purpose of Mask software into the following four functions: equipment, record, share and map. It also has access to the following classification information:

- Functional requirements of the record: through the acquisition of intelligent hardware data, send to the mobile phone using Bluetooth, visual display the mobile phone interface;
- View recent data by swiping and it can be used for data analysis and comparison;
- The shared content can be shared by screenshots to third-party software such as WeChat, QQ, Weibo;
- Dynamically share and chat with other users, friends can comment and like in real time;
- Uniform and rich visual style.

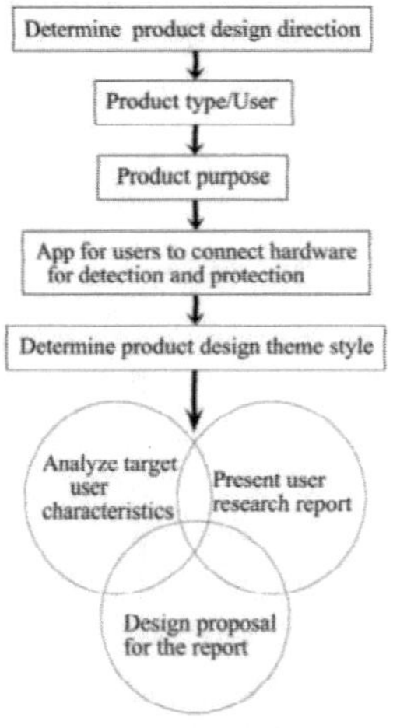

Fig.1. UI vision Design Process

Through the above analysis, four user demand dimensions can be set, including data analysis, my equipment, map, and personal center. The functional structure of the APP is divided into the following four aspects:

- The design of the data: guided by visualization and legibility, and provides users with statistical display of the data in a beautiful and concise view;
- Design connectivity: the display of the matching connection between the intelligent hardware and the APP removes the complicated function interface and forms a simple function display;
- The design of the map: after positioning, the icon shows the surrounding environment. Through the use of the official data from the Environmental Protection Agency, the user have a preliminary judgment;
- The design of the Personal center: Based on the function of displaying push friends and analyse personal activities. Increase the user's stickiness to the app through interactivity.

3.2. Identify target users

Mask users are not specifically defined by age or gender. Therefore, this product is mainly for users who are concerned about their own health. The target user's positioning can quickly convert the product concepts, ideas and functions required by the target user into a well-organized product. As shown in Figure 1, the UI visual design process for the product Foot notes

Foot notes should be avoided whenever possible. If required they should be used only for brief notes that do not fit conveniently into the text.

3.3. Mask information architecture and process design

In the early stage of design, the main functions of the product are roughly divided by means of information architecture diagrams. Use the text framework to list the features of the product for further analysis and supplementation. The program mainly focuses on the visual display and data sharing of data. According to the demand analysis, the information architecture diagram divides the products into four categories: equipment, data statistics, map, personal center. The information architecture diagram is composed as shown in Figure 2, from this functional information structure,further interactive flow chart analysis, the product function jump and switch to a detailed decomposition description, as shown in Figure 3. According to functional requirements, Divide product information into:

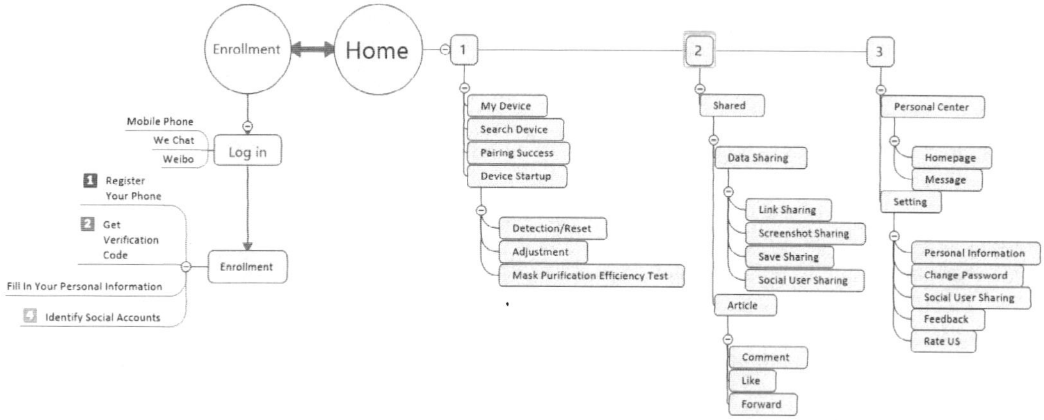

Figure2. APP information architecture diagram

- Login to register homepage: including mobile phone, email, third party login;
- Main interface section: including my equipment, shared, personal center, set;
- In addition to connecting the search function block, my equipment information framework also includes boot hardware for detection;

- The shared framework uses search neighbors and user repositories to increase user friends who use the same equipment. After joining the group, data sharing and information exchange can be performed;
- Personal Center: including homepage and messages, as well as comment forwarding for published articles;
- Setting up the framework: including data modification and viewing of other product information. According to the APP information architecture diagram, the detailed interaction flowchart shown in Figure 3 is completed, and the product function flow is completed through the graphic and text mode, and a clear logic idea is determined for the next APP design.

4.Design and display of UI interface design effects set by data visualization

The trend of the APP design interface is no longer the pursuit of beautiful, dazzling dynamic effects, but more attention to the user experience, visual comfort. The design further visualizes the visual read ability of the detected data as a design requirement in a data visualization manner. Eliminate complex functional interfaces, form simple data to check, hardware can be adjusted to the main function of the design. Mask in gray #4d5359 as the main background of the overall color with a

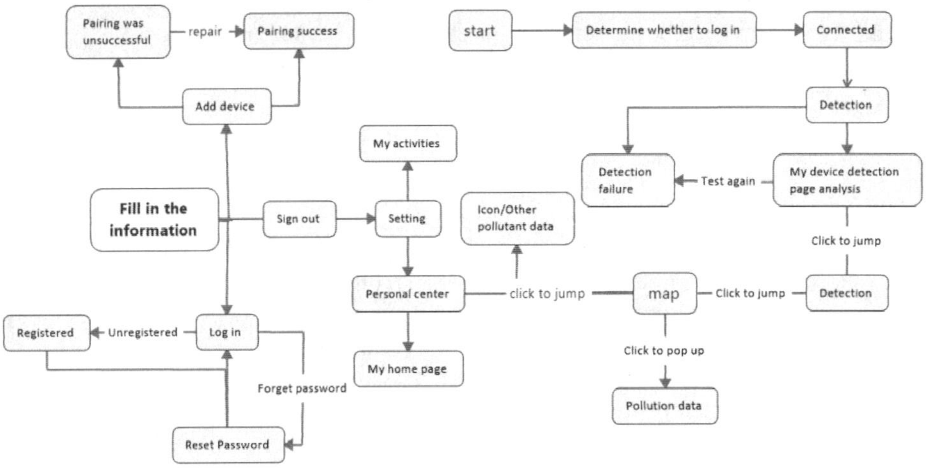

Figure 3. Interactive flow chart

transitional color, it looks more layered. The space is mostly based on ghost space, so that users can focus on the content. The overall color system is a cool color system, which mainly shows a visual perception of serious environmental pollution, which makes users pay more attention to environmental issues.

The shape of the Icon design is directly derived from the simplification of the mask, abstracted into a simple mask shape for identification. Gray-black background and low-saturation pattern have a warning effect on environmental degradation, thereby narrowing the distance between the product and the users. Reduce the burden of reading during user use. In each button, paragraph, form, you need to make a visual distinction by blanking. In the interface layout layout, it is necessary to understand the user's psychology and use a scientific visual method to visually guide the user. The bar and spacing, the spacing between the text and the line segment, and the proportion of the image select the golden ratio; in the global typography, by increasing the rhythm of the layout, the mask is used to control the tone of the image to make the interface complete and coordinated. According to the above functional analysis and usage requirements, the function pages of the APP design completed according to the user requirements as shown in Figure 4 to Figure 11 are completed. Each page also contains the details of the function page. Function allocation to meet the user's various needs.

5.Conclusion

At present, smart wearable devices and apps are mainly focused on the measurement of the number of motion steps, and the calculation of the calorie time is a basic function. Therefore, the combination of smart wearable devices and app applications will generate a large market demand. The paper will provide some reference and help for the research and development of similar or similar test products through the development of mask-based APP products, including interactive design and UI design. Nor does it pay more attention to the external environment and its own health problems. It proposed a certain persuasion method, the design schemes for easy operation, data sharing and visualization to solve such problems are given.

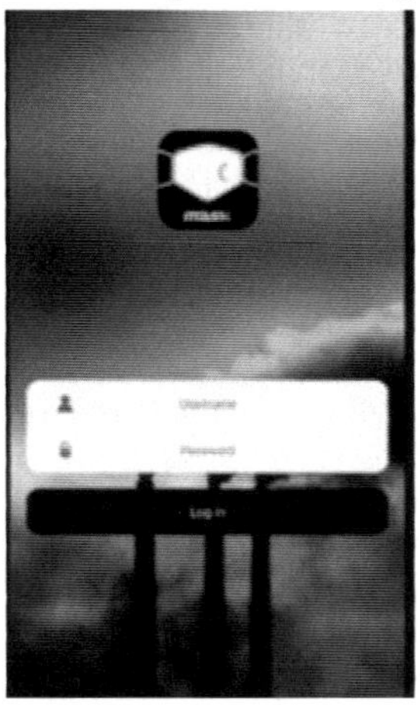

Fig. 4. Login Registration

Fig.5.Data Display

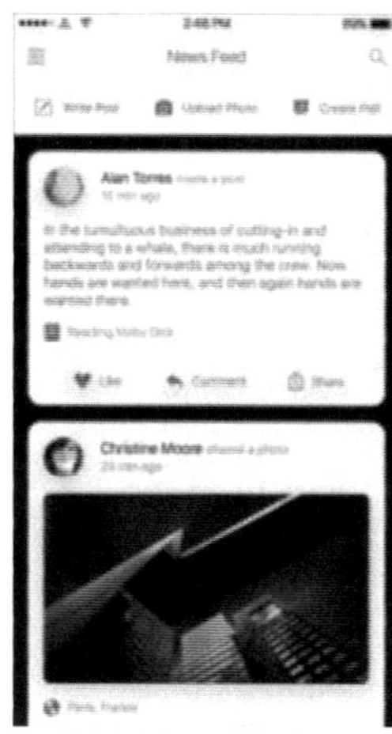

Fig. 6.Personal Home

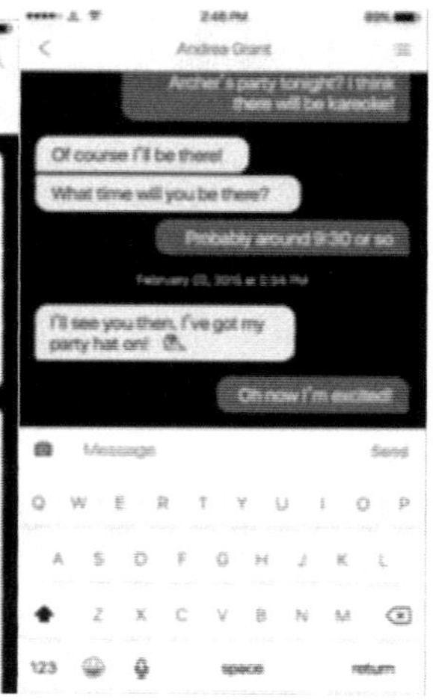

Fig.7. User Dialog

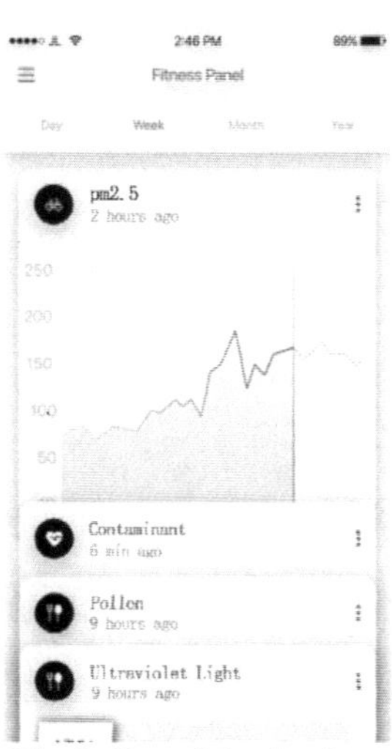

Fig.8. Data Monitoring

Fig.9.Pairing Connections

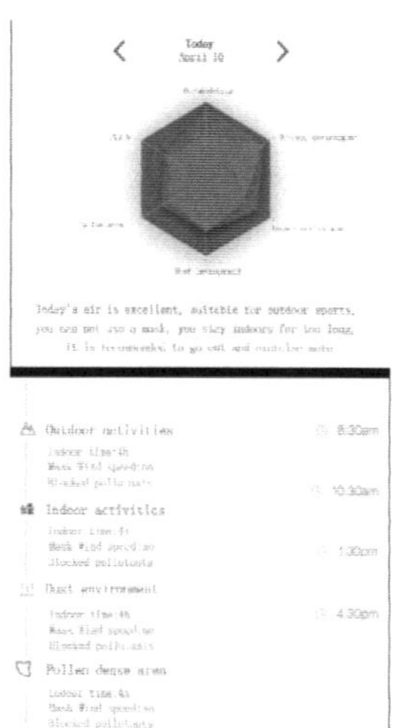

Fig.10.Personal Activity Detail

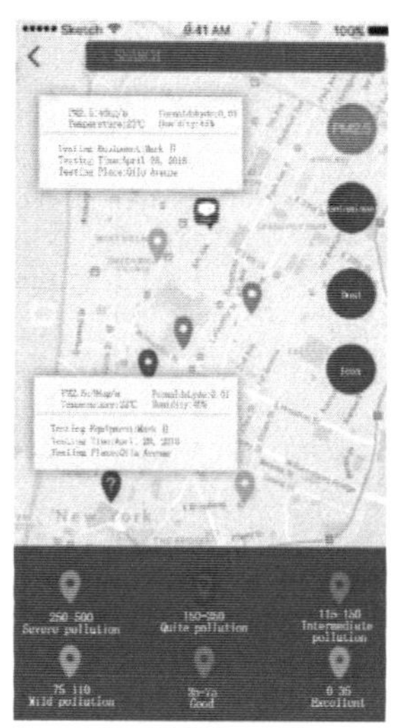

Fig.11.Map UI

Acknowledgments

First of all, I am very grateful to the organizers for providing a platform for communication and exchanges with their peers. Secondly, thanks to Ma Wangcheng and Liu Lifu for their help in the design and completion of the project, which is led by "the eight key industries in Yunnan Province" (51600630) "The project provides financial support, and I would like to express my sincere gratitude.

References

[1] YAO Jian Dong. (2011) Design and Application of a Universal Graphical Embedded Development Platform. Packaging Engineering, 28:20-21, 51.
[2] HAO Yi Xuan. (2012) Application and Prospect of Interactive Interface Design in Home Appliances. Scientific information, 29:147-148.
[3] XU Hong Quan. (2010) Research on the Perceptual Interface of Network Information Appliances (IA). Technology and Market, 17:9-10.
[4] Zhao Yuan. (2015) Based on the DOAS of ammonia discharge online monitoring technology research. Electronic Test, 15:98-99.
[5] Zhang Jia Qi. (2014) Research of user behavior persuasion design. Design, 9:188-189.

The Summary of Product Gene Network Theory

Lun Sun[1] *

[1] School of Media & Art Design, Wenzhou Business College, Wenzhou, Zhejiang, 325100, China

Abstract. Product gene network is a new research field of product morphology. This paper reviews the development and theoretical basis of product gene network and summarizes the method of the construction and application of product gene network. Borrowing some concepts of gene regulation network, product gene network takes product morphological gene as the node of network and the correlation of each gene as the edge of node to construct the complex network model of product morphological. On the basis of the product morphogenetic network, different node types in the product morphogenetic network are identified, and the node groups are obtained, which are ultimately transformed into design strategies. According to their own needs, designers can selectively combine the node types / node groups and other information to design the product form.

1. Introduction

These In recent years, the application of the concept of product genes in products has become a hot topic of morphological research. Many scholars have expanded the research field of product genes in product design from different perspectives, from the operation mechanism of DNA in product design to the mapping relationship between product genes and human perception images. The concept of gene network is formed by the microscopic study of genes contained in products. The purpose of product gene network is to form gene network by decomposing products, and to analyze and obtain gene nodes and node groups that need priority processing, so as to guide the actual product design.

2. Development of product gene network theory

Product gene network theory is derived from complex network theory. At the end of the 20th century, the research of complex networks began to be involved in many different disciplines such as engineering science, mathematical science and biomedical science. The complex network theory is to find out the commonality of different complex networks by analyzing the complex network system of these irrelevant disciplines, and to obtain the general rules for dealing with complex networks[1], then gradually form a number of classical models of complex networks. The complex network theory analyzes the characteristics of the whole network in these network models and studies the propagation path and variation of different information elements in the complex network. Therefore, the complex network has gradually become the main model to describe the principles of the organization of the objective world. By constructing the complex network, the internal topology is analyzed, and it has also become one of the main means to study the complex system in various fields.

The application of complex networks in the field of industrial design is still in its infancy, but the connection between complex networks and industrial design can be traced back to the 1980s. In 1981, Steward proposed the DSM model of design structure matrix to express the relationship between design elements[2], which first introduced the concept of complex network into the process of industrial design. With the establishment of complex network model and relevant statistical knowledge, scale-free characteristics of industrial product network are also confirmed[3]. Then a lot of research appeared in the field of mechanical products, the main theoretical application is mainly in the study of the relationship between the parts of the mechanical product family. It is found that the characteristic of typical complex network exists in mechanical products.

In recent years Xiaojian Liu[4] and other scholars used the theory of complex network research in the form of products, and found that the shape elements of product formed between gene networks, at the same time, defined the important characteristics of product gene nodes and node group which is formed by the identification method, and based on analysis of product gene networks results assist designer in product design. Since the concept of product gene network was proposed, the research objects involved water bottles, thermos cups, office chairs, product color matching and so on.

3.The theoretical basis of product gene network

Gene network is a complex network formed by genes. In biomedicine, genes are the basis of regulating cell behavior and completing biological operation. The interaction and expression of genes is the basis for the operation of biological systems. The interrelationship of genes can form gene networks by analyzing data and embody the characteristics of complex networks. Topological graph of the network is composed of many nodes and the edges between related nodes. If the nodes have no ownership relationship, the edges of the nodes are not directional. For gene network, the nodes of the network are genes, and the edges of the network are the graphical representation of the information transmission relationship between related genes.

The topological characteristics of complex networks are important indicators to determine the types of networks. In the existing complex network research, the clustering coefficient of nodes, the length of characteristic path, node degree and degree distribution of the network are analyzed. According to these three index characteristics, modern complex network theory forms complex network models such as regular network, random network, small-world network and scale-free network, and summarizes the characteristics of different network models.

Product gene network (PGN) applies the concept of gene network to products and abstractly represents product differentiation into a complex network form. As an artifact, the gene of a product is the smallest feature unit with the characteristics of heredity and variation, which determines the transmission of information in the iterative process of product morphology and determines the recombination and optimization of -morphology [5]. High recognition product morphological genes can be separated to further transfer and inherit excellent morphological genes, and more competitive product morphological genes can be formed through effective variation. The most basic components of a complex network are nodes and edges. The product gene network takes the modeling elements and parts of a product as gene nodes, and forms a network based on the correlation and dependency between genes as edges, and studies the topological characteristics of the network. Therefore, to construct the product gene network, the primary task is to determine the node and node edge of the product gene network, and then identify the node group and the central node through technical analysis, so as to guide the subsequent product design process.

4.The construction process of product gene network

4.1 Identification of product genetic nodes

The node of product gene network is the morphological gene obtained by product decomposition. Form is the visual characteristics of the external appearance of existence, and it is the overall cognition formed by observing the physical characteristics of the external appearance of existence and through psychological processing. Different scholars have proposed different classification dimensions for product genes, including the idea of life cycle gene classification, the idea of product gene classification based on information storage, and the classification of identifiable explicit and recessive genes. Product genes based on visual recognition should be objective, stable and operable, so recessive genes such as emotion, management genes and technology genes are not considered.

In different studies, different scholars have different decomposition and gene classification of the subjects. For basic morphological products, 3d modeling was carried out by parameterization method, and continuous numerical values were used for gene coding and subsequent research. Xiaojian Liu [4] studied the water bottle and established a parametric 3d model of the water bottle to obtain several continuity genes. Zhen Sheng defines the design parameters of automobile shape and establishes the continuous gene network model of automobile shape based on sample data statistics.

However, in the study of gene morphology with complex morphology, there are morphological contents that cannot be continuously encoded, so morphological decomposition is required. Product morphology is formed by the orderly combination of morphological genes. Therefore, the decomposition of product morphology can reasonably and orderly decompose the existing products in a reverse manner, and the morphological genes of the products can be obtained after peeling layer by layer. Based on functional logic analysis, a morphological analysis method of "mode module unit" is proposed in the genetic network research of office chair. Morphological analysis decomposes the product from the logic level and divides the product into single modules with different functions. Each module can be further subdivided into molecular modules according to functions. The basic elements of the product are mainly composed of three parts, namely, modeling, color and material. The product form is the ordered combination formed by the coordination and connection of these three parts. Therefore, each module is decomposed from the categories of modeling, material, color and combination to finally obtain product morphological genes and carry out operable coding for product genes. The partial discontinuity genes obtained by decomposition and digital coding of office chair morphology are shown in table 1.

Table 1. Part of the classification variable gene induction table.

Form of seat headrest	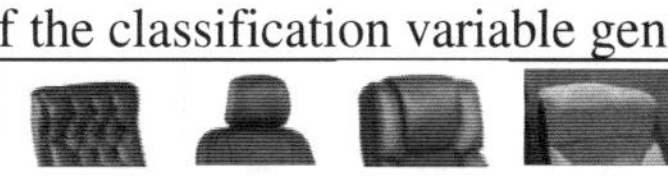 0 not obvious 1 separation 2 in the middle 3 The whole top

In the product gene network, the edge between gene nodes is the topological expression of the correlation between genes, that is, the existence of the edge between nodes means that the connected gene nodes have certain connections. The calculation of correlation between genes requires the numerical value of morphological gene coding and the retrograde calculation. For continuous genes, Pearson correlation coefficient was adopted for calculation, and Spearson correlation coefficient was adopted for calculation of genotype genes and the cross calculation of two genes. After the calculation is completed, the edges of the gene network under the corresponding threshold can be obtained by taking different thresholds.

4.2 Construction of product gene network

The purpose of building product morphogenetic network is to study the correlation between morphogenes from a micro perspective. By grasping the relationship between different morphogenes, it can be transformed into a design strategy that can be applied in the design to help designers carry out actual product design. The grasp of the correlation between different morphological genes is mainly from the two perspectives of node priority and node group in product gene network.

In bioinformatics, edge has directivity in biological gene network, and high-level gene node with regulation has important position. For product genes, the priority of gene nodes can also be identified through coding and mathematical processing, and transformed into the form of priority processing in design. The priority of a node is reflected by its centrality and sensitivity. The centrality of a node can be represented by its degree, that is, the number of other nodes associated with the node. Node sensitivity refers to the degree to which a gene is affected by other genes and is a feature of gene stability.

4.3 Node type and node group of gene network

The type of gene node can be identified by calculating its centrality and sensitivity[6]. (1) key nodes. (2) independent nodes. (3) passive node. (4) secondary node. Through the identification of genetic network nodes, it can effectively grasp the importance of the nodes in the genetic network, and tell the designer which morphological genes should be dealt with first.

Node group refers to the collection of nodes that interact and act together in the gene network, and the influence of these nodes on products has a synergistic effect. The identification of nodal group can help designers grasp the synergistic effect of morphological changes, and take several morphological features of the group into consideration in the design. The identification method of node group: by increasing the correlation threshold of the network, the low-correlation edges are screened and removed continuously, so as to obtain the node group with close relationship. By increasing the threshold (named of [r]) of clustering coefficient in the network, nodes with higher clustering coefficient and their connected node groups form node groups. As shown in Fig.1, it is the office chair gene network graph under different values of [r].

Through the identification of node types and node groups in the genetic network, the importance of nodes in the genetic network can be effectively grasped, and which morphological genes should be given priority to by the designer. It is an important process to transform product gene network into design strategy, and it is a key step to apply gene network to practical design.

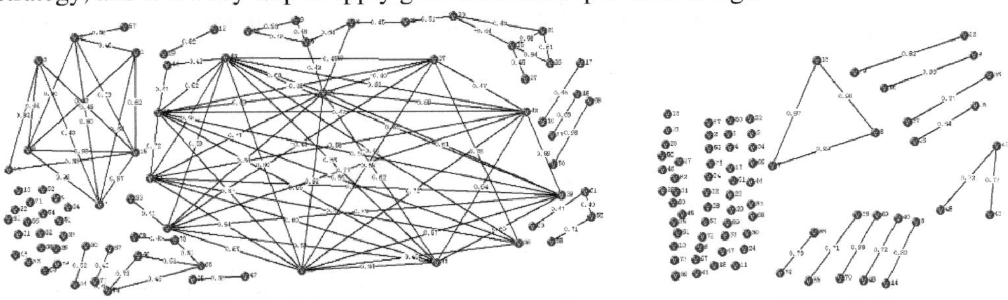

(a) [r]=0.4 (b) [r]=0. 7

Fig.1 gene network diagram in case of different values of [*r*]

4.4 Gene network data processing

After obtaining the product gene nodes and node groups that need priority processing in the gene network, the design strategy transformed into can be applied to the later design. The application of design strategy can be directly referred to the designer for design reference. In particular, for products containing subtype genetic networks, designers give priority to the design of important product forms while dealing with other forms. In the application of gene network to office chair, the evaluation of the design scheme obtained by the designer through the design strategy is superior to the control group in the design experiment.

For the products with continuous parametric modeling in the genetic network, interactive genetic algorithm can be used for further computer-aided design. Interactive genetic algorithm is an extension of genetic algorithm in product design. At the beginning of the interactive genetic algorithm, a number of model groups are generated by the software on the parameterized model, and then displayed to the user step by step. The user selects the displayed model, and the software generates a new model based on the user's choice and the design strategy obtained from the product gene network, and then gives the model to the user for selection until the optimal solution is obtained. By using interactive genetic algorithm, Xiaojian Liu et al. found that the addition of gene network design strategy can effectively reduce the number of iterations and quickly obtain the optimal scheme.

5. Summary

The research of product gene network is an exploratory practice of applying complex network to product gene field. Through the research of many scholars, the preliminary theoretical basis of product gene network and the construction and application methods of product gene network have been formed. The results show that product genes can form complex networks with effective features, and obtain gene nodes and groups in clusters, which provides new ideas for product shape design.

References

[1] Xiaofan Wang, Xing Li, Guanrong Chen. (2006) Complex network theory and application. Tsinghua University Press, Beijing.
[2] Steward D.V. (1981) The Design Structure System: A method for managing the design Of complex systems. IEEE Transactions on Engineering Management, 28 (3):71-74
[3] Yassien A, Dan B. (2003) Complex concurrent engineering and the design structure matrix method. IEEE Transactions on Engineering Management, 11 (3):165-176
[4] Xiaojian Liu,Yan Sun, Jianfeng Wu,Shiwei Cheng. (2013) Product's gene regulatory network model and its aiding to design process.Computer Integrated Manufacturing Systems,19(7):1463-1471
[5] Guanhua Hou. (2014) Establishment and application of product appearance gene intentional model. Journal of Machine Design, 31(3):105-109
[6] Lufang Zhang, Sheng Zhen, Lun Sun, Xiaojian Liu. (2016) Study of the product form design based on gene network. Journal of Zhejiang university of technology, 44(5):584-590

IOP Conf. Series: Materials Science and Engineering **573** (2019) 012032 doi:10.1088/1757-899X/573/1/012032

Researches on Museum Cultural Creative Products Based on Domains Crossing

Peng Hong[1]**, Shang Qianli**[2]

[1]Professor,Wuhan University of Science and Technology,Wuhan,Hubei, 430000,China

[2]Postgraduate,Wuhan University of Science and Technology,Wuhan,Hubei, 430000,China

395155952@qq.com, 3104237403@qq.com

Abstract. This paper carried out study on the domains-crossing cultural creative products of museum. Based on the analysis of existing domains-crossing products and the experience of existing domains-crossing successful cases in various fields, a new direction and new form of domains-crossing museum in the future are obtained. The author put forward two new forms as combining art museum with museum and combining with local characteristics, to explore a new direction of domains crossing, so as to facilitate the better development of museum cultural creation.

1.Introduction

Museum cultural and creative products rely on their own cultural resources, through creative transformation, science and technology, to promote the formation of various forms of culture as the core content of the product. In recent years, the state has given a lot of policy support, and a series of documents, such as *Guidance Opinions on Further Strengthening the Work of Cultural Relics* in 2016 and *Some Opinions on Promoting the Development of Cultural Creative Products in Cultural and Cultural Relics Units* in 2017, have encouraged museums to design cultural creative products.

2.Definition of Domain-crossing

Domain-crossing design means the mixing and integration of two domains. Domains-crossing design can inspire designers to create new inspiration and create new design environment.

The traditional culture represented by the museum continues to produce new ideas, combined with more new fields and new technologies, which will also promote the inheritance and continuation of the traditional culture and endow the traditional culture with a brand-new vitality of this era.

3.Museum's existing domains-crossing cultural creative products

The museums domains-crossing has become a major trend in the design of cultural creative products. "Museum +" has become the breakthrough point of the museum's cultural creation products.Domains-crossing cooperation not only presents the cultural content of the museum in different forms in different time and space, but also promotes the upgrading of the brand content and industry. The development of the museum's cultural creation products is an important means to attract the new public while retaining the core viewers. The characteristics of existing domains-crossing cooperation in museums are mainly focused on two aspects. Firstly, the dissemination of museum culture to the public, so that the public understand the cultural content behind the collection of resources. Secondly, it will go out, to cooperate with the brands with same attributes, therefore breaking the already-existed impression left on the museums from the public and creating the products that can attract the new consumers.

3.1 Domains-crossing cooperation based on the dissemination of culture

Domains-crossing cooperation based on the dissemination of culture means that museums disseminate museum culture through collaboration with other platforms or through the use of new technologies. Different forms of cultural communication can attract different groups of people. First of all.

Many museums in China offer museum courses. The Shanghai Museum has introduced a series of "Museum Youth Humanities Class" courses, inviting famous teachers to explain to students what "esthetic" is and what "aesthetic education" is. The "One lesson in Caotang" project, launched by Du Fu's Cursive hall in Sichuan, was included in the first national cases of social education. Courses are offered on weekends and holidays. On the one hand, it invites schools to organize students into the "One lesson in Caotang" at museums. On the other hand, the museum provides funds to the poverty-stricken areas for public welfare activities, which brings poetry culture out of the Hall.

The Shanghai Museum also launched "The Sound of All Things—Shanghai Museum's Parent-Child Concert". The concert included explanations of cultural relics, historical background, culture and art, letting the cultural relics and music to combine interestingly, materially, organically.

The museum also uses modern technology to display the museum. The Palace Museum has released its work using VR technology. restored the full extent of the imperial garden to a vibrant royal garden in a virtual reality world.

In addition, it made the publicity with multi-media. "National Treasure", "I repaired the cultural relics in the Palace Museum" and "If the national treasure can speak", using modern multimedia, showed the museum to viewers who couldn't get into the museum, which has set off a museum craze.

No matter the setting of curriculum, the application of modern science and technology, or the variety of arts, they are all the modes for museums to make the domains-crossing publicity. With the emergence of new-type consumers, museums are seeking for better ways for publicity.

3.2 Domains-crossing Cooperation based on Products

The domains-crossing cooperation based on products means that museums cooperate with third parties in the design of cultural creative products. The museum should choose brands that fit well and match well for cooperation, or the popular brands.

The popularity of reality TV variety has spawned "It's new, the Palace Museum", using the star effect to explain the story of the Forbidden City, rallying designers with the theme of cultural creative product design, and using crowdfunding and microblogging to promote the show, creating a heated discussion.

Most of the collaboration between the Palace Museum and Tencent has focused on the development of products in modern and more popular ways. Portable game, as a popular time-killing way, developed a Forbidden City Version named “Like Eliminating Everyday”. Combined with current popular music programs, it launched an "Ancient paintings can sing" music innovation competition, to create and sing songs according to ancient paintings, and call on the public to design the cultural creative products.

The theme restaurant Tianfu Jinxiu, which is co-established by the Dufu’s Hall Museum and KFC, features scenes of life in the Tang Dynasty. The table and walls are filled with Du Fu's classic poems and related calligraphy and painting works, as well as 3D holographic projection. The restaurant also sells cultural creative products from Du Fu's Hall.

They also launched products with local featured brands. Suzhou Museum, cooperating with the local intangible cultural heritage, the Forbidden City-exclusive Dessert, promoted the “Dessert Like 12 Chinese Zodiac” with the theme of “Edible Museum”.

There are,of course, many other products of domains-crossing cooperation. Domains-crossing-developed creative products need to be designed with cultural connotations, designed by combining with cultural relics themselves, rather than copying and pasting.

4. Enlightenment from the Successful Cases in Domains-crossing to the Development of Museum Cultural Creation Development

The design of domains-crossing cooperation in different fields has become mature after a long-term development. The museum domains-crossing is based on the early stage domain crossing. The domains crossing in different fields at the early stage has also contributed greatly to the domains-crossing development of the museum's creative products.

4.1 Domains-crossing Cooperation between Arts and Brands

Art and brand domains-crossing cooperation means to combine the art works together with the products. Companies choose to cooperate with well-known works of art, while artists transform their works according to the characteristics of their products. Such as the "XX" eye logo created by KAWS, which was very popular on the street and later worked with different companies, including UNIQLO, NIKE, Dior, etc. The Air Jordan 4, designed by KAWS in partnership with NIKE, is shown in Figure 1. The KAWS companion pattern is all over the body of the shoes. And the sole also has the personal logo of KAWS. Through the commercial cooperation, the "XX" eye mark was developed into a commercially valuable "brand".

By collaborating with well-known artists and the use of familiar works of art to design, on the one hand, products are easily to be accepted. on the other hand, works of art can be civilian.

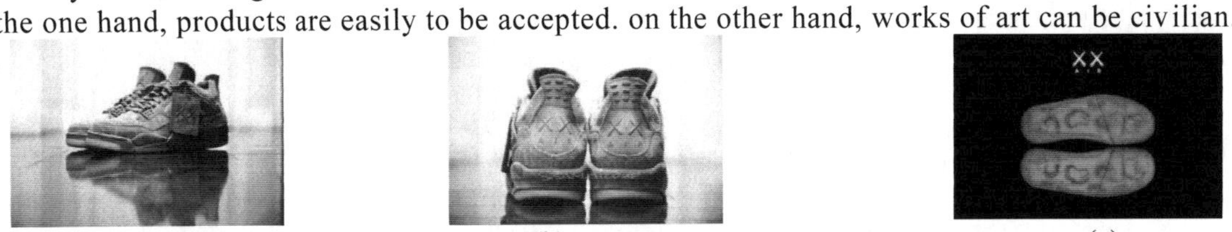

(a) (b) (c)

Figure 1. Air Jordan 4 from KAWS X NIKE

4.2 Domains-crossing Cooperation between Brands

The domains-crossing cooperation between brands is the cooperation between different industries, which is conducive to inter-industry exchanges. In this process, enterprises should pay attention to the fit and match of the two partners, add scores for the product. Besides, they should consider the preference of the target consumer group and their consumption demand.

The brand Nongfu Spring, as a traditional brand, is actively domains-crossing cooperation, to attract more young consumer groups. For example, Nongfushan spring fruit drinks cooperates with the portable game “Yin-Yang Master” in packaging design, cooperates with Alipay, Shared Bikes and Netease Yun music. Nongfu spring traditional brand has brand influence. But now the consumer structure changes, the original consumer has entered the old age. Therefore, to seize the new era of consumer psychology, Netease cloud music is now the largest user group of music platform. Nongfu Spring needs young consumer group, and Netease cloud music needs the brand influence of Nongfu Spring.

Museums represent the traditional culture and has the certain influence. Museums needs to be injected with the fresh blood to be able to follow the time step. Like Nongfu Spring, it could cooperate with the popular brands. However, the brands of cooperation shall be able to fit and match with museums.

5.Development Trend of the Museum's Domains-crossing Cultural Creative Products

An analysis of the development trend of museum domains-crossing cultural creative products will

IOP Conf. Series: Materials Science and Engineering **573** (2019) 012032 doi:10.1088/1757-899X/573/1/012032

help the museums to broaden their horizons and open a new road.

5.1 Cooperating with Artists

An artist expresses his/her needs or desires through his/her work. With his/her keen observation and artistic expression, the artist materializes some illusory unperceived consciousness. Cooperation with artists can be divided into two ways: First, the designer may use the works of art in the product design of cultural creative products, to enhance the value of the museum's communication and increase the added value of the product. Museums can build their own brand culture and brand personality through cooperation with artists. The second is the cooperation between museums and art museums in the creation of art works, and the exhibition of art works in museums, to experience the integration of contemporary art and ancient art, and discuss how to undertake tradition together.

Museums can learn from foreign museums, to integrate museums and local galleries, break the boundaries between ancient and modern art, and attract more visitors.

5.2 Cooperation with Local Featured Brands

China has a wide range of resources, rich and profound culture, each with its own local characteristics. Chinese museums are numerous and distributed all over the country, with different cultural features.

Museums could cooperate with local characteristics. Take the Sichuan Museum for example. It could choose the characteristic brand according to food, housing, travel, travel, shopping, entertainment, and, some ideas for the design are hereby proposed and a map has been drawn according to the characteristics.

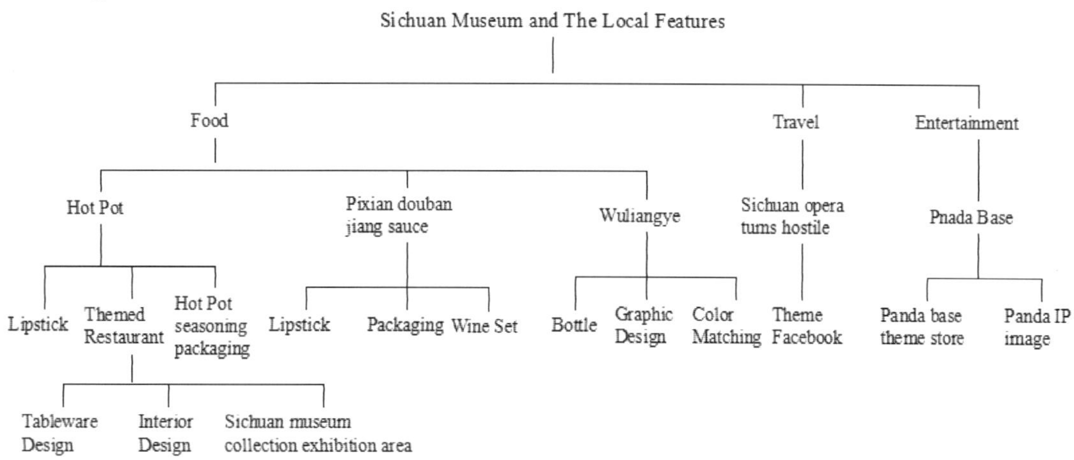

Figure2.A Thinking Map from Combining Sichuan Museum and The Local Features

6.Conclusions

The domains crossing in foreign museums has matured, but the situation in China is still in its infancy. As the pioneer of the museum, the Palace Museum is trying to cooperate across domains. Some other top-ranking museums are also trying to cross the domains. But currently speaking, the selection of brands, businesses and museum features is not matching enough. The author proposed to combine it with local art museums, to combine contemporary art with ancient art. Besides, the artists may create products or works of art based on the collections in the museums. The second one is to combine with local characteristics, to develop their own local characteristics, to deepen the impression of the museum. The museum is on its way to domains crossing. It can be counted on that it could have a bright future.

Acknowledgement
2017 National Social Science Fund of China of Art Project .No.160. Research and Development of Cultural and Creative Products Based on Museum-Level Assessment Criteria. Phased Research Results. Peng Hong. Wuhan University of Science and Technology.

References

[1] Yang,X.L.(2018)On the development of the cultural industry of the museum in the new era.China Economist,9:191-193.
[2] Shan,J.X.(2011)The social responsibility and the social development of the museum.Sichuan Cultural Relics,1:3-18.
[3] Wang,Y.(2015)On the realistic significance of art cross-border from the perspective of LV's art cooperation.Business Culture,12:43-44.
[4] Zhu,H.(2018)A cross-border study of brand in the internet age--a case study of NONGFU SPRING brand.Popular Literature,19:243-244.
[5] Wang,X.C.,Wang,S.Y.(2016)Research on the method of product design based on cross-border design.Packaging Engineering,4:148-151.

IOP Conf. Series: Materials Science and Engineering **573** (2019) 012033 doi:10.1088/1757-899X/573/1/012033

Research on Cloud Platform Service Based on Industrial Design Industrial Chains: Taking 'the research and develop demonstration platform of Sichuan Industrial Design Platform Service' as an example.

Yujia Jia[1], Renhua Zheng[1*]

[1] College of Nuclear and Automation Engineering, Chengdu University of Technology, Chengdu, Sichuan, 610059, China

[*]Corresponding author's e-mail: 236667991@qq.com

Abstract. In recent years, the government has vigorous carried out for the policy of industrial design industry, so that it has enabled the industry to develop rapidly. However, for the industrial design industry, the poor connection between the upstream and downstream industrial chains of industrial design has become an important factor which is restricting the development of this industry. This paper analysed why the industrial design industry chain cannot be linked closely through investigation and analysis as well as the pilot project in Sichuan. According to the pilot project of *Sichuan Industrial Design Cloud Service Platform (SCID Platform)*, it realized the self-circulation mode of industrial design industry chain, and play a substantial role in promoting economic operation.

1. Introduction

In recent years, the industrial design industry has made a considerable progress in China, and various regions have introduced industrial design through national guiding policies, which have made positive contributions to the transformation and upgrading of industrial economies. Industrial design has become an important support for economic growth in various regions. The industrial design industry has played a positive role in promoting the technological progress and competitiveness of local enterprises in various regions.

1.1. Industrial Design Industry Chain

The term "industry" is derived from the concept of economics. In the "Design Innovation Mechanism Research of Enterprises in China", Liu Guanzhong mentioned the concept of industrial design industry is : taking part in the process of production 、 circulation、 as well as increasing in products value, and final realization. [1]; Fan Jin also suggested that industry is a general term for economic sectors that can bring added value, and sometimes refers to all kinds of equipment that produce material of the products， and also contains the labour in this process, not just industrial design techniques [2]. One of the cores for judging whether an activity has industrial attributes is depending on whether it has "closely corporate economic activities among the enterprises" [3]. Therefore, the industrial design industry chain is not only a single part of designing or packaging link provided by the enterprise in the product development process. The enterprise providing the service in a certain link exists as a point of specialization in the industrial chain. The industrial design industry chain is the relationship between the upstream and downstream links related to industrial design in the whole product development process. Therefore, the industrial design industry chain is the concept from a social perspective. From this we can see that the industrial design industry should not be simply defined as the "design" process, but rather a product development and manufacturing process.

1.2. Industrial Design Cloud Platform and Product Family

"Cloud platform" is to consider the value of public information resources (enterprise information, processing information, exchange information, etc.) in the industrial design industry chain, and to share information related to industrial design to enhance the overall level and to promote the development of industrial design industry in that local region.

The product family is a kind of modular system, which can call shared components, modules, and subsystems within the product platform and make integration through standard interfaces, then, a sort of new specific products will be created [4]. It can generate diversified product clusters and provide personalized service. It can apply a small number of standardized modules to form a large number of products, and apply product commonality in each market segmentation as much as possible. [5]. The design idea of the product platform is to use the enterprise product platform to reduce the development cost of the enterprise, shorten the period of product development, launching configurable products that can satisfy the individual demands of customers efficiently and quickly, through customized development and combination with platform modules.

2. Industrial design cloud platform construction practice

2.1. Cloud platform construction background

Objectively, there are still some restrictive factors in the development of industrial design industry. One typical performance is the disconnection between the upstream and downstream of the design and manufacturing industry chain. This problem greatly reduces the efficiency of all links in the chain. As well as, it directly affects the development of industrial design companies.

For instance, as shown in Figure 1, after the design work finished, the new products of the design enterprise (product development enterprise) will usually entrust a third-party processing enterprise to carry out the OEM work, then they will face the next question: how to find Third-party foundries? Can the materials, technical specifications, process methods, process quality and process precision of the foundries meet the design and manufacturing requirements? So, how do design companies (development companies) find processing companies that meet the needs of the process? These situations are finally summed up as a question: how to find accurate technical information of the processing company? If the phenomenon of "link breaking" in such upstream and downstream enterprises cannot be solved in one-step, the design party (or developer) is in a blind state, it'll not only affects work efficiency but also the quality.

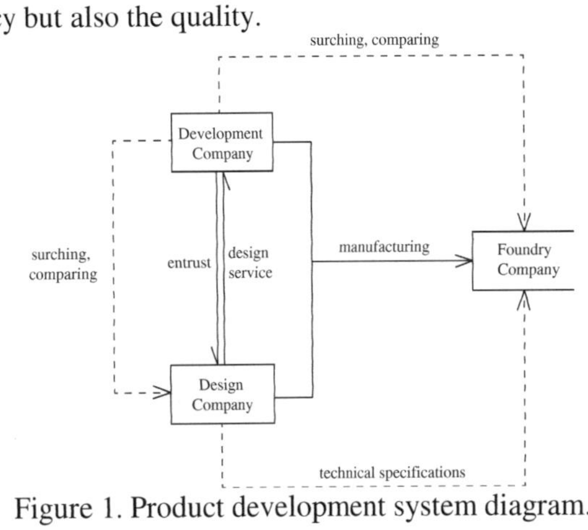

Figure 1. Product development system diagram.

However, the reality is that, despite the increasing development of the Internet, the information is very few and fragmented on the important technical specifications of processing enterprises, which is completely unable to meet the needs of the design party (or developer). At present, the main carrier of technical information of processing enterprises is the enterprise website. The technical information on the website is also limited to the processing method. For sheet metal processing companies, their websites generally show the type of process (such as bending, electroplating, etc.), and some companies will display their processing equipment. For example, the plastic model company website will generally display its type of process (such as injection modelling, blow modelling, etc.), but these companies almost haven't display of process precision, technical specifications, process equipment, but these data are very important. To further understand the detailed technical information, we can only communicate with the enterprise on the spot. But this will cost us plenty of time, furthermore, it'll affects the design efficiency (or developer's) work efficiency and even the quality of work. In fact, it also affects the processing quality of the processing enterprise itself, which is not conducive to develop the new customer resources. Therefore, it is necessary to establish a "one-step" information service platform.

2.2. Cloud platform construction method

The development of the cloud platform can widely help companies (R&D company – design service company – OEM) in the product development process to conduct information transparent and open information to exchange. This cloud platform can design for the regional "cloud platform" service model base on each regional development specificity. For example, in the southwest region, Sichuan Province is an important manufacturing district in China. There is a huge demand for industrial design in the fields of major equipment manufacturing, consumer electronics research and development, and light chemical industry. The development of industrial design is changing from "Made in Sichuan" to "Create in Sichuan". After experiencing the germination period and the cultivation period, the industrial design industry in Sichuan has now reached a key stage of growth and optimization.

The "SCID Platform", the basic content of the platform is to build a technical information platform based on processing enterprises of the local region, which is actually a processing The enterprise's technical information database, which collects no less than 100 processing technical information, the database contains the following information:

Technical information of plastic mold manufacturing enterprises;

(2) Technical information of metal processing enterprises;

(3) Technical information of wood processing enterprises;

(4) Technical information of ceramic processing enterprises.

Among them, the plastic mold manufacturing enterprises and metal processing enterprises are account for a great proportion in technical information, these two types of processing methods account for more than 95% of all manufacturing information. And through the following ways to make data collection and application:

(1) Data collection method: Through the network collection and data inquiry, gathering the contact information, business address, processing technology, process parameters, process quality, manufacturing cost and other related information of processing enterprises in Sichuan Province are collected in large quantities, and According to the four categories of plastic, metal, wood and ceramic lamps, to establish a detailed information database.

(2) Field investigation method: The project inspection team members have taken on-the-spot investigation to various processing enterprises that can carry out in-depth understanding. The processing scale, processing equipment, processing precision, processing quality, processing cost, and contact information of the processing enterprises are important. The information is confirmed one by one.

(3) Case study method: A small number of people in the industrial design industry and people from manufacturing companies are invited to attended seminars and survey visits to understand the actual needs related to the theme of the project.

For each processing and manufacturing enterprise, we strive to collect all relevant data, such as process methods, process precision, process specifications, process equipment (as far as possible), contact information, etc. The demand side (design or development company) can basically understand its technical information from this database, if anyone needs to know more deeply, he and can enquire it by phone or interview.

2.3. Cloud platform development significance

The construction goal of the industrial design cloud platform is to develop and build a "one-step" and "interconnected" demonstration platform based on social collaboration; and depend on the platform to participate in the innovative service system and service network. Through the close contact between online and offline, we will build an industrial design ecosystem by the process of "design resources + professional services + social collaboration", which chould effectively promote the cooperation of industry, education and research in industrial design, as well as promote cooperation and exchange of industrial design, and foster Innovative companies towards industrial design to help enterprises grow faster.

The technical information platform of the processing enterprise is built to solve the above problems. Building a platform for communication between design companies (or product development companies) and processing and manufacturing companies will be a very effective tool for connecting with design-manufacturing industry chain. This platform provides a processing (OEM) technical information database for a large number of design institutions, product development companies and design professional teachers and students, so that the can find the foundry partners easily, and also select the process methods in the design and development process. At the same time, it also provides important reference for process selection, process parameter determination, process quality verification, processing and manufacturing costs in the design and development process. In addition, the developer can also obtain important technical support and guidance from the foundry to improve design quality and design efficiency. It will also be an efficient integration of the design (development)-manufacturing industry chain for the local region, which will not only effectively promote the development of the industrial design industry, but also promote the economic development of the local region [6].

3. Industrial design cloud platform construction results

3.1. Technical goals for establishing a cloud service platform

The cloud service platform deploys software and design resources in the database. Users can select services according to their needs. According to the Information "Collection Stage" mentioned in the previous section, the cloud platform can be divided into two parts, named, "common platform" and "Lookup platform" as shown in Figure 2.

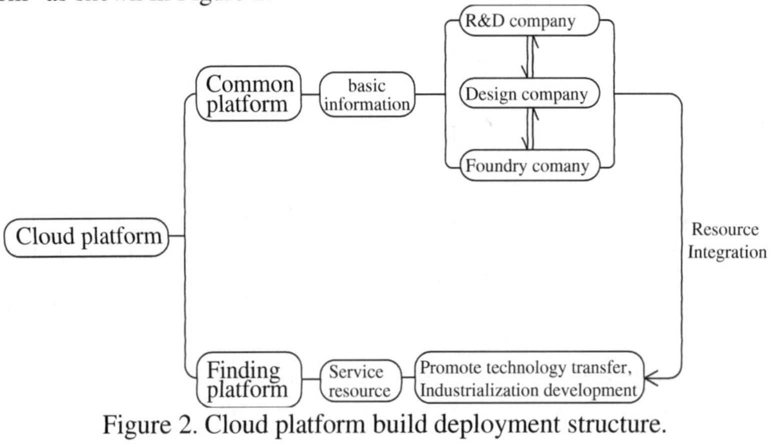

Figure 2. Cloud platform build deployment structure.

The "common platform" is mainly based on the first and second stages of the project, aiming to share the companies and technical information, so that, to open up the product development process of "R&D enterprise – design enterprise – foundry enterprise", to form the system of industrial design industry chain. The "find platform" mainly integrates resources based on enterprise interaction information in common platform, realizes network integration of product industrial design resources, integrates various professional support technical service function modules, and develops and constructs industrial design innovation cloud service platform. At the same time, it also cooperates with universities, scientific research institutions, professional technology service organizations and local excellent design companies in the region. Developing and building "one stop" by adopting "heterogeneous data and heterogeneous database integration in cloud service network environment", "service-oriented architecture and enterprise service bus", and "depth integration application of Web1.0 and Web2.0". The "interconnected and integrated" technology-based industrial design cloud service platform based on social collaboration forms an industrial design technology service innovation system and service network (including the entire platform mobile APP application).Therefore, the platform greatly gathers the resources which industrial design shared and demand, supporting the realization of industrial design services e-commerce, supporting the expansion of technical service function modules, the new service model combining online and offline provides services for product industrial design activities, and ultimately drives technology transfer and industrialization development.

3.2. *Technical process of building a cloud service platform*

In the process of establishing an industrial design cloud platform, the core technology of each different step will also be different. The technical key to the construction process of the industrial design cloud platform is how to accurately determine the technical indicators of each company.

There are two difficulties in obtaining accurate technical indicators. Firstly, the target of the project team is to obtain accurate technical indexes much as possible. However, from the perspective of processing enterprises, they tend to improve their technical indexes and report the fake parameters. Secondly, some companies' equipment has been used for a long time and it is impossible to confirm accurate indicators. Therefore, above this kinds of situation, the project team should know about the technical indexes on the one hand, and on the other hand, it is necessary to determine the confirmation technical parameters through various ways (such as the Internet) to confirm the information of the equipment brand and date of manufacture. This work is not too difficult, but it takes a plenty of time and requires a lot of work.

Therefore, after standardizing the construction process of the cloud product platform, it is mainly divided into the following stages, as shown in Figure 3.

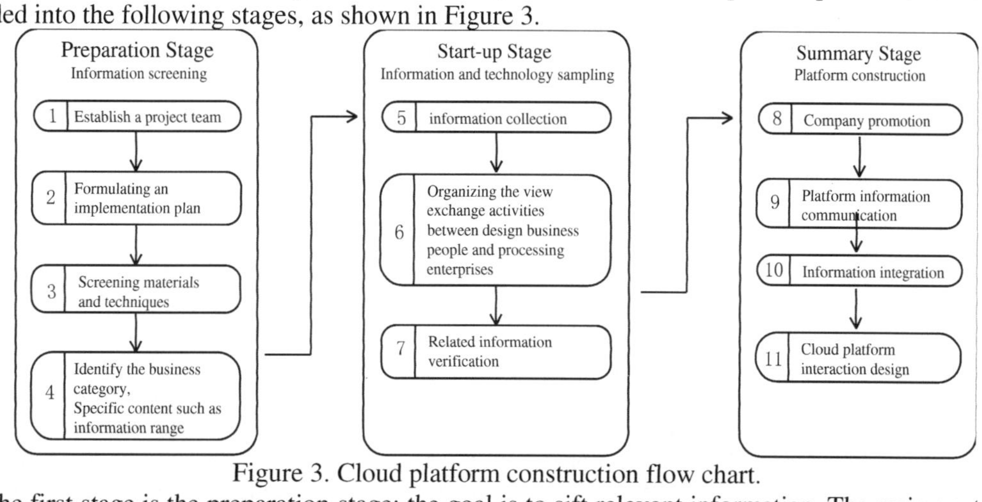

Figure 3. Cloud platform construction flow chart.

The first stage is the preparation stage: the goal is to sift relevant information. The main content of this phase is:

(1) Establish a project team and identify researchers;

(2) Formulating an implementation plan;

(3) Sifting material categories: first of all, to determine the direct relationship with industrial design, which will directly affect the type of process design of the product;

(4) Defining the specific direction of the type of companies which have been collected, as well as the type of materials, and the way of processing;

The second stage is the start-up stage, with the aim of identifying information of the preparation stage and sampling the information for relevant companies and technologies:

(5) Collecting a large number of relevant information in the first phase and gathering for a database, then verify and accept the acquired information;

(6) Organizing the view exchange activities between design business people and processing enterprises, inviting representatives of design companies and representatives from processing manufacturers, so that the processing factory can deeply understand the technical requirements of the design companies, also, it enables the design company to get further understanding about actual manufacturing process. This will enable both parties to get further interconnection with technical issues in the design work.

(7) Checking the survey information once again, in order to ensure the accuracy and practicability of the platform information, the members of the project team need to conduct a field visit to each processing enterprise that can have an in-depth understanding. Some companies have a strong awareness of privacy protection and are reluctant to disclose information about the company's equipment brand and processing details. In this case, it is necessary to persuade through the multiple communication of the group members, so that, to confirm the details of the information such as processing equipment, processing accuracy, processing quality, and processing costs.

The third stage is the summary stage, which aims to eventually build an industrial design cloud platform:

(8) The small group needs to visit the relevant enterprises, and obtain the information of the processing enterprises. In order to enable the design companies, designers and other design workers to understand the relevant processing information of these companies, the propaganda group has carried out the processing enterprises. The classification is divided into four categories according to plastics, metals, wood and ceramic lamps. The promotion includes four types of materials processing technology and common surface decoration technology. Designers can search for specific enterprises on the platform according to their own needs.

(9) Timely and effective communication of platform, so that design companies, designers and other design workers can not only find the contact information of processing enterprises in the “Processing Companies Technology Information Platform” in the “Industrial Design Cloud Platform” Information such as address, processing information, etc., but also can leave a message on the platform, asking relevant requirements and problems, project members will provide timely answers and help according to relevant issues, and establish a real-time communication on the platform for the design company to processing companies, as well as design a convenient and fast communication and cooperation platform for design workers and processing enterprises.

(10) Comprehensively organize the research technical data and the information that companies demanded, and upload data for modification and addition on this basis.

(11) Information platform interaction design. Project team members need to regularly organize relevant feedback problems and also the information requirements by companies, and update the cloud platform service system, at the same time, the platform’s interface design should base on user experience, to enable consumers could quickly find the information they need.

3.3. Industrial design cloud platform builds results

In fact, the establishment of the cloud platform information is a long-term through “injection-improvement-promotion” process. After the cloud platform project team has completed finish the planned tasks, they can continue to collect some valuable processing companies’ information, such as the new powerful emerging enterprises in plastic moulding and metal processing, as well as some minority companies focus on processing and manufacturing technology. Devoting to make it becomes a more complete database planform, and continue to link Industrial Design Industry Chain closely, enable the Industrial Design Industry Chain to operate healthily, Increasing the value of the original products, and finally realize the adjustment and upgrading of China's industrial structure.

Taking "Sichuan Industrial Design Cloud Service Platform" as an example in Figure 4.

The processing technology information platform as the bridges to communicate among the majority of design institutions and product development enterprises and processing (OEM) enterprises. In the construction of the "SCID Platform" module: a professional and effective information database is built, which includes the detailed technical information of 113 processing enterprises, and the tasks were exceeded than we planted. Among them, there are 27 plastic mold enterprises, 24 mechanical manufacturing enterprises, 22 sheet metal processing enterprises, 14 timber enterprises ,25 coating and printing enterprises. From the perspective of the type of these enterprises, that include our plants before: the 4 types of materials processing manufacturers in plastic molding (plastic production equipment), metal, wood, ceramics. Among those information, mainly in the plastic molding manufacturing enterprises and metal processing enterprises technical information, these two types of processing methods account for more than 95% of all manufacturing information. In the expected aspect of the "SCID Platform" project is: from the basic formation has been collected, the project team began to provide data to relevant enterprises and design practitioners (mainly for some industrial design enterprises in the local region and the industrial design school of universities), these groups are potential users of the database planform in the future. These data have played a very important role in the business work in related companies. In addition, after the technical information database of the processing enterprise went online, it has generated a large amount of traffic (page view) without any publicity, and gradually exerted its value. The project team plans to carry out appropriate publicity. It is believed that with the development of publicity work, the number of users will have a significant increase, and its value will be better utilized in industrial design of the local region.

Figure 4. "SCID Platform" page information.

4. Appendices

By analysing the existing "link breaking" problem in the industrial design industry chain, this paper proposes to build an industrial design cloud platform on network, which aims to make the industrial design developed to an "industrial chain", by increasing the connection between the upstream and downstream enterprises in the industrial design industry chain, not as an independent link. Meanwhile, taking the *"SCID Platform"* as an example, showing that establishing the industrial design cloud platform can not only directly enhance the efficiency of product design and development, as well as enhances the quality of design and development, but also brings opportunities of market to processing enterprises and makes contribution to the entire industry chain for developing. The value of this platform is mainly reflected in:

(1) It can greatly improve the efficiency of product research and development in industrial enterprises;

(2) Efficient and convenient supporting services could promote independent innovation;

(3) Effectively reduce the cost of innovation in enterprises;

(4) Enterprises obtain the cutting-edge in industrial design and product innovation methods through platform interaction.

Acknowledgments

This research is supported by the Science and Technology Project of Sichuan Province (Grand No. 2015GZ0080).

References

[1] Liu G.Z., (2014) Shaping the design innovation of manufacturing enterprises [J]. China new time, 10: 50-55.

[2] Fan J., Zheng Q.W., Mei J., (2004) Applied industrial economics[M]. Economy and management publishing house,.

[3] Li A., (2014) Design driven economic change-the rise and challenges of Chinese design industry [M]. China machine press.

[4] Luo J.Y., (2018) A modular method of adaptability-oriented product family [D]. East China Jiao Tong University.

[5] Ji H., (2012) Product platform design services based on cloud computing[D]. China academy of machinery science and technology group Co., Ltd.

[6] JS Fan, SH Yu, et. al. (2019) Multi-objective creative design evaluation method for industrial design cloud service platform[J], Computer integrated manufacturing system, 25(01):173-181.

Research on innovative design of lamp mix-and-match based on "sense of dissonance"

Cailian Wang[1], Jingran Wang[1], Lin Gan[1]and Zhe Gao [1]

[1] School of Art and Design, Wuhan University of Science and Technology, Wuhan, Hubei, 430065, China

E-mail: 5318973@qq.com

Abstract. With the development of the times, the lamp design bring forth the new through the old every year. "Sense of dissonance" is mentioned in the book of DaiSato, but the reason for its formation is not explored in depth. Based on the induction and analysis of DaiSato's related works, this paper defines the "sense of dissonance ", and studies the design trend of the lamps in the Milan design exhibition as the research object, so as to conclude that the main reasons for the "sense of dissonance" come from the mixing of materials, modelling, usage and experience. The design process of light fixture mix and match based on "sense of dissonance" is divided into the following four steps: defining the design content of "sense of dissonance ", determining the narrative design theme, determining the style, and completing the modelling. Following this step, a lamp design for 《over the rainbow》 is completed.

1. Introduction

The design of lamps has developed rapidly and is becoming more mature. The function of the lamp is simple, the form of the lamp is increasingly valued, and the lamp design tests the uniform processing of the form and function by the designer. The PH series lamps designed by Danish designer henningsen solve the unification problem of lamps in form and function, which is still regarded as a classic work nowadays. The tray table lamp designed by Japanese designer Naoto opens a new way of thinking in lamp design[1]. The design of lamps has become very mature. The design of lamps has begun to pay attention to the interactive relationship between lamps and the environment and people. More and more designs have started to transform from the level of function realization to the level of psychological experience.

Sato, the founder of Japanese design studio Nendo, once put forward that people tend to have a self-evident joy when they feel a little"sense of dissonance"[2]. The great abundance of lamps promotes people's aesthetic pursuit from the general inertial aesthetic development to the aesthetic surprise level. The innovation of lamp design has diversified perspectives, and product design with "sense of dissonance" has become a new direction of design. Therefore, it is of certain significance to study how to create "sense of dissonance" in design. This paper will carry out an innovative research on product design process from the way of creating "sense of dissonance" and taking lamps design as an example.

2. Definition and exploration of "sense of dissonance"

2.1 Definition of "sense of dissonance"

IOP Conf. Series: Materials Science and Engineering **573** (2019) 012034 doi:10.1088/1757-899X/573/1/012034

The design of "dissonance" is the result of mashing and merging the same or similar semantics when the original design semantics are unchanged. It will bring new experiences and surprise without affecting the use experience[3].

People generally think that the functional semantics of a product is as pure as possible, which is the characteristic of most products. Sato said that some "non-daily" things in daily life are rich and profound. In Sato's design, works with a sense of "dissonance" can be seen. Many of these works are mashed up and designed to mix and match the same or similar shapes or functional semantics. The "disappearance "brings a different sense of difference and can bring small surprises to the same old boring life

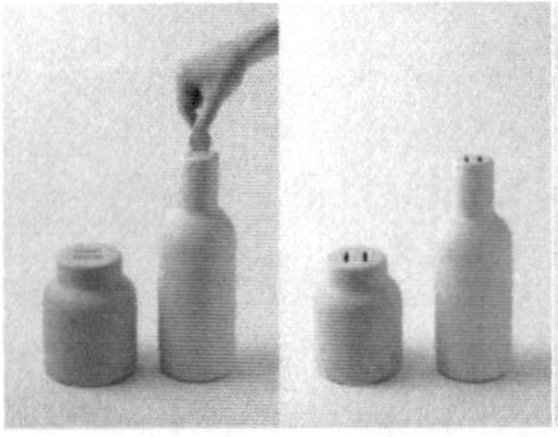

Figure1 Piggy Bank

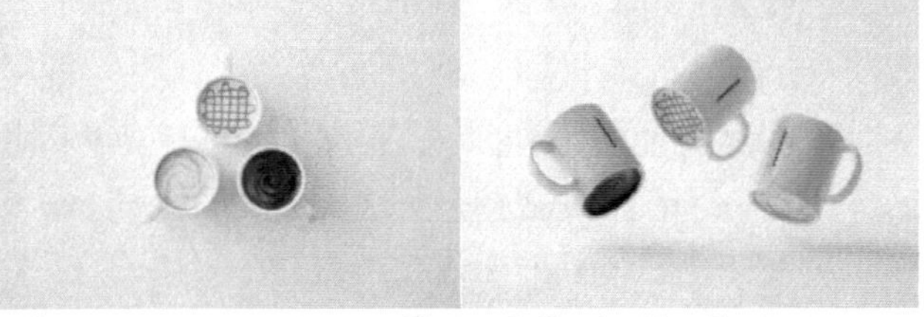

Figure2 Starbucks Cup

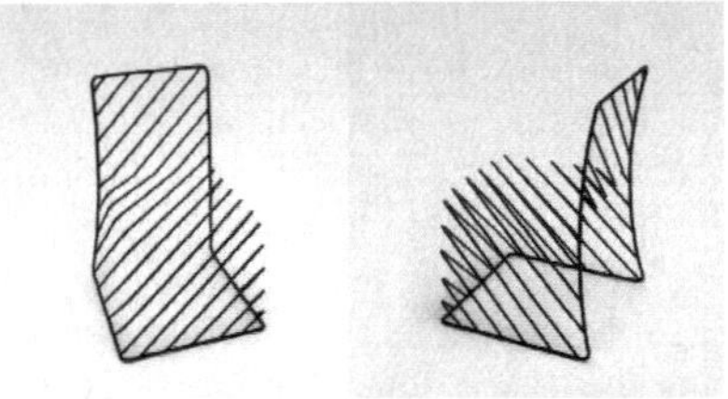

Figure3 "Two-dimensional"chair

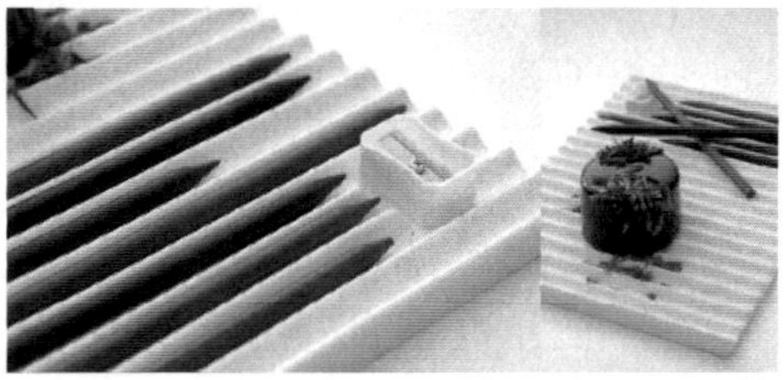

Figure4 Chocolate pencil

The "pig nostril" slot design in the piggy bank in Fig. 1 makes use of the similarity in the shape of the slot and the nostril, as well as the contrast in function and semantics, so that people will be surprised by the "sense of dissonance " every time they put money in the slot When the mug of Fig. 2 is inverted, the coffee pattern at the bottom of the mug can be mistaken as a cup of coffee. The surprise hidden in the bottom of the cup emphasizes the function of the coffee cup and also fully exerts its function. The "unused value" of the product can also reflect the visual beauty when not in use. The chair of Fig. 3 is formed by bending a set of parallel lines, and the planar vision of "two-dimensional is applied to the three-dimensional shape, so that the generated optical illusion also makes the product more playful than the practical function. The design of Fig. 4 is to apply the method of pencil sharpening to the action of cutting chocolate shavings. By using the 'mix and match' in this way, it is possible to give people a more interesting experience that is different from the traditional way.

2.2 "mix-and-match" creates "a sense of "dissonance"

Through the analysis of Nendo studio design works, it is necessary to find the same or similar factors in the same or different things in order to create a product with "sense of dissonance". Through the organic combination, the product semantics are not affected, the creation of "a sense of dissonance" can be achieved, so that the product has the possibility of creating "surprise".

The mix design creates new value by connecting two seemingly unrelated things. Mix-and-match is not just the mix of materials or external forms of styling, but also the mix of usage and experience[4]. The conditions for creating "mix and match" are not only common conditions with direct connection, but also non-common factors that seem irrelevant but have indirect relationships. Thus it can be seen that the mix design creates a "sense of dissonance". The mix-and-match design can be summarized into the following levels:

(1)Mix-and-match in material: It can be a crossover of materials in different fields, or a mix of two or more materials, exploring new possibilities of product forms, creating new product textures;

(2)Mix-and-match in modelling: natural or unnatural shape transplantation to achieve better human-computer interaction or fun experience; (3)Mix-and-match in using process: By changing the original way of using the product and using the new action to complete the product interaction, the purpose of more humanized or deeper experience can be achieved.(4)Mix-and-match in using experience: By changing the process of product experience, multi-level mixed experience is adopted to endow products with new experience, new cultural context and other higher-level design purposes.

3. Research on the mix-and-match design of lamps

At the annual weathervane Milan furniture design show, all sorts of lamps and lanterns design emerge in endlessly in order to achieve a more diverse and innovative ideas, crossover design method is adopted by many designers, In the design of lamps, not only the mixture of materials and shapes is adopted in the form, but also the way of use and the experience of use are mixed in the man-machine interaction relationship of products.

3.1 Mix-and-match in material

The shape and material of the lamp in different environments often directly affect the relationship between the lamp and the environment. The designers use different materials to mix and match or crossover to present the different textures of the lamp. The "Ghost" lamp (Fig. 1) designed by fashion designer Tsumura Yuki applies hair to the design of the lamp through the hair styling technology, which makes the lamp produce a new visual and tactile experience. The design of the "unexpected design" studio of the Chinese independent design brand, The half-mountain lamp (Fig. 2) combines the wood trim and the resin. After cutting into a sheet, each panel forms a different landscape-like picture. This difference also makes each product 'accidental'; Figure 3 is a lamp made by an Israeli designer using sand from the desert; it uses the sand that is considered useless or even harmful. Sand softens the light. In recent years, cement +X materials is also very popular. The cement and glass mixed in Figure 4 is also the same. Cement is a constructional material; its plasticity during forming and the stability after moulding are also favourable factors in product design. In the design of household items, various designs can be made of cement.

Figure 5 "Ghost "light

Figure 6 Half-mountain lamp light

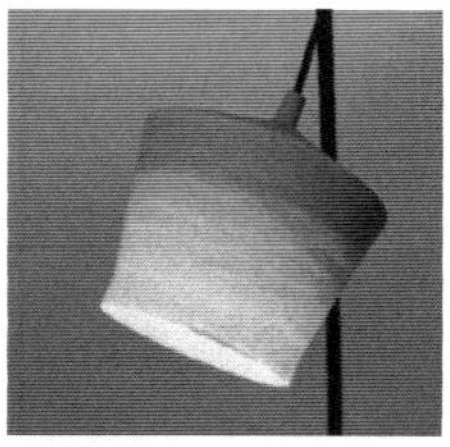

Figure 7 Sand light

Figure 8 Cement light

3.2 Mix-and-match in modelling

The diversified demands of the design have made the diversity of modelling of lamps. Many product shapes that were not originally included in the lamp have also been introduced into the lamp design by the designer ——it does not seem to be a lamp, It doesn't look like a lamp, but it can be illuminated, and the sense of "dissonance" brought by styling and mixing brings surprises. In Figures 9-11, the styling of books, balloons and water droplets is ported to the lamp shape for fun user experience.

Figure 9 Book lamp

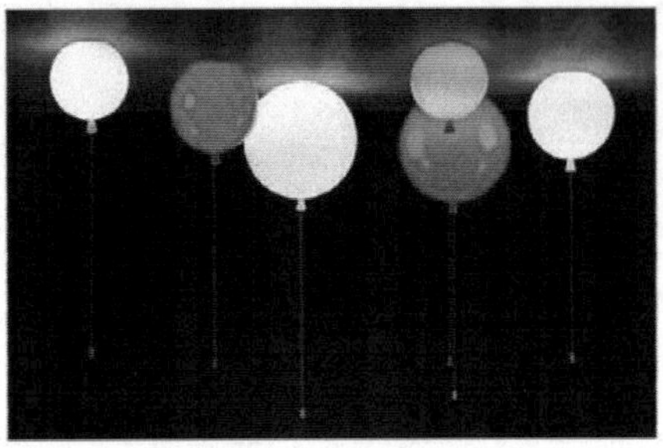

Figure 10 Balloon lamp

Figure 11 Water drop lamp

3.3 Mix-and-match of usage and experience

The usage and experience are a process of interaction. Changes in usage may lead to changes in the usage experience, and changes in the usage experience may also lead to changes in usage. This kind of mashup will make the lamp more interactive in use, and it is easier to reach Patrick's highest requirements in the three levels of design, that is 'happy'. A good design makes people smile[5].

The main function of lamps is to give off light, but if you change the way of turning on the lamp, it is possible to add a more interesting and more user-friendly interactive experience to the lamp design. The tray table lamp (Fig.12) is designed by Naoto Fukasawa. when you return home ,You may put the key into the tray, in the meanwhile, the light is on because of the gravity sensor to turn on the desk lamp. So it gives the user a smoother and more convenient experience. The "Heng" balance lamp of Figure 13 combines the concept of "balance" in Chinese culture to create a new way of opening the lamp. The two balls must be placed together to light up, and the balance of the ball also keeps people calmly and patiently. The "Inflation" lamp Figure 14 shifts the action of the inflation to the switch of the lamp, changing the way the lights are turned on, and increasing the fun of the usage.

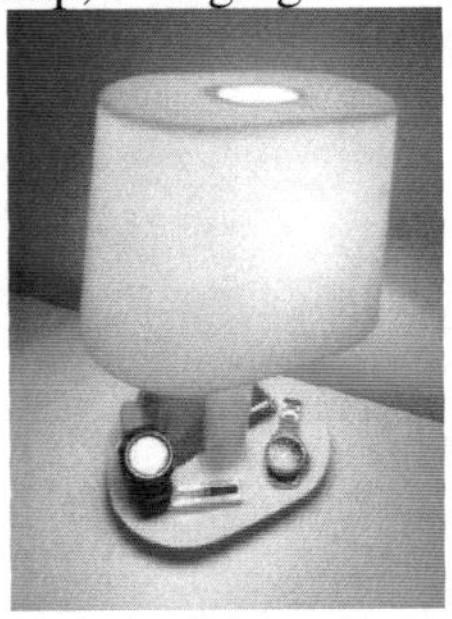

Figure 12 Tray table lamp

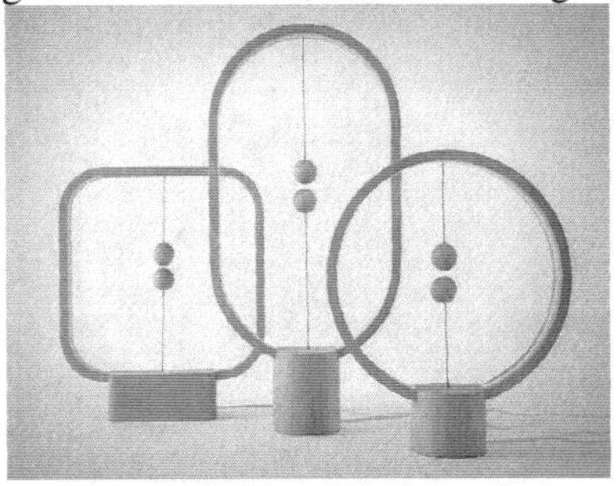

Figure 13 "Heng" balance lamp

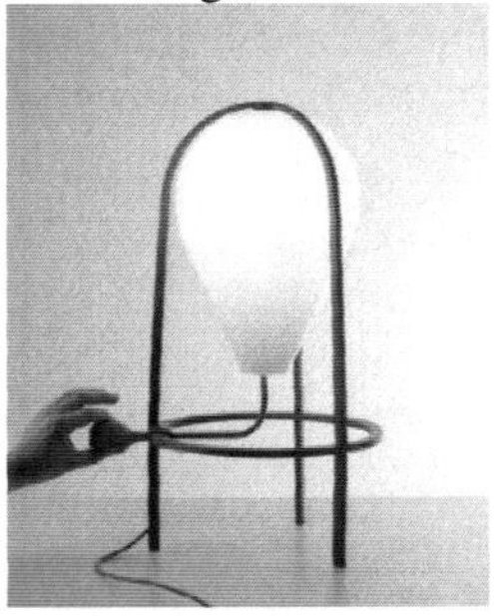

Figure 14 "Inflation" lamp

Lamps can not only provide lighting, but also create a sense of context by the shadow of it, which can arouse association of people. Lamp becomes the partner of the person. It is a lamp, with the minimalist appearance in figure 15. When the light is on, the shape of the light cast by a lamp is like a lampshade.

The collocation of virtual and real forms interesting visual completion, the fish shadow lamp in figure 16 makes the fish swimming in the light and shadow. It brings some Zen. The bulb in the "woops! " lamp[7] designed by the YUUE design studio falls down when you pull the switch rope. You perhaps think it is broken, but in fact, it is a great surprise of the experience.

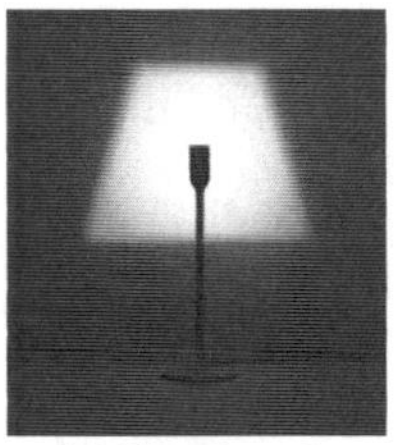

Figure15 Shadow lamp

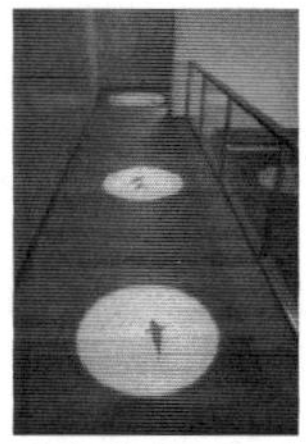

Figure16 Fish shadow lamp

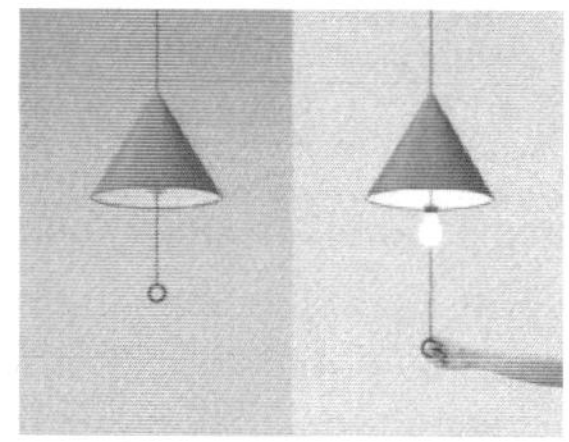

Figure17 woops! lamp

4. Innovative design process and examples of lamps based on "sense of dissonance"

In the innovative design process of lamps, four feasible design steps are designed from defining the content of "sense of dissonance ", determining the narrative design theme, determining the style and completing the shape. The following is the design process of lamps in 《over the rainbow》.

4.1 Looking for "sense of dissonance"

In daily life, the lights generally show a single colour of light, usually white, warm white, and cool white, it doesn't appear rich in colour or abnormal in colour. In physics, white light is a compound light which consists of seven light colours. White light forms a seven-colour spectrum through refracting by triangular prism;

In the design of lamp, the principle of dispersion in physics is used, and seven-colour light is produced by means of prism refraction. The seven-colour light and ordinary lights can produce a sense of dissonance (Fig. 18).

4.2 Design theme determination

After determining the expression form of "sense of dissonance", theme design should be carried out for this form to obtain emotional resonance through narration. Whenever the rain passes and the sky clears up, people always feel surprised and excited when they see a rainbow in the sky. All of them are exclamating the colourful nature and even the bad mood is swept away (Fig 19). A rainbow always brings warmth and strength. The theme song "Over the Rainbow "of the movie "The Wonderful Wizard of Oz" sings the imagination of the rainbow and gives the rainbow a deeper meaning[8]: "Somewhere above the rainbow, the sky is blue. As long as you dare to dream, it will be achieved".

Combining the artistic conception of rainbow given by this song, the image of rainbow is applied to the lamps that will be used every day, and such love and hope are injected into daily life, hoping that users can sweep away the haze every day and be full of power to move forward.

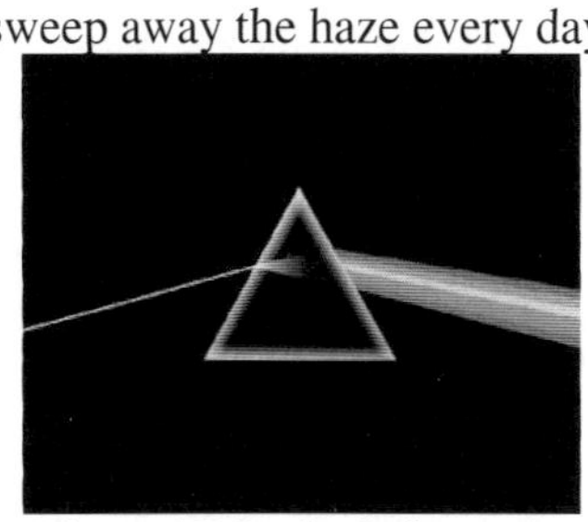

Figure18 Optical dispersion

Figure19 Rainbow

4.3 Lamp modelling style determination

According to the current design trend, combined with the warmth of the design theme and text, the design of the lamp should be simple, bright, and fresh and natural. In order to make the rainbow more clearly displayed, the overall shape is divided into two parts: the light box and the light blocking plate. The light blocking plate is the entire blank panel as the projection surface of the rainbow, but the volume and material of the panel need to be repeated in the specific design process. Resolving its coordination, the whole part of the styling is harmonious and unified.

4.4 Design process and results

Firstly, according to the concept of "dissonance" and the design method of "mix and match", we will find more design points in order to get a better way to achieve the design; then, according to the determined design positioning, the attempt and scrutiny of modeling will require simple geometry. The shape of the graphic, and finally the final shape and structural details are determined according to the shape and function.

The light box of the "Over the Rainbow" lamp is made of solid wood, giving it a fresh, natural and warm texture. The light source and the prism are hidden in the middle groove, and the light on one side of the groove is refracted by the prism to project the rainbow on the circular resin panel, On the other side is the white light used for lighting, which produces the contrast between the original colour light and the rainbow light, creating a "sense of dissonance", and connecting the "narrative" created and designed in the movie the wizard of Oz, creating an experience scene. The mix-and-match design on the lighting brings an interesting "sense of dissonance" and creates an emotional connection with the designer.

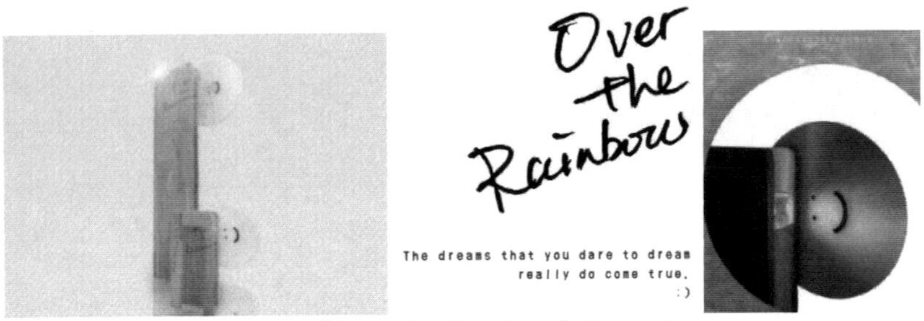

Figure20 《Over the Rainbow》 lamp design

5. Conclusions

From the perspective of "sense of dissonance", this paper puts forward that the sense of dissonance comes from different mixing ways of product elements, and makes an empirical analysis of lamp cases, combining with the characteristics and design trend of lamps, to complete the design of the example. Design innovative solutions for light fixtures. At the same time, the innovative process design can also provide a reference for other types of products.

Acknowledgements

Fund project： Innovation and entrepreneurship training program for college students in hubei province / General project of social science research of hubei provincial department of education
Project No.： 201510488046/2012Y088

References

[1] Huang Yanni. Return to the origin of product design -- a brief analysis of the design thought and practice of Naoto Fukasawa [J]. Decoration,2008(11)
[2] DaiSato, Looking at the world from the inside out [M]. Beijing: Beijing times Chinese book company.2015.3.2016.6
[3] DaiSato. DaiSato: Solving problems with design [M]. Beijing: Beijing times Chinese book company.2016.6
[4] He Jinghao. Discussion on the mixed use of materials in today's product design [J]. Art review, 2015(10):105-108.
[5] Yin Xianghua. Research on interest in interface design [J]. Fine arts, 2018, no.365 (05):104-105.
[6] The Red Dot award-winning work The Heng Balance Lamp[J]. Modern decoration, 2017(4):191
[7] PHOTO-YUUE. YUUE: explorer [J]. Design, 2016(10):46-53.
[8] Wu lan. Music in "the wizard of oz" [J]. Film review, 2015(9):96-98.

IOP Conf. Series: Materials Science and Engineering **573** (2019) 012035 doi:10.1088/1757-899X/573/1/012035

Sustainable Design of Cultural Creative Products Based on Museum Cultural Derivatives

Yuyao Zhang

College of engineering, Ocean University of China, Qingdao, Shandong, 266000, China

shizhangyuyao@163.com

Abstract. China's tourism industry has developed rapidly in recent years, but also promoted the rise of cultural and creative industries. However, due to the immature development of cultural and creative industries and inadequate market supervision, there are many deficiencies in cultural and creative products. At the same time, the quality of Museum souvenirs is uneven, and the utilization rate of resources is low. Given this phenomenon, the author puts forward the method of "PREACH March," and applies this method and sustainable design theory to the development of creative products in the Terracotta Warriors Museum. Taking the Terra Cotta Warriors and Horses Museum as an example, this paper analyses the defects of the existing souvenirs, refers to the design of the cultural creation products of the Palace Museum, redesigns the souvenirs of its museum sustainably, and solves the problem of how to make the museum souvenirs no longer be placed. "PREACH March" is applied to the design of the museum souvenirs of Terra Cotta Warriors and Horses, and a sustainable and recyclable bookmark is successfully designed. The purpose of this paper is to put forward a practical method for the design of cultural and creative products in the future, and to guide future design to follow the concept of sustainable development.

1. Introduction

1.1 Research Background

Chinese culture profoundly contains many precious treasures. In recent years, Chinese culture has caught worldwide attention, and gradually moved towards the world stage. No matter whether it was in the heritage art auction market, or the drawn films, it is gradually catching people's eye. From a product design view within the global competition market, the use of cultural features design application, not only improve the uniqueness of products, but also better provides an emotional consumer experience. It is evident in developed countries of the importance of highlighting the cultural and creative design innovations for a competitive advantage. In terms of cultural and creative industries mixed with the consumer process in modern culture, functional product development will gradually focus more on spiritual values (Chen, 2009).

Accompanied by the continuous improvement of people's living standards is that people's spiritual demand has been far exceeding their demand for food and clothing, which has triggered more concern and emphasis on culture, while the creative cultural products are also gradually known by the public. As an emerging industry in China, most of the products of the creative cultural industry are still demonstrating immaturity and incompleteness in spite of being immersed in thousands of years of cultural heritage and vast market.

1.2 Research Content and Scope
This paper mainly consists of four parts. Elaboration on the topic from different aspects is carried out in the first three chapters, while the specific design practice is conducted in the fourth chapter, in which the sustainable concept and existing museum souvenirs are combined and applied. The research drew on the background of design for sustainable behavior, a relatively new field of inquiry concerned with reducing the social and environmental impacts of products during their use. Despite the increasing development of theories to change user behavior through design, there is a lack of understanding of how different cultural contexts affect behavior.(Spencer, J,2014).

1.3 Research Purpose and Significance
Museums have been defined by the International Association of Museums as non-profit organizations. The core connotation of such characterization definition is to emphasize the non-profit essential nature of cultural and exposition units, instead of prohibiting such institutions from obtaining certain economic returns through opening up services, providing multi-cultural products and establishing other business entities to effectively apply to make up for the shortage of operational funds. The development of museum undertakings will also be promoted by profits out of the research and development of multi-category duplicates, audio-visual products, knowledge reading materials,and other cultural products by focusing on the historical information and aesthetic value of cultural relics. The successful design of museum cultural derivatives will for sure promote the exchange between museums and people’s cultural life at present, as well as attract more people's attention, visits and feelings to museums.

Therefore, this topic attempts to explore the development and design of museum cultural derivatives from the perspective of sustainable concept, with the view of constructing a sustainable recycling and utilization model of museum cultural derivatives.

1.4 Research Methods
Considering the unique cultural nature of museum cultural derivatives as well as their common aesthetics and commerciality with other cultural creative products, a set of "PREACH Methods" has been summarized in the research process of this paper, which shall be elaborated as follows:

The "PREACH method" includes six steps: "P", "R", "E", "A", "C" and "H", in which "P" stands for purport; "R" refers to "reside" for geographical location; "E" refers to "era" for age or time; "A" refers to "author" for author or person related to it, "C" refers to "culture" for cultural background; and "H" refers to "history" for the evolution of history. This method can be widely applied in design. The significance of design work should firstly be understood upon researching it, while the deep meaning of the work can be known through full investigation and understanding of its historical, social or cultural significance. The study of geographical location is mainly applicable to the architectural design, while product design does not require an in-depth understanding of the geographical location. At the same time, the era of the birth of this design work also has to be understood. As heroes are created by the times, so should design. Understanding the author of the work is conducive to the establishment of an understanding of design techniques and styles. Also, cultural background and historical development also play an essential role in design work. A complete understanding of a work can be achieved through the study of its significance, geographical location, era, author, cultural background and historical development.

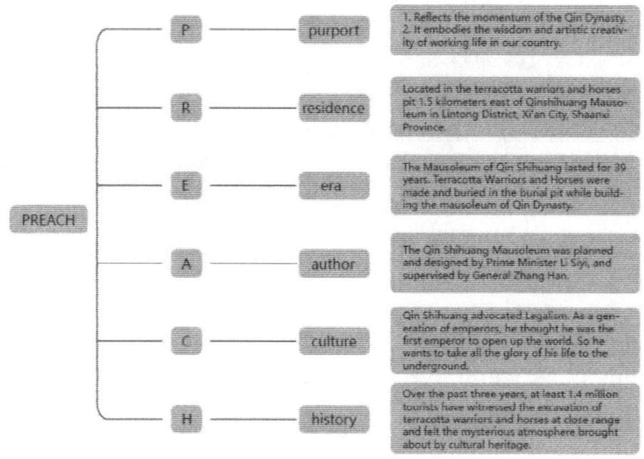

Fig. 1. The "PREACH method."

Other major research methods are as follows:

(1) Literature collection. Detailed study of the relevant content of this topic has been carried out and some literature materials have been sorted out and summarized by reading a large number of literature on domestic and foreign cultural creative products, museum cultural derivatives and sustainable design, so as to provides powerful arguments for the contention of this paper through the analysis of previous experience.

(2) Field research. The existing problems of cultural derivatives in Shaanxi Museum are summarized through the investigation of Shaanxi Museum in the early stage, to effectively promote the innovative redesign.

(3) Induction and deduction. The results of case analysis and field research are collated, summarized, analyzed and compared, to conclude the latter part of the paper.

(4) Design practice. The ideas presented in this paper are validated through the final innovative and redesign.

2. Research on the Current Situation of Museum Cultural Derivatives

2.1 Meaning and Categories of Museum Cultural Derivatives

Considered as powerful “messengers of meaning,” tourist souvenirs help consumers to maintain material links with cherished past experiences and to convey individual and cultural meanings to their broader existence. (Alain, D. and Julie, M.,2014)Cultural assets cover a wide scope. It can be summarized into two areas: tangible cultural and intangible cultural. The first part includes historic buildings, natural landscape, cultural landscape, floral, fauna, regional life objects, etc. The second part includes things such as Tang poetry, Sung lyrics, classical literature, religions, customs, festivals, ritual, and behaviors. Lee (1996) considers culture as referring to “products of human activities, including tools, social life used to maintain the laws and institutions that depend on spiritual life all the art products, but also includes many the human activity history in the creation process.” Highlighting the presence of the cultural characteristics, from the design side, the elements of national culture, such as symbols, lines, stories, shapes, and colors help to explore the product design and give information. Therefore, the consumers, in addition to the actual use of its functionality, can sense the designer through the commodity characteristics through desired shapes conveying the culture, to create a moving and common of emotional interaction(Chang, C.2015).

Compared with museum cultural products as a broad concept category, museum cultural derivatives are considered as a narrow concept category, while cultural derivatives belong to one of the various categories of museum cultural products. Cultural derivatives are an industry that plays a crucial role in the development of museums. Unlike ordinary souvenirs, cultural derivatives are

IOP Conf. Series: Materials Science and Engineering **573** (2019) 012035 doi:10.1088/1757-899X/573/1/012035

usually designed based on the tenet of museums, the concept of museums and the elements of exhibits. In spite of assuming the same function as souvenirs, they demonstrate a quite different nature from the pure commercialization of souvenirs and are more like artistic commodities due to the addition of cultural symbols and connotations. Museum cultural products in the popular sense refer to the unique cultural and memorial commodities related to the collection (or exhibition) of the museum which are displayed in the museum stores and the cultural and commemorative special commodities related to the cultural characteristics of the museums, which is regarded as a crucial part of the museum cultural industry and carries historical and cultural information related to the theme of the museum, while certain cultural connotations, local characteristics and artistic tastes are also endowed.

2.2 Commercial Model of Museum Cultural Derivatives
By the survey, the development of foreign museum stores has been far more developed than that of domestic museum stores. Standard museum stores are set up by most museums abroad, while except for the self-owned museum stores within the store, the corresponding shops online are also widely established on the Internet. In Singapore and the United States, for example, shopping platforms consisting of several different museum stores have even been constructed. Regardless of the size of stores or the degree of souvenir development, most foreign museum stores have been on the right track and demonstrating favorable operating conditions. In spite that several museums in China have also set up "Service Department," "Commodity Department" or "Cultural Relics Store," they are mostly engaged in small tourist commodities as well as food and beverages. Essentially, they can only be regarded as "snack stores" in terms of both interior decoration and products they are engaged, rather than museum stores in the real sense. Also, their operation is generally unsatisfactory. Except a few large museum stores (such as the Shanghai Museum and the Capital Museum which attract more visitors), most of the museum stores are in poor business. Some museum stores have not even been entered by even one visitor during the period of my observation because of their remote location and inadequate visible signs, which is in sharp contrast to a large number of visitors to the museums.

2.3 Existing Problems of Museum Cultural Derivatives
Tourist souvenirs are a typical example of special possessions that may convey deep meanings to tourists' lives(Alain, D. and Julie, M.,2017). Despite that the academic research results of museums are undeniable as the academic research institutions, they can only introduce local tourist souvenirs as substitutes due to the development of derivative souvenirs with the characteristics of the museum has encountered greater difficulties. Some tourist souvenirs are currently sold in museum stores, which present serious homogeneity. Although they may be attractive to foreign tourists who are new to China, it is ineffective for "knowledgeable" domestic tourists, which means that a huge market has been abandoned. By comparing the development of museum cultural products at home and abroad, it can be found that the development of derivatives in China's museum cultural products is rare, while they are mostly not innovative and assume a low level of design even if there are derivatives, such as simply adding the name or logo of the museum. Given that most of the existing museum cultural products in China are low-end products wholesaled from the small commodity market, they demonstrate rough appearance and poor quality, which not only fail to stimulate the purchase willingness of tourists, but also result in the impact of the national image to a certain extent.

3. Research on the sustainable design concept of cultural and creative products based on museum cultural derivatives

3.1 Definition of a sustainable design concept
The design is the process of understanding the world and articulating an alternative conception of how it should be shaped, according to the designer's intentions. Through design, we cause change and shape our environment. If we don't take sustainability into account when designing, no matter in which domain and for what purpose, we miss the opportunity to cause positive change. We recognize

that there is a rapidly increasing awareness of the fundamental need and desire for a more sustainable world, and a lot of genuine desire and goodwill - but this alone can be ineffective unless we come to understand that: There is a narrow perception of sustainability that frames it as protecting the environment or being able to maintain a business activity. Whereas as a systemic property, sustainability does not apply simply to the system we are designing, but most importantly to the environmental, economic, individual, technical and social contexts of that system, and the relationships between them(Becker, C et al.,2014).

From the concepts universally recognized in the world, we can sort out the rich connotation of sustainable development in the following aspects:

The earth is a complex giant system. Every country or region is an inseparable subsystem of the giant system. The most fundamental characteristic of the system is its integrity. Every subsystem interacts with other subsystems. As long as one system has problems, it will directly or indirectly affect the disorder of other systems, and even induce the overall mutation of the system, which is most prominent in the Earth's ecosystem. Therefore, sustainable development pursues overall development and coordinated development, that is, common development.The coordinated development includes the overall coordination of the three systems of economy, society and environment, the coordination of the world, the country and the region at three spatial levels, and the coordination of the economy and the population, resources, environment, society and all levels within a country or region. The sustainable development originates from the coordinated development.

3.2 The development of a sustainable design concept

Earlier, the concept of sustainability was first proposed by ecologists, which is called ecological sustainability. It aims to illustrate the balance between natural resources and their exploitation and utilization. In November 1991, the International Ecology Association (Intecol) and the International Federation of Biological Sciences (Cubs) jointly held a symposium on Sustainable development. The outcome of the seminar not only develops but also deepens the natural attributes of the concept of sustainable development, which is defined as protecting and strengthening the production and renewal capacity of environmental systems. Defining sustainable development from the concept of the biosphere is a representative of defining sustainable development from the aspect of natural attributes. That is to say; sustainable development is to seek the best ecosystem to support the ecological integrity and the realization of human aspirations, to make the human living environment sustainable. Culture creativity is a new factor for economic development; a country's development is not only dependent on its scientific and technological capability anymore, but on the ability to create local culture, lifestyle and taste and to add value with a combination of art, and business. Furthermore, in the era of an aesthetic economy, a product must possess charm, beauty and an emotional and pleasant element to stand out from the other competitors on the market(Chi-Hsien and Wang, C.,2015).

3.3 The application of the concept of sustainable design in cultural derivatives of museums

The application of the concept of sustainable development in creative products can be considered from the following aspects:

1. Environmental perspective. Tourist souvenirs should be strictly controlled in production to reduce environmental pollution and waste of materials. Also, in the process of transportation, environmental resources should also be saved, and souvenirs should be transported to museums in the most environmentally friendly way.

2. Practical angle. After purchasing souvenirs home, tourists should repeatedly use souvenirs to give full play to the value of products. If only as a handicraft placed at home as an appreciation, it loses the significance of sustainable development.

3. Economic Perspective. Tourism souvenirs in line with the concept of sustainable development should be recyclable. After buying the product home, tourists can not only use it for themselves, but also give it as a gift to others. When the product is used many times, it can still play its final value, that is, recycle it. Such souvenirs truly carry out the concept of sustainable development.

3.4 The significance of sustainable design concept

Sustainable development first advocates maintaining the progress and development of human society from the perspective of environmental protection. It calls on people to pay attention to the protection and improvement of the ecological environment while increasing production. As soon as the term "sustainable development" was put forward, it gradually gained recognition in the world and became one of the most frequently used words in mass media. This reflects the suspicion and abandonment of the development path that mankind has gone through before, and also reflects the longing and yearning for the development path and goal that mankind has chosen in the future (although there are still some ambiguities). It is gradually recognized that the path of development in the past is unsustainable, or at least unsustainable, and therefore undesirable. The only alternative is sustainable development. This reflection of human beings is profound, and the conclusions drawn from this reflection have epoch-making significance. This is the fundamental reason why the idea of sustainable development has gained consensus and universal recognition in countries with different economic levels and cultural backgrounds all over the world. Sustainable development is a goal that both developing and developed countries can strive to achieve. The vast majority of developing countries are actively engaged in the practice of sustainable development, which is also an important reason why the theory of sustainable development is popular all over the world.

4. Research on the sustainable design method of cultural and creative products based on museum cultural derivatives

4.1 Material selection for sustainable design

In the process of material extraction, preparation, production, use and waste, a large number of resources and energy are often consumed, and a large number of pollutants are discharged, resulting in environmental pollution, so the selection of recyclable materials should be worthy of attention by designers. When selecting materials, the following conditions shall be met:

(1) Energy saving: materials can reduce the energy consumption of a system. Through better performance to improve energy efficiency, that is to say, improve the material performance can reduce energy consumption, to achieve the purpose of energy saving.

(2) No pollution: the materials used shall be easy to process and free from pollution or minimum pollution during processing. Also, the selection of material types should be less the better in-depth analysis of the material processing process, the choice of easy processing and processing without pollution or less pollution material is to ensure the sustainability of the product requirements, reduce, control the manufacturing process pollution guarantee. That is to say; material selection should make products produce the least waste in the manufacturing process, and the least environmental pollution.

(3) Ease of disassembly and reuse: This is from of easy disposal of discarded products. At present, the disposal of waste commodities in the world has attracted more and more attention. Sustainable design can prevent the depletion of the earth's resources and effectively protect the environment. Easy recycling is crucial to the recovery of products.

4.2 Sustainable Design of Processing Technology

Sustainable design of processing technology can be considered from two aspects: the process of the product itself and the packaging process of the product. In the production process of products, the processing technology with lower cost and fewer consumables should be used. In packaging, the utilization and recovery of paper and the use of ink in the printing process should be taken into account.

4.3 Sustainability Design of Usage Mode

Most of the existing cultural derivatives are ornamental crafts or souvenirs. After purchasing, tourists either leave them at home or give them as gifts to others. However, whether placed at home or given to others, the end of the product is ashes or even thrown into the garbage can, no one cares about it.

Therefore, the use of creative products should play as many different roles as possible, rather than just as an ornamental. For example, in the cultural derivatives of the Palace Museum, there are many usable items, such as pillows, chopsticks, earphones and so on.

4.4 Sustainability Design of Business Model

Zukin (1995) emphasized that culture is not only capable for propelling the industries in terms of the economic aspect, but can also be used as the basis of the economy, providing product concepts for an entire line of products; the relationship between culture and economy has transformed from traditionally contradictory to mutually beneficial. Lash Urry (1994) had the same idea, indicating that a mutual beneficial relationship exists between culture and the economy; culture is presented as merchandise because of the economy, and economy is enhanced through culture in terms of beautification and embedded connotation, and elevates life quality and cultural recognition (Liang and Zhang 1995). Product design plays the role of cultural messenger in cultural economy, delivering cultural messages to the consumers through symbolic cultural codes and semantic con-version, such cultural merchandise also allows the consumers to foster affection toward local culture and boosts the value of culture(Chen, C.,2016). So the sustainability of the business model is particularly important.

5. Practical Application of Sustainable Design Method in Creative Products

5.1 Example analysis—the palace museum

The Palace Museum, as the representative of Chinese museums, has achieved great success in its creative products because it has found a way to develop creative products that conform to its characteristics. The design, production, and sale of creative products in the Palace Museum are based on the deep excavation of cultural connotations. Every process permeates the advanced concepts and painstaking efforts of the Palace Museum. Out. The development of creative products in the Palace Museum is far ahead of other domestic museums, but we must still maintain a serious attitude, keep pace with the times, innovate constantly, and achieve quality, quantity and sustainable long-term development in every link of research and development, production and sales.

5.2 Market Research on Tourist Souvenirs of Terracotta Warriors and Horses of Qin Shihuang

Shaanxi's tourism industry has developed rapidly in recent years. Terracotta Warriors and Horses, as one of the popular scenic spots in Xi'an, are very popular with tourists. Despite the increasingly perfect tourism industry, there are still many shortcomings in cultural derivatives. The author visited some merchants selling around the terracotta warriors and horses, and found that the existing cultural derivatives are mainly small copper alloy ornaments shaped by terracotta warriors and horses, followed by bookmarks, key chains and imitation bells. These derivatives are scarce in variety, single in form and lack of innovation. Also, in the processing technology, the quality of these derivatives is uneven, rough workmanship, there is no uniform price, and the market is chaotic.

Fig. 2. Existing terracotta museum tourist souvenirs.

5.3 Practical Application of Design Method for Tourist Souvenirs of Terracotta Warriors and Horses of Qin Shihuang

Tourist souvenirs not only reflect the local government's emphasis on culture, but also reflect the country's protection and respect for history. Therefore, by increasing the connotation of cultural derivatives to reflect the culture of the Qin Shihuang period, to improve national cohesion and national pride. To meet the concept of sustainable development, the impact of resources and environment in the production process is offset by increasing the use of products. For tourist souvenirs, the main purpose is to increase the types of practical tourist souvenirs, so that tourists can use the souvenirs to offset the impact of the production and transportation of the souvenirs, to achieve sustainable purposes.

Example: Bookmarks

After Qin Shihuang unified China, Xiaozhuan carried out the policy of "book with text, car with a track" and unified measurement and weights. It was under the responsibility of Prime Minister Li Si. Based on the original seal script used in Qin State, Xiaozhuan simplified, canceled the other six languages and created a unified writing form of Chinese characters. It was popular in China until the end of the Western Han Dynasty (about 8 A.D.) before it was gradually replaced by Li Shu. But because of its beautiful font, it has always been favored by calligraphers. Because of its complex strokes and ancient forms, and can add twists and turns at will, seal engraving, especially the need for anti-counterfeiting official seals, has been using seal books. The copper alloy is selected from the material, which has good chemical properties and corrosion resistance, is not easy to be damaged and has a long service cycle. The combination of a small seal and a bookmark not only enables the buyer to reuse the script, which is in line with the sustainable design, but also enables the visitor to have a deeper understanding of the qin culture.

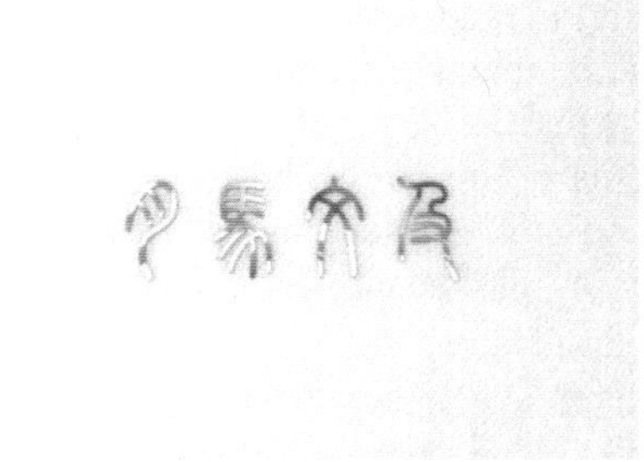

Fig. 3. Bookmark design.

6. summary

This paper summarizes the examples of cultural derivatives of Xi'an Terracotta Warriors and Horses Museum, and tries to apply them to other museums.

This paper mainly discusses the cultural and creative products based on the cultural derivatives of the Terracotta Warriors and Horses Museum in Xi'an. For other museums which may have different situations with the Terracotta Warriors and Horses, it can only be used as a case reference. Also, our on-site interviews and tests at the Terracotta Warriors and Horses Museum are limited, and the results do not represent the situation of all visitors to the Terracotta Warriors and Horses Museum. The concept of sustainable design is still in its infancy in our country. Enterprises don’t have a deep understanding of sustainability. However, in such circumstances, the souvenir market has less awareness of sustainability. It is hoped that the design of cultural derivatives will be of inspiration to other museums in the future. In the new era, the public's demand for material and spiritual consumption has become more and more intense. The purpose of visitors visiting museums is no longer limited to the view of exhibits and exhibitions. People's demand for consumption is derived from the preservation of museum visits and memories. As an important value-added component of Museum Cultural system, creative cultural products have the functions of disseminating culture, enhancing cultural confidence and classics. The important role of inheritance is similar to that of museums, which use their excellent cultural essence to gather advantages and make cultural innovations applied to the extension and expansion of museum functions, and provide better service for people with "sustainable design." While creating sales revenue, it also plays an important role in propaganda, inheritance, and promotion of traditional culture and Museum image, thus forming a virtuous circle for the whole development system of cultural and creative products(Ge, J et al.2018).

References

[1] Spencer, J.(2014).Exploring the implications of cultural context for design for sustainable behaviour.uk.bl.ethos.603027.
[2] Elizondo.(2011).Designing for sustainable behaviour in cross-cultural contexts : a design framework.uk.bl.ethos.556312.
[3] Wu, Q., Lv, J., Pan, W., Liu, D.(2017).Research on cultural and creative product design method based on case.Doi:10.3785/j.issn.1006-754X.2017.02.001.
[4] Chen, C., Lin, S.(2016).Message Delivery of Cultural and Creative Products Under Cultural Industries.Communications in Computer and Information Science book series.(CCIS, volume 617).Doi:10.1007/978-3-319-40548-3_3.
[5] Chi-Hsien and Wang, C.(2015).A Design Strategy of Cultural and Creative Products on the Global Market. Lecture Notes in Computer Science book series (LNCS, volume 9180).Doi:10.1007/978-3-319-20907-4_4.
[6] Chen, H.Y.(2009). A Study on the Information Design of Cultural Product. Department Of Product Design, Ming Chuan University. Master thesis.
[7] Chang, C.(2015).Next Step of Cultural and Creative Products - Embracing Users Creativity. Lecture Notes in Computer Science book series (LNCS, volume 9173).Doi:10.1007/978-3-319-20618-9_41.
[8] Alain, D. and Julie, M.(2017).A Journey Inside Tourist Souvenirs.Tourism on the Verge book series (TV).Doi:10.1007/978-3-319-44108-5_5.
[9] Alain, D. and Julie, M.(2014)."This is a piece of coral received from captain Bob": meanings and functions of tourist souvenirs.Doi:10.1108/IJCTHR-08-2013-0051.
[10] Becker, C et al.(2014).The Karlskrona manifesto for sustainability design.Version 1.0, May 2015.
[11] Li, Z.(2016).Research on the Value Characteristics and Development Strategies of Art Derivatives.Humanities World March 2016,No. 67.
[12] Wang, H.On the Research and Development of Creative Derivatives of Museums/Galleries.Unit Ten Art Window.
[13] Ge, J et al.(2018).Analysis of the Design of Cultural Creation Products in the Palace Museum and Its Reference Significance.Doi:： TB472,1003—0069(2018)03-0103—03.
[14] Li, N.(2017).Research and Application of Sustainable Design Concept in Tourist Souvenir Design.
[15] Wu, D.(2018).Sustainable Design of Furniture Products in Cultural and Creative Industries.TS664.01.
[16] Zhou, Z.(2016).iscussion on Cultural Products of Museums.Decoration Phase 276 2016, 04.
[17] Zhang, K.(2016).Cultural Relics Elements of Museum Cultural Derivatives Choosing Motivation and Strategy.Doi:1003-9481（2016）06-0052-04.
[18] Li, F., Guo, M.A Case Study of the Development Planning of Cultural Derivatives in Guizhou Museum New Museum.

IOP Conf. Series: Materials Science and Engineering **573** (2019) 012036 doi:10.1088/1757-899X/573/1/012036

Design of Intelligent Medicine Box for Elderly People Based on Context Awareness and FAST Theory

Siyuan Chen, Ming Chen and Jianhua Lyu*

Forestry College, Sichuan Agricultural University, Chengdu, Sichuan, 611130, China

*Corresponding author's e-mail: ljh@sicau.edu.cn

Abstract. Based on context awareness and FAST (Function Analysis System Technique) theory, this paper explored the needs of elderly users and designed an intelligent medicine box for them to provide more comprehensive health services. It defined context through context analysis method, combined four context factors to obtain human-machine interaction model, and built a functional tree through the FAST theory to analyze product design factors and functional transformation forms. Finally, this paper established a product shape matrix upon functional combination and human-machine relationship, and selected the best design scheme. Through a practical application to a case, the validity of context awareness and FAST theory hybrid approach and their technical feasibility and effectiveness was demonstrated, and an accurate mapping relationship between objective product performance and user needs was formed.

1. Introduction

China has entered aging society at the end of the 20th century, and by 2050 it will become a deeply aging country. According to the National Family Development Report 2015 issued by the National Health and Family Planning Commission, nearly 58.1% of the elderly have been diagnosed with chronic diseases and need to take medicine for a long time. In the case of accelerated aging and high incidence of chronic diseases, it is necessary to design better intelligent medicine box for the elderly[1]. Therefore, this paper obtained the needs of elderly people through immersive context awareness, and transformed user needs and integrated product functions by FAST theory.

2. Context awareness and FAST Theory

2.1. Context awareness

Context awareness was proposed by Schilit and Theimer in 1994. At present, the study of context awareness mainly includes two aspects, one is to study how people perceive, understand and feedback contexts, and the other is to build a context-aware system for users[2]. Hugh Beyer once pointed out that context awareness can collect users' behaviors, situations, and speeches, which can help designers obtain enough data information, and avoid personal subjective speculation[3]. Schilit et al. divided the situation into computing context, user context and physical context, and Chen and Kotz added time context to this[4]. Overall, the integrated context factor analysis assists design decisions and establishes a direct mapping between user requirements and product design requirements to ensure a high degree of consistency between them.

2.2. *FAST theory*

FAST theory was proposed by Bassevi in 1965. The theory follows a top-down analysis sequence, uses function trees to define and analyze product functions, and focuses on the important functions to increase product value[5]. Designers can use the function tree and the "How-Why" question and answer method to find the root node along the critical path, and determine the relationship between basic functions and sub-functions[6](Fig1).

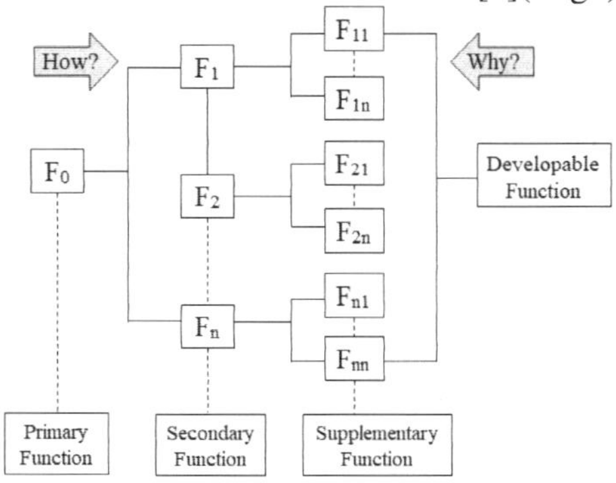

Fig1.FAST functional system model

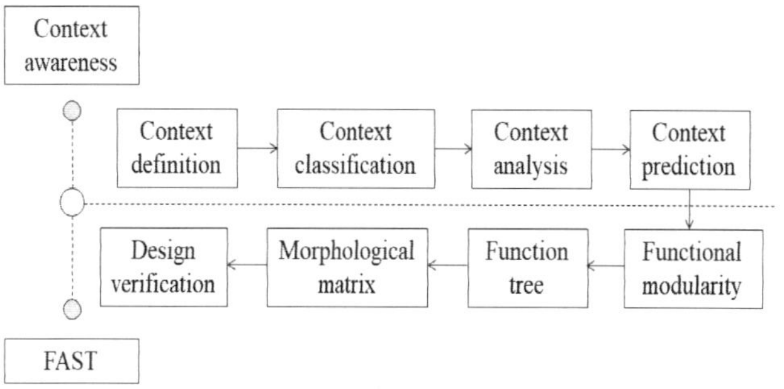

Fig2. The application process of context awareness and FAST theory

2.3. *Integration of context awareness and FAST theory*

Context awareness uses the context information of "human-machine-environment" to complete the classification of user cognition, behavior and experience, explore the invisible needs of users' sub consciousness, increase the pertinence and personalization of functional mapping, and solve the problem of "what to do". FAST theory can translate user requirements into design requirements, determine product core functions, develop extension functions, find reasonable functional implementation patterns, and solve the problem of "how to do". By combining context awareness with FAST theory, designers can complement each other's strengths, accurately capture user pain points, and make product function design clear(Fig2).

3. Analysis of the needs of the elderly based on context awareness

3.1. *Context definition*

A context includes all information about an entity's characteristics. Any information that interacts with the entity is a context[7]. Context factors include user context, social context, physical context, and temporal context. Traditional methods of analysis are difficult to adapt to the dynamics of user needs. Using context analysis to define contexts, designers can effectively organize and describe useful information to accommodate fuzzy user needs[8]. The context definition is shown in Table 1.

Table 1. Context definition

Level 1 context	Level 2 context
User	Target group: elderly people. Age: 60-74 years old. Characteristics: patients with chronic diseases, living alone, paying attention to physical health.
Product	Appearance: round and friendly. Color: mainly solid color. Function: modular classification. Structure: detachable.
Environment	Indoor or short outdoor activities.
Behaviour	Power on and off, interface interaction, scanning, APP interaction.

3.2. *Context factor analysis*

3.2.1. User context. The elderly aged 60-74 are called young seniors. In terms of health, they often suffer from chronic diseases such as hypertension, diabetes, and coronary heart disease, and have the characteristics of multiple diseases[1]. There are 6 common mistakes in elderly people during medication: don't take the medicine on time, taking the wrong dose, taking the metamorphic medicine, privately dispensing the medicine, don't pay attention to the medication precautions and don't supplement the lack of medicine.

3.2.2. Social context. At present, home-based care is still the choice of most elderly people in China. The weakening of family pension function makes the elderly have to improve their initiative in health care. The "Smart Pension" model introduces the Internet of Things and cloud technology into the elderly medical products, creating a variety of interactive modes[10].

3.2.3. Physical context. Through the analysis of the existing intelligent medicine boxes for the elderly, it is concluded that the main ways of interaction between users and products are voice interaction, visual interaction and behavioural interaction, which are realized through three ways: "human-machine", "machine-human" and "human-machine-human" (Fig3).

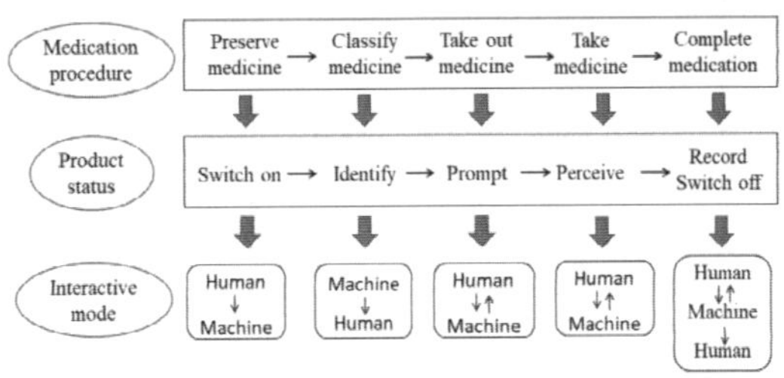

Fig3. Elderly intelligent medicine box "human-machine" interaction model

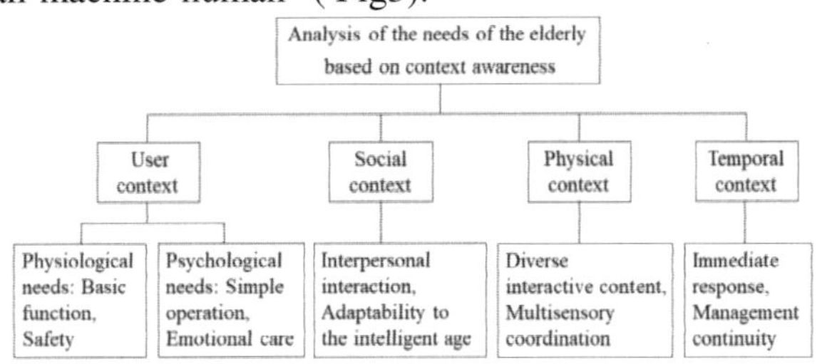

Fig4. Analysis of the needs of the elderly based on context awareness

3.3. Elderly needs analysis

Summarize the needs of elderly people based on context awareness, as shown in Figure 4.

4. Design scheme of the intelligent medicine box for elderly people

4.1. Functional structure construction

According to the function, the basic functional modules of the intelligent medicine box are divided into three categories: storage, prompt, and information service. Through the function tree, designers can explore the relationship between functions and make functions integrated (Fig5).

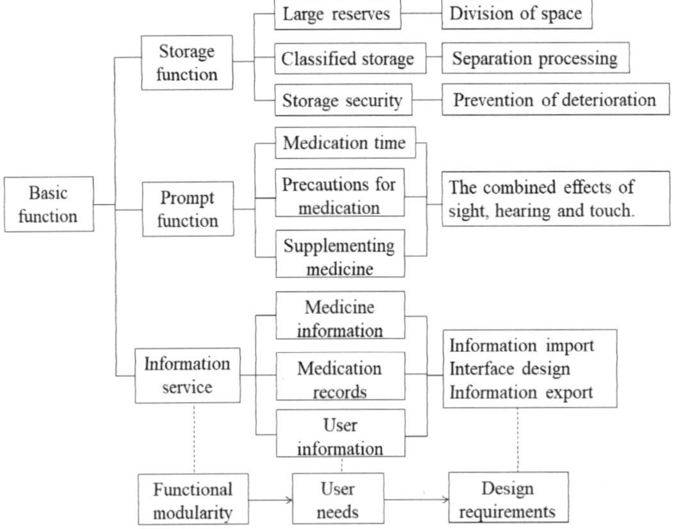

Fig5. Elderly intelligent medicine box FAST function tree

4.2. Morphological matrix construction

As shown in Table 2, the morphological matrix is established to further explore the functional implementation form and generate more technical solutions. For one of the functions, different technical processes can be selected to meet the design requirements, so it is necessary to select the best among several feasible solutions[5].

Table 2. The morphological matrix of elderly intelligent medicine box

Sub-function	Plan 1	Plan 2	Plan 3
A. Storage function	Drawer layering	Block partition	Monomer separation
B. Prompt function	Light	Buzzer	Speaker
C. Information import	Scanner	External keyboard	Touchscreen
D. Information export	Bluetooth	USB	Associated APP
E. Opening method	Twisted	Buckled	Sliding

According to the original understanding of the possible sub-functions in elderly intelligent medicine box listed in Table 2, there are 243 theoretically effective programs. By comparing various theoretical schemes, considering user characteristics, operability, production cost, use environment, and convenient maintenance, the design scheme of elderly intelligent medicine box is finally determined to be the combination of A2, A3, B1, B3, C1, C3, D3,E3.

5. Design verification

5.1. Appearance design

The design scheme is shown in Figure 6 and Figure 7. The main color of the product is white and gray, giving the elderly a relaxed and casual visual sense. In order to adapt to the vision of the elderly, only the "Power" and "OK" operating buttons are set. And the seven independent medicine bottles in the product configuration use high-purity color to enhance the discrimination.

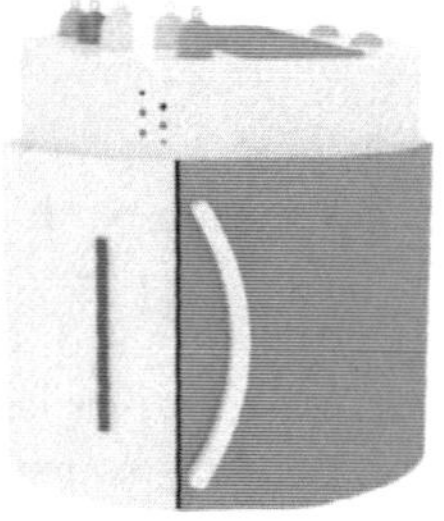

Fig6. Elderly intelligent medicine box rendering

Fig7. Monomer medicine bottle rendering

5.2. Material selection

In the material selection, the medicine box shell is made of medical PVC, which has low production cost, wide application, easy processing, excellent chemical resistance, mechanical properties and electrical properties. The handle and the sliding door surface are frosted to provide a non-slip and a better grip. The external medicine bottles are made of translucent PVC for easy access to medicines.

5.3. Functional analysis

This product consists of two parts. One part is the main medicine box, with storage area, scanner, touch screen and speaker. The other part is seven external medicine bottles with bases for the charging ports. The medicine bottle is divided into three parts. The top is a loop for easy carrying. The middle part is an LED light. The lower part is a storage space, and the internal separation structure can be modularized according to the number of times the elderly take the medicine. The functional analysis of the product is shown in Table 3.

Table 3. Elderly intelligent medicine box function analysis

Functional area	Function	Characteristic
Monomer medicine bottle	Dispense medicine	Packed by week, Easy to carry
Touch screen	Switch function, Display information	Tilted screen for easy operation, Indicative and concise icons
Scanner	Scan barcode	View the details of the medicine by scanning the code.
Speaker	Voice announcement	Alarm clock prompt, Synchronized voice prompt

6. Conclusion

Based on context awareness and FAST theory, this paper conducted research on elderly intelligent medicine box, explored the needs of elderly people through context factors, and discussed the interaction mode between users and products. Through design verification, the combination of context awareness and FAST theory can realize the effective transformation from the demand layer to the behaviour layer, making the product more suitable for the needs of elderly people. This method is feasible in the development of similar products and is expected to provide research ideas for related designs. Subsequent researches can continue with user demand dynamics and context-aware interaction design.

Acknowledgments

This research is supported by Research Project of Double Support Program for Art Special Project in Sichuan Agricultural University, 2018 and Opening Foundation for Industrial Design Industry Research Center, Key Research Base of Humanities and Social Sciences, Sichuan Education Department (GYSJ18-037).

References

[1] Jinjun Xia, Liu Yang, Zhiyuan Wu. (2016) Optimized design of intelligent medicine box for elderly people based on user experience principle. Packaging Engineering, 18:97-101.
[2] Zilei Zuo, Xiao Jiang. (2017) Design of mobile shopping application based on context aware. Packaging Engineering, 24:156-159.
[3] Jianbin Wu, Jingyuan Zhang, Linghao Zhang. (2018) Interaction design of in-vehicle infotainment based on context awareness. Packaging Engineering, 16:189-196.
[4] Jinling Liu, Xiao Jiang. (2018) Application research of context awareness in children's intelligent toy design. Design, 12:120-121.
[5] Chen Chen, Zhixue Sun, Le Zhang. (2017) Application of FAST theory in the conceptual design of domestic intelligent solid organic waste processor. Journal of Machine Design, 01:110-113.
[6] Xiaoli Wu, Yuxing Wang, Chengqi Xue. (2012) A Function Combined Children Bike Based on Design of FAST theory. Machine Design & Research, 01:98-101.
[7] Yuanyuan Chen, Zhengjie Liu. (2011) Research on mobile context awareness applications and interactions. Application Research of Computers, 12:4420-4425.
[8] Qing Zhang, Kaideng Chen, Suihuai Yu. (2018) User requirement analysis method based on FBS framework in product design. Journal of Machine Design, 07:119-123.
[9] Weihong Liao. (2013) Research on consumer behaviour in mobile internet. Science and Technology Management Research, 14:179-183.
[10] Qun Huang, Hanlu Tu. (2017) Research on the smart home medical product design under the background of intelligent pension system. Art and Design, 08:106-108.

IOP Conf. Series: Materials Science and Engineering **573** (2019) 012037 doi:10.1088/1757-899X/573/1/012037

Comprehensive evaluation of usability at the mobile end interface

Wang Sailan[1], Li Bei[2], Zhu Yiran[3]

[1 2]Architecture and Urban Planning College, Southwest Minzu University, Cheng Du, Si Chuan,610041, China

*Corresponding author's e-mail: 14795748@qq.com

Abstract. Research on interface usability has always been qualitative. This thesis conducted a quantitative research on the evaluation of usability of mobile terminal interface adopting Analytic Hierarchy Process in Management and Fuzzy Evaluation Theory in Fuzzy Mathematics. Firstly, according to the factors influencing users' evaluation of usability of mobile terminal, an evaluation index framework of mobile application was established. Then, according to the fuzzy comprehensive decision based on the data of questionnaire survey, a comprehensive evaluation of usability of mobile application was made. Finally, a detailed elaboration on the practice and application of this evaluation method was offered with the example of a tourism mobile application.

1. Introduction

The netizen number in China reached 0.829 billion and the popularization rate was 59.6% by the end of December 2018, with a 3.8 percentage rise from the end of 2017, and the number of new netizens in the whole 2018 was 56.53 million. The scale of mobile phone users in China is about 0.817 billion, of whom 98.6% connect to the Internet, and the number of new mobile phone users is up to 64.33 million. [1] The dependence on the mobile Internet in entertainment, social interactions and life services has been gradually intensified. The Internet big data company QuestMobile issued China Mobile Internet User Behavior Insight Report 2019, which showed the refined operation of users in 2019 was increasingly treasured. It is reasonable that in the near future, the Internet user refined operation will be targeted at the maximization of user values, and the user activeness, retention or payment index will be improved through the operation means.[2]

2. Research review

Donald Norman proposed in 1980s that usability and apprehensiveness were two parallel attributes of products and any product to possess these two attributes should meet two conditions. First, predictability: users can predict the results of any behavior; Second, visibility: users can see the operational effect and directly judge whether the operation is efficient. This viewpoint was based on his research conclusions made on the use of daily commodities. Donald Norman elaborated usability in his book. Davis proposed the technology acceptance model (TAM) in 1986 and considered usability as one key influence factor on the acceptance of information technology by users in addition to other influence factors of usefulness and satisfaction to techniques. [3]Hartson defined the usability of information systems in 1998 and defined usability as the interfacial interactive efficiency, learnability, and user satisfaction. Microsoft proposed the famous Microsoft usability guide (MUG) in 2006, which became a key index of web usability evaluation.[4]

The research on usability has always been synchronized with technical development. In recent years, experts and researchers concerning software and web design have diverted to the mobile end. Jakob Nielsen, an expert of web usability, targeted at mobile phones and panel computers for the first time in his Mobile Usability published in 2012, which offered foundation and guidance for small-screen design. [5]Lyndon Cerejo in his Mobile User Experience Elements in 2012 systematically studied the 12 elements of mobile end user experience, including usability. [6]So far, the research on mobile end usability gradually becomes deeper and systematic.

3. Mobile end usability evaluation index framework

In 2006, China released the national standard GB/T16260, which clarified the five aspects of usability, including understandability, learnability, operability, attractiveness and compliability, and this standard is suitable for diverse computer software products and relevant information systems.The MUG from Microsoft includes five factors of contents, usability, promotion, customized services, and emotions. Usability is thought as the endeavors made by users during the web cognition process and is decided by structures, goals and feedback. Structures refer to the overall organizational structure of the web; goals mean whether or not the theme of the web is definite and understandable; feedback is the operation and progression information given by the web to users.

Thus, the mobile App usability is affected by the vision usability, structure usability and interaction usability and is a psychical process of judging and balancing the three aspects and also a comprehensive result from the joint action of the three aspects. According to hierarchical analysis, the usability of mobile Apps is considered as the overall goal, and then vision usability, structure usability and interaction usability are regarded as the standard indices to be used to build a mobile App evaluation index framework (Table 1).

Table 1 .The evaluation index framework of mobile Apps

Overall goal *G*	Criterion index *C*	Plan index *P*	
Usability of mobile Apps	Vision usability (u_1)	Text structure u_{11} Picture quality u_{12} Metaphor of controls u_{13} Interface style u_{14} Text size u_{15}	P_1
	Structure usability (u_2)	Task flow u_{21} Guidance function u_{22} Information content u_{23} Newer guidance u_{24} Customization u_{25}	P_2
	Interaction usability (u_3)	Way of input u_{31} Way of output u_{32} Gesture operation u_{33} Feedback of errors u_{34} Help tips u_{35}	P_3

4. Evaluation methods of mobile App usability

4.1 Selection of evaluation methods

Since usability is one aspect of user experience and a psychological feeling, no quantitative method of usability can be absolutely precise. Due to the ambiguous relationship between the research targets, here the concept of fuzzy mathematics was introduced, which can comprehensively evaluate the mobile end usability.

4.2 Analytic hierarchy process (AHP) of fuzzy decision

The analytical hierarchy process (AHP) proposed by the American operational researcher T. L. Saaty 9 is a brief and effective method to make decision for complex problems. The tested problem from the overall goal G to the plan index P involved three layers: the target layer G, the criterion layer C and the plan layer P (Table 1). Then the importance of the decision-making plan, or namely the weights of decision plans P1, P2, …, Pn relative to the target layer G, was determined through pairwise comparison. Thereby, a very satisfactory decision was made.

AHP involves four steps:

1)Clarify the problem and build a hierarchical structure.

2) Create a pairwise comparison judgment matrix $A=(a_{ij})_{n\times n}$, which should be a positive reciprocal matrix and elements of which should meet:

$$a_{ii}=1,\ a_{ij}=1/a_{ji}>0,\ i,j=1,2,\ldots,n.$$

3) hierarchical ranking.

After the judgment matrix A is built, find its largest eigenvalue $\lambda_{\max}$ and then from the corresponding feature equation, find the eigenvector W. The normalized W is just the weights of importance of the elements in one layer relative those in the upper layer. This process is called hierarchical ranking.

4) consistency test.

In order to overcome the subjectivity and one-sidedness during the creation of the judgment matrix A, the consistency of A should be tested. When the random consistency test coefficient is $0 \le CR < 0.1$, the consistency of A and the corresponding eigenvector are acceptable; otherwise, A should be adjusted. The consistency test is divided into simple test and comprehensive test.

4.3 Steps of fuzzy evaluation

According to the rationale of fuzzy comprehensive decision-making,[9] the fuzzy evaluation of mobile App usability was finished as follows:

1) According to the mobile App evaluation index frame and the evaluation rationale, determine the factor set and evaluation set of the evaluation target.

The factor set is $G=\{u_1, u_2, u_3\}$; $u_i=\{u_{i1}, u_{i2}, u_{i3}, u_{i4}, u_{i5}\}(i=1,2,3)$, where G is the overall goal or namely the usability of mobile Apps; the primary indices u_1 to u_3 indicate the vision usability, structure usability and interactive usability respectively; the corresponding secondary indices are u_{ij} ($i=1,2,3; j=1,2,3,4,5$).

The evaluation set is $V=\{v_1, v_2, v_3, v_4, v_5\}$, where v_1 to v_5 correspond to the "very high, high, neutral, low, very low" degree of usability, respectively.

2)Build a fuzzy judgment matrix and use AHP to determine the weights of the indices.

3)Fuzzily evaluate each factor u_k of G according to the grade indices of the evaluation set V, forming a single-factor fuzzy evaluation matrix.

$$R^{(k)}=\begin{pmatrix} r_{11}^{(k)} & r_{12}^{(k)} & \cdots & r_{1m}^{(k)} \\ r_{21}^{(k)} & r_{22}^{(k)} & \cdots & r_{2m}^{(k)} \\ \cdots & \cdots & \cdots & \cdots \\ r_{n1}^{(k)} & r_{n2}^{(k)} & \cdots & r_{nm}^{(k)} \end{pmatrix}=(r_{ij}^{(k)})_{n\times m}$$

Suppose N experts participate in voting evaluation; $r_{ij}^{(k)}$ means the secondary index u_{ij} in factor u_k in the above matrix gets the frequency of grade v_j in V. Admittedly, $r_{ij}^{(k)}$ means the degree of u_k belonging to v_j (here $k=1,2,3$; $i\ j=1,2,3,4,5$; $n=m=5$).

4)Mark the weight matrix of u_k as w_k, and then the comprehensive evaluation result of u_k is:

$$b_k = w_k R^{(k)} = (w_{k1}, w_{k2}, \cdots, w_{k5}) \begin{pmatrix} r_{11}^{(k)} & r_{12}^{(k)} & r_{13}^{(k)} & r_{14}^{(k)} & r_{15}^{(k)} \\ r_{21}^{(k)} & r_{22}^{(k)} & r_{23}^{(k)} & r_{24}^{(k)} & r_{25}^{(k)} \\ r_{31}^{(k)} & r_{32}^{(k)} & r_{33}^{(k)} & r_{34}^{(k)} & r_{35}^{(k)} \\ r_{41}^{(k)} & r_{42}^{(k)} & r_{43}^{(k)} & r_{44}^{(k)} & r_{45}^{(k)} \\ r_{51}^{(k)} & r_{52}^{(k)} & r_{53}^{(k)} & r_{54}^{(k)} & r_{55}^{(k)} \end{pmatrix}$$

$$= (b_{k1}, b_{k2}, b_{k3}, b_{k4}, b_{k5})$$

Here b_{kj} means the corresponding weight of the sub-index in u_k in the evaluation set V (k = 1, 2, 3).

5) Comprehensively evaluate the 3 judgement elements u_1 to u_3 in G and its fuzzy evaluation matrix is:

$$R = \begin{pmatrix} b_{11} & b_{12} & b_{13} & b_{14} & b_{15} \\ b_{21} & b_{22} & b_{23} & b_{24} & b_{25} \\ b_{31} & b_{32} & b_{33} & b_{34} & b_{35} \end{pmatrix}$$

If the weight of the index set in G is W, then the comprehensive evaluation result is:

$$B = WR = (b_1, b_2, b_3, b_4, b_5)$$

where b_1,b_2,b_3,b_4,b_5 are the weights of corresponding grades in the evaluation set V of G, or namely the degrees of usability. According to the maximum membership principle, if b_2 is the largest, then this mobile system is the most usable; if b5 is the largest, then this mobile system is the least usable.

5. Case study

One tourism APP -- "bread trip" was used here and its usability was evaluated via questionnaire and analysis. In brief, 132 in-school students were asked to install and use this APP, and then were questionnaired. Questions of this questionnaire were set using a Likert5 scale (1 means "very much disagree" and 5 means "very much agree"). Then the indices of the plan layer P were measured. Finally 130 copies of questionnaire were returned, of which 125 copies were valid. The usability of this system was evaluated through comprehensive analysis of the fuzzy data set.

A judgment matrix concerning the importance of the criterion layer relative to the target layer was built:

$$G = \begin{pmatrix} 1 & 5 & 3 \\ 1/5 & 1 & 1/3 \\ 1/3 & 3 & 1 \end{pmatrix}$$

The eigenvector corresponding to the maximum eigenvalue λmax = 3.0387 was calculated by the summation method:

W = (0.6333, 0.1062, 0.2605)

This eigenvector was the weight vector of the primary factors u1 to u3 relative to the goal layer G. The random consistency test coefficient of CR = 0.0334 < 0.1 indicates the judgment matrix G meets the requirement of consistency test and this set of weights is acceptable.

Similarly, the weights concerning the importance of the secondary factors relative to the primary factors can be determined in the same way (Table 2).

Table 2 .Weight coefficients of factor u1

Secondary index	Judgment matrix	Weight vector w1	Consistency test coefficient CR
u11	1 5 3 4 5	0.4815	
u12	1/5 1 1/3 1/4 1/3	0.0580	
u13	1/3 3 1 1 2	0.1737	0.0369 < 0.1
u14	1/4 4 1 1 2	0.1778	
u15	1/5 3 1/2 1/2 1	0.1090	

Table 3. Weight coefficients of factor u2

Secondary index	Judgment matrix	Weight vector w2	Consistency test coefficient CR
u21	1 3 1/3 1/4 3	0.1551	
u22	1/3 1 1/3 1/4 1	0.0756	
u23	3 3 1 1/5 3	0.2150	0.0765 < 0.1
u24	4 4 5 1 5	0.4840	
u25	1/3 1 1/3 1/5 1	0.0703	

Table 4. Weight coefficients of factor u3

Secondary index	Judgment matrix	Weight vector w3	Consistency test coefficient CR
u31	1 1 1 5 3	0.2719	
u32	1 1 1 5 3	0.2720	
u33	1 1 1 5 7	0.3252	0.0189 < 0.1
u34	1/5 1/5 1/5 1 1	0.0597	
u35	1/3 1/3 1/7 1 1	0.0714	

For the vision usability factor u1, the fuzzy evaluation matrix of the secondary factors (namely the membership grade of the evaluation set V) can be determined from the questionnaire results and through statistical analysis.

$$R^{(1)}=\begin{pmatrix}0.2760 & 0.4984 & 0.7513 & 0.9593 & 0.8407\\ 0.6797 & 0.9597 & 0.2551 & 0.5472 & 0.2543\\ 0.6551 & 0.3404 & 0.5060 & 0.1386 & 0.8143\\ 0.1626 & 0.5853 & 0.6991 & 0.1493 & 0.2435\\ 0.1190 & 0.2238 & 0.8909 & 0.2575 & 0.9293\end{pmatrix}$$

Then from Table 2, the evaluation result of factor u1 is:

$$b_1 = w_1 R^{(1)}$$

$$=(0.4815, 0.0580, 0.1737, 0.1778, 0.1090)\begin{pmatrix}0.2760 & 0.4984 & 0.7513 & 0.9593 & 0.8407\\ 0.6797 & 0.9597 & 0.2551 & 0.5472 & 0.2543\\ 0.6551 & 0.3404 & 0.5060 & 0.1386 & 0.8143\\ 0.1626 & 0.5853 & 0.6991 & 0.1493 & 0.2435\\ 0.1190 & 0.2238 & 0.8909 & 0.2575 & 0.9293\end{pmatrix}$$

$$=(0.3280, 0.4832, 0.6859, 0.5723, 0.7056)$$

As for the structure usability factor u2 and the interaction usability factor u3, the fuzzy evaluation matrices of their secondary factors are listed in Table 5.

Table 5 .Evaluation set membership grade matrices of the secondary factors

Secondary factor	Weight index	Very high	High	Neutral	Low	Very low	Comprehensive consistency test coefficient C_R
u_{11}	0.4815	0.2760	0.4984	0.7513	0.9593	0.8407	0.0841 < 0.1
u_{12}	0.0580	0.6797	0.9597	0.2551	0.5472	0.2543	

u_{13}	0.1737	0.6551	0.3404	0.5060	0.1386	0.8143
u_{14}	0.1778	0.1626	0.5853	0.6991	0.1493	0.2435
u_{15}	0.1090	0.1190	0.2238	0.8909	0.2575	0.9293
u_{21}	0.1551	0.4173	0.4893	0.7803	0.1320	0.2348
u_{22}	0.0756	0.0497	0.3377	0.3897	0.9421	0.3532
u_{23}	0.2150	0.9027	0.9001	0.2417	0.9561	0.8212
u_{24}	0.4840	0.9448	0.3692	0.4039	0.5752	0.0154
u_{25}	0.0730	0.4909	0.1112	0.0965	0.0598	0.0430
u_{31}	0.2719	0.8687	0.4314	0.1361	0.8530	0.0760
u_{32}	0.2720	0.0844	0.9106	0.8693	0.6221	0.2399
u_{33}	0.3252	0.3998	0.1818	0.5797	0.3510	0.1233
u_{34}	0.0597	0.2599	0.2638	0.5499	0.5132	0.1839
u_{35}	0.0714	0.8001	0.1455	0.1450	0.4018	0.2400

The degree of usability was divided into five levels of "very high, high, neutral, low, very low" and thus was evaluated using a questionnaire. Users were asked to evaluate each secondary index according to five levels of "very high, high, neutral, low, very low". Finally, a decision was made via the fuzzy method.

The evaluation results of factors u2 and u3 are (Table 5):

$$b2 = w2R(2) = (0.7544, 0.4814, 0.4047, 0.5799, 0.2501)$$
$$b3 = w3R(3) = (0.4618, 0.4501, 0.5050, 0.5745, 0.1541)$$

Thus, the overall fuzzy evaluation matrix is:

$$R = \begin{pmatrix} 0.3280 & 0.4832 & 0.6859 & 0.5723 & 0.7056 \\ 0.7544 & 0.4814 & 0.4047 & 0.5799 & 0.2501 \\ 0.4618 & 0.4501 & 0.5050 & 0.5745 & 0.1541 \end{pmatrix}$$

The usability evaluation result of this system is:

$$B = WR =$$
$$(0.6333, 0.1062, 0.2605)\begin{pmatrix} 0.3280 & 0.4832 & 0.6859 & 0.5723 & 0.7056 \\ 0.7544 & 0.4814 & 0.4047 & 0.5799 & 0.2501 \\ 0.4618 & 0.4501 & 0.5050 & 0.5745 & 0.1541 \end{pmatrix}$$
$$= (0.4081, 0.4744, 0.6089, 0.5737, 0.5136)$$

The above results suggest the membership grade concerning the usability of this system to the evaluation set V = {v1, v2, v3, v4, v5} maximizes to 0.6089. According to the maximum membership principle, its usability is graded as v3, or namely the usability of this system is neutral. The comprehensive consistency test coefficient CR < 0.1 indicates this result is acceptable. The above computation was conducted on Matlab.

6. Conclusions

The mobile commerce is upsurging and shows a prospective future. Research on mobile commerce has never stopped, ranging from the commercial success and huge market to the popular discussion in the academia. The usability of interactive design becomes a key factor deciding the success or failure of a mobile end product, and usability evaluation is a very difficult process due to the complexity and subjectivity. This study was aimed to quantitatively evaluate usability. This study has two contributions from the theoretical and practical perspectives.

Theoretically, this study modestly enriches the theoretical research on usability of interactive design, especially at the level of quantitative research. The target was the popular mobile end interface, and the methods were analytical hierarchy process and fuzzy evaluation in the field of management. The usability of the mobile end was evaluated using a quantitative method, which offers a new method for quantitative research in the future.

From the practical perspective, research on usability is the key factor deciding the survival or failure of mobile products. Our findings can guide the management practice of mobile product

IOP Conf. Series: Materials Science and Engineering **573** (2019) 012037 doi:10.1088/1757-899X/573/1/012037

development and provide practical meaning for research & development institutions. Enterprises can evaluate the usability of products at the research & development stage or before release into the market, and thereby can modify and improve the usability problems in advance, which will reduce the loss of users due to usability problems.

As for limitations, the tourism APP was evaluated only from the user perspective, and the users were generally in-school students, which may have caused some bias. Nevertheless, since this study was aimed to investigate the evaluating method of mobile end usability, this case study only suggests the practical application of this evaluation method. In the future, more comprehensive and richer data can be collected in other evaluations. Thus, this preliminary study on the evaluation of mobile end usability should be further deepened.

Acknowledgments

Supported by 2019 The Fundamental Research Funds for the Central Universities of Southwest Minzu University 2019SQN03

References

[1] AppAnnie . (2019). https://www.appannie.com/cn/intelligence/
[2] QuestMobile.(2019).https://data.questmobile.com.cn/
[3] Cao Shengsheng., (2003) Preliminary study on product usability in the information society . Decoration., 135: 8-9.
[4] Davis F D. (1986)A technology acceptance model of empirically testing new end-user information systems: theory and results. Massachusetts: Sloan School of Management, Massachusetts Institute of Technology, :78-80.
[5] Hartson,RH.(1989) Human-Computer Interface Development: ConceptsAnd System for Its Management., ACM Computing Suvreys,, 1: 5-60.
[6] Viswanath Venkatesh V. Ramesh.(2006) Web and wireless site Usability: Understanding differences and Modeling use,MIS Quarterly.30: 81-206.

IOP Conf. Series: Materials Science and Engineering **573** (2019) 012038 doi:10.1088/1757-899X/573/1/012038

Greenhouse and Field: Two Field Domains that Cultivate Designers

Hu Sun, Yueqin Wu and Wei Li*

School of Art&Design, Xihua University, Chengdu, Sichuan, 610039, China

*Corresponding author's e-mail: 14344380@qq.com

Abstract. Based on the Ruskin's comparison of the designer, the design education and the national environment to the seed, soil and climate respectively, and combining with the background of industry and design development in our country at the present stage, this paper analyzes the internal and external causes of the designer's growth and development, and puts forward the comparison of greenhouse and field that the professional education and social practice in colleges and universities is the two field domains for the designer's growth, emphasizes the importance of integration of higher education and social practice, proposes that social practice is the main field for designers to grow up. The enlightenment to the development and growth of designers in our country at this stage is discussed.

1. Introduction

In 1895, John Ruskin used seeds, soil and climate to describe the growth of designers in Modern Manufacturing and Design. Soil and climate are the external factors for the growth of designers, which represent the design education and the national environment related to society, economy and technology. Seeds are the individual designers and the internal factors for the growth of designers. Ruskin's idea of artistic design is a specific product of his time, which expresses the artists and theorists' opinions and judgments criteria on the trend of artistic design in that time. The important role played by British creative industries in economy is closely related to the deep insights of early theorists and critics, the policy intervention and support at the medium-term national level, and the courageous exploration of design education. Ruskin's metaphor of "seed, soil and climate" in designer training are also adopted in this paper, and "greenhouse and field" are used to describe the designer's campus education and social growth. It not only looks at Ruskin's artistic thoughts from a historical perspective, but also reflects on the development needs of China's design industry, especially the development of the design industry and the reform of the training of design talents.

2. Seeds, Soil and Climate: Designers, Design Education and National Environment

In the late 19th century and early 20th century, the "happy and modern British life" described by Ruskin became a reality. Only the right soil and climate can seeds grow, germinate, branch, blossom and bear fruit.

Designer's artistic ability, thought and morality are the internal factors to cultivate designers. In Ruskin's opinion, only in a happy and modern British life can designers truly appreciate the needs of the people, keep innovating, so as to design for the masses. The power of art is a nobler and simpler power, which is the endogenous power to serve the masses. Designers should find a balance between hardship and luxury to achieve a moderate state of peace and reflection. Therefore, the peaceful art, happy family life and the general recognition of national culture are regarded not only as the premise and foundation for manufactured goods to meet the needs of the people, but also as the standard for good product design. At the same time, designers as individuals can also change the society, for instance, designers attract public attention, get decision-making power, and then produce meaningful works. This view has also made the government to pay more attention to the cultivation and creation of star designers. Art and design is an invisible force, which can permeate people's lives and cultures through works and concepts, "inform the soul with facts, touch the heart with sympathy, give comfort to the angry, and give civilization to the rude".

Design education is a direct external factor in training designers. Ruskin emphasizes the importance of perception in design learning by saying "Your design ability depends entirely on your own feelings and judgments. "Only when we have a deep and firm understanding of the design principles and put them into practice can it be meaningful".

Another external factor in training designers is the national environment, which is mainly manifested in the following aspects. 1) The slogan of "learning from nature" is clearly put forward, which considers that nature is the source of art and all creativity. Respecting for nature and learning from nature are the same as an old saying that "the art comes from nature but the inner feeling is the source" in ancient China. If we "ignore the natural form in pursuit of art, it will lead to the degeneration of temperament and intelligence". These expressions are directly related to Ruskin's passion for painting flowers, birds and natural scenery. 2) Although industrial development can promote economic prosperity and enhance national strength, there is also the destruction of the natural environment caused by industrial production and the demand for natural materials behind the seemingly booming economy. If the nature, which is the source of creativity nature is destroyed, people will lose their creativity. The occupation of agricultural farmland by industrial development has resulted in people moving passively or actively away from the countryside, initiatively entering or being forced to go into cities. As for designers and artists, they lost not only a good pastoral life, but also the spiritual home and spiritual sustenance. "The influences of modern industry on modern ornaments include industrial development, the rise of cities, the abandonment of rural houses in the Middle Ages, but people yearn for real rural life. Returning to the countryside and pastoral areas is the essence of Britain." 3) It confirms the positive effect of political rights on the development of art. The national-level political attitude and policy towards design art are also important factors in the development of design and the cultivation of design talents. In 21st Century Design, Woodham J.M. exemplifies the external environmental factors of the development of Ruskin's design industry, such as government policy support and national-level design attitude. In many exhibitions in the 1930s, the British exhibits and the focus of official publicity mean that the "happy and modern British life" expected by Ruskin has become a reality.

3. The Current Background of the Growth and Development of Designers

3.1. Design to Adapt to the Development of the Times and to Meet the Industrial Needs

Design industry is naturally linked with art culture, science and technology, and commercial economy. With the traditional industry development, transformation and the rise of new industries brought by the Industrial Revolution and the Information Technology Revolution, the government, enterprises and universities not only reflect on the past development model, but also construct the design industry from an all-round and multi-level framework. Design has undergone the evolution from traditional design (design 1.0) and modern design (design 2.0) to innovative design (design 3.0). Design 3.0 presents new features of green and low-carbon, network intelligence, open integration and co-creation and sharing. Different from the western countries with developed design, China has several different design times carried out at the same time in China due to China's special national conditions and complex historical development. The rapid development of economy and the industrial needs constantly raise new requirements for the knowledge structure and growth system of designers, therefore, the art design specialty in colleges and universities should meet the needs of the development of multiple industries and train innovative talents in many aspects necessary for society.

IOP Conf. Series: Materials Science and Engineering **573** (2019) 012038 doi:10.1088/1757-899X/573/1/012038

Under the background of economic globalization, technological innovation and network development, and with the constant industry optimization and upgrading, the new context of design industry is gradually taking shape and China's design education is facing new opportunities and challenges.

3.2. Responsible Design

It is constantly insisted that design theorists and educators should design for the masses, therefore, the design modernity has a natural and inherent democracy, which means the social responsibility of design, whether big or small. Along with the development of socialist ideology, this democratic artistic spirit has been influencing the Pioneer Design Movement throughout the entire twentieth century. Influenced by the founder Walter Gropius, BAUHAUS, a well-known modernist master, associated the dream of a "design power" proposed by Deutscher Werkbund with design, which changed from small responsibilities such as designers' practicability and reliability of their works, and the goal of not pursuing the short-term commercial success, into a great responsibility for the prosperity of the country. At present, the Craftsman's spirit strongly advocated by our country is also a kind of concern for occupational social responsibility, which is in line with the national development strategy of "making a country prosperous by manufacturing and a powerful country by designing ".

4. Greenhouse and Field: Two Field Domains that Cultivate Designers

The greenhouse shelters the weak seeds from the wind and rain. At the same time, carefully cared for in the greenhouse, the seeds are germinating, and growing with branches and leaves, and then transplanted into the appropriate fields, with the appropriate sunshine and rain, it is growing into the owners of the fields.

4.1. Greenhouse: Colleges and Universities are the Visiting Field to Train Designers

Excellent and qualified designers need not only theories and professional education from design universities, but also practical operation and vocational training of social industries. However, looking at the current situation of art design education in China, the current development speed of industry far exceeds the pace of the reform of university design education, so it is difficult for the cultivated talents to meet the needs of industry, that is to say, the university design education fails to meet the needs of industry for the cultivation of design talents. In 2015, the definition of industrial design was rewritten and updated, which further put forward new requirements for design education. However, " it takes ten years grow trees but a hundred to rear people", so talent training is not a problem that can be solved in a short time. At the same time, professional education lags behind the development of industry, which does not exist only in the design specialty. Almost all colleges and universities have this problem in their professional education, but the particularity of the design industry determines that the design is always related to the cutting-edge new technology, new materials, and new users' needs and so on, therefore, there is a clear distance between the two endpoints of talent output on campus and social input.

The change of ideas is equally important for educators and students. It must be clarified that the four-year college education is not equal to lifelong vocational education. Design innovation and change, not only refers to the transition and originality of design thinking, but also to the dynamic process of design practice constantly seeking the best solution because of the ever-changing market and challenges. Although students majoring in design are armed with design theories and methods in colleges and universities, their accumulated application skills and experience are difficult to fully deal with the actual work. Just as the result of ideal state and formula calculus, there is always a little practical error. Thus, colleges and universities are just an visiting field to train designers.

4.2. Fields: Society is the home field for designers to grow up

The words that "Genuine knowledge comes from practice" and "learning by doing" are the best generalizations of the professional characteristics of design. Before the Industrial Revolution, design was a craftsman's explicit or implicit workmanship and skills. The comprehensive activities of eyes, hands and hearts and their characteristics are also the common basis of art and design. Because of the practicality of design major, students need to enter the real social work environment to gradually release the energy accumulated in school. Even more simply and crudely, as for time and energy in the training of design talents, no famous school can guarantee that the four-year knowledge reserve can cope with more than 40 years' professional careers. Therefore, society is the home field for designers to grow up.

5. Enlightenment to the Growth and Development of Designers

Design industry is a service industry with the characteristics of "production relationship", which is facing the real industry. Design industry is related to any real industry, but it does not belong to any real industry. This "inter-" nature determines that design activities are the key to the design industry, and designers are the main body of activities. Design industry is naturally related to art culture, science and technology, and commercial economy, which makes it necessary for design education to enable designers to adapt to and actively "connect, integrate, cross and merge" among industries, disciplines, technologies and humanities, human knowledge and material knowledge.

To be an excellent designer, the concept of lifelong learning should be established and the habit of autonomous learning and active thinking should be developed from the time of students. For the students majoring in design, only with lifelong learning, self-learning, active thinking and daring to take the responsibility can they meet the future professional needs. It is suggested that designers take the initiative to shoulder the social responsibility and national responsibility, to serve life and society by design, to carry out design in a responsible manner, to lead life with a healthy and positive concept.

One of the external reasons for training designers is the transformation of university design education to meet the needs of industry. On the one hand, design universities strengthen their relationship with industry, such as establishing normal and two-way communication with enterprises, and penetrating enterprise projects and research contents into professional courses. Enterprises hold special design lectures from time to time in schools. Even career guidance courses consist of interviewing teams composed directly of human resources departments and design directors of enterprises, which can make it not only courses but also a real job interview drill. On the other hand, the industry's demand for design talents can also complement each other from the social education level to establish the professional growth education mechanism for designers. Only professional college education is not enough, the follow-up professional development and training for the designers, that is, design graduates, such as specialized design posts and highly targeted skills training. In addition, more and more enterprises have paid more attention to the training and growth of designers, and more and more industry designers have participated in the seminars of design conferences and workshops at home and abroad.

6. Conclusions

There are many issues of the growth and development of Chinese designers for industrial that should also be considered. How can greenhouses better let seeds take root, germinate, branch and blossom? How can greenhouse make seeds better adapt to the new environment, the field, after growing? What kind of soil is needed in the field to make the seeds grow strong and bear fruit? The similarities and differences in climate between greenhouse and field are also important determinants. The sustainable output of high-quality design talents and the knowledge management of professional designers are also issues that need to be considered in design education.

Acknowledgments

This work was supported by the Ministry of education of Humanities and Social Science Art Fund Project(19YJC760090) and Supported by Key Teaching Reform projects of Xihua University(XJJG2017074) and Supported by Project of Industrial Products Teaching Team of Xihua University(05050034).

IOP Conf. Series: Materials Science and Engineering **573** (2019) 012038 doi:10.1088/1757-899X/573/1/012038

References

[1] Jonathan M. W. (2012) Twentieth-Century Design. Shanghai People Publishing, Shanghai.
[2] Liu, H.X. (2012) Development Path of British Industry Design. Design Art, 12:45-47.
[3] Tang, L.T. (2014) Seed Soil and Climate- The Future Development of China's Design Industry. Zhuangshi, 254:06-10.
[4] Lu, Y.X. (2019) Research on Competitiveness of Innovative Design. Journal of Machine Design, 01:02-04.
[5] Zhang, Z., Li, B.Q.(2019) The Comparative Analysis and Enlightenment of National Design System Promoting Innovative Competitiveness. Scientific Management Research, 37:98-101.

Research on the Design of Henan Regional Characteristic Tourism Culture Creative Products——Taking the cultural and creative products of Chongdugou Scenic Spot in Henan as an example

Zhang Meng [1], Bo Meng Zhe [1], Li Wei [1*]

[1] School of Mechanical and Automotive, Southwest Forestry University, Kunming, Yunnan, 650224, China

[*]Corresponding author's e-mail: 1009858559@qq.com

Abstract. This paper studies how to design tourism culture creative products with Henan regional cultural characteristics and spiritual connotations. By analysing the advantages of Henan tourism product development and the main problems existing at present, summarize innovative ways to create distinctive brands, international, and sustainable design. Combining with actual cases of deepening design, trying to study Henan's regional tourism culture and creative products that have practical value, artistic value and cultural connotation value from various angles of design methods. I hope this article can be used as a reference for the innovative design of cultural and creative products.

1. Advantages of Henan Tourism Culture Creative Product Development

Henan is one of the main birthplaces of the Chinese nation and the Chinese civilization. In the long history of China's 5,000-year, more than 20 dynasties have been built in Henan. With rich historical and cultural atmosphere, rich tourism resources and convenient transportation, it has created superior conditions for the development of Henan tourism culture and creative products.

1.1. Strong historical and cultural atmosphere

The Central Plains culture can be called a diversified, typical, and heavy historical culture, and it can also be called a classic "Holy", "Fu" and "Soul" culture. In terms of content, the Central Plains culture contains 22 kinds of characteristic historical cultures such as Heluo culture, Chinese character culture and surname culture. In terms of spatial distribution, the Central Plains culture represents a cultural distribution pattern centred on Henan and spread to neighbouring provinces and autonomous regions. The Central Plains is not only the cradle of Chinese civilization, but also the birthplace of ancient political and economic centres and mainstream culture. It not only preserves precious spiritual and cultural wealth for future generations, but also provides a strong cultural connotation for the development of cultural and creative product design.

1.2. Henan is rich in tourism resources

Henan has abundant tourism resources and numerous historical sites. For example: Longmen Grottoes in Luoyang, Kaifeng, Shang he Garden in Qingming, and many other places of interest, as well as many amazing folk crafts, such as Luoyang Palace Lantern, Tang Sancai, China, and Zhuxian Town New Year. The development of tourism resources is closely related to tourist souvenirs. These rich natural geographical resources also provide unique elemental attributes for the development and design of tourism culture creative products. The combination of regional cultural symbols and products not only allows visitors to experience the baptism of Henan regional culture, but also to experience cultural differences.

1.3. Convenient transportation

Henan is located in the central part of China and is an important transportation hub with comprehensive transportation systems such as railways, highways, aviation and water transportation. The development of the transportation system has also provided advantages for the development of tourism in Henan, especially for the convenience of popular holiday tourism, which has promoted the rapid development of Henan tourism culture industry and increased the demand for tourism culture and creative products market.

2. Development Status of Henan Tourism Culture and Creative Products

Under the guidance of China's new era cultural protection policy, we pay more and more attention to the inheritance and innovation of culture, and at the same time enhance the attention to the cultural industry, especially tourism. With the reform and opening up and economic development, Henan's tourism and cultural industry has formed a certain scale. With the increasing participation of enterprises, Henan tourism has a series of cultural awareness of serialization and branding. However, compared with other major tourism provinces, it has not been able to maximize the unique advantages and benefits of tourism resources, and the design of regional cultural and creative products is still lacking.

2.1. Lack of innovation

The diverse, typical and heavy Central Plains culture is profound, and many traditional folk art cultures and craft products are bred. With the guidance of the policy, the government and enterprises pay more and more attention to the inheritance and development of the folk art culture. However, the inheritance of culture is not completely plagiarized. Cultural elements, innovative consciousness and production techniques need to keep pace with the times and innovation. Many traditional hand workshops and enterprises still do not know enough about this. For example, many enamel enterprises and handicraft workshops still use traditional production patterns and styling production. Although they have realized that they want to innovate their products, they have not changed the traditional model, just adding some decorations or making some model changes in the form of bottles, jars or bowls; Another example is the Tang Sancai handicrafts in Luoyang. The souvenirs sold in the market have always maintained their original form and lacked the awareness of recreating traditional cultural elements. Cultural inheritance is of course important, but blindly inheriting and not innovating will be out of touch with the development of modern society. If cultural and creative products simply add meaningless symbols of cultural elements, this will not only lose interest to the audience, but also reduce cultural identity.

2.2. Lack of regional characteristics, neglecting human care

The reason that tourists expect to travel is that tourists want to find and experience the differences in regional culture from the journey, and relax themselves in the experience, which is essentially a kind of cultural consumption, cultural enjoyment and cultural exchange [1]. Tourist souvenirs are commodities with regional cultural characteristics, with commemorative value and collection value, which can cause tourists to miss the tourist destination. Tourists are static for consumers, but tourist souvenirs are dynamic and can be collected and purchased. At present, although the development of tourism industry in Henan Province is consciously building brand culture, most of them are superficially formalized, and there are few things that can really impress tourists. Most tourists go to every tourist destination and they want to buy some local souvenirs. However, many souvenir sales places in the scenic spots are contracted by local residents. Souvenir sales show a state of stalls, serious homogenization of goods, lack of regional characteristics, will allow visitors to slowly lose interest in the lack of distinctive souvenirs from the initial expectations, thus reducing the curiosity and experience of the tourist destination.

3. The innovative strategy of Henan tourism culture creative product design

The main problem of Henan's regional tourism culture and creative products is the serious phenomenon of product homogenization and the lack of innovation in cultural elements. Based on the above viewpoints, the design and development of Henan cultural and creative products should be based on the emphasis on regional culture, incorporating modern elements to re-innovate culture and strengthen the combination of tradition and modernity.

3.1. Keep up with the times, strengthen innovation

Culture is not static, it has growth, and it can be said that all traditional art should be contemporary [2]. On the basis of grasping the essence of excellent culture, we must integrate design elements and new technology that conform to modern aesthetics, so that traditional handicrafts can develop in inheritance and innovate in development. Designing with culture can give products a soul with a new era of cultural significance. Applying the expressions and elements of excellent culture to the design of cultural and creative products can improve the quality and quality of products [3]. "The New Palace Museum" has received wide attention from the society since its launch. This brand perfectly combines the culture and products of the Forbidden City, making the Forbidden City culture "young". For example, the Forbidden City sleepwear series, the Forbidden City lipstick series (Figure 1-2), let it seem to travel from history to modernity. The key to the concept of cultural and creative product design lies in the word "cultural creativity", so we must consider how to modernize traditional culture and crafts and use it for today.

Figure 1. Forbidden City sleepwear series.

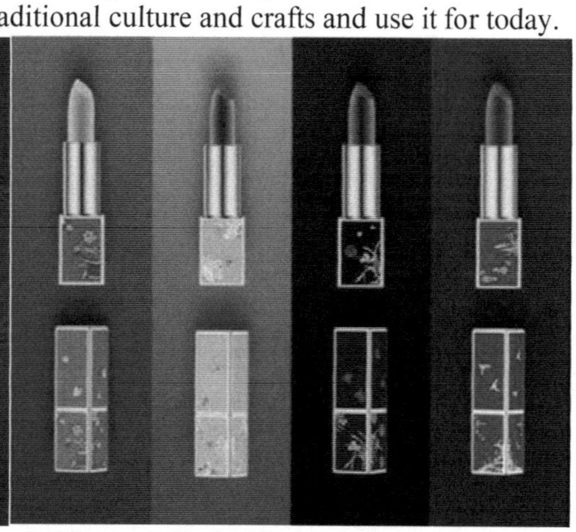

Figure 2. Forbidden City lipstick series.

3.2. Create a unique product brand image

The cultural brand value of the product is much greater than the value of the product itself [4]. In the face of the current homogenization of Henan tourism culture and creative products and the lack of characteristics and human concern, we must design unique cultural and creative products with regional attributes and selling points according to the regional natural characteristics and humanistic connotations. Creating a unique product brand image can enhance the brand identity of the product brand and enrich the cultural connotation and commemorative value of the product. By creating a unique product brand image and extending product categories and forms to form a series of branded goods sales model, not only can enhance the brand identity of the cultural brand, enrich the cultural connotation and commemorative value of the product, but also meet the diversified needs of tourists. Design a regional business card with texture and temperature for visitors.

3.3. Cultural and creative products go international

The inheritance and development of any culture is the result of cultural exchanges, and it is the collision and learning between historical civilizations. We must not only inherit and innovate traditional culture, but also carry forward our excellent culture [5]. Cultural exchanges between countries have become more frequent in the context of globalization, and Chinese brands have received increasing attention from all walks of life. We often encounter foreign friends in tourist attractions. Under this circumstance, tourism culture and creative products should focus on internationalization, let international friends interpret the profound Chinese culture through products, and let our excellent culture shine in the international arena.

3.4. Establish a sustainable development concept

The development of tourism resources will inevitably lead to the consumption and waste of resources. For example: the over-exploitation of natural resources mountains, forests and lakes; the development of beautiful and impractical tourist souvenirs; the uncivilized way of travel will cause the consumption of resources, so we need to apply the concept of sustainable development to the development of tourism. Learn from the excellent cases, use new materials, new methods, and new ideas to develop and design tourism culture and creative products in a sustainable way, so that they not only have the value of viewing and commemoration, but also have sustainable use value. For example, integrating modular design concepts, non-material design concepts, and green design concepts into cultural and creative product design to improve the recycling of product resources is also a heritage and continuation of product culture.

4. Taking the Chongdugou scenic spot as an example to carry out the innovative design of regional cultural creative products

Integrating the above-mentioned status quo of Henan tourism culture products and proposing innovative methods for designing tourism culture creative products. The author will use the above-mentioned innovative strategy to take a series of products with regional characteristics in the Chongdugou Scenic Area of Henan Province as an example to improve the whole process of product design to prove the feasibility of applying innovative research methods.

The innovative design of bamboo souvenirs in Henan Chongdugou Scenic Area is not only a reflection of the characteristic regional culture, but also a bold attempt to innovate the traditional bamboo product shape and function. In view of the current development status of bamboo products and the innovative methods of induction, we try to innovate and extend the series of traditional bamboo products to conduct in-depth research and exploration on the innovation of bamboo culture and creative products. In this way, we will try to design and apply the innovative methods of tourism culture and creative brands, and look forward to providing new design ideas for the innovative design of cultural and creative products.

Palace makeup - cosmetic products design kit (Figure 3) is the design element of the Chinese word "gong" as a jewellery box, and then blends the elements of the mountain and water of the water curtain fairy palace to make mirrors, combs, eyebrow pencils, makeup brush storage racks. In order to carry out product landscaping, through the patchwork distribution of the series products, the geographical landscape of the water curtain fairy palace is built. This series of products completely changes the original traditional product form, and is a product innovation design that integrates local elements, local culture and local raw materials under the guidance of modern design perspective. Its design is inspired by the iconic waterfall in the Chongdugou Scenic Area, the water curtain fairy palace. When the waterfall falls, it is divided by the moss on the mountain wall to form a white jade bead curtain made of silver wire jade beads. It is more feminine than the masculine character of the iconic venting cliff waterfall in the scenic area, like a fairy palace. This feminine style is complemented by the design style of women's jewellery products. The innovation of the product lies in the fact that its shape extracts the mountain and water symbol elements in the scenic spot for innovation, and presents the regional features of the local waterfalls and the bamboo forests. It has the characteristics of regional branding, modernization and internationalization, further sublimating the design connotation of product semantics.

IOP Conf. Series: Materials Science and Engineering **573** (2019) 012039 doi:10.1088/1757-899X/573/1/012039

At the same time, the use of environmentally friendly local bamboo as a raw material for the product not only increases the environmental friendliness and sustainability of the product, but also makes the whole set of products have a unique regional temperature to meet the needs of tourists, let visitors experience the romantic and unique scenery of the Chongdugou scenic spot.

Figure 3. Palace makeup - cosmetic products design kit.

5. Conclusion

The cultures of different ethnic groups in different countries and different ethnic groups in the same country vary widely, each with obvious regional characteristics, and the purpose of tourist tourism is to experience cultural differences and to feel the cultural characteristics of new, strange, heterogeneous and beautiful. Therefore, the author believes that in the innovative design of regional cultural and creative products, the effective expression of regional culture is the key to the deepening of cultural and creative product design. In the application research of the palace makeup - cosmetic products design kit, based on the uniqueness of the regional culture of the water curtain fairy palace, the rich cultural resources are fully explored, and the regional symbolic elements and the cultural connotations contained therein are extracted. The regional culture has penetrated into the design of tourism culture and creative products, which not only solves the problem of homogenization of tourism products in the current tourism market, but also solves the problems of lack of characteristics and innovation. It also brings the product to the road of branding, modernization, internationalization and environmental protection. Based on the above viewpoints, the design and development of Henan tourism culture creative products should establish a unique product brand image, pay attention to the combination of tradition and modernity, focus on internationalization, improve cultural differences experience and product sustainability design. Opening up new paths for innovative design of cultural and creative products.

Acknowledgments

First of all, I am very grateful to the organizers for providing this opportunity. Secondly, I would like to thank Bo M.Z. and Li W. for their help in the whole process of writing the topic. Finally, I would like to thank “the eight key industries in Yunnan Province to lead the brand professional (51600630)” I am sincerely grateful to the staff for the project and the staff of the Chongdugou Scenic Area.

References

[1] Gao, H. (2013) Tourism product design based on cultural demand psychology. Packaging Engineering, 34:34-36,40.
[2] Lu, B., Tan, X.D. (2018) Thinking of design culture and re-localization. Popular Literature - Art and Design, 84-85.
[3] Shen, S., Xu, Y.K. (2019) The application of traditional culture in the design of creative products. Packaging Engineering, 40:236-239
[4] Zhou, X.Y., Yu, B.Y. (2018) Research on the Intangible Design Innovation Strategy of Huizhou Cultural Brand under the Background of New Normal. Journal of Huaibei Vocational and Technical College, 17:91,92-95
[5] He, R.K. (2018) Industrial design under new opportunities. Design, 88-91.
[6] Han, X.J. (2013) Research on the Design and Development of Cultural Tourism Products in Henan Province. Urban Tourism Planning, 6:126
[7] Ye, D.H. (2012) Building a People-oriented Tourism Product Design System. Packaging Engineering, 33:133-135.
[8] Horman, D. (2014) The Development of Experienced Tourist product. Annals of Tourism Research, 15: 15-31.

IOP Conf. Series: Materials Science and Engineering **573** (2019) 012040 doi:10.1088/1757-899X/573/1/012040

The role of new industrial design training model in the transformation and upgrading of manufacturing industry

Bo Meng Zhe[1], ZhangMeng [1], Li Wei[1*]

[1] School of Mechanical and Automotive, Southwest Forestry University, Kunming, Yunnan, 650224, China

*Corresponding author's E-mail: 250974443@qq.com

Abstract. This paper studies how the new training mode of mechanical industry design promotes the transformation and upgrading of China's manufacturing industry. Through the analysis of the status quo of China's manufacturing industry design, the degree of industrial design engineering is adjusted to the mechanical category policy, and then from the perspective of the new model background, the industrial design complements the mechanical professional ability, the mechanical perfection of the industrial design professional ability, the complementary relationship between industrial design and mechanical skills, and analyzes the new model by improving the design ability of the new composite talents - improve market thinking, improve innovation capabilities, enrich professional skills, master technical theory, and enhance logical thinking to promote iterative upgrading of manufacturing products, thereby promoting the transformation and upgrading of China's manufacturing industry.

1. Introduction

Industrial design is an innovative cross-disciplinary subject, and it is also the primary process in the manufacturing industry. It provides an effective solution to meet the needs of consumers and to solve problems in industrial mass production. Industrial design is a sign of the level of scientific and technological development of a country and a region. Without advanced technology and methods, it is impossible to have a good and high-level design, and even develop new products [1].

China's manufacturing industry is big but not strong, and many problems are outstanding. At present, most of China's manufacturing industry is still only at the level of processing and production. The level of technology in manufacturing is low, lack of core technology and innovation, and product homogeneity is serious [2]. Nowadays, it is still in the catch-up phase of the strategy of "tracking, catching up, paralleling, and leading". In recent years, in order to speed up the pace of catching up, China has also proposed the development of manufacturing strategies such as "Made in China 2025" and "Building an Innovative Country". Today's design innovation has been listed as an important direction of the national innovation strategy. Under the background of the big environment and China's supply-side structural reform, the industrial design and the another core technical specialty of the manufacturing industry, the mechanical specialty, will coordinate and promote the product iterative upgrade and technology update to realize the transformation and upgrading of China's manufacturing industry. It is of great significance to improve the international competitiveness of China's manufacturing products.

2.Manufacturing industry design transformation and expansion of new models

2.1. Defects in product innovation and technology place higher demands on designers

The development of industrial design in China originated from the late 1970s. The industrial design in the early stage mainly relied on the introduction and imitation of foreign technology [3], and the design objects were mainly concentrated on small daily necessities. With the advancement of science and technology and the development of society, the scope and characteristics of industrial design have changed with the beginning of the establishment of China. Its scope extends to the design, interaction design and material selection design of large-scale products such as automobiles. The professional knowledge and skills that future industrial designers must possess are increasingly complex, professional and multi-directional.

However, in the international market, “Made in China” is still regarded as synonymous with “Chinese rough” and “Chinese counterfeiting” [4]. In recent years, in order to reduce the cost of design innovation, some manufacturers have exported many non-technical cores and even suspected high-imitation products to domestic and even international markets. The innovation of some products is the separation and reorganization of the color and function of the prototype products, and some products are used by designers to copy the original products by methods such as reverse engineering and reverse design. Finally, the low-quality "made in china" swept the world. It has lowered the competitiveness of China's manufacturing industry in the international market and limited the level of technology and design of China's manufacturing industry. The essence of design is that innovation is not plagiarism. Simple plagiarism cannot be copied to the essence. For example, plagiarism of foreign famous cars can only be copied to the style. There is no way to master the professional design point such as the optimum angle and position of the brake clutch throttle plate, the optimum curve curvature of the seat, and the optimal angle of inclination of the A-pillar. To change the status quo of domestic manufacturing, we must give product technology, charm and core, and fully coordinate the development of the two major manufacturing design professions - industrial design and mechanical design.

2.2. Professional Reform Cultivates New Compound Talents for the Development of Manufacturing Industry

Promoting the formation of the science and technology innovation industry group requires “knowledge-based universities, industries and governments to “overlap” and interact to promote the spiral of innovation” [5]. In terms of education and training, the education and industrial needs of engineering industrial design in some colleges and universities are not equal. The learning is not related to the practical application of product design innovation. Relevant technical disciplines do not pay attention to cross-learning. What they do is generally to improve the modeling problem, attach importance to innovative design, and despise technological innovation; In the design method, the design and technology are still stuck to the traditional sketch deduction. When a series of technical and theoretical contradictions arise, the ambiguity and instability between the problem and the solution cannot be avoided. In September 2018, the Academic Degrees Committee of the State Council of China issued the "Notice on the Corresponding Adjustment of the Authorized Points for Master's and Doctoral Degrees in Existing Engineering" - the degree of industrial design engineering is adjusted to mechanical. The policy officially opened up new models of industrial design and mechanical technology, ensuring complementary integration of mechanical and industrial design disciplines. The product design innovation of rational thinking and logical thinking is reasonably applied in the design process of two professional compound talents. The policy also responded to the slogan "Created by China to Transform by China and Enhance China's Quality" and implemented the "Made in China 2025" policy.

IOP Conf. Series: Materials Science and Engineering **573** (2019) 012040 doi:10.1088/1757-899X/573/1/012040

3.Adding and expanding mechanical design capabilities from the perspective of industrial design in the new model

When designing industrial products, most manufacturing companies separate the mechanical design and industrial design parts in the form of departments, resulting in less communication between the internal and external parts of the required design products, and there is a problem that the product form does not correspond to the function, and the product iteration upgrade is too slow. The author believes that after the training of new models based on mechanical industrial design, future designers of mechanical and industrial design professionals can solve problems in product design based on professional perspectives in the design process, reducing ideological conflicts. The design thinking, technical innovation and professional skills of mechanical designers can break through the original limitations of the profession.

- In the design thinking: the formation of the market, customer needs in the primary design thinking.

In the past, mechanical designers have often neglected the customer demand, market status and other factors in order to achieve the optimal function, structure, strength and other design points of the product, resulting in the result that the developed product does not meet the market demand. However, existing algorithms such as basic theoretical calculations, chip calculations, and artificial intelligence calculations are difficult to present the true demands of consumers. Under the new training mode, mechanical engineers can adjust the design, including the shape and structure, according to the user's needs and product functions at any time to improve the original defects and deficiencies. Thereby improving the ease of use, user-friendliness and user experience of the designed products, and achieving iterative upgrade of the product under the premise of reducing rework modification.

- In technological innovation: enrich innovative thinking and promote technological progress in line with innovation needs.

From the perspective of mechanical design, the innovative ideas of mechanical designers are often limited by the logic of their professional knowledge requirements and the rigor of the nature of the disciplines. Most of the designs tend to be old-fashioned and lack of creativity. It makes the domestic manufacturing industry progress slowly in innovation and iterative upgrading. The demand of the market makes innovation more traceable, and the technology is realized by innovative thinking. For example, the designer invented the rotatable screen to meet the needs of consumer mobile phone selfie; In order to meet the needs of consumers who want to answer calls more easily during sports, wireless Bluetooth headsets, etc. were invented. Therefore, innovative thinking needs to exceed the speed of technological development in order to promote the development of technology more quickly. On the international platform, breakthroughs in prototypes and innovation are key means for designing big countries to solve the problem of innovative design of manufacturing products. Therefore, the innovative thinking of industrial design and mechanical design can improve the level of innovative design, speed up the development of technology, and promote product iterative upgrade.

- In terms of professional skills: industrial design innovation theory and technology enrich the theory of mechanical design.

The traditional manufacturing industry has a single mode of production that restricts the mode and means of mechanical design. The manufacturing cycle of the manufacturing industry is longer, which also increases the risk of corporate investment. Design methods such as computer-aided design (CAID/CAM) and computer-aided engineering (CAE) in modern industrial design theory have played an important role in shortening the product development and design cycle, and it can promote new material technology in the selection of product materials; Reverse engineering design mode and concurrent engineering design mode can greatly improve design efficiency; 3D printing technology can design samples at low cost, improve the microstructure of the prototype several times before the production of physical prototype, greatly reduce the cost of design innovation; industrial design skills, such as ergonomics, green design, bionic design, optimized design, also provides more innovative thinking and theoretical basis for mechanical design.

IOP Conf. Series: Materials Science and Engineering **573** (2019) 012040 doi:10.1088/1757-899X/573/1/012040

4.The enrichment and extension of industrial design capabilities from the perspective of mechanical design in the new model

Joseph Binbit, the father of innovation theory, once said that "innovation is the fundamental source of enterprises to achieve uncertain growth." It is difficult to consider the deep content of product quality, function and structure only by divergent thinking innovation, and also make designers unable to grasp the whole process of life cycle such as product manufacturing and market circulation. In the past, the development of industrial design products relied mainly on market demand, ergonomic requirements, and practical and innovative professional perspectives to design products. The design process was mostly filled with research and free and divergent thinking; Product design research focuses on design styling, color and human-computer interaction, focusing on testing the user's visual and interactive experience with the product[6]; The product development process generally takes the prototype as the core, and the trial and error design cycle does not achieve satisfactory results in terms of development efficiency and cost saving. In the mode of industrial design engineering transferred to the mechanical category, the author believes that it can make up for the shortcomings of industrial design from the perspective of mechanical professional manufacturing technology and logical thinking.

- From the perspective of manufacturing technology: adding theoretical support for innovation and improving the technical and feasibility of the design.

In the new mode of teaching and training, the mechanical design theory course provides technical theoretical support for industrial design, which can effectively improve the future product design level, efficiency and reliability of industrial designers. In the product design process, the number of prototypes can be reduced to avoid falling into the repeated trial and error design cycle. Reduced design cycle while reducing costs. The mechanical calculation technology and other technologies also provide inspection reports for product materials, structures and shapes for industrial design, which not only facilitates the optimization of design schemes, but also ensures the feasibility of product design. In terms of material selection, a set of material optimization methods can be established. By using software such as CES Edupack, better materials are selected; Further introducing a compensation function that combines the goals of low cost and small quality, and provides designers with the most material-standard criteria [7]. The emergence of new materials and new technologies has also provided more possibilities for design. The free combination of various materials has expanded the design ideas [8], and promoted the development of manufacturing industry on the basis of promoting product iterative upgrade.

- From the perspective of logical thinking: make the innovative design logical, and ensure the feasibility of product design.

The thinking logic of industrial designers is nourished by the knowledge of mechanical theory. Every step and design point in the design process will be optimized by logical thinking in advance. A logical technical design process speeds up product design and development and productization, and promotes product performance and other features. Toyota Motor Corporation will avoid 80% design defects before trial production, and try to achieve the highest level of optimization design. If the domestic manufacturing industry compensates most of the design defects in the product design process and before the prototype production, it is strictly improved by logical thinking to avoid the judgment errors caused by the traditional design methods. Then we can save a lot of money, reduce waste of resources and improve manufacturing levels.

5.Industrial design and mechanical integration promote a new chapter in manufacturing

Internationally, the composite development of mechanical industry design has long been prevalent, such as the iPhone's original 4-inch mobile phone design: From an ergonomic point of view, since most users use a single handheld device, the thumb swipes the fan screen in the shape of a fan, which is about 4 inches in diameter, so 4 inches is the optimal size for a mobile phone screen design; button design takes into account the material knowledge of mechanical theory, the middle button uses metal buttons, compared to metal buttons in long-term friction Plastic buttons feel firmer, smoother and

IOP Conf. Series: Materials Science and Engineering **573** (2019) 012040 doi:10.1088/1757-899X/573/1/012040

more aesthetically pleasing. It can be seen that the design work of the composite industrial design can promote the iterative upgrade of the product.

Traditional industrial design pays attention to styling and despise technology. Mechanical design pays attention to function and ignoring creativity. Under the new mode of mechanical industrial design, designers will break through the original innovative thinking and complement each other in the process of product design and innovation, avoiding the stalemate between two different professional thinking. At the same time, innovative thinking and design theory and method work together to expand the design channels and improve the design level, which in turn promotes technological innovation. The reasonable and effective connection between product design and mechanical design can promote the design and practice of industrial design and mechanical design professionals to realize the harmonious integration of function and shape in design practice[9]. Creating design methods based on technical theory and logical reasoning will invisibly increase product innovation, accuracy and technology. In addition, it is also possible to integrate new technologies, develop new processes, and innovate materials according to the characteristics of enterprise development, demand, characteristics, etc., which is of great significance for enhancing product design quality, improving product technology, and thus promoting the development of national manufacturing. Accelerate the construction of a country with high quality and significantly enhance the quality advantage of China's real economy. It must not only complement the industry's "short board" in product quality supervision, but also comprehensively improve the technical content and technological innovation of export products, and further expand the manufacturing industry [10].

6.In conclusion

Today, with the rapid development of science and technology, in order to accelerate the development of China's manufacturing industry, universities and enterprises need to improve the design innovation model, concept and ability of design talents. The new mode of training in the direction of mechanical industry design is a major opportunity and challenge for traditional manufacturing. Under the new professional direction, the industrial design and mechanical design fields will be deeply integrated in the manufacturing industry, and the innovative thinking and design methods of new composite design talents will be developed to cultivate new full-link designers from the inside to the outside. At the same time, it can form a situation in which the theory promotes technological development and technological perfection of theoretical content, which is of great significance to the transformation and upgrading of China's manufacturing industry and the realization of developed countries in cornering overtaking manufacturing.

Acknowledgments

First of all, I would like to thank the organizers for providing an academic exchange platform for industrial design industry enthusiasts, which has created a good opportunity for the development of domestic industrial design. Secondly, I would like to thank Zhang Meng and Li Wei for their help in the completion of the project. The project was funded by the project "Leading Brands in the Eight Key Industries of Yunnan Province (51600630)". I would like to express my sincere gratitude and wish the meeting a complete success.

References

[1] Zhang, J.X., Shang, Y.N., Duan, J.J. (2008) Introduction to Industrial Design. China Ocean Press,Beijing.
[2] Liu, Y.H,. Liu, Q. (2018) The Effect of Industrial Design on the Transformation and Upgrade of Manufacturing Industry from the Perspective of Industry 4.0. Packaging Engineering, 39: 113-116.
[3] Jin, Y.Z. (2013) On Industrial Design and Innovation under the New Situation. Henan Technology, 3: 35.

[4] Liu, L.Y. (2016) Abandoning and Reflection:Chinese Industrial Design in the Post-Chinese Manufacturing Age. Design, 9: 86-87.
[5] Henry, E. (2016) Three-helix innovation model. Qinghua University Press, Beijing.
[6] Tang, B.B., Guo, G., Xia, J.J. (2017) Research on the Measurement Method of Industrial Design Material Based on User Visual/ Touch Experience. Journal Of Mechanical Engineering, 53: 162-173.
[7] Li, Y., Xue, Q., Li, M.Z. (2018) Design of new energy generation vehicle for the elderly facing lightweight target. Mechanical design, 35: 112-115.
[8] Yu, S.l., Mao, Y.M. (2016) The systematic influence of 3D printing technology on industrial design in the 4.0 era. Design, 19: 64-65.
[9] Liu, L.F. (2018) Reform and Exploration of the Course Teaching of "Basic Machinery Design" in Industrial Design Specialty. Mechanical design, 35: 413-414.
[10] Peng, S.T., Li, P.F. (2018) Quality Evaluation and Improvement Path of Chinese Manufacturing Industry Development. Research on socialism with Chinese characteristics, 5: 34-40.

IOP Conf. Series: Materials Science and Engineering **573** (2019) 012041 doi:10.1088/1757-899X/573/1/012041

Application and Expansion of Rationalism in Industrial Design Curriculum Teaching

Chunmeng Weng[1], Hong Yang[2]

[1] School of Arts and Design, Wuhan University of Science and Technology, Wuhan, Hubei, 430065, China

[2]School of Arts and Design, Wuhan University of Science and Technology, Wuhan, Hubei, 430065, China

[*]Corresponding author's e-mail: Hong Yang：420659302@qq.com

Abstract: At present, rationalism is still used in design education. Through analyzing the characteristics, advantages and teaching cases of rationalism in the course design of industrial design, the paper discusses the principles and practice directions of rationalism in the field of industrial design teaching, and promotes the further development of industrial design teaching in the information age.

1. Introduction

With the development of China's economy, industrial design has received unprecedented attention. It has been written into the outline of the five-year plan for the development of the national economy three times, and the Outline of the 13th Five-year Plan is even more explicit that it is necessary to "support the construction of the industrial design center." The establishment of the National Industrial Design and Research Institute has become an important driving force for the transformation from "made in China" to "Wisdom made in China". At the same time, China's design education has achieved rapidly. According to the statistics of the Chinese Ministry of Education, there were more than 690 colleges and universities have more than 45,000 students in industrial design (including product design) and related majors in 2017, which reflecting the huge demand for industrial design professionals. From 1980 to 2015, the official definition of industrial design continued to undergo major adjustments, and its connotation and extension continued to expand. How to strengthen the scientific and rational training in the process of industrial design teaching is also the requirement of the current era.

2. The Necessity of Rationalism to Industrial Design

Rationalism refers to a philosophical method based on the recognition that human reasoning can be used as a source of knowledge. It relies on logical reasoning and deductive argumentation to establish a vast system of knowledge. Rationalism in the teaching of industrial design refers to a set of methods based on logical reasoning and deductive argumentation in the teaching process.

Since the birth of industrial design, rationalism is indispensable. Mass production must depend on rationalism to achieve, which is the logical cause of industrial production, which is embodied in: maximization of benefits, division of labor and cooperation, repeated labor in batches, strict scientific management system, etc. The mode of mass production needs standardization, and the maximization of benefit also needs formula deduction and guarantee, these are the concrete embodiment of rationalism. Therefore, rationalism, according with the logic of industrial production, not only exists widely in the practice of industrial design, but also inevitably appears in the teaching link of industrial design.

Nowadays, experience economy and service design are popular, which does not mean rationalism will exit, and it still has positive practical significance. The experience economy starts from the life and the situation, molds the consumer's sensory experience and the thought identity, seeks the new development space, these choices and evaluation need through the accurate data, the scientific analysis to provide the guarantee and the support. Nothing is more typical than big data's analysis to help segment users, find new market growth points and rationally guide the behavior and process of industrial design.

The definition of industrial design has been adjusted with the progress of the times. The definition of industrial design given by WDO in 2015 is defined as a strategic problem solving process, which consists of four aspects: product, system, service and experience, these cannot be separated from rationalism.

3. The Application of Rationalism in the Teaching of Industrial Design

3.1. Origin of Rationalism

The design of rationalism can be traced back to the period of the Deutscher Werkbund (DWB), it clearly points out that in the standardization and mass production, the aesthetic should pay more attention to the functional and scientific nature of the design, so the rationalism highlights its role. In the 1920s, the establishment of Bauhaus in Germany promoted the application of rationalism in the field of design education. Among them, the most important contribution was the basic course, which was first founded by Johannes Itten and was a compulsory course for all students. In the teaching of the basic course, Johannes Itten makes students discover the law of vision through the rational visual training. It is different from the uncertainty and irrational teaching mode of traditional art teaching, and inspires students' potential talent and imagination on the basis of rationalism. Walter Gropius, the founder of Bauhaus, also emphasizes the rationality and function, logic and technology of design, and this rational design education view and measures are gradually accepted in the world.

This scientific, rational design thought has a great influence. From the teaching example of the composition of the layout in figure 1, we can see that the visual graphic design at that time has already had a mathematical expression, emphasizing the objective analysis of form and color, and paying attention to the relationship between point, line and surface, so as to guide students understand the composition of two-dimensional space and 3D space. There are also classic buildings such as the Bauhaus school building and the Gropius residence. Moreover, Ludwig Mies van der Rohe also proposes the slogan "Less is more", which is equivalent to the idea of anti-abuse of annexes and over-decoration, and its Le Corbusier's “Mechanical aesthetics” and the Modulor can be said to be a symbol of rationalism, which is used in the design of the Marseilles apartment, the Chandigarh and the Ronchamp chapel. The Modulor of figure 2 exerts its proportional control effect to varying degrees, which can be described as the ultimate manifestation of rationalism in mathematical planning and designing, which is the concrete embodiment of rationalism in the teaching of early industrial design.

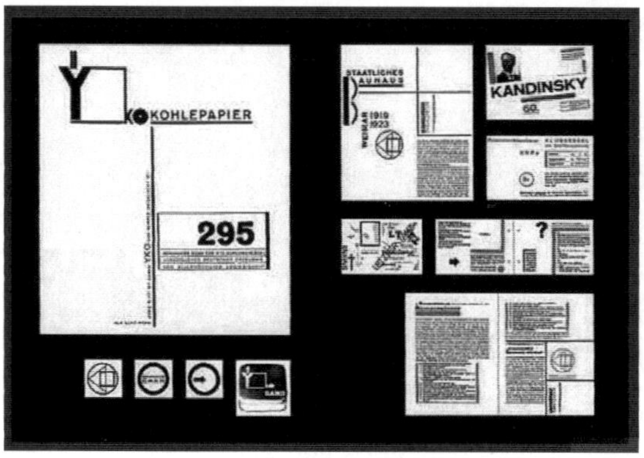

Figure 1. The layout of the Bauhaus School of Design constitutes an example of teaching.

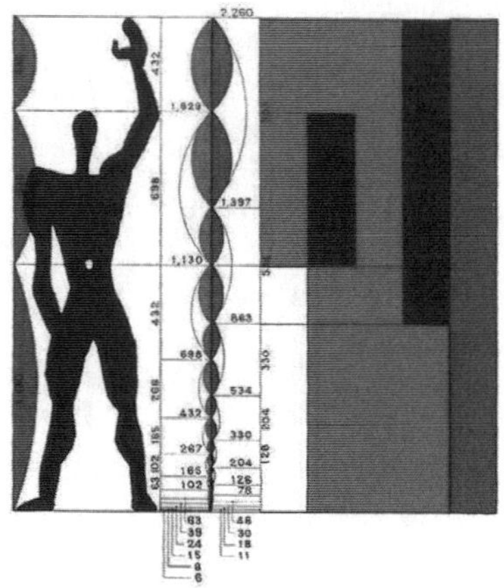

Figure 2. Le Corbusier's Modular System.

3.2. Application Examples of Rationalism in Industrial Design Teaching

At present, rationalism continues to play an important role in the teaching of industrial design, such as layout design, design expression, product design and so on. In the course of teaching, in addition to creating students' imagination from the perspective of sensibility, it is necessary to strengthen students' ability to master products from a rational perspective. For example, the “Layout Design” course is to arrange and combine the limited visual elements organically on the layout, and to present the rational thinking in a personalized way. In the teaching, it has its own design principles compared to other design courses. As shown in figure 3, the analysis of the overall layout design shows that in the design, we should pay attention to the design principles such as grid composition, primary and secondary distinction, classification and arrangement. There are also general basic forms for the layout design of text and graphics, such as the division between the upper and lower sides and the left and right sides, the repeated arrangement, frame arrangement, and so on.In this way, the designer is relatively easy to judge in the design process. If you do not consider any rules when designing a book or magazine, it will produce very messy results most of the time.

In the course teaching of “Product Form Design”, geometry is used to explore the laws of natural proportion, such as the golden section, perfect proportion and Fibonacci sequence, and so on. The quantitative mathematical proportion is the centralized reflection to the things with natural beauty, structure and law of change. In fact, there are strict mathematical rules in the modeling scale of many successful product designs. The positive cone pot designed by Aldo Rossi as shown in figure 4 is a cone formed by rotation of an equilateral triangle, which makes the whole shape easily decomposed into a 3×3 mesh system. The top of the kettle is just in the middle of the top three grids, the grid of the middle part comprises a spout and a handle, and the handle of the kettle can be regarded as an inverted right triangle, or a part of a square. This design makes the various basic geometries a part of its structure and also allows the bottom of the kettle to be in contact with the maximum area of the heat source. Strict mathematical relations make minimalist, generous, and observable products a classic existence.

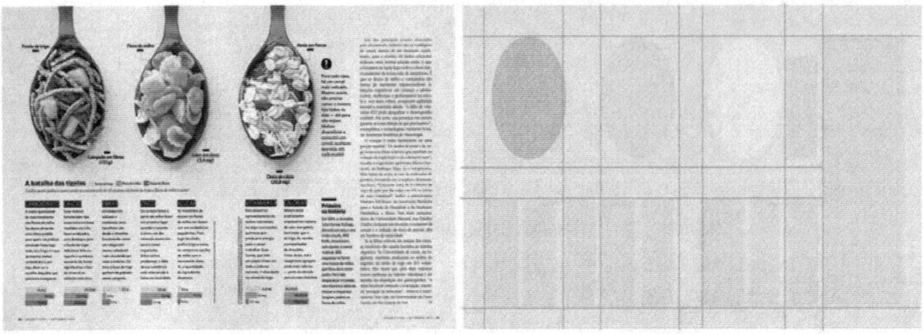

Figure 3. "Layout Design" course assignment analysis.

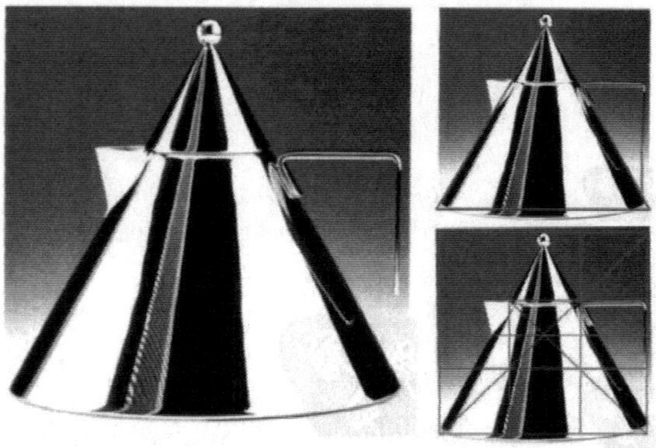

Figure 4. Positive cone pot.（Source: Author: Kimberly Elam Translator: Lee Leshan title: "Design Geometry" Press: China Water Power Press Publication date: 2003）.

In the course teaching of "Product Form Design", the application of rationalism can be reflected by the training of product control line. In the product form design, the form of the curved surface is becoming more complex, and the control line is required to organically link each visual element, the whole and the part through a certain relationship. For example, when designing the electric blanket controller, we first add and design the necessary parts of the product, such as display screen, display lamp, icon, button, etc., which are added to the surface of the product. In the process of design, it is found that more piecemeal parts are easy to lead to the disorder of the whole, so the control line design is needed in the surface modeling of the product. The long blue C-type hidden line in the diagram is tangent to the ellipse of the outer frame of the screen, and the upper end is tangent to the ellipse of the outer frame of the screen. The lower end gradually vanishes, the slide button track design is also the continuation of the blanking line, at the same time the other buttons oblique arrangement is also parallel to the lower end of the C type blanking line, so that all the elements are unified, the whole product is obvious more flexible and convenient. Through the use of these rational teaching links, students in the design process do not blindly rely on the feeling, they can be done to follow the evidence to make the design process more organized. At the same time, the results of the design guided by rationalism are more persuasive and acceptable.

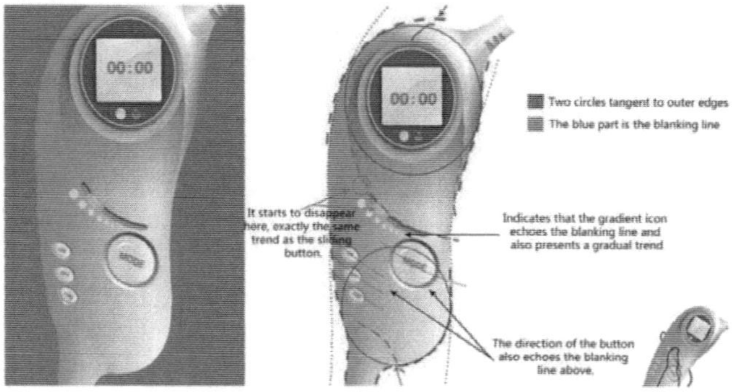

Figure 5. Electric blanket controller / designer: Fangfang Hao , Wuhan University of Science and Technology.

It can be seen that rationalism has been reflected in Bauhaus teaching as long ago, although rationalism training has always existed in the current industrial design teaching, the phenomenon that sensibility is greater than reason is still common in students' work. In the work of students, it is necessary to strengthen rational analysis and training in the usual teaching to ensure that the results of design output are more scientific, reasonable and reliable.

4. The Expansion of Rationalism in the Teaching of Industrial Design

Although the teaching of industrial design has been carried out with different degrees of reform, there are still a lot of teaching contents and links in traditional art teaching methods, and the proportion of art sensual teaching contents is very high. The strengthening of rationalism in the field of industrial design teaching promotes the development of industrial design teaching, and infuses the characteristics of logic, clarity, connection and continuity into the teaching of industrial design, which makes the teaching of industrial design more abundant.

4.1. The Expansion Principle of Rationalism

Nowadays, the mainstream teaching system of domestic industrial design can be roughly divided into four basic modules: basic course, technical foundation, specialized course, expansion part.

Every plate and every link should be fully involved in rationalism. We should be omni-directional, full-link expansion, each plate needs rationalism application. From the basic course, the students' rational thinking is cultivated, and the rational thinking of the technical foundation course is strengthened, so that the students can be applied in the professional courses, and then the students can use them well in the expansion course. Rationalism teaching should run through the whole course of industrial design teaching, so as to get better results. At the same time, we must also emphasize the expansion of the course link. For the design class subject, the emotional teaching link in the classroom is indispensable, but it is necessary to pay attention to the rational node control in the teaching process of the perceptual experience, so that students can feel the importance of rationalism in the classroom.

4.2. The Specific Expansion Direction and Thinking of Rationalism

Design education is carried out around the design process, and the design process, as shown in figure 6, requires three parts: design investigation, design conception and design and manufacture. In all aspects, the addition of rationalism can help students to design better.

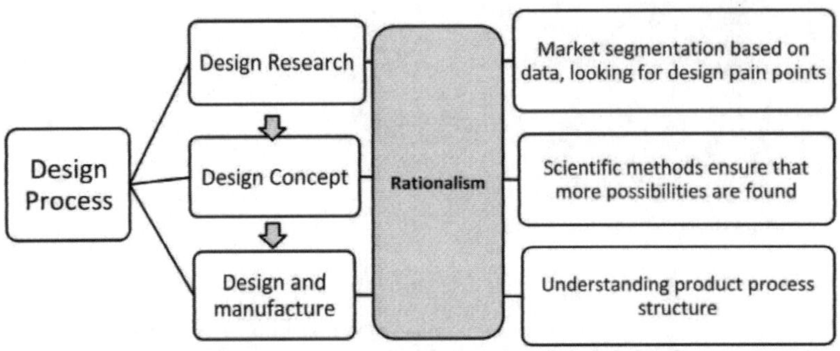

Figure 6. Design process(self-painting).

The preliminary research of design plays an important role, which determines whether a product has the meaning of design. Random selection is likely to conflict with the existing product in the market. With the development of science and technology and the application of intelligent products, the current industrial design has changed from "thing" to "person", from "product itself" to "product service", which also shows that people pay more attention to the psychology, behavior, demand and so on of "person". The current design is built around the user's behavior, which can be trained through rationalism. For example, in the process of product design and research, POEMS observation method can be used for recording. POEMS observation method is based on five aspects: people, objects, environment, messages and service. The establishment of POEMS framework in the course of investigation can guide the observers to observe what to observe, how to record and consider the later collation and analysis. Accurate and clear product design research can provide effective support for subsequent designs.

After completion the preliminary research of the product, the use of rational thinking in the form design of the product can increase the possibility of the product. The design is carried out from the plan view, and the inference process from two-dimensional to three-dimensional requires the use of rational thinking. As shown in figure 7, the plane curves that have been designed are named 1, 2, and their corresponding three-dimensional shapes are analyzed. When curves 1, 2 are projection of faces, the shape is simply extruded; When 1 is a projection of a line, but there is no turning point in the outer profile, and 2 is a projection of a face, it is a relatively rounded shape. In addition to these two shapes, there are many possibilities, as shown in figure 8. This is the "orthographic projection inverse deductive method", a basic training technique for deducing three-dimensional shapes from plan view sketches. It is not only clear in train of thought, easy to grasp, but also able to perform more relevant shapes according to the different combinations of the source of the projection. It is very conducive to the deduction of the form, and can cultivate students' ability to evolve the plan view sketch into a three-dimensional form.

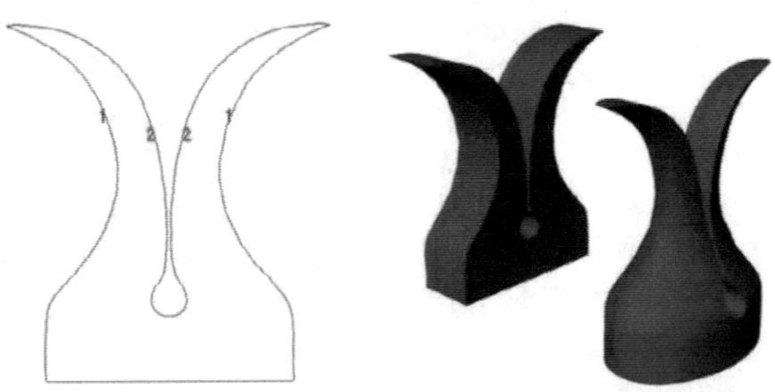

Figure 7. Positive projection reverse deduction method example1.(Source: Author: Chunmeng Weng, Xianfeng Ai title: "Product Form Design" Publisher: Peking University Press Publication date: 2016)

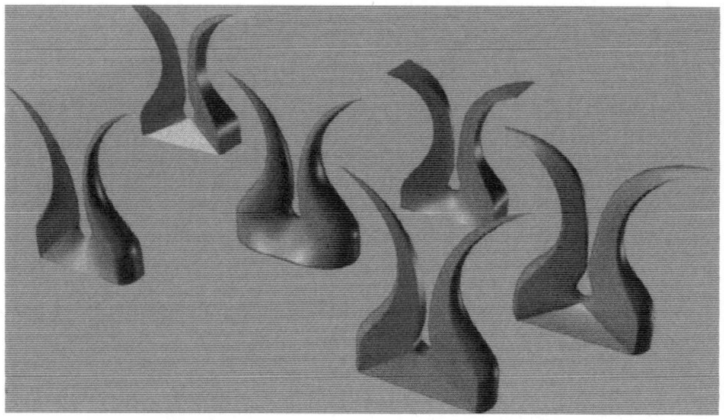

Figure 8. Positive projection reverse deduction method example2.(Source: Author: Chunmeng Weng, Xianfeng Ai title: "Product Form Design" Publisher: Peking University Press Publication date: 2016).

In the design process, it is necessary to fully consider the achievability of the molding process and functional structure of the manufacturing process, especially the function-oriented design. Therefore, it should pass certain in the teaching Methods and training to guide students to form rational derivations in design. As shown in figure 9, the function analysis of the product first determines the main function, determines the sub-function under its premise, and then expands the existing sub-function according to the second-level function module.In this derivation process, the necessary functions can be identified, the redundant functions can be identified, and the level of functionality that needs to be improved is clearly identified, making it easier to discover innovations in the design. As shown in figure 10, the student's shoe washing machine product display, through the analysis of the overall structural rationality of the product, the problems found in the entire product use process, and targeted solutions. For example, in the fixing aspect of the shoe, the retractable rotating bracket is used to fix the shoe, and the length of the bracket can be controlled according to the system; in the water sprinkling, three-dimensional spray water flow technology is used, and the shoes on the bracket are sprayed on the inner cylinder. And there are other aspects that have been improved and innovated. Similar training, methods and skills can be added in curriculum teaching, so rationalism can be expanded better in teaching.

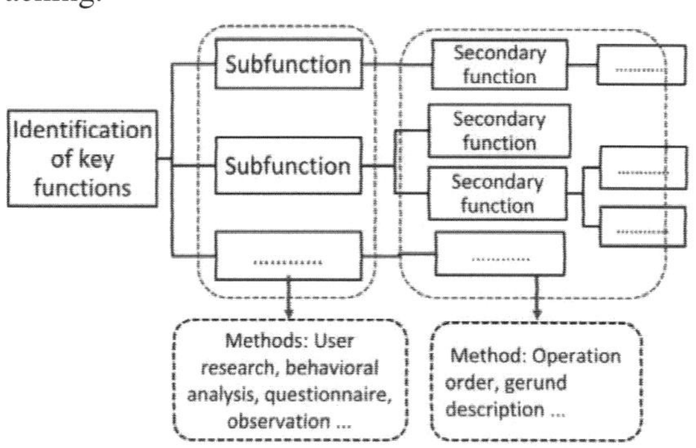

Figure 9. Product function design step diagram (self-painting).

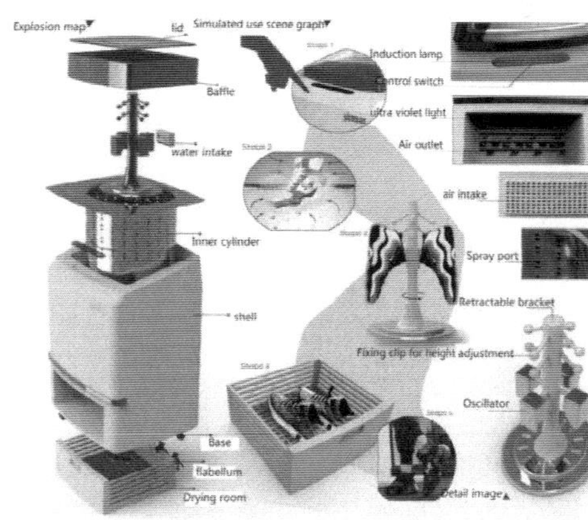

Figure 10. Household washing machine display map / designer: Cong Zhou.

5. Conclusion

With the development of society, the concept and method of industrial design teaching should also be changed accordingly. In the current courses of industrial design, the concept of rationalism has

IOP Conf. Series: Materials Science and Engineering **573** (2019) 012041 doi:10.1088/1757-899X/573/1/012041

gradually weakened. However, as far as the practice of industrial development is concerned, there is an urgent need to strengthen and expand it. Rationalism in industrial design teaching not only helps to cultivate industrial designers who meet the needs of the industry, comprehensive design talents with professional literacy such as design integration ability, but also promotes the integration of design education and market to adapt to the development of the information age.

References

[1] Han Jun. (2009) Development of the Rationalism Design Conception. Journal of Jilin College of the Arts, (2): 29–30.
[2] Zhao Yujie. (2010) From Perceptual to Rational: an Analysis of Bauhaus' Design Thought. Artistic Exploration, 24: 54–55.
[3] Zhu Bing. (2011) On the New Rational Design Thought of Bauhaus. Art Criticism, (09):133-136.
[4] Cai Shunxing. (2013) Contradictory Analysis of Rationalism Layout Design. Journal of Nanjing Arts Institute(Fine Arts & Design), (5): 175–177.
[5] Cheng Ling. (2014) The Development and Pedagogical Implication of Western Rationalism. Contemporary Education and Culture, 6: 48–53.
[6] Yin Hu. (2014) Exploration of Interdisciplinary Comprehensive Industrial Design Teaching Mode. Journal of Graphics, 35: 459–463.
[7] Zhang Hechen, Xiang Zhongxia, Yang Junyu. (2017) Explorations on the Cultivation Mode of Industrial Design Talents srom the Interdisciplinary Perspective . Design, (13): 85–87.
[8] Jiang Lin. (2017) Thoughts on the Current Situation and Strategy of Industrial Design Education in China. Education Modernization, (16): 106–107.
[9] Wu Lei. (2018) Teaching method and practice of industrial design specialty based on Emerging Engineering Education. Journal of Architectural Education in Institutions of Higher Learning, 27: 10–13.

IOP Conf. Series: Materials Science and Engineering **573** (2019) 012042 doi:10.1088/1757-899X/573/1/012042

General Design Education Practice: History of Design Materials

Jinbo Sun

School of Art, University of Sanya, Sanya, Hainan Provice, 572000, China

Corresponding author's e-mail: j.b.sun@qq.com

Abstract. Based on the universality, democracy and generality of design, this paper constructs the basic framework of general design education based on general education and design education. The general design education is an extension of the field of general education. The role and significance of designing general education are discussed in the paper based on the practice example of history of design material.

1. Introduction

The concept of general education has been widely accepted. General education originated in the United States in the 1950s. It has accumulated a lot of successful experiences and examples, and has a relatively comprehensive and far-reaching impact on education world widely. As the human economy and society enter the new stage of "Industry 4.0" and "Innovation 2.0", new demands are placed on comprehensive and innovative talents with diverse academic backgrounds. Therefore, as an important part of higher education, general education should also be expanded in accordance with the changes in the demand for talents and time changes.

2. General education

With the purpose of cultivating "total people", general education focuses on cultivating people with vision, understanding, liberal spirit, and beautiful emotions, not only on cultivating special or specialization in a certain professional field. General education requires students to comprehensively master the overall state of human knowledge achievement, including history, basic standpoint, and modes of thinking in major fields. They will have the ability of selecting or forming their own develop orientation or research area rationally based on existing knowledge reserves and learning experiences. With broad and solid professional foundation, clear and logical cognitive structure, complete knowledge structure, flexible ability structure, noble moral sentiment, and independent thinking and inquiry, students from general education can achieve mastery through a comprehensive study. They participate in social public affairs initiatively and effectively. They will be able to understand the most important issues in the current and future society. They will become citizens with a comprehensive personality quality, broad knowledge horizon, and social responsibility. [1]

Differences and links exist between general education and other training models. General education emphasizes combinations between multiple disciplines instead of universal genius education. It educates students to think and analysis from an interdisciplinary perspective, which can deal with problems from a broader perspective of cross-disciplinary and cultural exchange. Compared with Specialized Education, general education emphasizes the formation of a broad and solid knowledge base after extensive knowledge of the main areas of human knowledge, and then forms or selects a professional direction or field. Specialized education focuses on cultivating a particular field of expertise, knowledge and abilities. Compared with liberal education, general education is the inheritance and transcendence of liberal education in terms of knowledge function, social value and human development model. It is the integration and development of elite education and vocational education. Compared with humanities education, general education includes not only the content of humanities, art, and social sciences, but also the content of natural science and technology, which is broader and more inclusive.

3. General design education

3.1. Universality and democracy of design

Design is universal as a specific ability of human beings. As a verb, design is the intentional creation of a plan or specification, using a series of components, for the construction of an object or system or for the implementation of an activity or process. [2] As a generalized behavioural model of human beings, design is a comprehensive outcome of the three human talent abilities: skills of critical thinking, ability of creativity and ability of practice. Therefore, design is a behaviour based on human-specific abilities and is an ability that each of us has full potential to utilize. However, like other human beings, capabilities of design must be stimulated and cultivated. To a large extent, the existence and role of design capabilities depend to the circumstances in which the subject (whether individual or collective) live, and to stimulation and support provided by circumstances. Otherwise, design capabilities will be diminished or misguide to an invalid direction. [3]

The universality of design makes the activity democratic. As a unique problem-analysis and problem-solving ability of human beings, anyone with the motivation and opportunity can participate in design process, rather than the capabilities and activities of professionals engineers or designers. Compared with the professional design activities, the individual design activities in DIY (do it yourself) exist more frequently in amateur fields. They are self-driven and self-directed design or manufacture activities. The whole design process to the design object is more democratic and close to design subject. From the perspective of human history, this fruitful and creative activity enables human to actively participate in the design process at multiple levels and to express more individualized aesthetics through individual production, mass production or passive consumption. [4] Therefore, the democracy of design makes it into an interface and tool for humans to understand and transform nature and all things around. At present, the widespread popularity and application of the Internet has made the modern society more intellectual. The mass innovation design, open innovation design and user participation innovation design gradually emerge. In the future, the democratization of knowledge innovation in the whole society will be more remarkable. [5]

3.2. Construction of General Design Education

Branches of design encompasses a wide range of professions. In Chinese subject classification system, according to the *National Standard for the Quality of Undergraduate Professional Teaching in the General Colleges and Universities* promulgated by the Ministry of Education in 2017, the design category (professional code 1305) includes 9 majors in a narrow sense, including art design (130501), visual communication design (130502), environmental design (130503), product design (130504), clothing and apparel design (130505), public art (130506), crafts (130507), digital media art (130508), art and technology (130509T). In a broad sense, many disciplines, including mechanical design, architectural design, light industry, textiles, etc., can be included in the broad scope of design disciplines. A wide range of professional disciplines corroborates the universality, democracy, importance of design knowledge in human knowledge systems. It is precisely because of this, the construction of the general design education framework can not only start from the complicated disciplines, but from the nature and the commonality of design.

Design is the practical behaviour of human creative wisdom applying to material products and spiritual products. The whole process contains the subject, object, law and method. After analysing of the process model of human practice, we can map to the design discipline model with design thinking, engineering technology and design materials as the core. As shown in Figure 1, the whole process of practice is the process by which a subject (human) subjects the object (natural or man-made). The most important thing in the practice is the thinking of the subject. In the process of practice, the subject gradually accumulates and masters the laws and methods. And the object is the material. The thinking, laws, methods, and materials map into the framework of design, evolve into design thinking, engineering technology, and design materials, which construct the core model of design discipline. Differences exists between the narrow and broad design disciplines. These differences can't change the reality that, human beings utilize design thinking to construct design ideas and frameworks, and process design materials with engineering techniques, and finally create artificial objects or forms that meet the human requirements. Therefore, the core model can basically summarize the core knowledge structure of the design discipline.

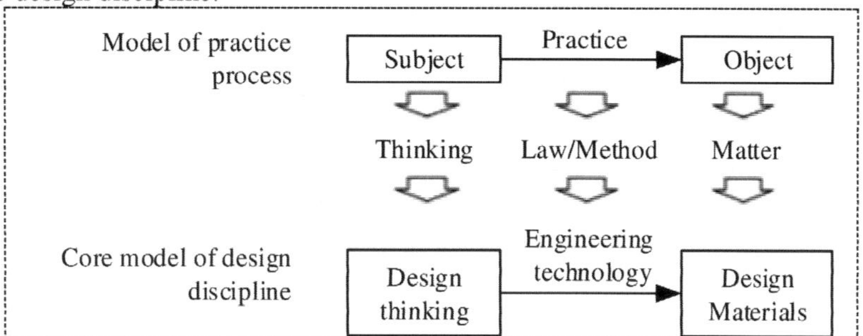

Figure 1. Core model of design discipline.

As the core knowledge of design discipline, design thinking, engineering technology, and design materials will be the main content of general design education. On the basis of the core model of the design discipline, the basic knowledge framework of the design discipline is summarized by combing the relevant knowledge of the discipline. As shown in Figure 2, in the design discipline knowledge system, design thinking, engineering technology and design materials are core, relevant knowledge categories have different focuses on application needs and practical issues in different fields, and can be gradually extended to design procedures and methods, innovative design methods, engineering design, and ergonomics, design cognition, materials and techniques, design expression. This system can basically cover the subject knowledge framework for entity design.

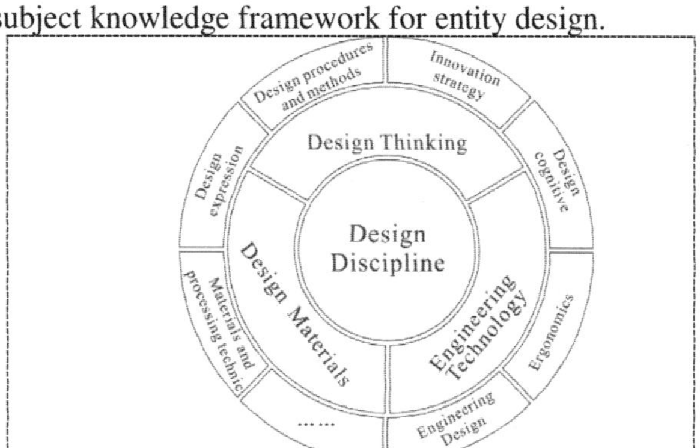

Figure 2. Basic knowledge structure of design.

3.3. The significance of designing general education

3.3.1. General design education provides individuals with more possibilities for development potential. The total development of the individual is the embodiment of social progress. The broader understanding of the progress of all aspects of human knowledge is an important basis for the overall development of individuals, which is in line with the fundamental principles of general education. In

China, higher education has been too focused on the cultivation of professional talents for a long time, and neglected the role of general education in the general knowledge structure.

This education model results the disconnection of art and human disciplines with engineering and technology disciplines. As a tool to popularize engineering thinking and knowledge and cultivate comprehensive creativity, design general education can bridge the intercommunication between different disciplines in human practice. Facing complex human social development trends and directions, design general education can promote knowledge construction to a more comprehensive and accommodating framework for individuals, and create more possibilities in knowledge for individual development.

3.3.2. General design education meets the needs of the comprehensive knowledge structure of talents in the era of "Industry 4.0" and "Innovation 2.0".

The requirements of the new era for the knowledge structure of talents are comprehensive, especially in the context of the "Industry 4.0" and "Innovation 2.0" era. "Industry 4.0" is a heterogeneous and customized industry through data flow driven automation technology, from economies of scale to economies of scope, and the cost of homogenization and scale. [6] The traditional science and technology innovation activities are oriented by technology development and scientific research, operated by research personnel in the laboratory or company. The "innovation 2.0" era innovations are user-cantered and society-oriented. In the knowledge-based society, "innovation 2.0" focus on people requirements. The challenge of this innovation model is gradually showing its vitality and potential value. [5,7]

The requirements for talents in the era of national innovation, or the requirements for people, are not the characteristics of traditional professionals. They need to comprehensively master all aspects of humanities, science and technology in a comprehensive knowledge framework, and have engineering thinking. On the basis of framework, they create innovation from different professions and perspective. General design education will play an important role in the cultivation of talents in the field of non-engineering. It is a feasible way to popularize and train engineering and systematic thinking, master the basic tools and methods of innovation, and understand the material basis of human production and living.

3.3.3. General design education will meet China's demand for the quality of compound talents in the manufacture upgrade.

The reform and opening up in the past forty years has pushed China to a point where deep transformation in manufacture is needed. China need to step into the next level to the manufacturing countries. In order to adapt to this development appeal, in 2015, China has proposed a series of ideas, plans and programs including the "Made in China 2025" plan, the "new engineering" construction, the "mass entrepreneurship, and the innovation". The core force that supports China's next step is the compound talents that adapt to the needs of development. The "Made in China 2025" document clearly puts forward the guiding ideology to meet these demands, including "taking talent as the foundation", "taking the development path led by talents", and improving the multi-level talent training system. [8]

At present, China's development needs not only a compound talented person with a comprehensive knowledge system, but also a variety of product life cycle related disciplines compound talent system. It includes the comprehensive "new engineering" training program, which cultivates engineering and scientific talents with the purpose of serving the needs of national strategic development and the construction of international competitive advantages. Also, human resource in design, management, human and art fields, with engineering thinking, understanding of design knowledge and integrated innovation capabilities are need. The multidisciplinary and complex talent cultivate system is the foundation of China's competitiveness and the basis for enhancing creativity and innovation. The cultivation of "new engineering" programs are more inclined to output high-level engineering and scientific talents, while the engineering thinking, design methods and innovative thinking need to be cultivated with general design education.

4. General education of design materials

Materials, information and energy are the core foundation of modern material society, they shape the current human lifestyle. As an important core of design general education courses, the design materials course is of great significance to the design subject knowledge framework in the current popularization, and will assist the development of innovation education.

4.1. The role of design materials

4.1.1. Design materials are the starting point for design activities.

The selection of design materials is the most basic part in the design activity process, and the quality of materials choices will affect the taste of products directly. In the selection of materials, a lot of influencing factors should be considered, including the inherent physical and chemical properties of the materials, the organic combination of materials, people and the environment [9], as well as the appeal of material emotional expression in Immaterial society era.

4.1.2. Design materials are the interaction medium between design and human.

Humans use materials to compose design, create the material-human relationship with the design procedure. The basic function of design materials is the realization of product function. In addition, design materials will impact human in sensory channels, such as vision, touch, taste, hearing and smell. Among them, vision and touch are especially important to human perception. Therefore, the research of physical properties and Kansei Engineering [10] properties to design materials, called CMF(colour, material and finishing), have been developed in recent years. [9]

4.1.3. The iteration of the design material drives the iteration of the design.

Throughout the history of human development, the materials used in different eras determine the state of production and living at that time. From the Stone Age, the Bronze Age, the Iron Age, the Iron Age, the Composite Material Age, to the Nanomaterials era, the distinctive features of each era are reflected in the design materials. Different eras and materials are in the one-to-one relationship. There are no material beyond time to much, only stable developments step by step.

4.2. The generalized value of design materials

Material is the fundamental basis of all human production and living activities. It has always been a symbol of productivity and a milestone in the progress of human society. The ability to comprehend and utilize materials determines the shape of society and the quality of human life. The world is made up of matter. The design material is the matter entity that people use to make certain properties, such as machines, devices, structures, etc. Material is the material basis of human social life. The development of materials has caused changes in the times and promoted human civilization and social progress. Materials, energy and information are listed as the three pillars of modern science and technology in the new era of knowledge economy, and their role and significance are particularly important now and in the future.

Beginning with the British Industrial Revolution, mass production of products has entered the human vision. Based on the low social productivity, rare product selectivity, insufficient consumer purchasing power, and the shortage of goods, consumers pay more attention to the function, structure, price and service life of products when purchasing goods. In the early stage of the function-oriented design thinking, in order to increase market profit and reduce production cost, designers had used materials at the minimal level to fit the function and structural rigidity requirements. At that time the colour and surface finishing process of products were as simple as possible. [11]

IOP Conf. Series: Materials Science and Engineering **573** (2019) 012042 doi:10.1088/1757-899X/573/1/012042

With the rapid development of social, economy, culture and information technology, the diversified evaluation criteria of consumers make the decision-making in purchasing no longer simply under the guidance of "function first". The characteristics of composite personality, aesthetics, experience and brand gradually become important references that influences purchasing decisions. The guidance of human beings' requirements has gradually moved from "material" level to "spiritual" level. After all, design materials compose all the physical products that in a "spiritual level". In the context of the human consumption system, whether as a producer, designer, consumer, communicator, bystander or user, all the system participators should understand hand master the material. This is the main significance of design materials in general design education.

4.3. The course of design materials

As the content of general design education, history of design material has two dimensions in knowledge system: the vertical dimension describes the design materials used in different eras, the horizontal dimension describes use of different materials at the same eras. With the two dimensions, the course of history of design material will construct a clear and complete development of design materials history.

The course is based on human history, design history, material technology and history, and lifestyle research. At the view of evolution in lifestyle, the course introduces the materials with the back ground information of society condition, economic characteristics and technology stage. In order to create a comprehensive image of materials, the possibilities and rationalities about materials will be analyzed and discussed in the course. The entire course is promoted layer by layer, according to the following sections:

- The relationship between human and material. This section introduces the development history of interweaving of human and design materials with special topics.
- Design materials of the Stone Age. This section introduces the selection and exploration of materials in the Stone Age, including the production, processing and use of stoneware, animal and plant materials.
- Design materials of farming and nomadic era. This section introduces the selection and use of materials in the farming and nomadic era, the changes happened in this era due to lifestyle characteristics, including the development and use of materials such as wood, metals, ceramics, and biology.
- Design materials of the industrial age. This section introduces the types and characteristics of design materials in the industrial age, including the origins of the industrial era, the discovery, development and use of metal materials, ceramics, polymer materials, new materials and other types of materials.
- Design materials of the information age. This section introduces changes in the use and development of materials in the information age, including the evolution of people's production and lifestyle in the information age, new materials and biological materials.
- Design materials in the post-industrial era. This section introduces changes in the use of materials in new context caused by social structure and social relationships in the post-industrial era, including materials for non-material design, wearable design, and emotional materials.
- Companionship - summary and discussion. This section summarizes materials in the process of the development and progress of human society. Design materials are the foundation of the era and the symbol of civilization. The history of human civilization is a history of using materials, making materials, and creating materials.

4.4. Significance of the history of design materials

This course introduces the historical process of human beings' discoveries, developments and utilizations of materials in the activities of designing creation. From the perspective of the development of design materials, the course describes richness and colourfulness of human development history. The meaningfulness, characteristics and significance of the material foundation in different stages of human society development will be the knowledge base in this course. And the course will help the participators to construct a scientific historical materialism and a sustainable development concept.

In the course, students from different majors have learned about the development history, physical and chemical properties, application scope, and twists and turns in development and use of design materials. They will understand engineering thinking and development. The content of thinking, iterative thinking, and historical materialism from course. Combining the original knowledge frameworks of different professional that students already has, the course will enrich their understanding of human society, material world and artificial world, and promote the improvement of their overall knowledge structure.

5. Conclusion

A history of human civilization can be understood as a history of human design, also known as the history of material development. Every new discovery and application of materials technology will promote design progress, which will bring about tremendous changes in social production and human life, and advancement of human society. The framework for designing general knowledge is to sort out the knowledge of design disciplines, and to enrich the general education. A more comprehensive general education will help more people build a more flexible knowledge system, strengthen thinking of design and innovative, and bring more possibilities for people's total development.

As the core of general design education, history of design materials course also serves as an early exploration of general design education. The course has achieved certain effects in improving the knowledge system and knowledge spread of design disciplines. Just like the material civilization and spiritual civilization of the human society moving forward, the history of design materials is constantly updated and iterative. The follow-up plan of this course will be gradually improved according to the development of the discipline and the needs of general education with scientization and, systematization.

Acknowledgments

Hainan Provincial Higher Education Educational Reform Research Funding Project (Project No. Hnjg2018-68)

References

[1] Xiangming. C. (2006) An analysis of some concepts about general education. Journal of Higher Education (03), 64–68.
[2] Wikipedia. (2019) Design. https://en.wikipedia.org/wiki/Design.
[3] Manzini, E. (2015) Design, when everybody designs: An introduction to design for social innovation. MIT press, Cambridge.
[4] Atkinson, P. (2006) Do it yourself: democracy and design. Journal of design history, 19(1), 1–10.
[5] Gang. S., Qiang. T., Rui. C., and etc. (2008) Technological Innovation from the Perspective of Complexity Science. Impact of Science on Society, 02: 28-33.
[6] Sendler, U. (2013). Industrie 4.0: Beherrschung der industriellen Komplexität mit SysLM. Springer-Verlag, Berlin.
[7] Gang. S., Nan. Z. (2009) Innovation 2.0: Democratizing Innovation in the Knowledge Society. China Soft Science, 10: 60-66.
[8] Ministry of Industry and Information Technology Planning Division (2015), Made in China 2025, http://www.miit.gov.cn/n973401/n1234620/n1234622/c4409653/content.html.
[9] Becerra, L., and R. Balla. (2016) CMF Design: The Fundamental Principles of Colour, Material and Finish Design. Frame Publishers, Amsterdam.
[10] Nagamachi, M. (1995) Kansei engineering: a new ergonomic consumer-oriented technology for product development. International Journal of industrial ergonomics, 15(1):3–11.

[11] Sanxin. L. (2014) CMF CREATE THE PERFECT PRODUCT USER EXPERIENCE. Design. (12): 114–116.

IOP Conf. Series: Materials Science and Engineering **573** (2019) 012043 doi:10.1088/1757-899X/573/1/012043

Product Design Decision Oriented by Hierarchical Relevance of User's First Perceptual Demand

Yi Wang *, **Siyin Lin**

Art and Design School, Xi'an university of Technology, 710054, China

*Corresponding author's e-mail:wymly@yeah.net

Abstract: The characteristics of "multi-dimensions, multi-hierarchy and correlation" of users towards the same product are shown in their perceptual demands. However, one or two perceptual expressions cannot precisely describe users' comprehensive perceptual demands for products. A semantic-associated multiple perceptual demand model is proposed in this paper, focusing on the multiple perceptual demand model oriented by users characteristics. Through feature analysis of users and perception evaluation test to product, different perceptual demands for products corresponding to the typical characteristics of different users are established. Then the *n*-dimensional perceptual demand information of users to product plan and the corresponding position in the semantic space with typical features of user groups can be comprehensively described. By taking the design of electric scooters as the example, the "Product Design Decision Oriented by Hierarchical Relevance of User's First Perceptual Demand" is verified and a hierarchical classification method for first perceptual demand is proposed.

1. Introduction

User's preference oriented personalized product design has become the key to attract purchase. Scholars established user emotional demand model to integrate user's requests into the early stage of design. Kansei engineering combines perceptional information of users with product modeling attributes through mathematical quantification. However, users often describe their emotional demands for products through language, while language has different meanings for different people and often shows obvious ambiguity and uncertainty. So it is difficult to predict users' perceptual demand for products and combine these demands with modeling features of products. Studies based on product conceptual design framework have proposed a user demand mathematical model to integrate user's perception into product design. Literature proposed that different fuzzy models can be used to determine product schemes and special demands of users will be transformed into different actual product schemes.Literature proposed that product design based on emotional space can comprehensively deal with emotional response of users to product scheme.

From the above studies, it can be seen that effective acquisition and processing of users' perceptual demand information is the basis of in-depth research. However, users' perceptual evaluation to products is described by language. In addition to ambiguity and uncertainty, their perceptual demands for the same product also show "multi-dimensional, multi-level and dynamic" features.Therefore, one or two perceptual expressions to product cannot cover the essential demand of users for product. Therefore, this paper proposes a multi-dimensional perceptual demand model construction method based on semantic associated hierarchy process. Through feature analysis of users and perception evaluation test to product, different perceptual demands for products corresponding to the typical

IOP Conf. Series: Materials Science and Engineering **573** (2019) 012043 doi:10.1088/1757-899X/573/1/012043

characteristics of different users are established. Based on the example, which has laid the foundation for personalized design of product based on semantic retrieval in future studies.

2. Analysis of users' perceptual demands and modeling

2.1. Analysis of perceptual demands

From the perspective of product design, users' demands are divided into demands for function, performance, structure, aesthetic perception, economy, reliability, maintenance, safety etc. According to the expression form, user demand information can be divided into numerical data (including discrete data and continuous data), semantic data (fuzzy data, redundant data, missing data) . In literature , according to the mapping relationship between user demand information and product design parameters, user demands are roughly divided into structured numerical demands and perceptual demands.

The structured numerical demand refers to user's ability to clearly express the characteristics of the product by numerical values. It is a numerical data and is directly mapped to the characteristic parameters corresponding to the product.

Perceptual demand is mainly the response generated by user's perceptual evaluation to product features. It is mostly qualitative semantic description data in the genre of language description, such as the aesthetic appearance of the product, style perception, product image and other evaluation. This information cannot be directly transformed and exhibited on product features, it must be analyzed, quantified and mapped with parameterized product features.User's perceptual demand for product is not unique. Even a single feature of the product will bring various demand characteristics from users, such as the demand for color, fashion, dynamic, feminine, nobility and so on.

2.2. Multidimensional perceptual demand model with semantic association

Demand modeling defines and analyzes the characteristics of user demands through mining, cluster analysis, etc., and models with abstract demand descriptions are established to aid product design decisions.The demand hierarchy model based on the association among the demand information and the topological relationship enables the product design to fully meet the needs of different levels of users. However, when establishing a multi-dimensional demand classification system, layer-by-layer mapping of different view models is required, and the process is too complicated to turn into the actual design. Therefore, it remains as a hot issue of current research.

Although users' perceptual demand for products is diverse, it also shows hierarchies to a certain degree, that is, the most important (primary) perceptual requests, secondary perceptual requests, weaker requests, etc. $R_i, \overline{R_i}, (i = 1,2 \cdots n)$ is set as an anti-sense semantic phrase for describing users' perceptual requests, such as "quiet-dynamic" etc. In addition, there is certain relevance among the semantic expressions used by users to describe their perceptual demand for a product. For example, in a user's perceptual demand for certain product, "elegance" has a stronger association with "simplicity and softness" than with "solidity and sports". "Elegance" and "nobility" are closely related in a higher level, while "sports" and "noble" are related in the middle level. On the contrary, "elegant" and "vulgar" are antonyms, the latter is in low level relation with "noble", and the antonym of "sports" is "quietness", which is also strongly associated with "noble" in the middle level. Therefore, "emotional semantic ball" model can be used to express a user's multidimensional perceptual demand, as shown in Figure 1. However, how to classify, refine, simplify and effectively map these sensible needs to product design features is the key to the decision-making of the personalized design of product oriented by users' visual perception.

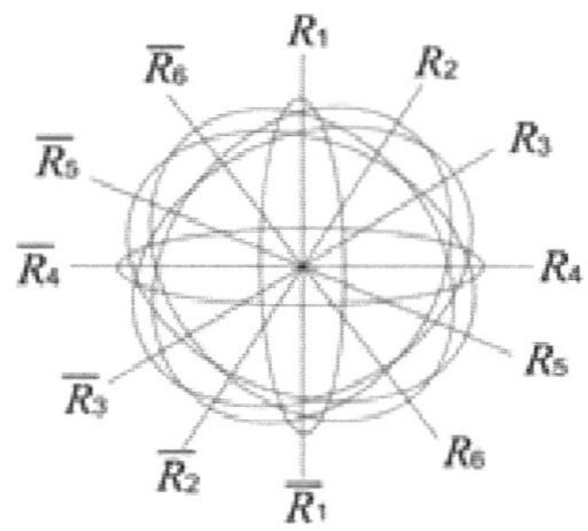

Figure 1. "Emotional semantic ball" model for expressing a user's multi-dimensional perceptual demands.

According to the "Emotional Semantic Ball" model of product perceptual demand, the multi-attributes of products and the diverse multi-user demands,the study proposes the relationship between the perceptual demand set and the semantic axis level to simplify the "emotional semantic ball" model, so as to better summarize the mainstream demand, improve the practical value of the model, and use the cluster analysis of the multi-dimensional perceptual demand of products. Here is the method:

Step 1: Use a statistical method to preliminarily extract the emotional words or phrases that a user frequently selects from a wide semantic vocabulary about perceptual demands to establish the user's perceptual information set about the product, $Y = \{Y_1, Y_2, \cdots, Y_n\}$;

Step 2: Establish the hierarchy of "demand R_i, $i = 1,2,\cdots,n$".

Step 3: Select the perceptual demands of higher hierarchy as the semantic axis MRj in the "perceptual demand semantic space". When $j = 1$, MR$_1$ is a linear semantic axis which can be used to define or quantify one-dimensional perceptual demands. When $j = 2$, the semantic axis MR$_2$ and MR$_1$ constitute a two-dimensional perceptual demand map, as shown in Figure 2, which is a perceptual design tool commonly used in current product designs. This two-dimensional perceptual demand cannot cover the user's multi-perceptual demand for a product, so the superposition of multiple two-dimensional demand maps often becomes necessary for the realization of the user's comprehensive perceptual demand for the product. However, this superposition process is too blunt and tends to cause the final solution to become an "empty set." Therefore, the two-dimensional perceptual demand map only provides reference and analysis in the design process. When j=3, the semantic axis MR$_3$ forms a three-dimensional inductive requirement semantic space with MR$_2$ and MR$_1$. As shown in Figure 3, the semantic axis MR$_3$ (popularity-elegance) are perpendicularly intersected with the MR$_2$ (dynamics-quietness) axis, with the focus being vertical with the main axis MR$_1$ (nobility-cheapness), constituting a "three-dimensional semantic correlation model of perceptual demands".

According to the complexity of the user's perceptual demand for products, when $j \geq 4$, it is relatively complicated to construct a multi-dimensional perceptual semantic association expression model. Literature proposed to use the rose diagram to express the multi-dimensional style, but how to express the inter-related characteristics of user's multi-dimensional perceptual demand require in-depth study. This paper proposes that there are different preference tendencies for users' perceptual needs of products that can be represented by multiple three-dimensional semantic spaces. By referring to different requirement semantics, products can be multi-stylized and serialized to meet differentiated needs of diverse consumer groups.

Step 4:The user's emotional information set is integrated into the association level between the product plan P_i and the demand semantic axis to determine the position of the product plan in the user demand semantic space.as shown in Figure 2 and Figure 3. The position of the product in the two-dimensional and three-dimensional perceptual semantic space can be found. Through the above four steps, the semantic hierarchical association clustering of the perceptual requirements of the user's multiple attributes is solved.

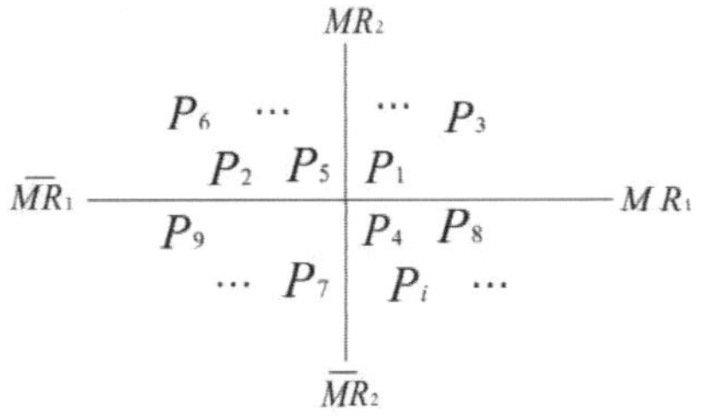

Figure 2. Location of products in the Two-dimensional semantic space.

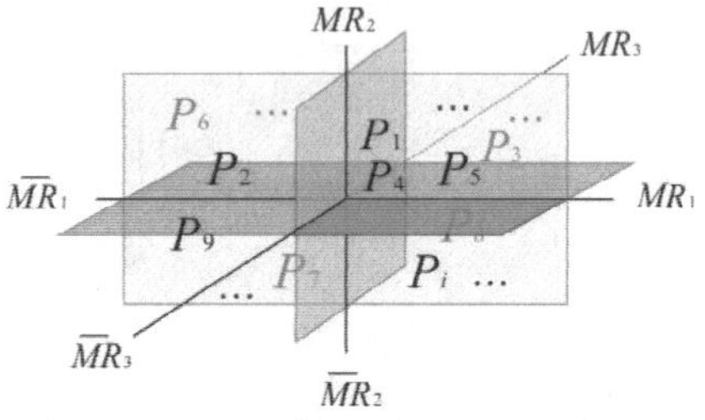

Figure 3. Location of products in the Three-dimensional semantic space.

2.3. Multi-dimensional perceptual demand model oriented by user characteristics

In Norman's "user-centered" product design, users' interests in products are more about the using experience than the product itself. Product development has shifted from functional, cost-based design to user-specific design.The multi-dimensional perceptual demand model oriented by users' characteristics extracts the typical features of users U_i, (i=1,2,...n) through survey and statistics at first, and uses two pairs of complementary adjective phrases $F_i, \overline{F_i}$ in its expression, forming a typical feature semantic space of the user group, as shown in Figure 4.

Through the product perception evaluation experiment, the user's "comprehensive perception" of the n-dimensional perceptual demand of the product plan Pi is related to the typical characteristics of the user group and to mark its position in the semantic space, such as the "multi-dimensional perceptual demand model of product oriented by typical semantic space of users" shown in Figure 5, it corresponds user's multi-dimensional perceptual information to the product plan.

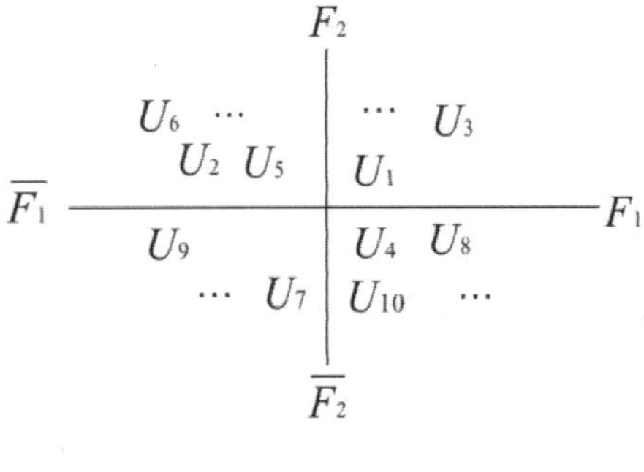

Figure 4.Typical semantic space of user group.

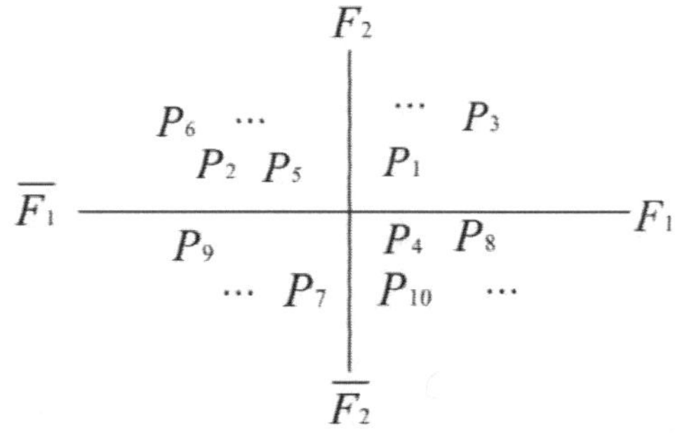

Figure 5. Multi-perceptual demands model of product oriented by typical semantic space of users.

3. Decision model of product design based on hierarchy correlation of first request

3.1. Feature analysis on target users

Product development directing at core users often serves as one of the key marketing plans of an enterprise. In order to find out the proper product plan satisfying the core user group, such target user group would be analyzed and subdivided into different sub-groups before making the final design decision. User group U_i is thereby divided into a core user group, a secondary core user group and a non-core user group.

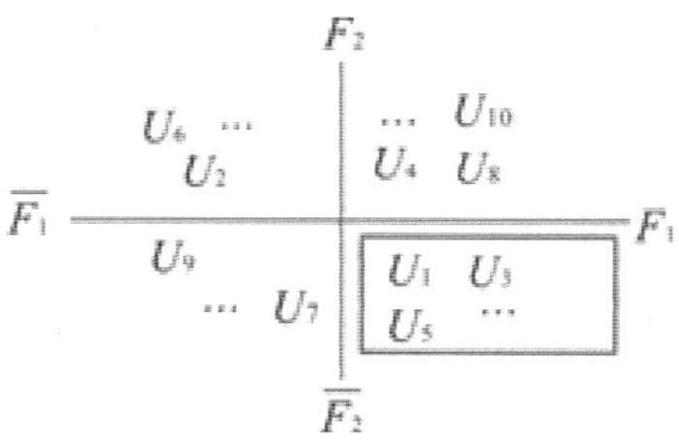

Figure 6. Typical semantic space of core users group.

Adhering to the same logic, two pairs of complementary adjective phrases F_i and $\overline{F}_i$ are selected to form a typical semantic space of users based on analysis on target users. Supposing that core user group is located at the lower right quadrant of typical semantic space of core users as shown in Figure 7, the rectangle clearly indicates the location and its characteristics of core users.

3.2. Establishment of user's first request

Mathematical statistics is adopted to rank the multi-dimensional perceptual demand of users for product. Top demands are selected as *FR* (First Request) and is divided into n hierarchies. When n=3, *FR* can be divided into "High, Medium and Low" sub-hierarchies; when n=5, it is divided into "High, Relatively High, Medium, Relatively Low, Low" sub-hierarchies. To improve the practicality of the decision model of product design based on hierarchy correlation of the first request, this paper proposed to express the multi-dimensional request of users to product with a visualized two-dimensional form rather than a three-dimensional model. Set n=3, *FR* is thus divided into high, medium and low sub-hierarchies and the "decision model of product design based on hierarchy correlation of the first request" will transform from that shown in Figure 7 to that in Figure 10, turning to be more intuitive and practical.

3.3. Design decision oriented by the hierarchy of users' first request

On the basis of the hierarchy of core users' first request, the n sub-hierarchy is selected for making final product design decision. If such semantic association of these perceptual demands is established based on big data of users' need, this model will be able to describe multi-attribute individual request of users more precisely. When users determine their characteristics with semantic retrieval, their perceptual demands for product and personalized product features can be obtained. Supposing that the final product planning is made focusing on the medium and the high sub-hierarchies, superimposing Figure 6 and Figure 7, the product planning in wire frame shown in Figure 8 will be the final decision.

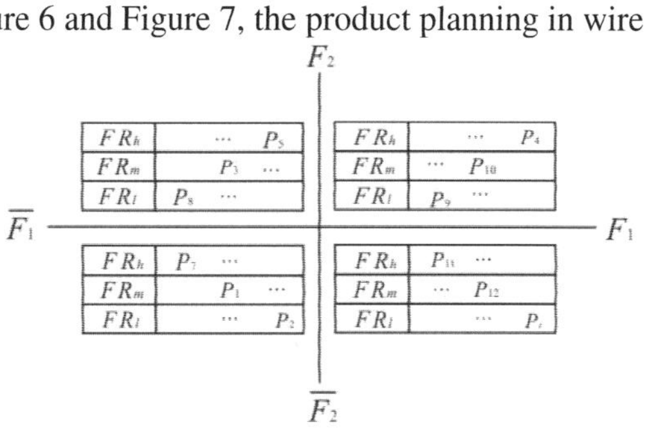

Figure 7. Decision model of product design based on hierarchy correlation of the first request

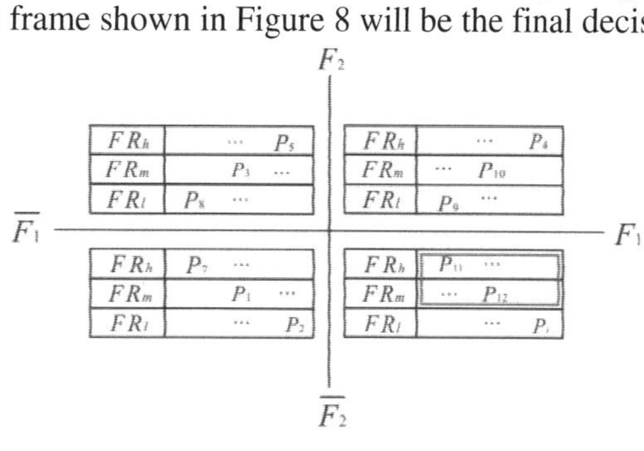

Figure 8. Decision model oriented by the hierarchy of users' first request.

4. Electric scooter design based on value hierarchy analysis

The design of 10 scooters of 3 kinds, i.e. "stand-riding scooter", "self-balancing scooter (the scooter using people's self-balancing to reduce the inertia force)", and "sit and stand-riding scooter", were selected in this research, as shown in Figure 9. Based on the analysis of the comprehensive perceptual demands of users on the product, such as the function and the style, (10 scooters were numbered as $P_i, i = 1,2,\cdots 10$ from left to right, top to bottom), this research aims at determining the product characteristics favored by target users and providing necessary data support and decision basis of personalized designing for product optimization in later period. First of all, the target consumers were determined as people between 12-45 years old, among which 25-35 years old ones were the core group while 12-24 years old ones were the secondary group, their characteristics were acquired through investigation. The target consumers are expressed as "elder -- younger, high-profile – low-profile", as shown in Figure 10.

Figure 9.Personal stand-riding electric scooter design samples

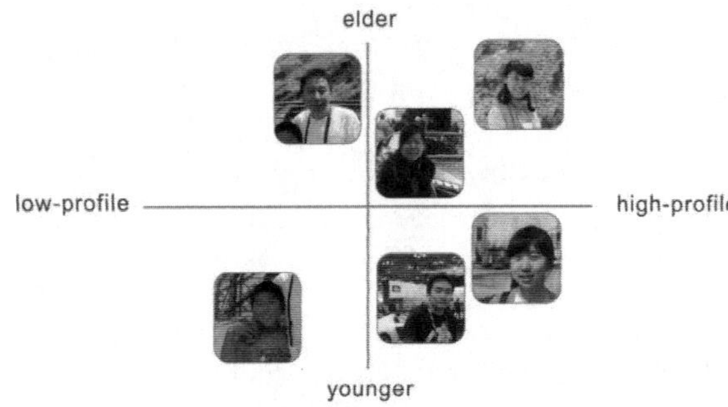

Figure 10.User characteristics analysis of personal stand-riding electric scooter

Through experiments and statistical analysis, user's sentimental demands of products were acquired primarily among many perceptual words, which were: (1)Y_1-value; (2) Y_2-prevalence;(3) Y_3-novelty; (4) Y_4-comfort;(5)Y_5-elegence;(6)Y_6-gender (refers to the style inclining to "masculine" or "feminine");(7)Y_7-sportiness. In order to simulate actual and accurate situation of design determination, the product development process underwent one year, including 3 parts: user's sentimental demands were acquired through market survey before designing, front-line salesman's sentimental demands were concluded through demand investigation after designing; and after the determination of the final plan, designer's sentimental demands were obtained through survey statistics analysis.

Hereby,through the experimental evaluation, the order of the perceptual demand information biased by the user's comprehensive perceptual evaluation can be ranked as: Y_1-value, Y_7-sportiness, Y_3-novelty, Y_2-prevalence, Y_6-gender, Y_4-comfort and Y_5-elegence.Specific experimental data and calculations can be found in the literature 12.

Since the primary demand of users is "value", it was select as the primary demand in this study. Three-dimension semantic level of product's perception image was constructed based on "the score of value". According to consumer's typical semantic space of "elder – younger, high-profile – low-profile", the corresponding plan in semantic space was divided into "high, medium, and low" level based on its "value" in the early stage, forming an "analysis chart at value level". With this chart, plans in the semantic space can be re-designed. The "analysis at value level" of "multi-dimension" sentimental plan decision of stand-riding scooters was completed accordingly (Figure 11). Several kinds of scooters favored by core user group can be selected. According to the value level, plan at medium and high value level were picked as chosen plans (plans framed by rectangle in Figure 3) for design optimization later.

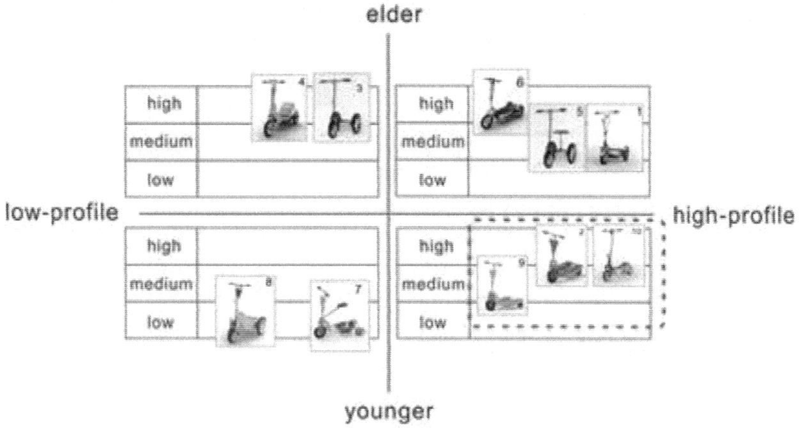

Figure 11.Value hierarchy analysis for the design of stand-riding electric scooters oriented by user characteristics.

5. Conclusion

“Emotional semantic ball” is adopted in this paper to describe perceptual demand features including “multiple dimensions, multi-hierarchy and correlation” of users towards the same product.Finding the user demand emotional semantic axis "MRj" from these tangible emotional needs.When j≥4, different preferences and tendencies of perceptual demands of users towards product are manifested by multiple three-dimensional semantic spaces. Stylized and serialized products can be achieved by referring to different primary request semantics to satisfy differentiated demands of various consumer groups. The approach provides a great solution to the dynamic trend of multi-attribute perceptual demand from users towards product.

Multiple perceptual demand model oriented by user characteristics proposed based on the above-mentioned approach comprehensively describes the position of n-dimension perceptual demand from multi-attribute perception evaluation of users towards product in typical semantic space of core users. The experiment results show that hierarchy of first request proposed in this paper is able to provide a simplified model, express multi-attribute comprehensive perception of users towards product intuitively and avoid one-sidedness, subjectivity and uncertainty in design decision. Such method can also improve the scientificity and availability of product style design decision and lay foundation for personalized innovative design theory based on users’ demands.

Acknowledgements

This research was financially supported by the National Social Science Art Fund of China(Grant No.18GB132)

References

[1] Jose A. Diego-Mas*.(2016) Jorge Alcaide-Marzal Single users' affective responses models for product form design. J. International Journal of Industrial Ergonomics,53: 102-114.

[2] Lu, W., Petiot, J.-F., (2014) Affective design of products using an audio-based protocol:application to eyeglass frame. J. International Journal of Industrial Ergonomics, 44: 83-394.

[3] Jerry M. Mendel(author),Zhang Qiye,Xie Wei(translator). (2013) Uncertain Rule-Based Fuzzy Logic Systems:Introduction and New Directions. Tsinghua University Press, Beijing.

[4] E. R. DECIU, E. OSTROSI, M. FERNEY and M. GHEORGHE. (2005) Configurable product design using multiple fuzzy models. J. Journal of Engineering Design, 16: 209–235.

[5] Miguel Angel Artachoa,Alfredo Ballesterb and Enrique Alcántarab. (2010) Analysis of the impact of slight changes in product formal attributes on user's emotions and configuration of an emotional space for successful design. J. Journal of Engineering Design, 21: 693–705.

[6] Wang Yi. (2016) Product Color Design. Chemical Industry Press,Beijing.

[7] Wang Liya. (2014) Study on Innovative Design of Product Platform Driven by Uncertain User Information.D. Hebei University of Technology.

[8] Wang yi, Wang Jia-min，Richard Wilfred Yelle, Hu gang. (2015) An approach to Extract Product Shape Features Using Variant Design of Curves and Surfaces . J. Mechanical Science and Technology for Aerospace engineering, 34: 107-112.

[9] Cathy Barnes, Stephen Paul Lillford.(2009）Decision support for the design of affective products. J. Journal of Engineering Design, 20: 477–492.

[10] Gao Zhu. (2010) Research on Form Innovation and assessment For Industrial Products. D.Xi'an University of Technology.

[11] K.Y. Chanb, C.K. Kwong, T.S. Dillon and K.Y. Fung. (2011) An intelligent fuzzy regression approach for affective product design that captures nonlinearity and fuzziness .J. Journal of Engineering Design, 22：523–542.

[12] Yi Wang.(2017) Research on User-visional Perception Oriented Personalized.D.Xi'an University of Technology.

IOP Conf. Series: Materials Science and Engineering **573** (2019) 012044 doi:10.1088/1757-899X/573/1/012044

Application of graphene and aerogel in winter running shoes products

Peng Hong[1] , Xia Xinqiao

[1] Wuhan University of Science and Technology, Wuhan, 430065

[2] Wuhan University of Science and Technology, Wuhan, 430065

421311535@qq.com

Abstract. In view of the functional requirements of winter running shoes, the characteristics of graphene and aerogel materials, the preparation process, and the feasibility of their application in products are explored. The results show that the application of new materials in winter running shoes products will make winter running shoes have better heat preservation, breathability, skid-proof, wind-proof, water-proof, deodorant, durability, wear-resistance, encapsulation, and so on, which greatly improves the overall performance of winter running shoes products.

1. The functional requirements of the inter running shores

Winters running shores is a necessary equipment for runners to run in winter or in other cold weather. The winter temperature is low, the temperature of the road surface is usually lower than 0 degrees, and there is often freezing of the road surface. Snow and ice remained on the upper of the shoe easy to melt, winter monsoon cold, early morning and evening light dim, set above the situation, Winter running shoes must have the function of keeping warm, anti-skid, wind-proof and waterproof, reflecting light. When runners wear running shoes for long periods in winter, the feet are lined with the shoes, and the insole produces friction, heat, sweat or moisture. Winter running shoes needs to remove excess heat and moisture in a timely manner, so as not to cause stuffy heat, bacteria, and foot health. In order to prevent snow particles, ice slag from the structural gap into the shoe, so the porosity of winter running shoes fabric needs to be greater than water vapor less than water droplets, and the irregular pores can also ensure the function of windproof, waterproof, and breathable at the same time. At the same time, running shoes should also have good encapsulation, stability and anti-skid performance when they encounter road slippage in winter.

2. Research status of winter running shoes

German brand PUMA is a running shoe designed for winter running _ PWRWARM (Figure 1). In order to reduce the weight of running shoes, multi-layer ultra-thin and ultra-light mesh cloth and traditional synthetic leather heat were used to fit the support, and the running shoes could not provide sufficient support when the actual weight of winter runners increased, and the running shoes could not provide sufficient support in the event of gravel or ice and snow on the road, and the upper of the shoes would strain quickly. In order to keep warm, a large number of fleece materials are used to make the air fluidity in the running shoes worse, so that the constant temperature technology of PWRWARM running shoes is the main part of the running shoes.——In terms of actual feelings, you will soon feel sweltering when you wear the ComfortTemp.

IOP Conf. Series: Materials Science and Engineering **573** (2019) 012044 doi:10.1088/1757-899X/573/1/012044

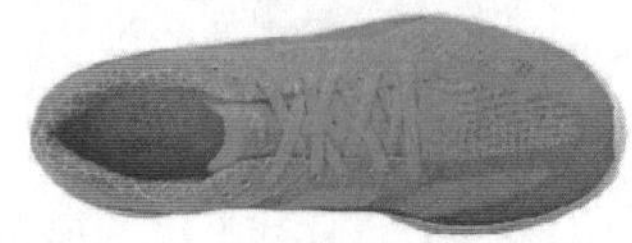

Figure 1:PWRWARM running shoes Figure 2: ComfortTemp thermostat technology in shoe tongue

Although the Japanese brand ASICS's GT-1000 7 SOLAR SHOWER running shoes (figure 3) can withstand the cold better, its use of multilayer mesh cloth and a large amount of towel materials increase the weight of the shoes.

Figure 3: GT-1000 7 SOLAR SHOWER running shoes

The use of traditional materials will inevitably increase the weight and breathability of running shoes when ensuring winter running shoes windproof, rain-proof, snow-proof and wear-resistant conditions. In pursuit of lightweight running shoes, the winter running shoes support strength is inevitably reduced, and under the premise of the pursuit of breathability, the performance of winter running shoes is often reduced. Therefore, some new materials need to be used to meet the necessary performance of winter running shoes at the same time.

3. Characteristics of graphene and aerogel

3.1 Characteristics of graphene

3.1.1 Graphene is a two-dimensional carbon nanomaterials with hexagonal honeycomb lattice composed of carbon atoms in sp ^{2}hybrid orbitals, which can be kept strong in a light and thin structure; Graphene has good thermal conductivity and good light transmittance; [6] In addition, it has excellent chemical stability and electrical conductivity, as well as wear resistance, graphene is plastic, can curl into a cylindrical shape, become one-dimensional carbon nanotubes, can also be made into spherical or ellipsoidal ball, to obtain zero-dimensional fullerene (figure 4). [6] Graphene blended with ordinary fiber can produce new yarn with antibacterial, antistatic, UV-resistant, low-temperature and far-infrared emission.; Add graphene to running shoe rubber to enhance outsole strength, elasticity, and wear resistance[11].

Compared with the materials in the above two cases, the upper made of graphene blended fiber is not easy to be damaged, it not only has good heat dissipation performance, but also has antibacterial property.

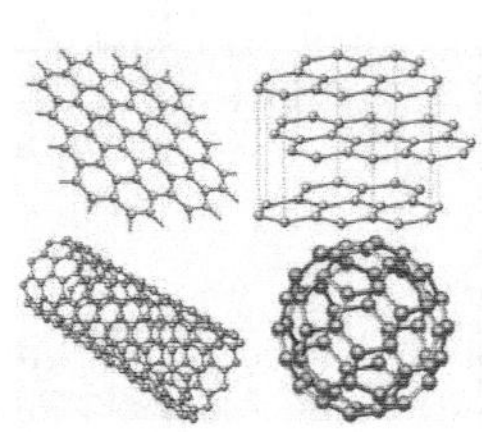

Figure 4: Structure of the graphene structure

4.Characteristics of aerogel

Aerogel is the world's lightest solid, with the characteristics of high specific surface area and low thermal conductivity, known as "one of the top 10 magical materials that change the world." At present, aerogel mainly refers to a solid material (figure 6), which is composed of nanometer-sized ultrafine particles. It has three-dimensional reticular structure, high porosity, large surface area and low density, and has the characteristics of high porosity, large surface area and low density. [4] The greatest advantage of polymer-

IOP Conf. Series: Materials Science and Engineering 573 (2019) 012044 doi:10.1088/1757-899X/573/1/012044

based organic aerogel in organic aerogel is its flexible molecular design, which makes it easier to control its performance. For example, flexible polyimide aerogel can be more convenient for shoes, clothing, fabric industry, easy transportation, cutting, sewing.

Aergoel is an environment-friendly material, does not contain toxic substances such as chloride, and is soft, easy to cut, difficult to deform and less in loss during the cutting process. [1] At the same time, aerogel belongs to ultra-light and ultra-thin material, and it needs thinner thickness when it reaches the same insulation effect as ordinary insulation material. [4]

Thanks to the development of textile-specific aerogel composite super-heating materials by the S.kistler team in 2017, aerogel has been used in the garment industry. The textile-specific aerogel composite binds each aerogel particle with no powder loss, softness, washability, breathability, and so on (figure 6).

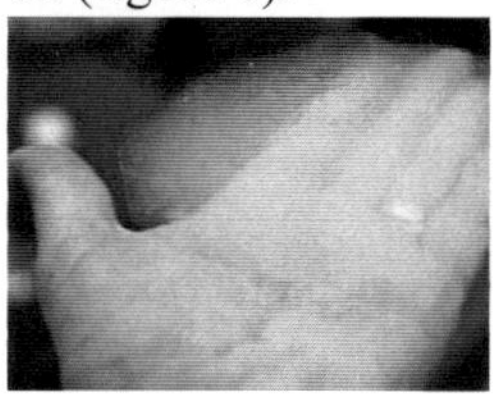
Figure 5: aerogel

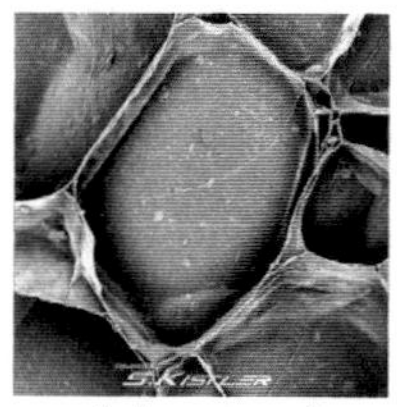
Figure 6; S.Kistler is a special aerogel composite for textile industry.

This material can achieve very strong thermal insulation performance, it is only 3mm thick, can resist-196 °C liquid nitrogen injection.

Compared with the previous two cases, the liner made of aerogel is lighter in weight, better in heat preservation and better in air permeability, and it is non-toxic and pollution-free.

5. Manufacturing process of graphene and aerogel

5.1. Graphene manufacturing process

The preparation method of transforming graphite oxide particles into graphene is as follows: firstly, carbon materials are obtained, and the carbon materials are rolled to form dispersed and comminuted graphite particles; Secondly, graphite particles are oxidized at high temperature above 600 °C to form graphite oxide particles. Then, use the graphene preparation system for microwave irradiation and tremor of graphite oxide particles, and then form the required graphene, for collection. The invention has the advantages of fast preparation speed, fast collection of graphene, avoiding deformation and warping of peeled graphene, and improving the efficiency of graphene preparation and the quality of graphene. [9]

5.2. Manufacturing process of aerogel

5.2.1 In 1931, Stanford University Kistler was prepared by ethanol supercritical drying using water glass as raw material. [4] In 1966, Peri synthesized SiO aerogel by sol-gel chemical method. [3]Organic aerogel was originally synthesized from American 1'ekala by high temperature carbonization of resorcinol and formaldehyde as raw materials. [4] Natural cellulose can be mechanically, ultrasonic, high-pressure homogeneous or chemically treated to form cellulose on a micron or nano-scale basis. These separated cellulose usually have a large specific surface area, and the groups on the surface can drive the micro-/ nano-fiber to self-assemble into a network structure by some interaction to achieve the prerequisite of preparing aerogel materials. By modifying other groups on cellulose, the interaction between cellulose molecules or cellulose microfibers can be further increased, and then cellulose aerogel with good properties can be obtained. [5]

5.3. Grapheneaerogel manufacturing process

The Chinese Academy of Sciences aerogel team spun oxidized graphene liquid crystals into a specific coagulation bath by wet spinning. After chemical reduction and supercritical drying, grapheneaerogel fibers with regular, continuous and porous structure can be obtained. Then, organic phase change materials (such as paraffin wax, polyethylene glycol, advanced fatty acids, etc.) were introduced into the porous network structure of aerogel fiber by impregnation. Grapheneaerogel phase change composite fibers were obtained: finally, fluorocarbon hydrophobic coating was coated on the composite fibers. A flexible grapheneaerogel intelligent fiber with self-cleaning function and multi-stimulus response behavior is obtained. [2]

6. Complementary characteristics of graphene and aerogel

Flexible grapheneaerogel intelligent fibers developed by the aerogel team of the Chinese Academy of Sciences achieve multi-functional integration in response to multiple stimuli, which can achieve fiber twist yarns and woven fabrics that cannot be achieved by single graphene and aerogel materials. [2] Aerogel materials have low strength, strong permeability, poor thermal conductivity, unable to provide support and high graphene strength, good thermal conductivity and release of far-infrared rays, graphene can be used in the intensity, heating and use, In addition, aerogel can also play a strong role in air permeability, heat preservation, the use of both materials can achieve complementary properties, such as the release of far-infrared graphene fiber, aerogel material locking heat to keep warm, a single graphene fiber thin. And aerogel can provide psychological thickness and aerogel can fill the crevice of graphene fabric and block the cold wind. Graphene fabric can avoid aerogel delineation, strengthen the solidity of upper, and increase the service life of composite materials. Graphene can also be prepared by hydrothermal reduction method as a three-dimensional cross-linked network structure, which has the characteristics of high elasticity, anti-fatigue, adsorption and so on. [12]

7. The advantage analysis of graphene and aerogel application in winter running shoes.

In the aspect of upper fabric, the preparation of Si/ cellulose composite aerogel, from cellulose and methoxytrimethylsilane precursors can provide good hydrophobic and adiabatic properties, which can enable winter running shoes to achieve the functions of keeping warm, preventing wind and waterproof. The blend of graphene and elastic fiber can provide good encapsulation and elasticity at the same time, and does not restrict the movement of the foot at the same time. For winter running shoes stability requirements, where running shoes need to provide support such as uppers, the use of graphene material to replace the traditional TPU material can enhance the support strength and reduce the weight of running shoes; Graphene material can also replace the traditional carbon plate, reducing weight while ensuring the original carbon board function. According to the requirements of winter running shoes comfort, the pores of aerogel are complicated and irregular, and the pores are larger than sweat steam. Combined with cotton or other skin-friendly material layer, the effect of absorbing sweat, keeping warm and breathable can be achieved. While keeping the feet warm, it can remove excess moisture and absorb excess sweat to make the running feet dry and comfortable. Graphene can also be prepared into grapheneaerogel, with high elasticity, fatigue resistance and adsorption characteristics, which can greatly enhance the life of aerogel, and absorb the smell inside the shoe bin to ensure the health of the user's feet. Dr. Aravind Vijayaraghavan, a material scientist at the University of Manchester, and the team added graphene to the rubber of the invo-8G series of running shoes. This increases the strength, elasticity and wear resistance of its outer base by 50% compared to industrial standard rubber, and that graphene rubber can bend and control more effectively on all surfaces, greatly reduce wear efficiency, and make grip more reliable and robust,Lasting. [11]

The synthetic gel materials have the functions of moisture absorption and air permeability, compressive water retention, heat preservation and cold storage. The aerogel fabric has the characteristics of water absorption and swelling, dehydration, deswelling, and a large number of pores on the aerogel under dry conditions can ensure the hot gas evaporating from the runner's feet. [4] It can provide a "breathable" function for winter running shoes products, if wet by water, the surface gel layer quickly swelling, the pores closed, the outside water can not soak through the aerogel material, thus "waterproof" or "anti-immersion" performance. Optimal solution for waterproofing and air permeability in ready-to-use shoes: "several microholes in the front palm of the outsole, and air-permeable waterproof cloth on the upper part of the opening...." It can replace breathable waterproof cloth with aerogel material, in addition to satisfying the function of breathable and waterproof, it also provides insulation and heat preservation function. Aerogel materials can reduce weight and thickness, reduce excess consumption during running, improve exercise efficiency, and aerogel materials are environmentally friendly materials, while achieving the same heat preservation effect as common winter running shoes, and it can reduce weight and thickness, reduce excess consumption during running, improve exercise efficiency, and aerogel materials are environmentally friendly. When the winter running shoes product is scrapped, the aerogel material can be directly landfill. Hydrophobic SiO / aerogel, polyimide aerogel has the advantages of lightweight, high strength, flexibility, durability and excellent high temperature resistance. The greatest advantage of polymer-based organic aerogel in organic aerogel is its flexible molecular design, which makes it easier to control its performance. For example, flexible polyimide aerogel can be more convenient for shoes, clothing, fabric industry, easy transportation, cutting, sewing.

At the same time, there is no winter running shoes among the world-famous shoe companies to use these two materials at the same time, if the two materials are used in a shoe at the same time, consumers will pay attention to the two materials at the same time.

8. Conclusion

The use of graphene and aerogel in winter running shoes products has great feasibility and commercial value. Its products can greatly enhance the elasticity, encapsulation, windproof, waterproof and breathability of winter running shoes upper, and enhance the winter running shoes shoe lining to keep warm and deodorize. Reduce the weight and service life of winter running shoes, enhance the strength of winter running shoes shoe body, wear resistance, enhance the wear resistance of winter running shoes outsole, toughness, anti-skid, which can improve the overall performance of winter running shoes. At the same time, graphene and aerogel belong to the environmental protection material, even if the product is scrapped, the recycling value of the waste is high, even if the waste can not be recycled and reused, it can also be directly landfill, which accords with the idea of green design of the product.

References

[1] Liu Yadi, Pei Ying, Zheng Xuejing, Tang Keyong. Preparation of cellulose-based aerogels and construction of functional materials [J]. Polymer Bulletin, 2018 (09): 8-22.
[2] New. Important progress has been made in the field of graphene aerogel smart fibers where Suzhou Nano is located [J]. New Chemical Materials, 2018, 46 (09): 275.
[3] Flexible Aerogel: From " Space Umbrella" to Clothing Warmth [J]. Textile Decoration Technology, 2018 (04): 5-6.
[4] Shen Xiaodong, Wu Xiaodong, Kong Yong, Cui Sheng. Research progress of aerogel nanomaterials [J]. Progress of Materials in China, 2018, 37 (09): 671-680+692.
[5] Chen Xiaoxing, Liu Xiaoming. Research progress of functional cellulose aerogels [J]. Green building materials, 2018 (10): 15.
[6] Zhao Liang Cheng, Li Bin, Li Qin, Wu Si Rui. Preparation and properties of three-dimensional graphene/polyurethane composites [J]. Fine Chemicals, 2018, 35 (11): 1848-1852+1858.
[7] http://iranshao.com/articles/2133-puma-pwrwarm-review
[8] Xu Meiji. Research progress on waterproof and moisture permeability of finished shoes [J]. Light Industry Science and Technology, 2012,28 (08): 92-93.
[9] Wu Chunliang, Gu Fang, Lu Ying, Niu Qianqian. Recent advances in patented technology for preparing graphene [J]. fine and specialty chemicals, 2018, 26 (11): 38-43.
[10] Xu Meiji. Research progress on waterproof and moisture permeability of finished shoes [J]. Light Industry Science and Technology, 2012,28 (08): 92-93.
[11] Graphene Running Shoes: Provides Strong Grip [J]. Universal Polyurethane, 2017 (12): 32.
[12] Wang Lina, Zhang Chidao, Wang Xinling, Zheng Zhen. Preparation and Properties of Superelastic Graphene Aerogels [J]. journal of functional polymers, 2018, 31 (05): 462-467.
[13] Qian Bozhang. New ultra-strong graphene material is expected to replace carbon fiber [J]. Aging

and Application of Synthetic Materials, 2018, 47 (05): 139.
[14] http://iranshao.com/articles/4374-2017-winter-shoes-review

IOP Conf. Series: Materials Science and Engineering **573** (2019) 012045 doi:10.1088/1757-899X/573/1/012045

Model design and analysis improvement of CNC Lathe Based on Ergonomics

Liu LiFu[1], LiWei[1], TanZhenHua[1], Zhang Meng [1], Bo MengZhe[1]

[1] School of Mechanical and Transportation, Southwest Forestry University, Kunming, Yunnan, 650224, China

[*] e-mail: 188561704@qq.com

ABSTRACT: As a typical mechatronics numerical control equipment, CNC lathes are widely used in various processing industries. To obtain better machining accuracy and effect of parts, there is a direct relationship between the structural configuration of the CNC lathe, the input and output mode of the machining information, the arrangement of the operating mechanism, the difficulty of the operation of the equipment, and the accuracy of the operator's operation. When designing a CNC lathe, we must consider the relationship between the human-machine-environment and apply the ergonomic methods and principles to the product design. This paper is based on the individual components of the appearance of the CY-PTC5655 CNC lathe. It proposes the ergonomic design points based on the appearance, creates a good human-computer interaction relationship, and effectively improves the operator's working environment. Reduce the operator's fatigue and improve the production efficiency of the product. The main contents and analysis methods of the thesis have certain reference for the designers of related majors.

1. Forword

CNC lathe is a typical mechatronic product and one of the most widely used CNC machine tools. The numerical control technology adopted by it is the core content of advanced manufacturing technology, and it is the basis for realizing China's "smart manufacturing" and "Made in China 2025". The performance of CNC lathes is not only reflected in the machining accuracy, production efficiency, processing capacity, reliability, etc., but also directly related to the operability, accuracy, ease of operation and comfort of the equipment. In order to obtain better processing results, it has a direct relationship with the structural configuration of the CNC lathe, the input and output mode of the processing information, the arrangement of the operating mechanism, the difficulty of the operation of the equipment, and the accuracy of the operator's operation of the equipment. Ergonomics is a discipline that studies the characteristics of human beings and the matching of working conditions with machines. At present, there are problems in the shape of CNC lathes that are not easy to operate, the display device is not easy to observe, the working space is limited, the color is not properly matched, and the body is indifferent. The use of ergonomics to guide the design of CNC lathes body can create a good human-computer interaction relationship, effectively improve the operator's working environment, and achieve the purpose of reducing operator fatigue and improve product production efficiency. This research has important practical significance and research value in the context of the current vigorous promotion of intelligent manufacturing in the country.

2. The development status of ergonomics in the design of CNC lathes.

The competitiveness of CNC lathe products depends and foremost on their own mechanical properties and processing accuracy. All major enterprises attach importance to the promotion of technology, for products whose performance and core technologies are not very different from each other, the convenience of their use and the comfort of operators have become the key to winning the market competition. The rapid development of CNC technology in China has also led to the improvement of the technology of CNC lathes. The automation level and user-friendly design of machine tools have gradually been applied to machine tool design, but compared with the application of CNC technology of similar foreign machine tools, there is still a large gap. As the operator of the CNC lathe, Not only should it have good physical strength, but also require the operator to have certain operational experience and familiarity with the equipment. The unreasonable design of the structure, shape and man-machine interface of CNC lathes will lead to lower operator's work efficiency, bad feelings to the operator's physiology and psychology, prone to fatigue and even cause operational accidents. Therefore, when designing a CNC lathe, we must consider the relationship between the human-machine-environment, apply the ergonomic methods and principles to the product design, and use the machine operator as the starting point for product design. The appearance, structure, performance and functions achieved are designed around the user's physical and mental experience. At present, China's machine tool equipment lacks in-depth research on the application and humanized design of ergonomics. The humanized design of the CNC lathe shape must be studied from the design elements of the machine's appearance, color, operational amenity, materials and so on.

3. The necessity and significance of ergonomics in the design of CNC lathes.

In the design of CNC machine tools, the relevant principles and methods of ergonomics will play an important role in the function, reliability and performance of machine tool design. The significance is mainly reflected in the following aspects:

- According to relevant data such as human body size and activity range provided by ergonomics, the relevant design can adapt to the physiological and psychological needs of human beings in design, and Meet the physiological and psychological emotional needs of the operator in long-term work; Applying the knowledge of ergonomics to the design of CNC machine tools can also improve the aptitude, convenience and safety of the equipment, reduce errors in the operator's operation, and improve processing efficiency and processing quality.
- In the design of CNC machine tools, the design of human-machine interface is an important part of the human-machine-environment design. It can guarantee the reliable transmission of processing information during processing. The man-machine interface of the CNC machine tool can not only provide timely information such as the position, speed and spindle speed of the machine tool, but also accurately transmit the information to the machine operator to provide the operation basis for the machine operator. In the design process, the relevant knowledge of ergonomics is fully considered and integrated into the design, which will effectively ensure the processing data of the CNC machine tool is accurately transmitted to the operator, Meet the operator's visual and tactile feel, ensuring the accuracy and convenience of the operation.
- Combined with the standard operating button graphic design features in the human-machine interface design, it satisfies the operator's operating habits and ensures that the CNC machine tool processing can be carried out accurately. According to mechanical design, electrical design features and standardization requirements, important buttons such as emergency stop switch, power start, stop button, cycle start of machining program, stop button, etc. must be selected according to its function. This is also the ergonomic design application of ergonomics in the design of CNC machine tool panel.
- In the design of CNC machine tools, combined with the psychological factors of ergonomics, optimize the shape, color, material and working space of CNC machine tools to improve the working environment and ensure the operator's good work mood and Convenience and comfort of operation to achieve the purpose of ensuring processing quality and improving work efficiency.

4. Ergonomic design of the appearance of CNC lathes.

CNC lathe is a metal cutting processing equipment. From the clamping of parts, the installation of tools, the preparation of machining programs, the setting of workpiece coordinate system, the operation of tool setting, every operation step requires the participation of machine operators to complete the machine tool. Every step of the machine requires the participation of the machine operator. The machine operator's vision, body size, hearing, touch, etc. are the main sources of reliable information during machine operation. Based on the principles and methods of ergonomics, using the knowledge of vision, human body size and hearing in ergonomics, the design of the appearance design of CNC lathe is completed to ensure the operator's pleasantness, safety and comfort in manipulating the numerical control equipment, it has a good practical value. The appearance of the CNC lathe is composed of multiple units protected by the CNC lathe, mainly including the surrounding panels, sliding doors, fixed covers, side panels, observation windows, CNC control boxes and other components. The appearance design of the CNC lathe should be based on the ergonomics theory from the appearance of the machine tool, from the observation window, handle, sliding door, CNC control box, machine tool color, each unit material and other parts of the analysis and research.

4.1. Problems in the appearance and design of CNC lathe CY-PTC56551

As shown in Figure 1, it is the appearance of the CY-PTC5655 CNC lathe. The sliding door is divided into a left sliding door and a right sliding door, and the left sliding door has an observation window; The CNC operation box consisting of the numerical control system, the numerical control system operation panel and the machine operation panel is mounted on the right sliding door of the machine tool. As shown in Figure 2, it is the size display of some unit parts related to operation in the design of CNC lathe. The height of the bottom of the CRT box from the ground is 1272mm, and the height of the top from the ground is 1706mm. The operation panel on the CRT control box is the operator's eye and manual operation are based. In the design, the humanized design of the unit such as the processing observation window and the numerical control operation panel is the key content of the appearance design of the CNC lathe. The performance of this part of ergonomic design will directly affect the user's evaluation of the performance of the equipment.

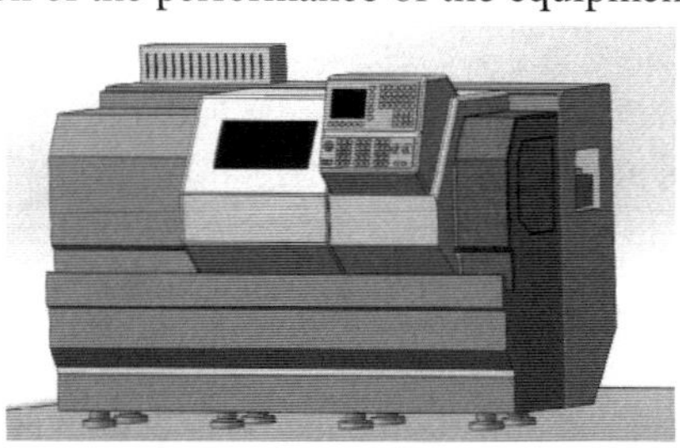

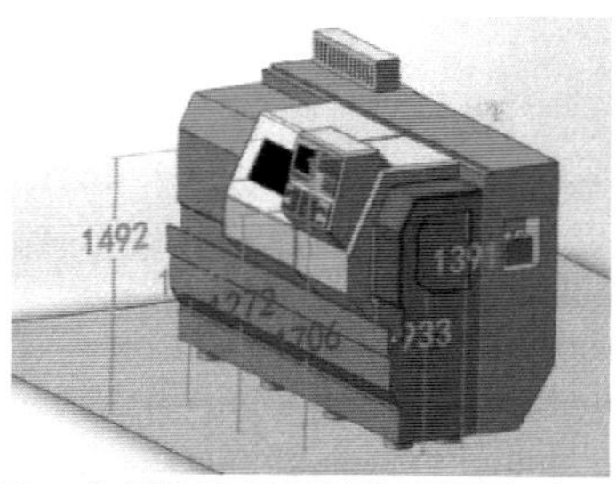

Fig. 1. Original appearance of CY-PTC5655 CNC lathe Fig. 2.CY-PTC5655 CNC lathe part size

Through the functional units such as the left sliding door, the right sliding door, the left door observation window, the door handle and the CNC control box on the right door of the CY-PTC5655 CNC lathe, the ergonomics analysis was carried out, and the appearance of the equipment was found. The following problems exist:

- The observation window on the left sliding door is viewed from a horizontal angle. The angle of inclination is too small, 63 degrees, and the size of the observation window is 540mm * 220mm. It is difficult for the operator to observe the workpiece processing under normal standing position, and it is necessary to close the observation window. It is very insecure and inconvenient to see the processing.
- The position of the numerical control system operation box on the right sliding door is 1272mm-1750mm from the ground. According to the human body height dimension analysis of the 50th percentile of Chinese adult males, the height position of the control box does not meet the requirements of human body height; meanwhile, The tilt angle of the CNC control box operation panel

is too small, 72 degrees from the horizontal direction, which is not convenient for the operator to operate and observe.

- The material used for the observation door on the right side is iron sheet. It is necessary to open it completely to observe the machining condition of the workpiece from the side. It is unsafe to open the door during the processing, which also causes inconvenience to observe the processing.
- As can be seen from Figure 1, there are no handles installed on the left and right sliding doors. The handle is a functional component that facilitates the operator's operation of the machine tool, and plays an important role in improving the operator's work efficiency and operational comfort. Without a handle, it is inconvenient for the operator to operate normally.

4.2. Ergonomic analysis and design modification of some unit parts of CY-PTC5655 CNC lathe

The ergonomic performance of the CNC lathe unit components will directly affect the operator's operational performance. Based on the problems found in the above-mentioned on-site investigation, the solution to the problem is given.

- Left sliding door.

The left sliding door is equipped with an observation window and a left door handle. In the process of machining parts, the operator stands in front of the left sliding door and observes the machining process of the workpiece through the observation window on the left sliding door. Therefore, the observer's observation line in front of the left sliding door is the main source of design parameters for the design of the lathe viewing windows.The design of the left sliding door should be matched with the fixed cover on the left side, and it must be ensured that after opening the sliding door, the opening distance is larger than the maximum workpiece size that can be processed by the lathe, and a certain margin is left, so that the operator can easy to install and disassemble workpieces. In order to ensure the smooth development of the above work, the distance dimension after the sliding door is opened is more than 100mm larger than the workpiece size. At the same time, the left sliding door should also cooperate with the right sliding door. When the workpiece is machined, the workpiece can be processed after locking each other. During processing, the operator observes the machining of the workpiece through the observation window. The optimal field of view of the operator in the horizontal direction is 15 degrees in the left and right direction, and the optimal field of view in the vertical direction is 25 degrees upward and 30 degrees downward. The height of the observation window should be designed according to the human body's apparent height dimension of the 50th percentile of Chinese adult men. Therefore, the observation window should be designed based on the two elements of the field of view and the height of the human body, ensuring that the height and size of the observation window on the left sliding door conform to the optimal field of view of the operator, ensuring the convenience of observation. The material of the observation window is made of plexiglass, and the window strength should be able to withstand the throwing of workpieces rotating at high speed.

After analysis, in order to facilitate the observation, in the improved design of the observation window on the left sliding door of the machine tool, the size of the observation window is increased to 580mm * 450mm, which satisfies the operator's optimal field of view range, and the inclination angle of the observation window is also observed. It is 73 degrees from the horizontal direction and meets the ergonomic design requirements, making it easy for operators to observe the machining process. The specific design is shown in Figure 3.

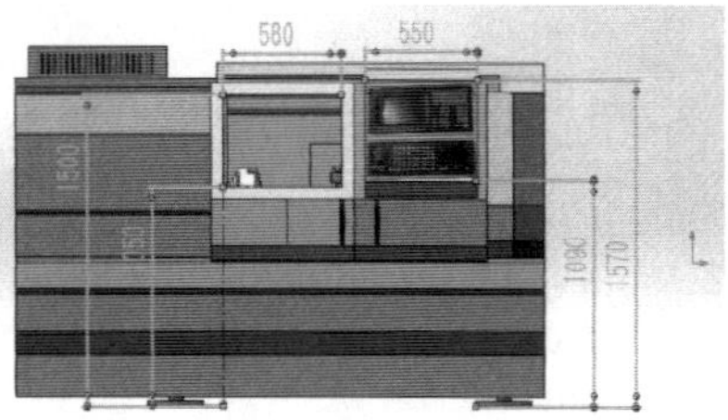

Figure 3. Location of the modified viewing window and CNC control box

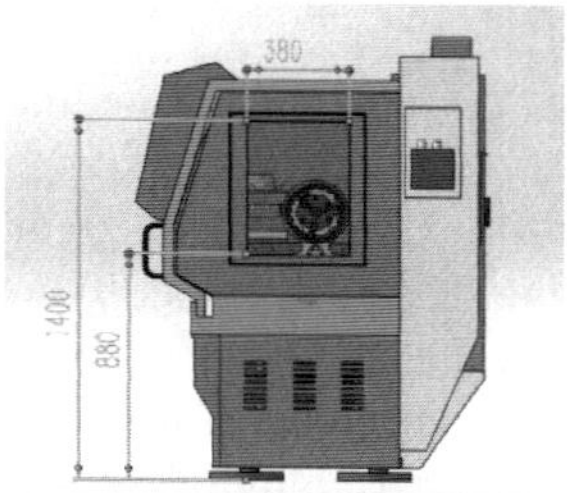

Figure 4. Modified CNC control box position

- Right sliding door and CNC control box

The CNC control box is mounted on the right sliding door. From the ergonomic point of view, in order to facilitate the operator to install and disassemble the workpiece, the knife and other operations, the left and right sliding doors should be locked to each other when machining the workpiece to ensure the safety of the machining process. The CNC control box on the door can be moved a distance with the right sliding door, which is convenient for the operator to clearly observe the condition of the tool and the workpiece when the knife is placed, and successfully complete the tool setting. As shown in Figure 4, it is the right side view of the CNC control box. The position of the CNC control box of the CNC lathe should be installed in the optimal field of view of the operator to meet the operator's eyes and manual operation to complete parameter setting, program entry, and operation during processing. Therefore, the optimal height of the CNC control box of the CNC lathe should be lower than the eye height in the standing position, higher than the elbow height in the standing position. When determining the design size and position of the operation panel of the CNC lathe, it is necessary to fully consider the compensation value of the eye height, elbow height and human body wearing shoes under the standing state of the operator. To ensure the comfort of the operator during the operation, it is necessary to design according to the 50-percent human body height of Chinese adult males. To adapt to different heights and genders of operators. As shown in Figure 5. The optimal size position of the operation panel on the CNC operation box should be lower than the eye height of the standing position and higher than the elbow height in the standing position. When designing, according to the size of the human body, consider the factors such as the eye height, elbow height, and the amount of correction of the human body wearing shoes in the standing position of the operator. During the operation, the operator's line of sight is required. The optimal field of view is about 10 degrees below the horizontal line of sight to ensure the convenience and comfort of the operator during operation.

The modified CNC control box is 1090mm-1570mm away from the ground, which is in line with the optimal height range of the CNC panel. When the operator observes the CRT screen, the line of sight is slightly downward; the manipulator operation panel on the CNC control box is modified from the horizontal direction. It is 78 degrees to ensure the best operating condition and meet the operational requirements of the operation panel on the CRT control box during ergonomic design. The specific design is shown in Figure 6.

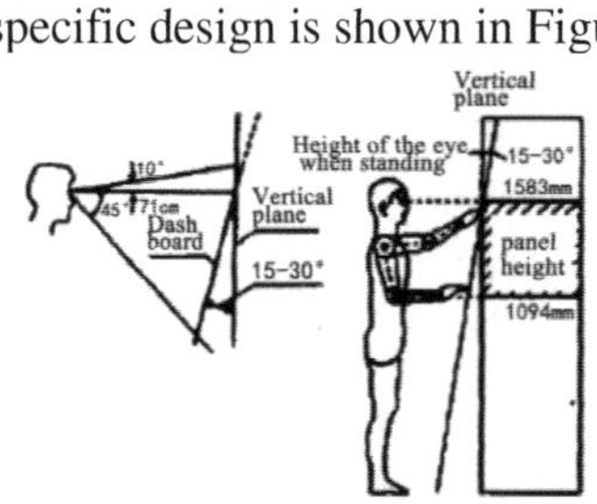

Fig.5. Ergonomic Operation Size Data Setting

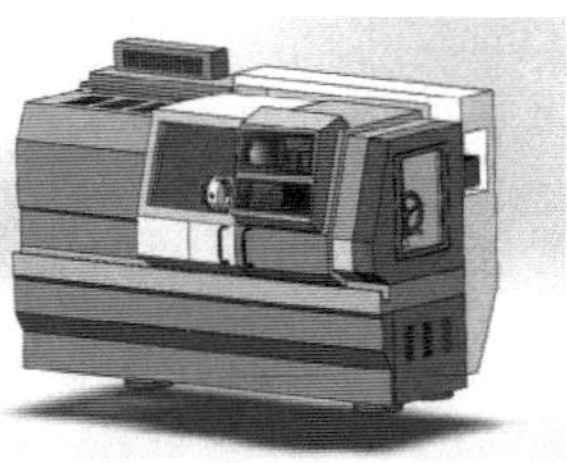

Fig.6.The Effect of The Modified Machine.

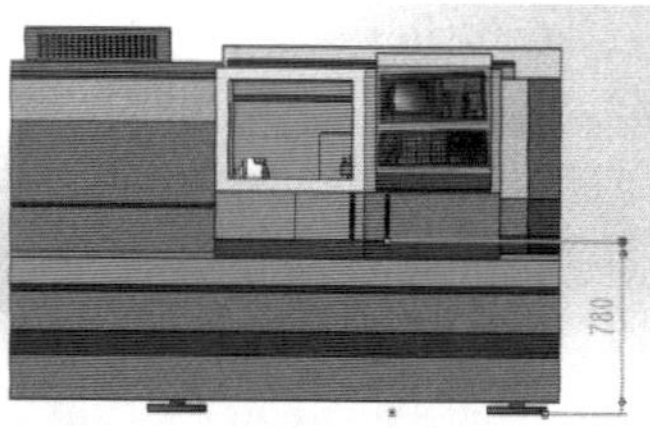

Fig.7. Handle Position.

- Handle .

In the design of the handle, the installation size of the handle should be considered. The shape, size and material of the handle should be selected from the tactile feel to ensure the operator's grip comfort and ease of operation. Through ergonomic analysis, the handles on the left and right sliding doors are mounted in the lower position on the left and right sliding doors. The design of the handle is based on the optimal height of 1050mm in human body operation. The cross-sectional area of the handle is rectangular with four corners and the length and width are 200mm*20mm. CY-PTC5655 CNC lathe, because the observation window of the left sliding door increases, the right CNC control box moves down, Therefore, the handles on the left and right sliding doors are installed at a distance of 780 mm from the ground, and their design meets the requirements of ergonomic design. The specific design and modified effects are shown in Figure 6 and Figure 7.

- Observation door on the right side.

As shown in Figure 3, On the right side of the observation door, an observation window made of plexiglass is provided, and the size is 520 mm*380 mm, which is convenient for observing the processing.

4.3. Color design of CNC lathe

Color is a very important visual information in addition to the appearance of CNC machine tools, the application of color must meet the physiological and psychological needs of the operator to operate the device for a long time, giving a comfortable psychological suggestion. Not only has the effect of increasing equipment aesthetics, reducing fatigue during operation etc., also ensure that the color of the CNC machine tool can comfort the operator's psychological feelings of working for a long time, played a role in function indication, regional division and improved observation accuracy. When the operator of the CNC machine tool faces the CNC machine tool for a long time in the processing process, the large-scale machine tool exterior color will largely affect the operator's mood and work efficiency. The appearance color of the machine tool is combined with different machine tool shapes and corporate culture, which will produce strong emotional color and performance characteristics, which will bring greater spiritual influence to the use of machine tools. The correct use of color will directly affect the processing quality and processing efficiency of the operator.

In the shape color design of CNC machine tools, the color configuration of the upper part of the lower part or the shallow part of the whole machine should be selected to give a stable and calm feeling. At the same time, because the machine tool is easy to produce a lot of chips, the chip removal device will transfer all kinds of chips downwards, and the bottom is designed to be darker to make the machine look cleaner and tidy. The upper part of the traditional CNC machine tool usually adopts white, but the operator is easy to produce a glare when facing a large area of white for a long time, so it is not suitable to select the white color of the whole machine shape. The CY-PTC5655 CNC lathe is entirely gray, which can calm people's minds and make people feel calm and meet the ergonomic color requirements of the equipment.

5. Conclusion

In order to better ensure the operator's convenience, comfort and safety, the ergonomic design principle should be applied to the product design of CNC machine tools to ensure a good man-machine relationship and effectively improve the operator's work. The environment achieves the goal of reducing the labor intensity of operators and increasing labor productivity. Starting from the appearance design of CNC lathe, based on the ergonomic design principle and method, the paper carries out the improved design of some functional components for the original equipment, which plays a certain role in promoting the appearance and appearance of CNC lathe.

Thanks

First of all, I am very grateful to the organizers for providing a platform for communication and exchange with peers. Secondly, I would like to thank Li Wei and Tan Zhenhua for their help in the design and completion of the project. The project is led by "the eight key industries in Yunnan Province (51600630)" and the Southwest Forestry University school-level education reform project provide funding, I would like to express my sincere gratitude.

Acknowledgments

Fund Project： "The eight key industries lead the brand profession - mechanical design and manufacturing and automation" (51600630)， Southwest Forestry University Teaching Reform Project

About the author:

Liu LiFu（1982—）,Male, SiChuan， master's degree, lecturer. Mainly engaged in teaching and research work in industrial design.

Reference

[1] Ding Y.L.（2005）Ergonomics [M]， Beijing Institute of Technology Press. Beijing.
[2] LiuTao,Chen Ming.（2014）Design and Analysis of CNC Machine Tools Based on Ergonomics [J]， Combined machine tools and automated machining technology,11: 49-52.
[3] Hu Ping.（2009）Comprehensive application of ergonomics in engineering machinery design [J],Quality cost management,47： 56-58.
[4] Zhang GaoMei.（2012）Application of Ergonomics in the Design of CNC Machine Tools [J],Mechanical research and application, 4： 131-132， 135.
[5] Cai XiaoYu.（2018）Application and Research of Ergonomics Principle in Control Mechanism of CNC Machine Tools [J],Southern Agricultural Machinery,49： 55-55.

Research on Image Modeling Design of Oxygenerator under the Guidance of Symbolic Semantic Thinking

Jianning Su[a], Jing Li, Yun Lv and Ruisheng Bai

School of Design Art, LanZhou University of Technology, Lanzhou, GanSu, 730050, China

[a]E-mail: sujn@lut.cn

Abstract. Based on the core theory of Semeiology, the transmission mechanism that product as an information carrier is simulated. From the perspective of symbolic semantics, on the basis of analyzing the signified and signifier of product image, the semantic and symbol of product image are excavated, and the fuzzy set analysis method in Kansei Engineering is used to establish the quantitative correlation mechanism for abstract image semantics and concrete image symbols. Taking the oxygenerator design of Lan Fei Co.,Ltd. as an example, it helps the enterprise to grasp consumer's image demand accurately, guides it to complete the image form design of the oxygenerator and enhances the competitiveness of the enterprise.

1. Introduction

The design concept of People-centered, such as Without Thought, Human-computer Interaction Design, Emotional Design and so on, is prevalent in the design community, which greatly reflect the direction after change of product design in the prevailing circumstance. At the same time, it also reflects the escalating demand of consumer which from rational demand to perceptual demand and from functional demand to spiritual demand[1]. Therefore, as the product developer, the enterprise should change the design strategy, deeply excavate the product image, fully understand the consumer's perceptual image demand, give play to the information transmission function of the product, and use it as the carrier of communication with users. The enterprise brand identification degree and the user loyalty are established on the basis of satisfying the perceptual image demand of users[2], as shown in figure 1.

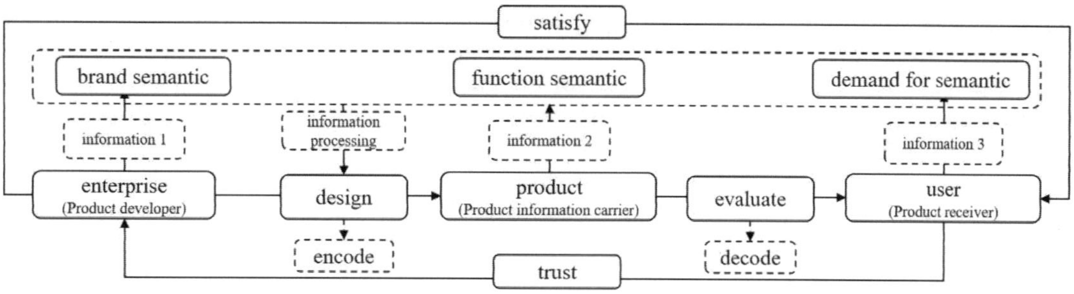

Figure 1. Information communication mechanism based on product.

By obtaining the brand semantic, functional semantic and user demand semantic information of the product, the enterprise carries on the comprehensive processing to the information in the design activity, instructs the product design, and then satisfies the user's demand. A series of activities such as aesthetic consciousness, perception, experience and so on are carried out by users before purchasing products, and the understanding and trust of product developer are gradually established. The product as the carrier of information has successfully made the communication between product developer (enterprise) and product receiver (user) become possible.

2. Symbolic semantic and product image

Each product has unique attributes, and only after interacting with the product can we understand its function, usage scenario, operation mode and so on. Thus, before using the product, it can only rely on visual perception to recognize the product, thus forming a preliminary impression on the product. These perceptible elements such as material, color, shape, structure, etc. are the image symbols that make up the image of the product[3]. Symbol is the beginning of conveying the meaning of information, which is also a kind of medium and means with ideographic function[4]. *F De saussure*, a semiotics researcher, believes that symbols are composed of signified and signifier. The signified is the meaning and purpose of the symbol, that is, the internal meaning of the signifier, while the signifier is the form used by the symbol, that is, the specific material image[5]. There are inevitable connections and similarities between symbols and its semantics, product image symbols and semantics, as shown in figure 2.

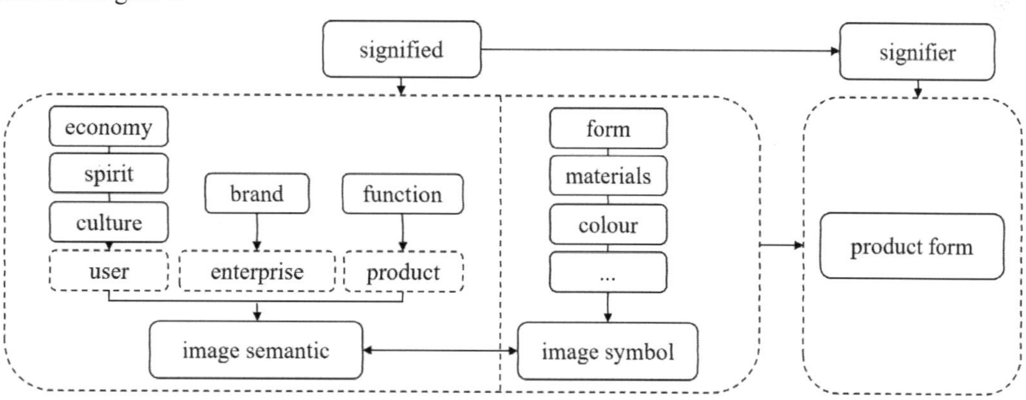

Figure 2. Signified and signifier of product.

Symbolic semantics not only make the symbolic symbols easier to understand and enrich the function of symbol communication, but also realize the expectation that users use symbols to express themselves. The product image semantic information also has the same status and function in the closed loop in which the product is used as the carrier of information so that the enterprise can communicate effectively with the user. In figure 2, the product signified contains the image semantics assigned by the user, product and enterprise, and image symbols that can reflect this semantic information. The signifier of the product refers to obtaining the image of product by coding the semantics and symbols through the design activity and this product image is the meaning and purpose of the product.

Hualiang Xiao[6], *Jiao Ma*[7], *Yuxiao Zhou*[8] and others studied product image form design based on Semeiology. *Chuang*[9], *Nagamachi*[10], *Yunfan Zhang*[11] and others obtained image semantics to guide product design based on Semeiology. On the whole, in the research of product image, scholars focus on the research of user's demand image, but lack of discussion on the relationship between image semantics and form elements. On the basis of the existing research results and the idea of symbolic semantics and, this paper will carry on the experiment and discussion to the image form design of oxygenerator, and excavate its multi-dimensional image in depth by using the method of Kansei Engineering. The correlation mechanism between image semantics and image symbol is established to guide the image form design of oxygenerator.

3. Analyzing image semantics of oxygenerator

The related data survey shows that with the change of human living environment and the improvement of health awareness of consumer, people's demand for this medical product, oxygenerator, has been actually increasing, and its consumers have been extended from hospital patients and elders to white-collar workers, students, pregnant women, outdoor players, drivers, etc. The universality of consumer and the diversity of image demand bring challenges to the design of oxygenerator. Because of the special function of oxygenerator, the product image becomes more and more important. Under the premise of completing the basic efficiency of the oxygenerator, the developer of the product needs to weaken the depression that the oxygenerator brings to the user, and then meet the consumer's demand for perceptual image of the oxygenerator. The oxygenerator as a carrier of symbolic information, refers to three major parts of its image, as shown in figure 3.

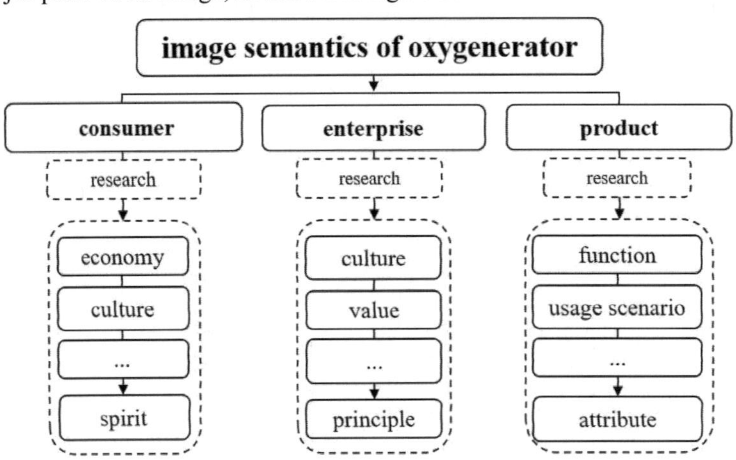

Figure 3. Image semantic analysis of oxygenerator.

The first part is about the semantic meaning of consumer demand image due to the different background of consumers, that is, different demand sources such as economic semantics, cultural semantics, spiritual semantics and so on. The second part is the image semantics of the products, such as corporate culture, enterprise value, enterprise concept and so on, and the third part is the product image semantics, such as the function, the use scene, the physical nature and so on. Based on the three directions of image semantic analysis, in-depth mining is carried out through oral interview, on-the-spot investigation, second-hand data investigation, network and literature review, etc. Then, according to the product semantic representation method based on natural language processing and by extracting the theme word (Subject Heading Terms, SHT)[12] which represents the image semantics that affect the shape of the oxygenerator, 195 words of the representative oxygenerator image semantics are

IOP Conf. Series: Materials Science and Engineering **573** (2019) 012046 doi:10.1088/1757-899X/573/1/012046

obtained. Based on the analysis of meanings, 17 typical image semantics were obtained by selecting the expert group members according to their experience. Finally, it is divided into several categories through expert discussion. The classification principle is the influence degree of category attribute and image semantics, and then four categories are obtained, as shown in table 1.

Table 1. Image semantic classification.

Classify	Meaning and quantity	Representative image semantic vocabulary					
1	material correlation(4)	lightweight	superior	environmental-friendly	stable		
2	performance correlation(4)	specialty	safe	durable	portable		
3	morphologic correlation(6)	like	exquisite	concise	symmetrical	round	cute
4	color correlation(3)	casual	personalized	technological			

4. Extracting image symbols of oxygenerator

The image symbol, which represents the image semantics of oxygenerator, is the design element of product image form design. Which is mainly reflected in the product image. As a result, product research is needed to collect mainstream oxygenerator brands through the Internet, magazines, periodicals, oxygenerator brochures, physical sales outlets, etc. (Haiyang House, Invikang, Kaiya, Haigui, Xinsong, etc.) and a number of niche but characteristic brands (such as Taishuai, Juren, etc.) product samples, a total of 90. Then, 30 initial samples were obtained by referring to the sales ranking and network clicks and analyzing 90 samples. Finally, the sample typicality is discussed by the members of the expert group, and the most representative product samples are selected, as shown in table 2.

Table 2. Representative samples of oxygenerator.

No.	Product sample	No.	Product sample	No.	Product sample	No.	Product sample
S1	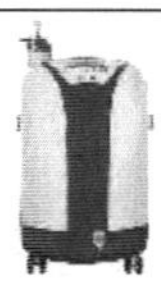	S2		S3		S4	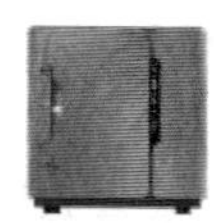
S5	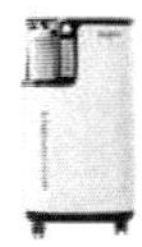	S6		S7		S8	
S9		S10		S11	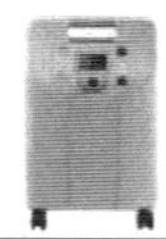		

In order to obtain image symbol elements, representative samples need to be deconstructed. At first, the image symbols of each sample are decomposed by means of morphological analysis[13] based on the model composition of the oxygenerator. Then, through the induction of image symbols of 11 samples, 15 image symbols, that is, the main design elements of oxygenerator are obtained. Finally, 11 image symbols are selected which have a great influence on the shape of oxygenerator, as shown in table 3.

Table 3. Image symbols of oxygenerator.

No.	Image symbol	Specific description	No.	Image symbol	Specific description
E1	plastic	polishing dull polish	E7	side profile	straight arc

		flame plating			Straight arc combination
		full touch screen			dot matrix
E2	touch screen operation	mainly touch,keys for assistance	E8	grille profile	Strip belt
		key operation			Segmented strip belt
E3	key operation	mainly key soperation touch for assistance			1:1:1
		concealed	E9	length-width-thickness ratio	1:(1/3-2/3):1
E4	handle	exposed			1:2/3:1/3
		folding			dominant tones: White
		rolling			dominant tones:black
E5	move mode	handheld	E10	monochrome	dominant tones:blue
		portable			dominant tones:grey
		straight			dominant tones:black+white
E6	front profile	arc	E11	multicolor	dominant tones:blue+white
		straight arc combination			dominant tones:grey+white

5. Extracting image symbols of enterprise

In order to ensure the integrity and consistency of the image form of the oxygenerator, the enterprise personality image form symbol was extracted and incorporated into the design of the oxygenerator, so that the enterprise personality was represented. Through in-depth investigation and analysis of enterprise characteristics and product image, the most representative sample (S12\ S13\ S14) is selected as the basis for extracting image form symbols (E12\ E13\ E14), as shown in table 4.

Table 4. Enterprise sample and Image symbol.

Enterprise sample			Image symbol	
S12	S13	S14		airplane
			E12 form	wings of flight
				military medal
			E13 colour	blue
				military green
			E14 material	mental

Through the analysis of table 4, we can see that the corporate image symbols mainly include material (metal), color (blue and green), form (aircraft, wings, medals), while the corporate personality symbol elements mainly depend on the form. Through the interpretation and symbolic transformation of the form, the erterprise image symbol is obtained, as shown in table 5.

Table 5. Image symbol transformation of enterprises.

The most symbolic image form	Image form sample	Image symbol transformation
airplane		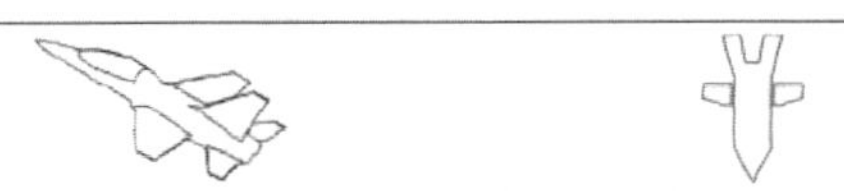
		simplification
military quality		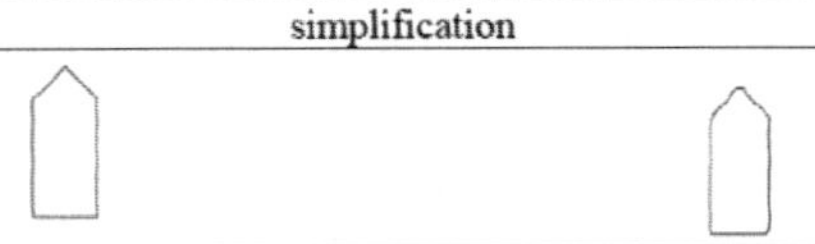
		deformation
logo		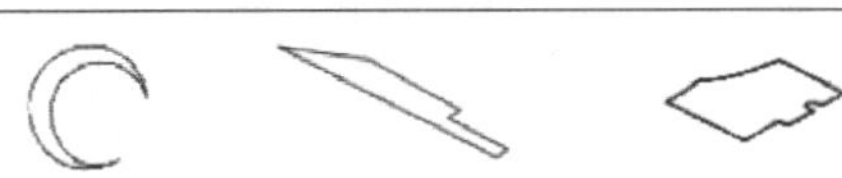
		scattered reorganization

The aircraft form is not only beautiful, but also has distinct individuality and recognition. Simplification of its shape can obtain visual form symbols as the morphological basis of enterprise image form; LanFei is a military enterprise, so the medal of honor, which represents the quality of military industry, is also the representative of its individual image symbol, which can be used in image form design. The third image symbol is the enterprise logo. LanFei's logo has a strong recognition, so the decomposition and reorganization of its elements can help people effectively use its morphology to strengthen brand identification.

6. Establishing quantifiable relevance model between samples and image symbols

In order to ensure the accuracy and readability of the image form of the oxygenerator, it is necessary to establish a quantitative correlation mechanism between the samples and the image symbols in order to guide the image form design of the oxygenerator accurately and effectively. The fuzzy set theory analysis method provides a systematic computing tool to express this kind of information in language. In this study, the fuzzy clustering analysis method[13] is used to establish the correlation model. Let M_1 be the matrix relation between oxygenerator sample and image semantic, M_2 the matrix relation between image semantics and image symbol, x the sample of oxygenerator, y the image semantic, z the image symbol, then the relation is expressed as:

$$M_1 = [uM_1(x,y) \mid x \in X, y \in Y] \tag{1}$$

$$M_2 = [uM_2(y,z) \mid y \in Y, z \in Z] \tag{2}$$

$$M = M_1 \bullet M_2 = \{[(x,z), \max\min(uM_1(x,y), uM_2(y,z))] \mid x \in X, y \in Y, z \in Z\} \tag{3}$$

The first step is to create the SD questionnaire of image semantics and representative oxygenerator samples. SD questionnaires were distributed to 10 designers, 15 consumers and 10 business employees. One invalid questionnaire was abandoned and 34 valid questionnaires were obtained. A uniform registration of the survey data is performed and the mean of each image semantics for each sample is calculated, as shown in table 6.

Table 6. The relationship between product samples and image semantics.

	S1	S2	S3	S4	S5	S6	S7	S8	S9	S10	S11
like	2.706	2.029	2.765	1.882	1.941	2.088	1.765	2.029	1.941	2.235	2.706
lightweight	5.441	1.706	2.382	5.353	5.441	6.118	3.059	5.412	1.853	3.824	5.059
specialty	4.176	1.971	3.765	3.529	2.206	2.118	1.824	2.000	5.382	2.647	3.235
exquisite	3.735	1.941	3.382	1.853	2.029	3.824	2.059	2.118	3.588	1.794	2.824
concise	2.765	1.794	2.206	2.000	2.147	2.059	5.059	3.235	3.676	1.735	3.882
casual	1.941	3.824	2.265	5.647	6.059	2.000	5.735	5.559	2.059	4.971	2.441
superior	4.059	2.324	4.029	2.029	2.029	1.912	1.647	1.882	4.735	2.324	3.971
round	2.147	3.824	4.353	6.059	5.647	1.912	5.176	2.941	2.029	2.088	4.735
safe	2.206	1.824	2.206	3.794	2.059	2.265	4.735	3.441	2.059	1.882	2.324
durable	3.412	2.471	1.971	2.088	2.118	2.059	4.000	2.088	4.353	1.824	2.647
Environmental-friendly	3.559	2.618	2.000	1.941	1.882	3.912	1.824	4.853	4.529	2.500	3.735
portable	5.500	1.941	1.971	6.088	5.971	5.235	5.441	5.676	2.265	3.029	3.941
technological	5.529	2.529	4.794	6.029	3.912	5.941	2.853	1.882	4.412	3.853	5.412
cute	4.118	4.000	4.765	5.971	6.059	5.294	6.088	5.412	1.971	3.882	5.471
personalized	5.588	2.382	2.500	2.029	2.853	5.353	4.147	2.382	3.294	3.206	4.500
stable	2.735	2.324	1.882	2.265	1.882	2.088	5.353	2.647	3.118	2.147	2.500
symmetrical	1.941	1.971	1.971	4.971	6.059	1.824	3.294	1.853	3.000	1.676	1.941

The image semantic mean value of each sample is less than 4, which indicates that the sample has the connotation of a certain image semantics. As in the mean statistics of S1 image semantics in the table, those<4 are like (2.706), exquisite (3.735), concise (2.765), casual (1.941), round (2.147), safe (2.206), durable (3.412), Environmental-friendly (3.559), stable (2.735), symmetrical (1.941). Through numerical comparison, the smallest number represents the strongest image semantics, so it can be concluded that the strongest image semantics of SI are casual (1.941) and symmetrical (1.941). On the horizontal side, we can get the smallest numerical value by comparing the numerical values, thus we can get the most representative one of the 11 samples in a certain image semantics. For example, S2 (1.706) is the least valued sample in lightweight, so it can be seen that the strongest consumer perception of lightweight is the sample 2. In order to import data into fuzzy mathematics calculation, table 6 data is normalized. Import the normalized data into the formula (1) get the fuzzy matrix M_1 between the sample and the image semantics, that is, the relationship between the sample and the image semantics:

$$M_1 = \begin{Bmatrix} 0.387 & 0.290 & 0.395 & 0.269 & 0.277 & 0.298 & 0.252 & 0.290 & 0.277 & 0.319 & 0.387 \\ 0.777 & 0.244 & 0.340 & 0.765 & 0.777 & 0.874 & 0.437 & 0.773 & 0.265 & 0.546 & 0.723 \\ 0.597 & 0.282 & 0.538 & 0.504 & 0.315 & 0.303 & 0.261 & 0.286 & 0.769 & 0.378 & 0.462 \\ 0.534 & 0.277 & 0.483 & 0.265 & 0.290 & 0.546 & 0.294 & 0.303 & 0.513 & 0.256 & 0.403 \\ \cdots & \cdots & \cdots & \cdots & \cdots & \cdots & \cdots & \cdots & \cdots & \cdots & \cdots \\ 0.391 & 0.332 & 0.269 & 0.324 & 0.269 & 0.298 & 0.765 & 0.378 & 0.445 & 0.307 & 0.357 \\ 0.277 & 0.282 & 0.282 & 0.710 & 0.866 & 0.261 & 0.471 & 0.265 & 0.429 & 0.239 & 0.277 \end{Bmatrix}$$

In the second step, the SD questionnaire of image semantics and image symbols of the oxygenerator is made.SD questionnaires were distributed to 10 designers, 15 consumers and 10 business employees. Four invalid questionnaires were abandoned and 31 valid questionnaires were obtained. The survey data are recorded and the mean values of each image semantics of each image symbol are calculated as shown in table 7(a) and 7(b).

Table 7 (a). The relationship between image semantics and image symbols.

	like	lightweight	specialty	exquisite	concise	casual	superior	round	safe
E1	1.806	1.387	4.129	4.903	5.548	2.000	4.323	3.871	1.935
E2	3.871	5.581	5.387	5.548	3.290	5.355	2.742	5.742	5.484
E3	2.000	5.258	2.032	3.581	3.516	3.161	4.516	3.516	1.452
E4	2.000	2.935	1.968	2.290	3.032	2.968	4.226	5.323	1.774
E5	1.774	2.032	2.226	5.839	3.161	2.613	1.903	5.968	2.419
E6	1.355	2.613	5.839	5.129	1.419	2.355	5.968	1.516	4.484
E7	1.839	2.484	5.968	2.806	2.226	4.677	6.323	2.194	3.129
E8	1.806	6.516	4.129	2.258	2.000	2.226	4.516	1.903	4.516
E9	1.516	1.452	4.516	5.194	6.419	4.806	6.387	4.516	5.065
E10	4.484	2.323	4.806	5.290	1.419	2.032	5.968	6.581	5.710
E11	4.097	5.065	5.581	5.645	6.452	5.935	5.968	6.419	5.968

Table 7 (b). The relationship between image semantics and image symbols.

	durable	Environmental friendly	portable	technological	cute	personalized	stable	symmetrical
E1	2.581	4.581	1.548	5.097	5.323	5.194	1.968	6.452
E2	5.774	4.097	3.355	2.355	5.419	5.032	5.419	5.581
E3	1.935	6.097	3.355	5.581	5.548	3.806	1.677	4.710
E4	2.032	5.355	1.581	6.097	3.935	2.935	4.323	1.806
E5	3.742	5.387	1.452	3.871	6.258	2.839	1.903	2.871
E6	5.742	6.484	3.742	6.484	1.903	1.935	4.161	1.548

E7	6.290	6.387	2.484	6.000	2.903	2.968	4.548	1.935
E8	5.452	6.516	6.419	3.806	2.581	2.000	5.968	1.935
E9	6.097	6.516	2.097	5.581	5.968	6.065	4.516	5.032
E10	2.645	3.065	6.484	5.742	6.484	6.097	5.032	6.419
E11	6.581	6.452	6.613	4.968	2.355	5.129	6.387	5.935

If the image semantic mean of the image symbol is less than 4, the image semantics of the image symbol can be determined horizontally. As in the table E1 mean value<4 are like (1.806), lightweight (1.387), casual (2), round (3.871), safe (1.935), durable (2.581), portable (1.548), stable (1.968). Through the comparison, the smallest value shows that the image semantics is the most significant. So the most obvious image semantics of E1 is lightweight (1.387). In the same way, we can get the smallest longitudinal value by comparing the numerical values. And then get the image symbols which contribute the most to a certain image semantics of 11 image symbols. Taking "lightweight" as an example, E1 (1.387) is the image symbol with the least numerical value, so consumers think that E1 (plastic) is the image symbol which has the greatest contribution to the image semantic "lightweight".In order to import the data into the calculation of fuzzy mathematics, the data in table 7 is normalized. The data is imported into the formula (2) the fuzzy matrix M_2 between image symbol and image semantics is obtained:

$$M_2 = \left\{ \begin{matrix} 0.258\ 0.198\ 0.590\ 0.700 & \cdots & 0.281\ 0.922 \\ 0.553\ 0.797\ 0.770\ 0.793 & \cdots & 0.774\ 0.797 \\ 0.286\ 0.751\ 0.290\ 0.512 & \cdots & 0.240\ 0.673 \\ 0.286\ 0.419\ 0.281\ 0.327 & \cdots & 0.618\ 0.258 \\ 0.253\ 0.290\ 0.318\ 0.834 & \cdots & 0.272\ 0.410 \\ 0.194\ 0.373\ 0.834\ 0.733 & \cdots & 0.594\ 0.221 \\ 0.263\ 0.355\ 0.835\ 0.401 & \cdots & 0.650\ 0.276 \\ 0.258\ 0.931\ 0.590\ 0.323 & \cdots & 0.853\ 0.276 \\ 0.217\ 0.207\ 0.645\ 0.742 & \cdots & 0.645\ 0.719 \\ 0.641\ 0.332\ 0.687\ 0.756 & \cdots & 0.719\ 0.917 \\ 0.585\ 0.724\ 0.797\ 0.806 & \cdots & 0.912\ 0.848 \end{matrix} \right\}$$

In the third step, the statistical operation of fuzzy clustering is carried out to establish the quantitative correlation between image symbols and samples. Firstly, the fuzzy relation matrix M_1 of oxygenerator samples and image semantics and the fuzzy matrix M_2 of image semantics and image symbols are introduced into formula (3). The contribution matrix of image symbol to sample is obtained. To facilitate observation and comparison of data, the matrix M is converted into a table of image symbols' contribution to product samples, as shown in table 8.

Table 8. The contribution of image symbols to samples.

	S1	S2	S3	S4	S5	S6	S7	S8	S9	S10	S11
E1	4.653	3.161	3.923	4.688	4.528	4.259	4.659	3.865	4.363	3.332	4.757
E2	5.734	4.044	4.646	6.156	5.922	5.256	6.317	5.374	5.069	4.355	5.749
E3	4.822	3.159	3.820	4.992	4.828	4.631	4.617	4.334	4.070	3.560	4.901
E4	4.105	2.887	3.470	4.292	3.971	3.837	4.188	3.643	3.668	3.095	4.324
E5	4.090	3.019	3.581	4.394	4.190	3.925	4.391	3.798	3.684	3.059	4.345
E6	4.886	3.017	3.793	4.489	3.953	4.288	4.323	3.958	4.635	3.376	4.584
E7	4.942	3.289	3.936	4.734	4.300	4.298	4.701	4.190	4.753	3.645	4.790
E8	5.067	3.077	3.604	4.900	4.510	4.600	4.871	4.595	4.348	3.557	4.790
E9	5.941	4.180	4.934	5.932	5.615	5.336	6.323	5.224	5.678	4.389	6.037
E10	6.077	4.062	5.038	6.440	6.054	5.562	6.227	5.132	5.181	4.338	6.041
E11	6.942	4.677	5.445	7.142	6.768	6.169	7.349	6.289	6.350	5.085	6.859

Through the systematic analysis of the data in table 8, the relationship between the sample and the image symbol can be obtained. Taking the whole column data of sample S2 as an example, through

IOP Conf. Series: Materials Science and Engineering **573** (2019) 012046 doi:10.1088/1757-899X/573/1/012046

numerical comparison, 2.887 is the smallest of the whole columns, indicating that the image symbol E4 (handle) is the largest contribution to sample S2 in accepting the evaluation of image semantics. As a result, when designing products such as lightweight oxygenerator, we should pay attention to the design of its handle.

7. Image modeling design of oxygenerator

Located in the actual project of oxygenerator design, the enterprise mainly develops a small oxygenerator with the image of "portable". First of all, the data analysis of 11 samples in table 6 shows that sample S2 (1.941) has the highest score. Secondly, the data of image semantics of sample S2 in longitudinal direction are sorted. In addition to "portable", the ahead three image semantics with higher scores are lightweight, concise and safe. Then, table 7 data is analyzed to obtain image symbols which have great semantic contribution to "portable", "lightweight", "concise" and "safe". Finally, based on the data from table 8 , the importance ranking of image semantics and corresponding image symbols can be obtained: handle > front profile >move mode > grille profile > key operation > plastic > side profile > touch screen operation > monochrome > length-width-thickness ratio.At the same time, the relationship between each image semantics and image symbols can be clearly defined, as shown in Table 9.

Table 9. The relationship between the image symbols with the image of "portable" and the related image semantics.

	lightweight	concise	safe	portable
E4	2.935	3.032	1.774	1.581
E6	2.613	1.419		3.742
E5	2.032	3.161	2.419	1.452
E8		2.000		
E3		3.516	1.452	3.355
E1	1.387		1.935	1.548
E7	2.484	2.226	3.129	2.484
E2		3.290		3.355
E10	2.323	1.419		
E9	1.452			2.097

We can see that the "portable" value (1.581) of the handle by referring to the ranking of the importance of image symbols in table 9 and combining the values of the image symbols in the relevant image semantics of the "portable" image is the smallest. So, during the handle design process, the portability should be mostly considered but the safety, lightweight and concise semantic might be slightly considered. Similarly, through analysis data can confirm that when designing an oxygenerator such as a "portable" one, we should focus on the simplicity of the side profile, the front profile, the grille profile, and the color matching elements, and the design of the material and the length-width-thickness ratio is mainly concerned with lightness, the most important thing about keystroke design is security, handle and moveable design should emphasize portability. The enterprise image symbols (aircraft, medal, logo) extracted from table 5 are integrated into the styling design by using the design aesthetics rule, and the design scheme of the oxygenerator is obtained. The 2D scheme is transformed into 3D model by solidworks software, and the keyshot software to complete the image styling design rendering picture of portable oxygenerator, as shown in figure 4.

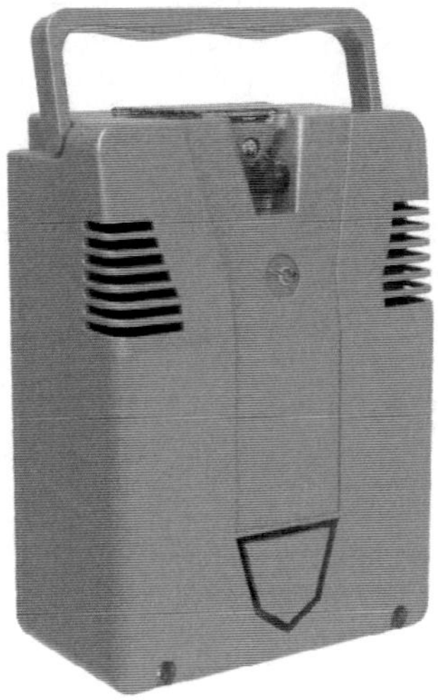

Figure 4. Rendering picture.

In the design process of the oxygenerator, the aircraft form with both aesthetic and enterprise personality is chosen as the styling element. Considering the simplicity of the handle, the side profile and the front profile, the aircraft shape is simplified and deformed to connect the oxygen outlet and the intake port, and the wing shape is naturally transformed into the grille contour line and extended to the side profile, which is harmonious and beautiful. The proportion is adjusted by the "Trisection" and "Golden Section Search" to meet the aesthetic needs of the users. Handle design is concise and in line with man-machine requirement. At the same time the whole oxygenerator uses blue plastic, light color and light texture, in line with the user's needs for portability. After expert evaluation, the design of the oxygenerator has met the demand for "portable" image in the overall styling, and directly reflects the brand identification of LanFei.

8. Conclusions

Product image styling design not only conforms to the development of the times, but also accurately explains the new requirements put forward by consumers to product developers and the new expectations for product design under the new consumer environment. When developing new products, enterprises not only pay attention to costs and technology, but also attentively explore products as a carrier of information synthesis. The discussion of product image is helpful to the development and design of product image styling. Under the guidance of the thought of symbolic semantics, through the deep probing and analysis of image semantics and image symbols of oxygenerator samples in this paper, the correlation model between oxygenerator samples and image symbols is established by fuzzy set analysis method. At last, referring to quantifiable data, the image styling design of portable oxygenerator of Lanzhou Flight Control Co., Ltd. was successfully completed by integrating the unique brand image symbol of enterprise, which proves the feasibility of the product image styling design process and the validity of the model about product image extraction and correlation.

References

[1] (us)Klaus Kripendorf. Design: a shift in semantics [M]. Beijing: China Construction Industry Press, 2017.

[2] Chun Ouyang. Product brand meaning continuation and innovation [J]. Art Education, 2015, (11): 142-143.

[3] Jianning Su, Zhijun Liu , Peng Wang. Research progress of product family form design method based on perceptual image [J]. Mechanical Design, 2017, 34 (11): 112-116.

[4] Yi Zhang, Jingwen Du, Ying Zhang. Research on the Construction of perceptual Image of Chinese Furniture from the Perspective of symbolic semantics [J]. Design, 2018, (21): 18-19.
[5] Linghao Zhang. Semiotic product design method [M]. Beijing: China Construction Industry Publishing House, 2011.
[6] Hualiang Xiao, Li Lin, Longhua Yan, Chen Zhiang. Tourism souvenir design based on design semiotics analysis and evolution of silver ornaments of Miao nationality [J]. Packaging Engineering, 2018, 39 (14): 221-226.
[7] Jiao Ma, Bin Zhang. Research on the extraction and innovation design of the color-paste paper-cut symbol of the city of china[j]. Packaging Engineering,2018, 39 (12): 249-253.
[8] Yuxiao Zhou, Shijian Luo, Chen Gencai. Icon design based on design semiotics [J]. Journal of computer-aided Design and Graphics, 2012, 24 (10): 1319-1328.
[9] Chuang M C, Ma Y C. Expressing the expected product images in product design of micro-electronic products[J]. International Journal of Industrial Ergonomics, 2001, 27(4): 233-245.
[10] Nagmachi M. Kansei engineering: a new ergonomic consumer-oriented technology for product development[J]. International Journal of Industrial Ergonomics, 1995, 15(1): 3-11.
[11] Yunfan Zhang, Zhuo Liu. Product family styling design based on brand image gene extraction [J]. Mechanical Design, 2018, 35 (3): 105-109.
[12] Zaolin Lu, Fritz Frenkler. Research on the design of new energy automobile model based on product semantic analysis[J]. Mechanical, 2017, 34 (3): 111-116.
[13] Meiyu Zhou. Sensibility and design[M]. Shanghai: Shanghai Science and Technology Press, 2011.

IOP Conf. Series: Materials Science and Engineering **573** (2019) 012047 doi:10.1088/1757-899X/573/1/012047

Application of Human-Machine Dialogue in Foreign Language Teaching at Universities

Ting Liu[1 2], Takaya Yuizono[1], Yifen Lu[2] and Zhisheng Wang[1 3*]

[1]Japan Advanced Institute of Science and Technology, Nomi, Ishikawa, 9231292, Japan

[2]School of Foreign Languages, Dalian Polytechnic University, Dalian, Liaoning, 116034, China

[3]Research Institute of Photonics, Dalian Polytechnic University, Dalian, Liaoning, 116034, China

[*]Corresponding author's e-mail: wangzs@dlpu.edu.cn

Abstract. With the development of internet technology, communication technology and artificial intelligence (AI), the versatile and convenient human-machine dialogue and interaction system has gradually penetrated into the education field and become a new type of teaching method. Focusing on the subject of the application of human-machine dialogue system in foreign language teaching at universities, this paper elaborated on how to apply the human-machine dialogue system in foreign language teaching, what effect can be achieved, and what application strategy can be adopted in human-machine dialogue interactive teaching. Combined with the problems existing in daily teaching, this paper presented suggestions for applying the human-machine dialogue system, which is expected to improve the effects of foreign language teaching.

1. Introduction

Human-machine dialogue is a kind of operation mode of computers, through which the user can command a computer to perform a certain task with dialogue through a console or a terminal display screen. The computer can display the calculated, processed and controlled results in time for the user to observe and understand, and the user can input various data and instructions into the machine through certain input devices to control by using commands or command processes. This is the process of dialogue between human and machine. Foreign language teaching at universities generally uses multimedia classrooms (such as video, projection, and synchronized slides, etc.) to simulate and represent real-life scenes for practicing listening and speaking skills, which has achieved impressive teaching results. With higher public requirement for listening and speaking level in foreign language teaching, the foreign language teaching mode at universities also needs to keep pace with the times. It is necessary to introduce or adopt new teaching equipment and new teaching methods. Human-machine dialogue expands AI into the field of education, and it perfectly integrates science and technology into teaching practice. The human-machine dialogue system can enrich teaching activities and teaching methods, build fresh atmosphere for foreign language learning, and reignite students' interest in learning foreign language.

Applying human-machine dialogue to foreign language teaching at universities is conducive to improving the teaching level of listening and speaking courses of foreign language. The human-machine dialogue teaching system include various scenes such as daily life conversation, travel conversation, and business conversation, etc. The conversation practice through role-play can help students independently practice conversations of different difficulty levels. At the same time, in the human-machine dialogue teaching mode that combines graphics and text, audio and video, students can receive various information through earphones, headphones and screens, so that their listening and speaking input and output abilities can be greatly improved. In such teaching mode, students' listening and conversational contents can be responded in a timely and objective manner, while their psychological factors in practicing speaking and listening skills can also be noted, which can improve their ability to use foreign language in communication. In addition, in the human-machine dialogue environment, the combined test of listening and speaking skills is conducted in a way that the results can more realistically reflect students' listening and speaking skills.

This paper examined the practical application effect of the human-machine dialogue teaching mode through its real application cases. Through listening and oral test of foreign language, learning motivation questionnaire and teachers' feedback, this paper evaluated the application effect of the human-machine dialogue mode in foreign language teaching at universities, and proposes suggestions for improving this teaching mode, with a view to improve the teaching effect of foreign language at universities.

2. Literature review

In the 1950s, Alan Turing proposed the famous "Turing test", which inaugurated the research on human-machine dialogue system; in the 1960s, MIT Artificial Intelligence Laboratory invented the first chatting robot-ELIZA dialogue system; In 2016, the United States released the "National Strategic Plan for Intelligence Research and Development" in preparation for the training of AI talents; subsequently, the Japanese government also promoted the staged development of AI industrialization; the "New-generation Artificial Intelligence Development Plan" released by China State Council in 2017 proposed a new education system for intelligent interactive learning under the background of AI development to better adapt to the teaching development in the AI era [1]. With the development of natural language processing technology, computer deep learning technology and artificial construction knowledge base, the AI era has been popularized in people's daily life. There are also a lot of research results and application methods for human-machine dialogue in the AI field [2-3]. At present, the teaching mode using human-machine dialogue has not been popularized in China, but the extensive use of human-machine dialogue in the process of foreign language teaching has become a trend. The research on the foreign language teaching mode of human-machine dialogue at universities has also increased [4]. We should adopt effective teaching strategies for improving students' spoken foreign language at universities, provide students with different course contents, combine traditional and new teaching methods, and promote human-machine dialogue teaching mode to organize foreign language listening and speaking activities and conduct spoken foreign language teaching in the form of dialogue and oral retelling, etc. [5]. There are many studies about the practical significance of the new oral teaching strategy in improving students' oral English level by investigating students who participated in the human-machine speaking test [6]. It means that the human-machine oral test has certain practical significance for the oral English teaching. In addition, the implementation of the human-machine dialogue test mode can scientifically test students' visual listening and speaking ability [7]. In order to apply human-machine dialogue to foreign language teaching at universities, it is necessary to further application strategy.

3. Practical application of human-machine dialogue in foreign language teaching

In listening and speaking teaching, teachers mainly guide students to understand the speaker's intention according to pronunciation and intonation, and extract key information from the familiar dialogue. For the unfamiliar conversation, students can skip new words and understand the general meaning of the dialogue, record simple information, and retell the dialogue close to the natural speed of speech, thereby applying it to daily life or business occasions. Through the human-machine dialogue system, we can create good conditions for dictation training and oral expression to give positive impetus to listening and speaking teaching. This paper is based on the practice of human-machine dialogue teaching for two classes of 47 students for a semester, using textbooks and some extracurricular resources to apply human-machine dialogue system for foreign language teaching.

3.1. Human-machine dialogue listening and speaking course (output by machine, acceptance and feedback by human)

A weekly listening and speaking teaching practice (90 minutes) was conducted by the human-machine dialogue system (Table 1). By setting the human-machine dialogue system and specifying the required listening and conversation materials, students can receive accurate audio and video materials, integrate into the language environment, practice voice morphemes, and improve understanding of foreign language through pre-study, listening and reading. By the means of human-machine dialogue, we can set specific scenarios and use network resources to conduct contextual dialogue examples. We can also provide templates for students to write conversations, exchange ideas and ask questions, thereby improving their ability for applying foreign language.

Table 1. Teaching practice of human-machine dialogue system.

Specific course design	Time (minutes)
Phonetic & morpheme practice	10
Vocabulary practice	10
Listening practice	35
Oral practice	35

3.2. Human-machine dialogue test (output by human, acceptance and feedback by machine)

This test used AI scoring function to test students' overall listening and speaking ability through the situational conversation, 1minute material reading and other forms. During the human-machine dialogue test, students' dynamic performance will be recorded throughout the process, so that good evidence can be referenced and the evaluation results are more convincing.

Natural language processing (computer linguistics, language computing) is a natural language processing technique that combines language science with information science by allowing computers to accept and understand human language skills. In the foreign language teaching test, the human-machine dialogue mode is used to randomly generate questions from the question bank and carry out IT-based test, automated scoring and scientific evaluation, thus reducing the intervention of human factors and creating a fair, objective, transparent test environment.

4. Application effect of human-machine dialogue in the practice of foreign language teaching

4.1. Listening and oral test based on Human-machine dialogue system

The listening and oral test based on human-machine dialogue system was conducted in stages. The automated computer test was adopted as the test method. Students completed the listening and oral test through the computer and the headset device in a human-machine dialogue manner, and the computer automatically scored students' spoken recording. The test was conducted at a fixed time on a weekly or monthly basis, and each test lasted for 60 minutes. As is shown in Table 2, questions and scoring standards for the listening and oral test were based on the human-machine dialogue system.

Table 2. Listening and oral test based on human-machine dialogue system.

	Amount	Corresponding scores	Scoring standard
Listen to the conversation and answer the question	10	30 points	If the student's answer is in line with the requirements of the question.
Listen to the passage and answer the question	10	30 points	

Read a short passage	2	20 points	If the student can read the passage fluently; If phonetic intonation and rhythm are correct; If listeners can understand the passage.
Situational conversation	2	10 points	If the student's answer is close to the purpose of communication, and can be scored accordingly.
Brief elaboration on the topic	2	10 points	

The researches on the influence of foreign language test on students mostly focus on the impact of different test modes on students' performance and learning motivation [8-9-10-11-12]. There are also some researches that explore the impact of students' cognitive level, personality traits, and physiological responses on their preferences for tests from their own perspective [13-14-15-16]. In order to make the foreign language oral test as an effective tool for checking the learning progress and improving the learning effect of students [17], this paper conducted a human-machine dialogue test on 47 students who received the human-machine dialogue teaching for a semester, and investigated their preferences for modes of foreign language oral test, such as "human-machine dialogue", "teacher-student dialogue", "group discussion". Based on a full scale of 10 points, the results are shown in Figure 1.

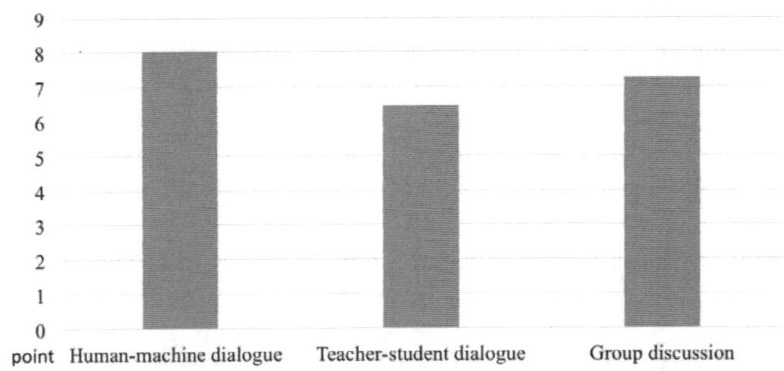

Figure 1. Preferences for modes of foreign language oral test.

It can be seen that students' preferences for human-machine dialogue, traditional teacher-student dialogue, and group discussion is roughly the same, with the proportion of preference for human-machine dialogue is slightly higher. It shows that students are accustomed to this type of test. Most students said that human-machine dialogue can relieve psychological stress and provide ease of mind during the test.

4.2. Motivation questionnaire

After the practical application of human-machine dialogue, a motivation questionnaire was administered to investigate students' basic attitude towards foreign language teaching based on human-machine dialogue. According to DÖrnyei (2001), motivation in foreign-language learning is modelled as a three-step process involving choice motivation (requires fostering a fun, supportive atmosphere in the lecture), executive motivation (requires instructors to make learning exciting and enjoyable), and increased motivation (requires increasing the learner's satisfaction) [18]. Based on this research, the motivation questionnaire in this paper consisted of three subscales, each with 2 items: choice motivation (two items: "this lecture was interesting" and "study in a fun atmosphere"), executive motivation (two items: "deepen understanding" and "leading to acquisition of new knowledge"), and increased motivation (two items: "sense of accomplishment" and "recommending human-machine dialogue to juniors"). All answers were evaluated on a scale of 5 (1. agree, 2. slightly agree, 3. neither agree nor disagree, 4. slightly disagree, and 5. disagree). There was also a free description section in the questionnaire in order to allow students to elaborate on their ideas about the exercises. The results are shown in Figure 2. It showed that about 80% of the students gave a good evaluation of all items.

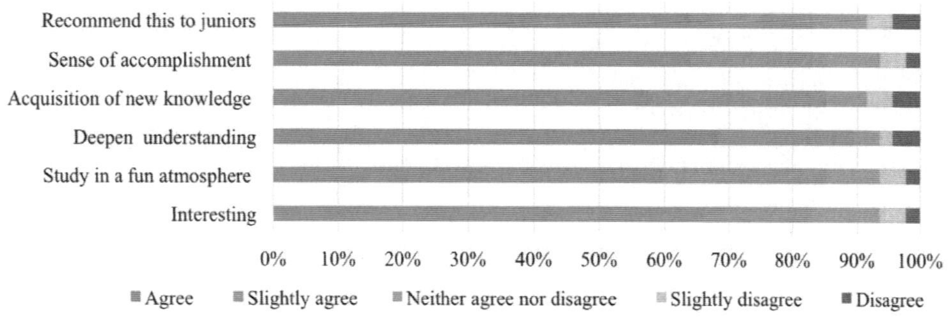

Figure 2. Results of motivation questionnaire.

In addition, the following comments by students were obtained from the free description section, which reflected a positive attitude towards this pedagogy: I think human-machine dialogue is very interesting; My listening and speaking skills were improved; I have a deeper understanding of the knowledge I have learned; I felt it was a fun learning atmosphere; My learning satisfaction was increased; I started to enjoy the foreign language listening and speaking course;

4.3. Teachers' feedback

The respondents are foreign language teachers who implemented human-machine dialogue teaching. They were posed about their opinions on foreign language teaching based on human-machine dialogue.

4.3.1. Human-machine dialogue teaching helps to create a foreign language environment

The multimedia software based on human-machine dialogue system can produce standard pronunciation and fluent conversation, which makes students more willing to practice listening and speaking skills. The human-machine dialogue system reads the standard textbooks and uses the pronunciation & intonation practice system to allow students to read materials and record them through the microphone. The speech recognition system enables students to clearly understand the gaps in their pronunciation & intonation, identify their own inadequacies by comparing with the recording, adjust their own pronunciation & intonation, thereby improving their foreign language level.

4.3.2. Human-machine dialogue teaching is conducive to improving students' in-class participation

The human-machine dialogue software is well illustrated and combines audio and video, which is conducive to cultivating students' interest in learning. By conducting listening and speaking training, using the voice laboratory, listening to and shadowing the recording, and conducting intensive or extensive listening, it can enrich teaching forms, improve the in-class efficiency, and demonstrate the course characteristics.

4.3.3. Human-machine dialogue teaching can expand the dimension of classroom listening and speaking training

Students can choose different learning materials for practicing foreign language according to their own interest. It can give full play to the active role of students in foreign language teaching to achieve the effect of human-machine dialogue system as an aid in teaching. At the same time, teachers can use the human-machine dialogue to provide personalized guidance, such as news, TV dramas, movies, lectures, reports, documentaries and other training materials, thus expanding the dimensions of in-class listening and speaking training.

4.3.4. Human-machine dialogue test can help students understand the learning outcomes

The scoring system of human-machine dialogue test can conduct effective and scientific evaluation and feedback, so that students can visually see their learning outcomes. This is conducive to helping students to develop interest in learning and form good habit of self-learning.

5. Discussion

The application of human-machine dialogue in foreign language teaching at universities is conducive to switching from traditional courses and improving the current situation of listening and speaking teaching. However, it has not reached a certain scale in terms of depth and breadth. As for the prospect of the human-machine dialogue development, we expect to promote its practical application in the field of foreign language education. The use of new scientific and technological means for foreign language teaching has put forward new requirements for teachers and students.

5.1. Teachers need to transform traditional teaching thinking mode

Although the human-machine dialogue system can improve the listening and speaking level in foreign language teaching in the absence of actual foreign language environment, it also puts forward certain requirements for teachers to use the human-machine dialogue system. How proficient can teachers use the system counts a lot for the efficiency and substance of the classroom teaching. Teachers should change the traditional teaching thinking mode, infiltrate the use of the human-machine dialogue system in the test phase in the normal teaching, introduce the skills of listening test and oral test based on human-machine dialogue, and actively play the guiding role in the classroom by combining human-machine dialogue with traditional teaching activities and organizing flexible and diverse activities to train students' listening and speaking skills.

5.2. Students need to receive adaptive training

At present, human-machine dialogue has not been popularized yet. There are still areas for improvement in terms of technology, as well as problems to be solved during the testing phase. The human-machine dialogue teaching mode entails communication with computers, so students may have subjective emotions in the learning process due to their poor proficiency in operating computers. Some students are still accustomed to paper reading and paper test, so they feel difficult to accept the electronic version based on human-machine dialogue. They are more inclined to traditional teaching and test methods, which may result in distraction and inability to make improvement. It is necessary to strengthen the adaptive training for students, so that they can adapt to the trend of the AI era as soon as possible.

6. Conclusion

In the context of AI era, the application of modern means, especially the realization of foreign language teaching by computers, is an inevitable trend in the development of data age today. As a new foreign language listening and speaking training method in absence of foreign language environment, human-machine dialogue plays an important role in creating a rich foreign language environment, increasing the time for students to practice and use foreign language, and improving students' learning motivation. Listening is input, and speaking is output. The combination of this two skills can improve the effect of learning foreign language. Studies have shown that the mechanism under the human-machine dialogue mode is conducive to stimulating students' interest in learning. The listening and speaking training course reasonably arranged based on human-machine dialogue can improve students' listening and speaking level, make the teaching more humanistic, scientific and pertinent, and comprehensively improve the quality of foreign language listening and speaking teaching. The teaching mode based on human-machine dialogue can help students to adapt to the intelligent era more quickly, and cultivate outstanding talents who can adapt to the intelligent society. Human-machine dialogue can accelerate education reform and promote educational development.

In addition, we still need to continuously explore and innovate teaching contents for foreign language teaching at universities. Through teaching and practice activities, teachers should learn from each other and pursue innovations to find the teaching methods that are most suitable for students in learning. In order to improve students' foreign language listening and speaking ability by using the listening and speaking test system based on human-machine dialogue, we should follow the trend of education reform in the human-machine dialogue teaching mode, and balance the key points of traditional education, traditional test methods and human-machine dialogue teaching mode. In order to improve students' foreign language listening and oral communication skills, we should make reasonable arrangements for its use, give full play to the advantages of human-machine dialogue teaching mode, improve the smooth progress of foreign language teaching reform at universities, and promote educational development and progress in the future.

Acknowledgments

This work was supported by the Humanities and Social Sciences Research Project of Liaoning Provincial Committee of Education in China (grant number 2017J068).

References

[1] Xu, F., Song, X. (2017) Research on the Acceptability of Oral English Test Based on Human-Machine Dialogue for Junior High School. Science & Technology Vision, 11: 29-30.
[2] Dai, Y.H., Xu, B., Chen, H.J. (2018) Promotion of Hybrid Teaching by AI and Construction of Ecosystem. Modern Distance Education Research, 02: 24-31.
[3] Li, K.F., Wang, Y.G. (2018) Artificial Intelligence. Hangzhou(Weekly), 20: 59.
[4] Wang, Y.B. (2018) Exploration of Role Orientation of Foreign Language Teachers in AI Era. Journal of Jilin TV & Radio University, 11: 82-83.
[5] Liu, J.J. (2002) Application and Practice of Human-Machine Dialogue in Modern Foreign Language Teaching .Technology Enhanced Foreign Language Education, 01: 15-17.
[6] Ma, L.M. (2012) Status Quo of Listening and Speaking English Teaching for Junior High Schools and Improvement Measures. New West (theoretical version), 14: 163-177.
[7] Yu, L.P. (2018) Practical Exploration of Listening and Speaking English Teaching Strategies Based on Human-Machine Dialogue for Senior High School Entrance Examination. English Teacher, 15: 112-115.
[8] Chapelle, C. (1998) Field Independence: A Source of Language Test Variance. Language Testing, 05: 62-82.
[9] Chapelle, C., Roberts, C. (1986) Ambiguity Tolerance and Field Dependence as Predictors of Proficiency in English as a Second Language. Language Learning, 36: 27-45.
[10] Hansen, J. (1984) Field Dependence-independence and LanguageTesting: Evidence from Six Pacific Island Cultures. TESOL Quarterly, 18: 311-324.
[11] Hansen, J., Stansfield, C. (1981) The Relationship Between Field Dependent = Independent Cognitive Styles and Foreign Language Achievement. Language Learning, 31: 349-367.
[12] Lu, C., Sucn, H. (1993) The Interaction Effect of Individual Characteristics and Assessment Format on the Result of Performance-based Assessment. Paper Presented at the Annual Meeting of the American Educational Research Association, Atlanta,GA, 43: 121-126.
[13] Birenbaum, M., Tatsuok, K. K. (1987) Open-ended Versus Multiple Choice Response Formats-It Does Make a Difference for Diagnostic Purposes. Applied Psychological Measurement, 11: 385-395.
[14] Birenbaum, M.,Tatsuoka, K. K., Gutvirtz, Y. (1992) Effects of Response Format on Diagnostic Assessment of Scholastic Achievement. Applied Psychological Measurement, 16: 353-362.
[15] Shohamy, E. (1984) Does the Testing Method Make a Difference? The Case of Reading Comprehension. Language Testing, 01: 147-170.
[16] Traub, R. E. (1993) On the Equivalence of the Traits Assessed by Multiple Choice and Constructed-Response tests.In R.E. Bennett,Ward.W.C., (Eds.), Construction Versus Choice in Cognitive Measurement. Hillsdale, NJ: Erlbaum, 13: 29-44.
[17] Fang, Y. (2015) A Research on Chinese Learners' Preference Degrees in Oral English Exam Modes: Based on SPSS Software Analysis. Foreign Language and Literature, 31: 151-157.
[18] DÖrnyei, Z. (2001) Motivational Strategies in the Language Classroom. University Press. Cambridge England.

INVESTIGATING THE ADVANTAGES AND DISADVANTAGES OF TAIWAN'S INDUSTRIAL DESIGN EDUCATION FROM DEPARTMENT EVALUATION DATA

Lin, Mei-Ting[1*], Chen, Xian-Wang[2], Huang, Liang-Peng[3] and Chen, Jheng-Han[4]

[123]School of Media & Art Design, Wenzhou Business College, Wenzhou, Zhejiang, 325035, China

[4]Department of Technology Application and Human Resource Development, National Taiwan Normal University, Taipei,10610, Taiwan

[1*]gua10035901@gmail.com

Abstract. With Taipei being nominated for the World Design Capital 2016, an event that emphasize "design thinking," issues such as identifying Taiwan's design education development trends and crucial factors influencing Taiwan's design industry development must be addressed for Taiwan's future design education. This study conducted a pilot study to investigate design department objectives, analysed designer abilities required by businesses, and referred to the design department evaluations. The results showed that for industrial design departments, issues that required most improvements were: (a) concerning those to better implement self-improvement mechanisms so that course learning objectives are adjusted to meet business expectations; and (b) Space and equipment should be upgraded. In addition to the upgrade, the actual requirements of the enterprise should be consulted for planning. In addition to making such changes, design departments should plan design education according to the actual requirements of businesses. The results of this study may be used as reference when engaging in department planning for the development of industrial design education in Taiwan.

1. Introduction

Taiwan's higher design education originated in 1964, the year when the Ming Chi Institute of Technology (current Ming Chi University of Technology) introduced the industrial design program. A period of five decades has passed since 1964, during which "design" has shown a clear intangible influence on Taiwan's economic development as well as various sectors such as national competitiveness and people's livelihood. A review of Taiwan's design education in practice shows significant developments. In fact, Ph.D. programs in design were further added in 2000 [1].

Under this premise, this study investigated the current situation and problems concerning the development of higher design education in Taiwan by analyzing the evaluation contents produced by various schools. The development of industrial design education was further explored by examining the educational objectives of design departments, and evaluation analyses. The analysis results are shown below.

2. Target Abilities Set by Design Department and Abilities that Business Expected from Industrial Designers

According to data compiled by the Statistics Department of Taiwan's Ministry of Education, the number of graduates from design-related departments in Taiwan amounts to more than 5,000 students per year. Similarly, students studying in undergraduate design-related courses such as visual communication design, product design, and interior design increased from over 30,000 during the 2007–2008 academic year to more than 40,000 during the 2010–2011 academic year, showing an increase of 40%. The number and quality of students graduating from such design departments have a direct effect on whether the personnel requirements of related industries can be satisfied as well as Taiwan's overall national competitiveness.

Previous design education mainly focused on practical training. The demands of related industries for designers were also relatively simple. However, as the overall environment evolved, the requirements diversified. Norman [2] proposed different views pertaining to future design education. For designers, they must not only possess the necessary professional skills, but also deal with work-related issues such as company organizational structure, social issues, interpersonal relationships, and service and design experiences. Following the government's policies to promote the cultural and creative industry, colleges and universities established related departments. However, whether the knowledge and skills taught in schools are able to meet current or future industry needs is an issue worthy of investigation. Therefore, we conducted a pilot study to investigate schools' objectives for founding cultural and creative design-related departments to assess students' acquisition of skills from higher education. The analysis results are shown below.

2.1. Analysis of the Objectives for Founding Design-Related Departments

Currently, design-related departments in Taiwan total more than 200. Competition between design schools remains fierce and schools are engaging in specialization and differentiation-related efforts to distinguish themselves from competition. Thus, an analysis of the objectives set by the departments as well as their unique characteristics will provide insight into the current trends in the abilities of designers under the new cultural and creative environment. For the first stage of the analysis, this study used a sample of 28 creative design-related departments. Such departments' name were divided into seven orientations or types: cultural vocation development, cultural and creative industry, cultural and creative industries management, cultural creativity and design, creative and digital media design, creative design, and others.

By combining the educational objectives of the 28 departments, we obtained a total of 49 ability items. Next, we separated these ability items into different groups and found that the abilities could be roughly divided into seven categories, which were "comprehensive abilities," "theory-based abilities," "digital technology-related abilities," "practical abilities," "property rights knowledge," "interior design abilities," and "marketing and management abilities." The top three most frequently mentioned abilities were "practical abilities," "marketing-based planning and management abilities," and "creative thinking abilities and cultural literacy." The analysis also showed that the focus of education differed between the seven department types, which are listed as follows: cultural vocation development departments focused on cultural literacy; cultural and creative industry departments emphasized on operational management; cultural and creative industry management departments paid attention to marketing-based planning and operational management; cultural creativity and design departments stressed the importance of theories, cultural literacy, and art literacy; digital media design departments focused on interdisciplinary integration abilities, digital technology-related abilities, and marketing-based planning; and creative design and lifestyle departments emphasized "creative thinking abilities," "product development," and "exhibition activity-based designs" (Table 1).

The 49 abilities were also divided into seven categories, of which comprehensive ability-related learning accounted for the highest percentage (i.e., 24.24%), followed by theoretical foundation ability-related learning (i.e., 22.44%) and digital technology ability-related learning (i.e., 16.32%). The results showed that these three abilities were the key abilities to be developed (Figure 1) [3].

IOP Conf. Series: Materials Science and Engineering **573** (2019) 012048 doi:10.1088/1757-899X/573/1/012048

An analysis of the educational objectives shows that cultural and creative design-related departments focused on practical abilities, marketing-based planning abilities, management abilities, creative thinking abilities, and cultural literacy as the key abilities to be developed. Concerning other departments, indicators for abilities to be developed also differed. Such ability indicators signified future abilities of the students. Therefore, whether these abilities meet the needs of employers is a topic worthy of further investigation.

Table 1. Key Abilities to be Learned According to Each Respective Department

Department types	**Focus of education**
cultural vocation development	cultural literacy
cultural and creative industry	operational management
cultural and creative industry management	marketing-based planning and operational management
cultural creativity and design	theories, cultural literacy, and art literacy
creative and digital media design	insufficient data for analysis due to low number of departments
creative design	
others	

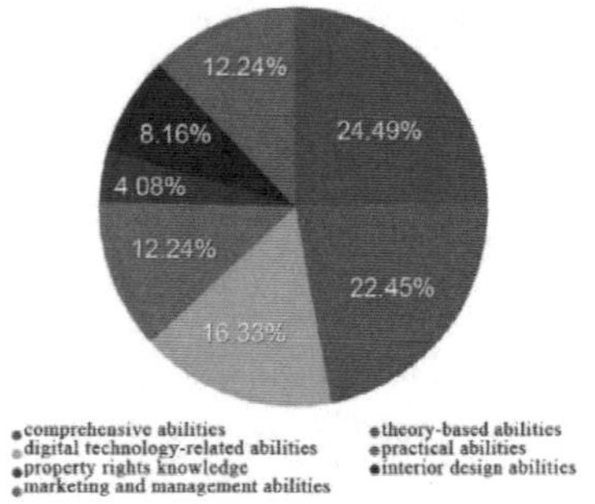

Figure 1. Abilities to be Learned as A Percentage of All the Abilities

2.2. Examining the Problems of the Current Era via Higher Education Evaluation

Taiwan's higher education evaluation has changed over time. Related evaluation systems have evolved from general school evaluation to professional education evaluation [4]. Most department evaluations in the second-stage further emphasized assessing students' learning results [5].

The Taiwan Ministry of Education has set clear targets for the two stages of higher education evaluation, in which the goals for the first-stage and second-stage department evaluations were to "ensure that students are provided a high-quality learning environment" and "promote and implement mechanisms ensuring favorable student learning results," respectively. The two-stage evaluation targets show how expectations and values of education have changed over time. In addition to the problem of sub-replacement fertility, imbalanced education quality, poor resource allocation, and globalization of education markets, challenges related to talent issues in the market has resulted in higher education resources becoming overly dispersed, defeating the intentions of the government when it rapidly increased the number of universities. The oversupply of education has led to incessant higher education crises and related problems such as low college and university competitiveness, low student enrollment rates, and poor student qualities [5].

Mechanisms to ensure the learning results of students were introduced to guide the abilities of students, emphasizing what they had learned and were capable of doing. The rapid increase in the number of universities resulted in poor student quality and the gap between students trained by schools and professional personnel required by businesses. Such an issue is a clear social problem and one that potentially jeopardizes national competitiveness.

3. Current Situation of Design Education Following Department Evaluations

This study first explored current situations and problems of higher industrial design education. Next, the advantages and disadvantages of Taiwan's current industrial design education were analyzed using evaluation content obtained by the Higher Education Evaluation & Accreditation Council of Taiwan and the Taiwan Assessment and Evaluation Association.

3.1. Sample Used for Industrial Design Evaluation and Analyses

This study selected only industrial design departments as the study sample and used department evaluation reports issued by evaluation centers on their websites as the basis of analyses. By searching through the online database of the evaluation centers, we obtained the evaluation data of six general university industrial design departments and four technical and vocational university industrial design departments. The evaluation and analysis sample thus comprised 10 university industrial design departments (Table 2).

Table 2. Sample of industrial design departments selected for evaluation and analysis

General Universities		**Technical and Vocational Universities**	
School	**Department**	**School**	**Department**
Tatung University	Department of Industrial Design	Ming Chi University of Technology	Department of Industrial Design
Dayeh University		Yunlin University of Technology	
Tunghai University		Taipei University of Technology	
Chang Gung University		Chaoyahg University of Technology	
Cheng Kung University			
Huafan University			

3.2. Evaluation of the Advantages and Disadvantages of General University Industrial Design Departments

Analysis showed that the primary advantages of general universities are as follows: characteristics developed by departments matched the education philosophies of the school. The long-term class teacher or mentor systems enabled teachers to fully understand students' personal characteristics. In addition, interdisciplinary development strategies could be adopted for resource integration, and joint work with the professional industry facilitated in finding cooperative education opportunities. Conversely, the disadvantages of general universities are listed below: curriculum planning and teacher expertise were unable to meet school's educational objectives; and lack of a systematic course plan, in which courses to be studied for each respective industry remained unclear. The advantages and disadvantages were compared and compiled into Table 3. Four items (emboldened in the table) featuring both advantages and disadvantages were: (1) the teacher and senior student guidance system helped guide junior students in making course selections. However, systematic course plans indicating courses to be studied for each respective industry were not available, indicating a need for improvement; (2) although the educational space and facilities were sufficient for learning, they were insufficient for meeting the departments' needs and educational objectives. To identify the reasons contributing to such problems, the departments' educational objectives and course directions must first be examined; (3) most departments were able to continue offering self-improvement courses. However, various self-evaluation mechanisms were ineffective; and (4) although alumni clubs were founded, they operated poorly and were thus unable to provide recommendations to departments for students' future developments.

3.3. Evaluation of the Advantages and Disadvantages of Technical and Vocational University Industrial Design Departments

The advantages of technical and vocational universities are as follows: teachers possessed practical and research experiences; teachers from the professional industry were employed, elevating academia–industrial connections; teachers possessed practical experiences that helped students connect with the professional industry; course syllabi were promptly uploaded online; multipurpose studios were established; and interdisciplinary counseling mechanisms and dual-class teacher/mentor systems were in place. Conversely, the disadvantages of technical and vocational universities are listed below: some classrooms failed to meet educational objectives and classroom F had to be done; inadequate employment of equipment maintenance personnel was observed; facilities were supervised by the school-level management, making usage inconvenient for departments; and computer and other equipment were insufficient in design workshops.

The advantages and disadvantages were compared and compiled into Table 4. Three items (emboldened in the table) featuring both advantages and disadvantages were: (1) complete management guidelines and systems for specialized classrooms were enforced. However, some classrooms failed to meet educational objectives and classroom replanning had to be done; (2) the schools supplied departments with sufficient funding and equipment. However, computer and other equipment were insufficient in design workshops, indicating the need for improved funding allocation; and (3) departments possessed interdisciplinary counseling mechanisms. However, no interdisciplinary learning mechanisms were in place to meet students' diverse learning demands.

3.4. Cross Analysis of the Advantages and Disadvantages Listed in the Industrial Design Department Evaluation Results of Both General Universities and Technical and Vocational Universities

By performing a cross analysis on the industrial design department evaluation results of both general universities and technical and vocational universities, this study identified the advantages of industrial design education in Taiwan, which are listed as follows: (1) students are successfully motivated to participate in design-related competitions; and (2) high ratio of graduates being employed. Conversely, the disadvantages of industrial design education in Taiwan are as follows: (1) improvements remain to be made in educational space and facilities; and (2) alumni club operation is inadequate and unable to facilitate department development.

Table 3. Evaluation of the advantages and disadvantages of general university industrial design departments

Item		Evaluation content of general university industrial design departments (advantages)	Evaluation content of general university industrial design departments (disadvantages)
Departments	Objectives	Clearly matched departments' educational objectives	
		Questionnaire surveys completed by graduating students showed high satisfaction, meeting departments' educational objectives	
	Characteristics	Characteristics developed by departments matched the education philosophies of the school	
		Long-term class teacher and mentor systems enabled teachers to fully understand students' personal characteristics	
		Interdisciplinary development strategies were adopted for resource integration	
		Worked with the professional industry to obtain cooperative education opportunities	
	Performance of school graduates	Performance of graduate students earned the recognition of their employers	
Education	Research	Research projects that received funding from the Ministry of Science and Technology showed outstanding results	Research projects and essays were completed by the same small group of teachers
		Various teacher rewards were set up as incentives to elevate their professional performance	
		Departments' research-related planning matched teachers' expertise and department development	
	Evaluation	Employed education evaluation systems through the use of online resources to reflect students' learning statuses	
		Education evaluation systems were built	
	Activities	Hosted international exchanges and seminars as well as related activities	
		Teachers and students actively participated inintracurricular and extracurricular activities (research competitions and exhibitions)	
	Courses	Teacher and senior student guidance system helped and guided junior students in making course selections	Curriculum planning and teacher expertise unable to meet schools' educational objectives
			Core curriculum failed to meet educational objectives
			Lack of a systematic course plan, in which courses to be studied for each respective industry was unclear.
	Education		Teachers' expertise failed to match courses and department development
	Facilities	Educational space and facilities for learning were sufficient	Educational space and facilities were insufficient for meeting the departments' needs and educational objectives
Development of abilities	Self-improvement	Self-evaluation improvement for courses were in place	Self-evaluation mechanisms were ineffective
		Curriculum committee operated favorably	
Students	Situation after graduation	Most graduate students worked in design-related jobs	Surveys on the employment or further education status of school graduates were missing
	Alumni clubs	Alumni tracking systems (alumni clubs) were founded	Alumni clubs operated poorly, failing to provide recommendations to departments concerning students' future developments

IOP Conf. Series: Materials Science and Engineering **573** (2019) 012048 doi:10.1088/1757-899X/573/1/012048

Table 4. Evaluation of the advantages and disadvantages of technical and vocational university industrial design departments

Item		Evaluation of technical and vocational university industrial design departments (advantages)	Evaluation of technical and vocational university industrial design departments (disadvantages)
Departments	Objectives		Educational objectives were unclear and students were trained only for limited fields
	Characteristics	Curriculum matched the needs of Taiwanese industries as well as international development trends	
		General and professional knowledge and abilities were stipulated for graduating students (i.e.,prerequisite for graduation)	
		Local characteristics developed and linked with the professional industry	
Education	Research	Number of teacher-conducted research increased	Low ratio of professor-level teachers
			Poor technology transfer
	Activities	Students' participation in design competitions showed great results	
	Courses	Promotion of comprehensive planning and learning portfolios in place	
		Courses matched students' career and professional developments	
	Education	Full-time teachers possessed favorable practical and research experiences	Insufficient number of teachers with a background in culture- related fields
		Part-time teachers from the professional industry hired to enhance academia–industrial exchanges	
		Course syllabus promptly uploaded online for students to check and learn	
		Interdisciplinary counseling mechanisms in place	Lack of interdisciplinary learning mechanisms to meet students' diverse learning demands
		Education based on multipurpose studios displayed concrete results	
		Teachers possessed practical experiences that helped students to connect with the professional industry	
		Dual-class teacher/mentor systems in place	
	Facilities	Complete management guidelines and systems for specialized classrooms were enforced	Some classrooms failed to meet educational objectives and classroom replanning had to be done
		A wealth of resources in libraries that helped students with their needs	Facilities were supervised by the school-level management, making usage inconvenient for departments
		Schools supplied departments with sufficient funding and equipment	Lack of computer and other equipment in design workshops
			Inadequate equipment maintenance personnel
Development of abilities	Self-improvement	Excellent department committee operation	
		Improvements were made according to recommendations made from previous visits	
		Regular teacher conference systems were in effect	
Students	Internship system	Comprehensive internship guidance system in place for off-campus internships	
	Situation after graduation	High ratio of graduates being employed	Lack of alumni follow up systems and personnel for long-term student status follow-up

4. Conclusion

Analyses of the industrial design departments of general and technical and vocational universities showed that compared with the industrial design departments of technical and vocational universities, those of general universities featured superior research potential and evaluation mechanisms. Conversely, industrial design departments of technical and vocational universities displayed more favorable education. In other words, the former better met the department evaluation expectations in research potential and evaluation mechanisms, whereas the latter better met business expectations in education. The analysis also proposed more suggestions in the area of education for general universities, showing that general universities required more improvements in such area.

General universities placed their focus on using ability indicators and considering abilities needed by industries in making course planning-related adjustments. Conversely, technical and vocational universities emphasized that teachers possess practical experiences and provide comprehensive learning portfolios to guide students' learning; most of the departments provided students with learning portfolios to help them select courses and attached great importance to related performance and competitions. However, alumni clubs operated poorly and were unable to provide department with resources effectively. In addition, education space and facilities were inadequate, preventing the departments from enhancing their education-related special features to improve their competiveness.

The analysis results are cross-analyzed with the goals of the Design Department, business surveys, and evaluation results, in which the results showed that for industrial design departments, the areas requiring most improvements were: (a) to better implement self-improvement mechanisms so that course learning objectives are better adjusted to meet business expectations; and (b) to upgrade design education space and facilities. In addition to making such changes, design departments should plan design education according to the actual needs of businesses. The results of this study may be used as reference when engaging in department planning for the development of industrial design education in Taiwan. Taipei is successful bid to host the 2016 World Design Capital, the spirit of main "Adaptive City- Design in Motion," that emphasized Taipei is a continuous improvement city, that related studies on design education and promotion are indispensable. This study should have an era representative and looking forward to providing the planning and development of recommendations for education of industrial design department in Taiwan.

IOP Conf. Series: Materials Science and Engineering **573** (2019) 012048 doi:10.1088/1757-899X/573/1/012048

References

[1] Lin, M. T. & Lin, R. (2013). The Analysis of Taiwan Design Changes from Industrial Design Magazine. Journal of National Taiwan College of Art's, 93,153-184.

[2] Norman, D. (2010). Why design education must change. Core 77 design magazine & resource. Viewed December 2015 http://www.core77.com/blog/columns/why_design_education_must_ change_17993.asp 7.

[3] Lin, M. T., Lin, R. & Lin, P. H. (2013). Analysis of Cultural and Creative Talent's Capability in the future from the Goal of Cultural and Creative Department and Institute. Journal of Education Research, 235, 33-45.

[4] Wu, C. S. (2010). Enhancing Evaluation Quality and Performance by Implementing Evaluation Committee Member Learning Courses. Evaluation Bimonthly, 24, 9-14.

[5] Hsieh, C. C. (2006). The Influence of Evaluation System on the Productivity for the Comprehensive Universities in Taiwan (Unpublished master's thesis). Ling Tung University, Taichung, Taiwan.

IOP Conf. Series: Materials Science and Engineering **573** (2019) 012049 doi:10.1088/1757-899X/573/1/012049

Research on the Attenuation Law of User Experience Satisfaction of Mobile Phone

Xuesheng Pei[1,2*], Yang Wang[1]

[1]School of Art and Design, Henan University of Science and Technology, Luoyang, Henan, 471000, China

[2]School of Mechatronic Engineering and Automation, Shanghai University, Shanghai, 200000, China

[*]Corresponding author's e-mail: pxsh12@126.com

Abstract. With the continuous development and widespread popularity of smart phones, people have been deeply studying mobile phones, and pay more attention to user experience. Based on literature research, this paper establishes a hierarchical chart of influencing factors of user experience satisfaction of mobile phone. A calculation method of user experience satisfaction is proposed. User interviews are used to determine the key factors, and user satisfaction model is constructed. Using questionnaire survey method to collect data and explore the process and rules of satisfaction changes in the process of purchasing, using and replacing mobile phones. The results show that user experience satisfaction of mobile phone decreases with the increase of mobile phone usage time, and battery loss, running stuck, damage and insufficient storage space are all key factors affecting the satisfaction of mobile phone users. Through data analysis, the change rule of user experience satisfaction of each key factor is obtained, and then guide relevant enterprises to upgrade and improve products pertinently.

1. Introduction

With the rapid development of mobile Internet technology, smart phones are rapidly popularized. However, the measurement and evaluation of mobile phones are often based on the evaluation of their hardware and software performance, but lack of attention to user experience. For a mobile phone, users will experience the process of purchase, use, (maintenance), replacement and so on. Especially in the use stage, user experience satisfaction change with the use of time. At present, there is still a lack of research on some issues, such as the factors leading to the change of satisfaction and the regularity of the change of satisfaction. Therefore, this study analyzes the influencing factors of mobile user experience satisfaction, establish a satisfaction model, and explore the reasons and rules of the change of mobile user experience satisfaction. This study will help mobile phone developers to improve the metrics of mobile phones in a targeted manner and improve user loyalty to the brand. At the same time, this study also provides some reference for the related research of user experience satisfaction.

2.Relevant research status

2.1. Summary of user experience evaluation

By consulting relevant literature, the current concept of user experience satisfaction mainly comes from the research of user experience measurement. Through reviewing the development and achievements of user experience at home and abroad, Ding Yi et al. put forward the elements of user experience, measurement methods, evaluation model, especially the evaluation method of integrating multiple evaluation indicators, which are the focus and difficulties in future research[1]. User experience has the characteristics of constantly changing with the process of using products by users. Therefore, it is necessary to collect data of user experience at different stages, including some measurements which aim at expected experience, immediate experience and long-term experience[2]. Law pointed out that the construction of user experience model is the basis of evaluating user experience, he believed that user experience evaluation should focus on the distribution of index weight and the evaluation of overall user experience[3].

2.2. Summary of mobile phone user experience research

For the user experience of mobile phone, Luo Shijian et al. put forward a user experience design method which based on usage scenarios, and validated the method by taking the design of mobile phone interface as an example[4]. By collecting the factors of satisfaction and dissatisfaction in the use experience of smart phone users, Si Yaqing et al. put forward that the results of user experience are reflected in emotions, satisfaction and intention of continuous use, and expounded the psychological changes of users when there is a difference between users psychological expectations and their actual use feelings[5]. Li Ding studied the modeling design of mobile phone based on user experience, pointed out the changing psychological needs of users, and promoted the innovation of mobile phone design concepts[6].

To summarize, although scholars' research on user experience has been relatively consummate, there is still a lack of research on the changing law of user experience satisfaction.

3.Brief analysis of the factors influencing the satisfaction of mobile user experience

3.1. Sorts and analysis of factors

User experience satisfaction mainly refers to the gap between users' psychological expectations of products or services and actual experience. The smaller the gap, the higher the satisfaction. Through investigation and analysis, this study divides the factors affecting the satisfaction of mobile user experience into four aspects: formal factors, economic factors, performance factors and humanistic factors. Formal factors refer to the form, color, quality and other external manifestations of the mobile phone, which is the most direct feeling of the product to the users; economic factors refer to the price, value of the goods before and after the purchase of the mobile phone and the impact of the relationship between the two on the user experience; performance factors mainly refer to the performance indicators provided by the mobile phone, reflecting the overall operation of the mobile phone; humanistic factors mainly refer to the cultural background and consumer psychology of consumers.

3.2. Establishment of hierarchical diagram

For mobile phones, formal factors mainly include the shape, color, material, weight and wear of the mobile phones; economic factors mainly include the brand, price and potential advantages and disadvantages of the mobile phones; performance factors mainly include the network performance, battery performance, storage space and operation speed of the mobile phones; and the humanistic factors include the user's cultural background, class background, consumer psychology and so on.

The hierarchical diagram is shown in Figure 1, it was established by assorting and extracting factors that affecting user experience satisfaction of mobile phone. Moreover, it can provide a preliminary theoretical support for further exploring the attenuation law of user satisfaction with mobile phones.

IOP Conf. Series: Materials Science and Engineering **573** (2019) 012049 doi:10.1088/1757-899X/573/1/012049

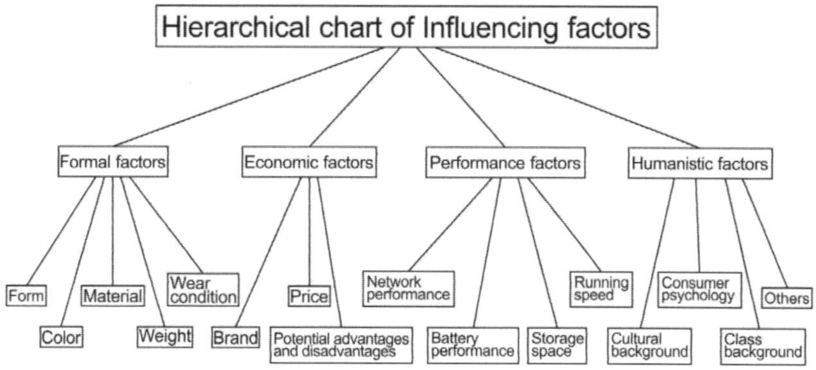

Figure 1. Hierarchical Chart of Influencing Factors of Mobile User Experience Satisfaction.

4.Construction and hypothesis of model

4.1. Proposal of satisfaction computing method

The products themselves has the characteristics of functionality, economy, humanity and aesthetics, which together determine the overall satisfaction of users for a product. However, the importance coefficients of each factor of different products are different. According to the use status of products, they can be divided into durable products, value-preserving products and general products. Furthermore, general products also include learning products, maintenance products and accidental damage products. Mobile phones belong to general products. At the same time, they have the characteristics of maintenance products and accidental damage products.

Based on the above analysis, this study proposes a calculation method for user experience satisfaction:

Assume each factor's satisfaction: performance factor's satisfaction is F, brand factor's satisfaction is B, price factor's satisfaction is P, and aesthetic factor's satisfaction is A.

Assume importance coefficient: performance factor is K_1, brand factor is K_2, price factor is K_3, and aesthetic factor is K_4.

Comprehensive cumulative satisfaction:

$$MP = K_1F + K_2B + K_3P + K_4A \qquad \text{Eq.1}$$

Considering the mutual influence of various factors, if any aspect is seriously dissatisfied, the product will not be selected. Then list the satisfaction product coefficient as:

$$MT = FBPA \times 10^{-8} \qquad \text{Eq.2}$$

Comprehensive satisfaction is the product of cumulative satisfaction and product coefficient:

$$M = MP \times MT \qquad \text{Eq.3}$$

Because of different users have different experiences on various factors of the product, their satisfaction is not the same. So the order of the results may also be different.

4.2. Mobile phone user satisfaction model construction

On the basis of a large amount of literature research, this study uses the main idea of the User Satisfaction Index Model (CSI)[7] to construct a mobile phone user satisfaction model, as shown in Figure 2.

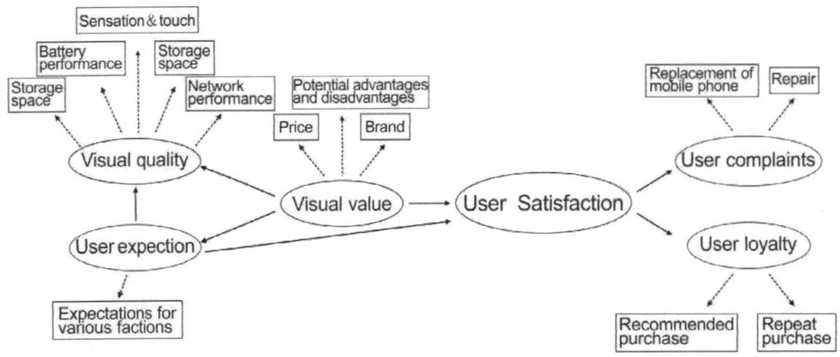

Figure 2. Mobile User Satisfaction Model.

4.3. Research hypothesis

After factor analysis, this study conducted user interviews with a number of users, selected the mobile phone usage time as an indicator to reflect the overall decline of mobile phone user experience satisfaction, and selected battery loss, damage degree, running stuck degree, storage space size as four key factors, and put forward research hypothesis.

4.3.1. Mobile phone usage time. With the increase of the use time of a mobile phone, the stability of various performance will change, and to a large extent affect the satisfaction of users. Just as the satisfaction of users for devices is the highest when new a mobile phone are purchased, with the passage of time, the satisfaction will vary. As a result, we propose the following assumptions:

Hypothesis 1 (H1) Mobile phone usage time is positively correlated with overall satisfaction decline of user experience.

4.3.2. Battery Loss. With the use of mobile phones, the total battery capacity will be irreversibly attenuated, which will have a great impact on the experience of mobile phone users. Whether the mobile phone's battery life is outstanding is also one of the reasons why users consider changing their mobile phones.As a result, we propose the following assumptions:

Hypothesis 2 (H2) Mobile phone battery loss is positively correlated with user experience satisfaction attenuation.

4.3.3. Degree of Damage. Daily use of mobile phone will inevitably occur wear and tear, different degrees of wear will have different degrees of impact on mobile phones: light to affect the appearance, serious to different parts of the damage directly affects the use of mobile phones. Therefore, the degree of wear and tear of mobile phone will affect the satisfaction of user experience to a great extent. As a result, we propose the following assumptions:

Hypothesis 3 (H3) The degree of mobile phone damage is positively correlated with the decline in user experience satisfaction.

4.3.4. Degree of running stuck. With the increase of mobile phone usage time, the hardware of the mobile phone will experience different degrees of loss. Application updates have higher requirements for mobile phone hardware performance. These will cause the phone to be stuck in different degrees. The slow running speed of mobile phones greatly affects the user's mood and seriously affects their satisfaction. As a result, we propose the following assumptions:

Hypothesis 4 (H4) The degree of mobile phone running stuck is positively correlated with the attenuation of user experience satisfaction.

4.3.5. Storage space size. Storage of photos, applications, system information and so on all shall occupy the storage space of mobile phones. With the increase of data and files, the storage space will gradually decrease. When the mobile phone space is too small to download applications and files, or the frequent prompt storage space is insufficient, it will seriously reduce user experience satisfaction. As a result, we propose the following assumptions:

Hypothesis 5 (H5) The size of storage space of mobile phone is positively correlated with the decline of user experience satisfaction.

5.Questionnaire design and data collection

5.1. Questionnaire design

This study uses a questionnaire survey method to guide users to fill out questionnaires based on their actual experience. From the purchase to the replacement of a mobile phone, the process of satisfaction change is a potential variable that cannot be measured visually. Therefore, it is necessary to obtain or estimate the useful data required for the study by observable and descriptive variables. The questionnaire mainly investigated and analyzed the factors such as mobile phone damage, battery capacity, running stuck, storage space, etc., and interviewed some users to get more data needed for this study.

5.2. Data collection

Questionnaires are conducted online and offline. Online surveys use the "Questionnaire Star" online survey website to collect questionnaires. Offline surveys are mainly completed through paper questionnaires. The subjects of the survey are mainly students, teachers and the general public. It is generally believed that the number of survey samples should be 5-10 times the number of survey items. There are 20 questions in this questionnaire, 164 questionnaires were collected, 160 valid questionnaires, the effective recovery rate is 97.6%, which meets the research requirements. The data of valid questionnaires and the corresponding question items of hypotheses are counted by Excel table. Some of the questions are listed in Table 1.

Table 1. Summary of Source Questions.

Hypothesis	Source issues
H1	Mobile phone usage time&Overall Satisfaction
H2	Standby time of mobile phone、Percentage of battery loss
H3	Damaged part、Mood after breakage
H4	Sorting the Importance of the Reasons for Changing Mobile Phones
H5	Sorting the Importance of the Reasons for Changing Mobile Phones

5.3. Data Reliability Test

Reliability refers to the degree of consistency of the results obtained when repeating measurements on the same thing, reflecting the stability and reliability of the measurement tool and the true degree of the measured features[8]. In this study, the SPSS 20.0 tool was used to test the reliability of the questionnaire data using Cronbach's Alpha coefficient and split-half reliability coefficient, which were shown in Table 2.

Table 2. Reliability test results.

Reliability index	General questionnaire
Cronbach's Alpha coefficient	0.722
Split-half reliability coefficient	0.762

After testing, the overall Cronbach's Alpha coefficient of the data in the questionnaire was 0.722, and the split-half reliability coefficient was 0.762, indicating that the questionnaire had good reliability.

6.Model Verification and Analysis of Satisfaction Attenuation Law

6.1. Hypothesis 1 (H1) Mobile phone usage time is positively correlated with overall satisfaction decline of user experience.

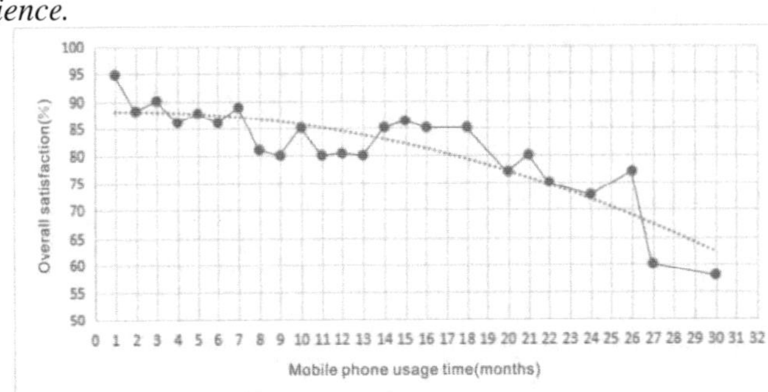

Figure 3. Attenuation curve of overall user satisfaction.

According to the data of the mobile phone usage time (x) and overall satisfaction (y) in the questionnaire, through the conversion and processing of some data, the coordinate scatter plot is drawn by Excel spreadsheet, and the trend line is generated, as shown in Figure 3. With the increase of mobile phone usage time, the overall satisfaction of users shows a gradual downward trend, and the decline of satisfaction shows a gradual upward trend. Therefore, the usage time of mobile phone is negatively correlated with the overall satisfaction of users, and the usage time of mobile phone is positively correlated with the decline of the overall satisfaction of users.

6.2. Hypothesis 2 (H2) Mobile phone battery loss is positively correlated with user experience satisfaction attenuation.

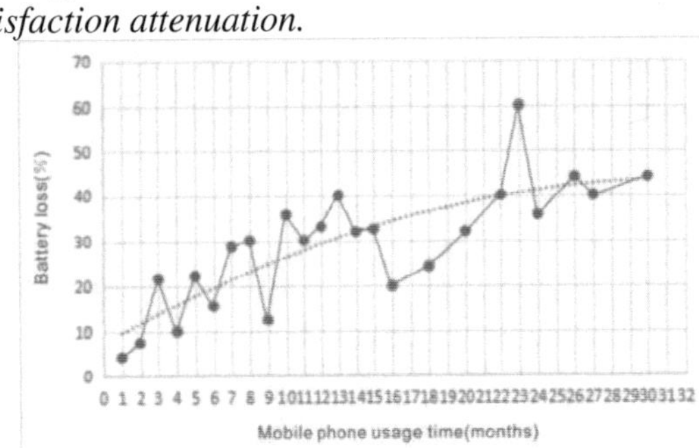

Figure 4. Percentage change curve of battery performance attenuation.

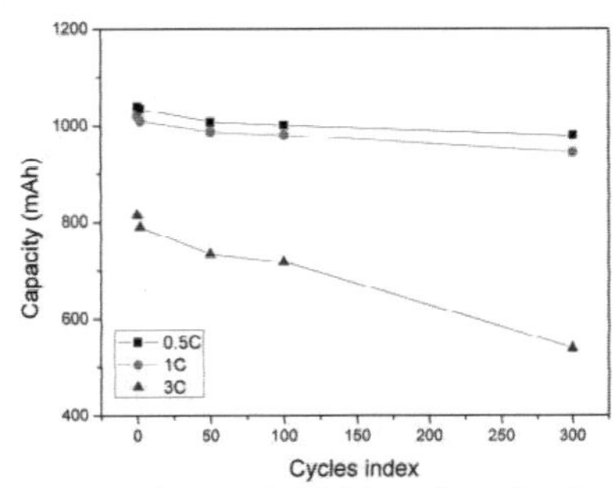

Figure 5. Lithium ion discharge capacity decay curve.

According to the data of mobile phone usage time (x) and the attenuation degree of mobile phone's standby time (y) in the questionnaire, the coordinate scatter plot is drawn by using Excel table through the transformation and processing of part of the data, and the trend line is generated, as shown in Figure 4. As can be seen from the figure, for different users, with the increase of mobile phone usage time, the battery life is shortening, battery performance is declining, the attenuation of standby time (%) of mobile phone shows a gradual upward trend, which reflects the continuous decline of user experience satisfaction. At the same time, according to Yang Peng's research on capacity decay of lithium-ion batteries, the capacity of lithium-ion batteries reduces with the increase of charge-discharge cycle numbers, as shown in Figure 5[9]. Therefore, the number of battery cycles is positively correlated with battery loss (%).

6.3. Hypothesis 3 (H3) The degree of mobile phone damage is positively correlated with the decline in user experience satisfaction.

This hypothesis involves four questions in the questionnaire: the wear and tear of mobile phone, the place of breaking, the impact on performance after breaking and the mood after breaking. The mood after breaking indicates the decline of user experience satisfaction from the side. According to the statistics of the broken part and the mood after the breaking, 42.31% and 30.77% of the users expressed "a little unhappy" and "relatively uncomfortable", while 26.92% of the users chose "indifferent" and "very uncomfortable". According to the analysis of the user's mood changes based on the broken part, 43.75% of the users chose "more uncomfortable" after the smash screen, which was the highest proportion. However, after the corner, border and back of phone are broken, the change of user's mood is "a little unhappy", which is the highest proportion. Therefore, mobile phone breakage leads to a decline in satisfaction, especially screen breakage, which has a greater impact. The data and results of this analysis are presented in the form of pie chart, as shown in Figure 6.

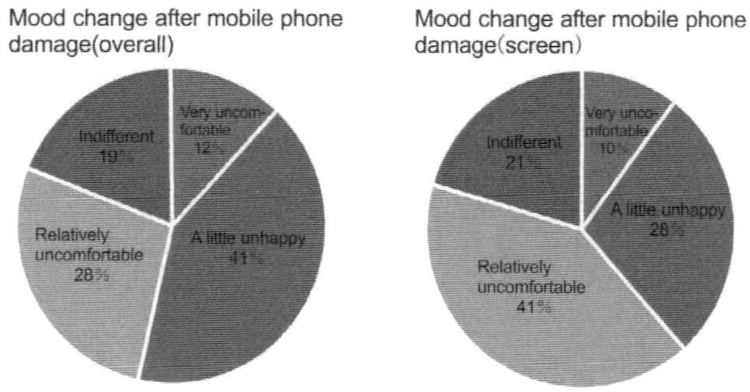

Figure 6. Pie chart of user's mood change after mobile phone breakage.

6.4. Hypothesis 4 (H4) & Hypothesis 5 (H5) The degree of mobile phone running stuck and the size of storage space are positively correlated with the attenuation of user experience satisfaction.

According to the question item of "importance ranking of reasons for changing mobile phones"in the questionnaire, after data integration and processing, the importance ranking "running stuck > insufficient storage space > broken down > battery durability > wear > slow network speed > other". The user chooses to change the mobile phone, which means that the experience satisfaction has reached or will reach the lowest point, and the data indicates that the slow running speed and insufficient storage space are the primary factors for the user to change the mobile phone. So the degree of mobile phone running stuck and the size of storage space are positively correlated with the attenuation of user experience satisfaction.

6.5. The general rule of mobile phone user experience satisfaction decline

It can be seen from the above analysis that the user experience satisfaction decreases with the increase of the use time of the mobile phone, and the key factors for the slow decline of satisfaction are battery loss, running stuck, damage, and insufficient storage space. At the same time, the sudden breakage and maintenance of mobile phones, which can also cause a sharp decrease or rise of the user experience satisfaction.

7.Conclusion

This study used the methods of user interviews and questionnaires to analyze the influencing factors of user experience satisfaction and establish the key influencing factors. Then the satisfaction calculation method was proposed, the satisfaction model was constructed and the model was validated by the questionnaire data. The results of this study can guide enterprises to improve products in a targeted manner, thereby improving user experience satisfaction and ensuring the sound development of products. At the same time, the satisfaction calculation method proposed in this paper will provide academic and theoretical support for researchers in related fields.

There are still some shortcomings in this study. Firstly, the sample size of the questionnaire survey is small, because the mobile phone user group is very large, 160 valid questionnaires are not enough to represent all users accurately; secondly, this study does not consider the impact of gender, age and occupation on the research results, and will continue to study in the future to obtain more accurate and complete research results[10].

References

[1] Ding, Y., Guo, F., Hu, M., et al. (2014) A review of user experience at home and abroad. Industrial Engineering and Management, 19: 92-114.

[2] Olsson, T. (2013) Concepts and Subjective Measures-for Evaluating User Experience of Mobile Augmented Reality Services. Human Factors in Augmented Reality Environments, 203-232.

[3] Law, E.L.C., Schaik, P.V. (2010) Modelling user experience-An agenda for research and practice. Interacting with Computers, 22: 313-322.

[4] Luo, S., Zhu, S., Ying, F., et al. (2010) Situational-based User Experience Design in Mobile Interface. Software, 16: 239-248.

[5] Si, Y., Liu, L. (2015) Research on the Influencing Factors of Smart Phone User Experience. Software, 3: 111-115.

[6] Li, D. (2013) Research on Mobile Phone Design Concept Based on User Experience. Xiangtan: Hunan University of Science and Technology, 10-30.

[7] Zhao, P. (1997) Customer Satisfaction Index (CSI). Management of Posts and Telecommunications Enterprises, 6: 33-35.

[8] Chai, H. (2010) Research on reliability and validity test methods in questionnaire design. World Science and Technology Research and Development, 32: 548-550.

[9] Yang, P. (2013) Study on capacity decay of lithium ion batteries. Shanghai: Shanghai Jiao Tong University, 23-26.

[10] Tang, X., Chen, F. (2015) Model and Empirical Study on Influencing Factors of Micro Credit Customer Satisfaction. Intelligence Journal, 34: 114-120.

IOP Conf. Series: Materials Science and Engineering **573** (2019) 012050 doi:10.1088/1757-899X/573/1/012050

Research on Internet Mutual Assistance Model of Art Education in Colleges and Universities

Xiaojing Liu [1] Ying Liu[1*] Peng Lin[1]

[1] Department of Architecture and Art, Central South University, Changsha City, Hunan Province, 410012, China

[*]Corresponding author's e-mail: [*] 81413450@ qq.com

Abstract. The art education of colleges and universities, which has a natural connection with various folk art forms and groups, undertakes the important mission of carrying forward the traditional culture confidence. Using Internet technologies to make different regions and groups absorb art nutrient from each other provides a new opportunity for the art education of colleges and universities with talents and innovations. Taking the art education of colleges and universities as the research subject, this essay tries to explore the characteristics, advantages, and formation mechanism of the mutual-assistance model between "Internet plus art education" of colleges and urban and rural areas by observing and analyzing the cultural demands and artistic resources of urban and rural areas, and provides a reference path for the construction of cultural confidence in colleges and universities' art education in a new era by discussing the adaptability through practical cases.

1. The significance and purpose of art education in colleges and universities

Art education, which cultivates people's perceptual ability, creative ability and expressive ability, subtly leads people into a more civilized way of life by showing the beauty of harmony between people, society and nature. Through art education, the government hopes to enhance the cultural literacy of the whole people and enhance the creativity and cohesion of the society, so as to build a harmonious society [1].

Colleges and universities undertake the functions of scientific research, talents training and social services. From the perspective of art education, scientific research refers to the study and enrichment of artistic expressions; talents training refers to the selection and cultivation of art professionals; and social services refer to the participation in the construction of social culture and art systems. With the advantages of the Internet, these three aspects can effectively bridge the gap between resources and demands to achieve mutual benefit and common prosperity.

2. The development status and opportunities of college art education

2.1 The development status of college art education

With a growing number of professionals and educated groups, China's college art education is developing rapidly, and the construction of relevant software and hardware is also continuously strengthened. At the same time, as The Opinions on Strengthening the Art Education in National Universities and Colleges is published, some *universities* and colleges have set up special art majors or public art courses. The development of activities such as national college student's art performances and the introduction of high art into campus has fully mobilized the enthusiasm and initiative of teachers and students to participate in art education, and the importance of art in college education has also been widely recognized.

At present, art education in colleges and universities is generally divided into two categories: the first category is art professional education, which is mainly for art students and creative groups in colleges and universities. It requires high artistic literacy and continuous creative materials from life. However, immersed in the ivory tower for four years, more and more students lack experience in observing life and practicing. Their art creation process also lacks imagination and judgment, and many art works are like copycat art after simple assembly and combination. The second is the art literacy education, and the audience, including the teachers and students and the public, is highly extensive. Under the comprehensive attention of the government and society, all colleges and universities have attached great importance to the construction of artistic literacy education, presenting a variety of forms of art education. But there have also been some obvious problems in the process of educational practice. First of all, the way of classroom teaching is obviously lack of openness. Many art clubs and literary activities are only limited in the campus, causing deficiency in the connectivity and communication with the community. This situation does not meet the requirement of openness of art education. Secondly, traditional art education is mainly carried out through classroom lectures, exhibitions and cultural performances, the forms of which have certain limitations in terms of time, frequency, interactive communication and detailed records [2].

2.2 The spring of art education

The advent of the Internet age has provided a wider space for art education and a richer means of presentation. By introducing the technological advantages and innovative ideas of the Internet into the art education of colleges and universities, we can innovate the traditional medium of educational communication to improve people's perception, creativity, imagination and understanding in a more flexible way, and it will become more convenient for us to serve audiences at different ages and levels and from different places.

*2.2.1 The model of "the Internet + art education" provides abundant inspirational resources.*The breadth and openness of the interaction of Internet is very suitable for colorful art education, and its infinite connection undoubtedly provides a huge resource base for art education and research.

2.2.2 The model of "the Internet + art education" removes the mysterious veil of art .. In the past, people received art education either in the exhibition hall, or in the classroom. The traditional way of art education not only has a sense of distance and passivity, but also lacks individual pertinence and interaction. In the Internet era, everyone has equal access to artistic resources and a platform to show their artistic abilities. Art education shows its popularity and becomes an activity that everyone can learn everywhere and learn all the time.

2.2.3 The model of "the Internet + art education" breaks the space-time barrier. In the Internet era, everyone has the equal ability to obtain and release information, and art education eliminates the influence of time and geographical division. A college art professor can conduct borderless, frequent real-time communication with a primary school student from a mountain village. For the students in the mountain villages, they can receive rich art knowledge. For the art professors, it is more convenient to collect and record art materials, which can stimulate the inspiration of artistic creation.

2.3 The mutual-aid demands among universities, communities and villages

2.3.1 Art education in colleges and universities has a wide range of needs for rich and colorful art resources in communities and villages. Tao Xingzhi, a renowned Chinese educator, put forward the theory that "life is education", emphasizing that art education in colleges and universities needs to guide teachers and students to leave their desks temporarily, get out of the ivory tower, and get close to and communicate with the source of art such as local music, wild scenery and folk customs and ceremonies. At the same time, urban communities and rural villages have great desire for various art forms, which are ideal exhibition stages for college art education achievements and works. Art comes from the conciseness of abstract life. Only when teachers lead students to embrace and understand life, can the artistic achievements be presented more resonatory to the public [3].

*2.3.2 The construction of community culture and art needs the participation and service of college art education resources.*Most colleges and universities not only have professional art talents and teams, but also have relatively complete venues and facilities, such as theater, dance hall, exhibition hall, piano room, studio, and so on. These personnel and material resources can serve not only the college teachers and students, but also the art education practice of community, reflecting the full use of educational resources and the equal rights of citizens to receive education.

2.3.3 The development of rural art education and the inheritance of folk art need the support and innovation of colleges and universities. At present, the developing trend of art education in rural areas is rather severe: art courses are generally neglected, and art teachers, as well as curriculum system need professional support. Simultaneously, art education in colleges and universities should protect and inherit the aesthetic law and art form of folk art, and explore the innovation path of characteristic art education nourished by traditional cultural heritage.

3. The mechanism of the mutual-aid model of "Internet + art education" in colleges and universities

The Ministry of Education has clarified the development idea of art education as exploring the ways and means of combining art education with other related disciplines, strengthening and improving the guarantee mechanism of art education, and narrowing the gap in the development of art education between urban and rural areas, different regions, and different schools. The art education in colleges and universities can combine its own advantages and environmental characteristics to build an "Internet + art education" mutual-aid model.

3.1 Mutual growth of college and community art education resources

Colleges and universities are a very special part of urban communities, including not only youth groups with strong vitality and cr88eativity, but also intellectual groups full of research spirit and practical motivation. Colleges and universities have natural merits and special advantages in participating in community culture and art construction.

3.1.1 We should make use of the interconnectivity of Internet information to realize the efficient utilization of software and hardware resources. Colleges and universities should participate in the construction of community culture and art, which can bring educational resources into the society and realize people's basic cultural rights and interests. Colleges and universities can provide art education courses, art creation places, rental of clothing and art tools and other relevant services to community residents through the Internet, so as to realize efficient utilization of art resources [4].

*3.1.2 We should make use of the flexibility of the Internet to meet individual needs.*People in the community with different artistic basic qualities like different artistic categories. Thus, it is difficult to meet the needs of everyone through collective teaching. But the Internet provides a technical channel for the spread of college art education. By spreading the course resources of college art education by means of micro video, MOOCs and interactive courses, we can fully show the characteristics of freedom and flexibility, meet individual differences in artistic perception and realize individualized teaching [5].

3.1.3 We should make use of the interactive feature of the Internet to provide opportunities and space for artistic creation..The geographical advantage of the community can provide more opportunities and broad space for art majors to carry out art practice. Through the Internet, the art education resources of the university can interact with the community cultural market in an efficient way, forming a long-term cooperation mechanism between supply and demand. Then the community becomes a fertile soil for students to stimulate artistic potential and show artistic talent. The teaching method of art has also changed from the simple in-class teaching to the combination of in-class and out-of-class teaching, focusing on the practical process and market feedback, so that students can have a deeper understanding of the multiple values of art. At the same time, the long-term output of art education by colleges and universities to the community can inspire the public's interest in and appreciation of art. If we make people in communities become closer to art, have the ability to appreciate art, and create their own artistic atmosphere, college art education will gradually cultivate potential energy for the market of popular art [6].

In 2015, ping99.com was reported by mainstream media as a successful case of online college art education, causing wide public attention. It provides a comprehensive service platform for online database, exhibition, transaction and communication, and provides online art education consultation and offline one-to-one guidance channels from more than 1,000 teachers of Academy of Fine Arts to the public users, so that more people can have access to professional college art education. At the same time, ping99.com also put forward the innovative service concept, namely "experiential art consumption". The users can first take the artworks home to get close to it, and then decide whether to buy after experiencing them for three months. The open Internet communication platform and professional art education atmosphere attract many potential college artists and people with artistic appreciation to carry out "socialized and interactive art education" online. This innovative mode based on the interaction between college art *education* and social resources on the Internet platform has also received business investment. At present, the platform has realized Series-A financing.

3.2 Mutual support and nourishment of college and rural art education resources

Rural community is also a form of community. Due to the diversity of regional environment and national personality, the characteristic cultural tradition and artistic form of rural community can be inherited and preserved. The folk custom in the countryside is a treasure house which can be excavated and developed by our college art education.

3.2.1 Since the Internet is not limited by time and space, we could utilize it to make up for the shortage of professional art education teachers. At present, the consciousness of traditional education in rural schools is relatively serious. Art courses are not given due weight; the overall quality of art teachers is not high; and only a few teachers graduated from art academies. After the implementation of the national targeted poverty alleviation policy, the Internet has entered the countryside. Art teachers and art courses from all over the world can enter the classrooms of rural schools, which has broadened the horizons of students and solved the problem of faculty for schools. [3]

3.2.2 We should use the Internet to reduce costs and establish a mechanism for tracking and exploring folk art. The establishment of a long-term mechanism to track and record rural folk art can provide a channel for the exploration and accumulation of artistic creation materials. For a long time, China's art education has been dominated by the western model, which makes the national art resources appear relatively weak and receive insufficient attention. The establishment of art education system with Chinese characteristics should be based on the consciousness of traditional culture. Most of the traditional culture is well preserved in underdeveloped villages, and is handed down from generation to generation mainly in the form of folk art. Internet technology can help college art teachers and students reduce the cost of field investigation and improve the efficiency of information dissemination.

3.2.3 We should use the extensive publicity of the Internet to form brand characteristic and create market value. Apart from the economic backwardness and the neglect of schools, the reason for the large development gap between urban and rural art education lies in the lack of understanding of the market value of characteristic folk art. After the folk art is condensed and sublimated into an art work, it will be understood and accepted by the public, and even go abroad to the world stage. We should make use of the extensive publicity of the Internet to build a unique brand image of folk-art works, and obtain commercial value through modern means such as cultural rural tourism and on-line sales, so as to improve the quality of life and inherit folk art. [7]

The "New Channel" project is an excellent case in which colleges enter the countryside to participate in art education, carry out joint research and development through the Internet and realize commercial operation, and promote national cultural brands to the world stage. The project, originating from Tongdao Dong autonomous county of Hunan province, is launched jointly by Hunan university, 16 internationally renowned universities, and some non-profit organizations, social enterprises and industrial research institutions. It aims to sort out minority cultures in central and western regions and explore innovative practical activities in the international society. The "New Channel" project brings together researchers from different art specialties and social disciplines to conduct research and produce works of art in the countryside, including video recording, product and environmental design, and art education. The convenience and openness of the Internet enable rural areas, universities and enterprises to interact with each other effectively, build a design innovation network and a social participation platform, and improve the cultural awareness of the urban and rural masses and the development of the art industry. The "New Channel" project has been exhibited at the Shenzhen International Cultural Industry Fair and the Milan Design Week successively, and won the approval of the art talent training project of China National Arts Fund and the special award of "G-mark" community research in Japan. Its value and significance have been widely recognized by domestic and foreign authorities.

4. Conclusion

The model of "Internet +" has created a new social environment and a new ecology. Faced with new environment, college art education should also renew its development idea, carry out initiative changes and reforms and fulfill the responsibilities and missions entrusted by the government. By constructing the model of "Internet + art education", colleges and universities can realize mutual assistance and benefits with communities and rural areas, which is the spark generated by the collision of emerging technologies and traditional art education system. The mutual-aid mode of "Internet + art education" is not only an effective way for colleges and universities to participate in the construction of cultural confidence, but also a new window for us to introduce the brand image of Chinese culture to the world with the help of new Internet technologies. The mutual-aid mode of "Internet + art education" has strong realistic significance and practical condition.

Funding Statement

The project of the Hunan Province philosophy and Social Science Fund "Research on value and performance evaluation of industrial design"， Number:16yba369

References

[1] The Ministry of Education. (2014) Some Opinions on Promoting the Development of Art Education in Schools. In: Jiaowailai. Beijing. pp. 2-5.
[2] Zheng, P. (2013) The Feasibility and Countermeasures of Art Colleges Participating in Community Culture Construction. Forum of Arts, 05: 108-10.
[3] Zhang, J. (2011) Current Situation and Strategies of Art Education in Mountainous Areas. Science and Technology Innovation Herald, 17: 143.
[4] Bao, H. (2012) The Research of Artistic Colleges Participating in Community Cultural Construction——The Case Study of Shanghai Conservatory of Music. Shanghai Jiao Tong

University, pp. 35-42.
[5] Yin, C & C, Ren. (2016) Discuss on the Characteristic of the Art and Design Education in the "Internet + " Times. Art and Design, 09:129-130
[6] Zhao, C & J, Li. (2016) On the Development of Art Education in the Era of "Internet + ". Journal of Xi'an Jiaotong University (Social Sciences), 05:128-130
[7] Yang, M. (2016) Research and Application of Art Education under the Background of "Internet + ". Asia-Pacific Education, 04: 253

IOP Conf. Series: Materials Science and Engineering **573** (2019) 012051 doi:10.1088/1757-899X/573/1/012051

Application of Industrial Design in the Development of Environmental Protection Quick-frozen Products

Mei Yun, Wang Fang, Jiang Haiyang, Lin Ying, Nie Qimeng

School of mechanical engineering and automation, University of Science and Technology Liaoning, Anshan, Liaoning, 114051, China

e-mail: meiyunliaoning@163.com

Abstract. With the change of the times, the industry design style and design concept of each period will continue to advance at a higher and higher level. Consequently, environmental protection, concise and practical design has been affirmed and respected by most people in the industry. Such as: environmental protection refrigeration equipment, nitrogen environmental protection quick-freezing refrigeration equipment for people's lives to provide fresh food design. With the gradual improvement of people's living standards, people's attention to food has shifted to the importance of nutrition and quality. Due to the diversification of current consumption patterns, people's requirements for food preservation have also increased a lot. With the extension of China's highway, highway transportation has also developed. The status of applications based on quick-freezing technology for fresh food will be the rapid development of fresh food refrigeration technology, and industrial design will play an important role in it.

1. Introduction

Research on the appearance design of environmental protection equipment and instrument products is an important manifestation of the improvement of independent innovation ability. Industrial designs can actively guide and support enterprises to develop more high-tech environmental protection technology and equipment with independent intellectual property rights. On the basis of mastering the core technology of environmental protection, more attention should be paid to product packaging design. In addition, we can also learn from the advanced new technologies at home and abroad, combine with the development needs of enterprises themselves, through the introduction, digestion, absorption and innovation, forward their core technology, thus it promoting the overall upgrading of the enterprise industrial chain.

In ancient times, people knew that cooling could prevent food from decaying. People began to use ice to refrigerate food. In order to transport food, ensure food safety and non-decay, people use ice and food loading together to reduce the temperature in the box and prevent food deterioration. In order to keep the temperature in the carriage at low temperature, people will make the material of carriage into heat insulation material. The heat insulation material is lighter. This is the quick-frozen food of ancient people. There was no foam in ancient times, but there were some kinds of thermal insulation materials such as cork, wood and so on. In contemporary society, the development of science and technology mainly uses chemical products to make thermal insulation materials, such as polystyrene foam and polyurethane foam.

In contemporary society, people have a higher pursuit of food freshness, which promotes the further development of food preservation technology. Vacuum freeze-drying of food is to freeze food materials below the eutectic point temperature to change water into solid ice, and then sublimate ice to water vapour at appropriate temperature and vacuum, and then condense water vapour with water vapour condenser of refrigeration system, so as to obtain drying technology of dried products. Vacuum freeze-drying technology can maintain the structure, nutrient components, shape and biological activity of food, which is incomparable with other food drying methods. But vacuum freeze-drying technology has its fatal disadvantage, that is, large investment in freeze-drying equipment and high energy consumption. With the localization of freeze-drying equipment, the one-time investment of the equipment has dropped dramatically, but its energy consumption is still the bottleneck restricting its development.

In order to ensure that the nutritional components of food are not lost, and that the colour of food is bright, food needs to be frozen to quick-frozen. Quick-freezing [1-4] can crystallize the internal ingredients of food. When food is eaten, thawing the food is to turn the crystallization into food juice, making the food return to the appearance before quick-freezing, and the food will have a very delicious taste and rich nutrition. Quick freezing is a very good freezing technology in today's society, which can ensure the nutrition and colour of food. Environmentally friendly energy promotes better development of quick-frozen industry. Diversified products and higher technology help industrial designers to broaden their horizons for quick-frozen product design [5-7].

2. Current Situation of Industrial Products Applied in Food Quick-freezing and Fresh-keeping

The present situation of fresh food preservation technology at home and abroad was analysed. The current situation in China is that domestic manufacturers have introduced advanced technology and mature equipment from abroad to establish multiple storage depots. At the same time, they have spent a lot of manpower and financial resources on food preservation technology, but the results are unsatisfactory. There is a big gap between the ideal and the result. The main reason is to follow the trend blindly, thus ignoring the domestic situation and the actual production situation of the current manufacturers. This will directly lead to the introduction of foreign technology and equipment cannot be quick-frozen, and reduce the efficiency of equipment. At the same time, there may be a combination of the original equipment and the introduction of new equipment, making fresh food preservation problems appear, such as: food loss, food deterioration and so on, which will cause inestimable losses to enterprises. It also poses a threat to food safety, which is not conducive to people's health. In addition, the state has few regulations on fresh food preservation technology, and there is a lack of innovation on fresh food preservation technology. In terms of food preservation technology, foreign capital is much better than domestic. Foreign capital forms a mode: purchase, production, processing, storage, transportation and sales. Fresh food should also pay attention to the process of refrigeration, sorting and transportation of fresh food while paying attention to preservation. In these processes, Different Fresh-keeping measures are adopted to keep fresh food. The national policy plays an important role in the domestic food preservation industry. Through the policy, the research on insurance food technology will be strengthened, which will make the food preservation technology more and more advanced [8-14].

In the 1990s, the scientific research staff of the National Fresh-keeping Engineering Canter began to work on the research and development of grape preservatives, adjusted the composition of sulphite preservatives, and added new technologies such as microcapsules, which kept the release rate of grape preservatives stable and sustained, greatly improved the freshness-keeping effect, and further developed two stages of release and recovery. The preservative of the mixture has a remarkable effect on preservation. At the same time, they have made great achievements in the development of new fresh-keeping film technology, invented a new type of fresh-keeping film with moisture-regulating function, thus forming a special fresh-keeping material for grape and professional and scientific fresh-keeping storage measures.

Preservation of biological preservatives is a new technology emerging in the world in recent years. Its principle is the method of producing bacteria by bacteria. At present, it has achieved good results in the preservation of fruits and vegetables. It is other mould that is used more in this method. This

IOP Conf. Series: Materials Science and Engineering **573** (2019) 012051 doi:10.1088/1757-899X/573/1/012051

fungus is a natural, broad-spectrum, efficient and safe filamentous fungal inhibitor. It can not only inhibit fungi, but also prevent the production of mycotoxins. The biggest advantage of this method is that will not be absorbed by the digestive tract of the human body, so it is a harmless fresh-keeping technology for the human body. Another advantage is that many microorganisms are difficult to resist, and because of their low solubility, they are generally used as food surface preservatives. It has been internationally recognized as a safe biological preservative for food storage and preservation.

3. Design and Analysis of Refrigeration Quick-freezing Equipment

In a narrow sense, industrial design mainly refers to product appearance design. Industrial design is at the front end of the whole product development process. Mainly participate in the design of exterior modelling programs, determine the artistic ideas. Only by defining the shape, can we carry out the next structural design (MD design). This is the reason why there is no hair attached to the skin. Industrial design in mechanical design and development can neither violate the principle of mechanical structure, nor design arbitrarily. They are interdependent and need to be adjusted and modified repeatedly. For example, when you design an appearance, the shape involved, the material used will make the die or welding impossible. That is meaningless appearance design.

3.1 Visual Design

The visual design of quick-frozen products mainly embodies the colour aspect. Environmental protection, green pollution-free is the national policy, but also we must do to protect the earth on which we live. Quick-freezing is mainly manifested in the speed, to quickly complete the food quick-freezing, so the use of streamlines and strong sharp symbols to express. The colour can be blue to express the feeling of freezing, or white or black to add ice and snow to the pattern. Some quick-frozen products are dynamic objects, moving objects, can use bright colours, attract people's attention, so can use blue contrast orange, yellow and other bright colours.

3.2 Structural Design

The structural design of quick-frozen products needs to consider the aesthetic design of the shape. The structure of refrigerated truck should conform to ergonomics, give people the feeling of common things in life, will not appear abrupt, absolutely will not have the feeling that it is difficult to integrate life. In line with the operation of people's common transport vehicles, easy to operate, so that people are pleased to accept this refrigerated vehicle.

3.3 Security Design

Safety design of quick-frozen products refers to the possibility of damage of quick-frozen equipment stored in open air or in secret places of people. Leakage of chemical gases or liquids used for preserving fresh food such as heat preservation will cause some unnecessary property losses and threaten the personal safety of people nearby. Complete design is to ensure the safety of products and eliminate these hidden points. Suffer from. So in product design, safety is a crucial link. Structural safety is to ensure the normal operation of the product, if there are problems and timely repair, timely reinforcement. As the name implies, the safety of colour is to let people see that the colour of this product will not be misguided.

3.4 Green Design

Green design originates from people's reflection on the destruction of environment and ecology caused by modern technology and culture. For industrial design, 3R principle is the core of green design. The so-called 3R refers to Reduce, Recycle and Reuse. Green design requires that every step of the product life cycle be taken into account: from the creative conception of the product to the manufacture and transportation, from the product being used by consumers to the recycling of the final product. Ensure that each link has the least impact on the environment. The most common manifestations are the following two aspects: Disassembly design. Disassembly design is a design method that makes products easy to disassemble and maximizes profits from material recovery and parts reuse. It is one of the main contents of green product design. For example, we can use threaded joints instead of welding and riveting in machine tool design. Thread connection is not only easy to disassemble and assemble, but also reliable, which has the characteristics of reuse. In order to reduce the number of parts, screw integrated structure can be used to replace ordinary screw and gasket, anti-loosening washer, etc. The design principles of disassembly are as follows: firstly, reduce the workload of disassembly; secondly, it is easy to disassemble and separate; thirdly, it is to use more standard parts. Recyclability design .Recyclability design is also one of the main contents of green design, which considers the possibility of parts recycling and reuse in the early stage of product design, that is, parts and materials in used or waste products can be used in other new products.

3.5 Humanized design

Humanized design is to design people-oriented. In view of the defect of the functionalist design idea, he proposed that the design should not be based on the function of machines, but on human beings. Based on people's use of products, the machine can serve people well. With the development of technology and the progress of society, people's requirement for humanization of products is getting higher and higher, especially for some designs related to people's safety in use, such as machine tool product design. Humanized design is to design people-oriented. In view of the defect of the functionalist design idea, he proposed that the design should not be based on the function of machines, but on human beings. Based on people's use of products, the machine can serve people well. With the development of technology and the progress of society, people's requirement for humanization of products is getting higher and higher, especially for some designs related to people's safety in use, such as machine tool product design.

4. Conclusions

The influence of industrial design on environmental protection quick-frozen products is getting deeper and deeper. Designers need study quick-frozen equipment, including its shape, colour, safety, emotion and so on. As long as they keep exploring, researching and designing, they will promote the development of this industry. Industrial design plays a very important role in it. Modern design concepts mainly refer to green design, humanized design, product semantics design and product system design. In terms of design, these main design concepts correspond to each other: in terms of environment, we should show our concern for the environment; in terms of people's use, we should show our concern for people; in terms of product semantics, we should reflect the local cultural characteristics; in terms of brand management, we should pay attention to the inherent quality of products, making design an effective means of brand marketing. There are two concepts: one is to conform to the human scale; the other is to have a good operating interface. Good operation interface can make the operation more convenient, fast, safe and reliable.

References

[1] ZHANG Yu-ning. Food Storage and Refrigerated Food Safety [J]. Jilin Agricultural，2016，3:18-19.
[2] HU Wen-zhong. Science and Technology of Fresh Cut Fruits and Vegetables [M]. Beijing: Chemical Industry Press, 2009.
[3] YANG Xiao-lou, XU Mao-zeng. Current Situation and Development Measures of Refrigerated Food Logistics Market in China [J]. Journal of Chongqing Jiaotong College (SOCIAL SCIENCES EDITION) ,2005,5(2):24-26.
[4] LI Zhi-jun. Application of Cold Chain Technology from Food Safety [J]. Modern Logistics Report, 2007,7: 36-37.
[5] RAYA Al-dadah, SAAD Mahmoud. Liquid Nitrogen Air Conditioning System for Domestic Application [C], 2016 International Conference for Students on Applied Engineering (ICSAE) , Mumbai, India,2016 .

[6] SU Zhao-kan. Discussion on the Application of Liquid Nitrogen Technology in Preservation of Cereals, Oils and Food[J]. Science and Technology of Cereals, Oils and Food. 1987,(6): 38-39.
[7] WANG Lan-zhen. Color Visual Beauty of Packaging Design [J]. Packaging Engineering, 2014, 35 (16): 110-113.
[8] YANG Ce-xun, WANG Zheng-gang. The Technology of Safe Production and Safe Use of Liquid Nitrogen [J]. Guizhou Animal Husbandry and Veterinary Medicine, 2014,2:15-17.
[9] Jun Hang, You-rui Huang, Li Li, Switched Reluctance Motors Direct Torque Control Research Based on RBF Neural Network and Fuzzy Adaptive PID Controller [C], In Proceeding of 2010 Second International Conference on Intelligent Human-Machine Systems and Cybernetics, Vol. 1, pp.135~138, 2010.
[10] Zhao Xiaomei, Jiang Ying, Wu Yupeng. Advances in ice-temperature storage technology of fruits and vegetables [J] Chinese fruits and vegetables, 2005 (3): 31-32.
[11] Qian Yong, Suggestions on Developing Highway Refrigerated Transportation in China [J].Transportation Operational Technology, 2009,10
[12] Zhao Pingyong. Design Colorology [M]. Beijing: China Media University Press, 2006
[13] Xu Yansheng, Li Yuchun, He Qinbo, Pan Shuqing. Design and application of control system for large-scale food vacuum freeze-drying equipment [J]. Journal of Vacuum Science and Technology. 2008 (04)
[14] Xu Yansheng, Long Jianyou, Li Yuchun, Xia Rui. Application of Secondary Steam Heating System in Vacuum Freeze Drying Equipment [J].Journal of Vacuum Science and Technology. 2007 (04)

User Research–Based Design Strategy for an Electric Water Heater and Its Application

Zhijian Liang[1], Yaqi Zheng[1], Ningchang Zhou[2*], Shifeng Liu[1]

[1]East Innovation Industrial Design Co., Ltd., No.1, Yaogang Rd, Shunde District, Foshan, Guangdong, China

[2]South China Agricultural University, No.483, Wushan Rd, Tianhe Distric, Guangzhou, Guangdong, China

[*]Corresponding author's e-mail: 93890290@qq.com

Abstract. To improve the performance of electric water heaters and innovate various product categories, this study primarily conducted user research and obtained data pertaining to the physiological and psychological characteristics, living habits, and user requirements of the target population as the starting point. Political, economic, sociocultural, and technological analysis, market analysis, household surveys, quantitative questionnaire surveys, and market visits were conducted to obtain the data for exploring the different user needs, integrating the needs of various users, conducting importance assessments of the needs, and developing design strategies for the electric water heaters. An industry breakthrough should be obtained by using the user experience–level data and by proposing new design strategies and corresponding design solutions.

1. Introduction

It has been increasingly noted that design of electric water heaters lack innovations. As the market demand for electric water heaters steadily increases, the new demands of Chinese families pertaining to water heaters and the user experience should be determined and improved, respectively. User-centered research, also known as people-oriented research, can help identify the target users of a product and accurately determine user demand. By conducting a qualitative analysis of user needs and the competitive environment, the precise status of a product in the early stage of development can be ascertained, the efficiency of the product upgrade process improved, and the product designed so as to achieve market success in the future.

2. Current state of the research in electric water heater design

On the basis of the type of energy consumed, water heaters are mainly divided into four types: electric-, solar-, gas-, and air-powered water heaters. Because electric water heaters have low installation costs and no regional limits and are eco-friendly and cost-effective, they have a large market share in China. However, in recent years, electric water heaters have been viewed as disadvantageous because of fundamental problems such as a low heating rate, less hot-water output, inconsistent water temperature control, and irregular water supply.

The use of clean energy is currently trending, and there has been a considerable increase in studies pertaining to air- and solar-powered water heaters. A linear analysis of water heater studies conducted from 1998 to 2014 indicates an uptrend in the number of solar- and air-powered water heater studies, whereas the number of studies pertaining to the electric water heater has exhibited a descending trend (Figure 1). Zhiqiang (2016) proposed a method for improving the water-storage-type electric water heater and designed a smart electric water heater based on a single-chip microcomputer control system for superior and safe operation. Lei (2017) conducted a correlation analysis on the factors affecting customer satisfaction of electric water heater installations and proposed methods and means to improve the service satisfaction in the industry of electric water heaters. Nel (2016) conducted a nationwide online survey in South Africa to evaluate the awareness of energy conservation measures with respect to electric water heaters and the key factors affecting behavioral willingness. Kapsalis (2018) proposed an optimization algorithm for exploring how to operate the electric water heater under dynamic electricity pricing. The main area of concern for existing studies has been the improvement of the energy efficiency, availability, and quality of installation services of products. However, to continuously upgrade product consumption, the homogenization of products in the electric water heater industry is crucial. We must understand the new requirements of Chinese households with respect to obtaining high-quality hot water and should meet their key requirements to improve user experience. This study proposes a breakthrough design innovation strategy for electric water heaters and presents the corresponding design solutions.

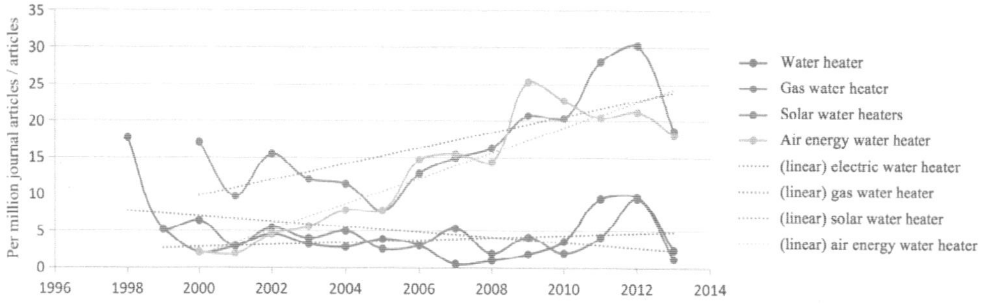

Figure 1. Linear analysis of water heater studies from 1998 to 2014
(Source: Wanfang Data, China National Knowledge Infrastructure)

3. Development of user research plans

The primary goal of this study was to identify a development direction for the design and technological innovation of electric water heaters, determine a road map for product technology development, find data and supportive logical points for product promotion and sales, improve the customer experience before purchase, enhance the sales and after-sales service experience, and establish a complete user requirement research system and related database. The secondary targets of this study were to determine the unmet requirements and potential problems associated with using electric water heaters and "accurately convert them into tangible as products or services based on their demand.

The research was divided into six phases:

(1) Information and data collection and analysis: Political, economic, sociocultural, and technological (PEST) analysis was conducted to determine comprehensive interventions at the policy level for sociocultural and technological development in China in terms of factors such as economic operation, changes in the population structure and culture. This study presents industry-related keywords as the core direction of the user research and the position of the target users.

(2) Development of the research plan: Using network information analysis and design, the "investment test questionnaire" was presented and demonstrated, and the household survey plan was formulated according to the demonstration results.

(3) Household observation: The "Daily Bathing Water Record Questionnaire" was issued during the initial 10 days. Then, the users' bathing recordings were obtained and analyzed, and the core

questions of the users' bathing life were explored by conducting in-depth interviews.

(4) Market visit: Interviews with sales staff and installers were conducted to understand the construction of the end user experience supply chain.

(5) Questionnaire: The users' pain points were refined, and then, qualitative research was conducted using the results of the quantitative questionnaire to verify the conclusions of the users' pain points. Finally, methods to solve the users' pain points and meet user requirements were developed.

(6) Conclusion Integration: According to the previous five phases, this step involved output product innovation development planning and product design strategy development.

4. Analysis of user requirements

4.1 Primary study

In the early stage of network information and data collection, we used PEST analysis–based research methods to collect information data from policies, the economy, the society, and technology and to summarize the information and attract social attention. A number of keywords were extracted and classified, as shown in Table 1.

Table 1. Primary perspective research keyword integration

policy economic situation fertility anticorrosion financial policy real estate subsidy policy event change accident urban/rural engineering	health safety quality maintenance	e-commerce industry disadvantage internet marketing solution	service industry status marketing GDP	heating method energy conservation efficiency multimode combination multienergy combination heat recovery surroundings environmental protection energy
	intelligent comfortable network control cloud/Internet of things intelligent control cross-border additional features	installation appearance		

On the basis of industrial attributes, five essential social concern–related keywords were extracted from the aforementioned keywords:

(1) Environmental protection and energy conservation: Although numerous energy-conservation studies exist, research attention has declined in the recent past. The number of dual theme and multiple theme studies has begun to increase, and studies on energy conservation and environmental protection have been stagnant as researchers have shifted in other study directions.

(2) Intelligent water heaters: The research attention on developing intelligent water heaters has been increasing at a steady pace. However, the existing technology is unable to meet customer expectations. Thus, the development process for creating a real intelligent water heater requires ongoing research.

(3) Health: In recent years, the emerging "keep healthy" selling point has received popular recognition.

(4) Installation: Installation is an inevitable process after purchasing and requires more attention.

(5) Environment and policies: Factors such as the economic situation of the country, real estate factors, and changes in fertility policies influence the industry. Companies have to seriously consider how to adapt to the environment and policies.

4.2 User research

In the user selection stage, we considered that the product is oriented toward the mass market. Because it caters to a high number of users, user categorization was conducted on the basis of three factors: city, family structure, and brand of heater. The cities were divided into Central China, East China, and South China. The following cities were selected for household assessment: Shenyang, Beijing, Changzhi, Nanjing, Guiyang, and Fuzhou. The family structure was divided into four categories: direct family, couple family, standard nuclear family, and single-person households. The mainstream brands of water heaters such as AO Smith, Wanhe, Midea, Haier, Wanjiale, Ariston, and Siemens were included. Eventually, the researchers recruited 80 households to record their life scenes and product usage process.

Here, Shenyang was considered an example. On the basis of the information regarding urban planning, population structure, income level, water quality, and energy cost in Shenyang (Table 2) and the full coverage principle of the three levels of city, family structure, and brand, ten households were selected for household observation. Among them, the questions in the user interview questionnaire pertained to "object," "person," and "environment" throughout the time before purchase, during usage, and after sales service processes. Recordings such as videos, photographs, and questionnaires were used, as shown in Figure 2.

Table 2. Overview of the urban economy, water quality, and energy cost in Shenyang (2014)

Overview of Shenyang	In second-tier cities, the main urban area is completely urbanized and the population is aging.
Economy	House price: more than 7,000 yuan/m^2; per capita income: 2,500 yuan, general consumption level
Water quality	Water quality is poor.
Energy price	Natural gas price: 3.30 yuan/m^3; electricity price: 0.50 yuan/kWh

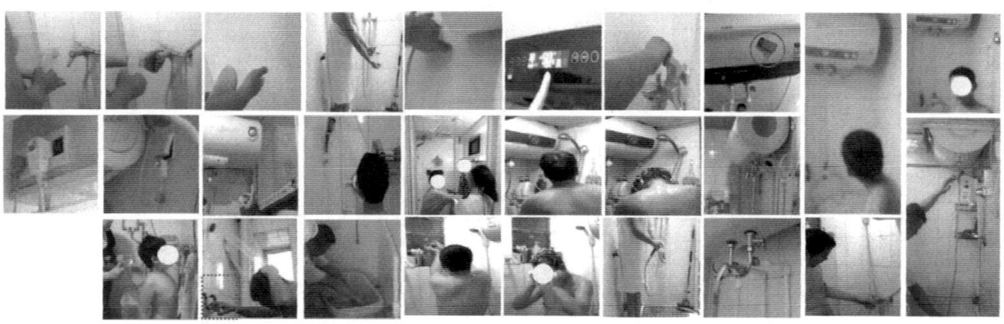

Figure 2. Home research video data finishing (operation section excerpt)

The aforementioned household observations, interviews, and questionnaires yielded a large amount of data. The demand data collected by conducting household research were optimized (Table 3), and the Kano model was applied. To effectively draw accurate conclusions, the analysis process was divided into five steps: the model integrates the demand, evaluates the importance of the demand, uses the quantitative data obtained from the questionnaires to conduct demand filtering, and finally optimizes the project output (Figure 3). After the demand filtering is completed, the data is sorted, as presented in Table 4, on the basis of three stages—before purchase, during usage, and after sales service.

Table 3. Household research data optimization (operational problem excerpt)

Operational problems	1. Generally, the function of the electric water heater control panel will not be operated. 2. The power switch of the outlet is pressed directly to start the machine without the operation panel process. 3. The installation height of the water heater is within the users' controllable range, and the remote control was ineffective. 4. For Beijing, a 24-hour boot state was observed in which only the mixing valve is used to adjust the water temperature.
	1. Users need to know the amount of hot water. 2. The user test time was approximately 10–15 s on average, and the test time for a

IOP Conf. Series: Materials Science and Engineering **573** (2019) 012052 doi:10.1088/1757-899X/573/1/012052

	woman was slightly longer than 30 s. 3. The user mixing valve adjusts the appropriate water temperature on the basis of the body temperature and repeats the operation approximately twice.
	2. The user turns off the mixing valve to use a shower gel while bathing (insufficient water or water wastage). 2. The user turns off the mixing valve and opens the valve again in the middle. The water temperature is unstable and thus a user can be burned easily. 3. The user uses a mouthwash while bathing. 4. Women do not understand the function of the water heater.
	1. Most users are in the state of rest while bathing and adjust their position. 2. There is no awareness pertaining to turning off the power of the water heater while bathing. 3. After the bath is complete, users does not change the position of the mixing valve. 4. The center knob is very high; thus, the elderly could not adjust the electric water heater. 5. When a child takes a shower, the shower head is removed, and the water pipe is placed vertically in the basin.
	1. It is difficult to adjust the water temperature by adjusting the mixing valve while washing.

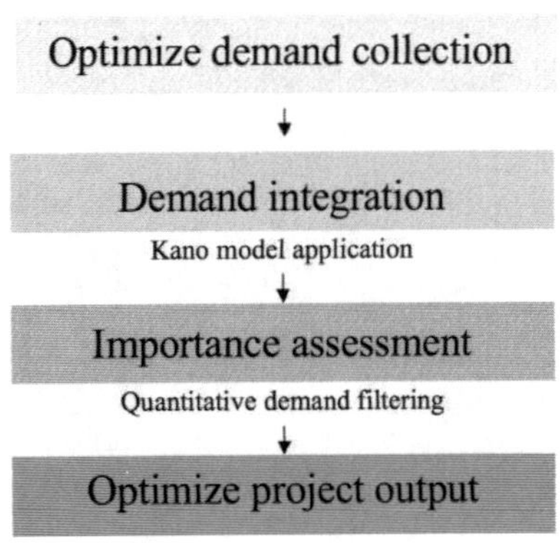

Figure 3. Process of demand mining, sorting, and exporting

Table 4. Demand export analysis

Before purchase	Users give special attention to the energy efficiency rating.	95%	Efficiency
	Users think that the water heater occupies a considerable amount of space.	85%	Installation space
	Users consider the hot water demand for the elderly, children, and women.	78%	Special population
	Users have concerns about the safety of electric heating.	85%	Safety
During	The water heater is kept powered on when the user is bathing.	60%	Usage habit intuitive operation
	Users conduct temperature adjustment.	93%	
	Users pay attention to the remaining amount of hot water and heating state of the water heater.	80%	
	The problem of water heater control is concentrated: the display		Prompt

	is not clear, the operation is not sensitive.		function
	The user indicates that the most expected function is product maintenance or water quality improvement.	50%	Water quality maintenance
	The user indicates that the most anticipated function is remote control.	42%	Remote control
	User does not receive maintenance activities and water change of the liner.	70%	
After sales	A user's dissatisfaction is concentrated in the slow response by the after-sales team and unclear charges.		After-sales specification

Analysis of the demand export results found that consumers pay attention to energy efficiency, installation space, water consumption and safety of special people before purchase; the user's pain points are easy to appear in the use habits, intuitive operation, residual water reminder function, water quality maintenance and Remote intelligent control of these nine aspects.

Judging from the data analysis results presented in Table 4, users' security concerns and usage habits conflict. Of the users, 85% worry about "electricity safety hazards" while using the appliance. The household observation results indicate that 60% of users unplug the heater before bathing (Figure 4). Because of this conflict, in-depth interviews were conducted, and it was determined that users are aware of the dangers of electricity because of their safety instinct. Moreover, some electricity leakage mishaps have circulated on the Internet. Thus, the impression that there is danger while using a "water heater" is deeply rooted in the hearts of the people. Two million results were obtained when the keyword "electric water heater leakage" was searched on Baidu. This demonstrated that electric water heater leakage accidents are not uncommon. Thus, users worry about safety while bathing when the electric water heater is in the on state, and this is not only a psychological factor.

The study of the functions of the six major brands of electric water heaters demonstrated that all major brands have improved their technology in terms of safety (Table 5 and Table 6). Most of the brands use the antielectricity technology of "three-pole break" and "anti-electric wall" for safety protection. The technical stage of the anti-electric wall cannot completely avoid the risk of electric leakage. The main causes of electric water heater leakage are the use of a machine for an extensive lifespan, failure of the protection technology due to the aging of the components, poor electrical environment, wrong connection of the circuit, aging of the wires, use of inferior sockets, and a user's wet hand touching the switch and plugs. The insecurity pertaining to the product safety causes a user to avoid use (Figure 4).

Figure 4. Some users plug in the power plug before bathing

IOP Conf. Series: Materials Science and Engineering **573** (2019) 012052 doi:10.1088/1757-899X/573/1/012052

Table 5. Technical finishing of the major mainstream electric water heater brand functions (1)

Brand	Heating energy efficiency	Protection safety	Liner technology	Energy saving	Healthy and clean	Intelligent humanization
Brand A	MAX rapid thermal capacity Separate double rod heating AES adaptive preheating	PS security system	Special protection liner	Medium temperature insulation AES adaptive energy saving system	IMM liner maintenance tips Patented liner cleaning technology Inner tank monitoring system	remote control One-click reservation Joint control shuttle system Real-time hot water tips
Brand B	3D+ rapid heating technology Rapid heat capacity Three-speed shift	Patented anti-electric wall Three-tiers power outages	Patented adamantine three-layer bile	Medium temperature insulation Inner insulation material (tight insulation layer)	Core magnesium rod	APP smart appointment Intelligent person washing play music
Brand C	E+4 times capacity increase heat Heating tube separation technology Water system - water cyclone technology Power - 3000W high power	M-SAFETY double earth leakage protection	"Blue diamond" liner	Thick insulation Peak Valley Nightingale Medium temperature insulation	Tech Water System One-button high temperature inhibition High purity activated carbon filter water core	remote control One-click reservation Washing Smart panel
Brand D	"Double engine" rapid heating technology Rapid heat capacity Submersible heating rod	Patented anti-electric wall Three-tiers power outages Hydroelectric separation	Three-layer "submersible tank" liner	Medium temperature insulation Inner insulation material (full latitude lock heat technology)	Fresh ion technology High purity magnesium rod fresh water	Smart butler function Schedule multiple timed heating
Brand E	Power - 3000W high power High nickel stainless steel heating tube Rapid heat capacity X4 Water system low mixing technology	Third generation double shield protection (anti-electric wall + three-tiers power outages)	Three-layer blue glazed adamantine liner	Liner insulation material - thickened energy foam layer Night energy saving	Extended magnesium rod	WIFI cloud intelligent interconnection system Remaining hot water display, reservation
Brand F	Titanium heating tube, increasing area Water system low mixing technology	Break gate Ten heavy security	Titanium liner	Intelligent - automatic energy saving	Silver net antibacterial	Intelligent WIFI Customized APP
Brand G	Navi Instant Heat Double heat expansion	ELCB Three-tiers power outages Hydroelectric separation heating tube	5A plasma liner	King Kong HoldTM overall foaming Smart Valley 5 day and night insulation	Easy to clean water function Smart liner maintenance tips	NAVI Reminiscent of rigorous intelligent tracking Smart booking water

Table 6. Technical finishing of the major mainstream electric water heater brand functions (2)

Brand	Heating energy efficiency	Protection safety	Liner technology	Energy saving	Healthy and clean	Intelligent humanization
Brand A	MAX rapid thermal capacity Separate double rod heating AES adaptive preheating	PS security system	Special protection liner	Medium temperature insulation AES adaptive energy saving system	IMM liner maintenance tips Patented liner cleaning technology Inner tank monitoring system	remote control One-click reservation Joint control shuttle system Real-time hot water tips
Brand B	3D+ rapid heating technology Rapid heat capacity Three-speed shift	**Anti-electric wall** Three-tiers power outages	Patented adamantine three-layer bile	Medium temperature insulation Inner insulation material (tight insulation layer)	Core magnesium rod	APP smart appointment Intelligent person washing play music
Brand C	E+4 times capacity increase heat Heating tube separation technology Water system - water cyclone technology Power - 3000W high power	M-SAFETY double earth leakage protection	"Blue diamond" liner	Thick insulation Peak Valley Nightingale Medium temperature insulation	Tech Water System One-button high temperature inhibition High purity activated carbon filter water core	remote control One-click reservation Washing Smart panel
Brand D	"Double engine" rapid heating technology Rapid heat capacity Submersible heating rod	Patented anti-electric wall Three-tiers power outages Hydroelectric separation	Three-layer "submersible tank" liner	Medium temperature insulation Inner insulation material (full latitude lock heat technology)	Fresh ion technology High purity magnesium rod fresh water	Smart butler function Schedule multiple timed heating
Brand E	Power - 3000W high power High nickel stainless steel heating tube Rapid heat capacity X4 Water system low mixing technology	**Double protection**	Three-layer blue glazed adamantine liner	Liner insulation material - thickened energy foam layer Night energy saving	Extended magnesium rod	WIFI cloud intelligent interconnection system Remaining hot water display, reservation
Brand F	Titanium heating tube, increasing area Water system low mixing technology	Break gate Ten heavy security	Titanium liner	Intelligent - automatic energy saving	Silver net antibacterial	Intelligent WIFI Customized APP
Brand G	Navi Instant Heat Double heat expansion	ELCB Three-tiers power outages Hydroelectric separation heating tube	5A plasma liner	King Kong HoldTM overall foaming Smart Valley 5 day and night insulation	Easy to clean water function Smart liner maintenance tips	NAVI Reminiscent of rigorous intelligent tracking Smart booking water

IOP Conf. Series: Materials Science and Engineering **573** (2019) 012052 doi:10.1088/1757-899X/573/1/012052

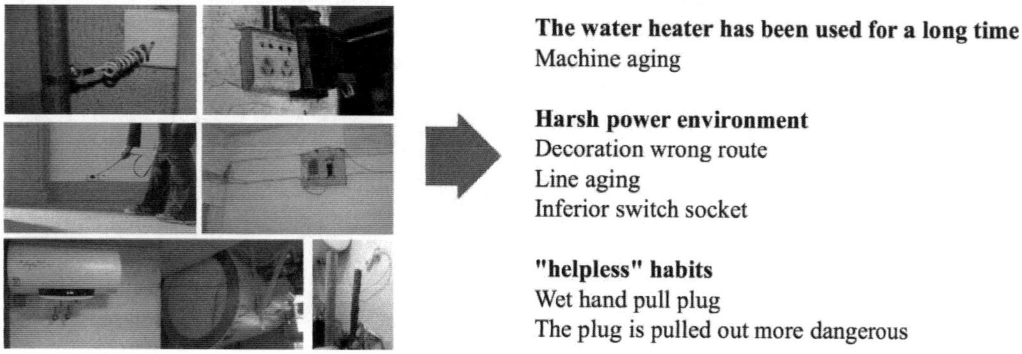

Figure 5. Reasons for electric water heater leakage

5. Redefine electric water heaters

In response to the users' security concerns, we redefined the electric water heater with respect to the function "shower with automatic power cut-out" and proposed a new design strategy in terms of the product, product design, functional technology improvement, user experience improvement, and promotion plan.

(1) Appearance design of the product

The control panels of common electric water heaters in the market are super layered and very rigid; thus, they cannot give people a visual sense of security. Based on the first level of Donald's three levels of emotional design, the visceral level, that is, products that produce the most direct primary emotions by the user, are instinctively stimulated by perceptual experience. The visceral level is also the level of attraction of the product's shape. In this study, the electric water heater control panel and the water deflector are combined, and the user is given a visual hint of the usage safety through a smooth and rounded curve and a soft and warm color (Figure 6).

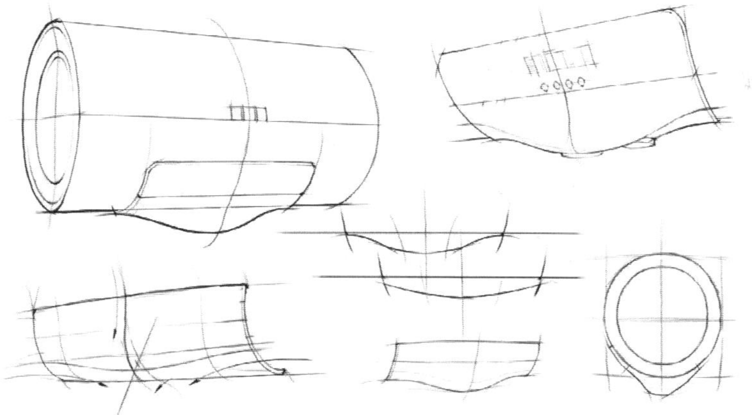

Figure 6. Design sketch

(2) Function and technology innovation strategy

This study proposed a "no electric washing" technology. When the electric water heater water valve is opened, the water flow causes the built-in sensor core to detect the signal, and the signal is transmitted to the broken core through the circuit board to realize the hot, ground, and neutral lines. All the lines are disconnected so that the whole water heater is in a state of no electricity, thus ensuring safety during a bath. After the user has finished bathing, the water heater is automatically reset after 5 min of closing the water valve.

(3) User experience design

To enable consumers to perceive that the product is in the safety state, when the water valve is opened, the display screen automatically turns black. The black screen accurately conveys the message that there is "no electricity." This innovation effectively eliminates user anxiety while taking a shower.

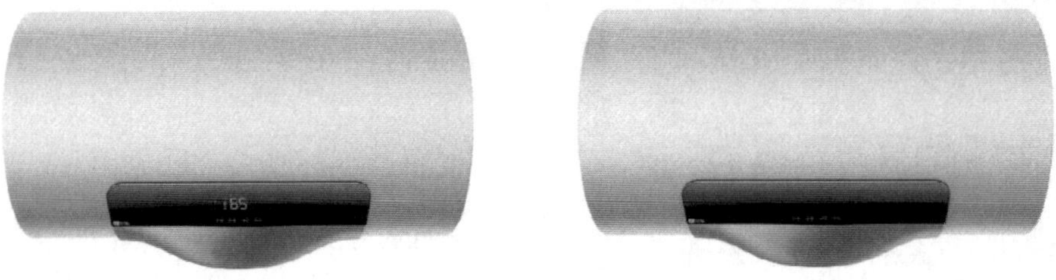

Figure 7. Design of the display interface based on the goal of eliminating user anxiety

(4) Developing product promotion plan

With the theme of "shower with automatic power cut-out," we developed a promotion framework by communicating with consumers and understanding their acceptable price for the product (Figure 8). By using the "smart water heaters, triple safety guarantee" as a slogan and the selling point, we continuously promoted the product online and offline. Consumer awareness of the existence of a safe water heater was achieved, and the advertising strategy strengthened the recognition of the product and redefined safe bathing. Therefore, the product secured a leading position in the market among electric water heaters.

Industry theme	Shower with automatic power cut-out		
Category value claim	Respect every life and lead the upgrade of electric water heater safety technology		
Representative series	S6 series products		
Product identity	S6 "Shower with automatic power cut-out" + 6 Safty functions		
Promotion theme	Without electricity, With safty		
Consumer pain points	Worried about leakage during bathing, psychological fear		
Reason	The decoration is connected to the wrong line, the line is aging, the inferior switch socket, the wet hand plug switch, the socket is easy to be pulled out, and the machine itself is aging.		
First-level selling point	Bathing without electricity		
First-level communication	Intelligent core technology, Triple security		
Core technical support	iSense sensor	iCut power off chip	iPlug switch chip
Communication language	Inlet induction transmission power off signal	The whole machine is powered off during bathing	Automatically reset power after stopping water for 5 minutes
Secondary selling point	Anti-electric wall	WI-FI reservation hot water	Patented antibacterial liner
Three-level selling point	Waterproof touch screen, fluid safety shape, anti-corrosion moisture-proof shell		

Figure 8. Product promotion framework

6. Conclusion

This study adopted user research as the core methodology to investigate solutions for electric water heater innovation breakthroughs from the perspective of user experience design. By analyzing the potential requirements of users in depth, it was found that some users were concerned about safety while using an electric water heater and removed the electric water heater plug from the socket before bathing. An in-depth study of this behavior revealed that a power cut-off function was required when users used the product. Thus, a product innovation design strategy was developed to eliminate user safety concerns and improve user experience, and a promotion framework was constructed. Guidelines were obtained to provide a reference for the electric water heater industry in terms of exploring product innovation.

Reference

[1] Cagan, J., Vogel, C.M. (2001) Creating Breakthrough Products: Innovation from Product Planning to Program Approval. FT Press, New Jersey.
[2] Tan, Z.Q., Yang, J.J., Lou, Z.B., Wei, C.J. (2016) Design of Intelligent Home Electric Water Heater Control System. Instrumentation Technology, 3: 1-4.
[3] Dong, L. (2017) Research on Customer Satisfaction of Electric Water Heater Installation Service Based on Structural Equation Model. Tianjin University of Finance and Economics.
[4] Nel, P., Booysen, M., Merwe, B.V. (2016) Energy Perceptions in South Africa: An analysis of behaviour and understanding of electric water heaters. Energy for Sustainable Development, 32: 62-70.
[5] Kapsalis, V., Hadellis, L. (2017) Optimal operation scheduling of electric water heaters under dynamic pricing. Sustainable Cities and Society, 31: 109-121.
[6] Joseph kim-keung ho, (2014) Formulation of a Systemic Pest Analysis for Strategic Analysis. European Academic Research, 5.
[7] Norman, D.A. (2005) Emotional Design: Why We Love (or Hate) Everyday Things. Basic Books, New York.
[8] Gu, R., Mu, B.N., Wang, G., Yang, Y.P., Li, K., Gu, Y.F. (2015) Analysis and Expression of Instinct,Behavior,Reflection in Emotional Design. Packaging Engineering, 36(16): 87-90.
[9] East Innovation Industrial Design Co., Ltd., (2018) Bathing Without Electricity. http://www.east-innovation.com/?p=11773.
[10] Yin, J.G., Wu, Z.J. (2012) Methods and Trends of Product Emotional Design. Journal of Hunan University, 16(1): 161-163.
[11] Zhang, L.H. (2006) Design Research Based on Cultural Anthropology. Art & Design, 2006(6): 14-15.
[12] Tang, L.T. (2006) Industrial Design Method. China Architecture & Building Press, Beijing.

Information aggregation of Qin Huai coloured-lanterns

Nannan Li[1], Zhangping Lu[2] and Mingzhu Li[3]

[1]Department of Industrial Design, Jiangsu University, Zhenjiang, 212013, China

763106949@qq.com

Abstract. The scattered information concerning Qin Huai coloured-lanterns was integrated from literature resources through information digging, organization and association. Firstly, The topic types about Qin Huai coloured-lanterns from literature resources were clarified on the software Cite Space and through card sorting. Then associations and semantic relation networks were constructed together with the semantic relations among topics. A literature resource information base on Qin Huai coloured-lanterns was established with the help of Ontopia topic maps. This study offers some convenient research methods for coloured-lantern researches and fans, improves the utilization rate of coloured-lantern resources, and promotes the transmission and sharing of coloured-lantern information.Finally, the usability of coloured-lantern information base was confirmed by designing a case study. The research methods and flowchart used here can be adopted by researchers from other fields.

1. Introduction

Qin Huai coloured-lanterns as a state-level intangible culture of China, intergrates both practical and esthetic art values and reflects the thinking ways and ideology of people living near Qin Huai River, China. Qin Huai coloured-lanterns have developed into a cultural symbolization and through appropriate use, these cultural connotations can significantly guide the living behaviors and works creation for modern people. After long-term baptism, Qin Huai coloured-lanterns have accumulated diverse resources, but like other intangible cultural heritages, coloured-lantern resources are under the scattered state as an information island without any association [1], leading to the inaccurate user information positioning and the low resource utilization rate. Thus, the scattered resources should be aggregated, so as to connect all information points through simple associated relations, forming a Qin Huai coloured-lantern information resource network, which will improve the utilization rate of coloured-lantern resources and promote the transmission of coloured-lantern cultures. In this study, the literature resources of Qin Huai coloured-lanterns were dug, organized and associated, and a coloured-lantern information base based on Ontopia topic maps as the information organization method was built.

2. Relevant research

2.1. Qin Huai coloured-lanterns projects

Qin Huai coloured lanterns also called "Jin Ling coloured lanterns", "Nanjing coloured lanterns" or "Qin Huai festive lanterns" (short for festive lanterns) originated from the Six Dynasties and were honored as "Qin Huai coloured lanterns are the best in the world" in Ming Dynasty. Qin Huai coloured lanterns are traditional coloured lanterns manufacture customs inherited along Qin Huai River and were included as the first batch of state-level intangible cultural heritages expanded projects --

traditional arts (folk arts) of China in 2008. Coloured lanterns integrate calligraphy, paper-cutting, coloured drawing and carving and possess profound cultural connotations. The various forms related with coloured-lanterns, including the making, put-up, watching, playing and enjoying of lanterns, are ubiquitous in life at specific periods and influence the ideology of the folks. The civilian coloured-lanterns are the ornamental products or decorations for physical and mental enjoyment and are the sustenance of yearning for living and working in peace and contentment at certain ages. Furthermore, Qin Huai coloured- lanterns to some extent reflect the social life, customs and cultures of Nanjing at different periods. During the long-term development, the cultural genes and innovations hidden in coloured-lanterns have gradually become cultural inspirits. Thus, Qin Huai coloured-lanterns are of significant values for us to study the humanistics and history of Nanjing.

However, the intensification of international cultural and economic competition and the expanding of modern industrial civilization have limited the development of festive lanterns. Thus, scholars and researchers from all walks of life are exploring the route of coloured-lantern development, and the information resources and the utilization of festive lanterns become increasingly diverse. So far, festive lantern resources are widely distributed in books, monographs, poetry, essays, newspapers, journals and news reports. The books include Nanjing Intangible Cultural Heritage: Qin Huai Lantern Festival by Zhou Anqing (2012), China Coloured Lanterns by Yan Dongdong (2013), and Nanjing Traditional Manual Arts by Sun Yiyi and Gao Anning. Poetry frequently mentions lantern raising and scenes on Lantern Festivals, such as Art and Literature Collections --80 Fires, Poetic Essay of Lanterns, and the popular works Romance of the Three Kingdoms and Water Margins. Newspapers include China Youth Daily, Nanjing Daily, and Yangtse Evening Post. The news websites include xinhuanet.com, Sohu.com, people.com.cn, sina.com.cn, cctv.com, js.ifeng.com, NetEase News, and qq.com. Traditional operas include Put up Dragon Lanterns and Nine Lotus Lanterns from Qu Opera, and Qin Huai festive lantern riddles, Qin Huai festive lantern ballads, and Qin Huai eight exclusive skills from Gathering of Heroes.

The current situations concerning the research and application of Qin Huai coloured- lanterns were analyzed from literature resources. Sun Lili from Zhejiang University analyzed in detail the theme selection, contents and nature of Qin Huai lantern festivals by utilizing four newspapers including Contemporary Express, Xinhua Daily, Wanjiang Evening News, and People's Daily, and summarized the mutual compliance between the media representation and cultural identity of Qin Huai lantern festivals [2]. Jiang Meiling from Nanjing University of the Arts analyzed the coloured-lantern crafts, design elements and artistic values, found the integrating point between traditional coloured-lanterns and modern lamps/lanterns and thereby redesigned Qin Huai coloured-lanterns [3]. Wang Juan from Jiangnan University comprehensively considered the idea sources, crafts, visual elements and other factors of Qin Huai coloured-lanterns and proposed new inheritance and developing methods for the theoretical innovation, design innovation, industrialization and branding [4]. Zhang Qun interviewed Gu Yeliang, a famous artist of Nanjing lantern-making, and introduced the details of coloured-lantern making and the current situations of coloured-lanterns [5]. Mu Yijun and Huang Tao collected and organized the materials of Qin Huai festive lanterns and through an animated interactive design, displayed the lantern festival scenes at critical periods during the development of Qin Huai lantern festivals and built VR/AR/MR virtual experience exhibitions in the animation to reproduce the coloured-lantern making process of craftsmen [6].

So far, much progress has been made in research on Qin Huai coloured-lanterns, but the affluent types of coloured-lantern resources are like isolated information islands without association. Researchers often have to consume huge time and labor to collect the resources of certain targets. Furthermore, the utilization rate of coloured-lantern resources cannot be maximized.

2.2. Literature information organization of Qin Huai coloured-lanterns with the help of topic maps

Topic maps are an information resource management technique originating in 1995, and mainly consist of three basic elements of TAO (namely Topics, Associations and Occurrence), which constitute the information network. Topic maps present information in the form of structures and are capable of information positioning and information connection. Topic maps are outstanding in information resource organization: (1) the massive and scattered information is gathered in an orderly way, which enhances the utilization rate of information resources; (2) information resources are logically linked by associations, which enhances the efficiency of information retrieval by users; (3) users search the matched information and acquire other associated information and concepts [7].

Research on topic maps in China focuses on two aspects: theoretical research on topic maps at the technical level (field of information management and computer application), and application research of topic maps in information management (in field of books information and archives management). Gui Sisi and Cheng Xiufeng introduced the XTM standards and XTM codes of XML grammar topic maps [8]. Liu Qian elaborated the interoperation between XML Topic Maps (XTM topic maps) and other information organization systems and proposed and validated the XTM-based interoperation models [9]. Li Qingmao et al. validated the accuracy and comprehensiveness of ontology information retrieval models of topic maps in the field of tourism [10]. Zhang Yunzhong et al. used topic maps into digital archives annotation systems [11]. Wang Meng built intangible cultural heritage topic map models and plotted topic maps in the case of Beijing Opera and Kunqu Opera. Shi Yi and Xiong Huixiang used topic maps into Huangmei Opera information organization [12, 13]. Li Qingmao et al. validated the feasibility of topic maps into tourism resources with Prefecture of Aba, Sichuan as example [14]. Li Qin explored the application of topic maps into information organization in the field of traditional Chinese medicine with the example of Treatise on Fevers. Moreover, topic maps have also been used to establish corporate and urban portal websites.

In this study targeting at coloured-lantern literature resources, the topic types of coloured- lantern were determined through information digging, a coloured-lantern literature information network was built according to the semantic relations among topics, and a Qin Huai coloured- lantern literature information base was constructed by using Ontopia topic maps.

3. Construction of Qin Huai coloured-lantern literature information base

3.1. Topic type analysis

3.1.1. Keyword classification .The topics were mainly classified by resources. Any things with the same property can be considered as one topic, such as concepts, characters and objectives, but the determination of topics should be comprehensive and accurate. Generally, a hierarchical method is used to first set the first-level topics and then the sub-topics or concrete contents. Shi Yi determined the archives resource topics of Huangmei Opera by referring to the professional terms, literature keywords and network resource theme words in the existing vocabularies of Huangmei Opera, but did not present any detailed introduction of steps [13]. Based on the thoughts of faceted classification by the Indian library scientist Ranganathan, Li Qingmao determined the topics of tourism culture information resources in Aba [14]. We think this qualitative method is still limited by subjectivity. In this study, we determined the topics by using a structured method involving 3 steps: (1) use Cite Space to extract the keywords from the literature; (2) analyze the keywords through card clustering; (3) extract and determine the types of topics according to the card sorting results.

The articles on Qin Huai coloured-lanterns from CNKI were imported into Cite Space for keyword exploration, which returned 80 keywords. These keywords were ranked by the occurring frequency from high to low (Table 1). Keywords are the frequently-occuring words and core contents of an article. The results of keywords from Cite Space are comprehensive and clear. Comprehensiveness means the literature of coloured-lanterns is comprehensively analyzed to involve all possible concepts in the literature resources. Clearness means the semantics of keywords are definite and have no different meanings.

Table 1. Cite Space keyword exploration (only a part).

Seq	Feq	Centralit	Year	Keyword
1	10	0.39	2012	Qin Huai coloured-lantern
2	8	0.22	2006	folk handicrafts
3	8	0.15	2006	festive lantern(art)
4	6	0.37	2015	intangible cultural heritage
5	6	0.28	2012	Qin Huai lantern show
6	6	0.08	2006	Confucius Temple
7	5	0.00	2007	festive lantern art
8	4	0.34	2011	Qin Huai festive lantern
9	4	0.28	2006	folk art
10	3	0.09	2009	industrialization
11	3	0.04	2007	Qin Huai River
12	2	0.17	2014	folk culture
13	2	0.22	2012	inheritor
14	1	0.00	2010	Inheritor of intangible cultural heritage
…	…	…	…	…
80	1	0.00	2017	Traditional folk culture

Card sorting is an effective and agile method that helps us to understand how to organize information and how to understand and describe different types of information and thereby to acquire inspirations and more effectively classify information into the first-, second- and third-level guidance types [15]. The keywords were classified via open card sorting. Firstly, the serial numbers and keywords were matched and made into 80 cards. Then 15 users (including 4 inheritors, 6 festive lantern researchers, 5 festive lanterns fans) were asked to sort the cards according to their personal information. The classifications of all users were recorded and imported into EZSort. Then tree-like plots by Complete, Single and Average were drawn, and through discussion with the users, the tree-like plot of the Complete algorithm was selected as the final basis for keyword classification (Figure 1).

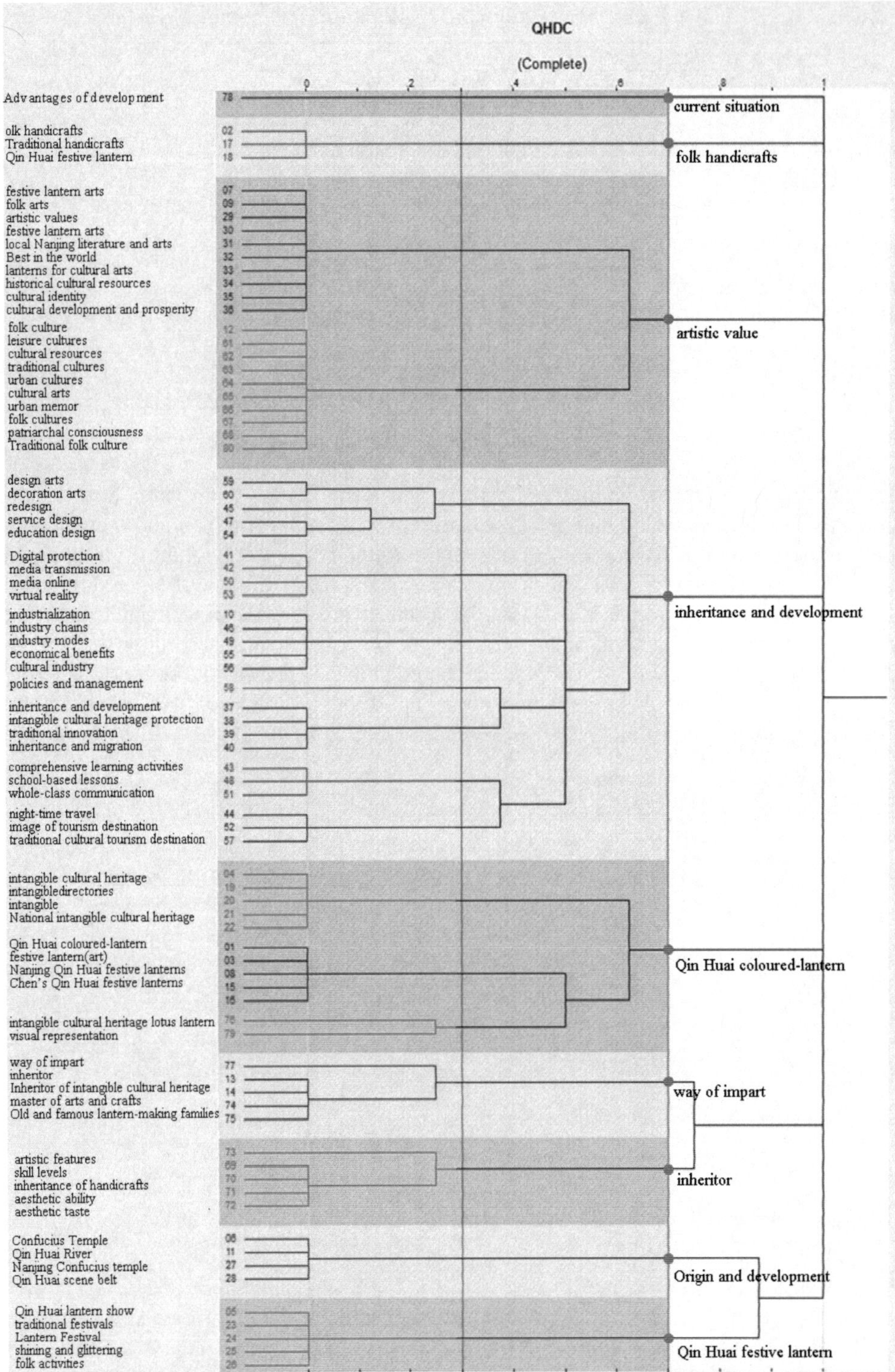

Figure 1. Tree-like plot of Qin Huai coloured-lantern keyword clusters .

The inheritors, researchers and fans of festive lanterns were all experienced users, which ensured the reliability of the classifications. On the Qin Huai coloured-lantern clustering tree-like plot, the boundaries of first-level (1), second-level (0.7) and third-level (0.3) information separated the keywords into 7, 9 and 18 groups, respectively. The 46 articles were analyzed by paragraphs with "article" as the unit, and the 9-group keyword classification was selected as the 9 types of topics. According to our preliminary investigation and the paragraphs of the literature, we selected 7 types of topics, including "05 Qin Huai lantern festivals", "02 folk handicrafts", "29 artistic values", "37 inheritance and development ", "13 inheritors", "01 Qin Huai coloured-lanterns" and "77 way of impart". Moreover, the keywords including Confucius Temple, Qin Huai River, Nanjing Confucius Temple Qin Huai scene belt, and Confucius Temple Qin Huai scene belt were summarized as "origin and development", while keyword development advantage was summarized as "current situations".

3.1.2. Determination of topics （1）Topic of "Qin Huai lantern festivals" Qin Huai lantern festivals are folk and custom activities with Qin Huai festive lanterns as the carrier. The keywords traditional festivals, Lantern Festival, and flaming are all associated with Qin Huai lantern festivals. The prevailing annual lantern festivals have gradually enlarged the diversity and scale of festive lanterns. From 2013 to 2018, the lantern festivals facilitated users to understand the categories and development of coloured lanterns at different periods.

（2）Topic of "folk handicrafts" The manufacture crafts are the core of Qin Huai coloured-lanterns. After baptism by ages, the materials of festive lanterns have slightly changed, but the lantern-making crafts are unchanged. After visiting festive lantern craftsmen and by reviewing literature, we summarized the lantern-making crafts into conceiving, frame preparing, dyeing, petal making, papering, and decorating. This topic offers users the steps of coloured lantern manufacture and incisively and vividly reflects the craftsman spirits of coloured-lantern handicraftsmen.

（3）Topic of "artistic values" Coloured-lanterns as the state-level intangible cultural heritage are known as "the best in the world" and can be used into festival decoration, entertainment, sacrifices to gods or ancestors, or daily use. The artistic values of coloured-lanterns are summarized as congratulation, praying and blessing, which offer users the way of macroscopic cognition of coloured-lanterns.

（4）Topic of "inheritance and development" Researchers have explored the developing ways of festive lanterns from different perspectives. The keywords of industrialization include industry chains, industry patterns and economic benefits, and the concrete ways of digitized protection are medium transmission, medium online and virtual reality. The keywords of festive lanterns inheritance from the design perspective include design arts, decoration arts and service design. Thus, the topic of inheritance and development is a summarization of coloured-lantern protection measures and development ways. This topic harbors 4 sub-topics, including industrialization, digitized protection, redesign, and others (including talent cultivation, brand establishment, school-based lessons). The aggregation of different inheritance methods is critical in understanding the developing direction of Qin Huai coloured-lanterns.

（5）Topic of "inheritors". Inheritors are indispensable roles in intangible cultural heritages. The keywords associated with inheritors include craft art masters, and old and well-known lantern-making families. Different factions have been formed during the development of coloured-lanterns. The topic of "inheritors" involves 9 sub-topics or factions, including Li faction, Cheng faction, Yin faction, Lu faction, Wang faction, Cao faction, Ke faction, Chen (Nanjing) faction and Chen (Jurong) faction. This topic offers users a convenient entrance to rapidly know the information of inheritors.

（6）Topic of "way of impart". Different styles or factions of coloured-lanterns have been formed during the development of "inheritance in old and well-known families". The way of inheritance differs slightly among different factions. Due to historical reasons, personal reasons of inheritors and other reasons, the limitations of "only imparting to males" or "only imparting within the family" have been gradually broken down during the inheritance process. The paragraphs concerning the way of coloured-lantern inheritance were grouped into the topic of "way of impart", so that users, from the way of impart, can discover the problems faced during the development of coloured-lanterns and indirectly understand the situations of coloured-lantern inheritors.

(7) Topic of "Qin Huai coloured-lanterns"Qin Huai coloured-lanterns developed from the Western Han Dynasty more than 1800 years ago and have accumulated different themes, ranging from small- and medium-sized lanterns for entertainment to large lanterns for decoration, with unique shapes and novel styles. The keyword "intangible cultural heritage lotus lanterns" reflects a classic type of festive lanterns. The paragraphs related to festive lanterns in the literature were summarized as the topic of "Qin Huai coloured-lanterns", which facilitate users to search for concrete works of coloured-lanterns.

(8) Topic of "origin and development" There are different views about the origin of coloured-lanterns. After prosperity and depression, coloured-lanterns have developed into the representative cultures in Qin Huai River areas. The keywords Confucius Temple, Qin Huai River, and Nanjing Confucius Temple Qin Huai scene belt correspond to the paragraphs describing the origin and development of coloured-lanterns in the literature and thus are summarized into the topic of "origin and development".

(9) Topic of "current situations". The keyword "development advantage" summarizes the advantageous conditions of coloured-lanterns, and accordingly, there must have been difficult times during the development of coloured-lanterns. The topic "current situations" is set with two sub-topics of advantages and problems, which correspond to the paragraphs elaborating the current situations of coloured-lanterns.

3.2. Topic association analysis

Association analysis is to connect the independent topics according to the relations among topics, forming a information network with a definite framework system. The same topic may have different associations, such as the association between "Qin Huai coloured-lanterns" and "folk handicrafts" is "coloured- lanterns/folk handicrafts", belonging to the association of phenomenon expressions, and the association between "Qin Huai coloured-lanterns" and "Qin Huai lantern festivals" is "inclusion/membership". To establish an efficient information network, we independently determined the types of associations according to the connotations of topics and the semantic relations between topics. As shown on the Qin Huai coloured-lanterns between-topic relation model in Figure 2, the dots represent different topics, where the large and small dots indicate the first-level and second-level topics respectively; the lines between topics indicate the between-topic relations. Seven types of associations were set according to the coloured-lantern between-topic relation model, including "coloured-lantern/folk handicrafts" association, "coloured-lantern/artistic value" association, "deciding" association, "creating" association, "including/affiliating" association, "master and apprentice" association, and "faction/way of impart" association.

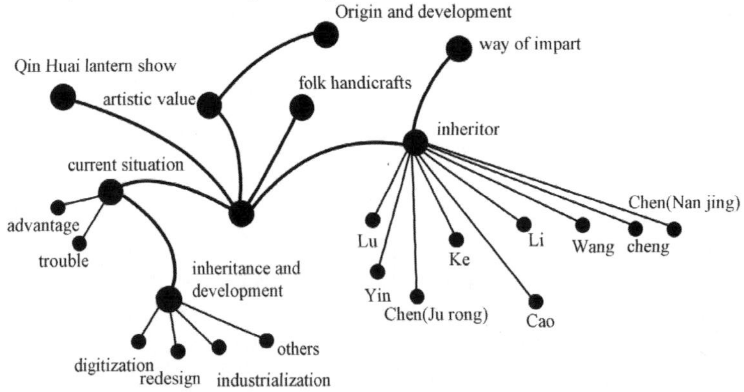

Figure 2 .Between-topic relation model of Qin Huai coloured-lanterns.

The "coloured-lantern/folk handicrafts" association and the "coloured- lantern/artistic value" association imply the crafts and artistic values of coloured-lantern works. For instance, the process flow of coloured-lantern works in Chinese Folk Coloured Lanterns is conceiving, frame preparing, dyeing, petal making, papering, and decorating, which convey the meaning of blessing. The "faction/way of impart" association shows the way of impart from inheritors. These three associations all belong to the type of phenomenon expression, but are analyzed separately here in order to clearly show the between-topic relations. The "deciding" association: topic (8) "origin and development" decides topic (7) "artistic values" to some extent, while topic (9) "current situations" decides topic (4) "inheritance and development". For instance, the view that coloured-lanterns originated from Taoism decides the meaning of praying for peace and health conveyed by coloured- lanterns. The "creating" association: topic (5) "inheritors" and topic (2) "Qin Huai coloured-lanterns" are under a relation of creation. For instance, Chen Bohua created the curled-up dragons, while Chen Youchang made lotus lanterns. The "including/affiliating" association: the affiliating relation is found under the same topic. For instance, topic (4) "inheritance and development" includes industrialization, digitized protection, redesign and other concrete ways of protection, while topic (5) Cao faction involves Cao Tingfa, Cao Zhengzhong, Cao Zhenrong and other inheritors. The "master and apprentice" association: In topic (5) "inheritors", some inheritors are masters and apprentices.

4. Construction of Qin Huai coloured-lantern literature information base

A coloured-lantern information base was constructed by using Otopia topic maps. Firstly, topic maps were compiled with the Ontopoly on Otopia. Then the topic maps were visualized on Vizigator. Finally, the target resources were reviewed and read on Omnigator.

4.1. Plotting coloured-lantern literature information topic maps

The editor Ontopoly on Otopia topic maps consists of four modules, including Description, Admin, Ontology and Instance. Firstly, the topic maps of Qin Huai coloured-lanterns were described as a whole on Description, and the content of Admin was set as "whole topic maps", which were outputted and locally stored in the form of XTM1.0. Ontology, as the core of the topic map ontology compilation, consists of 5 parts, including definition of topic types, resource types, association types, role types, and name types. The 9 topics of Qin Huai coloured-lanterns and the corresponding sub-topics and 7 associations were imported into the topic types (Figure 3) and association types. After two resource types (texts and pictures) and corresponding role types were selected, the concrete instances of the corresponding topics were compiled on Instance. Figure 4 shows the instances from the topic "inheritors" -- Ke faction inheritors.

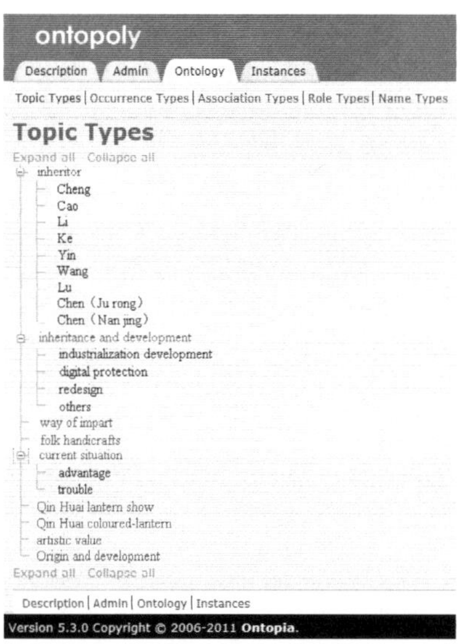

Figure 3. Qin Huai coloured-lanterns literature information topic types.

Figure 4. Qin Huai coloured-lanterns literature information with Ke faction as example.

4.2. visualization of coloured-lantern literature information topic maps

The Vizigator of topic maps consists of three inter-associated parts, including VizDesktop, VizLet and VizPlugin, which differ slightly in terms of functions. Here VizDesktop was used for visualization [16]. Visualization uncovers the associations between topic maps. Based on the between-topic relations, the topics can be presented in an affluent semantic relationship network. For simplification, Figure 5 only displays the semantic relation network of inheritors - Lu faction - Lu Youchang - lotus lanterns - Qin Huai coloured-lanterns. Clearly, the inheritors include Ke, Li and Lu factions, while the Lu faction involves Lu Youchang and Lu Min. Lu Youchang has created lotus lanterns and Chinese Folk Coloured Lanterns, which both belong to the representatives of Qin Huai coloured-lanterns. Users can select the topics of interest to trace relevant information. Figure 6 shows the visualized topic "Qin Huai coloured-lanterns". Due to the diversity of topics and the complex semantic relations and in order to prevent users from being lost in the huge relation network, VizDesktop offers a searching function, which ceaselessly shines the topic as-sought and thereby largely facilitates users to localize the target resources.

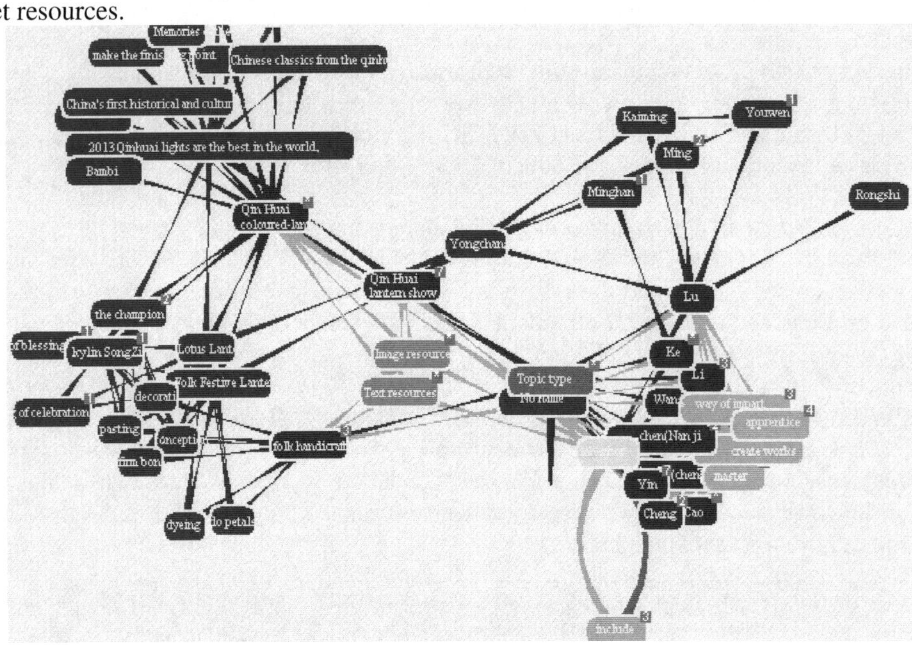

Figure 5. Visualized maps of some topics .

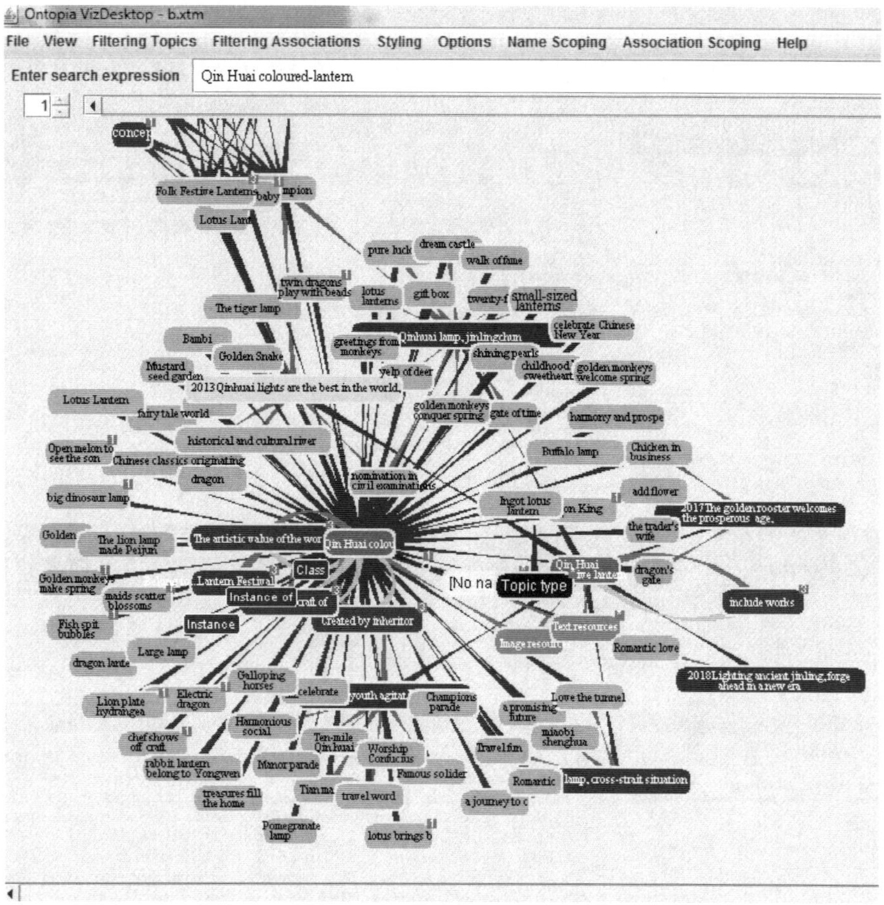

Figure 6 .Visualized map of the topic Qin Huai coloured-lanterns .

4.3. Reviewing coloured-lantern literature information topic maps

The Omnigator of topic maps can be used to review the details of a specific topic and facilitates text visualization. After a user selects the target information, he/she can directly read the internal resources on Omnigator or linking to external resources. With Qin Huai coloured-lanterns as example (Figure 7), the left side shows other information related to the topic "Qin Huai coloured-lanterns", such as the crafts and artistic values of coloured-lantern works, the year of being displayed, the inheritor of this work, which all help to expand the information of users. The right side shows the description of the topic and all the coloured-lantern works. For instance, users who are interested in Chinese Folk Coloured Lanterns can click and connect to Figure 8. On the browse interface of Chinese Folk Coloured Lanterns, users can see all resources related to this work. The left side shows the manufacture crafts, artistic values and creator of this work, while the right side shows all the texts and pictures in Chinese Folk Coloured Lanterns. In this study, the sources of texts are also noted, which help users to trace the corresponding resources.

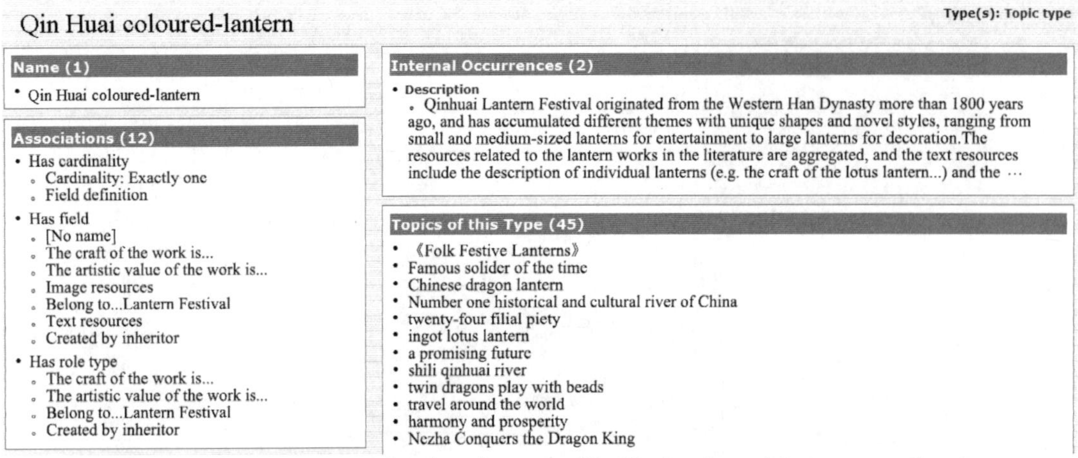

Figure 7. Browsing resources related to the topic Qin Huai coloured-lanterns on Omnigator .

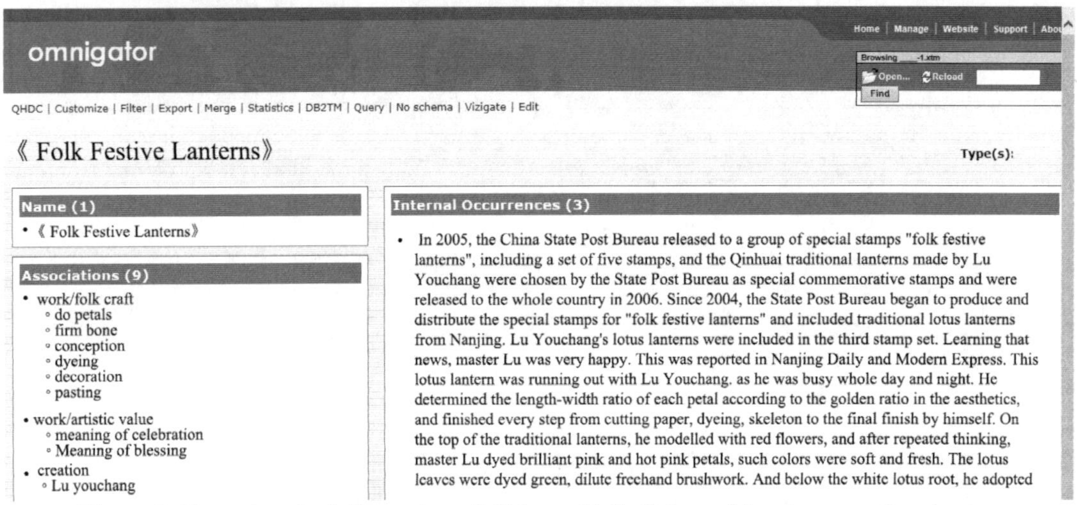

Figure 8 .Browsing the information of Chinese Folk Coloured Lanterns on Omnigator.

5. Application of the Qin Huai coloured-lanterns information base

The coloured-lantern literature information base is manifested in the form of topic maps, so that users can search the target information on Vizigator in a visualized way. When browsing any details on Omnigator, the users will not find information in a blind way thanks to t he coloured-lantern literature information resource base.

On the coloured-lantern information base, users can search the representative works of inheritors Lu Bohua and Lu Youchang through the topic "inheritors", and can be familiar with the crafts and feel the cultural connotations by reviewing the texts. Through picture screening and on-site investigation, we selected 35 middle- and small-sized coloured-lanterns and based on the topic of folk handicrafts, created an intelligent leisure game "coloured-lantern collection Qin Huai festive lanterns making crafts" (Figure 9a), where figure a is the home page of the game.

From the "current situations" of Qin Huai coloured-lanterns, users can know that there are large-size and small-size coloured-lanterns. The large-size coloured-lanterns are mainly customized for merchants and governments as ceremonial decorations, while small- and middle-size coloured-lanterns are created for children on holidays. The target of this game is a typical user group of festive lanterns -- children. Totally three levels are set, including easy, neutral and difficult (Figure 9b). The contents of the game are determined from two semantic relation networks of "inheritor - Chen Bohua -

representative works - folk handicrafts (artistic values)" and "inheritor - Lu Youchang - representative works - folk handicrafts (artistic values)". Figure 9c and d shows the easy-level papering craft interfaces of coloured-lanterns and rabbit lantern respectively.

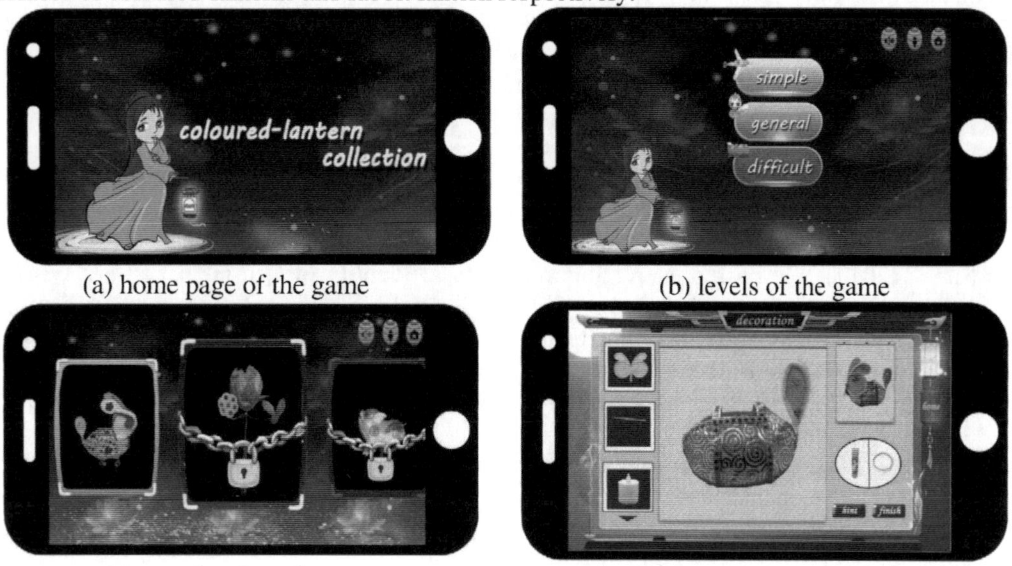

(a) home page of the game (b) levels of the game

(c) easy-level works (d)rabbit lantern "papering" craft

Figure 9. Design of game coloured-lantern collection Qin Huai festive lanterns making crafts.

The coloured-lantern information base can be applied in diverse ways. The game coloured-lantern collection Qin Huai festive lanterns making crafts, starting from inheritors, has innovated the making craft form of representative works from the inheritors. Moreover, specific festive lanterns can also be innovatively studied. With the representative lotus lanterns as example, researchers used to download papers one-by-one and read all the texts. Now, they can extract relevant contents of lotus lanterns and search for all pictures of lotus lanterns before theoretical innovation or design innovation. Through the associations of coloured-lantern literature information and the creation of coloured-lantern literature information base, researchers can directly search the topic maps, localize at lotus lanterns, read all texts about lotus lanterns on Omnigator, and browse all pictures of diverse lotus lanterns as well as relevant information, including the morals and crafts of lotus lanterns. Thereby, researchers can more comprehensively understand lotus lanterns before making one, which improves the work efficiency.

A coloured-lantern literature information base based on topic maps was constructed and stored in the computer system in a digitized form. In the future, we will design the auxiliary "coloured-lantern resource dictionary of the coloured-lantern information base, which will allow users to research information anywhere and anytime. The coloured-lantern resource dictionary will solve the cognition problems faced by special populations when using the emerging themed resource bases.

6.Conclusions

The Qin Huai coloured-lanterns are diverse in both types and forms, which complicate the organization and modeling of huge resources. Thus, in this study, the resources of coloured-lantern projects were first analyzed and then based on literature resources, the literature information was excavated, organized and associated. Then Ontopoly topic maps were plotted with the help of Ontopia and thereby a complete coloured-lantern information base was constructed. The topic association maps formed by VizDesktop were visualized and then browsed on the built-in Omnigator. Finally, the usability of the coloured-lantern information base was illustrated with the use of concrete instances. In the future, the Qin Huai coloured-lanterns information base will be improved and enriched through the superimposition or deletion with other types of resources of Qin Huai coloured-lanterns, which will offer convenient methods for subsequent research on Qin Huai coloured-lanterns and promote the information transmission and sharing of Qin Huai coloured-lanterns. This study offers a new method for the information organization, expression and discovery of Qin Huai coloured-lanterns. The research methods and flowchart used here can be referred to in research from other fields.

Acknowledgments
Thanks to the 17th batch of student scientific research projects of Jiangsu university(17C624).

References

[1] Meng Wang, Xin Xu . (2015) Research on the Application of Topic Maps in Intangible Cultural Heritage Information Resource Organization:Taking Beijing Opera and Kunqu Opera as Examples. Library and Information Service,59(14):15-21.
[2] Lili Sun . (2017) Media representation and cultural identity of Qin Huai Lantern Festival. Zhejiang University of Technology.
[3] Meiling Jiang . (2018) Modern lamps design based on Qin Huai coloured-lantern cultures . Nanjing University of the Arts.
[4] Juan Wang . (2012) Artistic values and innovative design of Qin Huai coloured-lanterns . Jiangnan University.
[5] Qun Zhang . (2007) Interview with Gu Yeliang,a famous artist of Nanjing lantern-making. Archives and Construction,(05): 34-35.
[6] Yijun Mu , Tao Huang . (2018) Animation of intangible cultural heritage -- Nanjing Qin Huai lantern festival . Art and Design,2(08): 91-92.
[7] Mingling Hu , Jiantao Wang . (2010) A Summary of Research on Topic Map Application in China. New Century Library, (02):33-35.
[8] Ping Wei . (2013) Topic map models in digital libraries. Ability and Wisdom, (12): 363-363.
[9] Qian Liu. (2003) XTM topic maps and information organization system interoperation. Sichuan University.
[10] Qingmao Li, Xingjiang Yang , Xiangbing Zhou , et al. (2010) Research on topic maps-based ontology information retrieval model. Journal of Computer Applications, 30(1).
[11] Yunzhong Zhang, Shuangshuan Fengg. (2018) Study on Resource Aggregation of Digital Archives Tagging System Based on Topic Maps. Library and Information Service, 62(14).
[12] Yi Shi, Huixiang Xiong, Yingying Lu. (2018) An Empirical Analysis on the Digital Resources of Intangible Cultural Heritage Based on Topic Map. Library and Information Service.
[13] Yi Shi, (2017) Archives resource aggregation of Intangible Cultural Heritage based on Topic Maps , Central China Normal University.
[14] Qingmao Li, (2007) Organization of tourism culture information resources based on Topic Maps. Sichuan University.
[15] Bella Martin, Bruce Hanington. (2013) Universal Methods of Design. Beijing: Central Compilation & Translation Press.
[16] http://localhost:8080/

Research on Teaching Design Based on BOPPPS Mode: A Case Study of the Course "Principles of Visual Communication Design"

Huang Liangpeng [1], **Lin Meiting** [2] **and Chen Xianwang**[3]

[123] School of Media & Art Design,Wenzhou Business College,Wenzhou City,Zhejiang Province，325035，China

*huanglp612@163.com

Abstract."Principles of visual communication design" is one of the important theoretical courses of visual communication design during undergraduate, it is also a basic course which combines tradition and modernity. In this paper, based on the phenomenon of " marginalization of theory courses "of art students in nowadays, we studied BOPPPS teaching mode of Canada and integrated it into teaching. We analyzed the problems that arised in the teaching process of this course optimized the teaching mode and method, explored more suitable methods for contemporary arts undergraduates to study basic theory course, it is showed that BOPPPS teaching mode has effectively improved the students' studing interests as well as the quality of homework. We hope this study could provide a reference for the education reform in the future.

1. Introduction

The Principles of Visual Communication Design is an important theoretical foundation course for visual communication design. The nature of the curriculum determines the need to export a large number of abstract concepts in the teaching process.

In the past, the traditional "blackboard + chalk" teaching, the one-way transfer of knowledge teaching mode has been unable to arouse the attention and enthusiasm of students' learning, resulting in the low participation of students in the class, which in turn affects students in practice. The theoretical knowledge learned in the classroom cannot be fully utilized.

Therefore, the traditional teaching mode has seriously affected the further improvement of teaching quality, failed to meet the need to pass knowledge in the classroom.

The BOPPPS teaching mode is a new teaching mode used by Canadian teacher skills training. The teaching mode emphasizes the student-centered, modular decomposition of the teaching process, and the strong practicality and operability make the classroom teaching arrangement more logic and rationality. This research takes the course "Visual Communication Design Principles" as a research sample and introduces the BOPPPS model to inject new possibilities for curriculum teaching innovation in the course teaching.

2. BOPPPS Model Overview

The BOPPPS teaching model is designed by Canadian Teacher Skills Training and is based on the qualifications of teachers in British Columbia, Canada. It is mainly used for the skills training of teachers. In the training, the teaching practice is used as the leading method to enhance the teaching of teachers. The effectiveness of skills and teaching (Yang Bo Tan, 2017). At present, the BOPPPS model has been

IOP Conf. Series: Materials Science and Engineering **573** (2019) 012054 doi:10.1088/1757-899X/573/1/012054

introduced and implemented in Chinese classroom teaching, which can effectively, efficiently and effectively promote students to actively participate in classroom learning while improving the quality of teaching.

The teaching design concept of BOPPPS mode is to divide the classroom teaching into Bridge in, Objective, Preassessment, Participatory learning, Post-assessment and Summary six links (Figure 1), the following describes each link.

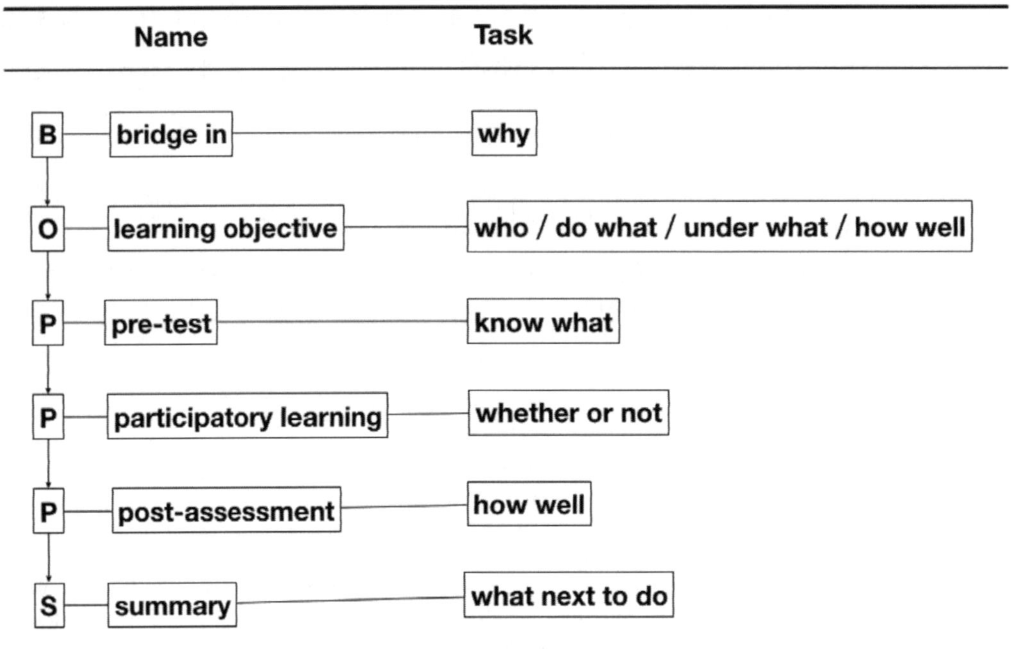

Figure 1. Boppps Model figure

2.1. Bridge-in

The bridge-in is the first link in the BOPPPS teaching mode. The main purpose is to enable visual communication design students to understand the main content of course learning and stimulate students' interest in learning. In the bridge-in stage, teachers need to focus on the design theory and value of the course, combined with design practice learning, to stimulate students' interest through a story closely related to the course learning content or successful cases, and put the cases with learned theoretical knowledge in a coherent manner, and at the same time, questions related to the teaching theme are proposed, and the students are allowed to enter the self-learning process of the core practical teaching content.

2.2. Object

The object emphasizes the student-centered teaching model. The purpose of design theory teaching is to enable students to translate basic theoretical knowledge into design skills and innovative creative ability through theoretical study. Through design theory learning, it is possible to obtain correct analysis objects and propose reasonable design solutions ability. Therefore, the design theory teaching objectives need to be more specific and clear. The object of a design course should be only for one or a few methods. Teachers should repeatedly emphasize and clearly explain the specific content points that should be mastered in this practical study, such as the creative performance of the design, the appreciation of the relevant design and the corresponding design methods. Obviously, the clear teaching goal is conducive to teachers in combination with the design of students encounter specific confusion in practice, better design of the corresponding teaching strategies and continuous improvement of teaching methods.

2.3. Pre-assessment

After clarifying the teaching objectives, it is necessary to pre-evaluate or explore the knowledge background and knowledge of the students. The bottoming can be carried out through various methods such as questioning, quiz or discussion. The main purpose is to quickly and accurately grasp the knowledge base of the students and the understanding of the teaching content. The teacher adjusts the teaching arrangement in time according to the needs of the course objectives and the actual situation of the students. (Danping Cao, Xingyao Yin 2016)By understanding the students' familiarity with the practical teaching content in a variety of ways, teachers adjust the curriculum in time, and could arrange more cases for students to explain or discuss according to the unfamiliar knowledge points; on the other hand, students can further understand the key contents to be mastered of the practical course through the teacher's pretest contents, which is benefit to clear purpose in the follow-up study process, and get better study results.

2.4. Participatory learning

Participatory learning is one of the other major links in the BOPPPS model. This stage is mainly through multiple forms of teacher-student interaction, repeated questions and answers, so that students can acquire new knowledge and master new skills in the interaction.

The main task of the teacher is to select the theme according to the goal of this teaching. After grouping the students, they will communicate with different groups and make basic plans for design through investigation, design and data reading. After each group completes the preliminary plan, each group of students will conduct detailed communication and report to discuss the design plan of the group. At this time, the teacher can raise questions according to the student's report situation and guide the students to think further. The arrangement of this method is completely different from the previous design theory lesson, which avoids the traditional cramming mode of teaching, making the students become learning subject of the class. In the process of communication, students not only exercise their ability to think, express, and cooperate, but also deepen students' understanding of the theory and their own unique understanding, so that knowledge can be more deeply impressed.

2.5. Post-assessment

The post-assessment is to test the students' mastery and understanding of the classroom knowledge through a targeted evaluation method, and fully understand whether the students really learn the knowledge and ability and achieve the set teaching goals. Through the feedback of learning effects, further adjustments to subsequent teaching content and methods are guided. The teaching of visual communication design principle can be to ask questions or short paper writing, to the classroom content as the topic for students to analyze and answer, to supplement the students ' learning missing or weak knowledge points.The main purpose of this test is to let students truly understand the omissions and deficiencies in their learning. From the point of view of teachers, teachers can continuously correct and improve the teaching content and methods through post-assessment results, enhance the effectiveness of teaching, and strive to achieve the best teaching objectives.

2.6. Summary

The final summary is a summary of the content in class, which plays as a connecting link between the preceding and the following class. The student's feedback on the mastery of the learning content, and the teacher also sorts out and summarizes the content of the teaching. Evaluation and feedback is the central link, can let teachers fully summarize the advantages and disadvantages in the teaching process, in order to improve the BOPPPS teaching model in the "Visual Communication design Principles" course of the application.

3. Teaching design of BOPPPS in visual communication design principle

This study introduces the BOPPPS teaching model into the course of visual communication design principles, and introduces the curriculum design through the bridge-in, learning objectives, pre-test, participatory learning, Post-assessment and Summary (Figure 2). The following sections explain each stage.

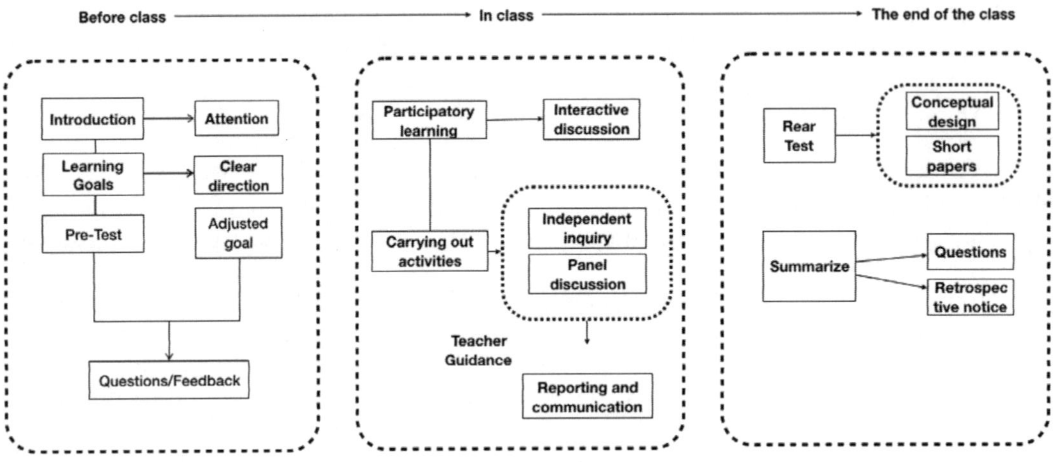

Figure 2. The design processes

（1）Introduction phase (Bridge-in). Played an advertising video, proposing that "eyes are active machines" to bring out teaching content, and to stimulate students' interest with a direct sensory experience.

（2）Learning Objective. The learning objective refers to the objective of student learning and the objective of teacher teaching, which can be used as an evaluation standard for testing learning effects (Weiwei Chen 2014). The learning objectives are reflected under what conditions, what will be learned and what level will be achieved.

The third chapter of the Principles of Visual Communication Design reflects the following objectives:

1 three elements of the visual process;

2 the process of producing visual phenomena;

3 visual psychological content.

（3）Pre-test. The class uses the following questions to test before class:

1.Why can one person see an object? (Graphic description, student induction, get feedback)

2 What is the process of visual production? (thinking about the production process)

3 Why do the same objects feel the same for everyone? (Thinking about visual psychology)

（4）Participatory learning. Participatory learning enables students to actively participate in the teaching process and change the way of passive learning in the past. The use of activity-driven teaching methods can effectively motivate students to learn.

Case 1: When talking about the light in the visual physiology, use the two master paintings to let the students separately express the expressions. On the left is the sketch of Georges Seurat, and on the right is the sketch of Anger, by watching the works of Georges Seurat. Feel the different results of the light in the transmission of information, clear light can see the clear image results of Angel's works, and can distinguish the details of the object, which is called resolution; and when the light is blurred or weak, only see When an object exists and cannot see the details of the object, it is called perception.

Through the comparison and appreciation of the works of the two artists, the students participated in the discussion of the difference between the two paintings and the influence of light on the vision. Through this kind of problem-driven approach, students will think independently, continue to study in depth, and fully grasp the knowledge points of light in visual physiology.

Case 2: In the visual psychological knowledge point, visual perception is a proactive visual activity, which includes grasping the basic characteristics-shape. At this knowledge point, a classroom activity is designed, the outline of the class teacher is depicted, and the live demonstration is performed. The evaluation draws the most similar one. Through this activity, the concept of visual experience is guided. The visual features of each person's visual experience are different.

(5) Post-assessment. Responding to the learning objectives, designing test questions for different goals. In this course, the effect evaluation is done by short essay writing. The short essay topic is combined with the knowledge points of the courses and the contents of the students' practical courses, such as "poster design". The application of visual psychology in the book, "Analysis of the application of optical illusion in poster design", etc., use visual communication to convey the principles of design principles to illustrate classic works or students' own works.

(6) Summary. Review the teaching content of the classroom, review the thinking methods to solve the difficulties and emphases, expand the practical cases of visual perception, and put the principle of visual perception into the four elements of communication mentioned in the previous course, which is the part of the subject. Finally, the content format tower theory of the next class is predicted, and the related extracurricular expansion reading is arranged.

4. Effect of contrast

Through the comparison of the teaching effect between the class using BOPPPS mode and the regular class, it is concluded that the class based on BOPPPS mode improves the students' enthusiasm, classroom atmosphere and teacher-student interaction

4.1. Students' evaluation on the application of this teaching method

The results the results of the questionnaire show that the teaching method based on BOPPPS mode is diversified, the case is properly applied, the writing is improved, the students' learning interest is improved, and the teaching efficiency is improved. At the same time, the teaching mode closely combines the theoretical part of the teaching content with the practical operation, corrects the previous teaching concept of paying more attention to theory and less attention to operation, and is recognized by most students.

Table 1. Students' evaluation of BOPPPS model teaching method.

Survey projects (29people)	Identity n %	Basic identity n %	Don't agree with n %	Recognition rate %
Boost your motivation	18 62.07	7 24.14	4 13.79	86.21
Teacher-Student interaction	19 65.51	7 24.14	3 10.34	89.66
Combination of theory and practice	20 68.97	3 10.34	6 20.69	79.31
Active classroom Atmosphere	22 75.86	5 17.24	2 6.90	93.10
Writing skills	20 68.97	6 20.69	3 10.34	89.66
Ability to think independently	23 79.31	3 10.34	3 10.34	89.66

4.2. Final examination results

In the final exam results, better data results can be obtained for the experimental class based on BOPPPS mode.

Table 2. Comparison of final grades.

Class	Score (100 Score)	Basic knowledge (60 score)	Thesis Writing (40 Score)
Classes	81.12±6.98	48.67±4.56	32.45±3.02
Control classes	76.53±7.65	45.92±5.79	30.61±3.84

5. Summation

This study uses multimedia courseware to assist classroom teaching. Through the creation of a situation, through a video advertisement, the concept is guided to the students, and then guide students to compare and analyze the visual content, using continuous heuristic and problem-driven teaching methods to promote Students actively think and learn actively. And can effectively guide students to sum up the role of discriminating visual perception in learning communication. The visual communication design principle is a relatively boring theoretical foundation course. This study initially introduces the BOPPPS model into the instructional design, and finds that the student's participation in learning is significantly improved, and the number of students including late arrival and absenteeism is also improved. The completion of the payment on time, the results show that the use of BOPPPS teaching mode can improve student participation, improve teaching quality, achieve effective teaching, and provide an effective teaching method for improving teaching.

References

[1] Cao danping and Yin xingyao (2016). Canadian BOPPPS teaching model and its implications for higher education reform. Laboratory research and exploration, 35(02), 196-200+249.
[2] Su Juan, guo feng, bai yu, zhao ge (2016). Application of BOPPPS model in heat treatment course of metallography. China modern educational equipment, 05, 92-93.
[3] Li qingqing (2017). The significance of micro-course guidance to the reform of college teaching mode -- taking visual communication design as an example. Research on fine arts education, 07,159 +161.
[4] Gong yanheng (2017). On the innovative development of multimedia technology and visual communication design teaching. Contemporary educational practice and teaching research, 04,16.
[5] Han xiaoyan, wang na, zhang shiyu (2016). Research on the teaching design of "chemical potential" based on BOPPPS model. Guangzhou chemical industry, 44(20), 180-181.

IOP Conf. Series: Materials Science and Engineering **573** (2019) 012055 doi:10.1088/1757-899X/573/1/012055

Application Research of AR Technology Based on Ergonomic Theory in Mining Industry

Haiyang Jiang, Yun Mei*, Ying Lin, Fang Wang, Qimeng Nie

School of Architecture and Art Design, University of Science and Technology Liaoning, AnShan, Liao Ning,114051, China

lnasmy@126.com

Abstract. In order to make the planning, design and production activities of mining parks more efficient, simple, safe and people-oriented, these processes are optimized and innovated from the perspective of workers. Based on the theory of ergonomics , combing a series of design and production process with AR technology in mining area, making the advantages of AR technology to reach its full potential. Using the ideas and methods of industrial design to establish the solving model of problems and proposing new design methods for mineral industry, so that advanced technology and the traditional industrial model can be highly connected, and in the end it must meet the consequence of human and machine with environment.

1.Necessity of introducing AR technology into mining Parks

AR, or Augmented Reality, is also known as Augmented Reality. By calculating the position and angle of the camera image in real time and adding the corresponding image technology, the physical information which is difficult to experience in a certain time and space in the real world is superimposed through computer science and technology simulation, so that the real environment and virtual objects appear in the same picture or space in real time and are perceived by human senses, and the superposition is realized. More realistic sensory experience. Augmented reality technology comes into being with the development of virtual reality technology, but there are great differences between them. First of all, they have different requirements for the sense of submergence and different emphasis in the application field. The development and design of mining parks is a comprehensive process. The project is huge and the task is arduous. There are many aspects to be considered. For example, to understand the topographic characteristics, to survey the geological environment of mining areas, to determine the location of main wells and auxiliary wells, to reasonably construct office areas, to do a good job of water resources reserve and ore transportation routes, .and to do these meticulous work, it is necessary to unite the strength and knowledge of the staff of various departments. A certain decision from the macro-coordination to the early design to the approval of execution must be through multiple parties. It takes a long time and there will be mistakes in the handover process. It consumes a lot of energy in the process of design and construction. The operation of the mining industry has not yet introduced products and technical means in the field of high-tech. There are many shortcomings in the current working mode. For example,1. There is no good overall working environment, including light environment and acoustic environment.2. It poses a great challenge to the reliability of people, because people's emotions and mental states have a high impact on their work reliability, thus affecting the production process [1]3. People's safety is not guaranteed to the maximum. However, the introduction of AR technology can bring new experiences, which can effectively improve people's reliability and increase work comfort. The design, implementation and management of mining areas can be greatly improved.(Fig. 1)

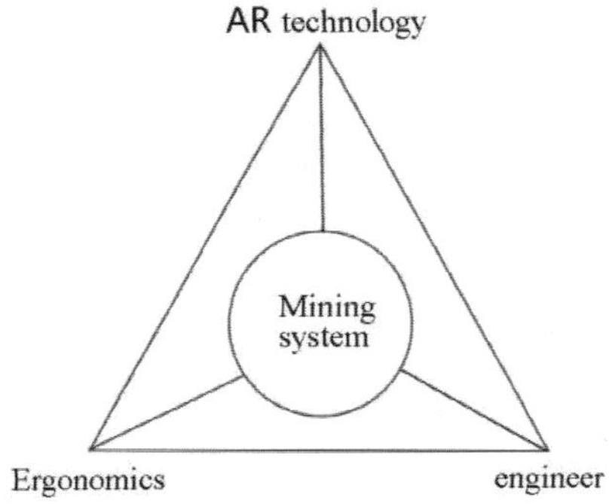

Figure 1 architecture diagram

2.Application of AR Technology in Mining Environment

Some scholars have studied the application prospects of combining AR technology with the coal mining industry. [2]From the perspective of high technology, this article will be based on ergonomics to push this application method to a wider area. On the other hand, it is applied to the entire mining area. Mining and rough processing is a very large project involving many details such as ground facilities and underground facilities. The following experimental explorations are proposed for the combination of certain aspects of the production process with AR technology.

2.1.AR technology effectively improves people's reliability

Human reliability refers to the possibility of a person's successful performance in a certain work or task under the given time parameter conditions.[3]For example, when transporting road wiring inside a mining area, it is inevitable that people will have omissions, the layout of the line is unreasonable, and the possibility of rework is very high. However, using AR technology to analyze the distance and relationship between the facilities of the site, processing the ground layout of the input computer, transforming the layout principle into its own design language, can present the best route, and the contact of transportation between the facilities is more convenient, the distance is shorter, the engineering quantity is smaller, the line is more compact, and with adequate room for development. The overall characteristics of AR technology can make the road and facilities along the site coordinate in distance and elevation according to their respective technical requirements. It can also be applied to railway transportation routing, so that workshops and warehouses with large freight volume can be arranged along the line. The line strives to be short and reduce the number of turnouts, and people will be greatly liberated.

In addition, the detection of all kinds of production and living pipe network lines in the mine is also very costly, because there are too many pipe routes, such as underground drainage pipes to the surface water tank, water supply pipes for underground production water supply, up and down pipes for domestic use, and hot water heating pipes for domestic use in northern mines. The damage or failure of underground pipeline route is inevitable. Once the fault occurs, it will cause the mine work to be unable to operate, resulting in property losses. AR technology can be used to test various pipelines and analyze the thickness of pipelines for timely repair and replacement to avoid leakage accidents. The use of AR technology is to prevent the occurrence of danger, rather than to save the accident has occurred.

2.2.AR technology improves auditory display design.

The application range of sound transmission information is very large, because people's auditory response is fast and the range of perception is wide.[4] We all know that the mine explosive storehouse is dangerous and the safety requirements are very strict, but some risk factors cannot be identified and identified with the naked eye in time, which will cause very serious consequences. It is a good choice to use the sound alarm device and AR technology for the safety detection of explosive magazine, because the magazine design supporting facilities should have fire pool and fire shed, fire ditch, power supply, communication and lightning protection facilities, walls, etc. The AR technology can detect the surrounding heat, humidity, and tightness coefficient, and is represented by different parameters. For example, the blue and red colors indicate the heat distribution(Fig. 2), and when the heat around the explosives exceeds the safety factor, prompts and warnings are given;

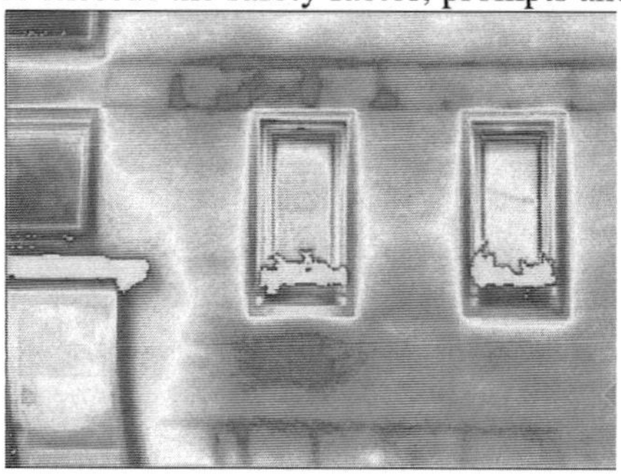

Figure 2 Heat distribution map

Figure 3 rigorous indication map

The rigor can be represented by a value of 0-10(Fig. 3); the degree of humidity is represented by a plane coordinate system, the abscissa is time, and the ordinate is humidity, so as to show the change of humidity. Assume that, as shown(Fig. 4), the humidity requirement is below 3, which is acceptable. No matter how the curve changes, it can be as long as it is below 3. A final alarm is given when the humidity is above 3. As shown(Fig. 5), if the humidity suddenly increases at 8:00, the initial alarm is given when the change is too fast, and the alarm sound is a beep. The sound pressure level and frequency of the ringtone are only slightly larger than the buzzer sound, mainly In the quiet environment attracts people's attention. When the safety factor exceeds the safety factor at 9:00, the sound of the alarm becomes the sound of the alarm. This sound is strong, the frequency is from low to high, and the tone is raised and lowered, which is used as an alarm for dangerous situations. From this model, it can be seen that 8:00 is a sudden change node. At this moment, the cause of the problem can be found, for example, it is raining, water leakage, and a valve is not closed, so as to get a solution to prevent it from happening. This is more intuitive, easier to manage, and safer to produce.

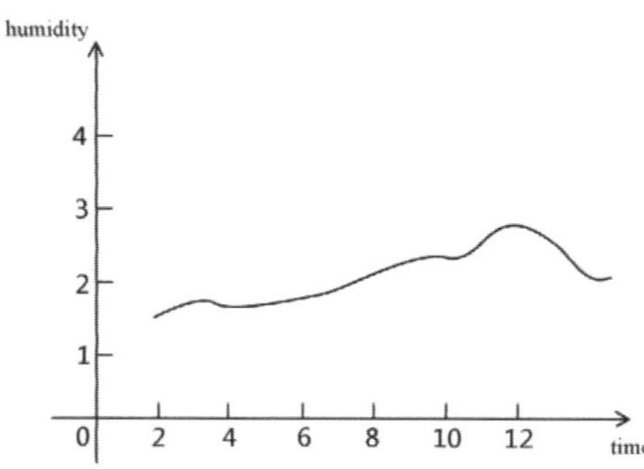

Fig. 4 Humidity Change Diagram (1)

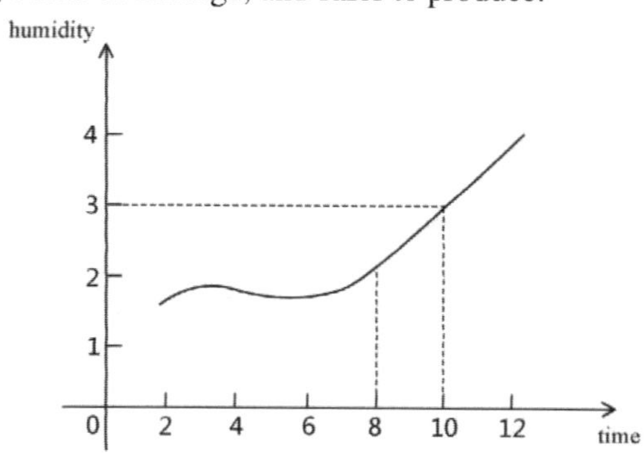

Fig. 5 Humidity Change Diagram (2)

This is the accident control strategy based on the key links and main methods of accident control. The above examples are only the technical principles in the 3E principle, because the technical countermeasures are one of the main measures to ensure safety. AR technology monitoring is a key technology based on careful research and analysis of potential hazards to predict, monitor and respond to hazards.

2.3.AR technology improves the working environment.

The light environment of underground operation is generally artificial lighting rather than natural lighting. Reasonable lighting and lighting in the work place are of great significance for the efficiency, safety and health in production. [5]Good light environment is the improvement of visual conditions on the one hand, on the other hand is the improvement of the environment. Wear AR glasses, you can create a reasonable level of illumination, so that the direction and diffusion of light is more reasonable, do not let the light directly to the eye.AR glasses can choose a reasonable light source color to coordinate lighting and hue, so that employees can concentrate more, reduce fatigue, improve labor enthusiasm and labor interest, thereby improving efficiency and attendance, and reducing labor accidents. The interactive experience is enhanced, the interaction between people and the environment, and the human-machine-environment form a coordinated system. In this system, the AR device improves the environment and enables the user to have a better working experience. For the enterprise, it improves the labor problem, improves management, reduces waste of materials and labor, improves production efficiency, and creates higher performance.

The mine air environment also needs to be monitored. The mine air mainly contains CO,SO2,H2S,H2,N2and so on. These toxic and harmful gases are extremely harmful to the health and safety of underground workers in the mine. AR technology can detect the gas index in the mine air in real time. When a certain harmful gas content is too large, the user will be prompted. This prompt is a classification suggestion. If the hydrogen sulfide content is too large, it should be represented by the color and sound of hydrogen sulfide gas. So that when we know which gas is more than the conventional value, we can use the corresponding means to reduce it, using physical methods such as opening the ventilating ventilation, use chemical methods such as neutralization reactions. In addition, records will be kept in the long-term detection process and a database will be formed. Under the database model, users can know which harmful gases are prone to anomalies and find the source of problems. As mentioned in engineering psychology, technology is designed to meet human needs, display data in an appropriate way, and ensure that representation and task requirements are displayed. Therefore, the design or technology that is really considered for users should be seriously explored and considered.

2.4.AR technology enhances feature recognition.

This is for the layout and design of the ground pipe network system. We all know that the mining park needs to set up many pipes and lines. If the various pipeline lines are crossed together, it will be particularly irregular. Even if the order is not reflected in its individual characteristics, Easy to repair and replace, so the pipe network system needs to be easily identified. Applying characterization and AR technology to this part of the project can eliminate a lot of unnecessary troubles. The ground pipe network system has water and sewage pipes, compressed gas pipelines, high and low voltage power lines, communication lines, etc. If this part of the pipeline is laid and laid The form only considers its own layout requirements without considering or not fully considering the layout of other pipelines, which may cause pipeline congestion and conflicts, so there must be a master plan to determine the plane coordinates and elevation of various pipelines.AR technology simplifies this part of the operation, allowing various pipelines to be coordinated on the plane and on the facade, and to maintain the required connection between buildings, structures, and transportation lines. The safety distance, erection mode, laying depth, laying sequence, groove specification of the various pipelines are all specified in the safety technical regulations. These requirements are converted into computer language, and virtual 3D models are generated. Different pipelines are used in different colors and The thickness is marked, for example, green represents the compressed gas pipeline, red represents the transmission line, blue represents the tailings pipeline, and orange represents the communication line. This makes it easy to identify and classify, and it is easy to distinguish the nature of the pipeline and the line by using different visual features, which greatly facilitates the design and engineering technicians.(Fig. 6).

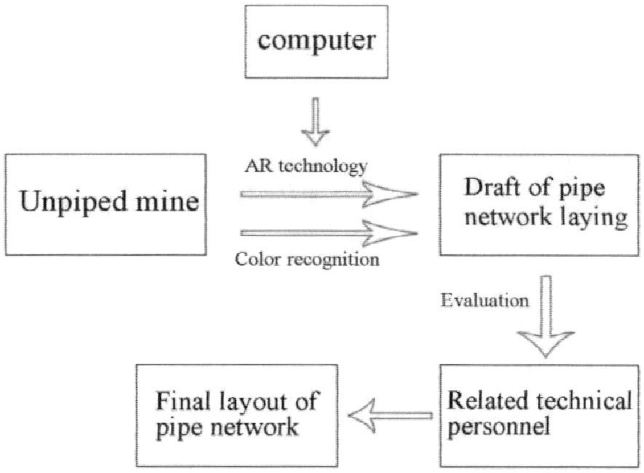

Figure 6 implementation model

3.Conclusion

Aiming at AR technology and mining industry, this paper puts forward a lot of applied research, and puts forward corresponding solutions for these application strategies. Because it is very difficult to combine high-tech with traditional industries, but there is a demand for this technology in the mining industry. Because these means and methods can be beneficial to the staff of the industry and the development of the industry, it is necessary to study the combination of the two from an academic perspective. In reverse engineering, there is a method called image reversal. It is based on the geometric dimension and color dimension information provided by the product digital image and video picture to realize the model reconstruction. Image reverse is a complex innovation process, generally using perspective transformation. And perspective projection, forming different perspectives, extracting geometric parameters from the information such as shape, size, proportion, color, etc., and then implementing the geometrical dimensions and the inverse of the color dimension through a professional system. This provides a solution method, through which the highway routing model and pipe network layout model mentioned above can be constructed, which is more conducive to construction. Although the industry of the future of augmented reality is optimistic, but after all, this is only a young industry into public view a few years. If these methods are implemented at home, we still need to look forward to the continuous improvement of AR technology. We believe that in the era of rapid development of science and technology, all methods can be effectively applied and all difficulties can be solved.[6]

References

[1] Zhang Z,Wu D. (2007) Ergonomics. M. Changsha Central South University Press. Changsha.
[2] Qu J.Cao Q.(2017).Application Research and Prospect of AR Technology in Coal Mine Industry.J.Technological innovation and application.(13)
[3] S.N.Chengelur,S.H.Rodgers,T.E.Bernard.Yang L's main translation(2006)Kodak practical ergonomics design.M. Beijing Chemical Industry Press.Beijing.
[4] Wickens,C.D.Translated by Zhang K.(2014)Engineering psychology and human homework.M. Beijing Machinery Industry Press. Beijing.
[5] Haik,Y.Shahin,T.M.Cao Y, Shi X, Yang L(2012)Engineering design process.M. Beijing Chemical Industry Press. Beijing.
[6] Cao Z(2018).New integration of VR and AR technologies in the field of industrial design.J.Fine art.90-91.

IOP Conf. Series: Materials Science and Engineering **573** (2019) 012056 doi:10.1088/1757-899X/573/1/012056

Smart TV Terminal Interface Design for the Elderly Based on User Experience

Yejun Chen[1], Mingzhu Li[2]

School of Arts, Jiangsu University, Zhenjiang, Jiangsu, 212013, China

[*]Corresponding author's e-mail: 1113634965@qq.com

Abstract. With the advent of aging society and intelligent era, designing smart product interfaces for the elderly is positively significant to improve their experience and life. Combined with the ideas and methods of user experience, an in-depth study on the elderly using smart TV terminal is conducted through interview, observation, questionnaire, etc. It constructs the personas and user experience map to excavate the pains and opportunities of the products. The current products and interfaces in the market are taken as the case to construct the relationship between the product opportunities and the five elements of user experience to analyze the terminal interface of smart TV and propose the optimal design scheme. It meets the needs of users, improves their experience and offers reference for the design of homogeneous interface.

1. Introduction

Smart TV is on the basis of the Internet technology. It has open system and chip together with the application platform that achieve two-way human-computer interaction. Moreover, it is an integration of audio, video, entertainment and data to meet diversified and personalized needs of users. [1]With the progress of the times, more and more terminal functions of smart TV are produced, and the operation interface becomes increasingly rich. These functions bring convenience for young people who are familiar with the Internet, while feelings of the old people are ignored. Faced with many uncommon functions and complex interfaces, intellectualization becomes the obstacle of old people and shake confidence of them using smart TV. As people have an increasing demand for the user interface, user-centered design is one of the trends of the interface design in recent years. Therefore, this research combines the design idea and method of user experience from the perspective of elderly users to enhance the availability of the smart TV interface, improve the user experience, and help the elderly users to use products more efficiently.

2. Smart TV terminal interface

Compared with traditional TV, smart TV terminal has more business content. With the operating system, users can not only use the video-watching function, but extend system functions by network operations and installing other plugins.

At present, most smart TV terminals are based on Android system, and the operating mode is similar to mobile phones. The majority of the TV screen is more than 42 inches. At the present stage, two common resolutions of the smart TV terminal are 1920×1080 and 1280×720.

There are few studies on the smart TV terminal interface for elderly users, and most of them are about the webpage and mobile APP. V.Mitchell comes up with the idea of keeping the user interface simple, avoiding multimodal keys and offering information portal appropriate for the elderly by analyzing their needs, physical and cognitive disorder. [2]Zhai Qianqian carries out the optimization design of smart TV interface on the basis of the use scenarios. [3]Zhang Ren combines the levels and elements of user experience, and creates the user experience design model of smart TV interface. [4]

3. Construction and analysis of the user experience map

User experience map is a visualization tool. From the view point of users, it describes the users' experience of using the products, discovers their pain in the experience, and summarizes the feasible product opportunity. It helps the designers to have a more intuitive impression on the experience of users and guide the follow-up design.

3.1. Constructing user roles and user experience map

Personas was proposed by Alan Cooper in 1999. [5]It analyzes large samples through a great number of interviews, observations and questionnaires to cluster and segment the users, refine their common features, summarize the comprehensive prototype of the target users, and carry out data visualization.

Constructing Personas can help to make the target users clear, understand the behavioral motivation and psychology, and assist the designers to design products. When determining the level of Personas, it is required to make clustering process for the data through qualitative research such as interview, observation and experience, as well as quantitative research such as questionnaire. [6] It is found that people over 65 years old seldom get access to smart products. As time goes by, old people will get more access to smart products in the future. Therefore, the target users in this paper are old people aged between 50 to 65. 12 people are observed and interviewed, in which 7 are men and 5 are women. The average age is between 50 to 65.

Generally, the user experience map includes the user need, behavior process, emotion experience, stage, contact, pain, opportunity, etc. On the basis of interview and questionnaire, Personas and experience map are constructed, which make it easier to discover the problems of the existing products and the potential demand of users. There are many differences between the survey users in free time and product-using proficiency, so two personas are constructed as shown in Figure 1. There are mainly five stages of the target users to use smart TV terminals: turn on the TV, read the content, choose the video, watch the video, and turn off the TV. By excavating the emotional satisfaction point and dissatisfaction point of the target users, it can be concluded that the major pains in the using process include: the interface level is complex, many videos are pushed without any purpose, and it is hard to find the History Records, Favourites and Search functions. The research excavates the existing pains and opportunities by further analyzing the users' behavior and emotional experience in the experience map to indicate the following design.

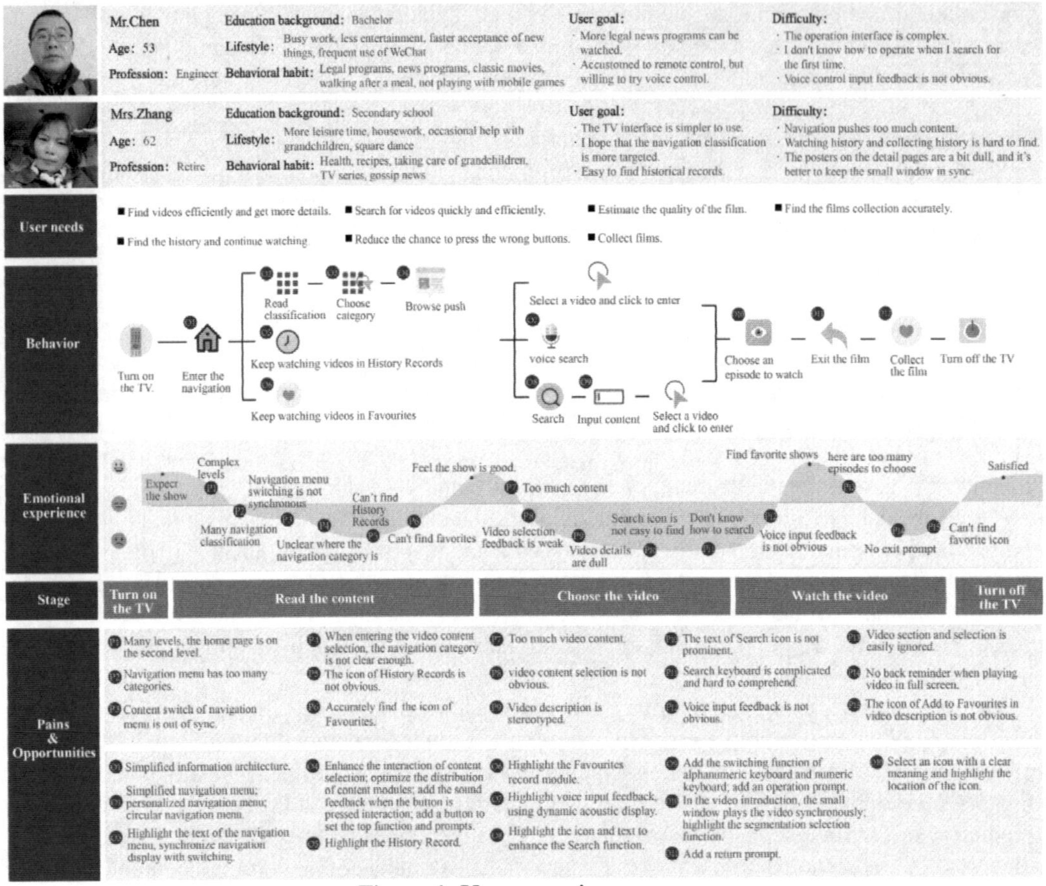

Figure 1. User experience map

4. A case study of smart TV terminal interface design

At present, there are rich smart TV terminal product brands in the market in different styles. According to online and offline surveys, sales volume of Tmall magic box 4A ranks the top. Moreover, it brings a good user experience and wins a high evaluation. Therefore, Tmall magic box 4A is taken as a case to optimize the interface.

4.1. Tmall magic box interface system problem analysis

In the preliminary research, interview and System Usability Scale (SUS) are taken to measure the users' satisfaction with Tmall magic box 4A. The SUS score of the original system is 60.5, and the level is D. In response to the issue that users feel difficult in operation, we focus on the confusing interface, complex content and unapparent icon instead of the problems such as network delay, frequent forced updates, advertisements and member limitation. Some interfaces of the original system are shown in Figure 2. By analyzing the framework and logic of Tmall magic box 4A together with the users' behavior and the research result, it is available to carry out a more definite design for interface optimization.

Jesse James Garrett [7] puts forward the concept of five elements of user experience, and divides them into strategy level, scope level, structure level, skeleton level and surface level. This model is widely recognized and applied to guide the Interaction design. It is found that the original system has solved the problems of P1, P3, P9, P12 and P15 in Figure 1 by adopting the five elements to analyze the above system, while there are still pain points in other aspects. By carefully analyzing the pains and exploring the problem description and performance, and connect it with the five elements of user experience, it is available to form Table 1.

a) Normal condition of TV series

b) Focus condition of TV series

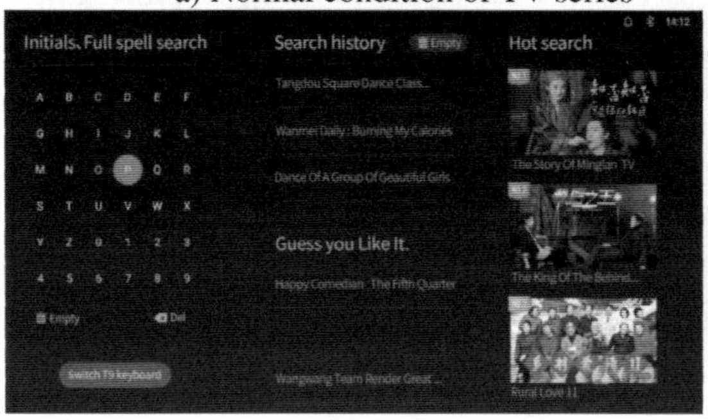

c) Search

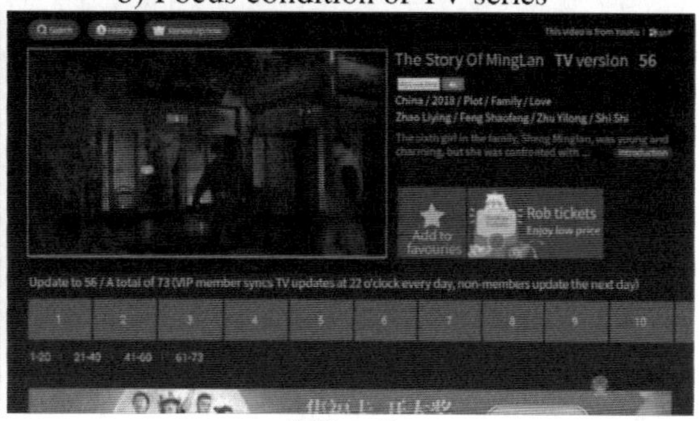

d) Details

Figure 2. Some interfaces of Tmall magic box 4A

Table 1. Main problems of Tmall magic box 4A

Problem levels	Original system has solved the pains	Original system has not solved the pains	Problem description	Specific performance
Strategy level		P2: Navigation menu has too many categories	Indefinite user types and needs	No distinction for different users Incomplete research on user targets and needs
Scope level	P3: Content switch of navigation menu is out of sync P9: Video description is stereotyped	P2: Navigation menu has too many categories	Some functions should be improved	Some functions are rarely used Display of some functions are not obvious Lack of interface guidance
Structure level	P1: Many levels, the home page is on the second level	P2: Navigation menu has too many categories P5: The icon of History Records is not obvious P6: Accurately find the icon of Favourites P10: The text of Search icon is not prominent	Categories of some functions are illogical Content feedback is unclear	The History module includes History Records, Favourites and Reservations The feedback of the selected image is not obvious
Skeleton level	P14: No back	P8: video content	Too many	Hints of 24 parts of the

	reminder when playing video in full screen	selection is not obvious P11. Search keyboard is complicated and hard to comprehend	navigation menus, lack of personalization Information tips are not obvious	navigation menus are not obvious
Surface level	P12: Voice input feedback is not obvious P15: The icon of Add to Favourites in video description is not obvious	P4: When entering the video content selection, the navigation category is not clear enough P5: The icon of History Records is not obvious P6: Accurately find the icon of Favourites P7: Too much video content P10: The text of Search icon is not prominent P13: Video section and selection is easily ignored	Crowded and flashy content	Interval of the video module is small with many colors favorites module and history record module are not obvious

4.2. Smart TV terminal interface design

An optimization design of the case's interface level is conducted on the basis of user research and analysis for competitive products.

4.2.1. Strategy level--clarify business objectives and user needs. Strategy level is the root and orientation of product design. Main content of this level is the user need and product objective. Category and characteristics of the target users in strategy level is not clear, and there is a lack of research on the needs of users. In the stage of user research, needs and goals of the elderly are analyzed constructing the user experience map to summarize the needs and pains of users, as is shown in Table 1. P2, the problem of too many categories in the navigation menu is due to the lack of in-depth analysis of user needs, which can be solved by deleting uncommon menus and adding customized navigation menus.

4.2.2. Scope level-- identify content requirements and feature requirements. The research studies the users, analyzes the original system, and utilizes Kano model to classify the categories of smart TV terminal interface needs to guide design, as shown in Figure 3.

The Kano model was developed by Noriaki Kano [8], It takes the influence of user needs on satisfaction as the basis to analyze the relationship between the product performance and user satisfaction. Noriaki Kano divides the model into five attributes: attractive attribute, one-dimensional attribute, must-be attribute, indifferent attribute and reverse attribute. He divides the categories of functional needs through interview and questionnaire. Here four attributes of the model are used to evaluate needs. The yellow part is the attractive attribute. It refers to the needs that won't be excessively expected by users, including voice control, multi-screen interaction, history records, basic information, add to favourites, favorites, etc. The green part is the one-dimensional attribute. It refers to the proportional need of the users' satisfaction and the satisfaction degree of demands, including synchronous jump. The blue part is indifferent attribute, which has no influence on the user experience no matter it is offered or not. It includes popular recommendation, go-to-top button, synchronized small window, etc. The red part is must-be attribute. It is generally the users' basic requirements for the product, which is the essential attribute or function, including the navigation menu, search function, etc.

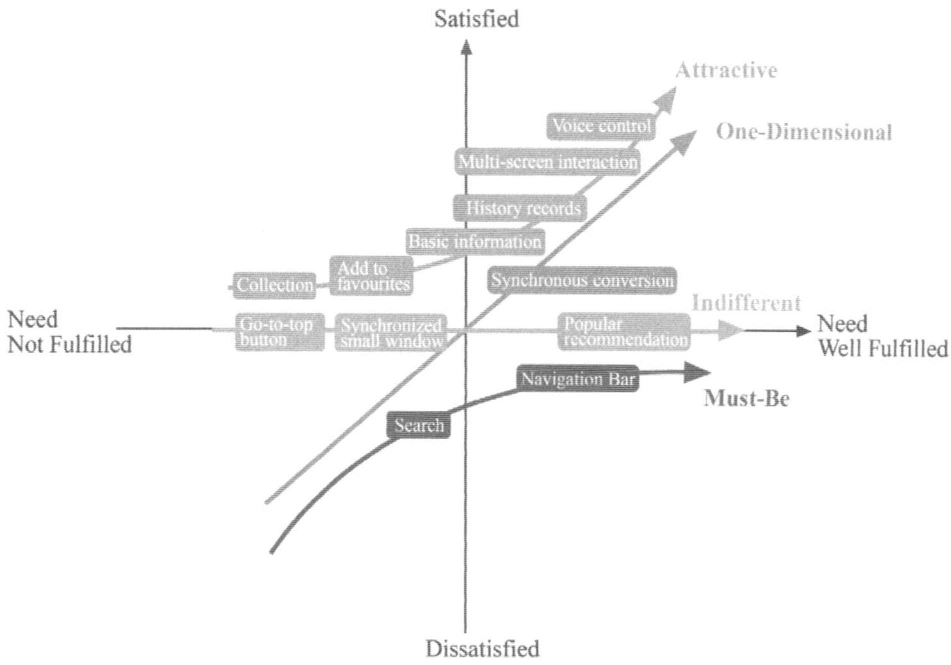

Figure 3. Kano model of Smart TV terminal interface needs

4.2.3. Structure level-- UX design and information architecture. Structure level mainly covers Interaction design and information architecture design. We carry out the information architecture and Interaction design of products according to the priority level of user needs. Interaction design pays attention to describing "possible user behavior", and meanwhile defines the behavior of "how can system cooperate and respond"[9]. Through analysis and organization, the information architecture of Tmall magic box 4A is shown in Figure 4. According to the user needs, the research optimizes the functional classification, interaction and structure. In the aspect of functional classification, it is required to adjust in accordance with the user needs and behaviors. For instance, simplifying the navigation menu function and adding personalized customization, taking circular navigation menu, adding the weather function missed in the original system, the switching function of alphanumeric keyboard and numeric keyboard, as well as the reminder of functional operation, etc. At the optimization function level, history and collection records are located at the next level of history. As the name is misleading, history and collection records are separated from the history and are improved to the higher level to make it more logical that enhances the understanding. According to the research, the functions of movie reservation and competition order can be canceled due to the low frequency of use. Additionally, recent watching record and history can be merged as well as the recent collection and collection history to avoid function repetition. Taking the home page as an example(as shown in Figure 5 a), b)), P2, P5, P6 and P10 in the user experience map are solved.

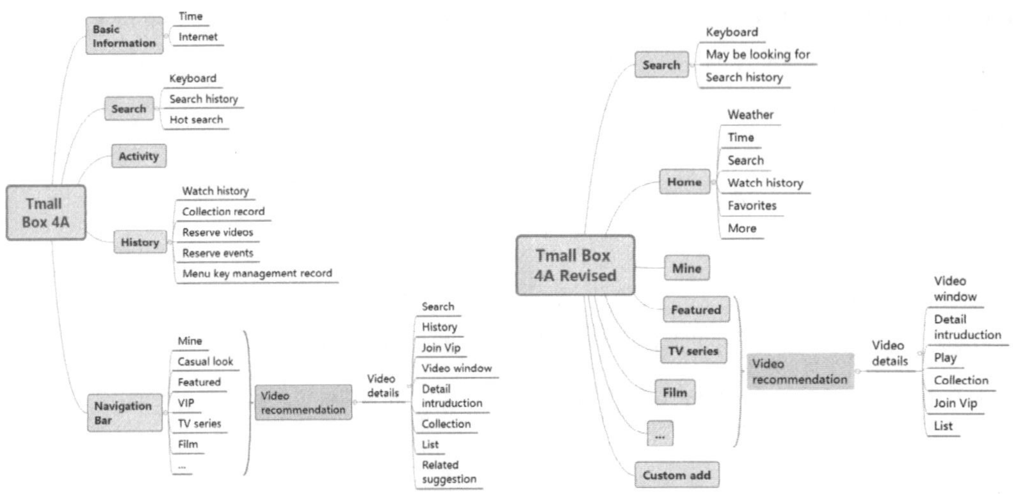

a) Before modification b) After modification

Figure 4. Changes of structure level of smart TV terminal interface

4.2.4. Skeleton level--interface design, navigation design, information design. Skeleton level mainly includes interface design, navigation design and information design. In the field of interface design, the relation and location between the elements and components should be determined in the skeleton level. Based on the previous research on user behavior, the integral interface layout is designed first. It gives priority to the card layout and top navigation menu, and content menu below the navigation menu is changed into horizontal stretch from the drop-down list, making it available to show more content while reducing the operating burden. Moreover, it establishes the operation logic that gray text only displays content, bold white text is the selectable option, and the selected text has a red frame effect. On the basis of keeping the original habits of users, the research strengthens the design of part of the information, such as adding the switching function of alphanumeric keyboard and numeric keyboard, as well as the keyboard operation reminder in the search page to solve the P11 in the experience map, as shown in Figure 5 e).

4.2.5. Surface level--visual design. Surface level is the last one of the five elements. It forms the final product design through satisfying the users' sensory feelings after meeting the other four levels. Moreover, it controls the product interface design from the visual level and conveys effective information, including the overall interface layout design, information configuration, icon design, graphic design, color collocation, visual style, etc. Visual presentation influence the intuitive feelings of the users. A good visual presentation can correctly guide the users, convey information and create a pleasant experience. The modified interface maintains the flat design of Tmall magic box 4A and the deep background. It takes a small amount of light gradient color that shows a light and delicate quality. By optimizing the visual element, it is available to help users operate more smoothly and improve their experience. For horizontal stretched menu, there is more blank space to make the information presentation more plentiful, the visual area extension broader, and the responsive interactive animation more delicate. The company color of Tmall magic box is red, so the blue frame of the selected text in the original system is changed into red. The modified interface darkens the unselected texts, strengthens the light floating effect of the selected texts to grasp the visual focus of users. For video content menu display, the menu control layout is too tight. Therefore, when entering the focus state of video content, the control layout will scatter, strengthen responsive interaction and make the visual effect of texts in the content navigation menu outstanding. Taking the normal and focus state of TV series menu as an example(as shown in Figure 5 c), d), P4, P7 and P8 are solved. Optimize favorites and history records module, and take simple icon to solve the surface level problems of P5 and P6, as shown in Figure 5 a). For P13, it is required to put the segment selection component on the top of the episode selection, and highlight the selection stage by bold text and red bottom line, as shown in Figure 5 f).

a) Home page

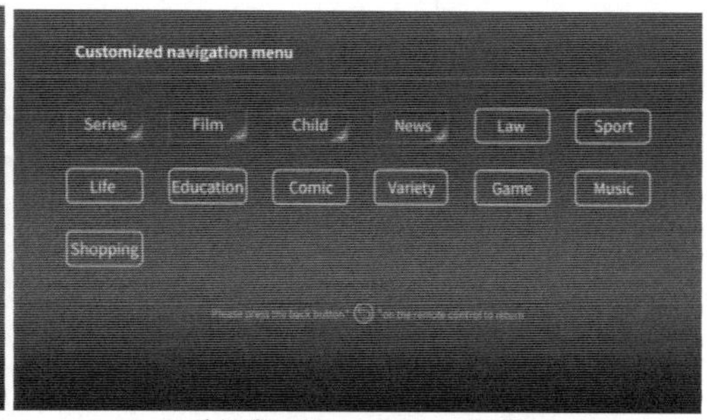

b) Custom navigation menu

c) Normal condition of TV series

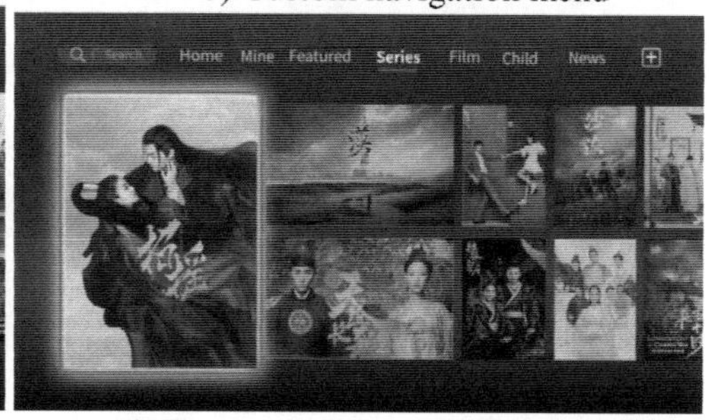

d) Focus condition of TV series

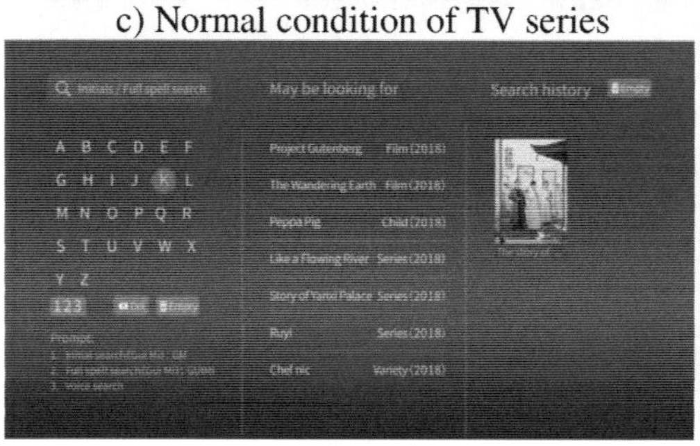

e) Search

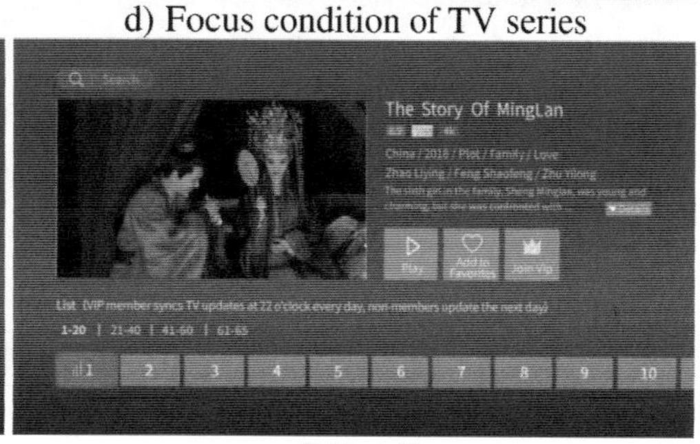

f) Details

Figure 5. Main interfaces after modification

5. Conclusion

At present, smart TV terminal interfaces and functions are complicated. This increases the burden and learning cost for the elderly to use such smart TV terminal. Therefore, it is required that interface design for the elderly should fully understand their needs and improve the operational efficiency. From the perspective of user experience, this study takes users as the center to excavate their needs and pains, and constructs the connection between the product opportunities and the five elements of user experience to analyze and optimize the smart TV terminal interfaces. Accordingly, it improves the user experience and helps the elderly users to use products more efficiently.

Acknowledgement

Thanks to the foundation of Jiangsu University's scientific research project（17C621）.

References

[1] Cui, B., Luo, S., Wei, K., et al. (2010) Analysis of key technologies of smart TV. Telecommunication network technology, (1): 51–59.

[2] MITCHEL, V., et al. (2007) Web-based interactive TV services for older users. Gerontechnology, 6 (1): 20–32.

[3] Zhai, Q.Q., Sui, Y. (2015) User experience design of smart TV interactive interface. Journal of Hefei Normal University, (16): 10–11.

[4] Zhang, R. (2015) Intelligent Product Packaging in the Information Age—Optimized Design Based on Smart TV Interface in Use Case. China Packaging Industry, 33 (6): 48–50.

[5] Cooper, A., Reimann, R. (2005) About Face: The Essentials of Interaction Design. Publishing House of Electronics Industry, Beijing.

[6] Dai, L.N., Xu, B.M. (2015) Research on Children's Furniture of Middle-class Families in Shanghai Based on Person Role Method. Packaging Engineering, 36 (16): 117–120.

[7] Garrett, J.J., Fan, X.Y. (2011) The Elements Of User Experience. Mechanical Industry Press, Beijing.

[8] KANO, N. (1984) The Journal of Japanese Society for Quality Control. Attractive Quality and Must be Quality, 14 (2): 147–156.

[9] Sternberg, J.R., Yang, B.J.,et al. (2006) Cognitive Psychology. China Light Industry Press, Beijing.

IOP Conf. Series: Materials Science and Engineering **573** (2019) 012057 doi:10.1088/1757-899X/573/1/012057

Research on Material Selection of Product Design under Environmental Awareness

Haokun Tian[1], Haifeng Zhang[2], Tingyu Liu[3]

[1] Northwestern Polytechnical University, Xi'an, Shanxi, 710072, China

[23]Heifei University of Technology, Hefei, Anhui, 230601, China

Abstract: Human demand for artificial objects is multifaceted and constantly changing. Therefore, product design system is becoming more and more complex, and there are various conflicts and constraints between design elements. From the requirement of product design with environmental awareness, this paper expounds the adverse effects of design materials on the environment. On the basis of analyzing the influencing factors of material selection in product design, the influencing factors model of material selection is constructed, and the environmental principles of material selection in product design are put forward to guide designers to realize the sustainable development of design. Based on analyzing the influencing factors of material selection in product design, this paper constructs a model of influencing factors in material selection, puts forward the environmental principles of material selection in product design, and guides designers to realize the sustainable development of design.

1.Introduction

With the development of the society, human demands on products have changed from only paying attention to using functions in the past to the requirement of all the properties such as using function, pleasure function and environmentally friendly function at present. Therefore, how to construct the relationship between the demands of human and the demands of environment to balance the system, improve design efficiency and guarantee product performance has become an issue of universal concern in the field of design at present. The life cycle of the products is getting shorter and shorter and the application of product molding materials not only affects the way of expression and the means of realizing the product modeling, but also directly relates to product recycling. With the rapid development of material science and technology today, it is a very important and urgent task for product designers to grasp the development trend of various new materials in a timely manner and apply the latest achievements of materials science correctly in modeling design to increase the added value of product modeling and reduce production and recovery cost. (As shown in Figure 1)

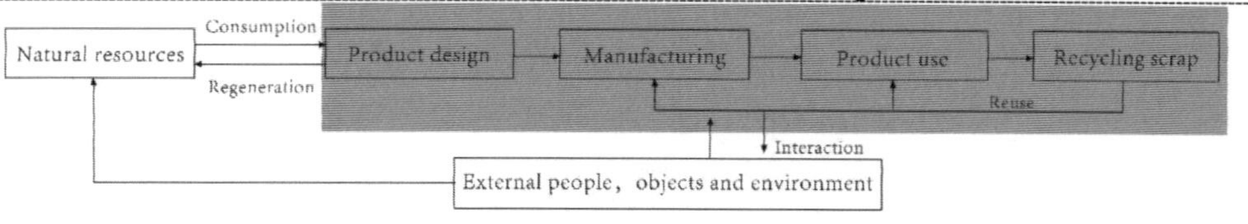

Figure 1. Green product basic system.

2. Product Design under Environmental Awareness

The concept of green products first appeared in the United States environmental pollution prevention and control regulations but the real meaning of green products was born in the former Federal Republic of Germany. Green Product (GP), also known as Environmental Coordination products (ECP) has the basic idea of incorporating environmental factors and pollution prevention measures into product design at the design stage, take the environmental performance as the design goal and starting point of the product, and strive to minimize the impact of the product on the environment. Generally, it means focusing on product environmental attributes (Natural resources and utilization, environmental impacts and detachability, recyclability, repeatability, etc.) during the product lifecycle and make it a design goal. While meeting the requirements of environmental objectives, the basic functions, service life, economy and quality of the product should be considered and guaranteed in parallel. Modern product design takes industrial product as the main object, and focuses on the comprehensive creative design of the function, structure, material, craft, visual communication, aesthetic foundation, agreeable nature and relation related to modeling. On this basis, we can build the basic system diagram of Green Product.

3. Application of materials and environmental problems

There is no doubt that material is the matter basis for product design. One of the core contents of material science is to study the relationship among the organization, structure and properties of materials. On the other hand, material is an applied science. Only through reasonable technological process can the material of practical value be prepared, and the material can be made into engineering material through mass production. The impact of the material on the environment consists of two aspects, like two sides of a coin, positive and negative. On the one hand, materials can repair the damage to the environment in varying degrees, control or reduce the environmental pollution and so on, but the effect is temporary, local and relative. On the other hand, the negative impact of materials on the environment mainly refers to the direct or indirect damage and damage to the environment caused by the production, preparation, processing, use and regeneration of materials.

In theory, the negative impact of materials on the environment has inevitability, irreversibility and universality. Environmental problems caused by materials arise and develop at the same time with human desires, economic development, and scientific and technological progress. From the point of view of natural environment, there are two limitations to all kinds of materials that human beings need. The first is that the amount of specific resources contained in the earth is relatively limited, and the resources of certain materials are not infinite or inexhaustible. The second is that the total amount of pollutants the earth can hold is limited, and its self-purification ability is extremely limited. Once the amount of pollutants goes beyond the bearing capacity of a Space-time Ecological Environment, there will be irreversible environmental pollution.

4.Influencing factors of material selection in product design

Design is a conscious modeling activity, which means that people consciously use tools and means in production to process materials into visible or tangible entities with certain shapes, making them useful or commodity names. Thus, design and materials are inseparable. To synthesize people's demand for green products at the present stage, we can construct the basic model of material selection from four aspects: functionality, aesthetics, economy and greenness. (as showed in **Figure 2**)

IOP Conf. Series: Materials Science and Engineering **573** (2019) 012057 doi:10.1088/1757-899X/573/1/012057

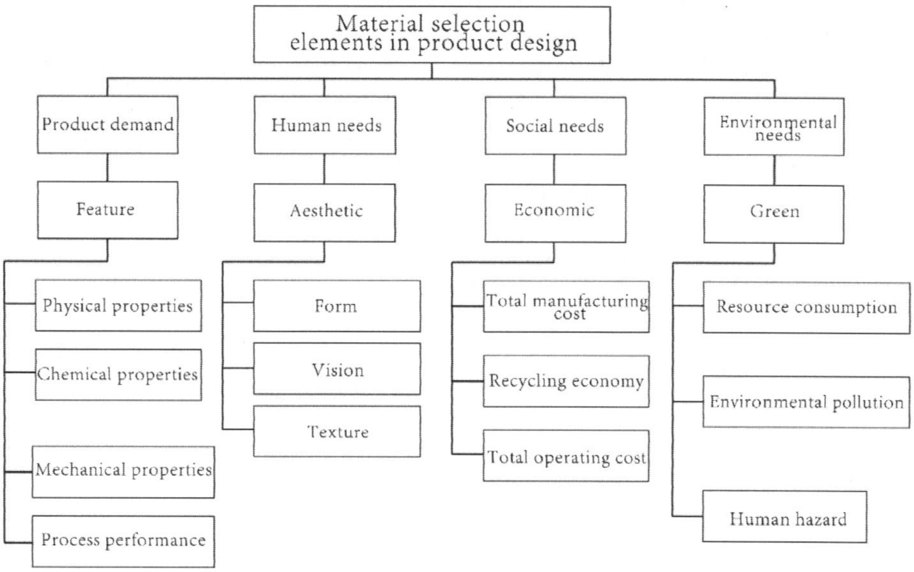

Figure 2. Material selection influence factor model

4.1 Functional

Function is the use and performance of industrial products and the design purpose of industrial products. Function plays a leading role in the structure of the product, and also has an impact on the appearance and shape of the product.

4.2 Aesthetic property

Aesthetic property, as a nonquantitative index of fuzziness, is often neglected in engineering design. In the design of industrial products, it is one aspect to realize the beautiful shape and orderly st ructure of products. Reasonable arrangement and application of the sensory and physical properties of materials can bring new features to product modeling.

4.3 Economy

The manufacturer and distributor of the product, while satisfying the needs of people or the market, part of the purpose is to obtain higher economic profits. Therefore, in product design, the economic content of the product is not a decisive index, but it has become one of the criteria for product evaluation.

4.4 Environmental Protection

The degree of influence of products on the outside world in the whole life cycle reflects the green degree of products. According to different design objectives and contents, the green performance of the product is manifested in material design, energy-saving design, human centered design, recycling and disassembly design, etc. It includes resource saving, energy saving, environmental impact and human hazards.

The above four parts embody the purpose and related principles of product green design, but they can not be defined simultaneously through unified technical indicators. These elements interact and restrict each other. Each aspect can adopt corresponding parameters and indicators to balance.

5. Environmental principles of product design material selection

The material selection of green product design should be based on material flow. In the concurrent engineering environment, environmental factors should be evaluated for each link in the whole design process, so that the product has the least adverse impact on the environment, and materials that meet the requirements of green products are obtained.

In addition to the principles of material usability principle, the process principle of material selection and the economics of material selection, the environmental principle of material selection should be paid more attention.

The environmental means the production, forthputting and scrapping of materials have a major impact, beneficial or harmful, on the environment. If the material cannot be reasonably selected and used, it will cause great waste of resources and energy, causing serious pollution to the environment. For product modelling, the following principles should be considered:

- Choose the green material. Green materials have environmental coordination, low energy consumption, low cost and low pollution, and have become a hot spot for global material science workers.
- Reduce the type of materials used. Try to avoid using a variety of different materials during design, especially for parts that are difficult to separate.
- Use a detachable structural design. This structure is used to facilitate product recovery or recycling, sorting and reuse after disposal when using a variety of materials.
- Try to use raw materials that are not coated or coated. The material has a "feeling of appearance", and through the surface characteristics of the organoleptic material, different psychological feelings are generated, and the rational use can enhance the unique artistic expression of the material. However, most of the paint will cause great pollution to the environment, so use it as little as possible.
- Use materials that can be decomposed and absorbed by nature. If the waste material of the product is not easily decomposed and difficult to be absorbed by the natural environment, it will cause great pollution to the environment. The high-scoring waste after processing and using the material is an example, especially the pollution caused by plastic packaging materials is extremely serious. In contrast, degradable materials are environmentally friendly. For example, starch toothpick is made from plant starch as a raw material, and is made into a disposable toothpick by curing, molding and grinding. It can be naturally decomposed when exposed to water, and is environmentally friendly. Compared with bamboo toothpicks, starch toothpicks have high smoothness, moderate hardness and hardness, and can protect teeth.

6. Summary

Through the analysis of material requirements of green design and product design, the article grasps the factors affecting material selection and summarizes the environmental principles of selecting materials. On the basis of considering the function, process and cost of the product, the environmental and sensory properties of the material are taken as important considerations, making the selected materials more in line with the needs of modern society, improving product performance and meeting the needs of sustainable development.

References

[1] Liu Zhifeng, Liu Guangfu. Green Design [M] Beijing: Mechanical Industry Press, 1999
[2] Chen Yanan. Research on the Innovative Application of Materials in Product Design[J]. Technology and Market, 2019, (1): 127-128.
[3] Zhang Huali. Rethinking the product design of environmentally friendly materials [J]. China Packaging Industry, 2015, (20): 145.
[4] Li. A new direction of product design based on low carbon environmental protection background [J]. Art Education Research, 2016, (15): 73.
[5] Liu Zhifeng. Green design method, technology and its application [M] Beijing: Mechanical Industry Press, 2008
[6] Gao Ruitao, Cao Yuhua, Zhang Xiaokai. The Transformation of Material Role in Contemporary Industrial Design[J]. Packaging Engineering 2006,27(6):315-317.

[7] Wang Xingwen. Research on the utilization and value of low carbon materials in product design [D]. Tianjin University of Science and Technology, 2016.

Research on the Design of the Elderly Wardrobe Based on Ergonomics

Chunling Gu[1], Jing Zeng[1, 2]

[1] Sichuan Agricultural University, College of Forest, Chendu, Sichuan; 611130, China

[2] Sichuan Agricultural University, College of Forest, Chendu, Sichuan; 611130, China

[*]1004674870@qq.com

Abstract. The elderly population of our country increases continuously, and the problem of aging population is becoming more and more serious. Based on the storage demand of the elderly, this paper analyses the problems existing in the use of the wardrobe by the elderly based on the ergonomics and the physiological, psychological and behavioural characteristics of the elderly. And from the perspective of safety, versatility and barrier-free design principles, the concepts of security, versatility and barrier-free design are used. It aims to improve the quality of life of the elderly by proposing a wardrobe design suitable for their elderly users.

1. Introduction

According to the latest population data released by the National Bureau of Statistics, by the end of 2018, China's population aged 60 and over was 249.49 million, accounting for 17.9% of the total population, of which the population aged 65 and over reached 166.58 million, accounting for 11.9% of the total population. Against the backdrop of China's upcoming aging population, China's elderly furniture is still in its infancy and has not yet formed a professional market and brand. Therefore, furniture suitable for the elderly will become a new consumption hotspot in China's future furniture market. The power of design will make life more secure and comfortable for older people.

2.Human factors analysis of the elderly

2.1 Physiological analysis

As we grow older, our body's physical function continues to decline, which has a certain degree of impact on our work and life. After entering the age of 65, the height of the elderly will shrink and the weight will continue to decline [1]. The bone content and muscle mass will decrease, and the tendons and ligaments will begin to shrink, which will directly affect the limb flexibility of the elderly [2]; arthritis, shoulder circumference The common diseases of the elderly such as inflammation and osteoporosis further reduce the activity of the limbs of the elderly.

2.2 Psychological analysis

When the elderly leave their jobs, they return to family life, social activities are gradually reduced, and the changes in social roles have also caused their lifestyles to change dramatically. There will be a huge gap in psychology, which may lead to some negative thoughts, often manifested as lack of Security, loneliness, sensitivity, etc. Affected by various factors, 421 families and two-child families have become the main form of modern family structure. As a result, more and more elderly people living alone have emerged. Their life circle has shrunk and no one has taken care of them. This makes the elderly feel very lonely.

3.Research on the use of wardrobes for the elderly

3.1 Research purpose and content

This research mainly analyzes the human-computer interaction process between the elderly and the wardrobe through interviews and observations of the use of the wardrobe by the elderly; selects the typical behavior decomposition action, and analyzes the behavior patterns and ability obstacles of the elderly. Through the research on elderly users and wardrobe products, to understand the actual needs of elderly users and the pain points of products, the design scheme of aging wardrobes that meet the actual needs of elderly users is proposed.

3.2 Research objects and methods

The sample of this survey was determined to be male and female citizens aged 60 and over, who are basically able to live on their own and have the habit of using the wardrobe for storage. First of all, we visited 13 elderly people, observed and interviewed the elderly using the wardrobe, and also learned about the living habits, behavioral characteristics, and health status of the elderly. Followed by the questionnaire survey, based on the survey, interviews and observations of households, combined with the relevant information to produce a questionnaire, through the questionnaire to investigate the main problems of the use of wardrobes for the elderly.

3.3 Research results

A total of 500 questionnaires were distributed for the elderly and 381 valid questionnaires were collected. Interviews and observations of 13 elderly people. According to the results of this survey, the dissatisfaction rate of the elderly in using the wardrobe reached 89.6%. The reasons are inconvenient use of the wardrobe, insufficient storage space, and unreasonable space division. The problems frequently encountered by the elderly in using the wardrobe are shown in Table 1.

Table 1. Common problems in the process of using the wardrobe for the elderly.

Behavior	Common problems
store clothes	insufficient storage space, wardrobe partition is too high, need to squat down, etc.
try on clothes	clothing storage disorder, dark light, forget where the clothes are placed, no mirror in the bedroom, etc.
drying the quilt	The quilt is placed too high, making it very inconvenient to take and store, and it needs to be done with tools. It is difficult for the elderly to complete alone.

4.Design principles

4.1 Safety design principle

Security is the first element of human survival. The safety design principle of aged furniture is to design and analyze from the physical and mental safety level of the elderly, to the greatest extent avoid safety hazards. The design of the elderly wardrobe should fully consider the physiological characteristics of the elderly, and try to avoid letting the elderly do the laborious actions. For example, in the wardrobe design, the hanger automatic lifting system is used to assist the elderly, and the inconvenient parts of the wardrobe function module are adjusted to ensure that the elderly can safely use the wardrobe.

4.2 Universal design principles
Universal design focuses on the special needs of the elderly while also paying attention to the needs of the young and middle-aged population; paying attention to the needs of the elderly at the age of 60, but also taking into account the needs of the elderly aged 80 or older. In addition to the elderly, the use of elderly wardrobes may also include family members, caregivers and housekeeping personnel. Therefore, it is impossible to metaphysically classify the design of the elderly wardrobe and the design of the ordinary wardrobe. Generally, furniture has a longer life than general industrial products. Older people may experience changes in physical function and living ability during the life of the furniture. Therefore, the design of the elderly wardrobe should also be designed for potential needs, so that the wardrobe can be adjusted as needed.

4.3 Accessibility design principles
Accessibility design is one of the main principles of elderly wardrobe design. The decline in motor function, sensory function and cognitive function of the elderly, resulting in audio-visual disorders, behavioral disorders, and disturbances of consciousness. The design of the elderly wardrobe is based on the study of the behavior, consciousness and movement of the elderly, making it a safe and convenient bedroom storage space for the elderly.

5.The design of aging wardrobe

5.1 The size design
Wardrobes are an important storage space for the elderly. If the size of the wardrobe is not suitable, it will cause inconvenience to the elderly. The overall size of the wardrobe is mainly determined by the size of the storage item and the size of the human body[3]. Therefore, the size of the aging wardrobe should be determined according to the size of the stored items and the size of the human body. The GB/T 3327-2016 "Main dimensions of furniture cabinets" [4] stipulates that the internal depth of the wardrobe should not be lower than 530mm, the height of long clothes is not less than 1400 mm, and the height of hanging short clothes is not less than 900 mm.

*5.1.1 Height dimension.*The bedroom and living room are the most frequent spaces in the home and are the place where large furniture is concentrated. The height of the residential floor, especially the height of the bedroom, has certain constraints and restrictions on the height of the wardrobe. In order to ensure the basic use requirements, according to the GB50096-2011 "Residential Design Code", the indoor net height of the bedroom and living room should not be lower than 2.40 m, and the local net height should not be lower than 2.10 m[5].

*5.1.2 Depth dimension.*According to the provisions of GB/T 3327-2016 "Main Sizes of Furniture Cabinets" and the common sizes of hanging and stacking clothes, according to the actual situation of the elderly using the wardrobe, the depth dimension of the old wardrobe is initially set to 550mm.

*5.1.3 Width.*There is currently no uniform standard for the width of the wardrobe, which is mainly limited by the width of the wall in the bedroom. In "Research on the Modular Design Method of Composite Wardrobe for Mass Customization", Dr. Lee rounded and normalized the width dimension data of 40 bedroom custom closets and positioned the size difference to 400. Then, according to the common size of the internal functional parts of the wardrobe, the width of the single door wardrobe of the artificial board and the solid wood composite wardrobe is determined to be 400 mm, and the width of the double door wardrobe is determined to be 800 mm. Finally, we can use a modular combination to match the 1200 mm, 1600 mm, 2000 mm wide composite wardrobe.

5.2 Functional space design
The internal functional space division of the aging wardrobe should be based on the principle of

clothing classification and storage, while taking care of the behavioral characteristics of the elderly group. We chose the two typical behaviors of drying quilt and trying clothes to decompose the action （as shown in Table 2） in order to be closer to the actual needs of the elderly.

5.2.1 Change the position where the quilt is stored. The top of the closet as the bedding storage area is the most common practice for wardrobe partitioning (as shown in Figure 1), but it will be more difficult for the elderly to use. From the action decomposition of Table 2, we can know that if the elderly want to take the bedding in the closet to the balcony to dry, it takes about 9 steps. However, most elderly people suffer from osteoporosis, so standing in a chair and taking a bedding is very dangerous for the elderly. If the elderly are required to complete this behavior alone, the difficulty coefficient and the risk factor are increased. We moved the position where the quilt was stored down to 0-600mm, which required space for squatting, and also designed a locker with pulley for the bedding. After adjustment, the action decomposition steps of the elderly quilt are reduced to 5 steps, and the dangerous action is cleverly avoided.

*5.2.2 Application of liftable clothes rail.*The top space of the covered wardrobe can be used to lift the clothes rail, and the hanging storage space is added in the wardrobe to reduce the stacking storage space. This form of design will guide older people to hang clothes for storage. Firstly, it can intuitively show the storage location of clothes to the elderly and solve the problem that the elderly forget the storage location due to poor memory. Second, a lot of clothes hang up space than stack less, the same size of space hanging can store more clothes than stack, increase the storage space of the wardrobe. Third, hanging and putting can better keep the wardrobe tidy, solve the problem of easy to mess up the wardrobe when storage clothes. In addition, clothes such as underwear and socks that are not suitable for hanging, drawers and stacking areas are also provided for storage.

Table 2. Decomposition of the action steps of the elderly using the wardrobe.

Behavior	Adjust	Action decomposition step
drying the quilt	before	stand on a chair—take a quilt—Stand on the floor—place the quilt—drying the quilt—take a quilt—stand on a chair—stand on a chair—put it in the closet—put the chair in position
	after	drag out the quilt box—drag it to the balcony—drying the quilt—take a quilt—put the box in position
trying on clothes	before	looking for a dress—get dressed—go to the room with the mirror—look in the mirror—return to bedroom—take off clothes—put the clothes in position
	after	pull down the adjustable hanger—take clothes—get dressed—open the mirror—look in the mirror—take off clothes—put the clothes in position

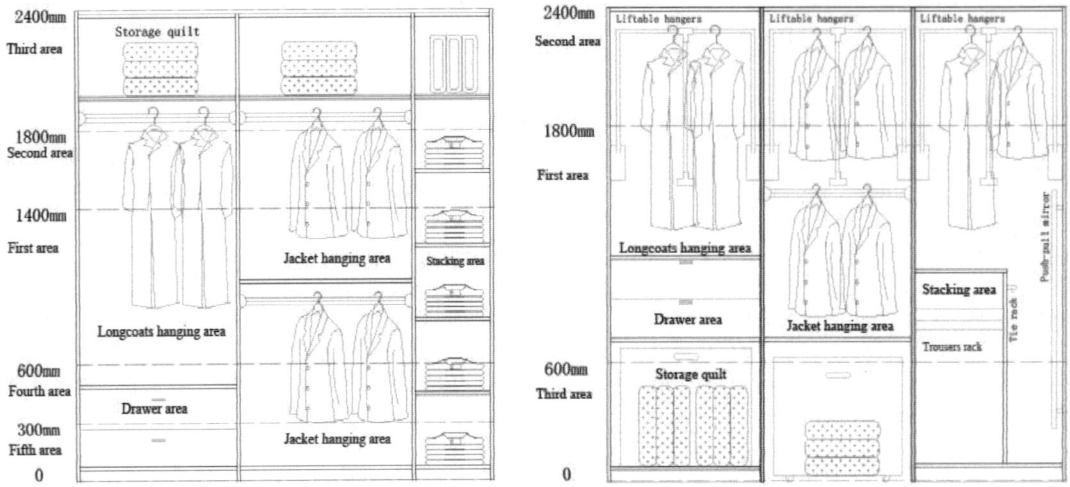

Figure 1. The common function modules of the wardrobe and the function modules of the wardrobe after adjustment for aging.

5.3 Detail design

More attention should be paid to details in the design of the wardrobe suitable for aging, so as to provide the elderly users with more intimate experience in vision, hearing and touch. According to the survey results, the elderly can't see when looking for things in the closet because the light is too dark. Therefore, we can install the sensor light in the closet. Second, placing a mirror inside the closet makes it easier for the elderly to use. Choose hydraulic buffer hinge, can avoid because drawer and move a door to close quickly and clip a hand, reduce wallop while still can accomplish mute. During the life of the furniture, you may experience changes in the physical function and living ability of the elderly. Considering the possibility of using a wheelchair in the future, leave a space of about 200mm in the bottom of the wardrobe and about 320mm in height to accommodate the wheelchair pedal. Even if the elderly use a wheelchair, it is easy to operate.

6.Conclusion

Many problems encountered by elderly users in the use of the wardrobe have greatly affected their experience. Reasonably adjust the functional partition of the wardrobe from the ergonomic point of view. Use the liftable hangers and increase the proportion of hanging storage to guide the storage habits of elderly users using the wardrobe. Propose targeted solutions such as insufficient storage space, inconvenient things, and confusion. With the continuous improvement of China's aging population, the demand for professional and aging furniture is also growing. This is an opportunity and a challenge, and it has placed higher demands on the market and practitioners. Only the practitioners of the furniture industry understand the actual needs of elderly users, solve problems in a targeted manner, and be people-oriented in order to gain a foothold in the elderly furniture market.

References

[1] Haitao Hu. (2005) Anthropometric measurement of The Elderly. Dissertation, Tsinghua University.
[2] Tao Tao, Huanyi Xu, Yanan Zhao. (2018) Design and application of old bedroom furniture based on physiological characteristics of the elderly. Forest Product Industry, 45: 47–49.
[3] Yingyi Xu, Zhihui Wu. (2018) Research and evaluation of ergonomic wardrobe products. Furniture, 39: 19–24.
[4] GB/T3327-2016, Main Dimensions of Furniture Cabinets.
[5] GB50096-2011, Code for Residential Design.

IOP Conf. Series: Materials Science and Engineering **573** (2019) 012059 doi:10.1088/1757-899X/573/1/012059

COMMUNICATION ABILITY IN THE DESIGN PROCESS OF THE IMPLEMENTATION PROCESS

Chen, Xian-Wang [1*], Lin, Mei-Ting [2] and Huang, Liang-Peng[3]

[123]School of Media & Art Design, Wenzhou Business College, Wenzhou, Zhejiang, 325035, China

[1*]mark1231@126.com

Abstract. In the design process, in order to make the whole design process more smoothly carried out, the article puts forward the concept of "flexible design", the core concept of flexible design is communication. From the analysis of basic communication ability to communicative ability, then from communicative ability to communicative ability, and ultimately to effective communication in the design process, it is concluded that good communication itself is an effective means of design. "Flexible design" is not only a concept, but also a way of thinking, reflecting the wisdom of design.

1. Introduction

Design is an interdisciplinary and comprehensive subject which contains many factors. Usually, a good design work is not done by one person, but depends on a solid design team. As far as this is concerned, there seems to be a big loophole in the personnel training system of design talents in domestic universities, that is, the lack of team consciousness. The lack of team consciousness leads to the drying up of "flexible design", which ignores the status of "communication ability" in the process of design organization and its role in the design process. In fact, the design of business processes starts not so much from collecting data as from interpersonal communication. Needless to say, "design" is a social industry with strong sense of service, and "market mechanism" and "customer demand" have become important considerations of commercial factors. How to grasp this entry point better depends not only on research and technical difficulties, but also on the strength of communication in the design process.

2. Presentation of Stream of Consciousness in Design Communication

2.1. Design communication is nature implement

Traditionally, excellent technicians are more or less craftsmanship, they pay attention to specialization. Today, the development of the times has catalyzed their rebirth. A good designer must be a generalist. They have changed from the original "craftsmanship" to the present "craftsmanship". If "modern design" is the perfect combination of science, technology and art, then "design communication" is a kind of ability. Human talent can be roughly classified into two categories, one is congenital, which is instinct. The second is to rely on the acquired experience and learning to acquire, which is ability. People with excellent innate linguistic communication skills can naturally add wings to design communication, but this does not mean that design communication skills can not be learned and exercised the day after tomorrow.

Design communication is an indispensable ability to engage in design work. Design in communication and communicate in design. Nowadays, design projects pay more and more attention to communication in detail. Of course, even small projects can not do without communication of big ideas. In short, large brand operation and project design need more innovative design in different details because their foothold is already in the high-grade brand scope, while small scheme design needs a strong design concept and cultural connotation urgently because of their weakness and many other factors. Implantation, thus rendering the finished product of the entire design. Having mastered this principle will help us design communication before taking over the project. The object of design communication often varies from person to person, so effective communication should actively find the meeting point of topic. The joint point of communication is to "create from nothing" and "excel from nothing", and so is the communication in design. In front of customers, we must show our unique design concept and creative thinking, reveal the rigor and professionalism of our design from time to time, and urge the other party to quickly establish information interaction and trust with us.

2.2. Design communication in several noteworthy design element

As information interaction matures, it can gradually intervene in psychological trust. During this period, we must have strong resources as the backing of the designers themselves, that is, based on the investigation and comparative analysis of the objective data of the design objects, the vertical design planning and the horizontal design comparison must be done. Specifically, it can be divided into the following four aspects: First, design communication should pay attention to variability. To communicate effectively, we need to pay attention to the way of language. Designers need to learn to adjust the state and change the way of communication. Diversified language helps to make the communicators and different people talk to each other deeply and achieve the purpose of communication. Second, design communication is a two-way behavior. One should be good at expressing, the other should be good at listening. Through the process of communication, listening, feedback and re-communication, listening and feedback, the theme and solution of communication are defined. Designing communication is a process of seeking mutual evidence. Only when the two sides cooperate actively, can the purpose of communication be realized. Furthermore, we should establish the preparatory work before design communication, and the communication content should be clear. Lack of preparation before communication will inevitably lead to confusion in the process of communication, which wastes the working time of both sides and is not conducive to the solution of problems. Finally, it is worth noting that the essence of "design" is the process of finding problems and solving them, and we can not only regard design as the production itself but also.

In a sense, designers need to play the role of customers and users from time to time, standing in the other side's position to peep at the rationality of their entire design. That's why it's always emphasized that the purpose of design is people, not the meaning of products. From experience to experience. Designer's greatest role is to help customers and users extract the most extreme and effective information, and then comb and integrate these information, combined with the "people-oriented" and "user-centered" design concept to convey the practical functions and formal aesthetics behind these information.

3. Communication ability runs through the whole design process

3.1. Establish the "3E → 3A" principle

The primary task of design communication is to develop the ability of "design expression". Design expression is a kind of presentation reflecting design. Its basic procedure is "observation expression elaboration", which is the "3E" principle. This principle refers to the design of the main body, the designer through language communication and non-verbal communication. Non-verbal communication, also known as body language, refers to people in the process of communication, do not use language as a tool to express opinions, but use other non-verbal ways to convey information. The function of non-verbal communication is to convey information, communicate ideas and exchange feelings. A deeper understanding of the design project will enable both parties to enter a reciprocal state of "Accept Agree Admire Appreciation". That is the "3A" principle. The promotion and evolution from "3E" to "3A" reflects the rational expression of design.

Design expression is to enrich the communication of the whole design. Therefore, as far as the external form is concerned, the ways of design expression can be varied. Computer media demonstration, entity model demonstration, case analysis and comparison, Market Research positioning and so on, according to specific design projects can choose specific programs to express. In addition, another important task of design expression naturally requires language communication to carry. Because the designer's language is very persuasive for customers, gradually penetrate the designer's own design concepts and design methods on the basis of gradually building trust with each other, and constantly remind themselves to think more from the standpoint of the design object, so as to achieve the most direct and effective design expression. Firstly, an invisible communication bridge is built between the designer and the design object, which will help to achieve a programmed realization in the overall design process.

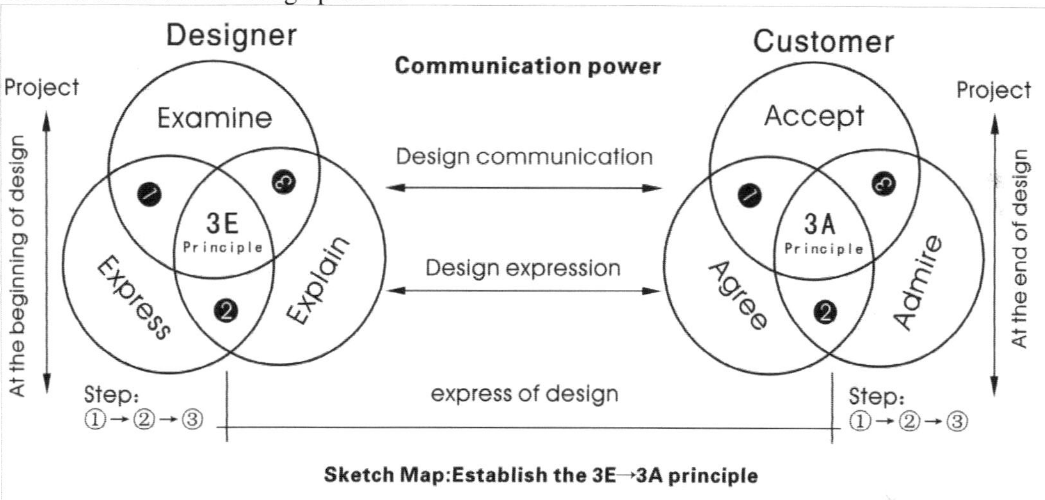

Figure 1. Establish the 3E→3A principle.

3.2. The design process of internal communication

Design refers to the design of communication in addition to outside customers and users, need to focus on the other hand, is designed for internal communication. Design is the design of internal communication in the organization or design team communication and exchange between higher and lower levels, between departments, are required between designers and designers of high-quality communication. As mentioned earlier are often a good design by a design team to complete, from the collection of material and market research, positioning creative ideas, artwork design, the project involves the design of the technical aspects, production technology, materials materials, post-manufacture and design feedback and so on. All aspects of the seemingly self-contained, but in reality is the need to complement each other to carry out the organic integration. Design processes in the 'stabbing outside must first be peace within "a sense still relies on" communication. " Communication design and creative process has become so passionate, so that the whole design process of communication becomes more smooth, more communication and exchange of the design in place.

Effective communication can eliminate a variety of interpersonal conflict, acts of good communication between people, so that the emotional interdependence between teams, a high degree of unity in the design concept, clear and bright on the project to achieve the smooth flow of information to change the team information barrier between the phenomenon of enhanced solidarity and cohesion among the team, team-building and lay a good inter-personal basis. I personally think that the design team's internal communication in the "parallel communication" is essential. Parallel communication and must be result-oriented. Designers and engineers, designers and technicians, designers and designers, between departments, etc., the most important thing is to communicate the results. The result is caused by a lack of information on the effectiveness of the attenuation, and finally affect the workload between different people. For example: a changes in demand, if not the first time to complete a thorough understanding of project team members, will lead a working part of the last member of a huge backlog of work unknown and uncertain changes to the content. Finally, internal communication in the design process should be noted that reducing the level of communication, because the transmission of information, the more participation, the greater the distortion of information. Therefore, it is best to communicate both direct interviews, in order to make the information timely and effectively with each other in passing, to the effectiveness of communication.

4. The Significance of Communication Ability for the Implementation of Design Process

4.1. Communicate ability within the conservation value of design

Generally speaking, the designer's communication skills develop from three aspects: First, to improve the ability to understand others; second is to improve skills; third is the systematic upgrade their professional level. Review of domestic design community as well as domestic design education, particularly in domestic design education, in terms of this point does seem stretched. Perhaps this will become the design of the education system in the development of a new focus.

China's industry design is not lack of design of processors, but full of wisdom and charisma of the innovative composite design talent. For a designer who Sublimation from the artisan type designers to "originality" of the designer, no doubt is to have a more lengthy process of development. Will work hard to go "do" rather than know how to extract from the practice of design thinking person that the "unimaginative", but only on paper mouth theory has never been to run the "front lines" of people said that the "pseudo-Carpenter "really good designer does not depend on a pile of thick books or a few will be able to take time on the practice. I do not exclude the need for both, but the key is whether the creative design thinking and commander of the overall communication skills. Without effective communication, even the world's greatest ideas will be wasted, good communication is a key step to achieve good design one.

4.2. Communication power of the external expression of the role in the design process

Mature design of communication is the client and designer mutual respect, mutual cooperation in a mutually interactive relationship. Customers must make their own requests and suggestions, that his style of intention to make the best in the design of the program. Such as in the entire product development and design process involves two major means of communication: interpersonal communication through the realization of linguistic signs, people and products to achieve through the form of symbolic communication. These two means of communication to establish a common basis of communication theory, while the performance of different media forms, the designer's focus is different. Communication not only in administration, marketing, business management and other industries has been a great attention but also in the modern design process is reflected in every possible way. Communication and information transfer between the two sides interact in the process, the purpose is to achieve mutual understanding, mutual recognition. Designers to communicate with these different objects constantly adjust the way to get on the feasibility of the product, technology, economy and other information. Information can be adjusted based on these design ideas to modify the design details.

For workers who engaged in the design is often just a fancy design and production of consistently, blindly emphasized design affair, has been designed to focus on the internal system, are often easy to overlook in the design of the body outside the "wisdom" and "strategy" . There is an urgent need for a rational approach for support, in particular about the rigorous and rational design of better than have to play its flexible side. How best to achieve the design of the finished product market acceptance, how to better play designs inherent charm of the extension of degree, how best to promote the audience's understanding of the degree of the design process, how we are able to travel more smoothly walking in the design both inside and outside it? In a word: communication. Communication ability in the design process to achieve the intensity of the program determines the design sense to start the business and development.

5. Thoughts on Future Design from Communication Ability

5.1. Training and Design Education of Communication Ability

Kenya HARA's book which named "Education of Desire - Creating the Future with Beauty Consciousness", mentioned that design is "education of desire", and the so-called commodity and environment are actually "fruits" harvested from the "soil" of people's desire. In order to produce high-quality commodities and create a good environment, fertile soil is indispensable, that is, high-level desire must be realized. And design is the essential factor that fundamentally affects desire. Carefully designed works can awaken the consciousness of sleeping, and then desire changes, which affect consumption patterns, resource utilization forms, and lifestyle changes.

There are undoubtedly two directions for the development trend of design in the future: one is based on the intelligent, systematic and scientific design after the rapid development of natural science; the other is based on the humanistic, emotional and sustainable design inspired by the humanities. This seems to be the two branches of the future development of "design", but in fact complements and promotes each other, because design is never a solo heroism, but a spirit of cooperation and teamwork. If we want to make the future "design" go further and better, besides the policy support at the national level, the whole society's design institutions, educational institutions, design organizations and designers themselves should devote themselves wholeheartedly to it as a great cause and tirelessly strive to build their own research paradigm and design alliance, which is bound to become a contemporary design alliance. Characteristic flag bearer of Chinese design. This is our rational understanding and attitude as a new generation of designers and design educators. The so-called "Wanzhang Tall Building" rises flat. From this, we can see that the future of China's design education is still a long way to go. Starting from a small point of view and from a small point of view is also our strong response and self-encouragement to the design education in this era.

In view of this, few courses such as Design Communication and Design Management are offered in the current domestic design teaching.From the perspective of design education, it is obvious that we have not done enough in this respect to cultivate the comprehensive qualities that future qualified designers should possess. We hope that in the future design education can be increased. Only by shaping and cultivating students'communication ability, strengthening the cooperation between schools and enterprises and professional practice, and reinforcing the idea of integration of production, teaching and research, can the study and cultivation of "communication ability" be truly realized. After all, design specialty is a highly applied specialty.

5.2. The Origin of Regression Design

With the advent of 5G era, new technologies such as big data science, cloud computing and artificial intelligence have sprung up, and people's desire for a better life seems particularly strong. All this seems to indicate that the design of the future will have a greater use. Then, in addition to professional design knowledge and skills, future designers should also know how to make good designs with both "demeanor" and "temperature" and human feelings.Design can be simple or complex.

There is no absolute gap between complex design and easy-to-understand design. That is to say, the boundary between design can be eliminated, because in essence, "design" is a materialized extension and Inspiration of human way of thinking, and with the continuous sublimation and innovation of people's way of life that is may be one of the contexts of the design language system with Chinese characteristics.Designing is nothing but often arousing people's unconscious emotional fluctuation or weak memory through different forms of expression to produce a "seemingly similar" cultural identity and behavior habits, so you are moved. Designing is nothing but just like picking up all kinds of shells on the endless beach, and then making them into all kinds of exquisite products locally, so that they have a "1 + 1 > 2" mode of thinking and behavior. This is the logic and presentation of design paradigm in the new era background, and also the self-repair and perfection of design connotation under the inspiration of the new era. Through design, we can make "unreasonable existence" become scientific and reasonable, make "existence is reasonable" become more humane, and make people's growing need for a better life and yearning for more possibilities.

References

[1] Reyner Banham. A Critic Writes:Essays by Reyner Banham. Los Angeles:University of Califomia Press,1999,P.5.
[2] Victor Papanek. Design for the Real World: Human Ecology and Social Change, New York, Pantheon Books,1971,p.4.
[3] Donald Albrecht.The Work of Charles and Ray Eames:A Legacy of Invention.New York:Harry N.Abrams Inc,1997,p.19.
[4] Liu Xihui. Design Management [M]. Beijing: Peking University Press, 2015,95-99.
[5] Yang Xianyi. Design Management Science [M]. Beijing: China Construction Industry Press, 2014,102-108.
[6] L.C. Deng design management: design of the organization, communication and collaboration [M] Taipei: Asia Pacific Press, 2003 books,108.
[7] G.Y. Liu Design Management [M] Shanghai: Shanghai Jiaotong University Press 2007, 111-119.
[8] H.Q. Chen Anquan Chen Designed to express the principle of effective communication [J] Wuhan University of Technology (Social Science Edition) No. 15 03 2002.
[9] Gavin •Ambrose Yingchun Zang, He Bei translation Design of thinking: the design of effective communication and creative strategy [M] Beijing: China Youth Press, 2010.
[10] Kenya HARA design in design [M] Zhu E, translation, Shandong: Shandong People's Publishing House, 2008,117-125.
[11] Kenya HARA Masayo AVE Dialogue in Design [M] Zhu E translation, Shandong: Shandong People's Publishing House, 2010,54-65.

IOP Conf. Series: Materials Science and Engineering **573** (2019) 012060 doi:10.1088/1757-899X/573/1/012060

The impact of design competition on design education

Shao-Feng Wang

Department of Product Design, XiaMen Academy of Arts and Design, FuZhou University, XiaMen, FuJian Province, 361024, China

262273724@qq.com

Abstract. Traditional design education needs to keep pace with the times when it comes to changing the ever-changing job market. It is the key to avoiding the pains and mistakes of transformation in the face of various opportunities and temptations. Taking the design competition as an example, this paper analyzes the game relationship between design education and design competition from the perspective of historical background and development trend of the design competition. Combine the case study method with the case method to summarize, analyze and discuss different design competitions. It is finally concluded how design education should objectively treat design competitions. Let design competitions better serve the strategy of design education.

1. Introduction

On April 7, 2018, the WeChat public account of the "Humanities Tsinghua Forum" published an article entitled "The Father of Chinese Industrial Design: Liu Guanzhong: The Red Dot Award is a commercial organization to deceive Chinese people" by Li Danyu [1]. Liu Guanzhong, a professor of industrial design at Tsinghua University, said in an interview in this article: "The progress in Chinese design does not require the use of red dot awards to recognize, even if you do not participate in the red dot award, Chinese design is still improving. Subsequently, the Red Dot Design Awards Operating Agency (Red Dot Limited) on April 12, 2018, Björn Steinhoff, Director of Communication and Public Relations, commissioned the Xiamen International Design Week to issue a statement containing the following three points: (1) The Red Dot Award has a long history. Over the years, it has developed itself by maintaining the professionalism and the objectivity of the judging; (2) The judging criteria are strict, and the award is difficult. In 2018, 6,300 works participated in the judging, and one of the "Red Dot Star Award" was awarded a prize rate of 0.015%; 45 of the "Red Dot Honorable Nomination Award" was awarded a winning rate of 0.71%; There are 69 works of the Best Design Award, and the winning percentage is 1.1%. The number of works that won the Red Dot Award is 1684, and the winning rate is 26.73%. A total of 1,798 works received different levels of awards, and the overall entrant's work was winning percentage was 28.54%. (3) Participants are voluntary, our entry fees and benefits are open and transparent, and our competition can promote their works well.

After understanding the views of both parties, it is not difficult to conclude that the contradiction between the two sides is on the surface of the red dot award high entry fee and higher winning rate, and the real contradiction lies in the lack of a credible design competition in the country leading to design The evaluation criteria are controlled by foreign awards. The invisible, design competition has a profound impact on traditional design education. As an education practitioner, on the one hand, we should make good use of the incentives of the design competition to improve students' enthusiasm; on the other hand, we must also prevent "teaching on behalf of the train." It is misleading of short-term

IOP Conf. Series: Materials Science and Engineering **573** (2019) 012060 doi:10.1088/1757-899X/573/1/012060

utilitarianism. This paper discusses the game relationship between design competition and design education from the perspective of design education. We are correcting the theory of competition and anti-competition, returning to the inner education-based thinking, establishing students' correct design values, and rationally screening and using external aids to promote the positive development of design education.

2. The development background of the design competitio

The following is a three-part discussion of the ins and outs of the design competition: (1) the background and definition of the birth of the design competition; (2) the status quo and growth of the design competition; and (3) the impact of the design competition on design education.

2.1 The background and definition of the birth of the design competition

1. The origin of the birth of the design competition

The design is commercial art. The design industry is a production collaboration method born after the industrial revolution. It not only drives the prosperity of the manufacturing industry but also promotes the growth of the surrounding affiliated sectors. The design competition is to grow in this environment. [2]

The early design competition was a public welfare act supported by the design industry. The main body of the undertaking was a government organization or an industry association. The first IF Design Award was held in 1953 at the iF Industrie Forum Design. In 1955, the German Design Association Design Zentrum Nordrhein Westfalen founded the Red dot design award; In 1957, the Japan Industrial Design Promotion Organization (JIDPO) founded the Good Design Award in Japan; in 1958, the Australian Industrial Design Association hosted the GOOD DESIGN AWARDS. 1968. BraunPrize, the German company, founded BraunPrize, Germany's first international design competition. In 1980, BusinessWeek and IDA (Industrial Designers Society of America) IDA (International Design Excellence Awards); in 1985, the Korean government organized the Korea Good Design Award (Good Design Products Selection) is designed to drive the development of the design industry to the high end of manufacturing. The high design competition witnessed the growth of the design industry and continues to be held to this day. Since then, various design competitions have appeared in multiple places at home and abroad, presenting a hot market(See Table 1).

Table 1: The establishment time of major industrial design awards

	IF	Red dot	Good Design Award	Australian Design Award	BraunPrize	IDEA
Established	1953	1955	1968	1958	1968	1980
Organization	IF Forum Design Company	German Design Association	JIDPO	Australian Industrial Design Association	Braun Company	IDSA&American Business Weekly
Commercial	Yes	Yes	No	No	No	Yes

The following design competition has both profitability and public welfare purposes. The government-sponsored design competition maintains the real significance of general welfare. Part of the design competition hosted by the company is the campaign for internal design elements, and the other part is the company that operates the award independently.

(1) Definition and classification of the design competition

The purpose of the design competition is to award awards in an open and fair manner. According to the Oxford Advanced English-Chinese Dictionary (98 editions, 84 pages), the award is defined as follows: "A judicial decision, esp. after arbitration," simultaneously edited by Gemser and Wijnberg. The book Economic Concept of Industrial Design Awards: A Conceptual Framework states that a prize consists of three elements: (1) the organizer of the competition: responsible for formulating the rules of the competition and providing substantial encouragement, for example, Money or honor. (2) The formulation and reviewers of the competition evaluation criteria: they are responsible for the regular screening of outstanding works according to the competition rules. (3) Participation in the design competition: the beneficiaries of the award [3] .

2.2 Development Status and Growth Mode of Design Competition

1. The status quo of the design competition

Due to the development of the economy and the popularization of design education, the people have a higher pursuit of quality of life, which indirectly promotes the growth of design competition. Regardless of the quantity and quality of the parametric works, or the improvement of the design business return, the facts have been proved, that is, the design competition is worthy of attention and input. [4]

Traditional design competitions have also evolved in the course of development for their respective purposes. The first category is a non-profit competition aimed at promoting the development of the design industry. They maintain their authority and brand power and are still organized by the government or industry associations. For example, "Good Design Award" and IDEA (International Design Excellence Awards).The other category is to achieve the profitability of the design brand commercialization. The design competition organizer has been transformed from an association structure to a corporate organization.

Whether it is a commercial or public welfare design competition, it is favorable for the progress of the business society and the growth of the human population. Therefore, the prosperity of the design competition is any way for the government to design the industry, the enterprise for the design return, the university for the design. Education has room for growth. This is a weather vane and deserves our attention and support.

3. The growth mode of the design competition

The design competition dominated by profit-oriented enterprises will be more adaptable and adaptable. The government-led design competition will lag in response to market development. The reason is mainly due to the lack of motivation for growth. The profitable race is always trying to promote it actively, while the non-profit design competition is relatively passive. At the same time, as a means of commercial promotion, it has a tremendous guiding effect on the consumer's purchasing ideology. [5]

Government personnel and industry associations mainly organize the non-profit design competition. Their vision is relatively macroscopic, and they plan investment and material from a strategic perspective. Once launched, it will generate greater design power for civil society organizations. For example, the 2018 Fujian Province Straits Cup Industrial Design (Jinjiang) Competition rewards the award-winning works as high as 2.05 million yuan, of which the most outstanding award is 300,000 yuan. The final entry reached 12,351, and this competition was only the fifth, showing that the government-driven design competition is growing very fast.

The growth of the for-profit design competition is more free and diversified, and their benefits are as follows:

(1) Phased registration fee, breaking the high registration fee of the one-time buyout in the past, starting from the lower initial review fee to attract more contestants, and then reviewing fees, final evaluation fees, award fees and receiving payments The variety of names also made the organizers profitable.

(2) Cooperate with the government or enterprises to outsource some of the awards and collect fees. For example, the IF Talent award for students and designers who graduated within two years from the previous year is changed from one year to one year. Twice. Each time there are about four design propositions, two of which are outsourced to companies or associations. The companies involved in the proposal include Samsung, Hansgrohe, and Haier.

(3) Join-style roving exhibition. In the presentation of award-winning works, on the one hand, you can get support from the local government. On the other hand, you can invite local enterprises to join the exhibition and charge fees. This is the value of the brand.

IOP Conf. Series: Materials Science and Engineering **573** (2019) 012060 doi:10.1088/1757-899X/573/1/012060

(4) Customize awards for specific countries or regions. For example, from 2003 to 2001, the IF Design Awards Organizing Committee customized the "IF China Design Award" for China1. Similarly, the German Red Dot Awards Agency is also optimistic about the booming Chinese market. In 2015, China Good Design (CGD) was established in China, which is also a new international design award for China. Of course, these are all commercial activities, which promote the development of the local design industry on the one hand and bring benefits to the organization on the other side.

4. The impact of design competition on design education

4.1 The design competition is in the ascendant

The emergence of the design competition is not only the meaning of the design of the popular science popularization but also the commercial return for the participating enterprises, which inevitably affects the design education of colleges and universities. [3] In 2010, Taiwanese players won one-tenth of the total number of awards in the IF Concept Design Award (later changed to the IF rookie award), ranking first in Asia. Even the main courses and evaluation standards of some universities. The main topic of this paper is to observe how design competition affects or changes the traditional teaching methods and rules from the perspective of the development of design education in colleges and universities. And hope to find the law from it, and take its essence to its trash, let the design competition actively promote the healthy development of design education. [6]

Analysis of the reasons for the design competition: [7]

(1) The design standard is vague. The ambiguity of the quality standard of design art education leads the society to pay more attention to explicit educational achievements, such as techniques and advertising, thus neglecting the core concept of design education, that is, the training of design thinking and the cultivation of design taste. [8]

(2) Pressure on college admissions promotion. There are too many colleges and universities to start designing, which leads to the survival of the students. Many colleges and universities use awards to promote their teaching achievements.

(3) Student employment advantage. There are some competitive competitions in the student's portfolio that have a competitive advantage. [9]

(4) Government promotion benefits. The local government's investment in manufacturing transformation and the concept of welcoming service upgrades have also actively promoted the development of peripheral design industries.

(5) There are three advantages to holding a design competition: (1) promoting corporate brand and image; (2) facilitating the search for talents and cultivating talents; and (3) harvesting the most programs at the lowest cost [10] .

4.2 The negative impact of the design competition

Design competitions are often not necessarily correct because they are limited by the needs of the market and the level of evaluation. Such guidance will affect social awareness and student creation methods. As Norman said: "Design competition is not only a dream machine but also a place for the birth and continuation of lies." [11]

(1) Many colleges and universities use the game to teach. This kind of destructive competition education has delayed students' basic theory and cultivated correct design values. [12]

(2) The transfer of design education standards. Because the level of the design competition is mixed, it is effortless for junior students to lose their design and focus on the design, and lack of independent judgment.

(3) The lack of protection of intellectual property. A large number of creative works are submitted without protection, allowing companies to get a lot of solutions at polar prices. More seriously, the indifference to intellectual property rights will affect the sustainable long-term development of the design industry. [13]

4.3 The positive impact of the design competition

The impact of the design competition on the rise:

(1) The transfer of the key points of the college education. Enrich the classroom teaching methods and materials, so that colleges and universities are closer to the market. At the same time, it also has a greater impact on the student's learning orientation. [14]

(2) Changes in student learning orientation. Let design students recognize the commercial and intellectual value of design earlier. Focusing on method guidance, such as the German Red Dot Award Chairman Peter Zec in an interview with Sanlian Life Weekly (65, pp. 63-165) in 2013, "We want to lead the way in product quality and want to share experiences with people. We have been giving back. For example, we have a Red Dot Research Center, research work. We will share these findings with the entire industry, and we are not focusing on one task. This is not the same as other design awards.

(3) Changes in corporate design awareness. Let the value of design become a selling point, while at the same time revitalizing the traditional handicraft market, on the other hand, let more designers invest in the cause of co-creation.

(4) Changes in government investment forecasts. Let the government realize the value of emerging industries and creation.

5. Corresponding strategies for design education

The design competition is more focused on economic benefits, and design education is more focused on social benefits. We cannot let design education standards adhere to design contest because of short-term profits.

(1) Return to the education standard, emphasize solid theoretical knowledge, establish correct design values, and cultivate students' independent thinking ability and judgment ability;

(2) Careful screening and custom entry. Let the design competition serve design education, not design education and design competition. For students of different grades, they should select events that suit their abilities. Conduct customized training.

(3) Adhere to the academic design standards. Don't prematurely give classroom evaluation criteria to business conduct. The design education focuses on strategic education, and the design competition focuses on tactical applications. Let students apply what they have learned and ultimately establish the right design values.

In short, the role of design competition can make design education more abundant, and the platform of design education can make the design competition develop faster.

Acknowledgement

Fund Project: "Inheritance and Development of Intangible Cultural Heritage under the Rural Revitalization Strategy" funded by the Xiamen Intellectual Property Office and the Fujian Provincial Socialist Theory Research Center (FJ2018B023);Fujian Provincial Department of Education funded "Internet +" background of Fujian Province "Industrial Design Intellectual Property Protection" platform mechanism research (JAS170064)

References

[1] Sohu News, Red Dot Award is used by commercial organizations to lie to Chinese people. 2018.
[2] Wu Pinjie, International Design Competition and the Value of Enterprise Entrepreneurs: A Study of Taiwan's Micro Design Brands, In Art. 2017, Successful University. p. 1-136.
[3] Huang Suzhen, Research on the Benefits of Product Design Awards - Take Taiwan Hand Tools Enterprise as an example, in the Master of Business Administration. 2011, Jinan University. p. 1-41.
[4] Nasar, J.L., Design by competition: Making design competition work. 1999: Cambridge University Press.
[5] Li Zhaoxiang, The Impact of Design Awards on Consumers' Willingness to Buy - Take the iF Design Award as an example, in Art. 2011, Taipei University of Technology. p. 1-94.

[6] Shi Hong, Yang Xiangdong, and Liu Fangwei, Reflections on the current domestic industrial design competition. 2009. 30(1): p. 216-218.

[7] Huang Li, "Packaging, and Design. Miscellaneous. J. Packaging and Design, Ten Years of Growth, Testimonials – Seeing the 10th Reluctant Design Award from the Design Competition. 2009 (2009 06): p. 74 -75.

[8] Lin Peiyu and Yang Qingtian, Image Analysis of Graphic Design Works of Taiwan Student Art Competition - Taking the winning works of the 90-97 academic year as an example. 2009(3): p. 43-59.

[9] Zhou Wanyi, The Impact of Product Layout Design on Attention—Taking Design Competition as an Example, In Art. 2011, Taipei University of Technology. p. 1-160.

[10] Zhang Juan, Brand Effect of Fashion Design Competition. 2010. 24(4): p. 84-85.

[11] Zhang Ying, et al., on the shortcomings and improvements of the design competition. 2012 (2012 05): p. 86-87.

[12] Zhong Yiting, Research on the Impact of Design Competition on Student Learning, in the Department of Industrial Design. 2011, Successful University. p. 1-136.

[13] Zhao Yongzhi and Tongji University School of Design and Art, Analysis of the Motivation of Enterprises to Establish Industrial Design Competition. 2012 (2012 02): p. 192-193.

[14] Xie Peixin, exploring design traits and thinking patterns with G-Mark's award-winning design, in Art. 2013, Chung Yuan University. p. 1-204.g Yuan University. p. 1-204.

IOP Conf. Series: Materials Science and Engineering **573** (2019) 012061 doi:10.1088/1757-899X/573/1/012061

Cross-border and Reconstruction: Exploration of Teaching Mode of Integration of Sculpture Art and Product Design

Zhao Aili[1*] Wang Liming [2], Yu Meng [3] and Qu Qingfeng [4]

[1, 2, 4] Industrial design department, Heilongjiang bayi agricultural university, daqing, heilongjiang, 163319, China

[3] Department of sculpture, Harbin normal university, Harbin, heilongjiang province, China,150000

Corresponding author's e-mail: aili_fiona@126.com

Abstract. Inter-disciplinary integrated teaching method can integrate the teaching staff of sculpture art and product design courses, complement knowledge in teaching, provide students with various learning experiences, develop students' cross-border design thinking, and carry out reconstruction and innovation of sculpture art or product design. This article starts from the collaborative participation of sculpture and product design teachers in teaching, intends to renovate the traditional teaching mode, explores inter-disciplinary teaching, and provides ideas for teaching reform.

1. Introduction

Sculpture is an ornamental and commemorative object with strong spiritual connotation. In fact, the development sculpture trend has been shown for a long time. Pluralistic, varied and fashionable sculpture forms are increasingly renovating and splitting [1]. Product design is a multidisciplinary design activity. This article intends to integrate sculpture art and product design across borders, and the results created can enable sculpture art to walk down the altar of art in the form of products, enter people's lives, increase the exchange experience between sculpture art and people, and explore the cross-border conversion mode between sculpture art and product design in the course. Usually, the teaching process of product design and sculpture design course is undertaken by the teachers of this major alone. In the teaching process, the traditional teaching mode and system of this major are often followed. In the teaching process of product design, emphasis is placed upon the evolution of form and innovative design. There is no more solid and in-depth research in the process of three-dimensional structure of form. In sculpture design course, only the form of product is emphasized while other comprehensive factors are ignored. This article will empower teachers and students with different professional backgrounds and knowledge structures to participate in teaching activities through cross-border integration teaching method. Through integration, students' creative thinking will be developed, and solid modeling ability, analysis and design ability will be improved.

2. Discussion on Cross-border Teaching Mode of Sculpture Art and Product Design

Before the teaching activities are carried out, teachers of product design and sculpture major will jointly form teacher teams. Through the integration of professional knowledge, collaborative development of teaching plans and methods, and clear the final teaching goal is to cultivate students' comprehensive design ability, form factors, functional characteristics, material application and other closely combined with innovative practice, and in the teaching process emphasizes the cultivation of design thinking ability, according to the knowledge structure and characteristics of the two professional students, according to the difference between their respective professional fields to set the teaching content, students through the form of homework to cross-border professional fields to carry out in-depth study, and the learning results will be presented in a new form. The teaching mode of the course mainly includes the following three ways: broadening the perspective of the course, changing the perspective of the course, promoting innovation through cross-border conversion mode, and training image thinking through cross-border themes. They act on every stage of the course and jointly realize the exploration of cross-border integration teaching mode.

2.1.Broaden the Angle of View and Change the Angle of Curriculum

In the process of teaching, first of all, through on-the-spot investigation, the acquired perceptual knowledge should be preserved and analyzed through various forms of records, and transformed into rational knowledge. The content of the course mainly includes basic theory teaching, field research, data analysis and display. Students who major in sculpture and product design are matched to form a group to implement the course content, this gives them a sense of teamwork on the basis of independent value evaluation system. Select cultural and commercial sites in the city for on-the-spot investigation, so that students can understand the application content and practical value of sculpture and product design, and discuss from different professional angles to form preliminary analysis results. The final analysis results will be reported in the form of PPT in class, so that students can share their research results and expand their knowledge. The specific content includes two aspects: one is " object" analysis, such as morphological element analysis, application environment analysis, material and process analysis, function analysis, etc. The other is " human" analysis, such as user demand analysis, display psychological analysis, interaction analysis between " human" and " object", etc. Through these two analysis methods, students can experience the deep feeling of the space or using the product, as well as the different psychological feelings that can be brought to people under different space environments and using scenes. These experiences can be displayed through the way of writing and picture recording. Finally, the comprehensive improvement of students' logical thinking ability and innovative thinking ability can be realized, enabling them to actively think about how to design.

2.2 Cross-border Transformation Model Promotes Innovation

In the process of product appearance modeling design, common modeling methods include stacking, wrapping, twisting, cutting, hollowing out, extruding, folding and others. This is consistent with the modeling methods used in sculpture design. This consistency also provides the possibility for its cross-border integration, and also enables us to clearly see the application of sculpture method in the field of product design and design. This application method can greatly increase students' design interest, expand students' design vision, and also promote the design works they create, bringing better aesthetic taste and artistic enjoyment to people. In the cross-border integration of sculpture art and product design, there are mainly the following ways:

The first is the cross-border conversion of sculpture into a product, that is, originally a sculpture product, which gives its product functions on the basis of retaining its original form, and realizes the feasibility of mass production in the design of materials and processing technology, and converts it into a product. Meanwhile, when sculpture art and design merge into a practical object that appears in people's life, what it gives to the object is not only a natural and romantic skin, but also a disguised handicraft with a special style. Its real value is the communication between sculptors and people through daily touching and using the object, and the transmission of personal experience and emotion [2]. For example, the sound of " sound sculpture" is a typical cross-border exhibit, and its sculptural shape highlights its good use function. See Figure 1.2

Figure 1.2.Sound sculpture product design

The second is that the product is transformed into sculpture across borders. Sculpture works absorb the advantages of product shape or product design elements to create sculpture, and can have the interactive experience process of " human" and " sculpture" to retain some of the functions of the product and form innovative sculpture works. Besides, for the product's own morphological characteristics, on the premise of not affecting the usage, the aesthetic appearance can be more favored by users, thus obtaining higher market evaluation. For example, chair sculpture is the sculpture works transformed from this product across borders. See Figure 3-4 for chair sculpture.

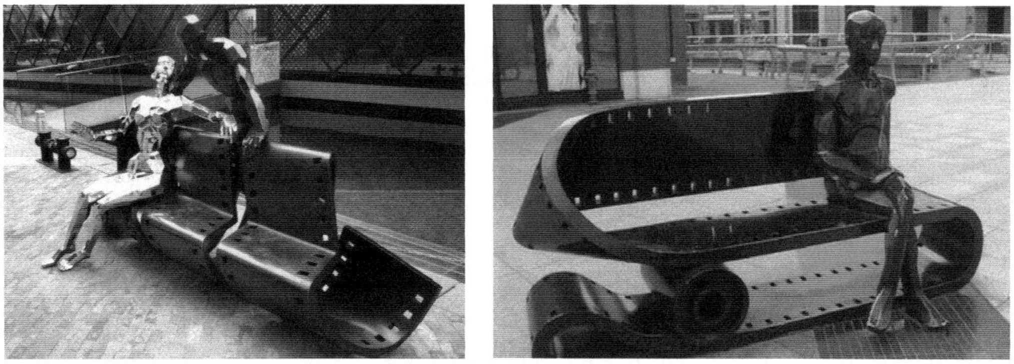

Figure 3.4. Chair Sculpture

The third is the integration and reconstruction of sculpture and products. With the development of new materials and technologies, the application of new materials, new technologies, new processes and even new science has brought plenty room for changes in product shape design. The application of new materials, new technologies, new processes and even new science also endows sculpture with new value connotation, expands the language expression form of sculpture itself, and makes its form more diversified and connotation richer. For example, the products designed by the famous female designer Zaha Hadid are very plastic in material selection and modeling design. The products designed by Zaha Hadid are shown in Figure 5-6.

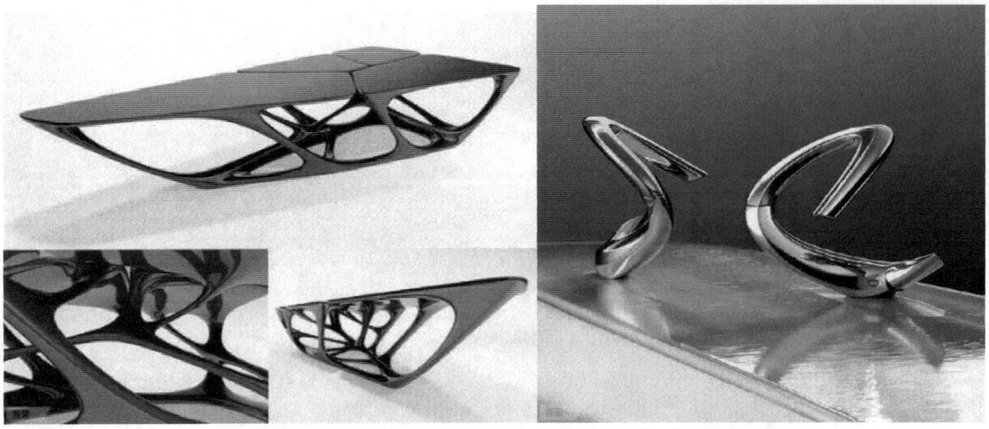

Figure 5.6. Products designed by zaha hadid

2.3 Design Theme to Implement Cross-border Innovation

In the practical part of the course, themes such as " distorted space" and " bionic trace" are selected. All cross-border design activities are carried out around the theme, which needs to be reflected by highly recognizable symbol elements.

First of all, students draw up specific design topics according to the given subject direction, use data analysis, brainstorming and other methods to find basic elements such as design routes and keywords, and then use graphics and diagrams to interpret the content expressed by the keywords. In the end, students form analysis reports based on the summarized data content. All the data content will pave the way for the future design.

The second is to draw the first draft of the next design around the theme and the information obtained in the previous link, and to tightly focus on the theme selected by the group in the design process. Teachers encourage students to try to use reverse thinking, reverse assumptions and other ways to show the form through exaggeration, distortion and other means of intention and comprehensive materials.

The third is the plan evaluation stage. According to the plan selection and evaluation activities among groups within the class in the first draft, the students show their design ideas to other group members and teachers in the form of PPT and design documents. In the process of showing, other group members and teachers give responding opinions, select the best plan, and show the group to record.

The fourth is the arrangement and revision of the plan, and in the subsequent courses, the final selected plan will be discussed and revised according to the questions raised by other groups and teachers in the previous stage. In the ultimate plan display, the characteristics of the product or sculpture should be fully reflected, and the whole picture should be displayed in the form of effect diagram, copy or video.

Through these four thematic trainings, the cross-border integration of sculpture art and product design can not only broaden students' vision, enable students to find new design methods and approaches through cross-border methods, but also broaden their professional perspectives, enabling students to create more excellent sculpture and product design works in communication and collision.

3. Inspiration from Cross-border Sculpture Art and Product Design

Integrating sculpture art into product design is not a modern design method. Many artifacts in ancient China are reflected. Stone Age tools are carved articles for daily life or labor [3]. Many modern industrial designers have well integrated sculpture into product design. For example, German designer Lugi Colani, known as the " design geek", had a solid modeling foundation in his early learning of sculpture. In his sculpture works, we can feel the extremely high modeling quality and strong sculpture artistic style. At present, we have introduced this kind of interactive cross-border exploration into the classroom. Focusing on how to carry out various attempts through integration and innovation, we have carried out design practice exploration on how to carry out design creativity and perfect the teaching method of design products. Teachers and students of the two majors jointly carry out theoretical study and practical work.

This kind of cross-border collaborative teaching can integrate and complement the professional skills of teachers, and at the same time open students' horizons, so that they can see different worlds outside their majors, and learn nutrition for their own use, thus achieving the purpose of improving their design thinking ability. This way of integrating the knowledge of many professional fields in the same course is very different from the traditional way of the whole class to carry out design practice according to the requirements of a teacher. This measure also provides a variety of references for the employment direction of students. Therefore, in the future teaching process, this cross-disciplinary learning mode will be continuously developed, instead of sticking to the traditional fixed teaching mode. In the teaching process, the purpose is to solve the needs of users and real life problems, and to cultivate students' comprehensive innovation ability that not only deeply understand their own professional characteristics, but also can integrate other design professional knowledge across borders.

4. Conclusion

Under this teaching mode, students are provided with diversified design thinking modes through investigation, discussion, design, evaluation and other links. They are no longer constrained by the original design field, but also have the spirit of pioneering and innovation. The final works also show good design results under this open thinking mode. This open cross-border design method greatly propels students' participation and interaction in the classroom, and at the same time it also greatly stimulates students' enthusiasm and initiative, encouraging students to create design works with unique artistic charm.

Acknowledgment

Fund Project: 2018 HeiLongJiang Province Art Science Planning Project -- Research on innovative design based on Sui Ling black pottery technology of heilongjiang province (Project Number: 2018B080)

References

[1] Peng De. (2017) Future Trend of Sculpture . Sculpture.
[2] Yu Ziye. (2015)On the Combination of Sculpture Art and Product Design-Creating a New " Communication Field" . Sculpture.
[3] Wu Shaolan, Yu Siying. (2015) Research on Product Design Combining Function Realization with Sculpture Art . Sculpture.
[4] Blanco Lukic, Barry Katz. (2012) NONOBJECT Design. Tsinghua University Publishing House. Beijing.

IOP Conf. Series: Materials Science and Engineering **573** (2019) 012062 doi:10.1088/1757-899X/573/1/012062

Application of Augmented Reality Technology in Industrial Design

Mei Yun, Nie Qimeng, Wang Fang, Lin Ying, Jiang Haiyang

School of mechanical engineering and automation, University of Science and Technology Liaoning, Anshan, Liaoning, 114051, China

e-mail: meiyunliaoning@163.com

Abstract. Augmented reality is an interactive experience of a real-world environment where the objects that reside in the real-world are augmented by computer-generated perceptual information, sometimes across multiple sensory modalities, including visual, auditory, and haptic. With the full integration of augmented reality technology and industrial design activities, new and breakthrough changes can be brought to industrial design. This paper takes augmented reality technology as the research object, expounds the development status of augmented reality technology, the related content of augmented reality technology, and the application related problems of augmented reality technology in industrial design. Finally, it describes the application strategy and direction of augmented reality technology.

1. Introduction

Augmented reality is an interactive experience of a real-world environment where the objects that reside in the real-world are augmented by computer-generated perceptual information, sometimes across multiple sensory modalities, including visual, auditory, haptic, somatosensory, and olfactory. The overlaid sensory information can be constructive or destructive and is seamlessly interwoven with the physical world such that it is perceived as an immersive aspect of the real environment. In this way, augmented reality alters one's ongoing perception of a real-world environment, whereas virtual reality completely replaces the user's real-world environment with a simulated one. Augmented reality is related to two largely synonymous terms: mixed reality and computer-mediated reality. Augmented reality, by calculating the position and angle of the camera image in real time and adding the corresponding image technology, the physical information which is difficult to experience in a certain time and space in the real world is superimposed through computer science and technology simulation, so that the real environment and virtual objects appear in the same picture or space in real time and are perceived by human senses, and achieve transcendence. Augmented reality technology comes into being with the development of augmented reality technology, but there are great differences between them. First of all, they have different requirements for the sense of submergence and different emphasis in the application field [1-7].

In the three-dimensional space simulated by computer technology, users can strongly feel the simulation of vision, hearing and touch. When applying augmented reality technology to the field of industrial design, it highlights the incomparable application advantages and performance of augmented reality technology. It can realize the observation and operation of things in three-dimensional space without obstacles and blockages, and vividly simulate. The actual industrial production scenario can be well applied in manufacturing, assembly, safety monitoring, architecture, interior design and other fields, which has extremely important practical functions and significance [8-12].

2. Development of Augmented Reality Technology

Based on computer network technology and modern information technology, augmented reality technology realizes the integration of virtual reality technology and modern advanced manufacturing technology [13-15]. It is a highly realistic simulation human-machine interface technology, which can achieve human-machine interaction system based on natural skills and real experience. Industrial design of virtual reality technology has its special performance and characteristics, which are embodied in the following aspects: In the concurrent design structure system supported by network technology, the unification and integration of design, engineering analysis and manufacturing are realized. It can quickly respond to market demand, generate digital prototype in virtual manufacturing environment, and predict and evaluate it. In the industrial design of virtual reality technology, the integration of design, engineering analysis and manufacturing is realized. Constructing three-dimensional virtual digital model can give people a real experience, modify and operate it in the sense of immersion, and better realize human-computer interaction. In the industrial design of virtual reality technology, it can better reduce the modification of design, save the design cost, and improve the design accuracy. Thus, virtual reality technology can be seen. Industrial design can greatly improve the efficiency of capital use.

The influence of augmented reality technology on industrial design is very obvious, mainly reflected in: Influences on the concept of industrial design. Once augmented reality technology is promoted in industrial design, industrial designers will participate more extensively in the whole process of enterprise production and provide services for their own design products. From the current definition of the concept of industrial design in academic circles, industrial design should include not only the shape and function design of traditional industrial products, but also man-machine design. Engineering, product function definition and other aspects, so in order to enable the final quality of industrial design products to achieve expectations, we need to define the application of virtual reality technology from the overall perspective, timely change the design quality evaluation criteria, pay more attention to product innovation and science. In traditional industrial design, designers always use two-dimensional plan to express their design ideas, and use specific computer three-dimensional software to obtain three-dimensional effect maps from different perspectives. In fact, the three-dimensional effect maps cannot fully express the designer's intentions; in contrast, virtual representation. The real technology can take the digital three-dimensional model as the core of the design idea, display the work perfectly in the way of multi-dimensional model, enlarge and rotate the product through software, and deepen the understanding of the relevant personnel to product design.

3. Application of Augmented Reality Technology in Industrial Design

Augmented reality technology can enhance user's visual perception and expand the function of user's visual system. Because of its combination of virtual and real advantages, augmented reality technology has good application prospects in medical, mechanical, design, entertainment, military and other fields. At present, the application of augmented reality technology in various fields mostly focuses on browsing and display functions, and lacks interaction with users. Augmented reality technology is applied to landscape design system. Real environment and virtual design objects interact with each other. Designers need to constantly modify and improve the design scheme according to the comprehensive effect of design objects in real scene. The application of augmented reality technology including： Virtual design, education and entertainment, safety training, advanced manufacturing field, architecture and art design.

The application of augmented reality technology in architecture and art design is in the field of architectural design. Augmented reality technology greatly expands the creative mode of graphic art design. Through the application of virtual roaming technology in architectural design, it can present three-dimensional simulation of urban landscape, residential landscape, interior design and historic building. It can make designers from multiple perspectives. The roaming technology of virtual reality in architectural display space has been greatly expanded to solve the dull and tedious communication defects in the design process. The virtual reality design of three-dimensional modelling and panoramic virtual reality display design can be used to fully demonstrate the innovative thinking of designers and flexibly apply it to different architectural design projects. Using virtual reality technology in interior design can create a virtual interior space of a building and visually express and transmit the designer's design intention and design thinking. It has more advantages than the traditional sand table model. It can realize the full-scale model design of the design space. Users can roam, modify and adjust in the virtual space at will to make it more effective. Humanization and fidelity greatly enhance and enhance the overall control and design level of interior space.

Augmented reality technology can be applied in automobile manufacturing system to find and solve the problems of automobile whole or parts in advance. Through the construction of three-dimensional model of automobile, designers can better experience the comfort degree, driving simulation degree, fault simulation of automobile interior, and improve the rationality and science of design index. Moreover, in the process of virtual development and design of automobiles, we can also realize the three-dimensional and virtualization of automotive modelling, die-making, stamping, welding and assembly processes. With the support and support of network data, we can realize the real-time design of virtual collaboration, and implement virtual experiment testing to predict the overall safety and dynamic performance of automobiles.

As augmented reality technology is more and more widely used in industrial design, it has been gradually applied in augmented reality racing simulator, aviation simulator and other related fields. Augmented reality racing simulator is a 3D racing simulator integrated with augmented reality technology. It uses VR and augmented reality technology in racing game, which makes the game have a new control mode and scene effect. In the mode of operation control, players no longer need to use keyboard to control the car, but to control the car through the specific logo pattern in the camera image; In the scene effect, augmented reality racing simulator is no longer in the virtual scene as traditional games, but the combination of real scene and virtual object, these characteristics will bring new interactive game experience to players. Such a change undoubtedly increases the authenticity and experience of the game process, which is no longer immersed in the virtual environment to experience, but to the direction of human-computer interaction. The injection of this interactive approach allows augmented reality to fully consider how to integrate with this technology when designing their appearance. This is the breakthrough point that the augmented reality racing simulator will ultimately be acceptable to everyone. This orientation, which originates from augmented reality technology and its use characteristics and specific environment, promotes the orderly development of creative activities between people and environment, between people and objects, and enables designers to explore more breadth of thinking. While creating and meeting new needs, it also creates a unique overall style of augmented reality ontology appearance. Feeling the most real three-dimensional scenes in the cognitive and experiential fusion of mind and object, making its concept more vividly interpreted, and integrating virtual into the real world, is precisely a variety of possible materialization processes in the context of human complexity and times.

The integration of augmented reality technology and aviation intelligent manufacturing fully demonstrates the status and effectiveness of the "intelligent window" of virtual reality technology. Based on the actual needs of aviation manufacturing in China, virtual reality technology can be applied to process design, workshop execution, management and other aspects of aviation manufacturing, which has extremely important practical significance. The core of aircraft manufacturing is process design. In this invisible hand, the standardization, controllability and expandable application of manufacturing process can be realized. Specific application items and contents include: immersion process audit. In the application of virtual reality technology in aircraft manufacturing, the first step is to carry out the audit of aircraft assembly process in order to design aircraft drawings. Data model, document and so on are the auditing contents, coordinating opinions and suggestions from all sides, and doing a good job of auditing and evaluating process design in an all-round way. Process designers can completely immerse themselves in the virtual and highly simulated environment without any interference and influence, and realize the interaction with virtual objects. With the application of virtual reality technology, aviation aircraft manufacturing has achieved a breakthrough in "scientific design mode based on Modelling and simulation". In the simulation virtual environment system of multi-source information fusion, simulation analysis of aviation manufacturing process design is realized, such as: digital prototype simulation of aircraft process, spatial analysis and roaming, virtual assembly, interactive virtual reality.

4. Conclusions

Supported by computer technology and network information technology, augmented reality technology realizes virtual and highly simulation of the real world through three-dimensional modelling. When this brand-new technology is grafted with industrial design, it can see the real function and significance of this "intelligent window". Under the integration of network, information and data, it can be clear. To present all-round details of products, realize the networking, systematization and intellectualization of industrial design, enhance the interaction of industrial design, make full use of industrial design resources, shorten the cycle of industrial design and improve the efficiency of industrial design. Augmented reality technology plays an important role in industrial design. By constructing the model structure of industrial design, this technology can further deepen the understanding of industrial design content of relevant personnel, and make relevant personnel further understand the content of industrial design in the way of model virtual presentation. Therefore, in the future work, it is necessary to further study the content of augmented reality technology. Finally, it lays a foundation for further improving the design quality.

References

[1] M.L.Yuan, et al. Augmented reality for assembly guidance using virtual interactive tool(J). International Journal of Production Research. 2008
[2] S.Nolle, et al. Augmented reality as a comparison tool in auto-motive industry(C). IEEE/ACM International Symposium on Mixed and Augmented Reality. 2006
[3] Zhao Luning. Research on the Development of Augmented Reality Technology and Interface Design [J]. Art and Technology. 2017 (03)
[4] Cao Weizhi. New integration of VR and AR technology in the field of industrial design [J]. Grand View of Fine Arts. 2018 (11)
[5] Sun Bing, Li Shenghui. Application of Augmented Reality Technology in Customized Packaging Products [J]. Printing Today. 2018 (02)
[6] Duranger. On the Impact and Innovation of Augmented Reality Technology on Industrial Designers [J]. Industrial Design. 2018 (02)
[7] Wang Jia. Research on product display based on VR-Platform platform of virtual reality technology [D]. Taiyuan University of Technology, 2013.
[8] He Chunguang. Exploration and Research of Virtual Reality Technology in Protecting the Manufacturing Technology of Mongolian saddle [D]. Inner Mongolia Agricultural University, 2012.
[9] Quan ning. Application of virtual reality technology in traditional houses in Kashgar [D]. Xinjiang Normal University, 2014.
[10] Wang Xuanlin. Application Analysis of Virtual Reality Technology in Architecture [J]. Computer Fan, 2016 (07).
[11] Lin Fengping, Chen Bichain, Zhang Yanding. Application of virtual reality technology in medical teaching [J]. Laboratory Science, 2016 (06).
[12] Sun Chao. Talking about the application of virtual reality technology in the field of smart cities [J]. China Public Security, 2017 (01).
[13] Lu Qin. Application of Virtual Reality Technology in Industrial Design [J]. Science and Technology Innovation Report, 2017, 14 (25): 115-116.
[14] Gao Yuanhua. On the Application of Virtual Reality Technology in Industrial Design [J]. Modern Decoration (Theory), 2016 (2): 119.

[15] Ranyang, Zhu Fei, Chen Kang. Application of Virtual Reality and Augmented Reality Technology in Industrial Design [J]. Laser Magazine, 2016, 31 (1): 4-6.

IOP Conf. Series: Materials Science and Engineering **573** (2019) 012063 doi:10.1088/1757-899X/573/1/012063

Application Research of Design Geometry in the Design of High Speed Trains

Ying Liu[1], Yi Chen[2], Chao Yuan[1], Xiaojing Liu[1]*

[1] Department of Architecture and Art, Central South University, Changsha City, Hunan Province, 410012, China

[2] Hunan Park Educational Consulting Co., Ltd, Changsha City, Hunan Province, 410083, China

Email: 81413450@qq.com, 48316365@qq.com , 1037303749@qq.com, 12856385@qq.com

* Corresponding Author: Xiaojing Liu; email: 12856385@qq.com; tel:+86-18974968823

Abstract: The visual characteristics of high-speed trains in China are comparatively not at an advantage, while vehicle serialization and product identification remains to be a major problem. From the perspective of design geometry, This paper analyzes the mathematical composition of modeling curves in foreign excellent models, summarizes the geometric proportional relationship of the vehicle system, innovates a design and operation method of the vehicle body shape, and establishes a new generation of model design based on the geometric law of the line shape of vehicle body so as to enhance the viability of vehicle serialization, and provide a basic premise for design strategy research.

1. Introduction of design current situation

Due to the continuous innovation of technology and the dynamic evolution of the aesthetic concept, the social environment and development strategy put forward higher aesthetic requirements for the design of high-speed trains. The new train is not only a high-tech carrier for the strong development of the national economy, but also an aesthetic materialization that highlights the soft power of culture. [1] After years of technology introduction and absorption innovation, China has realized the technical standardization of independent intellectual property rights of high-speed trains. However, in terms of the status quo of design, China's high-speed trains are still in a state of one shape for one car: The original models of CRH380 series (He'xie)-the largest number of operations-were imported from abroad, and its technical parameters and modeling structure are from three different countries. From the perspective of intuitive visual perception, the contours of each model are obviously different, and the commonality of form recognition is found lacking (*Fig.1*). The newly-launched "Fu'xing" series of CRRC Group comes from the original model designed by CSR and CNR, while CRH400A and CRH400B can't constitute brand recognition with different styles, different shapes, and lack of visual form uniformity. The design framework of the existing car system can't fully meet the image demand of a big country of high-speed train design and manufacture, which hinders the implementation of the national macro policy.

IOP Conf. Series: Materials Science and Engineering **573** (2019) 012063 doi:10.1088/1757-899X/573/1/012063

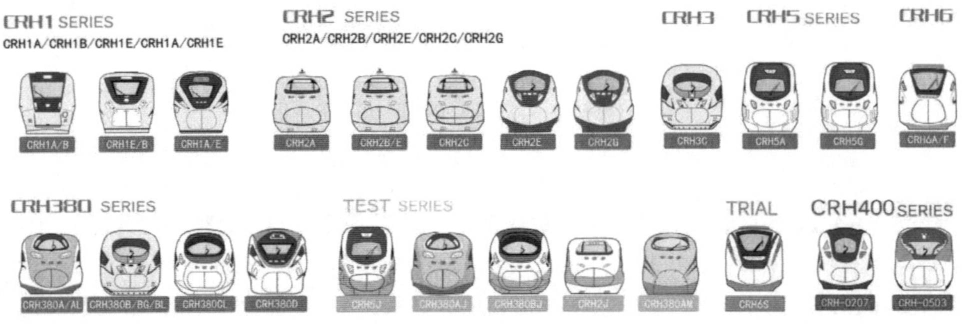

Figure 1. Current situation of CRH

In response to the problem of vehicle serialization and product identification, domestic academic circles have proposed design innovations and ideas from multiple research perspectives. On the whole, various research institutes focus on the design of high-speed trains on three major themes: the arrangement of design process, the metaphor of national culture, and the guidance of humanistic feelings. [2] Within the framework of this research, many design concepts of top-down qualitative studies are based on the theory of aesthetics, sociology, and psychology. [3] As mentioned above, the line shape of the existing models is quite different, and the visual identification system of the series is not formed. The overall qualitative research lacks the basis for conducting common analysis. Therefore, this paper starts with the design case study, analyzes the infrastructure composition of the line shape of foreign excellent models (series) from the top-down research route, explores the line shape rule of the model and the continuity of the serialized matrix features, with a view to making a contribution and forming the material basis of visual recognition of domestic models.

2. The theoretical concept of design geometry

2.1 Academic origins of design geometry

The research and development of design geometry is a process of exploration along with human transformation of the natural world. From the birth of civilization, people's artistic behaviors continue to accumulate aesthetic experience, and gradually form certain mathematical laws. The ancient Greek philosopher Pythagoras discovered that there is a very harmonious geometric proportional relationship in nature. He defines the geometric relationship between natural elements and proposes three kinds of qualitative differences: shape, order, position; If the relationship between the various parts of the object conforms to this law of division, then it has a strict proportionality, which can make people have a harmonious and pleasing visual experience.

In view of the aesthetic pleasure generated by the proportional order method, A large number of In-depth researches were conducted in the field of psychology research, from the perspective of visual balance and psychological balance to answer the psychological reasons for human beings to obtain sense of beauty. The Gestalt Psychology School regards the sensory response to order as the main research field. Gestalt psychologists believe that one of the most basic forms of our sense of order is the sense of balance. Our perception prefers simple structure, straight line, rounds and other simple orders. Such regular patterns are often the easiest ones to see in the complicated outside world, rather than cluttered shapes. The concept put forward by Fechner Gustav Theodora, a psychologist, is more concise: "Beauty is the order in complexity." Adolph Zeising, an esthetician, standardized mathematics of geometric proportion, and he defined the dominant ratio of "21:34" over the art and nature as the golden section. Later, Lalo C. conducted several cross-cultural experiments in a systematic approach to demonstrate the advantages of geometric proportion in human morphology judgment.

2.2 Research category of design geometry

Karl Heinrich Marx put forward in the Economic and Philosophical Manuscripts of 1844: "People know how to produce according to any kind of scale, and know how to apply scales to objects everywhere; therefore, people build following the laws of beauty." [4] Materialistic aesthetics regards scale as the materialized relationship of beauty, and scale is a universal and effective beauty rule formed by human creation of objects. The human creation activity is the exploration process of the scale law and the aesthetic law, and the design geometry is the discipline that studies the universal principle of the visual experience accumulated by people in the aesthetic activities.

Design geometry is not purely geometric in mathematics. Geometry studies mathematical relations and logical relationships; while design geometry focuses on finding visual aesthetics such as equilibrium, coordination, and rhythm. [5] Le Corbusier said in the book Towards New Architecture: "Geometry is the language of mankind... The geometry of art is not a sophisticated mathematics, but the morphological law everyone knows, and it is a standard of scale division that conforms to the form of human beings and nature. The design products that conform to this law will produce a harmonious aesthetic perception." [6] The design geometry research Covers three aspects: the scale of human and nature with the law of order, the basic principle of geometric proportion and the principle of segmentation of spatial vision. The main body of research is the visual relationship between proportion and composition. [7] Therefore, it is a combination of design art and geometry. Its purpose is to use geometric mathematical logic to achieve spatial separation to present harmonious perfect scales and sequences.

2.3 The application of geometry design

Based on the academic research category, design geometry is not simply a means of drawing that describes the visual composition, but an applied theory that makes objects that have geometric relations in space. It contains two-dimensional, three-dimensional spatial mathematical principles and segmentation principles, and standardizes the composition of scales in the morphological design, with the guiding principle of the basic principles. [8] In view of the combination of scientific rationality and intuitive image, the principle of proportional division of design geometry has played a positive role in the construction of modern practical aesthetics.

Today, design geometry has infiltrated into all areas of art design, and the construction of geometric language provides a guidance of many mathematical relationships for design activities. For example, irrational rectangles such as $\sqrt{2},\sqrt{3},\sqrt{5}$, and Φ(golden section) can be divided by horizontal lines and vertical lines to form a derivative rectangle, and a plurality of spiral curves can be generated by auxiliary lines such as tangents and extension lines. After a series of segmentation reconstruction, the various segmentation forms inside the rectangle are converted into new proportional relationships with each other to form a dynamic rectangle of a geometric cycle. Mathematical forms such as these operate the law of geometric relations of order aesthetics and become the paradigm of modern design. [9]

3. The application of design geometry in high-speed trains

The static image of the appearance of a high-speed train is an aesthetic form consisting of linear extension, surface articulation, and functional segmentation. This paper starts with the analysis of the scale law of the linear frame, and studies the most original structural relationship of the linear surface of the vehicle, which is conducive to grasping the basic principles of visual aesthetics.

Compared to other categories of product design, the geometric design elements in high-speed train design are difficult to be visually intuitive. The technical factors of aerodynamics and the consheafdts of high-precision mechanized production conditions make the geometric design language obscured in large-area of fluid surfaces. Therefore, the geometric relationship analysis of the design elements can only be deduced by the method of extracting the lines of the base surface, and the mathematical scale study can be carried out on the basis of simplifying the surface.

IOP Conf. Series: Materials Science and Engineering **573** (2019) 012063 doi:10.1088/1757-899X/573/1/012063

3.1 The geometric progression in the Korean KTX series

As a representative of the Pan-East Asian cultural circle, the segmentation characteristics of blue and white in the Korean KTX series of trains are very obvious. Take the Sea fog train as an example. It is the third generation of the KTX series, in which there are a lot of arc modeling elements, and large-span arc curvature is used in the cockpit and the color segmentation of vehicle body (*Fig.2*). If we deducted the partial arcs, a plurality of standard circles could be obtained. These circular circles of different sizes are arranged in order of small radius, the arc of the headlight is R1, the lower part of the opening and closing mechanism is R2, the upper part of the opening and closing mechanism is R3, and the contour of the window is R4. The blue and white coating split arc is R5. When the radius of the arc R1 is set to one length unit, R2 is 9 length units, R3 is 18 length units, R4 is 27 length units, and R5 is 36 length units; each arc radius constitutes geometric progression.

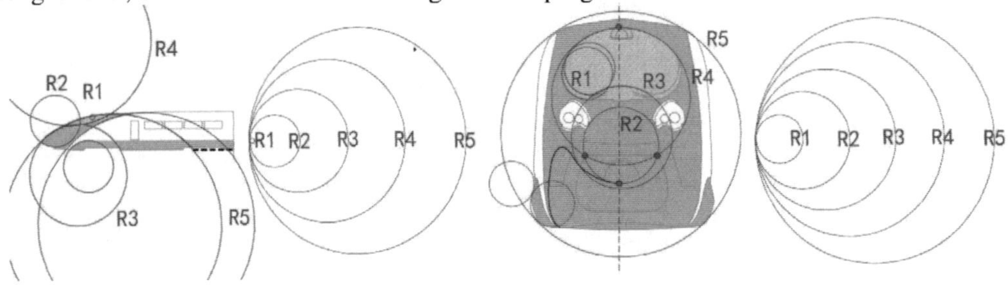

Figure 2. Views of sea fog

The front view of the sea fog leaves the visual impression of the cab windshield as an irregular shape. Following the simplification of the complex form, the contour curve is decomposed into four segments according to the law of curvature tangency. The standard circle of R1, R3, R4, etc. is obtained by the completion of the arc, and the inscribed circle R2 is obtained by using the two end points of the upper edge of the front opening and closing mechanism and the vertex of the front end of the vehicle locomotives and the inner edge of the light as the point of contact; The circumscribed circle R5 is obtained by using the four endpoints of the contour as point of contact. When the radius of the arc R1 is set to 1 length unit, R2 is 2 length units, R3 is 3 length units, R4 is 4 length units, and R5 is 5 length units; the radius of each arc also constitutes geometric progression. Although the dimensions of the various functional components of the train vary, and the curvature of the arc and the length of the arc are different, this geometric line that emphasizes scale and proportion can produce an effective visual response between the various modeling elements, forming a unified modeling paradigm.

3.2 $\sqrt{2}$ Proportional division of the Bombardier Zefiro series

The Zefiro model is a prototype of the CRH380D model introduced in the process of upstream technology. The contour curve of the vehicle has a large curvature and changes significantly (*Fig.3*). In view of the fact that the curvature of the front part of the headline is relatively prominent, the turning points in the side view can directly form a rectangular division of the geometrical progression according to the positional relationship of the vector coordinates. In this matrix, the first curvature change point A and the second curvature change point B are respectively located at 1/3 and 1/2 of the vertical height of the vehicle; the second curvature change is made from the end point of the vehicle locomotives to the third curvature change point C as the horizontal distance length. The point is located at 1/3 of the line in the horizontal coordinate direction.

In addition to the physical division of the geometric progression, the model uses more proportions of the $\sqrt{2}$ segmentation. The arc of the front opening and closing mechanism is divided into upper and lower curvature portions by the vertices of the front vertex D, and the vertical positioning points of the two sections are the lower curvature starting point E and the end point F of upper curvature respectively, and the two forms a rectangle together with the locomotives apex. Similarly, the $\sqrt{2}$ rectangle whose starting position of the light is coplanar with the apex of the front is divided by the upper end of the curvature of the opening and closing mechanism. The two $\sqrt{2}$ rectangles together establish the frame of front curve trend, which provides a basis for the curvature extension of subsequent lineshape such as cab windows and escape window contours.

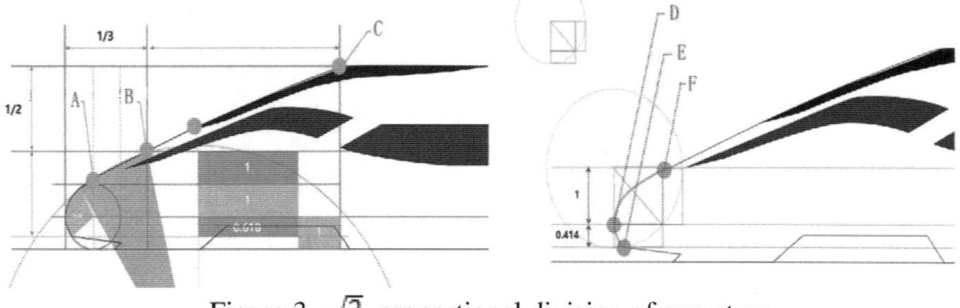

Figure 3. $\sqrt{2}$ proportional division of curvature

In addition, in the horizontal direction of the vehicle body, the front end point A of the window is located on the $\sqrt{2}$ proportional dividing line between the curvature change point B of the roof and the end point C of the front end of the light coating; in the vertical direction, after the end of the vehicle light is coated with the end point D, the car body height is divided into $\sqrt{2}$ proportions (*Fig.4*).

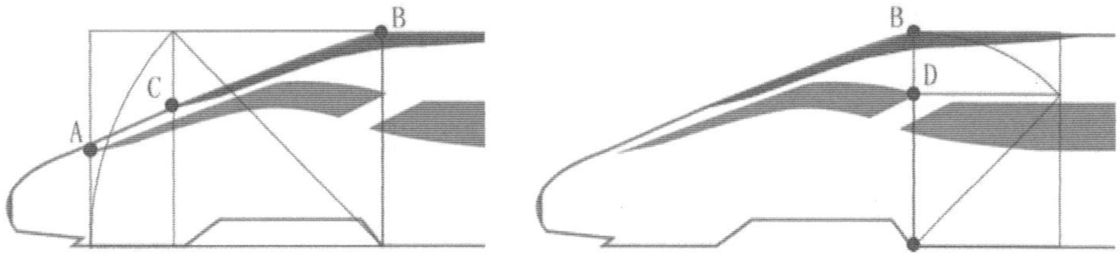

Figure 4. $\sqrt{2}$ proportional division in the right view

The research perspective is transferred to the front view, and the upper edge contour of the front opening and closing mechanism is at a $\sqrt{2}$ proportional division position of the vehicle body height; the lower edge contour of the driving window coating is located at a $\sqrt{2}$ proportional division position of the vertical height line from the roof to the apex of the front (*Fig.5*).

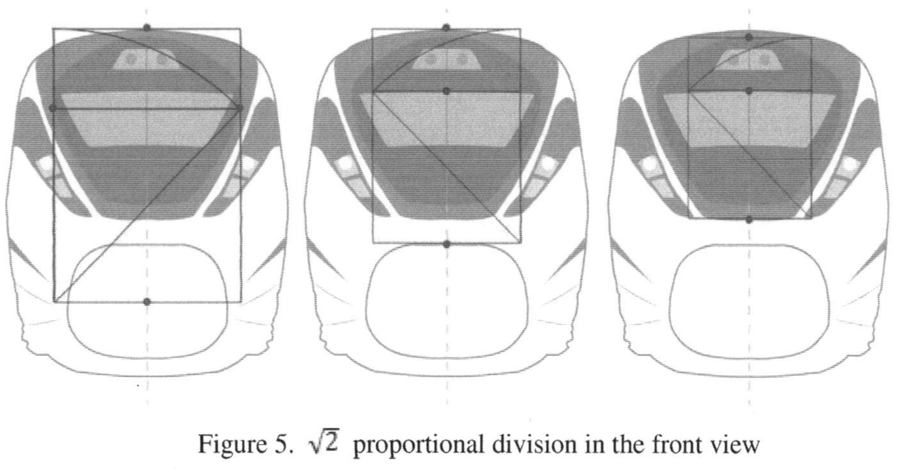

Figure 5. $\sqrt{2}$ proportional division in the front view

3.3 Golden section in German ICE series models

The ICE series train (or intercity multiple unit) is the fastest and highest train category in Germany. Its curve shape partially hides the golden section relationship. Observed and analyzed in the front view, the chamfering radius r1 of the upper edge contour of the second generation model is in golden section with the lower edge chamfering radius r2. This feature has been derived in the third and fourth generation models; The model radius r3 of the angle of chamfer of the lower edge and the upper edge chamfering radius r4 also constitute a golden section relationship (*Fig.6*).

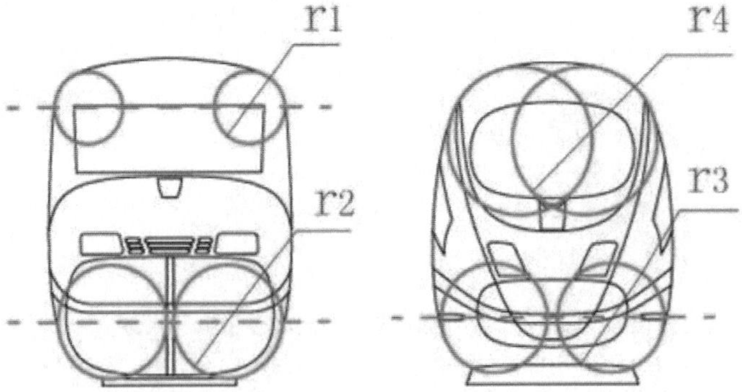

Figure 6. The front view of ICE series train

After reconstructing the streamline contour of ICE third-generation trains using the NURBS curve, four curvature control points on the curve can be obtained; the rectangles are formed according to the position of each point, and the ratio of the coordinate distance between each control point is the golden section. In addition, as the low-speed evolution version of the ICE three-generation model, the geometric proportional relationship of the modeling curves of the ICE four-generation model is more subtle and needs to be analyzed by means of auxiliary geometry. The specific method is as follows: on the side view of the vehicle body, the horizontal line (the extension line of the roof contour) is made with the vanishing point of the curvature as the end point, and the base is the edge of the train guard, and the parallelogram is constructed with the oblique parting line as the long side. The length ratio of the short side b to the long side a+b is a golden section, and the curvature change point is a golden point of the long side (*Fig.7*).

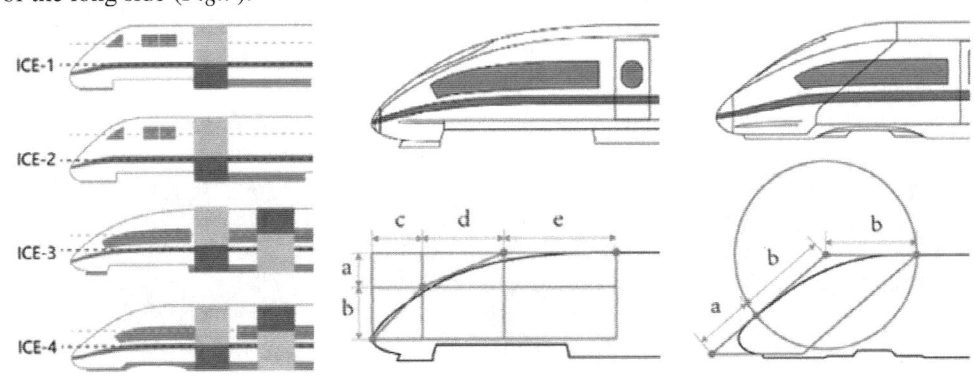

Figure 7. The right view of ICE series train

As the representative model of the German train, although the four trains of the ICE series have a quite different time-to-market, the unified light-colored body is decorated with long and thin red lines and has a strong overall visual recognition feature. From the geometric relationship, the height of the vehicle is divided by the red line into two parts according to the golden section. Its third-generation and fourth-generation models further expand the range of geometric ratios, from the height of the roof to the centerline of the window, and the ratio of the height of the vehicle to the centerline of the window is also the golden section. In the span of more than 20 years, the ICE series has always adhered to the geometric proportional division relationship, forming the distinctive style characteristics of the vehicle.

IOP Conf. Series: Materials Science and Engineering **573** (2019) 012063 doi:10.1088/1757-899X/573/1/012063

3.4 Research Summary

Based on the analysis of the above-mentioned models, the design of foreign high-speed trains uses a large number of rational induction methods to form geometric proportional relationship. All models have uniformity in order division and application. In the model change design, the geometric proportional feature was actively inherited, and finally formed the model gene of the model series. The main points of design of guiding significance are:

In the specific operation process of the design and implementation, the model designer optimizes the geometric relationship between the curvature nodes by enhancing the coordinate parameters of the curve control points to improve the smoothness of the vehicle body.

The functional parting lines and color dividing lines in all kinds of functional parts are all in accordance with the geometric proportional mode; these geometric lines with strict laws are used to derive the mathematical ratio by dividing the scale, and realize the visual equilibrium derivation in the constituent relationship.

In the design of the model evolution, the original proportional body is actively retained to form a visual continuity and eventually form a series.

4. The experimental argumentation

We try to verify the feasibility of the above three design elements through the virtual model, and comprehensively consider the application prospect of geometric design ideas in high-speed train design. The specific steps are: linear design - surface generation - structural line optimization - scale division - coating ratio - serial development - wind resistance calculation.

4.1 surface generation

We set the benchmark curve of the model according to geometric progression. Firstly, we establish the rational ratio of each CV control point of the NURBS curve, and then adjust the linear structure, the curve control point distribution, the order of the curve, and then rationally analyze changes of the curve slope. So the feasible line type is obtained, which is used as the modeling foundation of the newly designed train (*Fig.8*).

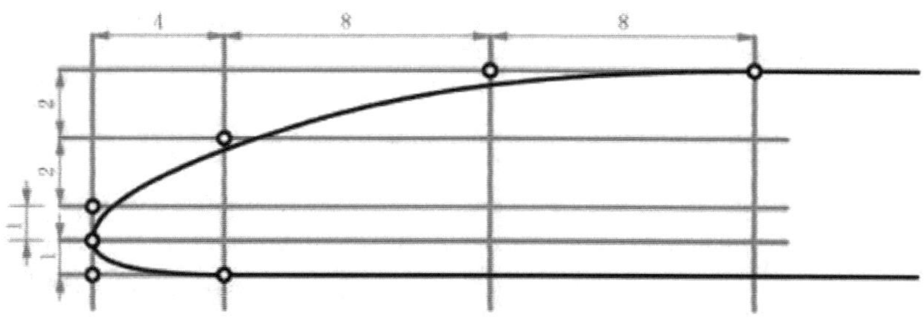

Figure 8. CV control point of the basic curve

In the design software, the closed surface is generated by curve lofting, and the original model of the new train model is obtained. These original surfaces are upgraded, the surface is adjusted, and the improved model is further optimized. As the order of the surface increases, the visual feature points of the shape gradually form, but the integrity of the reference curve must be guaranteed at any time, and any CV control points of the reference curve are not adjusted (*Fig.9*).

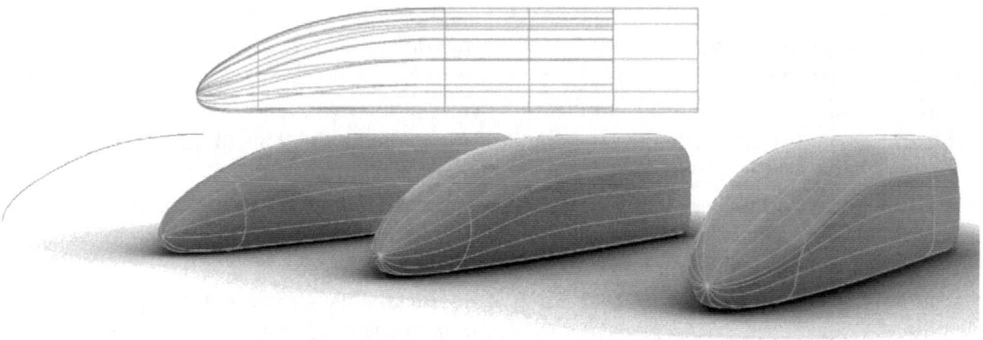

Figure 9. Surface increases

4.2 Proportional segmentation

After the solid model is generated, we divide the functional parts of the vehicle body according to the setting of the design target. In accordance with the technical requirements of mechanical design and ergonomics, these functional components must ensure the rationality of the scale (not to do technical analysis here). We use the vertices of the front as the starting point of the geometric segmentation. Based on the horizontal distance between the apex A of the front and the end B of the damper, as well as the positioning point C in $\sqrt{2}$ratio in the lower coating edge of the window in the cab; and then based on the horizontal distance between the point C and the end point D of the escape window, in turn, we establishes the front end point E of the cab window and the escape window coating position point F; finally, based on the horizontal distance between the point B and the vane curvature vanishing point G, a diagonal rectangle is formed, and we establishes the end point H in the cab window, lower edge end I in escape window, end point J of side window, and vehicle lower edge point K in proper order. After dividing the positional relationship of the front part of the vehicle, we keep the curvature of the curve in a smooth transition, and draw the golden section to make the body coating division (*Fig.10*).

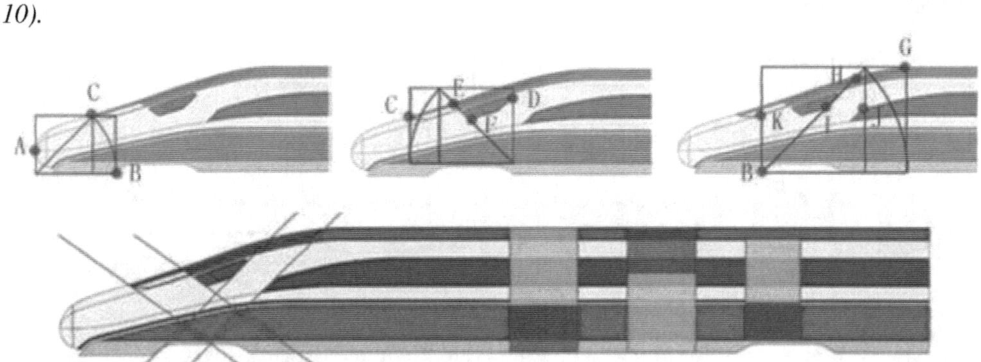

Figure 10. Division

Simulating the three-dimensional effect graph of the above-mentioned linear segmentation, we get the final design effect of the new model, (*Fig.11*). We reserve the proportional relationship between the window, the escape window, the driving window and the headlights, improve the geometric proportional relationship of the coating division, and adjust the coating connection of the window and the escape window to obtain the middle and right models in the picture. Due to their consistent division ratio, the three models effectively form a visual continuity under the changing form and color, forming a series of designs.

Figure 11. Series of new designs

4.3 Simulation calculation

The air resistance analysis of the optimized virtual model is carried out in the analysis software. The wind resistance coefficient of newly designed train locomotives is 0.2087, which is a good aerodynamic performance category (*Fig.12*). In view of the visual and data results of the above experiments, we believe that design geometry can be effectively applied to the design of high-speed trains.

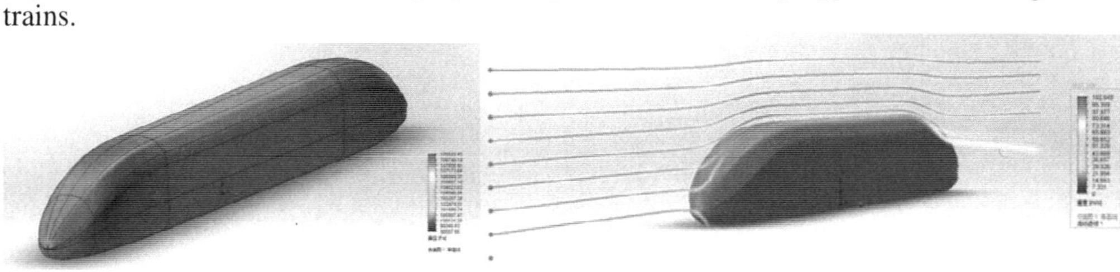

Figure 12. Air resistance analysis

5. Conclusion

Design geometry studies the relationship between geometric composition and visual perception, and studies the universal visual principle. It uses the mathematical law as the mediation of the order of the visual form, and interprets the morphological law originating from nature and the reason that its derivative style as a question of the human being's unconscious aesthetic choice [10]. With design geometry as an application guide, orderly structural relationships can effectively organize, design and express. Therefore, the design of high-speed trains requires reasonable data analysis, summarizing the linear rules behind perceptual thinking, and rationally planning the free curve by designers. This design method based on geometric and mathematical angle of view can rigorously establish the characteristics of the new generation of vehicle models; in addition, the geometrical visual proportion and rhythm law extends with the iterative process of the vehicle system, strengthening the homogenization identification system of the series models in the formation of the basis, and providing premise objects for design strategy research.

Acknowledgement

This research was financially supported by Hunan Provincial Philosophy and Social Science Fund (Grant No.: 18YBA436)

References

[1] Innovative Design Strategy Research Group. (2016) *China Innovation Design Roadmap*, China Science and Technology Press, Beijing, pp. 233-238.
[2] Xu, F. (2015) *China's High-Speed Train "Going out" Strategy: the main Theme, Strategy, Measures*, China Engineering Science, 2015, No. 4, pp. 4.
[3] Innovative Design Strategy Research Group. (2016) *China Innovation Design Roadmap*, China Science and Technology Press, Beijing, pp. 231-232.
[4] Marx: *The Complete Works of Marx and Engels*. (2008) People's Publishing House, Beijing, pp. 95-97.

IOP Conf. Series: Materials Science and Engineering **573** (2019) 012063 doi:10.1088/1757-899X/573/1/012063

[5] Kimberly, I. (2003) *Design Geometry*, translated by Li Leshan, China Water Resources and Hydropower Press, Beijing, pp. 76-78.
[6] Le Corbusier. (2014) *Towards New Architecture,* translated by Yang Zhide, Jiangsu Science and Technology Press, Nanjing, pp. 65-66.
[7] Wang W. (2010) *Research on the History, Theory and Application of Art Design Geometry*, Journal of Nanjing University of the Arts (Art and Design Edition), No. 11, pp. 168-171.
[8] Gail H. (2003) *Design Elements*, translated by Li Leshan, Han Qi, Chen Zhonghua, China Water Resources and Hydropower Press, Beijing, pp. 72-74.
[9] Wang W. (2010) *Research on the History, Theory and Application of Art Design Geometry*, Journal of Nanjing University of the Arts (Art and Design Edition), No. 11, pp. 168-171.
[10] Kimberly, I. (2003) *Design Geometry*, translated by Li Leshan, China Water Resources and Hydropower Press, Beijing, pp. 32-35.

IOP Conf. Series: Materials Science and Engineering **573** (2019) 012064 doi:10.1088/1757-899X/573/1/012064

The Incomplete Beauty of Visual Language

Ying Liu[1], Yi Chen[2], Xiaojing Liu[1]*

[1] Department of Architecture and Art, Central South University, Changsha City, Hunan Province, 410012, China

[2] Hunan Park Educational Consulting Co., Ltd， Changsha City, Hunan Province, 410083, China

Email: 81413450@qq.com， 48316365@qq.com ， 1037303749@qq.com， 12856385@qq.com

* Corresponding Author: Xiaojing Liu; email: 12856385@qq.com;tel:+86-18974968823

Abstract: The beauty of incompleteness in design is a kind of theory and a sublimation of incompleteness in cognition. It is a beauty that is opposite to perfection. It is a reversal of order and runs counter to the good wishes of people. But it can produce a strong visual and psychological impact, impressing people deeply. This paper illustrates the influence of incomplete design on visual perception through case studies, and discusses its application in the vocabulary, structure and pragmatics of visual language. Owing to the unique charm of the incomplete beauty in visual language, it can be concluded that the use of the incomplete design can achieve effective visual communication.

1. Introduction

As the Chinese saying goes, "men have sorrow and joy; they part or meet again; the moon is bright or dim and she may wax or wane". The incompleteness is a natural phenomenon, and it is the imperfectness that gives a lingering aftertaste. Venus de Milo with broken arms, the Roman Colosseum with crumbling fences and dilapidated walls, and the ruins of the stone carvings in Western Mansions area of The Old Summer Palace (known in Chinese as Yuanming Yuan) all showed a kind of residual beauty of defect. For contemporary design, this kind of incitement caused by the incompleteness cannot be replaced by perfection, and it is also a kind of spiritual realm constantly pursued by designers.

2.Visual language and visual psychology

2.1 Visual Language

Visual language is a set of norms or symbolic systems that convey meaning by two parts: visual basic elements and design principles. Among them, the basic elements include: lines, shapes, light and dark, color, texture, and space. They are the basis of a work and equal to the bricks, tiles, cement, steel and so on needed to build a house. It is also like characters and words in a written language. Design

IOP Conf. Series: Materials Science and Engineering **573** (2019) 012064 doi:10.1088/1757-899X/573/1/012064

principles include: layout, contrast, rhythm, balance, and unity. They are the principles and methods that artists use to organize and apply basic elements to convey meaning.

2.2 Visual perception and visual thinking

We often say "I saw something", or "I saw the world around me". What is the meaning of "seeing" here? For practical purposes in everyday life, to see or to watch is only the basic means of guiding the direction of practice. And Rudolf Arnheim explains it as a kind of elementary activity of confirming that something is in a certain location by one's eyes, which is what most people understand about "seeing". But then Arnheim puts forward that the content of "seeing" is more than that. What exactly does "seeing" include? Through numerous scientific experiments, Arnheim revealed the fact that vision is not only an activity of watching, but also a process of rational thinking. To this end, Arnheim proposes a concept that overturns traditional philosophy and general psychology: visual thinking. In traditional psychology, the distinction and boundary between perception and thinking have always been very clear. It is believed in traditional psychology that perception, belonging to the category of perceptual knowledge, is a direct reflection of an objective stimulus, a low-level cognitive psychological phenomenon of human psychological process. While thinking, belonging to the category of rational cognition, is an indirect reflection of objective things, a high-level cognitive psychological phenomenon in the psychological process with the characteristics of generality and abstractness. The concept of visual thinking put forward by Arnheim breaks the strict boundary between perception and thinking. He points out that perception, especially visual perception, has all the skills of thinking, and visual activity is completely a kind of positive activity, which can simplify, combine, abstract, and separate art and visual perception. It is not only selective, but also succinct and durable. In fact, in our daily life we have such experience: when we look at a three-dimensional sphere, the back of which is invisible to our eyes, we will still perceive it as a complete ball, that is the hidden back of the ball can still be "seen" by us; when we look at someone's face, even the hair flowing down his or her back becomes part of the overall picture we receive. It can be seen that human vision is by no means the same device as a camera with mechanical self-made parts. Because it is not a passive acceptance activity like a camera and the image of the outside world is not simply printed as accepted. Another famous contemporary psychologist, Sir E.H. Josef Gombrich of the United Kingdom points out that seeing is not recording, but the reaction of the whole body to those light patterns that stimulate our eyes. There is no unbiased eye in the world. Seeing is the explanation. The visual organs do not exist for the sake of understanding, but have evolved for survival. The result of this evolution leads to the special tendency of thinking when recognizing things. Visual activity is a creative activity of the human spirit.

2.3 The aesthetic principle of Gestalt psychology

According to Gestalt psychology, what is perceived is greater than what our eyes see; the conscious experience must be considered globally because the nature of the mind demands that each component, with its own characteristics, should be considered as part of a system of dynamic relationships. Thus, when an incomplete form is presented to people, incomplete individual elements will depend on the overall characteristics, causing a strong tendency in the vision to pursue perfection, symmetry, harmony and simplicity. In other words, it will provoke an impulse to supplement or restore individual elements to the complete state that they should have, thus greatly increasing the level of excitement of perception. This stimulating process of visual perception achieves the enhancement of visual language, which leads to the aesthetic sense of visual thinking.

3. The incomplete beauty

3.1 The definition of incomplete beauty

To ask what is incomplete beauty, we must first know what is beauty. British aesthetician Bernard Bosanquet wrote the definition of the beautiful in the first chapter of his book *A History of Aesthetic*:

"That which has characteristic or individual expressiveness for sense-perception or imagination, subject to the conditions of general or abstract expressiveness in the same medium." Alexander Gottlieb Baumgarte, a German philosopher and aesthetician, believes that sensory knowledge, namely, the completeness of perception or feeling, is called beauty. Although the definition of beauty among these ancient Western aestheticians are different, it is certain that beauty is the sensory feeling of human beings. Therefore, beauty is the process of human perception through thinking and the subjective feeling of the audience.

In *Shuowen Jiezi* (an early-2nd-century Chinese dictionary from the Han Dynasty), it wrote that "*can*" (a Chinese character), literally cruel injury, followed the meaning of its left lateral radical and phonetically resembled its right lateral radical; "*que*" (a Chinese character), literally the breach of a container, followed the meaning of its left lateral radical and phonetically resembled its right lateral radical. In the Chinese Dictionary, "*canque*" means "missing of limbs or organs caused by disability, etc.; incomplete or partially missing". Thus, "*canque*", or incompleteness, is a derogatory term, as opposed to perfectness. It is the destruction or imperfection of the appearance of materials, the destruction of things, the reversal of the order, and the psychological treason. The incompleteness in design refers to the defects, irregularities, oldness, involving the physical properties of the image's shape and contour, and textur. Interestingly, when incompleteness is applied to the art field, its imperfection has been greatly improved through the completeness resulted from human subjective experience.

Incompleteness can be divided into two types in the art field: narrowly defined incompleteness and generalized incompleteness. The narrowly defined incompleteness is the non-human or unintentional defects recognized by most people. The generalized incompleteness is the beauty of the incompleteness, that is, whether it is the incomplete part or the incomplete whole, it can be regarded as the beauty of incompleteness. It is the incomplete beauty referred to in this article.

3.2 The reasons behind the incomplete beauty

In daily life, people are often reluctant to see the fly in the ointment and are unwilling to experience things that are contrary to good ideals. But when people see the phenomenon of defect or experience imperfection, it is certain to cause imbalance of the mind, create conflicts, and trigger a strong physiological reaction. Therefore, the form of the incompleteness is likely to cause visual attention, which is more intense, prominent, and cohesive. At this point, the form of the incompleteness is not only the opposite of perfection, but a beauty that is shocking and awakening. This is why perfect things are easy to forget, but the incompleteness can make people unable to let go for a long time. Incompleteness can always cause people's compassion, and sometimes it is reflected on themselves.

In design, many defects do not affect our understanding. On the contrary, it is incompleteness that gives people a new visual experience. Human visual perception, a kind of visual thinking, has the initiative and gestalt consciousness. Although the graphics itself is incomplete, it can still obtain a complete perceptual object. Thus, the incomplete beauty becomes the beauty of artistic conception, quality, truth and imagination. The incomplete beauty of the design often uses this psychological impact on the viewer's innermost depths. Moreover, it is often the incomplete form that expresses more visual language and produces more visual feelings, resulting in a visual aesthetic, that is, the beauty of incompleteness.

4.The incomplete beauty in design

4.1 The incomplete form: a beauty of supplement

Regarding to many excellent graphic design works, we often find that some works are very prominent and compelling compared with other works. In fact, these works are finished based on breaking the organizational structure and visual perception in the general experience. This kind of graphics are not the most regular (symmetric, unified, or harmonious) shapes, but have their own unique tendencies

IOP Conf. Series: Materials Science and Engineering **573** (2019) 012064 doi:10.1088/1757-899X/573/1/012064

and laws. They may not look like perfect standard shapes, but they can produce intense excitement in the cerebral cortex through visual stimulation.

For example, in the second "Kan Tai-keung Design Award", the poster "Circle" (figure 1), which won the "Future Designer Award", is such a success of incomplete beauty. The theme of this poster is the reunification between both two sides of the Taiwan Straits. At the center of the white picture is a green jade bracelet. But the author He Shiyang has opened a small breach, representing Taiwan, on this perfect bracelet. This breach has become the visual center of the picture, quickly attracting our attention and triggering our thinking. People cannot help wondering why there is such a regrettable flaw on the perfect jade bracelet. Consequently, the atmosphere of patriotism has been heightened.

If figure 2 is three complete cubes, it will inevitably make people feel rigid and uninteresting. By subtracting a part of the shape, although the logo looks incomplete, it leaves a space for imagination. On the surface, it seems to be a kind of reduction, but it is an increase for the viewer, deepening the contrast among shapes.

In figure 3, the poster of "Fusion" by Yuan Youmin, an associate professor of the Department of Visual Communication Design at the Central Academy of Fine Arts, is a fragmented image formed by heterogeneous isomorphism. At the center of the picture is the image of an animal giraffe, but if you look carefully, you will find that the giraffe is the incomplete "Mona Lisa" of the famous Italian painter Leonardo da Vinci. This kind of incompleteness impresses a fresh feeling on the viewer. The fragmentation formed by this isomorphism produces a novel visual experience and strengthens the artistic conception of the picture.

Figure 1

Figure 2

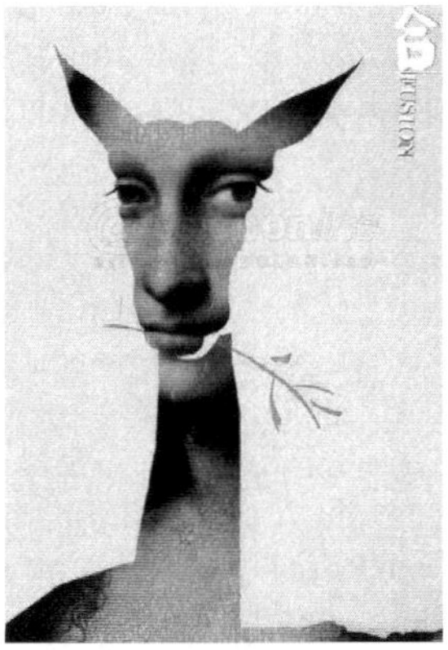

Figure 3

As shown in the above examples, the stronger the visual impact, the higher level of attention and memory index. Practice has proved that the incomplete image is more likely to attract visual attention. If applied reasonably, the defect can effectively enhance the attention of the vision, highlight the essence of objects, and expand the space of imagination.

4.2 The incomplete expression: a beauty of imagination

Paul Klee once said: "Art is not to depict something visible, but to create something invisible." This sentence shows that the real realm of art lies in the "invisible". In fact, as long as we pay attention, we will find that the "invisible" can be reflected in many art forms. For example, in traditional Chinese painting, there is an idea of "*liubai*", literally intended blank. In Chinese painting, we often find that only one theme is at the center of the painting, leaving a large space. In fact, this is a unique technique of traditional Chinese painting called "*liubai*".

Kan Tai-keung, the great designer in Hong Kong, is adept at using the technique of traditional Chinese painting, "*liubai*". With a grounding color of white, most of his posters (figure 4) inherit the sense of space of Chinese paintings, such as virtual reality and intended blank. There are not many elements in the whole picture, and rich and appropriate blurring is applied in some parts of the poster. It has reached a realm of "leaving enough space for viewers to imagine" and "producing limitless aesthetic enjoyment by limited ink paint" in traditional Chinese paintings. Whether it is tangible or intangible, virtual or real, it is all for the sake of accurately reflecting the artistic charm and ideological intention of the work.

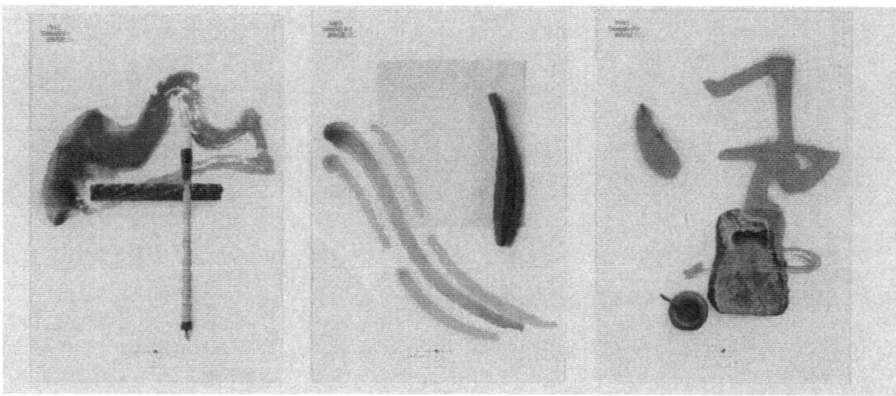

Figure 4

Figure 5 Figure 6

Figure 5 is the poster with the theme of "the return of Macao" designed by Li Shaobo, director of the design department of School of fine arts at Hunan Normal University. The incompleteness of the dragon pattern on the plate stimulates people's senses to complete the pattern. It is this complement that skillfully reflects the historic event of Macao's return to the Chinese mainland. To some extent, the imperfection of the plate is the utmost perfection of the inner idea.

We often see the slogan like "if you were her... " or " if you were here... ". Designers do not tell a story, but use a short half sentence with a sense of incompleteness to express thousands of possibilities, giving the audience enough space of imagination to put themselves in it. In ancient times, there was an old saying that "it is better to leave enough space for a writer when expressing his thoughts", which means that one should leave more leeway for writing and speaking, and leave space for the receiver's imagination and perception. As shown in figure 6, the same is true for the expression of visual language in design, which needs not to be perfect.

4.3 Summary

The form of incompleteness in visual art, as an art media, or as a linguistic feature, can satisfy the conditions of aesthetic pleasure. It is a concise but varied trace, a mixture of complexity and richness that implies a simplified trend and tension. It is an important visual element that is organic, seemingly random, but regulative. The texture of music is to put a simple theme into a rich, complex, various, repetitive movement, with a dense or elegant tune; while the texture of the art can be understood as the simplified structure or composition of the picture, with special processing, expansion, and variation of the media language to make it rich and colorful. The form of incompleteness, as one of the components of this texture, plays a role of deepening the theme in the communication of visual language.

5. The incomplete beauty in visual language

5.1 The design of visual grammar

5.1.1 Visual vocabulary in incomplete design

Figure 7

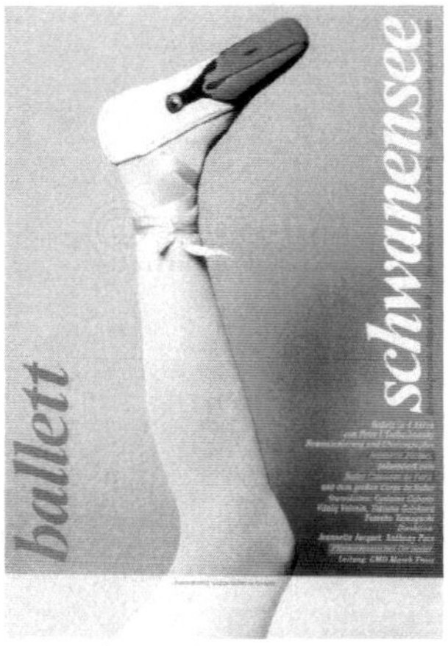

Figure 8

The vocabulary refers to the sum of words, that is, the aggregation of linguistic signs. From this, the visual vocabulary is derived, which is the basic elements of the visual language, like graphic patterns, colors, words, and so on.

Gombrich believes that regarding to the design elements, patterns that are easily perceived by people are tedious. When the expected pattern comes to sight, people's attention is not drawn to it. In information communication, the degree of surprise directly affects the memory of the audience. Thus, the simplification or even imperfection of the pattern naturally generate a kind of individual visual experience, which is against people's aesthetic will, so it can better attract the attention of the audience.

The aesthetician Kurt Beidirt interprets art as "the most intelligent organizational means that is based on the essence of insight. This essence is the essence of everything else that belongs to it". Therefore, in the case of grasping the essence of things, no matter how simplified, "the traces of the form of incompleteness contain rich connotation, which is the external manifestation of the 'intrinsic' simplified principle", in order to make better use of the incompleteness of the representation. Incorporating ideas into it, giving new meaning to the incomplete, and letting the incomplete image get the sublimation of art, thus creatively completing the concept and information communication. For example (figure 7), the designer skillfully uses the incomplete design to simplify the Chinese characters "*gong*" (literally national interests in Chinese) and "*si*" (literally personal interests in Chinese), with the lower lateral radical of "gong" removed to the right part of the character "*si*". And its copywriting of " it is glorious to honestly perform one's official duties but disgraceful to harm the interests of the country and the people" has profoundly revealed the theme of honesty and integrity.

5.1.2 visual language structure in incomplete design

Study of the structure of a language refers to the combination and arrangement of signs and symbols. The structure of visual language is not only the mechanical addition and subtraction of design elements, but also the expression of deeper connotation through the combination of various elements.

The change of composition can make the graphics incomplete and convey the visual language better than the complete graphics in the design. It is impossible for a picture of specified size to present all the images, and the choice itself is the design. The author must discard the superfluous elements according to the needs of the theme, and focus on the performance of some objectives. Thus, the imperfection of local graphics becomes inevitable. The composition of a picture is a noun at present, but in the Southern Dynasty, Xie He (the first Chinese painting theorist) called it "*jinyingweizhi*" (literally rules, composition and layout of a painting in Chinese). In the process of poster creation, designers should arrange the picture elements according to the needs, namely, what to show and what to give up.

For example, the poster of the ballet Swan Lake (figure 8) designed by German graphic designer Matthies Holger shows an incomplete image of a swan, whose head is a foot of a ballerina if viewed upside down. Other parts of ballerina's body are omitted for the reason of composition. The imperfection caused by this composition makes the visual impact of the figure stronger and the picture appears simpler and more concise.

5.2 The law of incomplete beauty of visual semantics

5.2.1 to highlight the subject as the fundamental

The subject of visual language is the main element and the main figure of design, which is the appeal point of poster itself. Once the semantics (theme) have been determined, the designer should focus on the subject. At this moment, semantic communication is particularly important, and all parts of the design should at the service of semantics. Any detail of the design must serve the design body and the design intention. And the main figure of design has the primary and secondary sequence, with some details. In order to be prominent in the ocean of information and attract the attention of the audience, it is necessary to highlight the important part of the main figure, so as to highlight the theme, namely, semantic communication. The incomplete design of the main graphics is one of the methods. The purpose is to highlight the focus and serve the main body, which is a clever incomplete on the overall effect, so that the audience can accept and remember the idea conveyed by the design.

For example, Japanese graphic designer Ikko Tanaka designed the poster of "the 27th Sankei Kanze Noh" (figure 9), in which the designer took the "Noh mask" as the main body and arranged it in the center of the poster. In order to highlight the main body of the picture, the author removed the secondary part, the white hat. After that, only the incomplete mask, namely the triangle in the center of the poster, was left, reaching a level of simplicity and refinement and producing a strong visual impact. It is just the integration of the white hat and the grounding color of white that makes the image of Noh become more vivid, prominent.

IOP Conf. Series: Materials Science and Engineering **573** (2019) 012064 doi:10.1088/1757-899X/573/1/012064

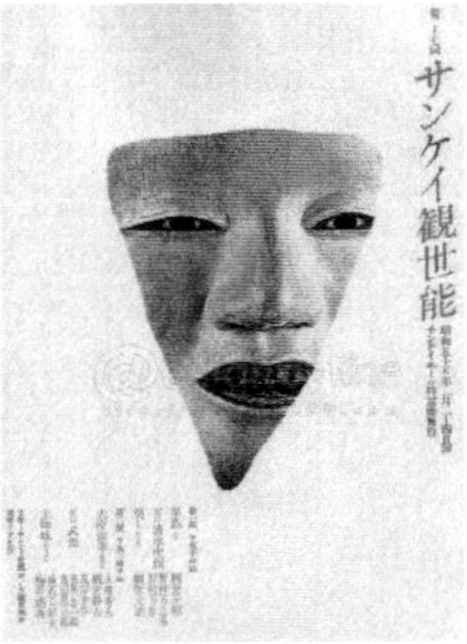

Figure 9

5.2.2 for the purpose of enhancing semantics

The meaning of design is to plan, regulate and visualize. That is to say, the purpose of design is to convey design ideas. Therefore, how to strengthen the communication of visual language is the goal of design activities.

Figure 10

For objective things, each person has an impression in his mind according to his own understanding. This is the "Gestalt" in everyone's mind. When the shape of an object we see is the same as the "Gestalt" in our mind, it will be too normal to be ignored. When the shape of an object we see conflicts with the "Gestalt" in our mind, it will cause visual attention. In principle, the stronger the conflict, the greater the degree of attention, and the higher the memory index. Thus, the audience always compares their predictions of changing external events with their inherent information framework. If these external events are consistent with their predictions, they will no longer be interested in it; if not, they will estimate the situation at that time.

The incompleteness itself is not beautiful, but it is precisely this kind of incompleteness that breaks some of the "Gestalt" in people's minds. It arouses people's visual vigilance and mobilizes brain activities, thereby rendering people to analyze why this object is illegal and incomplete. The biggest characteristic of the incomplete image is that it is easy to attract the attention of the vision, which is a highly fast, powerful and prominent process. If properly applied, it can achieve a shocking effect and produces aesthetic feeling at the same time.

Therefore, smart designers use the above-mentioned principle and apply it to the graphic design with his keen perceptive. They have better practiced the incomplete design of Chinese characters and enriched the expression and connotation of the works. For example (figure 10), the two points of the left lateral radical of the Chinese character "*you*"(literally oil in Chinese) are omitted, which constitutes the incomplete character. The information stored in the brain can be utilized by the viewer to form a complete character. This has greatly mobilized the viewer's interest and highlights the appeal theme of the advertisement.

5.2.3 with the intension of incomplete form but rich meaning

In visual language, the incompleteness of design elements does not represent the absence of semantics. On the contrary, the conscious incompleteness is aimed to produce a new meaning, having a clearer interpretation of the incomplete appearance of the subject figure and the inner thoughts that the designer wants to convey. This design process refers to the intension of incomplete form but rich meaning, that is, the author uses the incompleteness of the visual image to convey a perfect idea or a kind of unique emotion. The incomplete appearance is typically designed for the aim of incompleteness, but the perfect unity of perceptual knowledge and rational understanding, form and content. Under the incomplete external form, there is a perfect idea that matches it.

Professor Chen Fang designed a poster of victory (figure 11). There is no language in the picture but a broken hand with two figures made into the shape of V. It is the incompleteness of these three fingers that reminds people that victory is a painful price. A strong visual impact strikes people's heart and makes people feel desolate and bitter, prompting people to reflect on their own behavior.

Figure 11

5.3 The incompleteness of visual pragmatics

5.3.1 the breadth of incomplete design

In the design of fonts, modern design works with Chinese characters as the theme elements have begun to take shape and gradually developed, among which the incomplete design of Chinese characters is quite representative. From ancient times to today, the incomplete Chinese characters have left countless missing parts for people to guess and complete. Today's designers have found beauty in incompleteness and found rich design connotations. The incomplete design of Chinese characters not only reveals its formal beauty, but also is widely applied to graphic design, allowing viewers to feel the psychological collision of objective incompleteness and subjective pursuit of perfectness.

In graphic design, there are many different degrees of incompleteness in terms of the main figure of the poster. Graphic, with the original meaning of illustration, now refers to the illustrative visual carrier by means of painting, writing, engraving, printing, etc., and a way of information communication which can be copied. The main body in the poster has the characteristics of symbol, which means anything that can be used to deliberately replace another thing. Symbol has a clear and

IOP Conf. Series: Materials Science and Engineering **573** (2019) 012064 doi:10.1088/1757-899X/573/1/012064

simple connotation, so it has a universal sense of identity in the cognitive psychology. Similarly, the incomplete poster is also symbolic. After the intentional lack of the main body of the poster, its appearance has undergone partial or fundamental changes, and presented new visual effects with deepened language of design.

From the perspective of color, the defect of color in design is also a form of expression. It is an important principle to express emotions with the incomplete color. Incomplete color is an objective object, and emotions are something spiritually or vitally reflected in the object. The incomplete color is displayed in the picture, and emotions are perceived from the picture. The incompleteness of color always presents a fluctuant change with a modulated rhythm, and contains rich historic meaning. For example, in the posters expressing the theme of warfare, gray color and asymmetrical forms are often used. This kind of deficiency often has some special meaning and psychological suggestion. The hint is the spirit, the feelings and the vitality. The mottled color and broken shape give a shocking visual impact and successfully reflect the spiritual connotation of the works. When the little girl in red dress in *Schindler's List* passes through a black-and-white crowd, not only an emotion of sympathy, but also the reflection of history and life are presented. The application of incomplete color in design is the sublimation of the spirit.

5.3.2 The depth of application for incomplete design

Incompleteness does not always mean beauty. Only the incompleteness that accords with an order can produce aesthetic feeling. Thus, the designer must master the degree of incompleteness and adjust it flexibly.

In incomplete design, why can we achieve the complete expression of ideas but with incomplete form? The reason behind is the grasp of the integrity of the design object. "Perception is the process of selecting, organizing, and interpreting sensory stimuli to make it a meaningful and coherent image. The final product of this process is the perception of a particular object, forming a complete image of the object. "In the case of grasping the essence of things, no matter how simplified, "the traces of the form of incompleteness contain rich connotation, which is the external manifestation of the 'intrinsic' simplified principle." The famous painter Eugen Delacroix once said that in order to draw an object, "before drawing, the painter must clearly understand the main lines of the object, and the main lines is not the actual contour line, but the line that constitutes the structural skeleton of the visual object in many cases." Through the structural skeleton, that is, the grasp of the main axis skeleton, when the design element is partially missing, the audience can better utilize the incompleteness of the design, and complete the design element by the main structural skeleton.

Secondly, the unique correspondence between different parts determined by the main axis skeleton is of great significance. The fact that the structural skeleton determines the characteristics of a pattern is high important. If the structural skeleton of an image is similar to the structural skeleton of the viewer, this image can be recognized by the viewer effortlessly. Therefore, the charm of the incomplete design is reflected in the incomplete form, and its default or incomplete parts are supplemented by graphical hints and psychological associations.

6.Conclusions

The incompleteness in design is opposite to gestalt. However, the proper use of the incompleteness does not affect our understanding of things. On the contrary, it is incompleteness that gives people a new visual experience and highlights the essential part of things. It stimulates the audience's aesthetic tension and visual desire, which gives them more space for imagination. The contrast caused by the incomplete design can effectively enhance the memory index. What kind of visual objects are more likely to attract people's attentions? As designers, we should pay attention to this problem and use more theories of psychology to analyze the application of incomplete beauty in design. It is sure that incompleteness does not always mean beauty. Only the incompleteness that accords with an order can highlight the main figure, trigger imagination, and retain memory. Therefore, the optimal expression form of the Gestalt principle in visual language communication not only contains "incomplete shape, complete ideas", but also contains incomplete beauty.

Acknowledgement
This research was financially supported by Hunan Provincial Philosophy and Social Science Fund (Grant No.: 18YBA436)

References

[1] Arnheim, R. Art and Visual Perception[M]. Beijing: China Social Science Press, 1984
[2] Arnheim, R. Visual Thinking[M]. Beijing: Guangming Daily Press, 1986
[3] Zong, B. Artistic Conception[M]. Beijing: Peking University Press,1987
[4] Teng, S. Aesthetics Design and Art Education. Sichuan People 's Publishing House, 1998
[5] Yin, D. An Introduction to Design. Hunan Science and Technology Press, 2004
[6] Wang, F. A study on the incomplete beauty in graphic design[J]. Popular Literature and Art, 2009

Contemporary Packaging Design on the Elements of the Republic of China

Wenli Lu[1], Yijin Chen[1], and Xiangyang Bian[1*]

[1] College of fashion and design, Donghua University, Shanghai, 200051, China

[*]Corresponding author's e-mail: bianxy@dhu.edu.cn

Abstract. It was a rapid development period of Chinese packaging design in the Republic of China, and its distinctive design style was used to highlight the distinctive style of times. The study classified the styles of packaging applying and analyzed the packaging cases, which is about the elements of packaging design in the Republic of China. It explored the innovative redesigning in contemporary packaging design, such as Calendar Art, Meishuzi, trademark and decorative patterns. This paper generalized the elements of packaging design by analysing the contemporary packaging design and provided reference in design.

1. Introduction

China's modern packaging design started in the late 19th century. Packaging as a foreign word, only made a clear definition of packaging in national standard until 1983 [1]. During the Republic of China, Chinese packaging design developed at a relatively fast stage. As a large-scale industry at that time, packaging design demonstrated its distinctive style with unique technology and craftsmanship. Packaging design was an important part of commercial art when commercial art flourished greatly in the Republic of China. It was also glowing and leaving a large number of excellent packaging designs. The national enterprises began to pay attention to and improve the packaging design of commodities, in order to seize the market with western enterprises and promote their own products. As a result, packaging design has formed a new fusion of Chinese Art and Western Art. Based on the Chinese geography culture concept, the combination of Chinese and Western patterns has created countless exquisite packaging, and has also become a special era symbol. As a gem in the history of packaging design art in China, it still provides reference for contemporary art design.

2. Elements of the Republic of China for directly applying

The elements of the Republic of China in packaging design is mainly divided into two categories. One is applying directly, the other one is redesigning. Direct application in Minguo-style design uses extensive pattern externalization, especially using some landmark items, such as cheongsam and old buildings which were in the Republic of China is a more common design method [2].

In 2015, Disney's Minguo-style Mooncake [3] has a traditional graphic design combination of typical things, such as cheongsam beauty, Shikumen, oil umbrella and rickshaw which are all in the Republic of China. It forms a brand new picture with classical charm and modernity.

IOP Conf. Series: Materials Science and Engineering **573** (2019) 012065 doi:10.1088/1757-899X/573/1/012065

Figure 1. Minguo-style Mooncake by Disney

3. Elements of the Republic of China for redesigning

3.1. The pattern of Calendar Art

The Calendar Art can be the earliest advertising paintings in the history of Chinese commercial art [4]. It is the product of the continuous integration of traditional Chinese painting and Western painting. It has a long tradition, distinctive nationality and wide practicality. It is also the advertising art of China and the most brilliant achievement in the development process. This is the Calendar Art advertisement of Guangshenghang [5] in 1928 (Figure 2). The center of the picture is the realistic character image of the company's well-known brand, SHANGHAI VIVE. It is the most common that the products around the company are arranged neatly in rectangular rules in the Calendar Art. The form is that Chinese characters are on both sides of the pattern border, and the product is at the bottom.

Figure 2. The Calendar Art advertisement of Guangshenghang

Figure 3. The post of Tongrentang

In contrast to the imitation of the Calendar Art [6] in modern design (Figure 3), it can be clearly seen that the style of the rectangular border of the picture is completely preserved. The border is either decorative graphic pattern or Chinese font pairing. The inside of the rectangular border is the main content of the advertisement. The decorative patterns in the frame are simplified in the modern design. It is a straight line curve to replace the complex border decoration of the Republic of China. The entire picture becomes clear and clear as it simplifies the design language. The content depicted on the screen is also more targeted. The purpose of the promotion is clear. The theme of the picture is clear at a glance, with a vintage flavor and good artistic effects.

3.2. Modern application of Meishuzi

Meishuzi is the art character in the Republic of China. The revival of art characters is not a simple copying [7]. It is a modern aesthetic transformation and is suitable for modern Chinese character design. As designers pay more and more attention to the traditional Chinese character design in China, the art of the Republic of China is naturally attracted by the designers. The classical art design is gradually emerging in modern design and out of its essence.

The Sans-serif of the *Literature Weekly*'s art characters in the cover of the book which was created by Qian Junyi in 1925 (Figure 4), had evolved from the black body. The glyphs are squeezed and stretched, and they are all in the whole picture. It looks full, full of tension, and the font is tough and straight. Based on the profound understanding of Chinese traditional books, Qian Junyi's flexible art design has opened up the modernist design style from the sensitivity of Chinese characters.

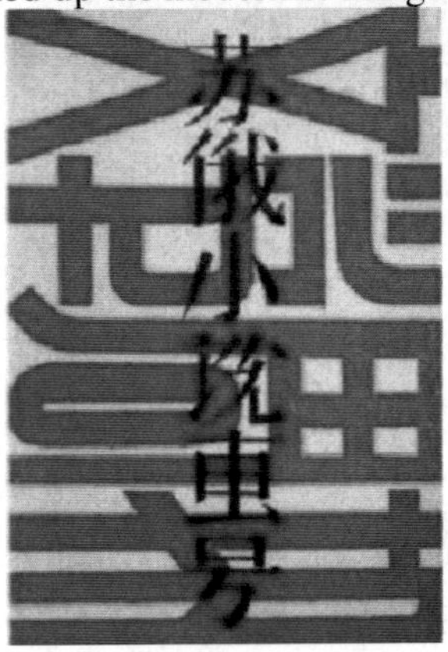

Figure 4. "Literature Weekly" cover by Qian Junyi （1925）

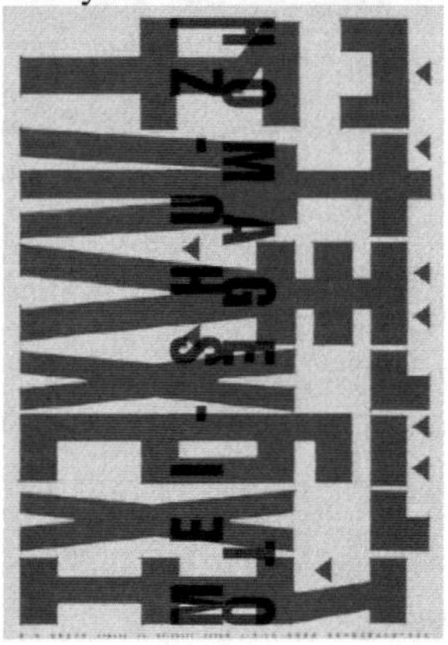

Figure 5. The post of Homage to Meishuzi by Jiang Hua （2005）

Jiang Hua, a famous modern designer, designed the poster "Homage to Meishuzi" in 2005 for the Chinese Book Design Forum. He changed the vertical text to English letters (Figure 5). Compared with the 1925 version of the design, it can be seen that although the typographic style is similar, the overall style is completely different. In the poster design of Jiang Hua, the triangular and linear stroke design gives a more rigid and sharp vision. It is full of modernity and structuralism. In this way, the layout of the bottom text in the middle of the text is also used in modern design.

These imitations of the modern design's typography and fonts have fully proved that the design art, in the Republic of China, has gradually received the attention of the public and returned to the sight of the modern art world. It also shows that the new generation of Chinese designers began to re-examine China. The text design explores a design path that integrates Chinese traditional culture into a modern and internationalized world.

3.3. Contemporary application of Logo and decorative pattern in the Republic of China

Contemporary Mingguo-style design, its edge decoration and decorative elements have been simplified. And it shows color block flat wind design. However, it can be clearly seen that compared to the complex and decorative trademark design of the Republic of China.

There are two aspects of design elements that are retained and widely used. The first is to retain the border design of the trademark frame (Figure 6). Whether it is a square or a circular frame design, it plays a decorative role and defines the theme content. In addition, contemporary design style is compared with the period of the Republic of China. It is more realistic than before. The trim design condenses into a more abstract line border, which is more succinct. Secondly, it retains the design method of using the trim design around the font to highlight the text content (Figure 7), and decorates the words in the picture with different border designs such as the shape of the circle, ribbon, heart, and fan.

Figure 6. Fonts decorated with ribbon patterns in the design of trademarks in the Republic of China

Figure 7. The design of the trademark in the Republic of China using an oval frame to highlight the subject

Zhiyaoju Studio [8] designed this group of labels (Figure 8). It uses simple wireframe to outline graphics, fills colors into flat shapes, and uses three or less colors in a picture. Therefore, design elements are diverse, but it still appears disorderly. The design of the text in the picture is either in the form of a banner across the whole picture, or placed on the top of the picture like a tag. It's also not placed in the upper left corner of the picture in the form of an embellishment, decorated in a continuous circle. The decorative techniques of the characters are various, and the different typesetting styles in the pictures also show different design styles. There are Chinese traditional elements (Guanyin and Koi carp), as well as western design themes (wstern doctors and angels wearing white hexagrams) in the pictures. The unified design language constitutes a series of complete Mingguo-style designs. The theme and content are humorous and interesting, which are deeply loved by young people.

Figure 8. The products of Mingguo-style by Zhiyaoju Studio

4. Conclusion

With regard to the contemporary application of packaging design elements in the Republic of China, this paper summarizes two common design methods and trends in modern retro design. First, the image of beauty in cheongsam was used more in packaging design in the Republic of China, and the figure image was usually simplified by design. Second, by imitating the style of contemporary design, there are many cases of five-color printing (yellow, eosin, turquoise, blue, black) in the period of the Republic of China, which can also adjust the brightness and purity of color. It is worth mentioning that purple is one of the commonly used colors in contemporary cosmetics design. Third, the design of Calendar Art was more and more widely used in contemporary application, and it was used for reference and imitation from layout structure to pattern style. Fourth, art characters, in the Republic of China, were widely used in contemporary packaging design, such as logo design, posters, books and other theme fonts design. Generally speaking, based on the sustainable development strategy of the times, the designers go to learn the inspiration of innovative design from the packaging design for a long time by using the elements in the Republic of China, and to find the fitness between Minguo-style design and the contemporary packaging design.

Acknowledgments
Project Fashion Culture, History and Heritage supported by Shanghai Summit Discipline in Design.

References

[1] Wei, JB. (2017). Research on the Origin and Development of Modern Packaging Design. China Packaging. pp. 24-26.
[2] Li F. (2014). Application of Minguo Style in Modern Packaging Design. Packaging Engineering. pp. 93-97.
[3] Minguo-style Mooncake by Disney (2015). Minguo-style Mooncake by Disney. http://www.shinerayad.com/news_info.asp?id=1289.
[4] Ding Jieqiong(2011). Research on the official month of the Republic of China. Zhejiang Institute of Technology. Hangzhou. pp.1.
[5] The Calendar Art advertisement of Guangshenghang.(2018) The Calendar Art advertisement of Guangshenghang. http://www.sohu.com/a/279728743_656524.
[6] The post of Tongrentang.(2018) Tmall old-fashioned GIF poster. https://www.digitaling.com/projects/25574.html.
[7] Yuan Fang(2015). Exploration beauty of Meishuzi in the republic of China. Zhejiang Institute of Technology. Hangzhou. pp.6.
[8] The products of Mingguo-style.(2018) Extra cash by Zhiyaoju. http://www.gtn9.com/index.aspx.

Research on Index System of Industrial Design and Development in Manufacturing Enterprises

Xiaojing Liu[1], Mengmeng Li[1] and Ying Liu[1*]

[1] Department of Architecture and Art, Central South University, Changsha City, Hunan Province, 410012, China

[*]Corresponding author's e-mail: 81413450@qq.com

Abstract. In recent years, the central leaders and relevant ministries and commissions have issued guidelines for the development of China's industrial design, and the development of industrial design in manufacturing enterprises has attracted more attention. At present, the development level of industrial design in various regions and industries in China is uneven. No institution or organization can accurately provide nationwide survey data, and the existing regional or industry-related survey data lack uniformity and continuity, which cannot provide sufficient reference for the country to formulate relevant policies. This paper mainly describes the building of index system of industrial design and development in manufacturing enterprise. The paper builds the framework based on literature at home and abroad, two rounds of questionnaire, expert interview and field research, selects indexes through Delphi method, analytic method and frequency analysis approach and then revises index system according to the expert evaluation of regional pilot. It attempts to connect with national statistical framework, so as to grasp the problems and trends of the development of industrial design. The establishment of index system of industrial design and development in manufacturing enterprise plays an important role in the future development of China's industrial design and the formulation of economic transformation policies.

1. Introduction

In recent years, Sino-US trade friction has been continuously upgraded, with the focus on the strategic layout of "Made in China 2025" aimed to upgrade China's manufacturing industry. As is known to all, industrial design is the primary link in the chain of productive service industry. It is of great strategic significance for the transformation and upgrading of manufacturing enterprises. It is an important indicator of the level of industrialization. The central leaders have emphasized to encourage and support the development of industrial design, and the governments at all levels have also intended to formulate supportive policies in hope of the rapid development of the industrial design. [1] However, at present, the level of industrial design and development in various regions and industries in China is uneven. No institution or organization can accurately provide nationwide survey data. The existing regional or industry-related research data lack uniformity and continuity, which cannot provide a sufficient reference for the government to formulate relevant policies [2] .

Under such a background, how to scientifically describe the development mode of industrial design of different types of manufacturing enterprises and how to accurately evaluate their development level has become an inevitable issue in the development of industrial design industry. We have deeply probed into the basic situation of industrial design in the development of manufacturing enterprises based on the analysis of the existing research results at home and abroad, field research of typical enterprise in different areas, questionnaire and expert interviews. We hope to build index system of industrial design and development in manufacturing enterprise, and effectively reflect the basic situation of China's industrial design and regional development.

2. Research process of index system

2.1. The design principle of index system

In order to highlight the basic state and industry characteristics of industrial design in manufacturing enterprises, and to ensure the objective and systematic evaluation of industrial design applications and development status of manufacturing enterprises, the design of index system of industrial design and development in manufacturing enterprise mainly follows five principles: reasonability, system optimization, universal comparability, practicality, and goal-oriented principle.

2.1.1. Reasonability. Reasonability is embodied in the combination of theory and actual situation and the scientific methods adopted. The index system can not only meet the theoretical requirements, but also reflect the objective reality of the evaluation object. Theoretical guidance can make the index system rigorous, reasonable and more targeted in the basic concept and logical structure.

2.1.2. System optimization. System optimization means that the evaluation object must be measured by several indicators, which are interrelated and mutually restrictive. There are horizontal linkages between some indicators, reflecting the mutual constraint relationship from different aspects, and vertical linkages between some indicators, reflecting the inclusion relationship at different levels. Meanwhile, the boundaries of indicators at the same level are as clear as possible and the indicators are designed to avoid internal connection in the same group. Hence, these indicators are highly systematic.

2.1.3. Universal comparability. Universal comparability refers to the comparison between periods and objects, namely vertical comparison and horizontal comparison. Vertical comparison means the comparison of the same object in different periods. This requires that the index system and the connotation of various indicators remain stable, and the reference values (standard values) for calculating indicators remain unchanged. Horizontal comparison means comparing different objects, so as to find common points and design the index system according to them.

2.1.4. Practicality. Practicality refers to be practical, feasible and operable. Firstly, indicators should be simplified; the method should be easy enough. And the index system should be as simple as possible on the premise of an objective and comprehensive evaluation result. Secondly, the data should be easy to obtain and collect. Whether for a qualitative evaluation indicator or a quantitative evaluation indicator, the source of information must be reliable and easy to obtain. In addition, all evaluation indicators and their corresponding calculation methods must be regulated and standardized. Thirdly, the data must be accurate and reliable. [3]

2.1.5. Goal-oriented principle. The goal-oriented principle embodies the purpose of the evaluation, namely, to guide and encourage the evaluated object to develop in the right direction. The purpose of the indicator system is to promote the better development of industrial design in manufacturing enterprises. Therefore, both design indicators and data collection should be oriented towards this purpose.

2.2. Research process of the indicator system

2.2.1. Source of indicators. With the confirmation of the definition and extension range of industrial design, we have conducted research on the management model, evaluation indicators and data of industrial design in developed countries such as Norway, Germany, the United States, Japan and South

Korea. We also have referred to relevant domestic research on the development of industrial design in some industries. With reference to the conventional statistical index system of developed countries and above-mentioned researches, we have initially summarized and sorted out 109 indicators of industrial design and development in China. Through subsequent expert interviews and field research, some necessary deletion and supplementation are made to the indicators. [4-12]

2.2.2. System construction and index screening. The construction of the index system is a thinking process from the concrete to the abstract and to specific dialectical logic. It is also the process that people gradually deepened understanding of the overall quantitative characteristics of the phenomenon. The screening methods adopted in the index system of industrial design and development in manufacturing enterprises are the Delphi method, comprehensive analysis method and frequency analysis approach. [13-15] The Delphi method is based on a series of systematic procedures, with the way of speaking on condition of anonymity anonymous methods of opinion. In other words, experts should not discuss with each other, but could only communicate with the investigators. After repeated consultation, induction and modification, the opinions of experts on the questions raised in the questionnaire are collected into basically consistent opinions, which can be regarded as the results of prediction. As a subjective and qualitative method, Delphi method, on the one hand, integrates the knowledge, experience and judgment of experts, which can help to get a conclusion reflecting the real state of evaluation objects. On the other hand, it also expands the screening scope of indicators, and can effectively deal with the subjective qualitative indicators that are fuzzy, uncertain and subjective. With a wider range of indicators screening, the system is much more applicable. Delphi method can not only be used in forecasting, but also be widely used in the establishment of various evaluation index systems and the determination of specific indicators. Comprehensive analysis method is to divide the overall goal of evaluation into several different components or different aspects (namely subsystems), and gradually subdivide, until each part and aspect can be described and realized with specific statistical indicators. This is one of the most basic and commonly used methods to construct the evaluation index system. Frequency analysis approach is the study of frequency of relevant reports and papers concerning industrial design at present. Hence, we can select those indexes with high frequency, so as to make the index system more comprehensive and accurate.

2.2.3. Working procedures. According to the working procedure of the Delphi method, the construction of the index system of industrial design and development in manufacturing enterprise mainly goes through the following procedures, as shown in Figure 1:

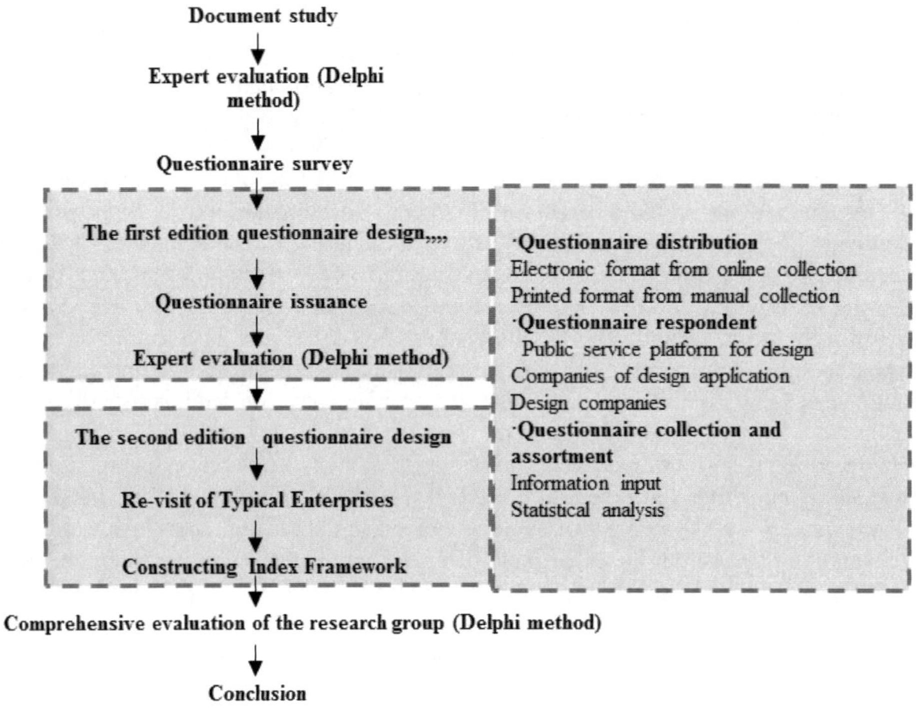

Figure 1 Working procedures of the index system of industrial design and development in manufacturing enterprises

The first procedure is literature review. Through domestic and international journals and magazines, 152 monographs and papers on manufacturing enterprise and index system of industrial design are collected. More than 50 papers are selected and 25 websites are selected from 63 relevant websites. With the method of frequency analysis, we have summarized 159 indicators as the design material for the first edition of the questionnaire from relevant literatures and opinions put forward by experts. Then, we have conducted a small-scale trial in Beijing and made adjustments to the content. At last, the first edition of the questionnaire is determined.

The first round of questionnaire distribution is in the form of electronic format and printed format. The questionnaires are distributed to experts in industrial design across the country, including the heads of industrial design related departments of manufacturing companies, the heads of industrial design companies, freelance designers, leaders from industrial design associations and relevant government departments, heads of industrial design related institutes, heads of design park, and experts from colleges and universities. A total of 119 valid samples are obtained. Valid questionnaires are screened for information entry and a preliminary screening of indicators. In this round of Delphi expert consultation, the boundary value method is mainly used to screen the evaluation indicators. The frequency of full score, arithmetic mean and variable coefficient are calculated according to the importance of each indicator. For inappropriate indicators, the experts of the research group will make adjustments based on the principles of comprehensiveness, reasonability, and feasibility. At the same time, the adjustment of the indicators also fully considers the additional opinions put forward by the experts. According to the results of the first round of surveys and the expert evaluation, the overall model of the index system of China's industrial design is initially determined. With reference to domestic and international design-related index system of industry evaluation and statistical habits in manufacturing enterprises, we demonstrate the overall status of evaluated object from three aspects: basic status, input and output. [16]

The second round of survey is mainly carried out by field research and face-to-face interview, with the key enterprises and institutions in the industries concerned by the ministry of industry and information technology as the research objects. Meanwhile, the deficiencies in the industries and the number of respondents in the previous round of questionnaire are balanced. In this survey, a total of 39 people from 21 enterprises or institutions are interviewed, including senior managers of manufacturing enterprises, directors of industrial design departments, senior designers, and managers of financial and personnel departments. Through three rounds of research and three rounds of expert evaluation, we have revised the overall framework and three-level index system of China's industrial design and development. Based on simplicity, comparability, manipulability, verifiability and sustainability, and the final index system of industrial design and development in manufacturing enterprises is completed, with the assortment of indicators.

3. Establishment of index system

Through the previous rounds of questionnaire and expert evaluation, we have preliminarily designed the index system of industrial design and development in manufacturing enterprises. The index system is divided into three levels, of which the first level contains the basic situation, input and output. The basic situation reflects the most distinct characteristics of the development of industrial design, and the input and output reflect the annual change of the development of industrial design. [17] The specific indicators are shown in table 1.

Table 1. Index system of industrial design and development in manufacturing enterprises.

Level 1 index	Level 2 index	Level 3 index	Details
Basic situation (A1)	Research and development needs (B1)	Design requirement (C1)	The demand for industrial design in the development of new products and upgrade products
	Organization structure (B2)	Department setup (C2)	Independent industrial design department setting
		Department level (C3)	The organizational structure of an industrial design department in an enterprise
			The decision-making power of the industrial design department in the development of new products
		Department structure (C4)	Settings in the industrial design department
	Personnel (B3)	Personnel quantity (C5)	Number of personnel in the R&D department and the industrial design department
		Personnel structure (C6)	The ratio of technicians, designers and researchers in the R&D department to the total number of R&D personnel
		Personnel qualifications (C7)	Age composition of industrial designers
			Titles of industrial designers
			Education backgrounds of industrial designers
	Equipment	R&D assets (C8)	Total amount of fixed assets for R&D department

	and facilities (B4)	Office environment (C9)	Environment and equipment of industrial design department
		Professional equipment (C10)	Investment amount of high-tech equipment for product development (such as rapid prototyping machine)
			Investment amount of design equipment for product development (such as computer, graphics tablet)
			Investment amount of equipment for product development (such as eye tracker, audio instrument)
Input (A2)	Capital investment (B5)	R&D investment (C11)	Annual total investment in product R&D
		Project proportion (C12)	Proportion of annual industrial design application project investment
			Proportion of annual industrial design pre-research project investment
		Cooperative research and development (C13)	Annual outsourcing status of industrial design project
			Research and development cooperation with other enterprises or research institutes
		Innovation protection (C14)	Protection cost and method of innovate design
	Talent investment (B6)	Personnel training (C15)	On-the-job training of industrial designers
			Design course training of non-designer
		Learning & exchanging (C16)	Participation and discussion in the design exhibitions
		Personnel income (C17)	Average annual labor compensation for employees in the industrial design department
		Incentive mechanism (C18)	The internal incentive forms in the company
	Equipment and trial production input (B7)	R&D investment equipment (C19)	Annual total investment in product development and professional equipment
		Trial production investment (C20)	Annual total model investment for product development
			Annual total mold investment for product development
	Policy support (B8)	Capital support (C21)	Relevant support fund from the government
		Policy support (C22)	Whether enjoy relevant support policies
Output (A3)	Innovation ability (B9)	New product development (C23)	The growth rate of new product development

IOP Conf. Series: Materials Science and Engineering **573** (2019) 012066 doi:10.1088/1757-899X/573/1/012066

	Achievement transformation (C24)	The growth rate of the transformation of patents
	Patent application (C25)	The patents produced through the industrial design
	Awards (C26)	Important industrial design awards
Economic benefits (B10)	Profit growth ratio (C27)	Enterprise profit growth
	Value ratio of new products (C28)	New product annual sales revenue share of product sales
Social benefits (B11)	Brand competitiveness (C29)	The influence of industrial design on product differentiation positioning
		The influence of industrial design on brand recognition
		The influence of industrial design on consumer loyalty
	Environmental benefits (C30)	The influence of industrial design on environmental protection

4. Description and discussion
The index system of industrial design and development in manufacturing enterprises is divided into three levels. The first level is to summarize the indicators of industrial design in manufacturing enterprises into three aspects: basic situation, input and output. The second level is the eleven major influence factors, which are further refined into 30 third-level indicators. [18-19]

4.1. The basic situation.
The first level is the basic information and living status reflecting the application of industrial design in manufacturing enterprises, including five factors: R&D needs, organizational structure, personnel status, equipment and facilities. The research and development needs are the demands of manufacturing enterprises for industrial design, which is mainly divided into two cases of new product development and product upgrade. The organizational structure presents the departmental structures of manufacturing companies, checking whether they have independent industrial design departments and internal structures, and the decision-making capabilities in product development. The personnel status shows the number, structure and qualifications of personnel in the design and R&D department as well as the proportion and experience of the R&D personnel. Equipment and facilities show the fixed assets, office environment and professional equipment of the industrial design department in the manufacturing enterprise and the various types of equipment and office facilities that have been invested.

4.2. Input.
The input reflects the annual investment of manufacturing enterprises in industrial design, which mainly includes capital investment, talent investment, equipment and trial production investment, and support policies. The capital investment demonstrates the R&D investment, project proportion, cooperative R&D and innovation protection of the manufacturing enterprise, reflecting the degree of emphasis on product R&D. It also shows the types of projects, the amount and quantity of project cooperation and the cost and method of innovation protection. The talent investment shows the learning and exchanging, personnel training, personnel income and incentive mechanism in

manufacturing enterprises, and on-the-job training of designers, design course training of non-designer, the design-related exhibitions and seminars as well. Equipment and trial production investment is to investigate the investment in R&D equipment and trial production of manufacturing enterprises, and to have the knowledge of the annual investment in equipment, model, and mold in product R&D of manufacturing company. Support policies shows the extent of government support for manufacturing enterprises and the policies and the amount of money for industrial design innovation.

4.3. Output.

The output reflects the annual output of industrial design in manufacturing enterprises, including innovation ability, economic and social benefits. Innovation ability shows the new product development, achievement transformation, patent application and awards in design of manufacturing enterprises, reflecting the important role of industrial design in the development of new products and patents application and the recognition of those high-level design competitions. The economic benefits, namely, profit growth ratio and value ratio of new products when applying industrial design demonstrate the changes of profits brought by new product research and development. Social benefits, namely, brand competitiveness and social environmental benefits, are implicit and difficult to quantify. Social benefits demonstrate that industrial design can promote product differentiation positioning, brand identification, consumer loyalty and environmental protection.

5. Conclusions

The index system of industrial design and development in manufacturing enterprises is a visual scale to measure the development level of industrial design between enterprises and the development characteristics between regions. It can reflect the corresponding relationship between industrial design and relevant concepts in national statistical system, and clarify the indicators of input and output and the way to assume value in industrial design of enterprises in product development process. The main features of the index system of industrial design and development in manufacturing enterprises constructed in this paper are as follows.

Firstly, the setting of the index system is highly objective. It is mainly based on the annual statistical data of the enterprise. The subjective factors and human factors are excluded as much as possible, and the industrial design and development of manufacturing enterprises is objectively reflected.

Secondly, the logic is simple and clear and the indicators are divided into progressive layers. It is of strong relevance. Thus, it is easy to accept and easy to analyse.

Thirdly, comprehensiveness. It reflects the overall situation of industrial design and development of manufacturing enterprises from multiple angles and through multiple indicators.

At the same time, we have seen some shortcomings in this research. First of all, this is the first nationwide evaluation of manufacturing enterprises. The indicators cover a wide range and need a large amount of statistical data. Secondly, some qualitative indicators are difficult to express in the form of a unified data. Finally, the weight determination of the index system and the calculation of mathematical model should be supplemented and improved in the following research.

Acknowledgment

The project of the Hunan Province philosophy and Social Science Fund "Research on value and performance evaluation of industrial design"， Number:16yba369

References

[1] Ministry of Industry and Information Technology. Core Index System for Government Website Development Evaluation (Trial) [S]. 2009.
[2] Social Science and Technology Statistics Department. Research on Index System Beijing Informatization Technology [R]. 2011,15-23.
[3] Zhao, Y, Hu, F. Principles and Structure of Establishing Index System of China's Sustainable

Development [J]. Jinan: China Population, Resources and Environment, 1997, (4): 54-59.
[4] Policy-making Group of the Manufacturing Industry Bureau of the Ministry of Economy, Trade and Industry. The Report of Sminar on How to Strategically and Effectively Use the Design Research [R]. Japan: Ministry of Economy, Trade and Industry, 2003-06-10.
[5] Kazunori, I. The History of Japan Institute of Design Promotion [EB/OL]. http://www.jidpo.or.jp/about/history.html.
[6] Wang, H. Research on the Selection Mechanism of Industrial Design Excellence Awards [D]. Beijing: Central Academy of Fine Arts, 2010.
[7] Wen, Z, Chen, M. Innovation Value Chain: New Thinking about Governmental Innovation Policies——The Case of Taiwan Innovation Policies [J]. Business Review, 2008,08:03-09.
[8] Li, Y. Research on the Situation and Improvement of Taiwan Innovation Policies [J]. Shanghai: Journal of Economics, 2008,12:02-11.
[9] Korea Institute of Design Promotion. Kiponet Brochure [M]. Korea Institute of Design Promotion,2008:02-13.
[10] Culture and creative industries in Germany. Research report No 577. Summary. [R].2009. Berlin, Germany
[11] Cho, D-S. Design, Economic Development, and National Policy [M]. Design Management Review,2004:10-20. [12]
[12] Nils, H.S., Marit, H. Design_Norway [M]. Nordic Innovation Centre,2008:01-24.
[13] Beijing Industrial Design Center. The Framework of Application Sample for World Design Capital of International Design Alliance [R]. 2010.
[14] Information and Statistics Department of China National Light Industry Council. Evaluation data of the informatization and industrialization level of househould appliance industry [EB/OL]. http://www.clii.com.cn/news/information/content-330277.aspx
[15] Qin, Q. Design on Index System of Evaluation for Competitive Capacity of Creative Industry in Haidian District [D]. Beijing Forestry University,2008.
[16] Wu, H, Hua, H. Empirical Study on the Efficiency Evaluation of Scientific and Technological Activities of Industrial Enterprises [M]. Science Press，2010：10-34.
[17] Wang, F. Study on Appraisal Index System of Sustainable Development of Mining Resource--Typed City [J]. Inner Mongolia Petrochemical Industry, 2008(8)
[18] Beijing Industrial Design Center. Development Situation and Existing Problems of Key Branches of Beijing Design Service Industry [R]. 2010
[19] The Impact of Industrial Design Effectiveness on Corporate Financial Performance》Julie H. Hertenstein, Marjorie B. Platt, Robert W. Veryzer Journal of Product Innovation Management，Volume 22, Issue 1, pages 3–21, January 2005.

Analysis on Advantages and Influencing Factors of Design and Development of Museum Cultural & Creative Products

Shi Bai[*]**, Guangyun Zhou**

School of Art and Design, Bengbu University, 233000 Bengbu, Anhui, China

[*]Corresponding author's e-mail: artbaishi@163.com

Abstract. The design and development of museum cultural & creative products originated from the needs of museum management and development, the promotion of Cultural and Creative Industry Park and the rise of new museology. Museum resources are cultural resources that can be utilized and developed, which provides the basic conditions for the design of cultural products. Through in-depth excavation and utilization of its own cultural advantages, museums can design and develop products that are popular with the public. The design and development of cultural and creative products in museums are affected by various factors, but actively seek strategies and ways to solve problems, they still have broad research and development prospects.

1. Introduction

Museums are embraced with a large number of resources and contain rich historical and cultural connotations. Cultural and creative products with high added value can be developed by transforming museum resources into effective elements through creative design. The design and development of museum cultural & creative products originated from the needs of museum management and development, the promotion of Cultural and Creative Industry Park and the rise of new museology. Museum cultural & creative products enjoy the characteristics of contemporary design language, which not only facilitate the dissemination of the spiritual connotation of museum collections to the public, but also can create certain social and economic values and effectively promote the sustainable development of cultural and recreational undertakings. Museum resources are cultural resources that can be utilized and developed, which provides the basic conditions for the design of cultural products. The uniqueness of museum cultural resources creates the uniqueness of cultural products, thus making them different from ordinary commodities. It is of great practical significance to study and analyze the advantages and influencing factors of the design and development of museum cultural and creative products.

2. Advantages of Design and Development of Museum Cultural & Creative Products

2.1. Museum Resources Provide Inspiration and Elements for Design and Development

Compared with ordinary enterprise organizations, museums have more abundant collection resources for the development and utilization of cultural and creative industries. All kinds of collection resources not only have the characteristics and significance of different times, but also contain many different factors caused by changes in history, environment, humanistic spirit, social consciousness, etc., which provide a lot of inspiration for the design and development of cultural and creative products. Thorough analysis of museum collection resources and full investigation of the needs of the public and the market are the key to the success of cultural and creative products. The museum's collection resources can provide creative sources of design from two aspects: first, select some objects and images in the museum's collection that have the most public influence and can best meet the aesthetic needs of the public, extract interesting and identifiable elements from them, and then transform them into commercially attractive designs; Second, the design, form, implication and process of the museum collection are refined to convey some feelings and feelings to the public through in-depth excavation of the museum collection information, so as to show the cultural characteristics of the museum collection[1].

Museum resources provide inspiration and elements for the design and development of creative products, and the design and development of related products also enable the value of collection resources to be better extended and spread. As for the traditional mode, the ownership and dissemination right of museum resources are often confined to the interior of the museum, which makes it difficult for the public to obtain resource information and the channels of resource utilization and dissemination are relatively blocked. Nowadays, more and more museums use advanced technical means to collect digitization of resource information, which is convenient for sharing pictures, text materials, videos and educational materials, etc. of the collections. This provides more powerful conditions for the design and development of cultural and creative products. For designers of museum cultural and creative products, they must first be familiar with and understand the museum's historical context and the connotation of the collection, and then they can reasonably and fully use museum resources to provide inspiration and elements for creativity, so as to establish design situations and strengthen design tension. On the basis of the above, it will finally realize the matching of product design expression methods and strategies to stimulate consumers' emotional resonance, so that museum cultural and creative products will convey the core cultural connotation of museum resources to the public.

2.2. Academic Resources Provide Basis for Design and Development of Cultural & Creative Products

The museum enjoys not only rich collection resources, but also professional scholars, experts and researchers. By integrating academic resources, the accuracy and foresight of the museum's dissemination of culture are ensured, and academic basis is also provided for the design and development of related cultural and creative products. The Palace Museum in Beijing has a strong academic research team. In recent years, the Palace Museum has launched a number of creative products App, such as "The Emperor's Day" (Figure 1). Through the role-playing of lions outside Gate of Heavenly Purity, the Palace Museum has an in-depth understanding of the emperor's food, clothing, daily life, work and entertainment 12 hours a day. It also has as many as 200 interactive points full of interest to lead users to complete game tasks, visit the Palace Museum buildings, appreciate the Palace Museum's cultural relics and learn historical allusions. Each interactive link is equipped with written explanations, which not only show vivid palace life, but also restore the misread historical truth. Another example is App "Han Xizai Banquet Map" (Figure 2). In the form of a whole long scroll, you can preview the overall picture and the partial picture of the painting by zooming in and out. You can get a detailed explanation of the figure information by clicking on the figure. The figure in the painting can be displayed in the form of animation. App inside's high-definition cultural relic images, professional academic materials, rich media content and innovative interactive design are combined to fully mobilize the user's sensory experience[2].

Figure 1. App "The Emperor's Day "

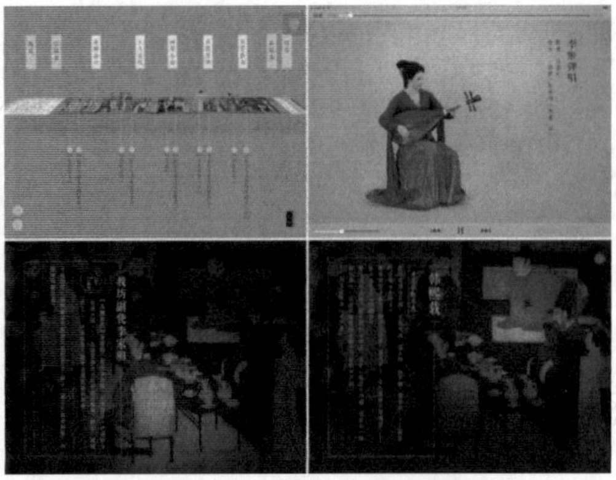

Figure 2. App "Han Xizai Banquet Map"

The content of App inside, a cultural and creative product of this kind, is produced through in-depth research and careful textual research by museum scholars, experts and designers. It is full of interest and knowledge for users. Thus, professional academic resources ensure the accurate interpretation of the museum's history and culture by cultural and creative products and provide all-round academic support for product design and development.

2.3. The Construction of High-quality Platform Provides Basic Guarantee for the Marketing of Cultural & Creative Products

As an information collection and storage institution, museums have natural advantages in the richness and popularity of their cultural resources. With the help of museum platforms, official shops are set up to sell their cultural and creative products so that the public can recognize the originality and authority of products from the heart. With the popularization of the Internet, the combination of museums and the Internet is increasingly close. More and more museums are starting to set up official product online shops to sell cultural and creative products online, and more actively take various modes to display collections to the public and sell related cultural and creative products. The promotion effect of museum platform on the marketing of cultural and creative products is mainly reflected in the following two aspects:

（1）It provides a platform for display and sales

Both the museum physical store and the online store provide a platform for the display and sale of their cultural and creative products, carrying a very rich amount of information. Taking the Tmall flagship store of the National Museum of China as an example, the website classifies all cultural and creative products in detail, which is convenient for consumers to search for relevant products clearly and quickly, and provides detailed introduction of creative sources, historical allusions, design elements and other aspects of the products, deepening consumers' cultural awareness of the products, thus promoting product sales.

（2）It provides a platform for communication and discussion

Many museums release the latest design and development of cultural and creative products through official social media, thus causing public concern and hot discussion. At the same time, they open Fansite for everyone to exchange and discuss. For example, the "Pearl Headset" and "Emperor Folding Fan" in the Imperial Palace in Beijing, the "I Know" adhesive tape in Inside The Emperor's Treasure, and the "Yue secret color porcelain Lotus Bowl Cookie" in Suzhou Museum have all once aroused public discussion on social media and thus become Internet celebrity products. Originally solemn and serious museum resources have become the topic of people's daily life, which has played an effective role in promoting the museum itself and cultural and creative products. The identity of online social media has also changed imperceptibly during this period, from "information disseminator" to "content creator". The public can freely discuss and express their views on social media and give a large amount of feedback information to cultural and creative products, which provides a big data reference for the museum's subsequent product design and development.

3. Factors Affecting the Design and Development of Cultural & Creative Products in Museums in China

3.1. The Influence of Museum System and Management Mechanism

The government's support in policy and finance is an important factor in the development of cultural and creative industries in museums. In recent years, the state has strongly supported the development of cultural and creative industries in museums. Public museums belong to public welfare institutions. On the one hand, the financial management system of public museums is that all operating funds of museums are fully allocated by finance, and state funds are not allowed to be used for operation, development and investment[3]. On the other hand, all museum revenues must be turned in. Because of the influence of the system and mechanism, although the museum's cultural and creative products have broad prospects for development, they cannot give full play to and release the advantages of talents and resources, which will definitely reduce the motivation and enthusiasm of the museum in developing cultural and creative products.

3.2. Impact of Adverse Protection of Intellectual Property Rights

Nowadays, with the rapid development of Internet technology, it is very convenient to spread and obtain information in the network. In addition, China's current intellectual property protection mechanism needs to be strengthened, which makes the entire cultural and creative industry, including museum cultural and creative products, vulnerable to infringement. Excellent Product innovation is likely to be quickly copied and imitated after it is highly praised by consumers and recognized by the market. For example, after Sichuan Sanxingdui Museum first introduced "Bronze Mask Biscuits", other museums have followed up the development of similar cultural relic-themed biscuits, resulting in serious homogenization of products. At the same time, museums will follow each other's product types, performance and even marketing methods, thus unable to concentrate all their efforts to create products with unique charm of the museum. The innovative research and development of cultural and creative products requires a great deal of manpower, material resources and financial resources. If it is copied and imitated by others without authorization once it is launched into the market, it will inevitably cause the creative commercial value of the originators to be difficult to fully realize and affect their Creative Power Entertaining. Without the protection of intellectual property rights, the legitimate interests of creative subjects cannot be protected. The government should improve the protection system of intellectual property rights and strengthen the protection of intellectual property rights.

3.3. Impact of Technological Development

Funds and talents are the basic guarantee to support the design and development of museum cultural and creative products, while technical support is the only way to realize the transformation from Creative Concept to real products. More and more museums integrate all kinds of collection information, systematically carry out digital design and establish multimedia data databases to facilitate Internet visitors from all over the world, which all benefit from the development of digital equipment technology. At the same time, the establishment of the digital museum makes the collection resources open to the public anytime and anywhere, which provides a new development opportunity for the design and development of cultural and creative products. On the other hand, through the application of technology, natural history, culture and creative products are more interactive and interesting in form, so that the public can understand and know the collection in a rich way of user experience[4]. For example, the high-tech art performance " Riverside Scene at Qingming Festival 3.0" , jointly designed and produced by the Beijing Palace Museum and Phoenix Satellite TV Group, perfectly combines the application of digital technology with the contents of national treasures. Through excavating the artistic verve and cultural connotation of the original work, it integrates the double 8K ultra-high definition projection technology to show the various living beings and social features of Bianjing, the capital city of the Northern Song Dynasty, in a dynamic effect, and constructs the immersion experience of real and virtual people in the painting. It can be seen that the development of technology can enrich the design methods of cultural and creative products, and the application of technology can create ideal cultural and creative products, so that the public can be satisfied with the experience behavior, and finally realize the conversion of cultural and artistic values to cultural and artistic prices.

3.4. The Influence of Museum Collection Quantity and Value

The number of museum collections is directly proportional to the number of materials that can be used for reference in the design and development of cultural and creative products, and the value of the collections directly affects the connotation and market attention of their cultural and creative products. The interaction between the museum and the collection takes the museum platform as the carrier, and the museum will be famous for its precious collection. For example, Mausoleum of the First Qin Emperor Museum's Terracotta Warriors, Inside The Emperor's Treasure's The Green Jade Cabbage, Shanghai Museum's Big Grams Tripod, Neues Museum's bust of Egyptian Queen Nefertiti and so on. There are many kinds of related creative products designed and developed based on these precious cultural relics, which are loved by people all over the world. For the design and development of museum cultural & creative products, we should seriously explore the collections that can show the value of the museum as creative materials, create cultural products that are symbolic and can be integrated into contemporary life, and activate the collections, so as to give full play to the educational, collectible, cultural and commercial nature of museum cultural and creative products.

3.5. The Influence of Museum Professionals

Although the museum is extremely rich in cultural resources, if there is no relevant professional personnel to work, the museum's cultural and creative products cannot well show the charm of culture. China's cultural and creative industries started late, especially in the museum's internal institutions, which generally lack talents in creative design, management and other aspects. Therefore, lack of experience in the design and development of related cultural and creative products has led to a series of problems, as follows:

（1）Products lack innovation.

Most museums in China have insufficient investment in the design and research of cultural and creative products, lack their own team of design talents, and have yet to form original design and research capabilities and competitiveness. The design of most products is only a simple copy of the cultural relic resources in the library or a copy of the excellent literary creation products at home and abroad, ignoring the originality of the design and lacking the extraction of the cultural essence of the library. For example, the calligraphy themes developed by the Shanghai Museum, Nanjing Museum and Suzhou Museum have no difference except for the color matching of the products and the calligraphy content printed on the surface. Consumers can hardly identify the source of the products. Cultural & creative products of various museums should display their characteristics to the greatest extent. Imitation of each other will hinder the development of cultural and creative industries. From the consumer's point of view, the lack of creative and similar products affects the consumption choice space and cannot meet the growing material and spiritual needs, thus making the museum cultural and creative products less attractive to consumers.

（2）Products lack brand awareness

Brand is an intangible asset with economic value, occupying a certain position in the consumer's consciousness, covering specific expressions such as product packaging and visual image. At present, the cultural & creative products of most museums lack brand difference and brand identification, and the unique cultural brand image of museums has not yet been established. In order to gain an advantage in the market for museum cultural and creative products, we must stick to original design and brand promotion, which are also the two ends of the "Smiling curve" of the cultural and creative industry with high added value (Figure 3), seize every opportunity to establish Brand Identity, demonstrate the uniqueness and quality of its products, and enhance consumers' trust in the museum brand[5]. The museum avoid consuming too much investment in the middle and lower parts of the curve, and strive to exert power at the left and right ends of the curve with high added value, i.e. pay attention to design research and development and brand promotion.

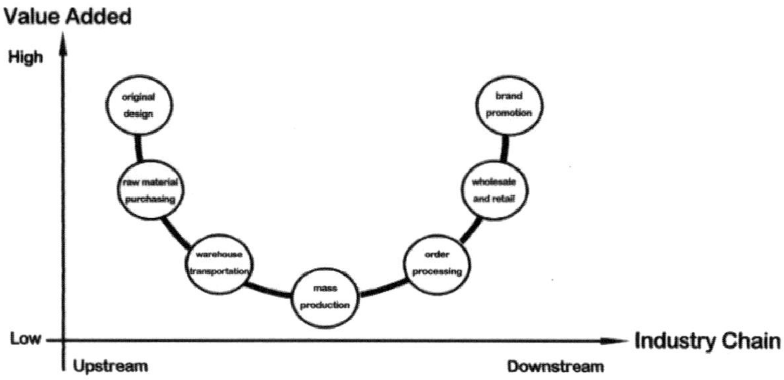

Figure 3. Smile Curve of Cultural and Creative Industry Chain

4. Conclusion

The richness and uniqueness of museum resources are an important source of the form and connotation of cultural and creative products. Through in-depth excavation and utilization of its own cultural advantages, museums can design and develop products that are popular with the public. Although the design and development of cultural and creative products in museums are affected by various factors, they still have broad research and development prospects. While making good use of its rich collection resources, academic resources, high-quality platforms and other advantages, museums should also consider the limitations of product design and development, actively seek strategies and ways to solve problems, and integrate various forces to provide better conditions for the design and development of museum cultural and creative products, thus providing broad space for the sustainable development of cultural and creative industries.

Acknowledgment

Supported by Social Science Key Project of Bengbu University (No.2017SK04zd), Quality Engineering Project of Anhui Higher Education (No. 2018jyxm0704)

References

[1] Xinmeng, L. (2018) Research on the Development of Cultural Creative Products in Xi 'an Museum. (Master Thesis) Northwest University, China.

[2] Dongrun, Y., Xiang, M. (2018) Research on Art Information APP Interaction Design from the Perspective of Semiotics. J. Media. Commun., 01: 82–84.

[3] Ning, L. (2011) A Research on the Development of Chinese Museum in the Past Decade. (Master Thesis) China University of Political science and Law, China.

[4] Ting, ZH., Xinwei, T. (2018) The Enlightenment of Japan's Digitalization of Cultural Heritage. J. Studies in Culture & Art. Commun., 04: 19–27.

[5] Xiuwei, W. (2016) The Research on Museun Cultural Licensing in the Horizon of Cultural Creative Industries. (Doctoral Thesis) University of Science and Technology of China, China.

Experimental Study on the Preference Characteristics of children with ASD for Product Design Materials

Xiaowen Liu[1*], Wanqing Zhang[2], Ni Liu[3]

[1]School of Art and Design, Xi'an University of Technology, Xi'an 710054, China

[2]Engineering Research Centre of Rubber and Plastic Sealing, Xi 'an Aerospace Propulsion Institute, Xi'an 710100, China

[3]Rehabilitation Centre for Special Children, Xi'an Children's Hospital, Xi'an 710003, China

[*]Corresponding author's e-mail:631039377@qq.com

Abstract. In recent years, individuals with ASD have attracted increasingly great attention from all circles of the society, but the studies concentrating on the life product design for these special users are rare. This study adopted the paired comparison method and involved three groups of Chinese children aged 3-6 years (children with ASD, children with mental retardation and normal children), with seven of the most common materials in life products as the evaluation objects, to investigate the preference characteristics of children with ASD for product design materials. The results indicated that the preference order and preference degree of children with ASD were both highly different from those normal children at the same age. The preference order of children with ASD was similar to the children with mental retardation, while the preference degree was different. This study provides designers and researchers with a more comprehensive understanding of the psychological features of children with ASD, and it can be the theoretical basis for designers to develop further guidelines encouraging upgraded humanized product design for this special population.

1. Introduction

Autism pedigree disorder (ASD), known as autism or autism in a broad sense, is one of children's extensive series of complex neurodevelopmental disorders [1]. The individuals with ASD are associated with disorders of social, behavioural and communication abilities. In 2016, a report released by the National Centre for Health Statistics showed that the incidence of autism among children aged 3 to 17 was estimated at 1/45. Additionally, according to the latest data published in the Report on the Development of China's Autism Education and Rehabilitation Industry in 2017, it is conservatively estimates that there are at least 10 million autistic people and 2 million autistic children in the 1.3 billion populations in China, and the annual growth rate is nearly 200,000[2]. Therefore, it is particularly important to provide innovative life products that really meet the special needs of ASD children in view of their intellectual features and living conditions.

On the basis of the current findings, although the related comments from individuals with ASD were broad to some extent; few valuable insights of product design in this field were obtained. There are almost no specific products developed for this group in the market, and most of the products still originate from the concept of normal children's product design. The current research on products for children with ASD is mostly confined to product appearance and entertainment. Specifically, it is rare to explore the material preference of this special group so far, which is an important content related to their physical and mental health development.

In industrial design, materials are substances used to form products that exist objectively and independently of human consciousness. Whether traditional or modern materials, natural or artificial materials, single or composite materials, they are all the material basis of industrial design and the basic elements for products to satisfy functional requirements and reflect structures[3]. Thus, it is necessary for designers to consider not only the materials performance characteristics, the corresponding process conditions, costs and material resources, but also the psychological impacts on users in the selection of materials for product design. Furthermore, with the continuous advancement of technology, the improvement of people's living standards, the combination of materials and design are combined more closely, as the user's demand for products has gradually improved from material satisfaction to spiritual and psychological satisfaction [4].

The particularity of children with ASD results in the particularity of the products they use. Material selections of products for this group need to satisfy their special visual and psychological preferences in order to design life products that truly meet their physiological and psychological demands, so as to create a comfortable and meaningful living environment for them, which will promote their physical, intellectual and spiritual development.

In view of this, this study adopted the paired comparison method to study the material preferences of three groups of children aged 3-6: children with ASD, children with mental retardation with matching physical age and psychological age and normal children with matching physiological age. This study attempts to explore the following two problems:

- What are the material preference characteristics of children with ASD, and what is the order of their material preference?
- Whether there are significant differences of material preference between children with ASD and the other two types of children, and what are the results?

2. Data Collection

2.1. Participants

Basic characteristics of the subjects are shown in Table 1. The experimental group consisted of 15 children with ASD from the rehabilitation centre of Xi'an Children's Hospital. They were all diagnosed as autistic children by Xi'an Children's Hospital according to DSM-IV criteria, and were further identified by the convergent evaluation of doctors and researchers. The control group consisted of 15 children with mental retardation who came from the same institution and had no characteristics of autism. The matching discrepancy of verbal and operational intelligence between the experimental group and control group was +/-10 percentage points [5]. Another 15 normal children came from ordinary kindergartens in Xi'an, matching the physiological age of the experimental group and control group. All three groups of children were excluded from visual and auditory diseases.

Table 1. Basic characteristics of group with ASD, control group and ordinary group

	Sex Ratio	**Age (years)**		**Verbal IQ (WISC-CR)**		**Performance IQ (WISC-CR)**		**Full IQ (WISC-CR)**	
	Male/Female Ratio	Mean	Range	Mean	Range	Mean	Range	Mean	Range
ASD Group (n=15)	10:05	4.88	3-6	60.43	46-74	66.65	48-100	60.22	45-85
Control Group (n=15)	9:06	5.08	3.5-6	66.88	42-87	66.23	35-96	63.02	45-90
Normal Group (n=15)	9:06	4.2	4-6	—	—	—	—	—	—

2.2. Experimental materials and tasks

A great variety of materials are used in actual products, and seven of the most commonly used materials for actual children's products are selected as evaluation objects in this experiment : glass, wood, metal, plastic, rubber, silicone and felt. Each material is presented as a sphere with a diameter of 4 cm, as shown in figure 1. Seven material balls are combined in two pairs, a total of 21 combinations. Considering that most of the children with ASD have language function barrier but strong operation abilities, the operation of the entire experimental system is effective. The subjects were asked to select one of the two material balls presented in the experiment and put them into a black foam board with a hole in the middle. The surface of the foam board is black, with a circular hole in its Centre (white around, so that the hole will be found easily). The diameter of the foam hole is 4 centimetres, which is available to hold a material ball.

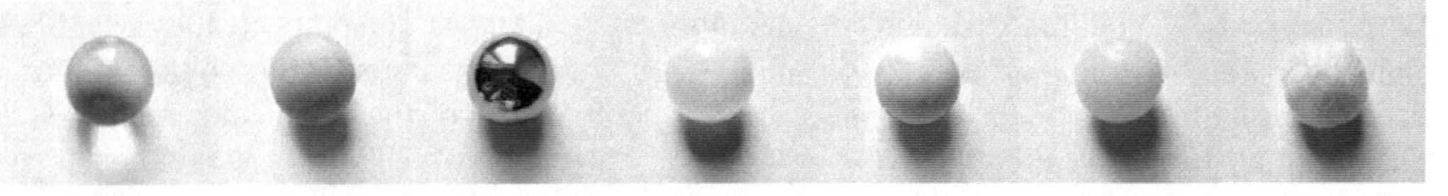

Figure 1. Evaluation objects: seven material balls

These seven evaluating material balls were presented in two groups of presentation sequence (group A and group B). Group A: glass-wood, glass-metal, wood-metal, wood-plastic, metal-plastic, metal-rubber, plastic-rubber, plastic-silicone, rubber-silicone, rubber-felt, silicone-felt, glass-plastic, glass-rubber, wood-rubber, wood-silicone, metal-silicone, metal-felt, plastic-felt, Glass-silicone Glue, glass-felt, wood-felt. Group B: felt-wood, felt-glass, silicone-glass, felt-plastic, felt-metal, silicone-metal, metal, silicone-metal, silicone-metal, silicone-wood, rubber-wood, rubber-glass, plastic-glass, felt-silicone-rubber, silicone-plastic, rubber-plastic, metal-wood, wood, metal-glass, wood-glass. In order to balance the sequential order presented by the 15 combinations, stepwise approach was used to arrange the presenting order of material balls, as shown in Table 2.

Table 2. Material presentation sequence

Material	Glass	Wood	Metal	Plastic	Rubber	Silicone	Felt
Glass	—						
Wood	1	—					
Metal	2	3	—				
Plastic	12	4	5	—			
Rubber	13	14	6	7	—		
Silicone	19	15	16	8	9	—	
Felt	20	21	17	18	10	11	—

2.3. Experimental Procedure

In order to ensure the quietness of the surrounding environment and the emotional stability of the subjects, all experiments need to be conducted in different rooms. Firstly, the experimenter explained the demonstration method in detail to the subjects: choose a preferred one from the two stimuli presented and put it into the test board. The formal experiment was started until the understanding and operation of the subjects were confirmed after several demonstrations. The experimenter presented two material balls in a preset order each time, and then the subjects are asked to select one of them to put into the black foam board. After placing one material ball, means a selection is completed. Each participant completed 42 judgments in Group A and Group B.

3. Data Analysis

Table 3. Average Number of Material Selection for Three Groups of Children

	Order	**1**	**2**	**3**	**4**	**5**	**6**	**7**
Group with ASD	Material	Glass	Metal	Felt	Silicone	Plastic	Rubber	Wood
	M	8.53	8.33	8.13	6.93	5.47	2.47	1.93
Control Group	Material	Silicone	Glass	Metal	Felt	Plastic	Wood	Rubber
	M	8.93	8.27	7.47	6.87	4.00	3.33	2.80
Normal Group	Material	Felt	Silicone	Rubber	Metal	Wood	Plastic	Glass
	M	8.27	7.20	6.07	5.67	5.67	4.20	3.93

(Note: Average M = cumulative number of times each material ball is selected / number of children in the group)

Table 3 shows the orders and the degrees of material preference of three groups of children. Firstly, as listed in table 3, the children with ASD in this experiment preferred glass and metal most, followed by felt, silicone and plastic, and rubber and wood at last. They preferred smooth and bright materials to rough materials. The material preference order of children with retarded intellectual development in the control group was silicone, glass, metal, felt, plastic, wood and rubber. By contrast, the most favourite materials for normal children were felt, silicone and other soft materials, which was consistent with recent research results. Nevertheless, there was a higher preference for metal materials in normal group children, and it was possible because that the gender of the majority of participants in normal group was male, which was in order to match the ASD group. Then, the preference order comparisons of the three groups were analysed using the Pearson's correlation coefficient method, as in equation (1). The results indicated that the material preference order of ASD group and control group were similar, with the rank correlation coefficient $r = 0.878$, $P < 0.01$. However, the order differences between ASD group and normal group were highly significant, with rank correlation coefficient $r = 0.077$, $P > 0.05$.

$$r = \frac{\sum\left(X-\overline{X}\right)\left(Y-\overline{Y}\right)}{\sqrt{\sum\left(X-\overline{X}\right)^2\sum\left(Y-\overline{Y}\right)^2}} = \frac{l_{XY}}{\sqrt{l_{XX}l_{YY}}} \tag{1}$$

Secondly, the degree of material preference was expressed by the mean value M. For example, the glass score of group with ASD was 8.53, which means that there were 8.53 person-times choosing glass when it was compared with other materials in that group. Then, the preference degree comparisons of the three groups were analysed using the chi-square test, and the following conclusions had been obtained:

- Overall, the numbers of material selection among the three groups were significantly different, with $X^2(12) = 37.07$, $P < 0.01$;
- There were large differences in selection times between ASD group and control group, with $X^2(6) = 14.098$, $p < 0.05$;
- There were highly significant differences between ASD group and normal group, with $X^2(6) = 30.062$, $p < 0.01$;
- There are relatively remarkable differences between control group and normal group, with $X^2(6) = 13.373$, $p < 0.05$.

As shown in the analysis of the above, children with ASD aged 3-6 were similar to children with mental retardation in material preference order, but their material preference degree (selection times) were significantly different, for example, rubber and wood were all at the end, but the ASD group made fewer choices of them, with the average of 2.47 and 1.93 respectively, while the control group with the average of 2.80 and 3.33 respectively. Furthermore, it also indicated that there were great remarkable differences both in material preference order and preference degree between children with ASD aged 3-6 and normal children of the same age. Overall, children with ASD preferred bright and smooth materials, while ordinary children preferred soft and warm materials.

4.The Pedagogical Implications of This Study

For educational application, the materials preferred by children with ASD can be used in design products for early interference treatment. Amotivational syndrome is one of the important characteristics of children with ASD, especially learning new information and executing new tasks [6].Appropriate products will not only attract their attention and stimulate their intrinsic interest, but also maintain their motivation to participate in activities, rather than relying on external reinforcements or rewards. Barker believes that paying attention to children's interests and choices will improve their social motivation and promote their common attention. [7]. Robert L. Koegel also found that low motivation level of children with autism is usually manifested in the form of grumpy temper, refusal to cooperate, inability to concentrate and avoidance of problems [8]. Laski's research shows that stimulating the motivation of children with ASD in natural situations will not only improve their language and game skills, but also create effective and positive emotional effects. Therefore, motivation has become a critical area [9], and it is one of the important measures for effective intervention in children with ASD.

In addition, the intervention method of motivation stimulation strategy is the core content of Pivotal Response Treatment (PRT) [10], which was proposed by Robert L. Koegel and Lynn Kern Koegel, and firstly published in the 1980s as a form of naturalistic behavioural intervention based on Applied Behaviour Analysis (ABA) [11]. PRT is child-led, emphasizing natural situations and improving motivation to help children with ASD learn key skills, conducting multiple clue analysis, and integrating feedback on natural events. It aims at guiding children to promote their own abilities in an active way, in order to promote generalization to cope with the real living environment [12]. PRT has been recognized gradually in theory and practice, and has been promoted in various fields all over the world in recent years. The National Project Standards of the American Centre for Autism Professional Development and the National Centre for Autism Research have repeatedly identified PRT as the most scientific and empirical method of autism intervention[13]. Therefore, keeping insight into the preferences and interests of children with ASD and design products that will satisfy their physiological and psychological needs in the real sense will helpful for their effective rehabilitation promotions.

5.Conclusions

Autism Spectrum Disorders (ASD) brings tremendous mental and economic pressure to the society and families. Children with ASD are usually hardly be integrated into normal social life, and long-term training and care will be necessary [14]. As a result, products in living environment play an important role in the psychological development for children with ASD. This paper explored the material preference of children with ASD aged 3-6years in China by method of paired comparison. As shown in the results that children with ASD generally prefer bright and smooth materials, and the material preference order is: glass, metal, felt, silica gel, plastic, rubber and wood. Their material preference order is significantly different from that of normal children, but there is no significant difference with the retarded children matched by age and IQ. In addition, the material preference degree of ASD group children is remarkably different from the other two groups of children.

Overall, the results provide a theoretical basis for researchers and designers to develop further guidelines to encourage products that satisfy the special needs of children with ASD in real sense. Nevertheless, continuous investigation on different aspects of their preference characteristics is lacking. Further research should investigate more effective and diversified innovation approaches for product design practices.

Acknowledgments

This project is supported by National Social Science Foundation Art Project of China (Grant No. 18CG199)

References

[1] Mikle, S., Jacqui, R. (2017) Sensory, emotional and cognitive contributions to anxiety in autism spectrum disorders. Frontiers in Human Neuroscience., 11:1-7.

[2] Han, Y., Ma, X. (2018) A review of autism case studies in China. A Journal of Modern Special Education,. 329 (02): 32-37.

[3] Su, J. N., Li, H. Q. (2005) Study on Sensory Properties of Materials in Industrial Design. Mechanical Design and Research., 21 (3): 12-14.

[4] Tambouratzis, T. , Karalekas, D., Moustakas, N., (2014) Special Issue on Sustainability in Manufacturing: The Role of Life Cycle Engineering. Journal of Industrial Ecology., 18(4):508-516.

[5] Cao, S. Q., Fei, W., Sun, A. J. (2012) An experimental study on color preference of autistic children aged 4-6.5. Chinese Journal of Special Education., 5: 49-52.

[6] Deanna, S., Angela S., Susan, W., Elizabeth, L. (2005) Emotion Dysregulation and Anxiety in Adults with ASD: Does Social Motivation Play a Role?. Journal of Autism and Developmental Disorders., 45(12):3971-3977.

[7] Baker, M. J. (2000) Incorporating the Thematic Ritualistic Behaviors of Children with Autism into Games: Increasing Social Play Interactions with Siblings. Journal of Positive Behavior Interventions., 2(2):66-84.

[8] Koegel, R. L., Koegel, L. K. (2006) Pivotal response treatments for autism: communication, social and academic development. Brookes Publishing Co., New York. pp. 3-30.

[9] Laski, K. E., Charlop, M. H., Schreibman, L ., (1988) Training parents to use the natural language paradigm to increase their autistic children's speech. Journal of Applied Behavior Analysis., 21(4):391-400.

[10] Huang, W. H., Chen, X .Y., Li, D. (2010) Key Skills Training Method: A New Direction of ABA Applied to the Education and Intervention of Autistic Children. Chinese Journal of Special Education., 124 (10): 63-68.

[11] Koegel, R. L., Koegel, L. K., Koegel, B. L., et al. (2014) Children and Youth with Autism Spectrum Disorder (ASD): Recent Advances and Innovations in Assessment, Education, and Intervention. Oxford University Press. The Oxford.

[12] Hardan, A., Gengoux, G., Berquist, K., et a1. (2015) A Randomized Controlled Trial of Pivotal Response Treatment Group for Parents of children. Child Psychol Psychiatry., 56 (8)：884—892.

[13] Whalen, C., Schreibman, L., Ngersoll, B. (2006) The Collateral Effects of Joint Attention Training on Social Initiations, Positive Affect, Imitation, and Spontaneous Speech for Young Children with Autism. Journal of Autism and Developmental Disorders., 36 (5): 655-664.

[14] Simpson, R. L. (2005) Evidence-Based Practices and Students With Autism Spectrum Disorders. Focus on Autism and Other Developmental Disabilities., 20 (3):140-149.

[15] Yafei,Q., Liang, L., Shuliang, M. et al. (2016) Autistic Children's Visual Attention to Restrictive Interest Stimulation: An Eye Movement Study. Chinese Journal of Special Education., (5): 29-34.

[16] Luling, S. (2017) Retrospect and Analysis of the Research on the Toy Preference of Autistic Children. A Journal of Modern Special Education., 16: 38-41.

IOP Conf. Series: Materials Science and Engineering **573** (2019) 012069 doi:10.1088/1757-899X/573/1/012069

Exploring the Application of Serious Game Based on Augmented Reality: A Case Study on Tsingtao Beer Museum

Xueai Li[1], Miantao Xie[1], Junjie Chu[1*]

[1] Departments of Industrial design, Ocean University of China, Qingdao, Shandong, 266100, China

[*]Corresponding author's e-mail: Chujunjie@ouc.edu.cn

Abstract. This work aims to explore the feasibility and significance of serious game based on augmented reality in museums, and approaches its effect on the experience of visitors. Depending on the status of museums and the characteristics and advantages of AR serious game, the article demonstrates that the experience and the follow-up purchase intention of visitors can be enhanced by improving the interestingness, learning and interaction of visiting museums, and setting up personalized mechanisms. Taking Tsingtao Beer Museum as an example, we designed a game concept prototype and described the method of game development. Finally, the article proposes the prospect and significance of further study on the application of AR in museums.

1. Introduction

As a significant part of the national economy, tourism has unlimited potential while developing vigorously. The data issued by National Bureau of Statistics [1] has shown an annual trend of growth in the number of visitors of museums in China. In 2017, the number of visitors of historical relics institutions in China has reached more than 1.1 billion, including more than 0.9 billion of museum visitors. "We're heading to a golden age of relics and museology," said Dr. Jiang Hu, the vice curator of Shanghai Museum. [2] Meanwhile, the 21st century poses some challenges to the relevance of museums and museums have to strive hard to avoid coming across as inaccessible ivory-towers [3]. AR technology creates a blend between the real world and the digital, bringing the digital 3D objects literally in our hands [4].

Taking Tsingtao Beer Museum as an example, in this paper, the authors reviewed the previous studies in the field of cultural institutions to understand the application and development of AR in museums, investigated whether the use of serious games in museums can engage the experience of visitors transfer cultural knowledge as side effects to the visitors in the process. On this basis, combined with augmented reality technology, we designed a serious game prototype in Tsingtao Beer Museum, analysed and summarized the level of the game prototype to solve the problems in traditional museums and the ability to enhance visitor's experience. Finally, we proposed the plan of game development and future work.

The remainder of the paper is organized as follows: section 2 provides a literature review of the application of AR in museums, the related experience and serious game based on AR; section 3 describes the investigation of Tsingtao Beer Museum; section 4 mainly introduces the flow and the development of our game; section 5 discusses the differences between traditional visit and visit with the game in 6 factors; section 6 outlines the conclusion and section 7 discusses the future work.

IOP Conf. Series: Materials Science and Engineering **573** (2019) 012069 doi:10.1088/1757-899X/573/1/012069

2. Literature review

In recent years, the construction of museums in China has developed rapidly. In 2017, the Cultural Technology Innovation Plan [5] presents that "fully facilitate the integration of scientific technology into the cultural sector". In order to conform with the trend of times and alleviate problems in the construction of museums, an increasing number of museums begin to use new media to reform and innovate [6]. AR is the practice of dis-playing digital information over people's real-time view of objects, people, or spaces in the physical world [7]. It has become an efficient, automatic, and playful method towards the appreciation and understandings of tangible and intangible cultural heritage [8]. The use of AR was found to not only bring history to life, but also actively engages visitors and facilitates the gathering of new information [9]. However, there are also some important restrictions in the use of AR be found. For instance, the special conditions of museums (e.g. artificial markers cannot be used, small, crowded rooms, etc.) and the limited computational capacity of smartphones [10]. Thus, there is a need to study areas relating to the application of AR in museums and find a way to combine museums with AR.

Museum tourism, like heritage tourism, is viewed to a great extent as an experiential consumption [11]. Though the combination of AR and museums has been put into practice, study indicated that museums in China are overall in an early transitional period in terms of the adoption of, and the development of cultural technology appropriate for visitor access [12]. Study showed that AR has the potential to influence the experience of muse-um visitors significantly and positively [13] Besides, in terms of the subsequent impacts, AR-enriched user experience produces higher perceived value [11], and user satisfaction [14]. Thus, Chen et al. [11] establish a relationship "experience quality / perceived value / satisfaction / behavioral intentions". It is noticeable that individuals have different approaches in the way they forage, retrieve, process, store and recall information, so the adaptive and personalizing mechanisms should be proposed [15].

Investigations reveals that participants often have a preference for, and expectations of "interactive games" and "Interactive Exhibit Design" [12]. In addition, serious games have been proven to be a source of enjoyment and, if well-made, are a powerful tool to communicate knowledge [16]. The study pointed out that the game facilitated a wider exploration of the museum [17]. Therefore, serious games may be a good way to combine AR with museums, which can also enhance the experience of visitors.

3. Investigation of Tsingtao Beer Museum

3.1. Tsingtao Beer Museum

The investigation is based on the case of Tsingtao Beer Museum, which is a comprehensive topography museum of history and art. The museum is the only beer museum in China. Its exhibition area is more than 6000 square meters which is divided into three parts: history and culture, production technology and multi-functional area. The museum is located in the old workshop and with old equipment of Tsingtao Beer a hundred years ago. It concentrates the historical development of beer industry in China, based on the centuries' history and techno-logical process of Tsingtao Beer, integrating cultural history, production process, entertainment, creative shop-ping and catering, which is intellectual, entertaining and engaging.

For matching up the vising process of production technology, we understand the current information and process of the museum with the guide map (see Fig. 1). There are two topographical problem can be summarized from the map: some spaces are narrow, and the processes are too many and confusing.

IOP Conf. Series: Materials Science and Engineering **573** (2019) 012069 doi:10.1088/1757-899X/573/1/012069

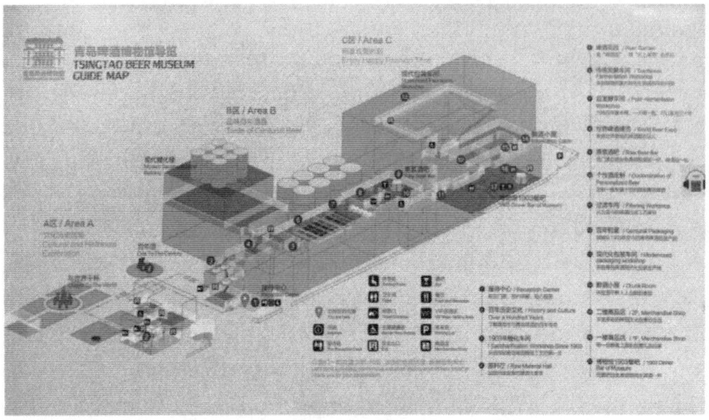

Figure 1. Tsingtao Beer Museum guide map.

3.2. Online comments

In order to gain a preliminary understanding of visitors' experience in Tsingtao Beer Museum, the online comments from several digital travel platforms were collected and translated from Chinese to English. Here are some of the online comments on the touring experience in Tsingtao Beer Museum which can represent many other similar comments (see Table 1).

Table 1. Typical Comments of Tsingtao Beer Museum Extracted online.

No.	Comment
1	"We were deeply impressed by the cold beer and beer beans in Tsingtao Beer museum, everyone received and ate them. The museum provided beer twice, but the second time we don't want to drink without beer beans. But the beer crafts and exhibits were a little boring and the arrangements are jumbled."
2	"Visitors rarely stopped to understand and appreciate the beer history or equipment. And there were many people so that it's crowded. I can't remember an exhibit that brightened my eyes in the journey. However, it was the beer that made me feel fresh."
3	"I like travelling, especially the ones about history and this is my second time to Tsingtao Beer Museum. I'm here to have a glass of beer at 10 degrees Celsius, and to make some like-minded friends. The Tsingtao Beer Museum introduces a lot of knowledge about the production of beer. It's great, but I haven't found difference from the first time I came here. I think it's time to change."
4	"The introduction of cultural relics in the museum is too rigid. At last, I want to buy some souvenirs, but the expensive price stopped me. It's a pity that I haven't understood the whole production flow of beer after visiting."
5	"Tsingtao Beer Museum is famous for the production of beers. And the most happiness thing is that my family came here to have a glass of beer. We understood the raw material and process of beer together, and the instructions of the exhibits were easy to understand, but the words are so boring that I didn't want to read."

Summarizing these online comments, we can extract some common words that the visitors have used to describe Tsingtao Beer Museum: crowded, boring, narrow space, cannot remember, jumbled arrangements and expensive price (if the visitor pay for it). It is worth noting that most visitors mentioned the free beer pleased them. Therefore, there are both advantages and frequently mentioned problems on the experience of Tsingtao Beer Museum.

3.3. On-site investigation

In order to examine the needs of visitors in Tsingtao Beer Museum further, we went there to interview both the visitors and the stakeholders of the museum (see Fig. 2). A questionnaire is designed to investigate visitors' needs and their experience in Tsingtao Beer Museum.

Figure 2. The on-site interview with several visitors and an on-site staff in Tsingtao Beer Museum.

100 questionnaires (adapted from Fenu and Pittarello [18]) were sent out in total at the exit of the museum (where is a bar for visitors to drink), and 91 questionnaires were finally collected.

From the analysis of the survey we can obtain an overall idea of the profiles of the users who engaged in the experimentation. Most of the visitors (70%) were between 18 and 45. Many people (about 55%) visit the museum for entertainment and viewing, 75% of them spent 2 hours or less visiting the museum. Based on existing exhibition form, visitors showed little interests in history and technology.

The latter questions were designed on problems in most museums, and the feedback has been collected. The experience of visitors in the museum has been analyzed, and based on these data, factor analysis is carried out with SPSS. The result of factor analysis is shown in the table. (see Table 2)

Table 2. Factor analysis of elements of experience in the museum.

Factors	Communality	Mean value of factor load	Name	Variance contribution rate (%)	Cumulative contribution rate (%)
F1	0.625~0.684	0.527~0.729	Interestingness	14.78	14.78
F2	0.567~0.632	0.554~0.735	Transmission efficiency of knowledge	12.19	26.97
F3	0.568~0.649	0.583~0.676	Satisfaction degree of personalized service	11.85	38.82
F4	0.577~0.610	0.566~0.763	The science and technology in exhibition	9.56	48.38
F5	0.516~0.565	0.558~0.692	Immersion	7.63	56.01

According to the results of factor analysis, the cumulative contribution rate of these five factors is 56.01%. Therefore, we can conclude that the main problems of visitors exist in these aspects. The main requirements of visitors for the exhibition in museums are interestingness, transmission efficiency of knowledge, satisfaction degree of personalized service, the science and technology in exhibition and the immersion. The results provide the data basis for the next design.

3.4. Exploration of technology
Based on the literature review and the actuality of Tsingtao Beer Museum, the authors find three key problems.

Firstly, the difference between Tsingtao Beer Museum and other historical museums is that visitors can observe closely as well as touch the tools to understand the production flow, however, most visitors cannot emerge with a full-fledged concept of the production flow after visiting. Visitors can't experience a complete process because of the limitations in the museum. However, the problem can be solved by using AR. Therefore, it is necessary to present the production flow that these visitors want to know in an appropriate context, and recreate a cognitive and emotional link with the author and the place they visit [19].

Secondly, the exhibition in the museum lacks personalized service. Due to the museum is rebuilt from the old factory building of Tsingtao Beer, it is difficult to solve the problems about the settings of exhibits and topography in the museum by existing technology. Tsingtao Beer Museum has a large number of visitors, it is necessary to set the display in the open, but the practical halls are narrow (because the equipment was not built for visiting, but for the production in the old days). Thus, there is a possibility to design personalized tour route by using AR to solve the problem.

Thirdly, the guide in the museum is not clear enough. Visitors are difficult to know their positions as well as the progress of the visit. In this situation, visitors can't remember and classify the contents in the museum. Thus, it is in accordance with the rules of people's memory if location information can be added in the process.

3.5. Conclusion
In conclusion, here are the needs of visitors in Tsingtao Beer Museum which are most important: better interaction and more interesting, learning in an enjoyable and relaxing process, a clear and appropriate system for guiding.

4. A serious game based on AR
Aiming at the result of our investigation, we designed an augmented reality serious game prototype based on mobile devices in order to find a way to address the needs of visitors in Tsingtao Beer Museum and improve the overall touring experience.

4.1. Game prototype flow
"Finding" is a serious game prototype based on augmented reality. The equipment and tools displayed in the production flow hall can be recognized by the user's camera to carry out the game, which encourages them to collect relevant clues and knowledge from the exhibition. After getting the definite information, visitors can interact with the game and get real-time feedback. After their tour, they can get some coupons for souvenirs which engage their experience and intentions to purchase. Hence, the game meets the needs of visitors for interestingness, as well as learning effect.

Visitors should choose the tags they are interested in first to determine an appropriate game them (there are five different game routes in total, system will match the latest similar one with the tags according to the user's choice). After that, visitors can explore the museum according to the hints, which gives them a lot of space where both interestingness and knowledge can be reflected. What's more, visitors may work together when meet problems which creates opportunities to interact with others in reality.

In order to design a complete AR game, the logical flow between the main pages is shown as follows. (see Fig. 3)

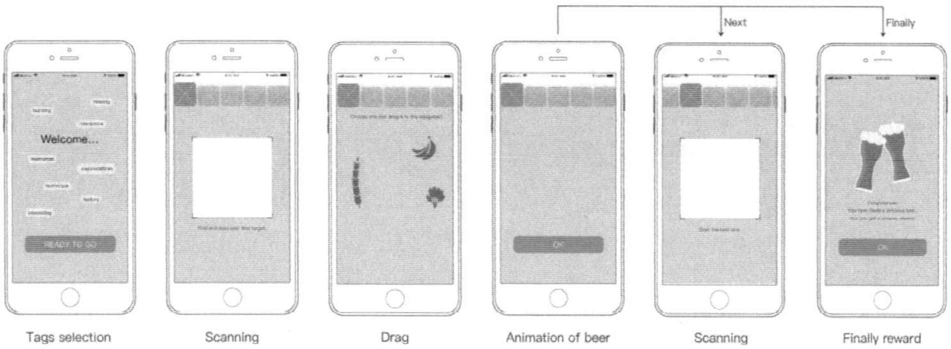

Fig 3. Low-fi prototype illustrating the game flow of the mobile augmented reality game Finding.

Tags selection: In order to meet to preferences and needs of different visitors, five different game routes in the prototype are designed in the background (For example, there are differences in the number of checkpoints and the content displayed on each page between users who choose history and learning and users who choose technology.), and a most appropriate game will be chosen by system according to the tags that visitors choose.

Scanning: After selecting the interest tags, users can visit the museum and enter the scanning page, which will prompt users to find the target. According to the real production process of beer, visitors need to complete the game items which are set in each process step (Scanning in each exhibition hall). Each step in the overall process is shown at the top of the page. Users can know their real-time location and progress according to the status bar, and enjoy the game and memory better.

Dragging: A pop-up tip will be displayed if users scan the specific equipment, which shows the required material or conditions of the corresponding tool. Users can complete the task by dragging and pulling the correct material (barley is the material needed in fig. 3) or conditions onto the equipment. Visitors can experience the pleasure of brewing and create an environment-friendly interactive experience by throwing in the mock raw materials.

Animation of beer: When user drags the materials or conditions onto the corresponding equipment, the collider on the equipment will be triggered, then corresponding animation will be displayed. The wort rolling in the equipment is showed in fig. 3. At the end of each operation, users will gain real-time feedback, which simulates the changes of raw materials in beer brewing process.

Scanning: After the displaying the animation, users will receive the next task (scanning the next equipment).

Finally reward: Ultimately, users can get some reward when they complete all the tasks based on their completion accuracy.

4.2. User journey map

A user journey map illustrating the visitor's experience with the mobile augmented reality game in Tsingtao Beer Museum was drawn (see Fig. 4). With the game, visitors can learn in a relaxing and interesting way. Due to the personalization design, they can visit with their own touring route and interest. Besides, the different experience engages and rewards can influence their intentions to purchase.

IOP Conf. Series: Materials Science and Engineering **573** (2019) 012069 doi:10.1088/1757-899X/573/1/012069

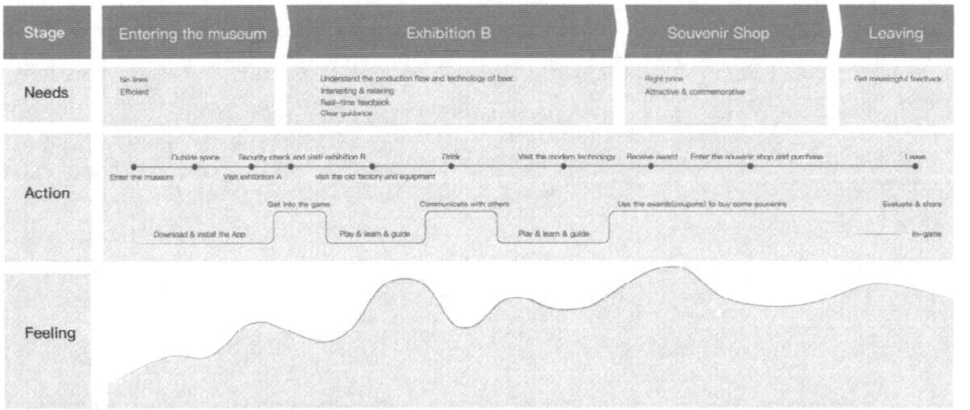

Fig. 4. User journey map of the experience with or without the mobile augmented reality game.

4.3. Game prototype development mode

In order to develop our design into a mobile application in the future, we have learnt and mastered the necessary technology, and then chose EasyAR SDK Pro (an augmented reality engine that supports 3D object tracking and generates dynamical tracking targets from standard .obj model files) to achieve AR features. We need to obtain the .obj model files of the equipment firstly to identify the equipment in the museum as a 3D Object Target (the object that be detected and tracked). And the models are got by Maya. What's more, at least one material image should be imported to the models so that the 3D objects can be identified the tracked in the hall, which is different from most existing image Target recognitions. Ultimately, we'll develop demos in Unity in the future, the model is as follow (see fig. 5).

Fig. 5. High-fi prototype illustrating a main part of the game flow.

4.3.1. Interaction. We have made many comparisons on how to realize the interaction between AR objects and visitors. And we chose to use the Lean Touch plug-in finally, which allows visitors to control the position, size and rotation of the objects in the game better. In this way, the game elements can interact with visitors well and have rich interestingness. Visitors can achieve the goal easily by habitual gestures in daily life. In Blender, we will build the prototypes (such as hops, barley and so on) which we need to use as sub-objects of 3D Object Target. What's more, we need to add some C# scripts to realize some UI attributes such as drag and zoom to be used by visitors.

4.3.2. Award mechanism and data collection mechanism. After the game, according to the setting of award mechanism, system will compare the materials and equipment chosen by users with the standard process. Different number of coupons (from 20% to 50%, which will be obtained from the museum) in souvenir shop will be presented to users based on the consistency of the data. The aim is to test the intentions of visitors to purchase the cultural and creative products, for the psychology to take advantages. Every visitor's game data will be transmitted to us through mobile network or wireless network, so as to iterate and update in the later stage.

5. Discussion

We extracted and analysed the factors related to visitors in the game prototype. They are: 1. Degree of knowledge about craft exhibits; 2 The interestingness and entertainment of visit; 3 Degree of personalization; 4 Interaction experience with exhibits; 5 Intention to purchase souvenirs; 6 Desire to advertise (share with others) the museum. Based on these factors, comparing with the existing museum visiting mode, it can be found that visitors can interact with exhibitions joyfully through AR game. Besides, the immersion and experience of visitors will be greatly improved. For the benefits of museums, the application of AR games will increase the sales of souvenirs and attract a number of visitors. In the future, the application of AR games will have a better situation.

6. Conclusion

This paper describes the design and evaluation of a serious game prototype in museums based on augmented reality. Depending on the results of previous research in the field of cultural institutes and the investigation of Tsingtao Beer Museum, we discussed the key problems that can be solved by AR and the effect on visitors' experience with serious games based on AR.

In the game prototype we designed, the traditional visiting mode in museums has changed, which has opportunity to enhance the visitor's experience in many ways. Visitors can explore while learning knowledge in an interesting process (different from imparting knowledge). What's more, visitors can reach the peak of emotional experience in stages (getting real-time feedback when interacting with exhibits). In this process, visitors play with their own personalized routes to ensure that each visitor gets the best experience as far as possible. People can also communicate and cooperate with others at the same level, which can promote potential social interaction.

Nevertheless, there are some key factors that we must pay attention to. For example, the limited time and energy to engage in virtual games to some visitors. Hence, in order to avoid the unpleasant misunderstandings and troubles which may reduce the travel experience when using AR, the design should be more intuitive and relatively simple to attract more visitors.

7. Future work

Due to the constraints of resources, we have only designed the concept prototype of the game for the time being. In the future, we will continue to explore technology and develop the game as a complete augmented reality game based on mobile devices. Because of the uniqueness of each museum, this paper can only be used as a reference for other museums. The future game mechanism will be more intuitive and easy to understand. The on-site test will be conducted to bring greater improvement to the experience of visitors. At the same time, the con-tent of Tsingtao Beer Museum will be better combined with augmented reality games, so that to play the role of theory and games in other fields. Next-generation technologies such as Augmented Reality are fast permeating many society sectors. Their market is projected to reach $95 billion by 2025, representing a large portion of the economy within the next decade [20]. With the coming of 5G era, high-speed and low-latency data transmission methods will give rise to more applications of augmented reality. Thus, it is of great significance to study augmented reality in museums.

References

[1] Lu L. Q., (2018) Development status and trend analysis of Chinese museums in 2018, https://www.qianzhan.com/analyst/detail/220/180621-935f24ed.html.
[2] 360doc, (2017) The needs and pain spots of technology application in museum. http://www.360doc.com/content/17/0815/13/35024114_679348636.shtml.

[3] Neuburger, L., Egger, R. (2017). An Afternoon at the Museum: Through the Lens of Augmented Reality. Information and Communication Technologies in Tourism, 241-254. Doi: 10.1007/978-3-319-51168-9_18.

[4] Kyriakou, P., Hermon, S. (2018). Can I touch this? Using Natural Interaction in a Museum Augmented Reality System. Digital Applications in Archaeology and Cultural Heritage, 12, e00088. http://dx.doi.org/10.1016/j.daach.2018.e00088.

[5] Ministry of Culture of the People's Republic of China. (2017). Cultural Technology Innovation Plan during 13th Five Year period from Ministry of Culture.

[6] Liu, N. (2018). The Application and Future development countermeasures of New media of Chinese museums. Master. Hunan University.

[7] Scholz, J., Smith, A. N. (2016) Augmented reality: Designing immersive experiences that maximize consumer engagement. Business Horizons, 59, 149—161.

[8] Clini, P., Frontoni, E., Quattrini, R., Pierdicca, R. (2014). Augmented Reality Experience: From High-Resolution Acquisition to Real Time Augmented Contents. Advances in Multimedia, 597476. http://dx.doi.org/10.1155/2014/597476.

[9] tom Dieck, M. C., Jung, T. H., Rauschnabel, P. A. (2018). Determining visitor engagement through augmented reality at science festivals: An experience economy perspective. Computers in Human Behavior, 82, 44-53.

[10] Gimeno, J., Portalés, C., Coma, I., Fernández, M., Martínez, B. (2017). Combining traditional and indirect augmented reality for indoor crowded environments. A case study on the Casa Batlló museum. Computers & Graphics, 69, 94–105.

[11] Chen, C.-F., Chen, F.-S. (2010). Experience quality, perceived value, satisfaction and behavioral intentions for heritage tourists. Tourism Management, 31, 29–35.

[12] Ch'ng, E., Cai, S., Leow, F.-T., Zhang, T. E. (2018). Adoption and use of emerging cultural technologies in China's museums. Journal of Cultural Heritage, CULHER-3519; No. of Pages 11. https://doi.org/10.1016/j.culher.2018.11.016.

[13] Mortara, M., Catalano, C. E., Bellotti, F., Fiucci, G., Houry-Panchetti, M., Petridis, P. (2014). Learning cultural heritage by serious games. Journal of Cultural Heritage, 15, 318–325.

[14] Poushneh, A., Vasquez-Parraga, A. Z. (2017). Discernible impact of augmented reality on retail customer's experience, crossmark satisfaction and willingness to buy. Journal of Retailing and Consumer Services, 34, 229–234.

[15] Raptis, G. E., Fidas, C. A., Avouris, N. M. (2016). Differences of Field Dependent/Independent Gamers on Cultural Heritage Playing: Preliminary Findings of an Eye–Tracking Study. EuroMed 2016, Part II, LNCS 10059, pp. 199–206. Doi: 10.1007/978-3-319-48974-2 22.

[16] Froschauer, J., Arends, M., Goldfarb, D., Merkl, D. (2012). A Serious Heritage Game for Art History: Design and Evaluation of ThIATRO. 2012 18th International Conference on Virtual Systems and Multimedia, 283-290.

[17] Rubino, I., Barberis, C., Xhembulla, J., Malnati, G. (2015). Integrating a Location-Based Mobile Game in the Museum Visit: Evaluating Visitors' Behaviour and Learning. ACM J. Comput. Cult. Herit. 8, 3, Article 15 (May 2015), 18 pages. Doi: http://dx.doi.org/10.1145/2724723.

[18] Fenu, C., Pittarello, F. (2018). Svevo tour: The design and the experimentation of an augmented reality application for engaging visitors of a literary museum. Int. J. Human-Computer Studies,114, 20–35.

[19] Bak, D. (2016). Literary museums today: definition, status, prospects. In: Proceedings of the ICOM International Committee for Literary Museums (ICLM) Annual Conference. ICOM, Paris, pp. 66–76.

[20] Li, H., Gupta, A., Zhang, J., Flor, N. (2018). Who will use augmented reality? An integrated approach based on text analytics and field survey. European Journal of Operational Research. https://doi.org/10.1016/j.ejor.2018.10.019.

Analysis on the Establishment of Bionic Design Thinking Model in Design Innovation

Meina Wang[1*]

[1]Departments of Art, Sichuan University Jinjiang college, 610000, China

[*]Email: cimona@126.com

Abstract. Our purpose was to evaluate the usefulness of the Bionic Design Thinking Model. On the basis of previous studies, a set of innovative product design thinking models have formed, based on the theoretical basis of product modeling design thinking and the theories of bionics, design psychology and thinking science.This paper calls for a high degree of harmony and unity between human society and nature through bionic philosophy of design, and calls for a reasonable symbiotic dialogue platform between human and machine, ecological nature and artificial nature.

1. INTRODUCTION

In recent years, the research on the thinking mode and method of product innovation design has become a hot spot in foreign design research, but the domestic research on this is still relatively insufficient. LIU studied the psychological phenomenon of the designer when designing the product shape, and initially established a mapping relationship between the product shape and the psychological response. GERO et al. analyzed the problem-solving process of innovative design of modeling, and pointed out that analogy reasoning is an important means to generate innovative design of modeling. SAKOL et al. put forward a creative design method of modeling based on design knowledge and user knowledge, and established a knowledge transformation model in modeling design. At home, Zhao Jianghong, Su Jianning et al. have mainly put forward image-based design methods. These studies have expounded the innovative design thinking of products from various angles, but have not yet formed a simple and effective design thinking model system.

Therefore, on the basis of previous studies, I have formed a set of innovative product design thinking model taking bionic design as an example based on the theoretical basis of product modeling design thinking and the theories of bionics, design psychology and thinking science, which will be described in detail below.

IOP Conf. Series: Materials Science and Engineering **573** (2019) 012070 doi:10.1088/1757-899X/573/1/012070

2. MATERIAL AND METHODS

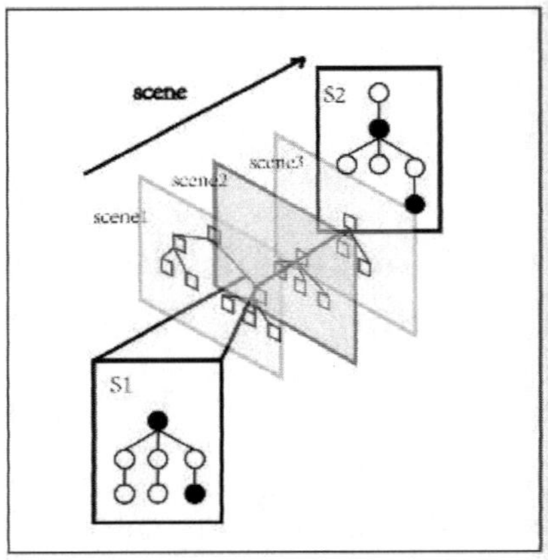

Figure 1. The different answers of one problem and their problem scenes
The answer S1 correspond with the problem P1, the answer S2 correspond with the problem P2, S1 and S2 are different because of the different scene

From the perspective of logic, product innovation design is a generalized problem-solving process, and the result of design may not be a specific final solution of an actual problem, but rather an "intermediate" solution from the initial problem state to the final solution. As the problem scene changes, the solution is adjusted and changed accordingly. As shown in **Figure 1**, for the same design problem, the solution may also change due to different scenes.

At the same time, the problem domain of product innovation design is a typical pathological domain. Under such circumstance, the initial state of the problem (design requirements), the target state of the problem (design results) and the operation mode of the problem (design rules) are not clear. So, it is difficult to solve the product modeling design problem through rule reasoning, and the solution process of the problem needs to be completed through the expert knowledge and experience of the designer. However, compared with the expert knowledge and experience of other fields (such as product reliability, manufacturability, etc.), the expert knowledge and experience of innovative design have greater uncertainty and fuzziness. The strategy adopted by many innovative designs is to generate new design ideas through analysis of past or other design cases. Therefore, the problem solving of product innovation design is a typical analogical reasoning process, as shown in Figure 2, from solution s in scene 1 to solution in scene 2.

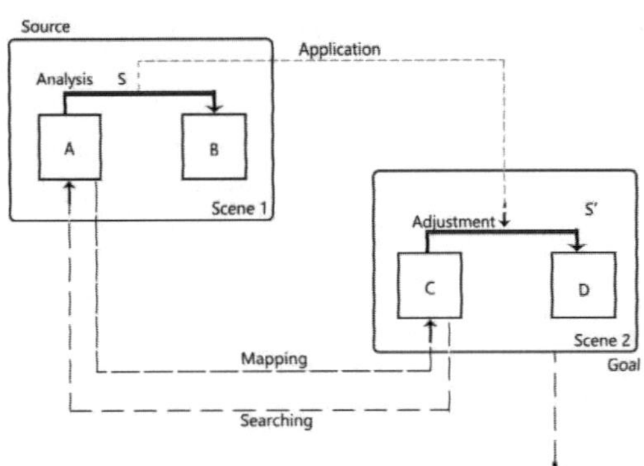

Figure 2. The analogical reasoning process of product design
A——primeval state in scene 1 B——objective state in scene 1
C——primeval state in scene 2 S ——solution of the problem in scene 1
D——objective state in scene 2 S ' ——solution of the problem in scene 2

2.1 Design Thinking Dominates the Problem-Solving Process

Product innovative design requires the designer's expert knowledge and experience to complete the problem-solving process, while knowledge and experience form the designer's design thinking, in other words, design thinking plays a leading role in the solution process of innovative design.

2.2 Using Bionic Design Thinking to Solve Problems

The process of bionic design of common products can be simply shown in **Figure 3**:

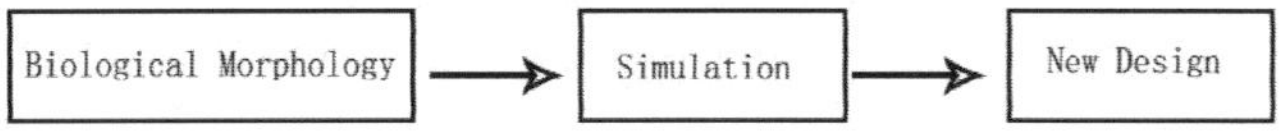

Figure 3. The normal bionic design thinking model

However, starting from the "quasi-materialized" design and using bionic design thinking to carry out product innovation design process, it can be shown in **Figure 4**:

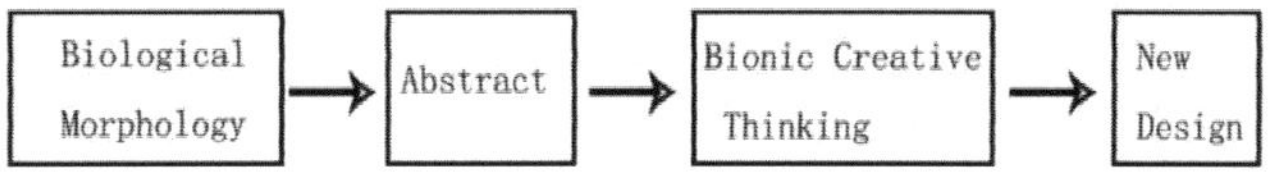

Figure 4. The product design thinking model based on bionic design thinking

It is worth pointing out here that bionic design is not an exact imitation of biological prototype, but based on biological prototype, through creative thinking, carries out secondary or even multiple innovations to obtain the design form--this is the essence of bionic design thinking. The model is shown in Figure 1.5.

Therefore, if people want to make the "enlightenment" of bionic design enter into the advanced stage, they need to abandon the pure natural form (i.e. biological prototype), use its principle to carry out the "quasi-materialized" design, and enter the "transformation" from "quasi", so as to create a real new design.

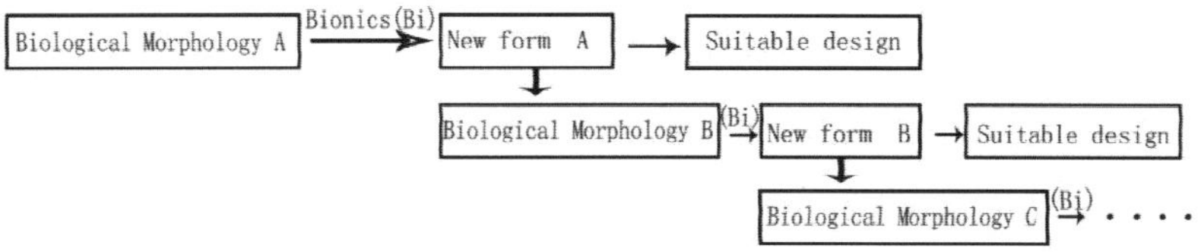

Figure 5. The flow chart of the bionic design thinking model

3. RESULTS

3.1 Establishment and perfection of bionic design thinking model in product innovation design

Product design is a systematic project, so it needs to be carried out in an orderly way. According to the design's own regular program, it reflects the different links of design behavior, and each link should show clear stage goals.

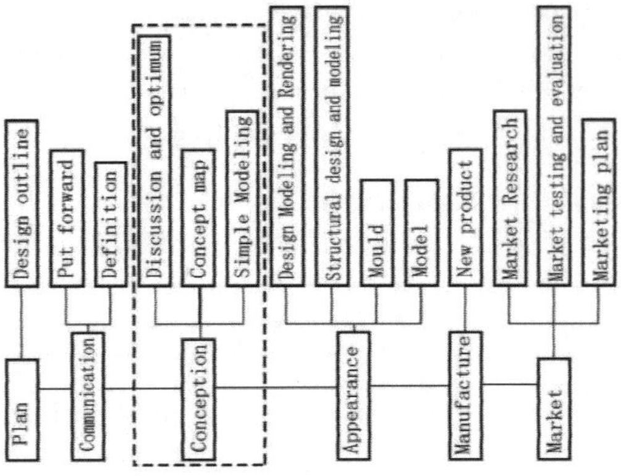

Figure 6. The flow chart of product design

As shown in **Figure 6**, this is a basic design process. Each stage in the process has its own method and theoretical guidance, and this paper mainly studies the part outlined by dotted lines in the figure--the conception stage, which is of course an important link before a good design work come out.

From the decomposition part of the figure, we can know that the "conception" part includes three parts: discussion and revision, conceptual diagram and simple modeling. Through these three tasks, it is undoubtedly the designer's thinking, that is, the interpretation and understanding of this design task in his mind. From this point of view, it is the next work that we have to complete--form a set of effective bionic design thinking models, which will provide designers with broader thinking space and more references in the conception stage.

3.2 Establishment Process of Design Thinking Model

3.2.1 Research on the Difference between Designers' Rational and Perceptual Thinking Modes

1) Cognitive Test of Product Innovation Design

Innovative design of products is mainly to code and reproduce the modeling elements (symbol information, semantic information and performance information) of products. In this process, we can use interviews, oral analysis and design concept sketches to study the thinking characteristics and performance methods of designers, to externalize and transfer their perceived images, and to discover their innovative design rules. This paper uses the experimental method of "process research" in psychological research--oral analysis method 1, and takes a typical household product--seat as the design object to analyze the rules in the designer's thinking mode.

2) Test process

In the test, a total of 30 industrial design-related staff or university teachers and students were selected to provide design requirements and reference materials with seats as design objects, and the subjects were required to carry out bionic modeling design work. The test uses real-time oral report, requiring the subjects to speak out their thinking process and focus of attention in words and record them in the whole process.

3) Test analysis

After the test is completed, the collected oral materials are classified and summarized. Through analysis, it is found that in the process of product modeling design with the use of bionic design thinking, there are mainly two different modeling thinking models according to the designer's thinking mode: Designers with rational thinking mode, after being familiar with the design task, try to decompose the design task into several subtasks, and then answer several subtasks respectively. Finally, they synthesize the final design scheme as shown in **Figure 7**(left). For example, when designing a seat shape, they will decompose the task into subtasks such as side contour shape, seat back shape, armrest layout, man-machine relationship and color matching, and then synthesize them after answering respectively. Designers with perceptual thinking are more inclined to describe their design plan in the design process through semantic concepts, such as curved surface, straight face, unified and changed words, and then carry out step-by-step optimization design with unique ability to grasp modeling semantics and artistry, combined with product design positioning. Based on the analysis of modeling design tasks and questionnaires, the design thinking process flow chart was established. As shown in **Figure 7** (right):

Through comparison and summary, it is found that designers with rational thinking mode follow the characteristic of designing from low abstract level to high abstract level, which is a bottom-up reasoning process of design thinking. This way of thinking analyzes the design tasks hierarchically according to the modeling characteristics, decomposes them into several subtasks, then seeks the solutions of the subtasks, and finally integrates the results. However, designers with perceptual thinking mode follow a characteristic of designing from high abstract level to low abstract level, which is a top-down thinking process. It is guided by concept to process the information of the design content, starting from the overall form, through the design semantics, looking for a breakthrough, combined with their own design knowledge, put forward creative design scheme. As shown in **Figure 8.**

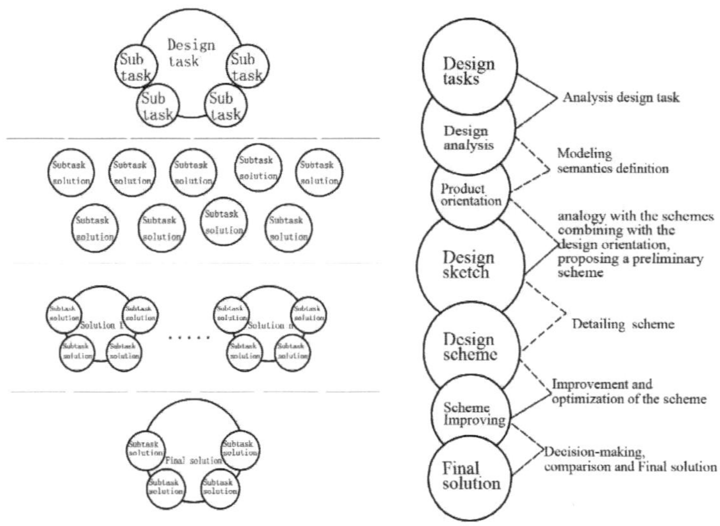

Figure 7. The rational thinking mode (left) and the perceptual thinking mode (right)

To sum up, the design thinking mode is to follow a bottom-up or top-down design information processing process, in which the designer's knowledge is externalized into product modeling. This is the premise of establishing bionic design thinking model. Through this step, designers can be helped to find their own way of thinking more quickly and apply it comprehensively to open the door of creative inspiration.

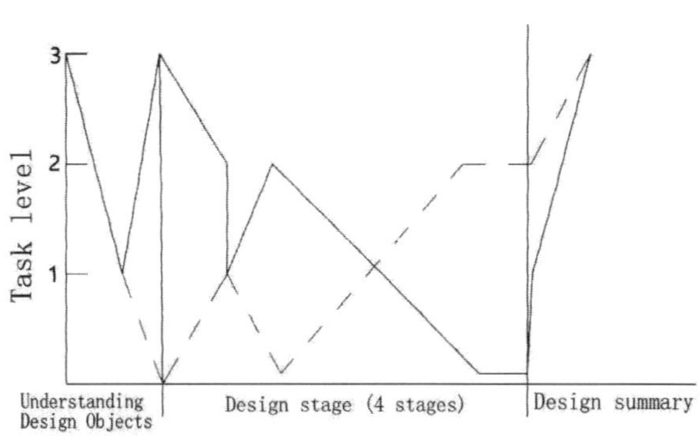

Figure 8. The different designer based on different thinking way have different thinking mode
———perceptual thinking ---------------rational thinking

IOP Conf. Series: Materials Science and Engineering **573** (2019) 012070 doi:10.1088/1757-899X/573/1/012070

3.2.2 Designer's Perceptual Image Model

Whether it is a bottom-up or top-down thinking mode, the application of bionic design thinking means a process from concrete to abstract. It's just that the designers with rational thinking mode ask questions first, then seek corresponding "answers" in natural objects according to the questions asked, i.e. corresponding objects, and then abstract them. However, most designers with perceptual thinking mode first discover their excellent characteristics in the nature, and then abstract the design concepts they need according to the inspiration of such characteristics, and then create a concept of new product. This process from concrete to abstract requires the designer's perception and idea model to guide it.

First, look at Figure 4. The thinking mode model diagram mentioned earlier is already a basic framework, but it is still too simple. We need to further refine and perfect it to form a set of thinking model that can be generalized, that is, specifically discuss how to use creative thinking to carry out bionic design product innovation. After investigation and summary, bionic innovation can be summarized as fixed-point method, the association method, combination method, extraordinary method and Imitation. These skills can effectively help designers divergent thinking and put forward more suitable design schemes. These methods and theories have been mentioned in previous studies and will not be repeated here. To sum up, it can be represented by the following figure. As shown in **Figure 9**.

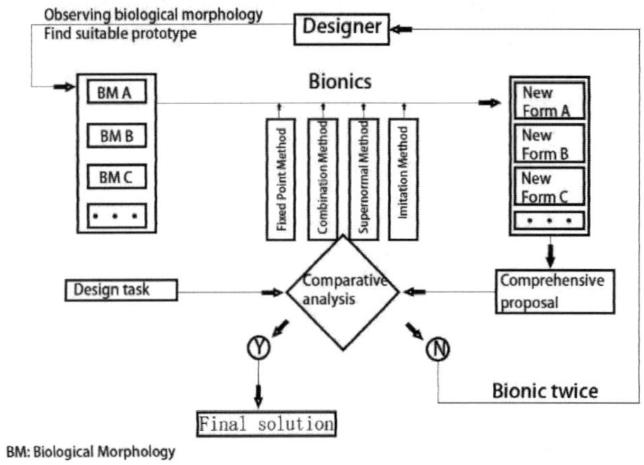

Figure 9. The thinking model based on bionic design thinking for designer

In short, it is based on the requirements of product design tasks, seeking all kinds of suitable biological prototypes in nature, and then through various innovative techniques to spread and sum up, forming new design proposals. If the proposal is not passed, it will be returned to carry out secondary or even tertiary bionic innovation to form the sketch scheme that best meets the design task requirements.

The establishment of this model is mainly to help designers expand their thinking and accelerate the formation of concepts. However, in order to strengthen the design scheme and form the final design product, the following perceptual image matching model between the user and the designer needs to be used to achieve the ultimate goal of user demand.

3.2.3 Perceived Image Matching Model between Users and Designers

At the conception stage, we should not only pay attention to the designer's way of thinking, but also the user is a group we cannot ignore, after all, they are the ultimate users of the product. Therefore, it is necessary to study the user's perception image before the design starts, standardize the design knowledge and make it match with the user's perception intention. On this basis, we proposed a perceptual image matching model between users and designers as shown in **Figure 10**.

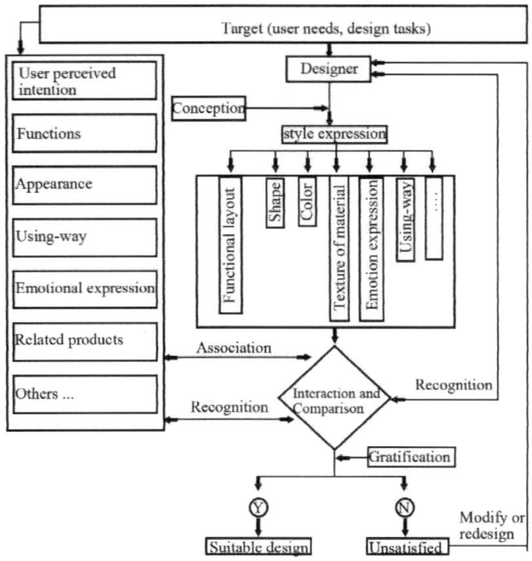

Figure 10. The thinking model based on bionic design thinking for matching between the designer and the user

Designers combine design goals (user requirements) with their own experience and skill, transform ideas into the appearance of products (including interfaces, symbols and colors, etc.), giving users a kind of induction. Users, with association way of perception to their own experience and needs, have an interaction with the product shape designed by the designer, using the help of the previous product usage. When the interaction is consistent or some kind of coupling, the design of this form meets the needs of users to a certain extent. When the interaction is not consistent, the satisfaction of the user cannot be achieved, and even the user's behavior may be misled, requiring modification or redesign.

So far, we have come up with an interlocking product innovation design thinking model based on bionics and creative thinking. The model firstly obtains two basic thinking modes of sensibility and rationality through investigation and research, which can be used as references for designers according to different people. After that, the designer's perceptual image model was consulted to carry out innovative thinking, and the design inspiration was drawn from biology to form a preliminary sketch plan. In the end, the user's needs were further adapted through the user and designer's perceptual image matching model, and the design plan was screened and re-innovated, so that the design plan was "Right the first time". The specific flow chart is shown in **Figure 11**.

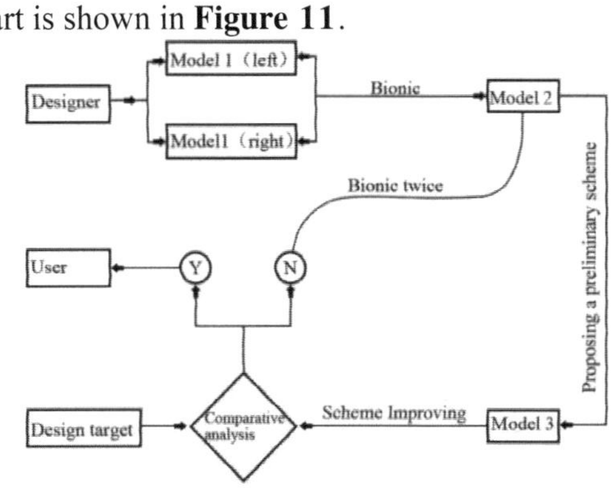

Figure 11. The general thinking model based on the bionic design thinking
Model 1 is Figure.7, Model 2 is Figure.9, Model 3 is Figure.10

IOP Conf. Series: Materials Science and Engineering **573** (2019) 012070 doi:10.1088/1757-899X/573/1/012070

4. CONCLUSIONS

The establishment of the bionic design thinking model has the following obvious advantages:

1. It can explore in the nature with infinite charm and stimulate the designers' inspiration.

2. From a psychological point of view, the launch of a new product is a new stimulus with a certain intensity for consumers. The process of being recognized is actually a process in which consumers are not accustomed to it and become accustomed to it. The design works produced under the guidance of bionic design thinking model are easily recognized by consumers because they are drawn from familiar biological forms, and the adaptation time will be shortened accordingly.

3. Focusing on breaking the traditional thinking set and innovating concepts, it focuses on the theme of harmony and unity between human and nature. It advocates to study the laws and functions of the relationship among human, machine and environment while designing products, to reconcile the contradiction between function and creativity, and to adapt engineering technology to the functional aesthetics of products, to the needs of human beings, to the physiological and psychological structures of human beings and to the development of the surrounding environment.

4. The source of design comes from nature, and the design results are projected to nature, thus achieving a balance point in the transformation. It not only makes the human-machine-environment system dynamically balanced, coordinated and consistent, but also can make people obtain physical comfort and psychological pleasure, thus winning the most, the largest and the highest working efficiency and economic benefits with the least, the smallest and the lowest cost.

Barry Shepard, one of the founders of SHR Perceptual Management Company, which designed Beetle for Volkswagen, pointed out: "Manufacturers realize that consumers are not only pursuing the function of a product, an important product should reflect the personality of its owner." Using bionic techniques, the product is endowed with the tension of life, and a large number of bionic elements full of ideal colors and realistic personalities, which makes the product stand out from the dazzling variety of shelves. The beauty of unfettered and unabashed personality is irresistible. Moreover, the morphological structure of living organisms evolved and developed in the long-term survival principle of survival of the fittest, which can make people feel a kind of life and vitality of self-consciousness, arouse people's potential consciousness of cherishing life and merging with nature. This paper calls for a high degree of harmony and unity between human society and nature through bionic philosophy of design, and calls for a reasonable symbiotic dialogue platform between human and machine, ecological nature and artificial nature.

References:

[1] Li Leshan. Industrial Design Psychology [M]. Beijing: Higher Education Press. 2004. 1

[2] Gao Longchang, Lu Shuhe, Li Zongfang. An Introduction to Thinking Science [M]. Chengdu: Southwest Jiaotong University Press. October 2004. 10

[3] Gao Nan. Methods and Cases of Industrial Design Innovation [M]. Beijing: Chemical Industry Press. 2005. 10

[4] Su Jianning, Li Heqi. Research on Product Shape Design Method Based on Perceptual Ideas [J]. Journal of Mechanical Engineering. 2004.40 (4): 164-167

[5] Tan Hao, Zhao Jianghong, Wang Wei, Zhang Jun. The Thinking Model and Application of Product Modeling Design. Journal of Mechanical Engineering. 2006. 5: Volume 42 supplement

[6] Research on Construction of Multidimensional Perceptual Semantic Recognition Model [J]. Chai Chunlei, Li Dan, Bao Defu, Tong Shang. Mechanical Design. 2017(03)

[7] Liu Jihong, Zhi Jinyi, Zhi Yu. Research on User Cognition Based on Adaptive Thinking Model in Product Design. Mechanical Design. 2018(02)

[8] Kantowitz B H, Rodrigue H L, Ellmers D G. Experimental Psychology [M]. Translated by Guo Xiuyan. Shanghai: East China Normal University Press. 2001

Analysis of the Design of Cultural and Creative Products from the Perspective of Regional Culture

Li Zhang*, Fei Hu

School of Art and Design, Bengbu University, 233000 Bengbu, Anhui, China

*Corresponding author's e-mail: 844940950@qq.com

Abstract. Under the context of the rapid development of regional culture, the highly distinctive regional culture has become the important resource for the cultural and creative products. Therefore, it is necessary to study the regional culture and its features, derive and consolidate the regional cultural resource and include the extraction principles of the regional cultural symbols into the research. Meanwhile, because the regional cultural can only be transmitted culturally through the cultural and creative products, the two of them are in a mutual penetration and influence relationship. Therefore, the study starts from the relationship between regional cultural and the design of cultural and creative products to explore the concepts, forms and nature of cultural and creative products. Also, the application of regional cultural in the design of cultural and creative products is illustrated to provide strategies for the design of regional cultural and creative products and offer effective path for the long-term development of regional culture.

1. Introduction

Because the nationality and territoriality of the culture mutually reflect each other[1], finding common grounds with others and all-embracing are the sources of the design inspiration for the national culture[2].For one thing, the application of national cultural symbols has increased the cultural connotation of the product; for another, people's sense of belonging and pride can be enhanced with the national cultural embedded in the design[3]. Therefore, it has become increasingly important to demonstrate local cultural features in the cultural and creative product. As a result, current design academic filed focuses on reflecting national spirituality and regional cultural characteristics in the cultural and creative products. As the soul of cultural and creative business, cultural and creative products is not only copying the cultural treasures or the product with traditional craftsmanship. Rather, it is a product of regional culture, product innovation, and technology industry guided by the needs of modern consumers. The cultural and creative products designed from the perspective of regional cultural should integrate regional culture into the products, make product based on the soil of regional culture, imprint the cultural stigma on the products and put on the regional culture outfit for the product. In this way, the unique national spirit and regional characteristics of cultural and creative products can be demonstrated.

2. Regional Culture

Regional culture refers to the cultural tradition with long-standing and unique features in a region. It integrates with the environment within certain territorial scope and remains distinctive[4]. Because of the different behavioral patterns and thinking modes of people in different area, together with the diversified regional cultural elements in different regions, they have become the root of the distinctiveness for a regional culture. At the same time, regional culture is featured with the characteristics of universality, group, inheritance and permeability. For example, historical sites, cultural features, social customs, methods of productions, etc. On the material level, it should reflect local people's diet, architecture, clothing, utensils, etc.; on the spiritual level, it should reflect their values, customs, etiquette, law and religion.

3. Cultural and Creative Products

Cultural and creative products refer to the products taken cultural connotation as the inspiration, and convey the cultural connotation in the design factors to the consumers through design. With the prerequisite of satisfying consumers' utilization demand, cultural and creative products should maximize their cultural demand in the product. In short, it refers to the recognition of cultural value from the consumers in the cultural and creative products, so as to gain recognition from more consumers and arouse cultural sympathy from the users in regard to the product design. For example, when a consumer sees a notebook with hometown features, it can trigger his memory of hometown and arouse cultural sympathy of homesick. Also, when you see a souvenir with local characters, it can reflect the function and features of cultural and creative products.

Derrick de Kerckhove believes that design is the external form of the visible and audible manmade products with texture in the cultural sense. Therefore, design is the "skin of culture" [5]. Regional culture is the rich soil to create cultural and creative products, which is at the center in the cultural and creative products design, the leading factor in forming "cultural identity" of the product, the material source of the product design and the important measure to improve the product's additional value.The cultural and creative products use design as a principle to transform regional culture into products, so that products are ultimately presented to consumers in a certain cultural form that convey and spread cultural meanings themselves. This is the cultural attribute of cultural and creative product design. For example, the exquisite design of Japanese company Sony comes from the reflection on the national resources and the oriental philosophical expressions of "small, subtle and emptiness", which is based on the deep understanding of regional culture and is reflected by the design. On the other hand, if the product design only considers technical skill and is in lack of the cultural and spiritual exploration, it will fall into the formality and the vicious cycle of imitation. Therefore, it is shown that regional culture should studied to realize the industrial upgrade of the cultural and creative products and render indistinctive cultural character to the products.

While study the relationship between regional culture and cultural and creative product design, the specific culture of a regional or a nationality (group of people) should be taken as the starting point to study the design application in the cultural and creative products, which is to transform the regional culture into a symbol to convey the product's cultural attribute. From the realization method, cultural and creative products mainly transmit the hidden factors contained in culture into explicit design elements, using design as a cultural factor to seek a new language that conforms to the modern life form, and explore the satisfaction of the human spirit after using the product[6].

4. Regional Culture Symbols

4.1 Concept and Features of Regional Culture Symbols

To select regional cultural phenomena such as natural environment, cultural heritage, customs and habits as objects and extract the material as well as ideological elements that represent regional cultural features are the combination of regional cultural symbols. Also, the regional cultural symbols cover the elements of shapes and patterns visually, which are of unique features, visuality and rich regional cultural connotation. Regional cultural symbols have representativeness and symbolic meaning. The representativeness means that the symbols can take people to have cultural imagination and arouse cultural sympathy.

4.2 Extraction of Regional Cultural Symbols

4.2.1Abstraction
Because the entity of the regional culture is specific, such as the historical sites, architecture, humanities and folk culture, which are the fixed form of the objects that cannot make consumers have more profound cultural interpretation. Therefore, the extraction of cultural symbols must be summarized abstractly. With abstraction as prerequisite, the simple language can generalize regional culture and extract the symbolic modelling. According to the visual abstraction and the morphological beauty of the form, it is abstracted and classified into several core key elements, and the symbolic prototype for the design of the literary product is obtained. For example, picture 1 below is the design of cultural and creative products in Yiju Temple. In this case, we can see that the designer has optimized the design symbols with Buddha statues, niches and embossments on top as the representative design symbols. The symbols that can represent Buddhist cultures are selected to demonstrate regional culture. Also, the abstract extraction principle of cultural symbols is put in the first place.

4.2.2 Reconstruction
Secondly, the extraction of regional culture should follow the principle of reconstruction, and adopts design methodologies including appropriate forms, deformations, dislocations and metaphors to analyze and reconstruct the cultural symbols. In the process of transforming specific regional cultural symbols to the abstract design symbols, the procedures of extracting classic symbols, optimizing the abstraction of symbols and reconstructing the symbols, so as to realize the artistry of the symbols and realize the reconstruction and sublimation of the symbols. For example, picture 2 shows the design case of the public chair design in American Grand Canyon Area. Because American Indian culture is generated here, and the railway has played important role in the western development in American history. To maximize the historical background of the region and the unique geographical culture, the designer utilizes the reconstruction principle in the geographical cultural symbols to reflect the hidden culture in the public chairs. The design approaches enable the communication between the product and the user, and highlight the historical meaning conveyed in the geographical cultural symbols.

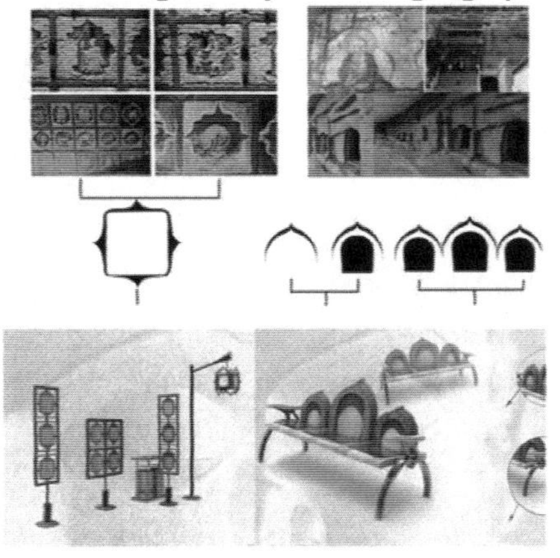

Figure 1. Case Study of Cultural and Creative Products in Yiju Temple

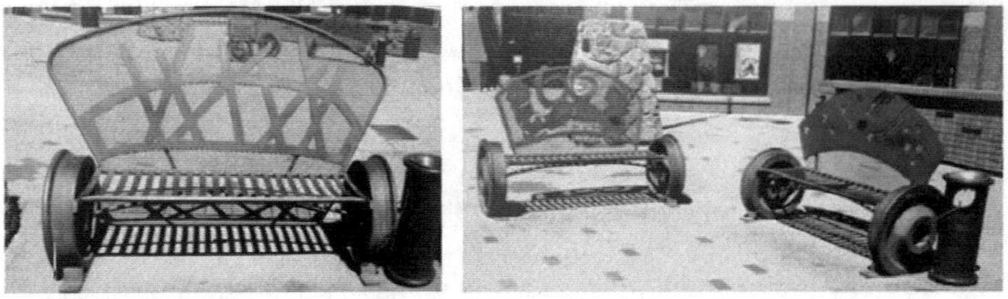

Figure 2. Design of Cultural and Creative Public Seats in the Grand Canyon Area of the United States

4.2.3 Integration of Tradition and Modernity

The cultural and creative products based on the regional culture carries the past and flows into the future. Therefore, the regional cultural tradition should integrate into modern life and realize the regional cultural symbol "rebirth". The mode of life and thinking pattern of modern people should be included in the research. The optimization and extraction of design symbols can be realized through the modern design approaches to convert them into design elements and use them reasonably and appropriately in the product. The new forms should be created to satisfy the design concepts of the cultural and creative products and rive the cultural promotion. For example, the design of torches in 2008 Beijing Olympic Games is a classic design case to integrate tradition and modernity together. It successfully demonstrates the paper in traditional Chinese culture to domestic and overseas tourists. As it is known to all, the application of paper is transmitted with the development of human society. The cultural and creative design symbols of the Olympic torches come from the traditional Chinese paper scroll (shown as picture 3). The paper scroll conveys the cultural sensation to the audience that paper has made tremendous contribution to the record of culture. Also, the adoption of paper which is a representative regional cultural signal can better express the cultural concept. At the same time, the sports spirit is embedded in the product form of torch. The product uses traditional Chinese auspicious clouds as decoration, which is of distinctive cultural representativeness and visual identifiability. Also, the proper and vivid application of regional cultural symbol is a self-evident way to show the cultural connotation in the product.

Figure 3. The design of Olympic Torch

5. Case Study of Regional Cultural and Creative Product Design

Faced with the identical and homogeneous issue in the market currently, it is especially vital to integrate regional cultural characters into the product design to win in the competition. Meanwhile, it is also important to add cultural additional value in the product. Next, faced with numerous cultural creative materials, the expression way must be distinctive, artistic and meeting market demand no matter it is the specific patter or the condensed abstract symbol. The cultural element or symbol cannot simply exist as a "surface", it should carry more significant meaning and function except for the visual effects. The development direction of cultural and creative product design lies in how to maximize the regional cultural features, symbolic meaning and value of the product. Therefore, the problem can only be solved by illustrating and analyzing the regional culture.

The designer can only find the direction of design through understanding the geographical cultural uniqueness and conducting research on the local people. For example, more than 70 procedures of making Yuhang traditional Chinese umbrella are disassembled in reference to the manufacturing skills, materials and inspirations of the regional culture. Also, they are utilized in the design of chair. "From Yuhang" integrates the regional handcraft and Yuhang cultural together to reconstruct Yuhang culture and blends them into modern design. The design is the interpretation to the traditional culture of handcraft in Yuhang region, which is of important impact to its transmission and development. Extracting design element from the unique cultural object and use them into the product design improves the cultural additional value of the product, and transmits the culture in daily life through the usage of the product has achieved dual effect in using and inheriting the culture[7].

Picture 5 below is the innovative design of red clay exorcising dance pottery. It uses Zhanjiang red exorcising dance as the classic representative of regional cultural symbol. Through processing the cultural symbols, the signals including the design modelling, pattern, ornamentation and color are converted into design element to achieve the regional culture feature that the product wishes to convey. Another example in picture 6 is the design of tea set of swimming fish in clear water. The swimming fish, sea and the wave on the surface of the water are used as the design elements by the designer. With the combination of product and sea culture, the distinctive and rich ocean culture features are reflected. The two cultural and creative products in picture 5 and picture 6 are used properly in the culture and creative products, which accurately demonstrate exorcising culture and sea culture. It is shown that reflecting regional cultural features can accurately and effectively demonstrate local features, highlight the characteristics of cultural and creative products, and improve the product's cultural taste.

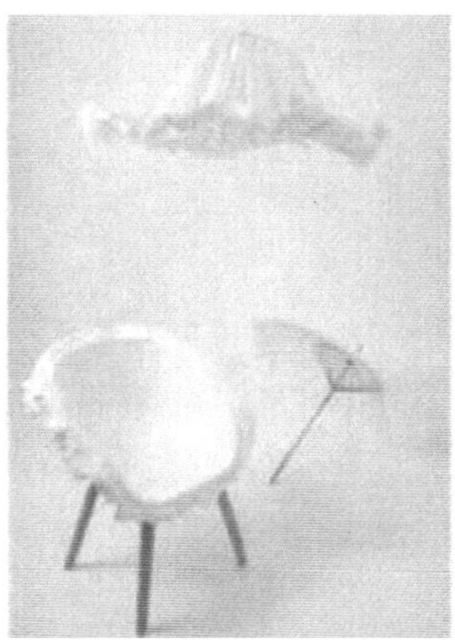

Figure 4. The future of Yuhang paper umbrella exhibition

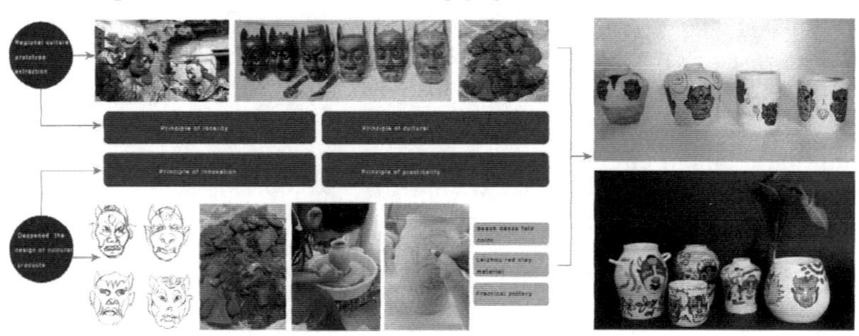

Figure 5. Clay potteries with Luowu

IOP Conf. Series: Materials Science and Engineering **573** (2019) 012071 doi:10.1088/1757-899X/573/1/012071

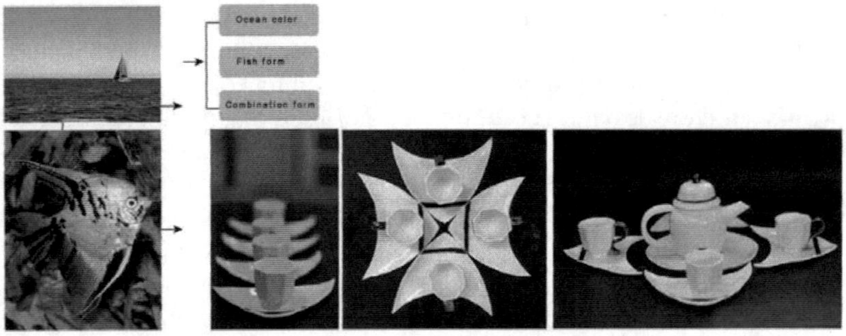

Figure 6. The tea set with fish in blue waves

Picture 7 below shows the cultural and creative product of earphone design using court beads. It is designed by a team of designers from the Palace Museum and has a strong Chinese cultural identity. As a design symbol, the bead string in the Qing Dynasty is smartly combined with the earphone products to perfectly show the cultural connotation behind the product.

Picture 8 below is the cultural and creative product with Changzhou regional cultural features. The designer uses a new bookmark modelling to increase the sense of un in the product. Also, the architectural symbols with local features are converted into the form of bookmark. While functioning as bookmark, it is of full of interactive meaning while reading with strong Changzhou regional cultural feature of the product.

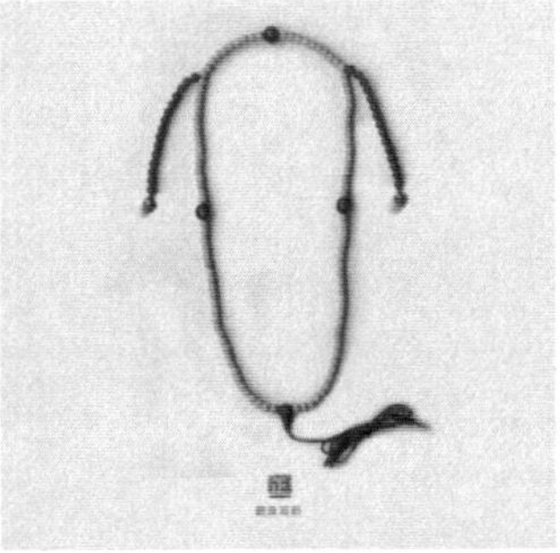
Figure 7. The Pearl form of earphone

Figure 8. Oriental Salt Lake City Bookmark Form

6. Conclusion

As a great cultural country boasts rich geographical and cultural resources, which provides diversified inspirations for the designers, it is of high value for their attention and utilization. The design and development of cultural and creative products centering around regional culture cannot only better express the cultural connotation of the cultural and creative products, but also inject vitality into the products. It is capable to drive the promotion of regional culture and build new regional cultural brand. Through analyzing the regional culture and cultural and creative products together with interpreting the extracting principle of regional and cultural symbols, the corresponding design strategies are put forward. Also, the real cases are studied for the application of regional culture in cultural and creative products to provide beneficial reference to the product design and promote the interference and innovation of regional culture.

Acknowledgment

Supported by Quality Engineering Project of Bengbu University (No. 2018JYXM17)

References

[1] Chang,LIU.,Qiangling,HUANG.(2010) A Brief Introduction for Tourism Souvenirs Design and Western Sichuan National Culture Protection. J. Packaging Engineering. Commun., 31(14): 129-131.

[2] Pengbin,LI.(2014) Application of Traditional Cultural Elements in Modern Packaging Design. J. Packaging Engineering. Commun., 35(8): 18-20.
[3] Jinguang,WANG.(2013) Research on the Design of the Tourist Souvenirs Based on the Hainan Culture. J. Packaging Engineering. Commun., 34(20): 99-102.
[4] Yanan,LAI(2014) Cultural Value of Design and Cultural System of Society. J. Packaging Engineering. Commun., 35(14): 137-140.
[5] Derrick,D.K(1998) The Skin of Culture. Hebei University Press, Baoding, China.
[6] Qinyan,GAO.(2012) The Application of Creative Elements in the Packing of Creative Products. J. Packaging Engineering. Commun., 32(3):107-111.
[7] Mei,MEI.,Huaiwang,YUE.(2011) The Modern Product Design Method Based on the Cultural Object of Minorities in Yunnan. J. Packaging Engineering. Commun., 32(3):47-50.

Research on the Development Strategy of Shared Appliances under the Low Tide of China's Shared Economic

Yinqiu Yang *, Qiang Sun and Shi Ba

[1]School of Art and Design, Bengbu University, Bengbu 233030, China;

[*]Corresponding author's e-mail: 419121349@qq.com

Abstract. In recent years, the sharing economy has not only become a new economic model that has received widespread attention, but also created enormous economic value. Many traditional industries have entered the sharing economy, including home appliance companies. However, under the business policy of gaining profits through manufacturing increments, many shared economic enterprises have encountered difficulties. Based on the resources of idle household appliances, targeting the huge second-hand home appliance market, based on big data, Internet of Things and block chain technologies, and propose feasible home appliance production technology improvement programs, it can become the main resource for shared home appliances. This paper establish a new shared appliance operation mode from a technical perspective, and provide reference to the horizontal development of home appliance enterprises and new enterprise positioning.

1. Introduction

The concept of the shared economy was first proposed in 1978 by Marcus Felson, a sociology professor at Texas State University, and Joel. Spaeth, a professor of sociology at the University of Illinois[1].The sharing economy is a different economic model. It generally refers to the main purpose of obtaining certain remuneration. The two sides of the transaction are between strangers, and there is a new economic model in which the right to use goods is temporarily transferred. A system for the direct exchange of goods and services.

Although the concept of the sharing economy has been around for a long time, it has been a prevailing economic phenomenon for nearly a decade. In the fall of 2008, Airbnb was introduced to the market as the world's first travel rental platform. In the ensuing decade, the sharing economy became a high-frequency hot word in people's lives. The sharing economy is not only attracting attention from the business, political, and academic circles, but also affecting people's lives.

According to the forecast of China's Annual Report on Shared Economic Development (2018), by 2025, the global shared economy market is expected to increase from US$15 billion in 2015 to US$335 billion. As of the end of 2017, the transaction volume of China's shared economic market was approximately 492.05 billion yuan, an increase of 47.2% over the previous year. Among them, the transaction volume in the non-financial sharing sector was 2,094.1 billion yuan, an increase of 66.8% over the previous year[2].

But time is up to 2018, China's sharing economy faced a huge crisis, and many shared companies had operational problems: Mobike bicycles were acquired, OFO shared bicycles dying, and the TOGO shared cars were frozen property. Everything has seriously weakened people's confidence in the sharing economy. There have been many voices in the society that are not optimistic about the sharing economy, and the sharing economy has fallen into a low tide.

2. The operating form and current status of China's shared home appliances

In the period of sharing the hot economy, sharing has also become the focus of the home appliance industry. During the period, there were a variety of shared appliances on the market, and the way they operate was different.

2.1. Shared TV

In the TV field, LeTV has launched a TV that can be shared: users do not need to pay for TV, but they need to pay for TV packages, watch TV bundled advertisements and programs, and purchase related products, so that users can gain revenue by participating in TV operations, and Return the user in a way that becomes a member of a particular service.

2.2. Shared washing machine

In the field of white electricity, a washing machine with laundry and drying functions has appeared in Shanghai, and users can use the washing machine by cash or mobile payment. But this public washing machine is no longer a new business model. In the early years, coin-operated washing machines were commonly used in collective dormitories and apartments. However, the way to convert coins into a payment method for scanning QR codes has not much changed in essence, and there is no qualitative improvement in technology and user experience.

2.3. Shared air conditioner

On August 8, 2018, a company in Shunde, Guangdong Province, first issued a shared air-conditioning public beta version. The shared air-conditioning system adopts the "deposit + on-time charging" mode. Each air-conditioning deposit is 3,000 yuan, which is charged at 1 yuan per hour. The electricity fee is paid by the user. Users can place orders through the mobile phone app. Lease shared air conditioning, and install and debug the enterprise. After the use expires, it will be removed and transported by the company.

At present, there are many different ways of operating shared appliances. Most of the shared home appliances are still the enterprises that retain the ownership of home appliances and lease their rights to use. However, the problems faced are concentrated on high operating costs and difficulties In profitability[3].

3. Research on the existing problems of leased shared appliances based on the concept of shared economy

From the perspective of shared resource objects, the ideal resource for sharing objects should be resources already being in the society as a whole, such as goods, labor, public facilities and so on. The existing shared resources of shared appliances are not social idle resources, but incremental resources. At the same time, sharing by renting also makes the sharing method became sole, which brings a series of problems.

3.1. The method of paying the deposit is not enough to restrain the leased user

In the lease-type sharing behavior, the item holder retains the ownership of the item and transfers the right to use the item. The user of the item obtains the right to use the item within a certain period of time by paying money or other benefits. In this way, the item holder revitalizes the idle item, and the item user's demand can be met by paying a small amount of money.

However, at the same time as the transfer of the right to use the item, the management of the item has also been transferred. In the process of using the item, it is inevitable that the item will be damaged. In order to cover the losses that may occur during the use of the goods, the user of the goods will usually be required to pay a certain amount of funds as a deposit to serve as a credit guarantee for the safety of the goods.

For items of lower value, a lower amount of deposit is acceptable to the item user. However, for shared items with higher value, the following problems exist: when the deposit is equal to or higher than the value of the item itself, the willingness of the user to participate in the sharing behavior will decrease; the deposit is lower than or even lower than the value of the item itself， it loses the binding force on the user of the item and the protection of the item.

3.2. The company bears high management costs

The existing business entity of the shared economy is the enterprise, and the ownership of the shared goods is also in the enterprise. The timeshare lease amount for shared items is lower. Therefore, in order to obtain the income, the enterprise must hold a large number of shared items and ensure that the shared items can operate efficiently. This makes the shared enterprise become a real heavy asset enterprise, and the difficulty and cost of asset management are much increased.

Taking the OFO shared bicycle as an example, the public data shows that the manufacturing cost of an OFO shared bicycle is about 300 yuan [4], and its volume is more than 17 million. The huge amount brings great management difficulty and loss. The data shows that the OFO damage rate is up to 20-40%. According to this figure, the loss caused by OFO sharing bicycles is only 900-1.8 billion yuan due to damage[5].

3.3. Home appliances have high privacy

The home appliance itself is distinct from the shared items such as cars and bicycles. Its use place is fixed and its privacy attribute is strong, which is not suitable for public use. Moreover, the sanitary requirements of household appliances such as washing machines and refrigerators are equally high. Therefore, for shared washing machines in public places, time-consuming rental of air conditioners and other products, the market reaction is cold.

4. Ignored idle home appliance resource sharing mode: second-hand home appliance trading

In the home appliance market, another de facto sharing economy model is second-hand home appliance trading.

China's home appliance industry has participated in the third stage, which focuses on updating demand. There are fewer and fewer claims that “it must be bought” and “electric appliances are broken and replaced”, and home appliances are changing from necessities to consumer electronics. The reasons for people to change their home appliances are more subjective factors such as "uncomfortable use", "outdated" and "not good-looking". There are even consumers who do not have a demand for use, but only because of the promotion and impulse consumption. The new consumer industry has produced a large number of old appliances that are in good condition or repairable.

According to "China's waste electrical and electronic products recycling and comprehensive utilization industry white paper 2017", in 2017 China's TV theory scrap amounted to 32.16 million units, refrigerator theoretical scrap amounted to 24.39 million units, washing machine theory scrap amounted to 16.2 million units, air conditioning The theoretical scrapped amount reached 27.23 million units. However, China's existing green recycling, dismantling and digestion capacity is less than 50% of the total amount of used household appliances[6]. More old appliances are once again flowing into the home appliance market. The number of idle home appliances is huge and the market potential is high.

However, the current second-hand appliance trading market still lacks effective supervision. Most of the second-hand home appliances are traded by C2C. Most of the old home appliances are purchased by SMEs and even individual operators, and then repaired and sold. There is no uniform maintenance and quality inspection standard throughout the entire process from acquisition to resale. The corresponding parts sales and purchases have also loose supervision, and the overall lack of effective regulatory mechanisms. This has significantly affected the quality of second-hand home appliances and reduced consumers' willingness to purchase.

The second-hand home appliance market has huge potential and development significance, but there are also many problems. Addressing these problems should be the driving point of the home appliance sharing economy.

IOP Conf. Series: Materials Science and Engineering **573** (2019) 012072 doi:10.1088/1757-899X/573/1/012072

5. The future development strategy of shared appliances: second-hand home appliance trading and light asset leasing system under big data monitoring

The home appliance industry is an important component of the social economy. In the future, sharing home appliances is the only way to reinvigorate the idle inventory of home appliances and promote the development and upgrading of the industry. In the future, shared home appliances should focus on upgrading and developing the second-hand home appliance sharing industry, and at the same time to establish a reasonable and feasible lease-type shared home appliance industry model. At the same time, before establishing a new type of shared home appliance industry, it is necessary in order to solve the following problems.

5.1. Establish new standards for home appliance production

At present, the core problem of the quality of second-hand home appliances is the lack of effective supervision means. Consumers do not have professional knowledge and skills. Pre-set monitoring modules on home appliances, use sensor technology to monitor and record the operation of electrical appliances, fault conditions, maintenance history. , accessories information, sales records and usage records, etc. This information is recorded by the server through the Internet system, and the home appliance can be effectively monitored, and the user can also obtain the information through the inquiry software. The pre-installed unified monitoring module as a standard for home appliance production enables all home appliances to be effectively monitored. For one thing, the quality of home appliances is transparent and open, and for another, relevant data records can be used as an important reference for maintenance. Figure 1. In the intelligent temperature control regulator introduced by Nest in the United States by 2015, this technology has been applied, and the technical foundation of the wide-scale promotion has matured.

After the user data are recorded, combined with big data and block chain technology, the home appliance monitoring module can be networked, and many new services can be provided to the user. With the rapid accumulation and continuous updating of user data, traditional home appliance enterprises will also acquire the second attribute of information enterprises and complete industrial upgrading.

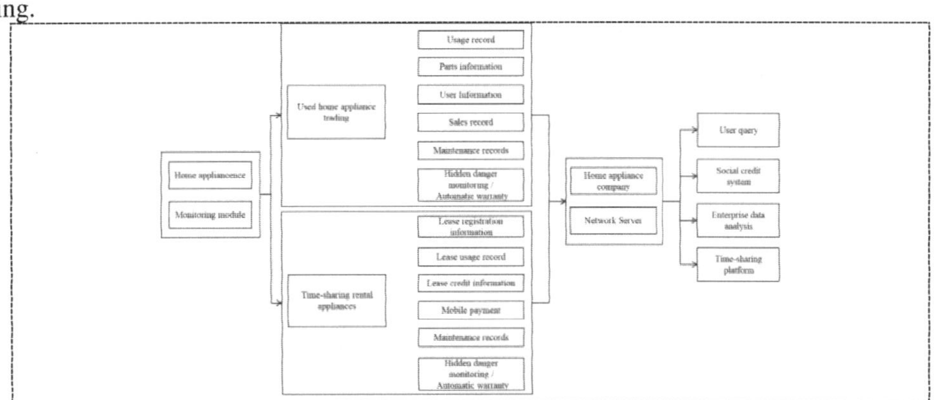

Figure 1. Used home appliance trading and light asset leasing big data system.

In the long run, the intelligent module is an essential accessory for home appliances, and the establishment of new home appliance production standards is also conducive to the establishment of the Internet of Things in the whole society, which will become an important part of the future intelligent society

5.2. Second-hand home appliance acquisitions and maintenance enterprise scale

When the product performance monitoring module becomes the electrical production standard, the standardized second-hand appliance repair and acquisition efficiency will be greatly improved, and the cost will be reduced. The conditions for establishing a unified second-hand home appliance purchase standard and the system will be mature: including quality standards and maintenance standards.

Relevant and pricing standards will be transparent and relevant, and relevant systems and processes will be standardized. It will become possible to create a large-scale second-hand home appliance acquisition and maintenance enterprise, and a large and standardized second-hand home appliance market will be established.

5.3. Improve relevant laws and regulations and social credit system

In addition to the transfer of ownership of the second-hand home appliance industry, collective housing, shared apartment and other public housing, shared apartment and other public housing space for public appliances rental is also an important part of shared appliances. The management of property rights of shared home appliances that only rent the right to use depends to a large extent on the user's consciousness and the norms of laws and regulations on users. This requires the establishment of viable technical monitoring tools and credit system specifications[7].In addition to the relatively low deposit system, the strict and perfect credit reporting system has a considerable deterrent effect on violations, and serious violations of laws and regulations must bear corresponding legal responsibilities.

6. Conclusion

Although the sharing economy represented by industries such as shared bicycles has encountered a low tide in the moment, in the long run, the sharing economy still has good development prospects. From an economic point of view, the sharing economy is the only way to revitalize the stock of social resources. From a historical perspective, the sharing economy can promote the construction of a big data intelligent society and accelerate the integration of social information. The development of shared home appliances requires further investment, and identify the key points to solve fundamental problems.

Acknowledgments

This work was supported by a grant from Anhui Provincial Federation of Social Sciences, Innovation and Development Research Project (NO.2017CX017) and Social Science Foundation of Bengbu University (NO. 2017SK07zd).

References

[1] Felson M , Spaeth J L . Community Structure and Collaborative Consumption: A Routine Activity Approach[J]. American Behavioral Scientist, 1978, 21(4):23.
[2] China's annual report on sharing economic development [M]. National Information Center Sharing Economic Research Center, China Internet Association Sharing Economic Work Committee.
[3] Cheaa (2017) With the rise of home appliance rental and shared appliances, do you really buy it?.http://news.cheaa.com/2017/0612/510263.shtml.
[4] Lao suji (2017) ow much is the manufacturing cost of shared bicycles? http://news.cheaa.com/2017/0612/510263.shtml.
[5] Cai Hui , Zhong Jianting . (2017) Who is sharing the bicycle? "Radish + Big Stick" can save 1% damage rate. Southern Metropolis Daily.
[6] China Waste Electrical and Electronic Products Recycling and Comprehensive Utilization Industry White Paper 2017(2018).China Household Electric Appliance Research Institute.
[7] Cai Chaolin. (2017) The rise of the sharing economy and government regulatory innovation[J] South China Journal of Economics No.3, 2017.

IOP Conf. Series: Materials Science and Engineering **573** (2019) 012073 doi:10.1088/1757-899X/573/1/012073

Study and Practice on Skills Competition Help Vocational Education Students Growing up

Congquan Li[1*] Ji Huang[2] Jinbing Wang[1] Jun Yang[3] Gang Tian[1]

[1]College of Mechanical and Electrical Engineering, Tianmen Vocational College, Tianmen City, Hubei Province,431701, China

[2]Department of Neurology, Tianmen People's Hospital, Tianmen City, Hubei Province,431701, China

[3]School of Nursing, Tianmen Vocational College, Tianmen City, Hubei Province,431701, China

[*]Corresponding author's e-mail: licongquan@sina.com

Abstract. In order to research the influence of skill contest on students'growth, a questionnaire survey was conducted among students of Tianmen Vocational College. Research shows that 100% of students want to be respected. 92% percent of students care about their advantage. This brings good enlightenment to teaching. Competition events should not be single. Only in the way that competition promotes each other can lead promoting learning through competition. The purpose of the competition is not to win the prize. Competition is learning, finding gaps and shaping students. Competition avoids the fictitious elements contrast with E-learning. Through the competition students could endeavour to study , self-respect can be developed, psychological quality can be enhanced and the sense of social responsibility can be enhanced. Skills competition pursues frontier ideas. Only by reform can achieve good results. Teachers play a key role in the course of the competition.

1.Introduction

Vocational education students are divided into two categories: higher vocational and secondary vocational. Some students have the ability to study undergraduate course, but like vocational technology, so they come to vocational schools. However, more students do not develop good learning habits, and only their grades are poor, can they study in vocational and technical schools.The sources of secondary vocational students are more complex, some have participated in the entrance examination of senior high school, some have not participated in it, and some have returned to school after many years of working. Some excellent secondary vocational school students are very excellent, but many students lack in-depth thinking and collective concept.Most of them are left-behind teenagers and lack communication skills, which has brought a lot of troubles in teaching. Vocational colleges aim at training high-quality skilled talents based on a certain theoretical basis and can carry out creative work. Skills competition fully carry forward the glory of labor, the value of skills and the great spirit of creation[1].Skills Competition can promote students growth

2.The premise of competition helping students developing

2.1. Promoting competition by competition

There should be more competition events. Management should not favour one more than another.Otherwise, it is difficult to play the role of competition in promoting learning. Only by promoting competition through competition can we do well in promoting learning through competition. Now general colleges and universities have formed a four-level competition pattern, namely: school level, municipal level, provincial level and national level.

School-level competitions are the foundation. There are many students taking part in such competitions and win a wide range of prizes. Competitions organized by government departments, by national societies and associations and selection competitions for international disciplines are called "National competition", which have a wide range of influence. In response to national competitions, provinces and municipalities carry out corresponding types of competitions.

College competitions should be diversified, which gives teachers and students more choices, so that students can be guided according to their situation and it is conducive to competitive among the competitions themselves. The atmosphere of competition, learning, catching up and surpassing can improve the level of competition. Tianmen Vocational College has actively launched competitions in recent two years.The competition is based on theoretical learning, highlighting practical learning, skills training. Training method is shown in figure 1. In the course of competition, we have grasped the key of glorious labor and precious skills. Tianmen Vocational College College support participating in national vocational college skill contest, provincial skill contest, provincial tea industry skill contest and world skill contest. Vocational education students can not only focus on skills training, They will meet new problems in work, need new methods to solve.So they require innovative thinking. In developing skills training, college carries out the cultivation of innovative ability. Schools support innovation competitions, such as college students' mechanical design innovation competitions, digital innovation design competitions.In this way, we can cultivate the spirit of innovation.

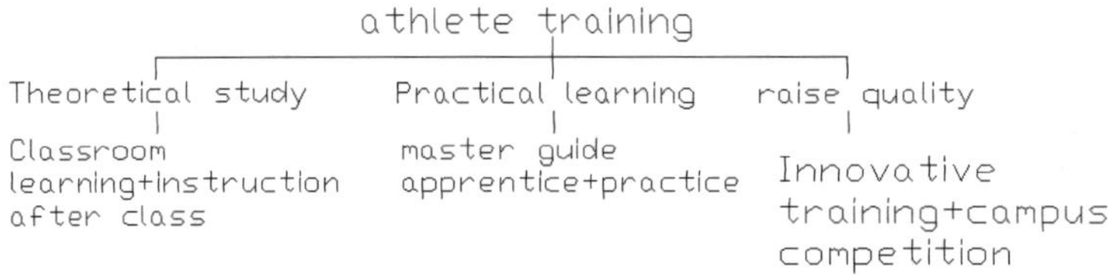

Figure 1. Training mode

2.2. Not aiming at winning a prize

Vocational students have done well in primary and junior middle schools. At a certain stage, parents and schools have not managed well, and their grades plummeted, which turning them not to strive for progress. After arriving at vocational schools, about half of them are either isolated, extreme or lazy.Tianmen Vocational College has obvious local characteristics. Its students mainly come from this city. Its language expression and living habits have great similarities. Butterfly effect is easy to form in class and after class. One student dozed off, some of the students next to him followed. One student talked, the students next to him talked. It's obviously not a good idea to aim for winning a prize in competition.

Gu Wenhao argues that students' competitions "can not be **directly** aimed at winning awards" [2]. This view is correct in economically developed areas and in large cities. Many places are underdeveloped, teachers are weak, facilities are backward, and students' comprehensive quality is weak.In this case, training students is a difficult thing.They go into arena is a good thing, and should be praised. To be exact, competition is not for winning prizes. Competition is for learning, finding gaps, and shaping students.We should be responsible to the society and the students when carry out competition .

students see the navigation lights on their way .Through the competition competitors feel that I can learn, I can do. If teachers and students are aiming at the grand prize to compete, they will not be able

IOP Conf. Series: Materials Science and Engineering **573** (2019) 012073 doi:10.1088/1757-899X/573/1/012073

to understand the original intention of the contest. In this case, they are often disappointed , the learning is difficult to sustain.

3.Internal Causes helping students grow up

3.1. Developing self-esteem

Often many secondary vocational school students are late for class, or leaving class early, or be tired and disobedient to advice.

In order to understand the psychological activities of secondary vocational school students, the research group conducted a questionnaire survey on 62 students, whose majoring is mould in 2017. The result is shown in Table 1.

Table 1. Questionnaire for secondary vocational school students

problem	Answer sheet	
Do you care about being respected	Yes (62 students)	No (0 students)
Do you care about your advantage?	Yes (57 students)	No (5 students)

The answer, 100% of students want others to respect themselves, 92% secondary vocational school students are very concerned about their own advantages.They believe they have the advantages.Although considered to be rather poor the students have a better idea, just like ordinary people. Valuing oneself is a psychological activity of self-esteem, that is, self-esteem sentiment. British psychologist McDuhu believes that self-esteem is not only the key to understanding volitional activities, but also the basis of self-respect and moral cultivation. Vocational teachers not only should teach professional skills, but also should interpret the inner world of vocational students, persistently and carefully cultivate a healthy self-esteem of students.

Competition is a good platform to cultivate students' self-esteem. Excellent people will win the honor of the competition and become learning models. They will be widely publicized through the network and newspapers. Where is the standard of higher excellence for students? Where is the right direction?Other students will think about it.

When you see others being respected, you also have the idea of being respected. Gao Xihua and Huang Ruijuan investigated the influence of skill contest on nurses' self-esteem.

The self-esteem score of the competition students was 30.1639, but the nurses who had not participated in the competition was 30.1631+ [3]. Carry out nursing skills competition has obviously improved the advantages of nurses' self-esteem. Competition platform has brought positive energy to vocational school students. Once the shadow was neglected in primary or secondary school has disappeared. Through the competition, they find the dignity of being a man.

3.2. Strengthen Psychological Quality

Repeated alternating action can cause fatigue. The damage caused by repeated alternating action is called fatigue damage. Pavlov School believes that fatigue is the result of the development of brain protective inhibition [4].When an action is repeated, the signal impulse action is used in the nerve center. The long-term repetitive excitation of the central nervous system will consume the energy in the cell.In order to avoid more consumption, it will produce protective inhibition, resulting in the decline of competition training ability. Skills training has a dull and tedious process. After a long period of intensive training, there will be psychological resistance, which is manifested by lack of confidence and lack of interest in things.But at the end of the game, after a period of rest, the nerve center has no training impulse stimulation, and habitual sensation is missing again. The spontaneous training needs stimulate the brain nerve center again when seeing or hearing the winning information of the competition. This is the performance of the operational conditioning reflex [5].

Yao Ying and Cui Jinggui of Qinghai Normal University conducted a comparative study on 65 awarded elite athletes and 295 general skill athletes. The data cited is shown in Table 2 [6]

"t" is the test value, "p" is the probability of the sample observation results when the original hypothesis is true. The smaller the p value, the more significant it is. $p < 0.002$, so there are significant differences in player' emotional perception, emotional regulation and emotional use. It can be seen from the table that the psychological quality of excellent players is obviously higher than that of ordinary players.

Table 2. The contrast of psychological quality

	excellent player	general player	t	p
emotion perception	4.109±0.490	3.759±0.504	5.097	0.000
emotion regulation	4.257±0.520	3.976±0.514	3.987	0.000
Using emotions	4.398±0.423	4.007±0.51	5.689	0.002

Skills competition can not only improve their knowledge, but also enhance their hands-on ability and innovation consciousness. More importantly, students who have participated in competitions are accustomed to the pressure of skill training and can bear more similar pressure.Competition matures the players psychology, makes them can think more comprehensively, and correctly treats praise, criticism, can correctly face mistakes and doubts.

3.3. Enhancing the sense of social responsibility

Not paying attention to things outside the window, only reading page after page.This method of learning is not related to practice, it is not a good one.Students need professional practice and social research.The purpose of practice can deepen the understanding of this major and firm the correct direction of learning.Students whose objectives are not clear will grow slowly , and they will not work hard enough.Students who lack social practice are easy to become selfish, become extreme and lack of respecting for others.Some students with weak self-control will do things against the law and discipline.

Competition skills mostly come from enterprise production and people's needs.Students should combine professional learning with enterprise needs, or combine with people's life, which solve the difficulties for the society , solve the difficulties for people.In this way students take on social responsibility and can achieved the promotion of comprehensive quality.

4.External Causes Helping Students Growing

4.1. The vanguard of teaching reform

The form of textbooks is fixed and some knowledge lags behind.[7][8].It takes a long time to compile a textbook.There are strict procedures for examining and revising textbooks.The general process is three times revision, three times modifying and one time reading must be done before using a new textbook.The content of textbooks can not keep up with the pace of scientific and technological development. Traditional teaching must be reformed.Educators should understand what level of vocational education has developed and what are the needs of enterprises.Yu Zhibin believes that the World Skills Competition embodies the real life or work scene. As a professional, he should have higher quality and skill level.Nanjing Communications Vocational and Technical College carries out the core curriculum teaching reform aiming at skill competition.The college make contest content as part of students' learning content.The integration of "teaching", "learning", "doing", "training" and "examination" has achieved good results.Skill competition leads the direction of skill development, and skill competition is always at the forefront of teaching reform.Tianmen Vocational College attaches great importance to competition teaching, actively participates in seminars, exchanges competition experience and discusses teaching methods.Seize the opportunity to improve teaching methods.Learning theory in practical, teaching is highly pertinent.It is easy to understand theory by theory teaching with practice.The teaching situation of "promoting learning by competition and improving by practice" has been formed.

4.2. Not idle practice

There are certain difficulties in skills competition, which requires higher physical strength, skills and theoretical level .For example, World Skills Competition CAD Machinery Design Project lasts two days, it takes 2.5 hours every morning and afternoon in competition.Player want to achieve good results in competition, physical training is necessary.Learning-by-doing is a effective way in studying, which cooperated with the brain, eyes and hands.It's difficult to make a desertion in training.Modern classroom is a classroom with network as its feature. Teachers use the network to check on work. Teachers and students use network to communicate and use network resources to teach.The malpractice of network is that students are addicted in it. The network classroom has a larger fictitious component.Skills competition training adopts the method of teaching, demonstration and actual operation.In order to achieve good results, the skills training mode of work-integrated learning, school-enterprise cooperation and work on regular post has gradually been formed.Study in the right direction and study on a firm footing.Competition training extends the classroom space, which taps the potential of team.Students can not only think independently, but also discuss with each other. Creative thinking bursts out naturally.

4.3. Promoting teachers development

Jia Guiling believes that who guide the competition should not only learn new professional technologies, but also know the requirements of enterprises.Teachers benefit teaching as well as students in the competition[10][11].The step-by-step teaching procedure is: preparing lessons, teaching in classes and correcting exercises.The content of other professional is less involved , and teaching and examination are not separated.So many teachers' abilities have been improved slowly without the competitive pressure.Students have weak sense of competition and innovation. This is far from excellent skilled personnel.

In the competition, the students are the subjects to be tested, but in essence, the tutor is the driving force.The tutor needs to clarify three questions: Can I teach? What should I teach? How should I teach?Teaching concepts can not be copied blindly, teaching methods must be updated.The competition will set some new technology problems.We should use professional theory to solve them creatively .Competitors can accurately judge the cause of the problem and quickly find a reasonable solution, which is the key to achieve good results.If we break through this point, we can cultivate more excellent talents and promote teachers continuous developing.

5.Putting into practice

Tianmen Vocational College is a local vocational college built on the basis of local secondary vocational schools.After ten years of hard work, we have formed a competitive situation from scratch to more.Constantly explore, create our own competition methods, which is effective.In 2018, 21 prizes were won in the student competition, including one second prize in the national contest, one first prize and three second prizes in the provincial contest.Teachers win 15 awards in teaching competition.

6.Summary

As long as the proper method is found and the teachers work hard, the students in vocational colleges can be changed from mediocre students to excellent students.Competition is a good education. In order to do well in competition, students should have a solid theoretical basis, strengthen practical skills training, and develop the docking between teachers and apprentices.Skills Competition, as a fair competition platform, has the opportunity to learn and show, in which students regain their self-esteem.There should be many types of competition project for students. Management should not favour one more than another.Secondary vocational school students have weak foundation. They should be encouraged to participate in competitions and not be encouraged to win prizes.Competition can expand the scope of learning, clarify the direction of learning, and accelerate the growth rate.Skills contest can avoid the fictitious network, overcome difficulties and make progress.Competition brings positive energy for students' growth. Vocational colleges should actively study and participate in competitions according to their own characteristics.

Acknowledgement
This research on skills competition help vocational education students growing up was financially supported by investigation project of China Vocational Education Association, Hubei(document:[2018]13, HBZJ2018028), and supported by Teaching Research Project of Tianmen Vocational College (document: [2018] 32 , tzy2018a004) Tianmen Vocational College.So I express my sincere gratitude for this. Also I express my sincere gratitude to all friends who care about and support the research.

References
[1] Xue Yongwu.(2018)Talent Strategic Thought in Xi Jinping's Report at the 19th National Congress of CPC and its Enlightenment.Journal of Shanghai Normal University(Philosophy & Social Sciences Edition), 47:40-46.
[2] Gu Wenhao, Qi Yuansheng, Zhao Shiying. （2013） On the Dialectical Development of Students'Practical.Innovation Ability and the Platform of Discipline Competition.University Education，3:19-20.
[3] Gao Xihua, Huang Ruijuan, Zhang Xia. （2012） Discussion on the influence of nursing skill competition on nurses'self-esteem.Chin Med J Metal Indus, 29(5): 601- 602.
[4] Niu E, Yang Li, Wan Xupeng. （2017） A Summary of Research on Sports Fatigue. Sport World (Scholarly)，1：178-180.
[5] Zhang Hua Jun. （2009） Operational Conditional Reflection and Active Component Knowledge: Rational Exchange of the Roles of "Teaching" and "Learning".Capital Medical University Journal (Social Science Edition)，00:492-494.
[6] Yao Ying.（2017）The Relationship between Emotional Intelligence, Psychological Resilience and Stress Coping of Vocational School Skills Competition Players.Qinghai Normal University Master's thesis.36-38.
[7] Guo Zhaosong, Wen Aimin. （2019） Reform of Higher Vocational Automobile Operation and Maintenance Technology Specialty Course Based on Skills Competition. China Management Informationization，1:207-208.
[8] Ji Hui Li. （2012） Research on the Lagging of Textbook Construction in Higher Vocational Education.China Electric Power Education，3：92-93.
[9] Yu Zhi bin.（2018）Influence of World Skills Competition on China′ s Vocational Education Ｒ eform.Journal of Beijing Polytechnic College，4:60-64.
[10] Jia Gui Ling.(2018)Analysis and Consideration on Skills Competition in Vocational Colleges.Chinese Vocational and Technical Education, 1:55-60.
[11] Jiao Yongfang, Pan Jian.(2014) On the Promotion of Professional Skills Competition to the Construction of Professional Teachers. Course Education Research, 33:237-238.

Electric car design based on wheel motor drive

Xin Cao[1], Weimin Yang[2*] and Yangfan Yao[3]

[1,2,3] School of Mechanical and Electrical Engineering, Beijing University of Chemical Technology, Beijing, Beijing, 100029, China

[*]Corresponding author's e-mail: 2018310050@mail.buct.edu.cn

Abstract. Under the increasingly severe energy and environmental issues, new energy vehicles are receiving more and more attention. In the new energy vehicle series, new energy vehicles with wheel hub motors as the driving technology are unique. The new energy vehicles of the hub motor have the characteristics of compact structure, independent drive and high integration of the power system. These characteristics make the hub motor have unparalleled advantages in the development of new energy vehicles. By analyzing the principle of the driving mode of the hub motor, the paper draws the limitations of the hub motor itself, and starts from this defect, combined with its own advantages, combining the advantages and disadvantages, thus making a new breakthrough in the car's shape. After the design of the styling was completed by modeling, the appearance of the car was evaluated by the fuzzy evaluation method, and the result was good. This paper combines the design examples with the scientific and theoretical basis to enhance the car's styling image and provide reference and basis for the future design of the wheel hub motor.

1. Introduction

With the diversification of energy sources and the deteriorating environment, people are increasingly paying attention to electric vehicles. This type of car has the characteristics of quiet driving, environmentally friendly and low emissions, high flexibility in body design, and easy weight reduction of the car body. In the car design, most electric cars are more conservative in shape, and the basic model of the design is still the traditional diesel locomotive. This design is not innovative and does not conform to the technological sense of electric vehicles.In addition, the appearance of automotive products has a high-speed update,which will shorten the life cycle of a product, and it will cause waste of various resources and materials.Consequently,this way is not conform to the development concept of green ecology. The hub motor integrates the motor, drive train and brake system into the hub, thus making the vehicle structure simpler and providing a possible split design. Split-type design is a more forward-looking design idea, which not only greatly changes the phenomenon of homogenization of existing models, but also provides new possibilities for future car design [1].

2. Wheel hub motor defect analysis

2.1. Cooling method of the hub motor

*2.1.1. Problems.*The hub motor technology puts the power unit, transmission and brake device together into the hub, which can greatly simplify the original mechanical parts of the car. However, the installation of these components in a small space will inevitably lead to poor air circulation. The problem of not being able to dissipate heat in time. During the operation of the motor, due to the high electrical load and magnetic load, huge heat will be generated, and at the same time, it will be affected by heat conduction and heat radiation of other components. Therefore, the motor cannot be cooled only by the flow and air cooling. Excessive temperature will have a huge impact on the overall performance of the motor [2], if this heat can not be dissipated in time, it may damage the insulation performance of the winding, reduce the service life of the motor, and may even lead to permanent Irreversible demagnetization of magnets [3].

*2.1.2. Solution.*At present, there are three main heat dissipation schemes for the hub motor, namely air cooling, liquid cooling and gas-liquid mixed cooling. Among them, air cooling includes natural air cooling and forced air cooling. Because of the poor heat transfer capacity of the air, natural cooling is suitable for small hub motors. This method is an ideal method without additional cost. However, for a powerful high-power hub motor, the amount of heat generated by natural air cooling alone is not sufficient to cool the high-temperature gas. Therefore, forced air cooling and liquid cooling are required for timely heat dissipation [4].Forced air cooling is divided into open air cooling and closed air cooling. Open air cooling refers to the use of special ventilation mechanisms such as air slots to achieve internal and external air circulation of the hub motor. According to different ventilation directions, it can be divided into axial type and radial type. The axial type is often used for small motors. In axial ventilation, the heat of the motor is taken away by the wind in the axial direction, which can directly dissipate the stator and the rotor [4]. This method is simple and easy to implement. Although the open air cooling cost is relatively low, it is easy to blow external dust into the motor and it is noisy during operation. Closed air cooling means that air flow is realized inside the motor casing. The advantages of closed air cooling are mainly two points. The motor is relatively clean and the noise is small, but the manufacturing cost is high.

Air cooling has limited heat dissipation capacity. In comparison, the cooling capacity of liquid cooling is much better. Liquid cooling is suitable for high-power permanent magnet motors. Liquid cooling mainly includes oil cooling and water cooling, and oil cooling usually uses transformer oil. Oil cooling is divided into direct cooling and indirect cooling, and direct cooling is divided into oil-injection and fuel-injection [4].

*2.1.3. Selected plan to determine design ideas.*The front face of a conventional diesel locomotive often has an iconic intake grille designed primarily to cool the cooling water tank. Now, the air intake grille on the front of the car has become an important element in the design of the car, and each car company has a relatively stable style. For electric vehicles, there is no need for a cooling water tank, and a large area of the intake grille is no longer needed. The hub motor on an electric vehicle usually uses a high-power motor, and the heat dissipation requirement is very high. Therefore, in the design process, forced air cooling or liquid cooling is adopted to control the temperature, so for the electric vehicle, a small area of the intake grille is still needed to direct the airflow for forced air cooling.

Congxi Jiang, one person at key laboratory for aerodynamic and thermal environment simulation of ground vehicles in Shanghai, has studied the heat dissipation law of electric vehicle hub motors. He found that if the front compartment parts of the car can be arranged in a regular order, the high-temperature cooling gas is When the chassis is out, it can reduce its influence on the motor temperature.

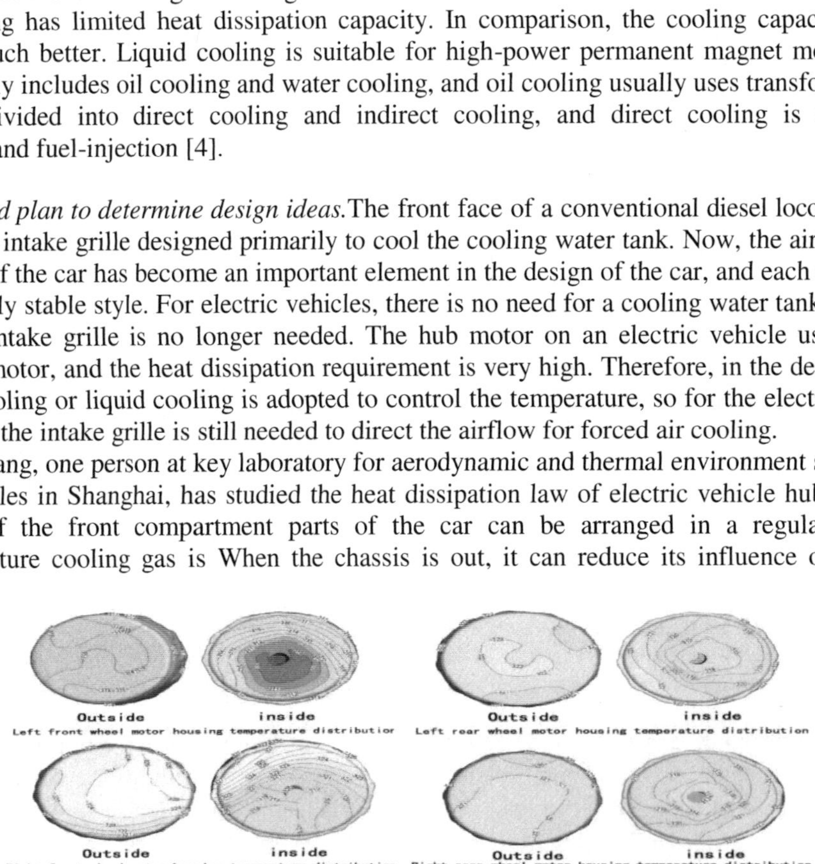

Figure 1. Motor case temperature distribution

As can be seen from Figure 1, the surface temperature of the left front wheel motor is lower than that of the rear wheel. As can be seen from Figure 2, the surface air mass flow rate near the front wheel motor is higher than that of the rear wheel, so that better ventilation and heat dissipation conditions can be obtained. In Figure 2, the maximum mass air flow in the lateral air mass flow at different vehicle speeds is the right front wheel. The inflow in the front end is a high temperature gas, and most of this gas will flow to the area of the right front wheel, in which case the temperature of the motor rises. In the low speed range, the air flow rates of the left front wheel and the left rear wheel are similar, but due to the presence of the inflow, the intake air amount of the left front wheel increases significantly as the vehicle speed increases. It can be seen that the inflow has an important influence on the wheel flow field and thermal management.

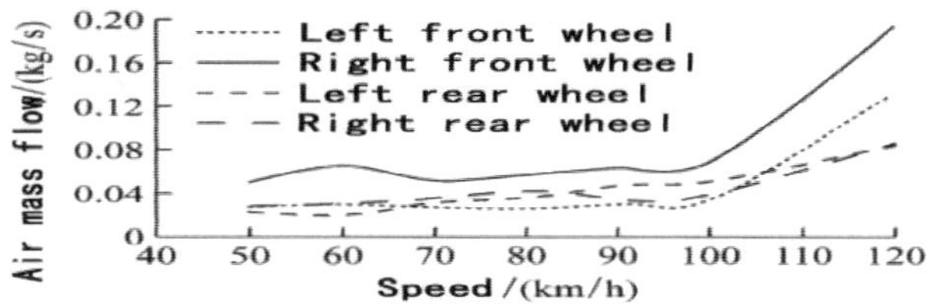

Figure 2. Horizontal air mass flow of the wheel at different speeds

Considering the above factors, when designing the appearance of the car, the shape should be aerodynamic service, so that the front wheel of the electric car is moving forward, so that the front wheel has sufficient air intake to cool. There are two main methods commonly used: the design method of opening holes in the front part of the car or the design structure and design mode of the spokes and wheel covers are improved, and the air flow is directed to the motor surface to increase the intake air amount of the cooling air of the wheel.

2.2. *suspension optimization of hub motor*

*2.2.1. Problems.*The hub motor and its accessory parts are complex in structure, large in size and high in quality, and are significantly different from conventional diesel locomotives. The main performances are: the hard points of the suspension and steering system components change, the wheel span becomes wider, the unsprung mass increases significantly, the vehicle mass distribution changes significantly, and the steering characteristics are also affected. The large increase in the quality of the unsprungs will bring great inconvenience to the comfort and handling of the car, and also cause safety hazards in the process of driving the car [5].

*2.2.2. Solution.*In order to meet the requirements of the hub motor system in layout, four-wheel independent drive and the driving performance of the electric vehicle, combined with the structural characteristics of various independent suspensions, it is proposed to use a double wishbone torsion bar spring independent suspension. Shi Tianze of Jilin University has optimized the traditional suspension system for loading hub motors and designed a new double-arm suspension system. The system is hinged to the steering knuckle by two lower control arms, while the intersection of the two control arm extension lines forms a "virtual hinge point". The virtual hinge point and the connection point on the mounting point on the suspension form a suspension kingpin. With the properly designed double-arm suspension structure, the virtual hinge point can be located in the wheel, and the actual hinge points of the two control arms and the steering knuckle are still outside the wheel, so that the position of the suspension kingpin is not limited to the physical structure. Constraint, thus providing a larger space inside the wheel, meeting the installation requirements of the hub motor, and not changing the overall parameters such as the wheel track of the vehicle under the premise of ensuring the parameters of the kingpin.

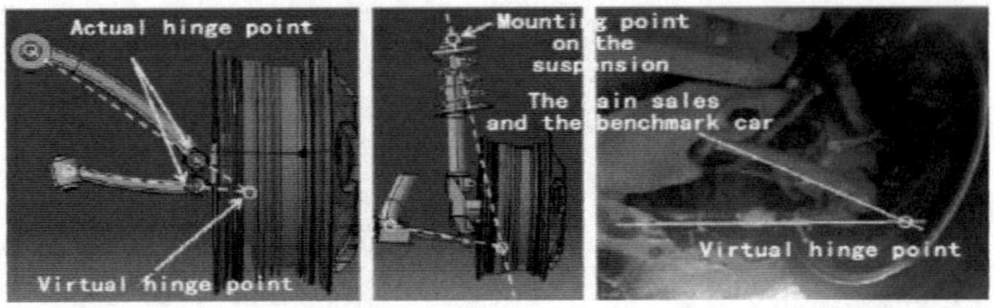

Figure 3. Double-section arm suspension and virtual hinge point

Xinbo Chen and others from Tongji University have designed a new suspension of "double wishbone suspension - torsion bar spring - electric wheel module". The suspension structure is relatively compact and consists of a double wishbone suspension, a torsion bar spring, a shock absorber, a rotational speed sensor, a brake disc (clamp), an outer rotor DC brushless hub motor and wheels. The outstanding feature of the suspension is that it can realize the flat generalization of the chassis structure. For the electric vehicle, solving the problem of the chassis is equivalent to providing a split design for the electric vehicle [5].

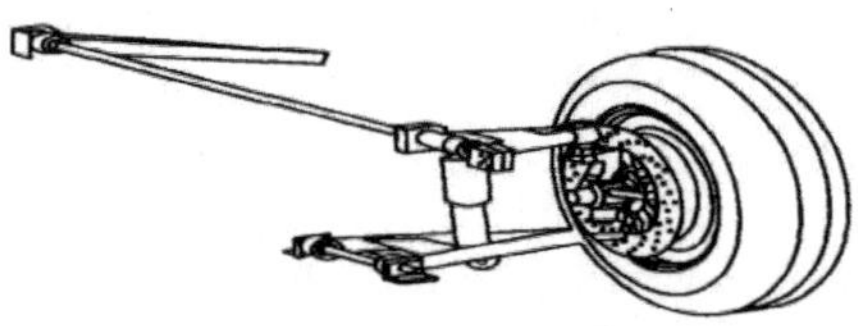
Figure4. Torsion Bar Spring - Electric Wheel Module Structure

*2.2.3. design plan.*The elastic, damping and unsprung mass of the suspension are the few relevant parameters that have the greatest impact on ride comfort. The roll center and roll stiffness are the main indicators for evaluating the handling stability of the car [7]. At the same time, it is also necessary to refer to the influence of the change of the camber angle of the wheel on the ride comfort. The double wishbone independent suspension is selected for the car design. The main positioning parameters include: the camber angle and the toe angle of the wheel, the caster angle and the camber angle of the kingpin [8].In the design, in order to optimize the layout of the shock absorber and reduce the non-suspended mass, a longitudinally arranged torsion bar spring can be used as the elastic element, and the energy per unit mass is high. The shock absorber uses a single cylinder hydraulic shock absorber. In the specific structural design, it is necessary to study the positioning parameters of the upper and lower control arms of the suspension.

2.3. Aerodynamic design of the hub motor

*2.3.1. Problems.*The aerodynamics and body shape of the car are closely connected [9]. Aerodynamics is an important branch of mechanics, which not only directly affects the design of an electric car, but also has inextricably linked with air resistance [10]. At present, the cruising range of electric vehicles is generally not high, so under the existing battery technology conditions, electric vehicles must be as energy-saving as possible. In the power-saving measures, it is most effective to reduce the running resistance of electric vehicles and design a low-resistance shape. The design of the car is a split-type design, which is a brand-new field of car design. It has a breakthrough significance in car styling, but one of the most prominent problems in this type of design is that the wind resistance is usually higher than that of ordinary cars. Therefore, how to achieve a low-resistance design of the split-type car design is an urgent problem to be solved.

*2.3.2. Solution.*When designing a car, consider the aerodynamic factors. At the time of design, the designer needs to consider two criteria: first, through the streamline design of the car shape to reduce the air resistance when the car is running, and secondly, the car will be subjected to lift and side force when driving. The role of the designer, therefore, the designer should focus on reducing the role of

IOP Conf. Series: Materials Science and Engineering **573** (2019) 012074 doi:10.1088/1757-899X/573/1/012074

these two forces to ensure the stability of the car. Usually when designing a car, designers will follow a line of thinking: adopt an overall optimization method, starting with a basic shape with low resistance, and gradually designing features such as cabin, cabin and luggage compartment in combination with the practical requirements of daily life. In this process, the drag coefficient of the car gradually reaches the requirement [11].

The reason why the car is subjected to various forces and moments during the driving process is mainly due to the influence of the external flow field of the car [12]. Nowadays, when people study automotive aerodynamics, they use fluid state analysis, and through flow analysis, people can recognize important flow processes [13].

2.3.3. design plan. In the design of the hub spoke of the vehicle body, on the one hand, the heat dissipation of the motor can be carried out through a reasonable and effective wheel spoke shape, and the trapezoidal opening design is mainly adopted, so that the cooling capacity of the motor can be improved. On the other hand, it is also possible to dissipate heat by adding a braking force feedback system. Excessive wheel hub spoke openings have a large impact on the degree of freedom of design. Therefore, in the design process of the automobile, the combination of the braking force feedback system and the opening method can be adopted, and under the heat dissipation of the braking force feedback system, the size of the opening of the hub spoke is appropriately reduced, so that in the design process There will be a higher degree of design freedom.

In the exterior design of the car body, if you want to achieve the smallest wind resistance of the car, you need to design in a streamlined shape. Figure 5. is based on the design principle of reducing the resistance, based on the original appearance of the car. When designing the front end of the car, avoid the bumps in the small parts [13].

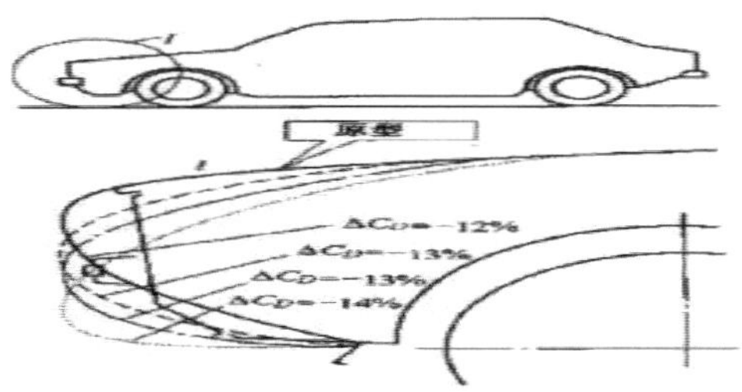

Figure 5. Example of changing the front end shape to reduce aerodynamic drag

2.4. Layout and modeling of electric vehicles loaded with hub motors

2.4.1. Arrangement of internal combustion engine vehicles. Conventional internal combustion engine vehicles have great limitations in the design of automobiles because of the necessary mechanical mechanisms. Nowadays, the traditional diesel locomotives produced by the car companies not only occupy a considerable space for the engine, but other accessory components also have an impact on the car space. However, in new energy vehicles that load hub motors, these traditional automotive components will be replaced by mechanical changes, which will bring greater interior space and design possibilities for new energy vehicles.

2.4.2. Arrangement of electric vehicles for loading hub motors. The hub motor electric vehicle no longer uses the previously inherent mechanical structure as the power source and transmission mechanism, and the power system is integrated into the wheel. Because the engine and transmission mechanism were simplified in the design, a major change occurred in the overall layout of the vehicle.

Wei Zhang of Beijing Hyundai Motor Technology Center has proposed an innovative design of electric chassis - skateboard chassis, which has many advantages, including: high degree of freedom of design, chassis and body independent, strong handling, safety Good performance and other aspects. The front and rear weight of the chassis is 1:1. If the car collides with an accident during driving, the chassis can absorb a large number of impacts in a short period of time to prevent deformation of the passenger compartment. Finally, the skateboard chassis is highly integrated, reducing the complexity of the manufacturing and handling process.

Tongji University's "Chunhui No.3" is the first four-wheel independent fuel cell vehicle with wire-steering in China. The chassis of the car adopts a double-layer frame structure, which can make the power battery, fuel cell and drive control. The system is placed in it. This flat generalized structure and its separation from the body structure make it suitable for the development of various models of different shapes and uses, which is conducive to shortening the development cycle of new automotive products.

*2.4.3. Design plan.*In the design of the chassis, the arrangement of the battery pack is the first problem to be considered. In electric vehicles, the tiling method is usually adopted, but this method has drawbacks and has a relatively large impact on the passenger's seating space. Therefore, in the design process, the designer needs to pay attention to compensate for the defects by coordinating the layout of the car. Secondly, the battery pack should be placed in a distributed manner as much as possible in the process of placing, so that the space in the car can be enlarged. In addition, there are many outstanding advantages in the dispersion of the battery: reducing the friction of the battery. The risk of collision, maintaining global stability, etc.

In order to improve the performance and fuel economy of electric vehicles, designers need to consider the lightweight work of the vehicle chassis. In the design process, the chassis should adopt advanced technology and materials. The materials are mainly made of aluminum alloy and magnesium alloy, and the structure is mainly hollow structure.

3. New energy vehicle design based on hub motor drive

3.1. Inspirational exploration and hand-painted design

Designed with the target user in mind, the design inspiration can come from any item that the user likes, extracting the elements and shapes of the design from the user's favorite items, and using this to explore the sketch. In the process of sketching, combined with various defects and compensation measures in the electric motor driven by the hub motor, the shape of the car is continuously optimized. The design of electric vehicles is a multi-culture, multi-disciplinary and multi-functional combination, which not only requires aesthetics, but also aesthetics, but also meets the requirements of practicality and other conditions. Sometimes it is necessary to combine the unique culture of the region where the designer is located. Art form [16]. Therefore, after a deep analysis of the practical requirements of automotive design and the aesthetics of Chinese residents, after many sketches and modifications, the sketches of electric vehicles are shown in Figures 6 .

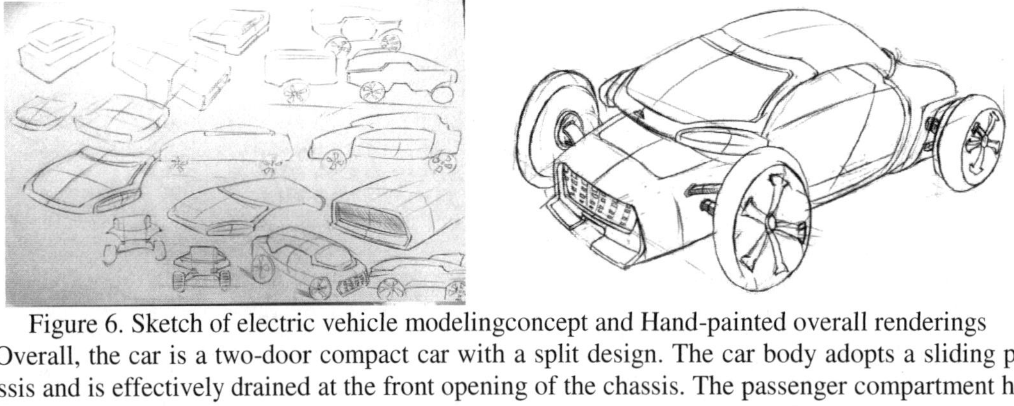

Figure 6. Sketch of electric vehicle modelingconcept and Hand-painted overall renderings

Overall, the car is a two-door compact car with a split design. The car body adopts a sliding plate chassis and is effectively drained at the front opening of the chassis. The passenger compartment has a sleek design, which can reduce the air resistance when the car is running, and can also give a smooth and silky design. Guohua Wang of Jilin University pointed out in the paper "Wheel Air Drag Reduction and In-Wheel Heat Transfer" that the aerodynamic drag coefficient of the trapezoidal opening is generally lower than that of other shape openings, that is, the piston with trapezoidal opening is aerodynamically reduced. Resistance can have a better effect. The spokes are of a five-spoke design with a trapezoidal vent at the end of the opening.

3.2. Modeling through modeling software

After collecting enough inspiration, we can paints by divergent thinking. In the process of hand-painting, the general outline of the car shape needs to be outlined first, and there is no need to consider engineering problems such as manufacturing process, cost and technical difficulty. In the design process, a process of gradual transformation from whimsical to realistic, in the initial stage, the designer needs to conceive the shape, proportion, structure and other elements of the car body. After that, the designer needs to carry out various design constraints according to the previous article. Corrected. After completing the preliminary hand-drawn draft, the design was put into 3D effects through modeling software. In this design, the body mainly adopts black as the main color, and the black gives a calm and steady feeling, which meets the psychological aesthetic needs of most people. The car is designed with a separate chassis, and the chassis and passenger compartments are mounted above and below. The headlights are horizontally stripped. This design can illuminate the road at night to ensure basic driving safety.

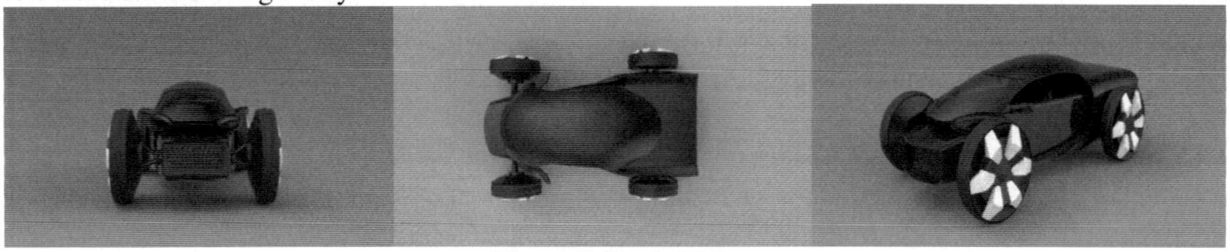

Figure 7. Body View

The design adopts a small slip-back shape in the design. The main body line is the contour curve of the car body from the front face to the rear of the car. The overall shape gives a sense of impact, and the visual effect is given in the movement. A steady and steady atmosphere, the business combines the rhythm of youthful speed. The vehicle uses four huge wheels, which improves the passability of the car. At the same time, the spokes are designed with a trapezoidal reference on the opening according to aerodynamics. They are decorated with silver and are very visually stunning. The body is made of black metallic paint, which shines under the sunlight, giving people a very smart feeling and good flow. The chassis of the vehicle is separated from the passenger compartment. The chassis is provided with an opening at the front end. On the one hand, it is designed to meet the heat dissipation requirements of the battery pack and the motor. On the other hand, it also plays a decorative role to avoid the structure that does not occur because the electric vehicle does not have an engine compartment. Coordination phenomenon.

In the basic determination stage of the sketch form, the heat dissipation form required for the pure electric vehicle driven by the hub motor is adopted. In this design, the cooling of the hub motor can provide a variety of cooling methods, including the air inlet of the chassis part of the front part of the chassis, the guide channel of the car bottom, the liquid cooling and the air intake of the hub for cooling; suspension and shock absorption The system uses double wishbone shock absorption.

4. Fuzzy evaluation method evaluation design

4.1. Determine the indicator set U

The indicator set sorts and sorts the factors to be evaluated. First, collect sensible image vocabulary about electric vehicles. Perceptual image vocabulary refers to the comprehensive performance of the user's perception of the appearance and color of the car.

In order to collect representative vocabulary, it can be done in the following ways: (1) access to relevant literature; (2) through structured access with consumers and use of self-administered questionnaires; (3) through network search Related terms. In the end, we collected a total of 112 related vocabulary, and after the vocabulary screening, we got 64 sensible vocabulary words. The KJ method and the cluster analysis method are used to screen the vocabulary again. Finally, 16 image semantic space tables are obtained. As shown in Table 3, they are sorted into four categories, which are the contents of the evaluation index set U.

Table 1. Sensual Image Semantics Glossary

Sensual imagery	Style vocabulary
Lovely, warm, friendly, at home	Nearly human - annoying
Muscular, speedy, offensive, smooth	Dynamic - still
Simple, generous, beautiful, clean	Simple—complex
Personal, novel, trendy, unique	Innovative - stale

4.1.1. Establish evaluation matrix and calculation decision matrix. Invite 30 electric car owners to participate in structured visits and self-administered questionnaires. The comment set is set at 5 levels, which are very good, good, average, poor, and very poor. The evaluation index set U and the comment set V are combined to price each evaluation, and the percentage of the number of the reviews in each indicator is counted to form an evaluation matrix R. 30 evaluators were invited to give weights to the evaluation indicators according to the survey data, which is 0.3 for the near person, 0.3 for the dynamic, 0.3 for the innovation, and 0.1 for the simple one, then W={0.3,0.3,0.3,0.1}; therefore, the decision matrix scheme has :

$$B = WR = [0.3 \quad 0.3 \quad 0.3 \quad 0.1]\begin{bmatrix} 0.41 & 0.33 & 0.06 & 0.04 & 0.01 \\ 0.45 & 0.38 & 0.02 & 0.01 & 0.00 \\ 0.43 & 0.37 & 0.05 & 0.03 & 0.01 \\ 0.52 & 0.34 & 0.04 & 0.01 & 0.00 \end{bmatrix}$$

$$= [0.423 \quad 0.423 \quad 0.085 \quad 0.056 \quad 0.014]$$

B and the V^{t} vector are subjected to a point multiplication operation, and the comprehensive evaluation value is B· V^{t} =7.9675.

It can be seen from the calculation results that the final score of the design scheme is 7.9695, which is a relatively good score. Therefore, the scheme has market value and has certain feasibility, and can be subsequently optimized.

5. summary

With its independent and integrated features, the hub motor drive will become the best driving method for electric vehicles in the future. Although there are still many technical obstacles to be overcome in the current hub motor, such as the increase in unsprung mass, sealing problems, heat dissipation problems, etc., with the advancement of technology, the optimization of the automobile structure, in the electric vehicle, the hub motor will be fully development of. Through rational analysis, this paper systematically proposes the modeling method of the electric motor driven by the hub motor, and designs the suspension system of the double-arm suspension-torsion spring-electric wheel module based on the sliding plate chassis. The electric vehicle with split design was used, and then the fuzzy evaluation method was used to evaluate the design of the car. With the increasing number of automobiles, the continuous consumption of natural resources, and the increasing environmental pollution, vehicles powered by clean energy such as electricity are receiving more and more attention from countries and enterprises. The electric motor driven by the hub motor has provided a higher degree of design freedom for the electric vehicle because of its new technology, which has brought about tremendous changes in the inherent body design of the diesel locomotive. Through researching this topic, it aims to provide reference for the appearance design of electric vehicles driven by current hub motors, and also hopes to provide more reference value for automotive shape design research.

References

[1] Chen Haiyang. Research on the modeling of pure electric vehicles for the driving mode of hub motors [D]. Jilin University, 2017.

[2] Jiang Congxi, Zhao Lanping, Miao Lu, Yang Zhigang. Heat dissipation law of electric vehicle hub motor[J]. Computer Aided Engineering, 2018, 27(02): 47-52.

[3] Wang Yazhen. Design and analysis of permanent magnet motor with fractional slot concentrated winding for electric vehicles [D]. Huazhong University of Science and Technology, 2015.

[4] Liang Peixin. Research on heating and heat dissipation of permanent magnet synchronous hub motor [D]. Harbin Institute of Technology, 2013.

[5] Shi Tianze. Design and performance matching of hub motor driven electric vehicle suspension and steering system [D]. Jilin University, 2015.

[6] Chen Xinbo, Wan Gang, Li Wei, Wang Wei. Development and Application of Double Wishbone Suspension-Torsion Spring-Electric Wheel Module[J]. Journal of Mechanical Engineering, 2005(12): 92-95.

[7] YE Yuchen. Research on the vibration-supporting pillar of the electronically controlled suspension of the new gas-hydraulic coupling of passenger cars [D]. Jiangsu University, 2013.

[8] WANG Shukun, FANG Liang, ZHENG Long.Design and Research of Wheel Drive System for Electric Vehicles[J].Machinery Design & Manufacture,2012(06):146-148.

[9] Xu Fei, Li Yanlong, Zhang Chenming, Yang Zhigang. Design Features of Future Electric Vehicles Based on Aerodynamics[J].Beijing Auto,2012(05):6-9.

[10] Yang Qianli. Shape design of new energy single-drive electric vehicle [D]. Beijing University of Chemical Technology, 2017.

[11] Li Yanlong, Tan Jiawei, Li Qiliang, Wang Fan. Model Design of Low Wind Resistance Split Electric Vehicle[J]. Journal of Tongji University(Natural Science), 2017, 45(09): 1359-1365.

[12] Hu Jiehua. Research and improvement of turbulence model of automobile external flow field based on rational shape [D]. Hunan University, 2009.

[13] Zhang Haifeng. Research on aerodynamic modeling optimization design of electric vehicle body based on Fluent [D]. Yangzhou University, 2012.

[14] Gao Yongwei. Design and Research of Electric Vehicles [D]. Qingdao University, 2017.

IOP Conf. Series: Materials Science and Engineering **573** (2019) 012075 doi:10.1088/1757-899X/573/1/012075

Research on Service Design of Community Medical Facilities Based on Aging-appropriate and Elderly-centered

Yangfan Yao[1], Xiaoru Zhou[2] and Tianyang Zhu[1*]

[1]Departments of Mechanical and Electrical Engineering, University of Beijing University of Chemical Technology, 100029,china

[2]Departments of Mechanical and Electrical Engineering, University of Beijing University of Chemical Technology, 100029,china

[*]Corresponding author's e-mail: zhuty@mail.buct.edu.cn

Abstract. Research on urban community medical service facilities, from the perspective of service design methods, combined with the community medical service system, according to the needs of the community elderly, analyze the needs of community elderly for community medical service facilities, and seek contact points between community elderly and medical facilities. Conceive community health services and build medical service facilities that meet the needs of the elderly in the community. Through the optimal design of the urban community medical facilities service system, the effectiveness of community medical service facilities will be further improved, thereby improving the satisfaction of community elderly people on community public medical service facilities, and providing reference for the research of community medical facilities service system and service mode design.

1. Introduction

According to the results of the sixth national census in 2011, China has officially entered an aging society [1]. The degree of urban aging is far greater than that of rural old age. With the increase of age, the living obstacles accompanying the elderly are gradually increasing, and medical care is increasingly becoming the biggest problem for the elderly. As a place where the elderly often travel, the community lacks medical facilities that serve the elderly and cannot meet the daily needs of the elderly in the community. Further optimize the community medical service system by analyzing the growing demand for medical services for the elderly in the community.

2.The need for community health care for the elderly under the service model

The definition of service design concept is the basic component of the service design strategic advantage seeking process. It improves the service usability, satisfaction and loyalty by starting from the user's needs, service planning, product design, visual design and environmental design. And efficiency, to provide users with a better experience, so that users get satisfaction and create value [2]. The service model is a new model based on service design. From the perspective of users, user value, contact points and value-added activities are the main research directions of service design. Through the analysis of the daily life of the elderly in the community, combined with the specific content of the service design, the design thinking of service design is applied to the optimization process of the community-based medical service system centered on the elderly, and the effective design scheme is explored through the guidance of service design.

According to the survey, the number of chronically diagnosed patients in China has exceeded 260 million, and the number of patients aged 60 years and older is 2.3 to 3.2 times that of patients under 60 years old, and 60%. The above elderly patients have two or more chronic diseases at the same time [3]. Chronic patients rely on drugs for a long time, and they need to go to the hospital regularly to get medicine.

2.1 Contact points of the community elderly in taking medicine

Contact points occupy an important position in the service system and are important nodes that connect the various parts of the service process. Mastering the contact points is the most important for controlling the entire service system. Each contact point can have different effects on the sensory experience of the community elderly, and through the analysis and processing of appropriate contact points, the satisfaction of the elderly users in the community can be improved. By analyzing the process of taking medicines for the elderly in the community, as shown in Figure 1, it is found that the elderly in the community go out, take medicines, and go home for three periods of contact, thus further analyzing the needs of community elderly for community service medical facilities.

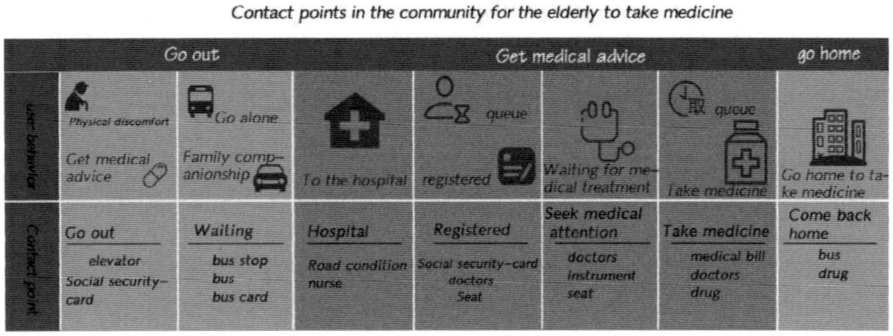

Figure. 1 Contact Diagram of Medication for Elderly People in Community

2.2 Integrate service design into community health care

The essence of service design is to design an effective model for organizing, planning people in the service system, infrastructure, communication, and the various components of tangible matter [4]. Service design integrates the design concept into the service planning and process itself, focusing on user needs and user experience, and by matching resources and requirements in specific environments, formulating system solutions for service optimization, thereby improving service quality and improving The service experience of the target user [5].

In the process of community medical service design, this model of user participation is introduced into the service design research to carry out participatory service design. By using users as the design leader, emphasizing the initiative and participation of users in research, to explore user demand information and support the development of design activities [6]. Through the participation of community seniors in the design, potential medical needs and service experiences can be further discovered, and more systematic and more targeted design solutions can be provided for the improvement of community medical service facilities.

3.Analysis of community medical service needs of the elderly

The semi-structured interview method is used to set the direction of the interview questions in advance, such as the analysis of the contact points of the elderly in the community when taking medicines, and record the sensory analysis of the contact points in the process of taking medicines for the elderly in the community, as shown in Figure 2. Except for a few major issues set in advance, other questions were randomly asked. Finding problems and solving problems in free conversations makes interviews more dynamic and makes it easier to discover the feelings of respondents.

Figure. 2 Sensory analysis of contacts

In the process of community medical experience analysis, the research is carried out around the sick elderly in an open way, summarizing many factors affecting the user experience in the medical service process, and selecting key people to construct the character role model, as shown in Figure 3. From the role model card, all factors affecting the patient can be summarized from the patient's point of view, and all formal and informal contact points can be found to gain more patient-centered experience [7].

Through the sensory analysis of the contact and the construction of the key persona model card, it is found that the community elderly people take medicine from the community to the hospital for a long time to wait and walk, which brings great to the elderly. inconvenient. As the memory of the elderly declines, the memory of the doctor in the disease, there is very little memory when returning home. The community health care in the community is insufficient to meet the needs of the elderly. In daily life, the elderly in the community lack knowledge about the life of controlling their own diseases, and do not know how to control their living habits. The elderly in the community will not have a correct lifestyle to alleviate the disease. Through the analysis of the process of taking medicines for the elderly in the community, the emotional changes of the elderly at the contact points are analyzed. Design a medical facility that serves the elderly in the community by optimizing service processes.

4.Community medical facilities in line with the community's elderly under the service design—automatic drug sales machine

Through previous research, it was found that most of the elderly in the community suffer from various chronic diseases and need to take drugs for a long time to maintain. In the process of going to the hospital to buy medicine, the distance from the hospital to the community, the waiting time, the time of the number, the distance of taking the medicine is very inconvenient for the elderly, so through the user experience map, the old people take the medicine The contact points in the middle, optimize the elderly drug collection process, and design an automatic drug dispenser that serves the elderly in the community.

4.1 Based on the design of the automatic drug dispenser

The design of the vending machine includes the shape and color of the product, and the two determine the user's first impression of the product. The shape mainly adopts the geometric form. The appearance of the product is simple and generous, the technology is strong, the operation mode is direct, and the soft chamfer is applied around the product, which can bring security and trust to the elderly users. The color is blue and white. From the perspective of color psychology, blue is the color of the sky and the ocean, which can bring refreshing and relaxing feelings to the elderly.

The feelings and feelings are conducive to the patient's emotional stability, while the gray gives people a sense of weight, increasing the patient's sense of security and trust. The body of the elderly begins to decline. In the process of design, the problems of vision and touch should be fully considered. The styling strives to combine traditional aesthetic concepts, simple and generous, and is suitable for the characteristics of the product itself. Abandoning the traditional monotonous colors in color, but it can not be made into a colorful look, which is easy to stimulate the senses of the elderly.

IOP Conf. Series: Materials Science and Engineering **573** (2019) 012075 doi:10.1088/1757-899X/573/1/012075

4.2 Functional design based on automatic drug dispenser

As a community-based medical service facility centered on the elderly, it mainly serves the elderly. Through preliminary research, it optimizes the process of taking medicine for the elderly and designs a self-selling machine for the elderly. Older people in the community are more likely to suffer from chronic diseases. The drugs needed for chronic diseases are mainly prescription drugs, and they cannot be purchased at the pharmacy. When purchasing drugs, you can purchase them through the social security card. The social security card records the chronic diseases of the elderly. The social security card can get the prescription drugs for the elderly. On the interface of the automatic drug dispenser, you can also inform the elderly through the video. The precautions for the disease in the daily life can be those preventive measures. The automatic vending machine is open at 24 o'clock, and in an emergency, the elderly can take medicine at any time. And the big ones reduce the unnecessary travel and waiting of the elderly in the process of taking medicine.

According to the investigation of shape and color in the early stage, the shape of the product is determined by the simple geometry. The color is mainly blue, supplemented by gray. The function is mainly designed to serve the elderly in the community. Two automatic drug dispensers were designed, as shown in Figure 4. Both designs are blue in color, gray in color, and slightly different in form. The shape of the sketch is relatively short, and the overall look is relatively thick. The sketch 2 is mainly slender, and the overall look is slightly lean. In the function sketch 2, there is a first aid kit below, which is convenient for emergency situations in the emergency first aid community.

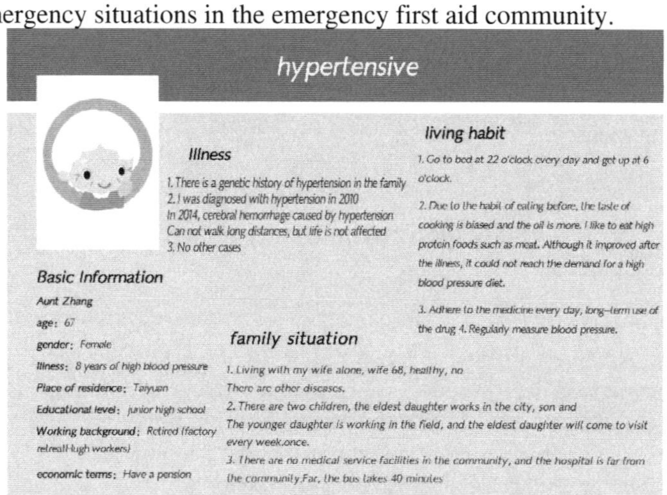

Figure. 3 Role model card

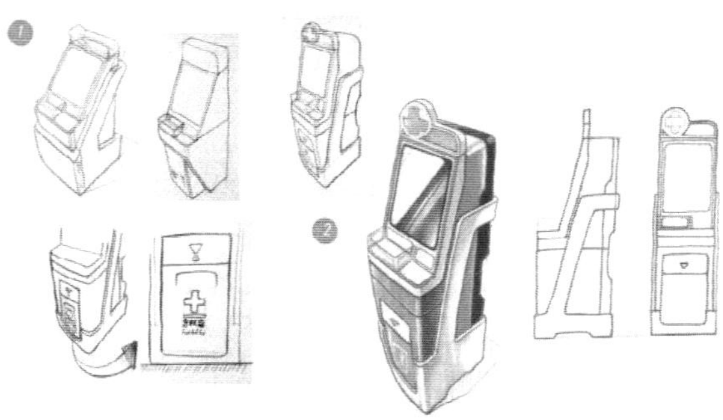

Figure. 4 Product sketch

5.Automatic drug dispenser design evaluation

5.1 Construction of Fuzzy Comprehensive Evaluation Model

In view of the fact that the product shape is difficult to objectively evaluate due to various factors, the fuzzy comprehensive evaluation method can be used to comprehensively evaluate the design scheme and achieve the selection of the best scheme [8]. The fuzzy evaluation method combines qualitative and quantitative measurement methods, and uses the idea of fuzzy mathematics to make a total evaluation of things restricted by various factors, which is characterized by clear results and strong systemicity. Through the user's perception of the shape of the automatic drug dispenser, the importance and the degree of affection of the evaluation index of the program are characterized, and the evaluation score of the evaluation index is obtained. The calculation formula of the index evaluation value is as follows:

$$b_i = \sum_{i=1}^{n} r_{ij} w_j$$

In the fuzzy comprehensive evaluation method, U, V, R are the three most important elements, U is the evaluation index set, and the sensory image pairs to be evaluated are integrated and sorted, and the evaluation index set can be obtained. U=$\{u_1, u_2, ..., u_n\}$; V is a collection of reviews, visual senses are very good, good, general, poor, very poor and other fuzzy concepts as the main filling vocabulary of the comment set, can get the comment set V=$\{v_1, v_2, ..., v_m\}$; The evaluation matrix R of the indicator is formed by the evaluation value obtained by the i-th index evaluation index on the j-th comment set, and the requirement is $\sum_{j=1}^{m} r_{ij} = 1$。 The evaluation index is assigned to the corresponding weight coefficient, and the index weight matrix W=$\{w_1, w_2, ..., w_n\}$, Claim $\sum_{i=1}^{n} w_i = 1$。 The decision matrix is calculated from the weight matrix and the evaluation matrix.The ecision matrix can be expressed as:

$$B = WR = [w_1 \quad w_2 \quad \cdots \quad w_n] \begin{bmatrix} r_{11} & r_{12} & \cdots & r_{1m} \\ r_{21} & r_{22} & \cdots & r_{2m} \\ \vdots & \vdots & \cdots & \vdots \\ r_{n1} & r_{n2} & \cdots & r_{nm} \end{bmatrix} [b_1 \quad b_2 \quad \cdots \quad b_m] =$$

Where V={10, 7.5, 5, 2.5, 0}, the decision matrix B is multiplied by the vector to obtain a comprehensive evaluation index.

5.2 Determine the indicator set U

In order to collect representative vocabulary, it can be done in the following ways: (1) access to relevant literature; (2) through structured access with consumers and use of self-administered questionnaires; (3) through network search Related terms. In the end, we collected a total of 112 related words, and after completing the vocabulary screening, we got 55 words. The KJ method and the cluster analysis method were used to screen the vocabulary again. Finally, 16 vocabulary words about the automatic drug dispenser were obtained, as shown in Table 1, and they were sorted into four categories, which are the evaluation index set U. content.

Table. 1 Glossary of the Appearance of Drug Vending Machines

Sensual imagery	Style vocabulary
Delicate, natural, dexterous, friendly	Conspicuous - neglected
Comfortable, beautiful, full, smooth	Comfortable - stimulating
Convenient, precise, beautiful, fast	Simple—complex
Personal, novel, direct, creative	Simple—complex

5.3 Establish evaluation matrix and calculation decision matrix

Invited 30 community seniors to participate in structured interviews and self-administered questionnaires. The comment set is set at 5 levels, which are very good, good, average, poor, and very poor. The evaluation index set U and the comment set V are combined to price each evaluation, and the percentage of the number of the reviews in each indicator is counted to form an evaluation matrix R. Invite 30 evaluators to give weights to the evaluation indicators based on the survey data, with a conspicuous of 0.3, a comfortable of 0.3, and a simple of 0.3. Innovative is 0.1, then there is W= {0.3,0.3,0.3,0.1}; The fuzzy comprehensive evaluation method matrix algorithm is different from the ordinary matrix algorithm. It follows the principle of multiplying and multiplying, adding and maximizing. After the matrix is obtained, the sum of the data is not equal to 1, so it needs to be normalized to obtain a scheme. The final decision matrix. The purpose of data mining is to obtain the relationship between perceptual reviews and samples, and to obtain the best solution from the order of preference, so as to deepen the design.

Decision matrix of scenario one:

$$B = WR = [0.3 \quad 0.3 \quad 0.3 \quad 0.3]\begin{bmatrix} 0.45 & 0.34 & 0.05 & 0.04 & 0.01 \\ 0.43 & 0.33 & 0.03 & 0.02 & 0.00 \\ 0.33 & 0.42 & 0.02 & 0.01 & 0.01 \\ 0.40 & 0.38 & 0.06 & 0.04 & 0.02 \end{bmatrix}$$

$$= [0.417 \quad 0.417 \quad 0.086 \quad 0.056 \quad 0.028]$$

B and the V^t vector are subjected to a point multiplication operation to obtain a comprehensive evaluation value of B·=7.868.

Decision matrix of scenario 2:

$$B = WR = [0.3 \quad 0.3 \quad 0.3 \quad 0.1]\begin{bmatrix} 0.28 & 0.24 & 0.20 & 0.05 & 0.01 \\ 0.45 & 0.25 & 0.20 & 0.02 & 0.01 \\ 0.25 & 0.45 & 0.15 & 0.03 & 0.01 \\ 0.40 & 0.25 & 0.35 & 0.04 & 0.03 \end{bmatrix}$$

$$= [0.361 \quad 0.361 \quad 0.181 \quad 0.060 \quad 0.036]$$

B and the V^t vector are subjected to a point multiplication operation to obtain a comprehensive evaluation value of B·=7.375.

It can be seen from the calculation results that the final score of the design scheme is 7.868, which is a relatively good score. Therefore, the scheme has market value and has certain feasibility.

5.4 Product Description

The first option is a medical facility that is more suitable for the elderly in the community, as shown in Figure 5. In designing this device, the reasons for the mobility of the audience are taken into consideration. Therefore, in order to avoid the person from squatting and bending down, the medicine outlet is designed to be easily accessible to the elderly, and the medicine port is blocked by the box to prevent The drug fell. There is an emergency first aid function in the function. In the community, there may be sudden illness in the elderly. In this case, proper first aid measures and necessary first aid tools are very important, if there are no suitable drugs and props. After the first aid, then the difficulty of the hospital's rescue work will be greatly increased, and this automatic drug dispenser uses the space below to place the first aid kit to deal with similar emergencies.

Figure. 5 Product Scenario Diagram

IOP Conf. Series: Materials Science and Engineering **573** (2019) 012075 doi:10.1088/1757-899X/573/1/012075

6.Conclusion

This paper examines community medical care from the perspective of service design, and conducts research on the process of medical treatment for the elderly in the community. It analyzes the contact points of the elderly when the community takes medicine, studies the sensory changes of the elderly contacts in the community, and constructs the character model card. Conduct a specific analysis. It is found that the elderly in the community spend most of their time traveling and waiting during the process of taking medicine, which brings great inconvenience to the elderly. The community lacks the medical facilities that serve the elderly with long-term medication. To this end, in response to the discovery of the previous problems, propose a solution strategy, reduce the process of buying drugs for the elderly, facilitate the elderly to buy medicines, and design a service and community automatic drug sales machine for the elderly. The preliminary product plan uses the fuzzy evaluation method, constructs the program evaluation model, collects the vocabulary about the product, conducts the questionnaire analysis and focus interview on the user, and calculates the data and matrix calculation to obtain the comprehensive evaluation value of the two programs. Screening for the program. The fuzzy evaluation method is used to comprehensively evaluate the program, and the automatic drug selling machine serving the elderly in the community is obtained, and new ideas for community medical service facilities are provided in the future.

References:

[1] The Office of Population Census of the State Council. Main Data of the Sixth National Population Census in 2011 [M], China Statistics Publishing House, 200, 11.

[2] Mali. Study on the Design Model of User-Centered Urban Community Public Facilities Service [J]. Packaging Engineering, 2017,38(12): 253-258.

[3] XUE Li-mei. The Role Redirection of CommunityHealth Service Center in the Nursing Care of ElderlyPatients with Chronic Diseases[J]. Laodong BaozhangShijie, 2015(S2): 201—203.

[4] VARGO S L, MAGLIO P P, AKAKA M A. On Value and Value Co-creation: a Service Systems and Service Logic Perspective[J]. European Management Journal, 2008, 26(3): 145—152.

[5] Zhang Xiao, Xu Xiaoyun, Li Jie, Yang Aihui and Myrica rubra. Research on participatory community medical service design methods for elderly chronic diseases [J]. Packaging Engineering, 2018, 39 (12): 197-202.

[6] Li Xueliang. Participatory User Research for Health Medical Design [D]. Wuxi: Jiangnan University, 2016.

[7] Gao Jiawei. Research on Virtual Display Design from the Perspective of Service Design [J]. Packaging Engineering, 2014, 35 (14): 89-93.

[8] Li Fuxing, Sun Jian. Evaluation method of product design scheme [J]. Mechanical design, 2017,34 (04): 104-108.

IOP Conf. Series: Materials Science and Engineering **573** (2019) 012076 doi:10.1088/1757-899X/573/1/012076

How does interaction design affect user experience throughonline shopping interfaces?

Haibo Cheng [1*]

[1] Departments of Art Design & Architecture University of Monash, VIC 3145, Australia

[*]Corresponding author's e-mail: 1249176012@qq.com

Abstract. Since the birth of eBay, the world's first online shopping platform in 1995, there have been more and more shopping websites. The interfaces of online shopping websites have also changed dramatically. And user experience of online shopping is influenced by interaction design factors of the web interfaces. Thus, this article discusses how interactive design affects people's shopping experience through online shopping interfaces. Rosenzweig (2015) points out being easy to use and effectiveness are two important principles in user experience design. Therefore, the purpose of this article is to illustrate how different interactive design factors affect users' shopping experience by analyzing the usability and efficiency of the examples of eBay's and Amazon's interfaces. The analysis of the two online shopping interfaces indicates that user experience is influenced by the feedback and Fitts' law factor in terms of usability. Besides, it is affected by the navigation and time factor based on efficiency. According to the research of this article, designers can optimize people's shopping experience by improving the usability and efficiency of websites in interface design.

1. Introduction

With the widespread application of e-commerce websites, a growing number of people choose online shopping. Ranganathan and Ganapathy (2002) suggest that good interface design is the key factor to meet the needs of consumers and enterprises to survive in competitive online sales. In addition, Kurniawan (2003) contends that interaction design should focus on the interaction between people and products, taking into account users' backgrounds, experiences, and feelings during the operation. Therefore, user experience is critical in interaction design. Moreover, Flavia´n and Gurrea (2008) propose that an attractive design can evoke feelings of pleasure in the use of a website. This means a well-designed shopping website will not only provide users with a quick and convenient way of shopping but also make people enjoy this convenient purchase. Upon all these points, there is the main research question. ------How does interaction design affect user experience through online shopping interfaces? Meanwhile, Kolko and Connors (2010) add that usability and effectiveness are of great importance in interactive design. As a result, user experience in the online shopping interface is closely related to interaction design. For this reason, there are three sub-questions in this report

------How does interactive design affect people's shopping experience in terms of the efficiency of shopping interfaces?

------How does interactive design affect people's shopping experience in terms of the usability of shopping interfaces?

------What shortcomings in the interface can be improved to make people's user experience friendlier?

Based on the three questions, this article analyzes the effectiveness and usability of different elements in the interface design of shopping websites to explain how different interactive design factors affect shopping experience and makes recommendations for interface improvement through the analysis and comparison of different forms of similar functional interfaces of different websites.

2. Material and Methods

2.1. Venn Diagram

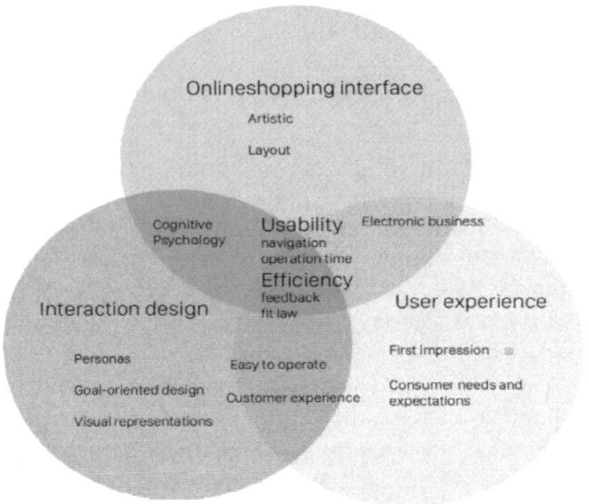

Figure 1. Venn diagram

2.1.1. Online Shopping Interface. The online shopping interface is a display platform that allows consumers to use a shopping search engine to find products of interest and buy goods directly from sellers over the Internet. Customers can use a variety of electronic devices to access the platform, such as desktop computers, laptops, tablets, and smartphones.

2.1.2. Interaction Design. Interactive design is a field of design in which people interact with products and services to achieve a common purpose. Its core is to focus on user needs and then meet them.

2.1.3. User Experience. User experience is a kind of subjective feeling that the user builds up in the process of using the product. User experience occurs in every step of the operation when users use an online shopping app. Therefore, a good user experience is crucial to selling goods online.

2.1.4. Relationships among Interface, Interaction Design and User Experience. When a user opens an online shopping website, there is interaction in every step of the operation. As a result, the interface of an e-commerce website is more interactive than other web pages. For this reason, interactivity is an important factor to be considered in the interface design of online shopping websites. Whether consumers are attracted to the interface for the first time, and whether consumers can first find the product they need, these user experiences are all about interactive design. The intersection of the interface, interaction design and user experience in the Venn diagram is usability and efficiency, which means that these two principles can affect the three parts simultaneously. Thus, usability and efficiency are the central focus area of this report.

2.2. Literature review

2.2.1. Influence of Navigation on Efficiency. Simple navigation is essential in the online shopping interface. According to Cooper et al (2000), the definition of navigation is a feature that can help users understand a new interface or help them to locate objects in the system. Moreover, they propose that the poorly designed navigation presents the most common problems in interactive design and difficult navigation is a major frustration to the user experience. Meanwhile, Kurniawan and Panayiotis (2007) give a new explanation of the concept of navigation: Web interface navigation is similar to spatial navigation. The navigation bar across the top of a website which points to the top-level categories of the ontology is called global navigation, which is equivalent to a map used in space navigation. Sub categories under global navigation are called local navigation, which is equivalent to the landmarks of every place on the map. Good web navigation can help users reach their destinations effectively. Furthermore, the authors point out a good design principle to improve the efficiency of users is that each page should have the global navigation bar at the same location so that users can easily jump back to the home page and maintain a certain operational memory.

2.2.2. Influence of Time Factors on Efficiency. In terms of interface efficiency, another factor that affects people's shopping experience is the time. Middlebrooks, Murayama, and Castel (2016) propose that people's memory and attention are limited, and short-term memory information loss will bring psychological frustration to users. Therefore, complex interface elements and too many operations will not only affect the speed of page loading but also affect people's shopping experience. Furthermore, Schneider and Anderson's study (citing Hick's Law from 1952) suggests that the response time of users making decisions is positively correlated with the number of options. This means the more options the user faces, the longer response time is required. Consequently, the application of Hick's Law helps designers improve users' efficiency and users will spend less time on their choices. When the user is given fewer choices in the design, the user's decision time cost will be lower. Shortening the user's selection time is a solution to improving the efficiency of the interface; however, it does not mean that the shorter time for the system is better. Kolko and Connors (2011) argue that an interface that changes information too quickly can cause information loss, which means the user may forget the information that has just appeared. For this reason, the designer must evaluate the time read of the interface and keep the user's operating time within a reasonable range.

2.2.3. Influence of Fitts'law on Usability. The usability of an interactive design interface also affects people's user experience, and Fitts' law is widely used in n design. Fitt (1954) proposes that the time required to rapidly move to a target area is a function of the ratio between the distance to the target and the width of the target. This law has developed into a human movement prediction model that is mainly used in human-computer interaction and ergonomics. Mackenzie (1992) adds that Fitts' law is also applicable to the interaction interface. In Fitts' law, the location time of the mouse cursor is determined by the distance from the cursor's position to the object and the size or width of the object's area. That is, positioning time is a function of cursor moving distance and object area. The closer the cursor is to the object or the larger the object area is, the shorter the positioning time is. Therefore, the result of the user's operation time is often different in the interactive interface. The larger the clickable button area within a reasonable range is, the easier it is to click, while the smaller the clickable area is, the less easy it is to operate. Furthermore, the findings from a study conducted by Mcguffin and Balakrishnan (2005) indicate that target expansion contributes to selection, even if the user has moved 90% distance to the target. It can be argued that this is the reason why the size of the option button increases when the mouse moves to a certain area on many shopping sites. Consequently, making full use of Fitts law in web design can reduce the user's operation burden and improve the user's shopping experience.

IOP Conf. Series: Materials Science and Engineering **573** (2019) 012076 doi:10.1088/1757-899X/573/1/012076

2.2.4. Influence of Feedback on Usability. Another aspect of usability is reflected in the feedback of interaction design. Eick (1995) argues that interactive visualization tools (such as animation) can create better matches between tasks and decision environments, improving the quality of information and reducing the amount of work required. Moreover, Rogers, Preece and Sharp (2011) claim that feedback on progress is one of the key elements of flow and does not prevent users from the normal flow of activities and interaction. Feedback must be easy to see and understand without obscuring or interfering with the user's actions. Nevertheless, there is little evidence that all feedback on the interactive design page is desirable. Cooper et al (2000) suggest that when an error occurs, the sound generated by the interface is called negative acoustic feedback, and negative auditory feedback should be avoided in the design of the interface, because the sound is a public statement of users' operation failures.

2.3. Mind Map

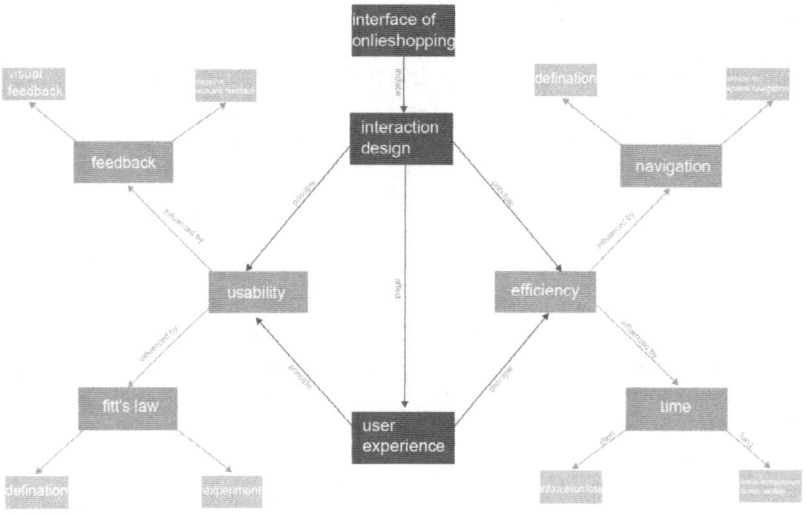

Figure 2. Mind map

3. Results

3.1. Design precedents

As the world's first online shopping platform, ebay's interface has changed dramatically in more than 20 years. In fact, the original interface has been able to meet people's shopping needs, but its interface is still continuously improved to optimize people's user experience.

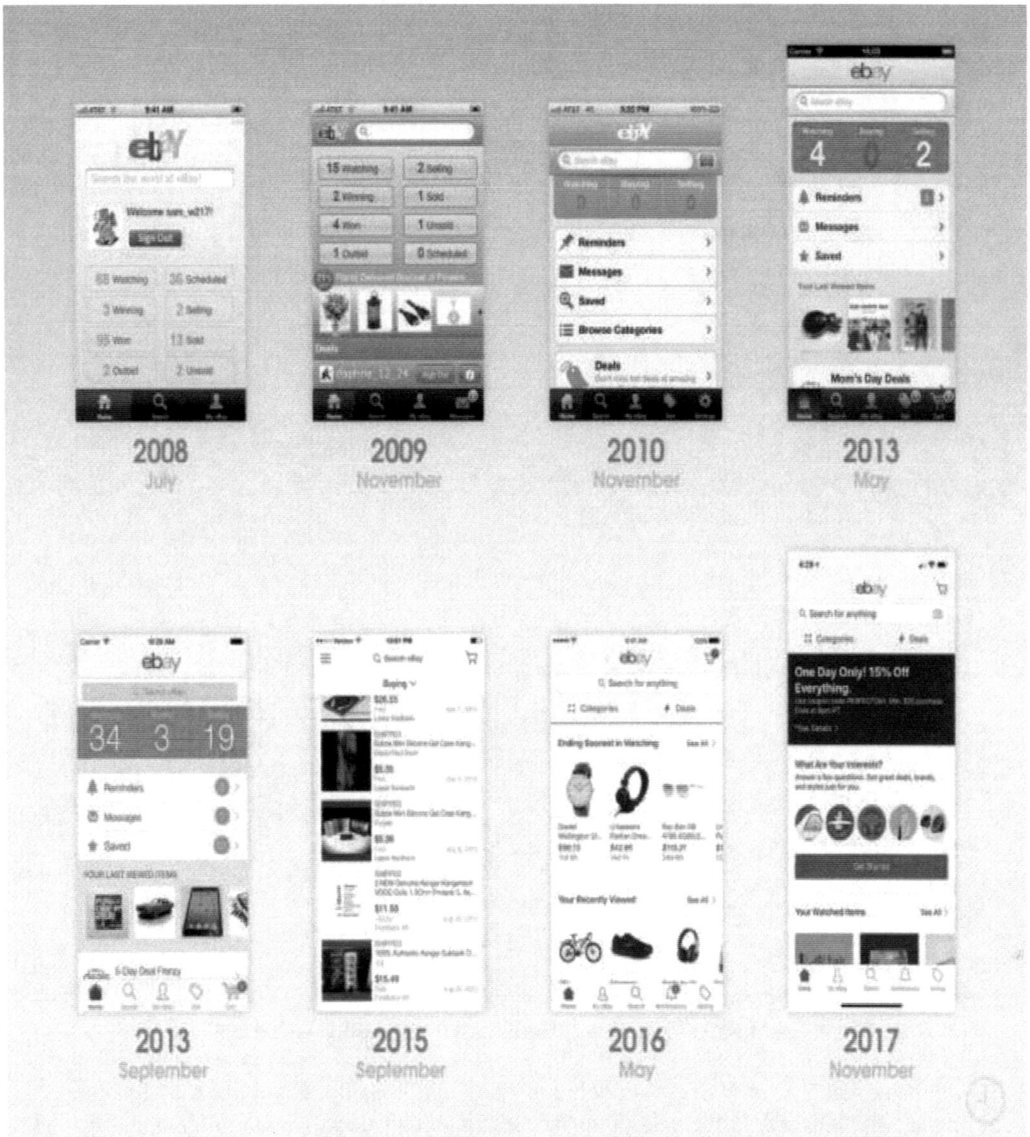

Figure 3. Evolution of eBay Interface

3.2. Case studies

3.2.1. Comparative Analysis of Registration Interfaces between eBay and Amazon. When customers first log in to the two sites as new users, they need to create an account first. By comparison, it is found that the user only needs to select "create an account" from the login and create account options in the Amazon interface, while the user needs to distinguish the button of "create an account" from four similar buttons and then proceed with creating an account in the eBay interface. According to Hick's Law, the more choices the system provides, the longer time it takes for the user to make a choice. Such interface settings not only prolong the user's operating time but also affect user experience in terms of the efficiency of the web interface.

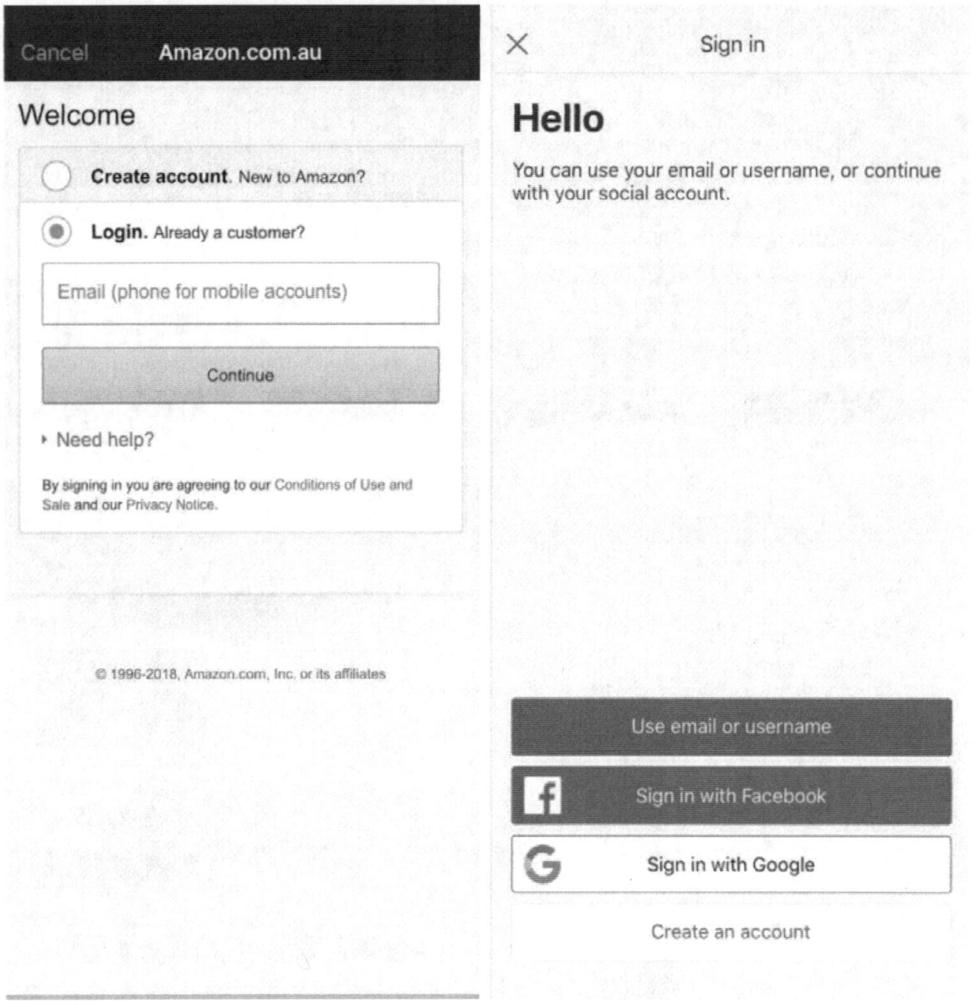

Figure 4. Registration interfaces for eBay and Amazon

3.2.2. Comparative Analysis of Navigation between eBay and Amazon. Navigation performance is also different on the two websites. In the eBay interface, the user can find the categories option in the main interface effortlessly, while in the Amazon interface, users need to click on the left folded navigation bar and select items according to different categories. This means that if the user wants to find the kind of product he needs, he needs an additional step. Besides, the navigation bar under eBay's every interface allows customers to return to any page interface rapidly. By contrast, the user needs additional option to jump to another page in the Amazon interface. Consequently, the simple operation of eBay can greatly improve the efficiency of user operations, make the operation process smoother, and optimize users' shopping experience. Whereas, Amazon's customers must spend more time locating and looking for goods. Thus, the navigation bar is a critical part of the interactive interface.

IOP Conf. Series: Materials Science and Engineering **573** (2019) 012076 doi:10.1088/1757-899X/573/1/012076

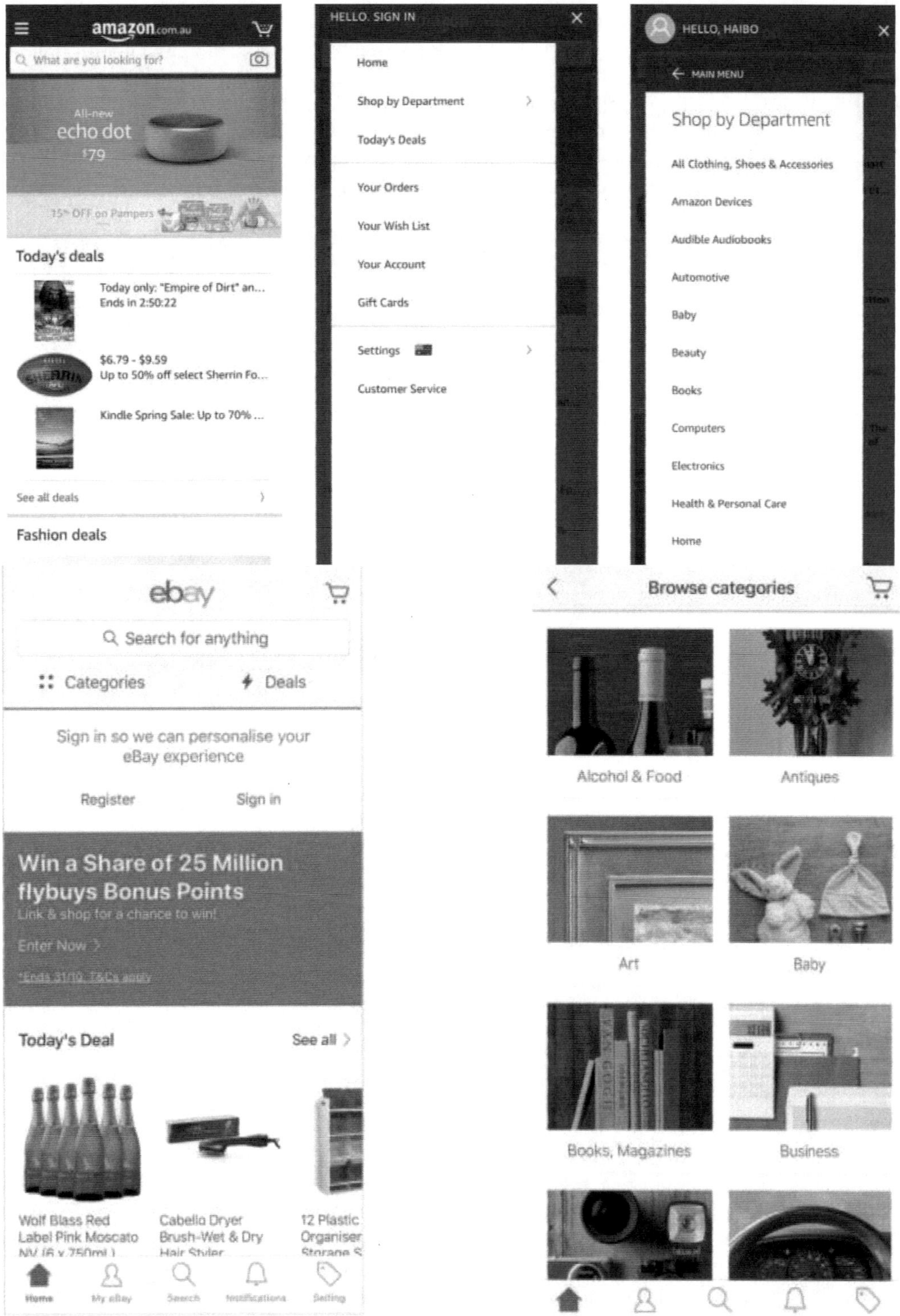

Figure 5. Navigation for eBay and Amazon

IOP Conf. Series: Materials Science and Engineering **573** (2019) 012076 doi:10.1088/1757-899X/573/1/012076

3.2.3. Comparative Analysis of Sign- in Interfaces between eBay and Amazon. The login buttons for the two interfaces are significantly diverse. In the Amazon interface, the login button is most prominent in the main interface, and the orange button background gives users a larger area to click on. However, in the eBay interface, the login button is at the top of the interface, and the clickable area is just a small arrow. According to Fitt's law, the larger the button is, the easier it is for the user to click and the less time it takes. Therefore, the login button in the Amazon interface is of higher usability, and the user spends less time to complete an operation and has a superior user experience compared with eBay.

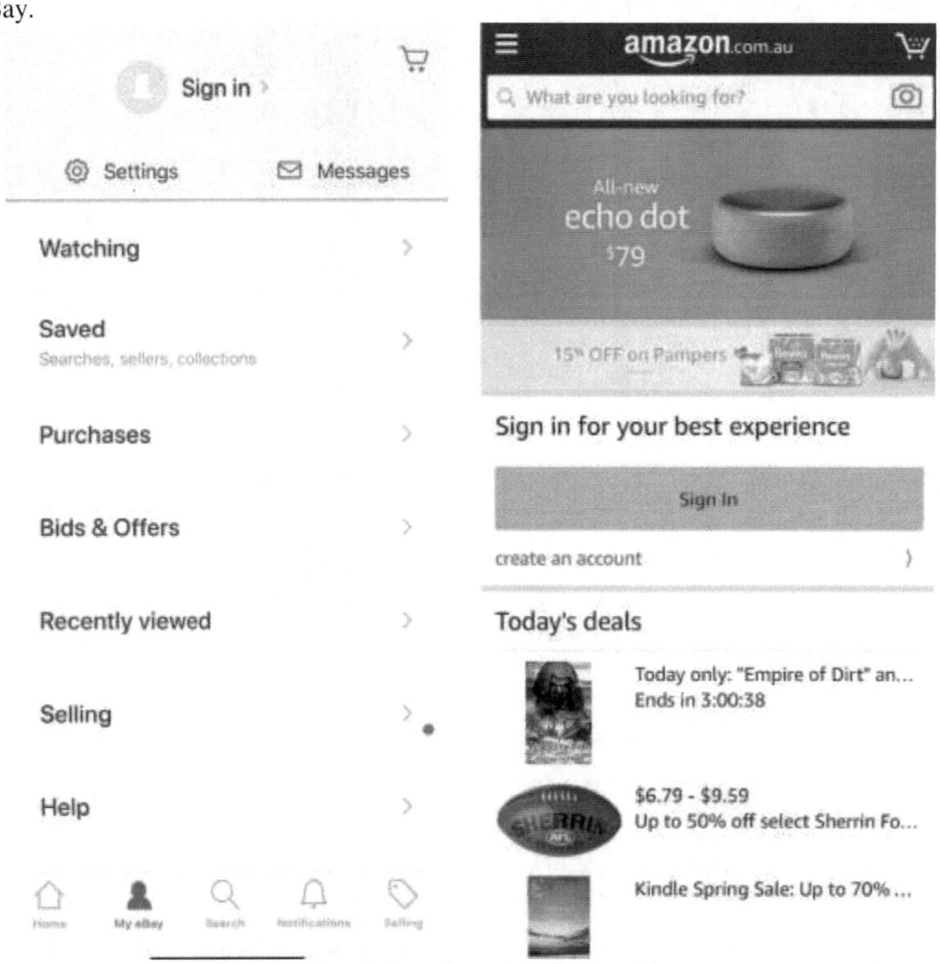

Figure 6. Sign-in interfaces for eBay and Amazon

3.2.4. Comparative Analysis of Feedback between eBay and Amazon. Both sites give positive feedback when customers add items to the shopping cart. In the Amazon interface, in addition to a text hint, there is addition vibration that reminds users of adding commodities successfully, and the top right shopping cart also appears in the number of new items added to the shopping cart. In terms of eBay, a new interface for adding commodities to shopping carts appears, and a hint for adding quantities appears on the shopping cart at the upper right side. Rich visual feedback in the two interfaces gives users the most objective and direct experience suggestions through the guidance of the text and clearly presents the status and effect of these manipulations to users.

IOP Conf. Series: Materials Science and Engineering **573** (2019) 012076 doi:10.1088/1757-899X/573/1/012076

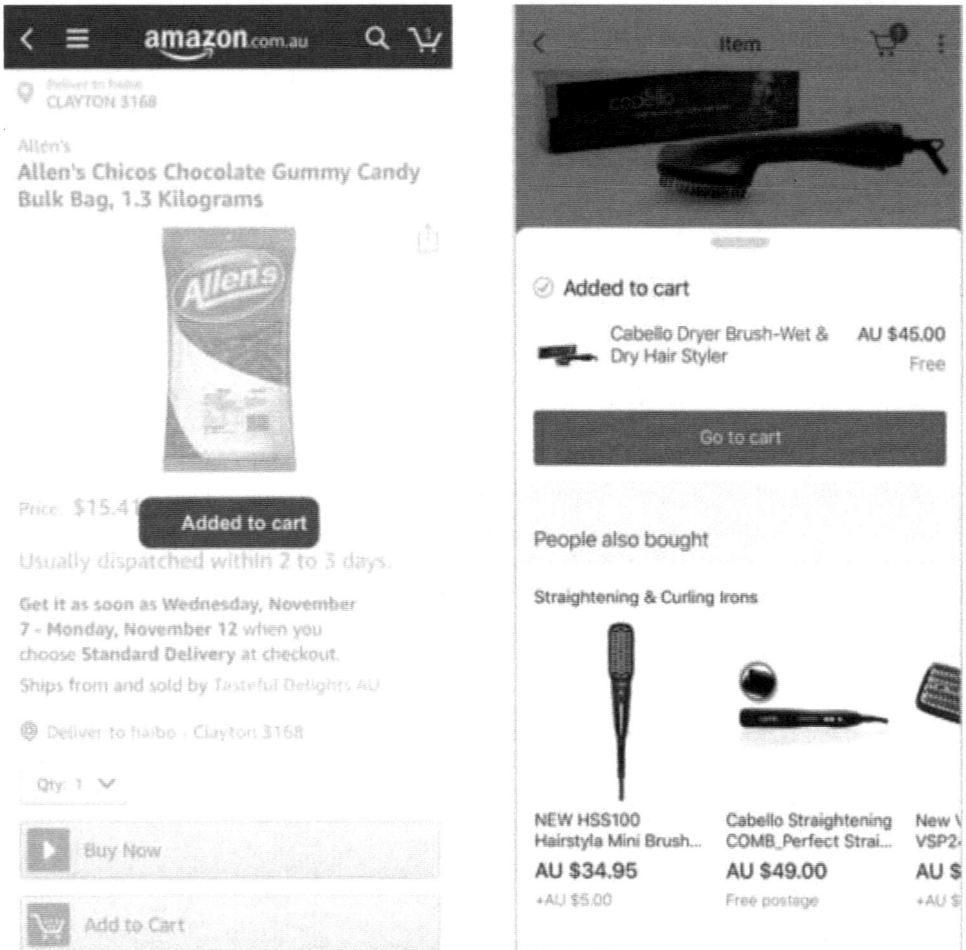

Figure 7. Feedback for eBay and Amazon

4. Design methods

4.1. Web analytics
Hanington and Martin (2012) argue that web analytics is a gateway for your organization to become deeply invested in what your customers are doing online. In website analysis, a large number of data must be collected, such as the number of visits to the site, the click rate of different items, the time customers stay at different interfaces, and the probability of the customer's second repurchase. The data of the analysis can help verify the research results clearly. Moreover, website usability tests can be carried out and the website can be optimized according to the data.

4.2. Usability testing
Hanington and Martin (2012) point out usability testing is an evaluative method that allows teams to observe an individual's experience with a digital application as he or she walks through the steps of a given task. Through the previous data analysis, the prototype for usability testing can be improved. This process can further improve user experience by testing whether the improved interface is easy to operation for average users.

5. Conclusions

This article analyzes the influence of different interaction factors on the usability and efficiency of online shopping interfaces and points out that the design of navigation and the time factor are two main factors influencing the efficiency of an interface. In terms of effectiveness, the rational use of Fitts' law and appropriate feedback can help users to simplify operations and improve the user experience. This article takes user experience as the leading direction for design and compare the interaction factors in ebay and amazon interface. Interaction design connects consumers and Internet products closely, effectively increases the usability and efficiency of online shopping interfaces, and forms the most natural interaction between users and interface.

References

[1] G. Eason, B. Noble, and I.N. Sneddon, "On certain integrals of Lipschitz-Hankel type involving products of Bessel functions," Phil. Trans. Roy. Soc. London, vol. A247, pp. 529-551, April 1955. (references).

[2] J. Clerk Maxwell, A Treatise on Electricity and Magnetism, 3rd ed., vol. 2. Oxford: Clarendon, 1892, pp.68-73.

[3] I.S. Jacobs and C.P. Bean, "Fine particles, thin films and exchange anisotropy," in Magnetism, vol. III, G.T. Rado and H. Suhl, Eds. New York: Academic, 1963, pp. 271-350.

[4] K. Elissa, "Title of paper if known," unpublished.

[5] R. Nicole, "Title of paper with only first word capitalized," J. Name Stand. Abbrev., in press.

[6] Y. Yorozu, M. Hirano, K. Oka, and Y. Tagawa, "Electron spectroscopy studies on magneto-optical media and plastic substrate interface," IEEE Transl. J. Magn. Japan, vol. 2, pp. 740-741, August 1987 [Digests 9th Annual Conf. Magnetics Japan, p. 301, 1982].

[7] M. Young, The Technical Writer's Handbook. Mill Valley, CA.

IOP Conf. Series: Materials Science and Engineering **573** (2019) 012077 doi:10.1088/1757-899X/573/1/012077

The Rebellious Journey: Archizoom Associati and Andrea Branzi

Yeqiu Yang[1*], Zhuolin Xie[2]

[1] Department of Design, Politecnico di Milano, Milan, Lombardia, 20158, Italy

[2] School of Architecture Urban Planning Constriction Engineering, Politecnico di Milano, Milan, Lombardia, 20133, Italy

[*]Corresponding author's e-mail: Yeqiu.Yang@polimi.it

Abstract. Archizoom Associati design studio is one of the founding groups of the Radical Movement in Florence dating back to the 1960s. Their design activity is extremely significant in understanding the philosophy of the movement and their design works challenged functionalist design aesthetics and the mainstream ideology at the time with sarcasm, subversion, irony, and a playful design language. Andrea Branzi, which was the leader of Archizoom Associati, together with his 50-year reflection on Italian architecture and design, has witnessed this rebellious design philosophy. Based on the historical background, practical activities of Archizoom Associati and the design teachings of Andrea Branzi, this article aims to review the theoretical consciousness into architecture sphere and the empirical research into contemporary design practice.

1. Introduction

Whenever the modernist design movement is mentioned, it is inevitable to think of the German Bauhaus, a period when the post-modernist phase is scrutinized, the Italian radical design is a case study that undoubtedly contributed to our notion of understanding design today. Alessandro Mendini, a legendary Italian designer and architect, in the article "The Art of Italian Design" sums up in five significant movements the history of Italian design: Futurism, Bel Design, Radical Design, Alchimia, and Memphis[1]. He believes that in the history of world design, Italian design has always been characterized by a particular style, concerning both the aesthetics and the innovation of industrial and architectural design. The artistic sense of the Italian design is the direct inheritance of the Italian history and culture, particularly of the Renaissance period, while the design innovation is the result of the spontaneous and continuous experimentation from the "design studios"[1]. Italians consider design as a fundamental value rather than an added value. Although the "design studios" differ from one another in terms of "design" methodology and ideology, their emotional drive and dramatic playfulness seem to be a tacit trait.

2. Radical atmosphere and environment: History of Archizoom Associati

After the Second World War, despite the rapid recovery of the economy, in Italy, social reforms and urban planning progress have been hindered by political interference. Urban plans and architectural projects are often used as bargaining chips in political struggles, and most of the architects had set their political orientation[2]. During the1968 protests that broke out in France, its influence had spread throughout Italian students, affecting as well the architecture students. Architecture was used as an instrument for political and ideological criticism. Architectural design was also seen as a mean to regain the cultural rights that were stripped away by the capitalist society and its labor division, a way to freely express their ideology to the mass. In Florence, during the 1960s, the newly graduated architecture students were frustrated by the general stagnation and depression of the construction industry. These students spontaneously formed design groups and self-proclaimed themselves as "Italian radicals". Those groups included UFO, Gruppo 999, Superstudioetc. [3], amongst which the most far-reaching one is Archizoom Associati.

In December 1966, four new graduate students from the School of Architecture of the University of Florence, Andrea Branzi, Gilberto Corretti, Paolo Deganello, and Massimo Morozzi gathered together and established a common goal together, an ambition to found the Archizoom Associati design studio [4]. The name Archizoom comes by mixing the prefix of the British architecture group "Archigram" of the 1960s, with the name of "Zoom" magazine [5]. In 1968, Dario Bartolini and Lucia Bartolini joined the group. Inspired by the British Utopian graphic architectural language and pop art, they co-hosted the "Superarchitettura" exhibition with the Superstudio architecture group. In this occasion, they also declared the establishment of "Paper Architecture" experimental research study [5]. At the same time, some architecture students of the University of Florence decided to occupy Florences city square by putting up installations. The action of this time was done as a protest against the conservative and traditionalist way of urban planning that does not make use of the architectural tools and knowledge, that they have learned to enact in a real urban transformation. At that time, Branzi, as a student, proposed a fundamental change in the concept of the city: The idea of an inexpressive, catatonic architecture, the outcome of the expansive forms of the logic of the system and its class antagonists, was the only form of modern architecture of interest to us. A society freed from its own alienation, emancipated from the rhetorical forms of humanitarian socialism and rhetorical progressivism: an architecture which took a fearless look at the logic of grey, atheistic and de-dramatized industrialism, where mass production produced infinite urban decors [6].

Archizoom was one of the main protagonists of the "Radical Architecture" movement of the 1960s, which opposed the modernist architectural ideology. They proposed the design concept of "function above all." Instead, they use imaginative storyboards, collage images, visual materials, etc. to add an emotional dimension to the design. By doing so, they proposed an innovative and pioneering design for city, building, landscape, furniture, and clothing. Their creation is an open criticism to the capitalistic mode of production, mass consumerism, and the anarchism of their period [3]. Like many other radical architects and designers from the late 1960s and early 1970s, the group's design philosophy was greatly influenced by the "Cold War" period's political turmoil and anti-cultural spirit. Any new innovative ideas characterized this subversive as "Anti-Design". By using new materials, new techniques, and new forms, they challenged the traditional aesthetics and the mainstream ideologies. After the disbanding of the group, their rebellious, radical spirit was inherited by the later Alchimia group and the Memphis group, making their influence lasting until today.

3. Sarcasm and subversion: Design practice of Archizoom Associati

Pop Art has deeply influenced Archizoom, the members of the group were still full of provocative and innovative ideas. Through the use of waves, curves, spots, stripes, and discarding colors, they reorganized and subverted established daily habits (Figure 1). In 1968, Archizoom participated in the 14th edition of Milan Triennial with a series of revolutionary works such as "Centro di Cospirazione Eclettica". In 1972, they participated in the exhibition "Italy: the New Domestic Landscape" [7] at MoMA, New York, a milestone in the development of Italian design of those years; In 1973, alongside other key representatives of radical design movement, they created the "Global Tools" teaching workshops. From 1966 to 1973, they designed all sorts of provocative and satirical furniture for the "Poltronova" brand, which has two classic examples : the first piece is the Safari chair (Figure 1), which is covered with skins from exotic animals, whose impactful color scheme and shapes challenged the mainstream middle-class common sense of "Good Tastes"; the second piece is the Mies van der Rohe armchair, named after the famous architect (Figure 2), in which they combine Mies'

IOP Conf. Series: Materials Science and Engineering **573** (2019) 012077 doi:10.1088/1757-899X/573/1/012077

signature chrome tubes with Le Corbusier's leather pillow materials: a design that was used as a mockery, a form of a protest against the 20th century society, and an extreme criticism of the functionalism design approach of "less is more". In 1968, they created the Gazebo series, published in the first issue of Ettore Sottsass's "Pianeta Fresco" magazine, in which their pop sensibility was abandoned in order to allow space for "elementary compositional operations" thus initiating an essentially theoretical process of rethinking architecture as a discipline and also in terms of its cultural role in relation to reality[8]; In 1970 the studies of Archizoom reached its peak, carrying out the study on the "No-stop city", wherein taking into consideration the state of decline of the modern cities, they considered the design as the basic tool to improve people's life quality and the planning of the urban space. From 1971 to 1973 their main focus was on "Dressing Design", covering the full process from the element to the finished garment, and expanding that concept to the surrounding environment (from the design of the furniture or to the large-scale city planning) that matched their garment design. In 1974 the group disbanded, but most of its members still worked with other radical design organizations, passing down the design ideas and passions of this singular generation of Italian design radicalists. They believe that the liberation of unconstrained architecture and design is a fearless cultural concept and lifestyle, a way of moving towards rebellion and freedom[9].

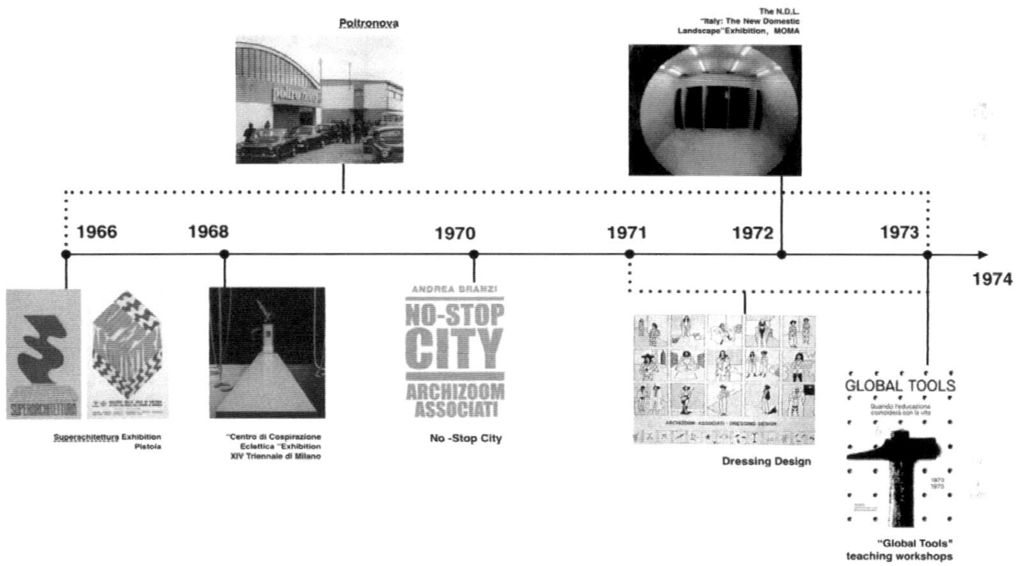

Figure 1. The timeline of work of Archizoom Associati

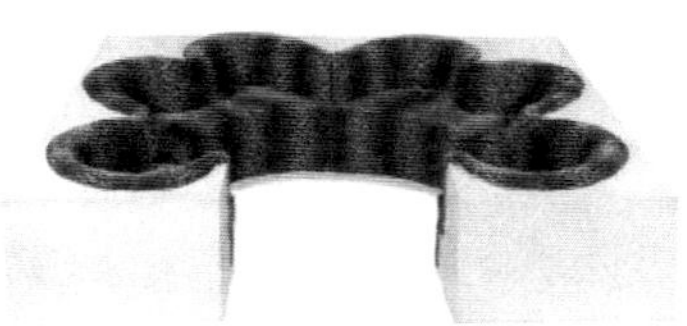

Figure 2. The Safari Sofa

Figure 3. The Mies chair

4. Utopia Magician: Andrea Branzi with the No-stop city

Andrea Branzi, with his unique reflection on more than 50 years of Italian architecture and design was considered a great witness to his peers [10]. He graduated from the School of Architecture of the University of Florence in 1966; Between 1964 and 1974, he was one of the founding members of

IOP Conf. Series: Materials Science and Engineering **573** (2019) 012077 doi:10.1088/1757-899X/573/1/012077

Archizoom; In 1976, he was a leading member of the Alchimia design group; In 1980, he became a principal member of the Memphis design group; In 1982, he founded the first international design graduate school, Domus Academy; Until 2017, he held the role of full professor and director of interior design program at the School of Design of Politecnico di Milano, which is the best design university in Italy. In 1979, 1987 and 1995, as an individual and taking part of a team, he awarded the Compasso d'Oro Award for with the highest honor in Italian design. In 2008, he was appointed honorary royal designer in the UK; In 2018, he won the Rolf Schock Prize in the Visual Arts category, from the Royal Swedish Academy of Arts. In his long designer career, he reached the peak of his experimental design during the period when with his colleagues in Archizoom Associati, expanded his research fields through urban, environmental and popular culture design, culminating in the conception of the utopian project of "No-Stop City" [11] [12].

In 1970, the Italian magazine Casabella published an article entitled "City, assembly line of social issues, ideology, and theory of the metropolis," in which the "No-Stop City" project was presented to the public for the first time: it is a model of utopic global urbanization [13], in the form of and radical project-on-paper (Figure 4). Like any other radical architectural groups of the 1960s, they opposed the pragmatism of modernist architecture and advocated the dilution of practical, imaginative, science fiction-like creative methods[3]. They rejected the traditional approach of scales, functions, materials, or needs as essential criteria to start the architectural design. Instead, under the influence of the pop culture, they made extensive use of paintings and collage paper schemes, using emotion as a medium to emphasize their critics to the society and the material reality, re-affirming of the autonomy of architecture.

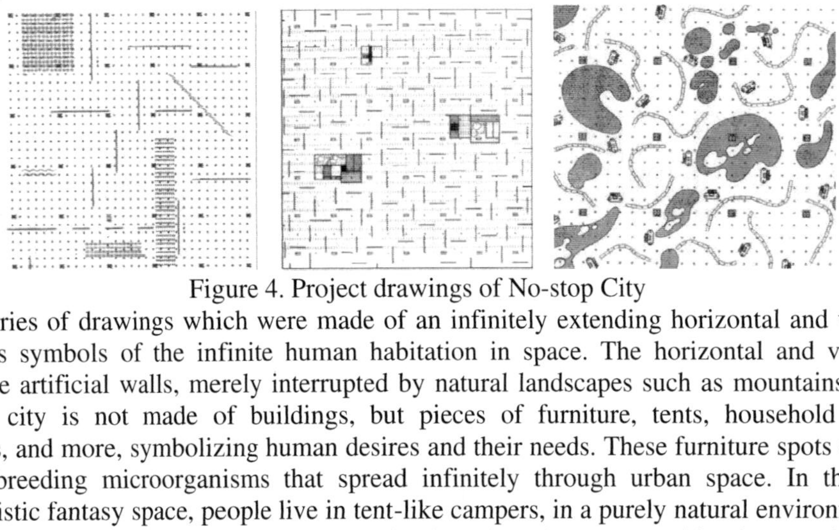

Figure 4. Project drawings of No-stop City

These series of drawings which were made of an infinitely extending horizontal and vertical grid are taken as symbols of the infinite human habitation in space. The horizontal and vertical lines represent the artificial walls, merely interrupted by natural landscapes such as mountains and rivers. The whole city is not made of buildings, but pieces of furniture, tents, household appliances, motorcycles, and more, symbolizing human desires and their needs. These furniture spots on the maps seem like breeding microorganisms that spread infinitely through urban space. In this free and uncharacteristic fantasy space, people live in tent-like campers, in a purely natural environment, where all their necessities are met by just adding more furniture into the map, resulting in a continuous urban space expansion[10]. In this fantasy city without centers and suburbs, people can choose where they want to live. A satire to the nowadays urban living: ultra-modern cities filled with desires and needs, and at the same time transpiring despair, decadence, and tragedy. As Branzi said: "The mass-production has generated the infinite spreading of the urban art. The city gives us this blank, uncharacteristic identity that allows us to be anyone anywhere [6]."

Branzi jokingly defined himself as a child who betrayed rationalism. Through the "No-Stop City" project, he aimed at mocking and criticizing the future of urban factories that capitalism helped to develop, and the ideology of the modernist architecture[9][10]. In his utopia, there are no boundaries between the natural and the artificial, no difference between the city and the countryside; the development of the city will be completely industry orientated, to satisfy the needs of the consumers. In this way, the world will become a "capital of desire"[6][11][14]. Instead of the ability to provide viable solutions, this project aims to change the world with the power of criticism, or at least awaken

IOP Conf. Series: Materials Science and Engineering **573** (2019) 012077 doi:10.1088/1757-899X/573/1/012077

the backward mindset of people who are so deeply immersed in this world, that they have lost perception of it.

5. Legacy of the rebellion: Branzi's design teachings

In the last period of his teaching at the Politecnico di Milano (2011-2017), Andrea Branzi alongside with Michele De Lucchi (Co-founder of Memphis) and Francesca Balena Arista, has carried out a research/teaching activity within the first-year design studio at Master level. Following anthropological studies, the design studio adopts topic-oriented "keywords" as a guide to project development. The outcomes of the design research are usually divided into two parts: the first one part includes analysis and interpretation of a theme, after which students are asked to make a short video expressing their understanding of the topic. The second part, is about the designing and making a smaller scale model of an interior based upon their research. During the lessons, students are committed to three main tasks: exploring bibliographic resources to research for the keyword; making use of video language to explain the meaning of this keyword, and finally, employing the space as mean to express the purpose and the meaning of the keyword in reality. The significant of the studio is "Transformation": From the "Keyword" to the "Space"; From the"Artistic thinking" to the "Design language"; From the"Understanding" to the "Expression".

One of the authors of this article had the chance to participate in 2015-2016 master's interior design course, with the assigned keyword "Chaos". The author's group took contemporary "Hikikomori" [14] (From Japanese, meaning reclusive adolescents or adults who withdraw from society and seek extreme degrees of isolation and confinement) and their living environment as a model. This problem exists in Japan, Hong Kong China, as well as in Italy. The living environment of the "Hikikomori" is a chaotic and messy one, mostly because those people refuse to communicate with the outside world. To prevent the outside world entering in their life, they lock themselves in their rooms, and, at the same time, make themselves prisoners of their own space. "Abyss" is the name of the space project; it shows the contradictions and conflicts of this inhabited environment. The viewer can have a view of the "Hikikomori" living space through a hole on the top (Figure 5). The computer, the only bright object in the room, is the daily tool inhabitant most frequently work with. The only physical connection of this space to the outside is a ladder on the wall. Inevitably, the project makes us think about: where does the "Hikikomori" come from? Where are they headed? Have their lives always been like this? Will our society keep ignoring them? Are they willing to climb out of this abyss?

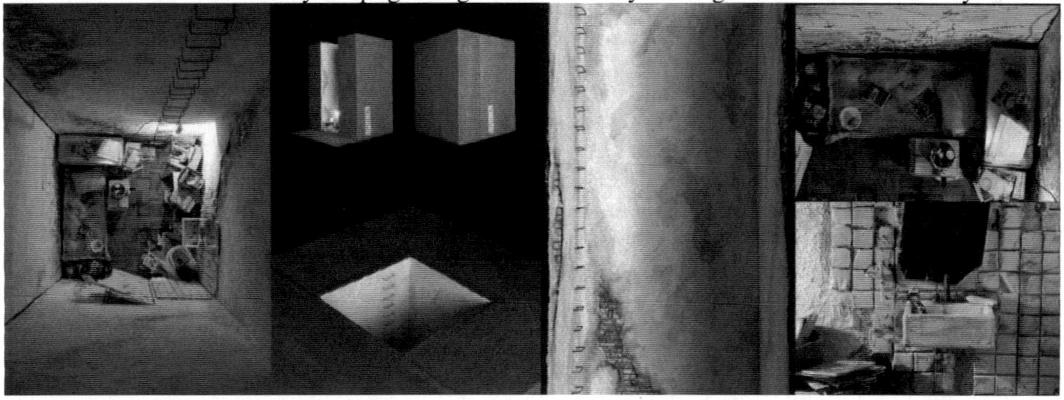

Figure 5.The project "Abyss", completed by the author and directed by Andrea Branzi (This project was selected in the Polimi Design X Designer in Milan Design Week 2017)

6. Conclusion

The intelligence of Archizoom Associati design studio lies in their ability to awe the world with a revolutionary new approach to criticism. Paying tribute to the Beatles, winking an eye at Pop artists, bowing in solution to Karl Marx, they pushed the life to ultimate aesthetical freedom[4][15]. As once in the Italian magazine Domus wrote: "We want to introduce to you everything that remains out of the door: the fabricated banality, intentional vulgarity, urban furniture, voracious dogs. We prefer a postcard-like horizon complete with rainbow. We also want to say: we can't be found where you are looking for us, don't trust the way we greet you. And then, the air smells of dead roses, a smell which we don't like that much[16]". As a rebel, Branzi is not only a designer, an architect, a theoretician, an educator, is also a philosopher, he criticized modernism without hesitation by "No-stop city", which challenged the functionalist mythology and popular aesthetic in Italy, with the striking manifesto of "Anti-Design" and actively participated in activism activities between the 1960s and 1970s. During the span of his career over 50 years, he has been proceeding with architecture, design, and craftwork projects that provide important significance and lessons for post-modernism design.

Wang Shu, the first person with a Chinese nationality who was awarded the Pritzker Architecture Prize, has always been known as a "Rebellious Architect". He even named his studio "Amateur" in order to resist and satirize the contemporary Chinese urbanization buildings, which he described "junk". In his work, he uses a lot of historical materials and Chinese traditional construction techniques to revive the Chinese traditional literati spirit in contemporary architecture. Obviously, Wang's construction method is incompatible with this era: Compared with technology, he prefers the craftsmanship; Compared with architecture, he prefers to build the house; Compared with the architect, he believes he is a literati first[17]. It could be said that this is why the Pritzker nominated Wang Shu - for the critical rebellious thinking, especially in China, where mediocre buildings are mass-produced during rapid urbanization.

Therefore, both Italian and Chinese architects or designers, can spot signs of "Rebellion", which represents a strong, unconventional and an initiative spirit, which is often mandatory for our contemporary practice and research.

Acknowledgments

This article is supported by the China Scholarship Council (CSC, File No. 201807820027), which sponsored the first author's Ph.D. study at the Politecnico di Milano.
In 2017, Prof. Andrea Branzi was honored to retire from the Politecnico di Milano. As disciples of him, the authors will keep the spirit of rebellion alive in the future and writes this article to pay tribute to him.

References

[1] Mendini, A., Mendini, F. (2005) Art of Italian Design. http://www.ateliermendini.it.
[2] Wolf, A. (2012) Discorsi per Immagini: Of Political and Architectural Experimentation. California Italian Studies, 3(2):1-20.
[3] Penny, S. (1988) Italian Design: 1870 to the Present. Thames & Hudson, New York.
[4] Gargiani, R. (2007) 1966-1974: Dall'Onda Pop Alla Superficie Neutra. Mondadori Electa, Milan.
[5] Warren, C., Dennis, C., Peter, C. (1964) Amazing Archigram 4: Zoom Issue.Archigram, London.
[6] Branzi, A. (2006) No-Stop City: Archizoom Associati. Hyx, Orleans.
[7] Ambasz,E. (1972) Italy: the New Domestic Landscape. Exhibition catalogue. MoMA, New York.
[8] Poltronova. (1969) Archizoom Associati. https://www.poltronova.it/1969/01/01/archizoom-associati/
[9] Annicchiarico, S., Branzi, A., McGilvray.B., Venerella.J. (2008) Che Cosa è il Design Italiano? Le Sette Ossessioni del Design Italiano. Mondadori Electa, Milan.
[10] Annicchiarico, S. (2007) The new italian design. Il paesaggio mobile del nuovo design italiano Milano. Triennale di Milano Servizi, Milan.
[11] Archizoom Associati. (1971) No-Stop City. Residential Parkings, Climatic Universal System. Domus, 496: 49-55.
[12] Branzi, A. (2006) Andrea Branzi. Thames & Hudson, New York.
[13] Archizoom, Superstudio. (1969) Discorsi per immagini. Domus 481: 46-48.
[14] Kremer, W., Hammond, C. (2003) "Hikikomori: Why are so many Japanese men refusing to leave their rooms?". BBC News, London.

IOP Conf. Series: Materials Science and Engineering **573** (2019) 012077 doi:10.1088/1757-899X/573/1/012077

[15] Branzi, A. (2006) Weak and Diffuse Modernity: The World of Projects at the beginning of the 21st Century. Skira, Lausanne.
[16] Sottsass, E. (1965) Archizoom, good fellas. Domus, 455: 51–59.
[17] Wang,S. (2016) Building the house. Hunan Fine Arts Publishing House, Hu Nan.

IOP Conf. Series: Materials Science and Engineering **573** (2019) 012078 doi:10.1088/1757-899X/573/1/012078

Realizing product serialization by Grasshopper parametric design

Binbin Sun[1*], Shangyou Huang[2]

[1]School of Design, Beijing Normal University, Zhuhai, Zhuhai, Guangdong, 519085, China

[2]School of Design and Art, Beijing Institute of Technology, Zhuhai, Zhuhai, Guangdong, 519085, China

[*]Corresponding author's e-mail: sunbinbin100@126.com

Abstract. The purpose of this paper is to explore how to realize serialization with parameterization by studying the application of Grasshopper parameterization modeling technology in the modelling design of serialization products. With the method of formulating the basic parameter series and setting the parameter algorithm, combined with the inheritance relationship of digital evolution of product characteristics, to reflects the series sense of product modelling, and demonstrates the feasibility of the method with practical cases. The conclusion is that in the background of high-speed development of information age and computer numerical control technology, parameterization provides a new way to realize product serialization design, promotes the combination of art design and digital manufacturing, maximizes design efficiency, flexibly updates products, and timely meets market demand.

1. Introduction

With the continuous advancement of economic globalization and the increasingly fierce market competition, the class diversity of consumers and the personalized and diversified demands of growing young consumers have caused irreconcilable contradictions between standardization and customization in large-scale industrial production. Product serialization is a systematic design method that adapts to the new economy and new business conditions, which can meet the needs of users to the greatest extent and improve production efficiency. Parametric technology is a very effective way to develop series products, which can quickly generate various design scheme by developing a series of basic parameters and dynamically editable logic programs. The following part 2 of the article introduces the concept of product serialization, design flow, and design principles. Part 3 introduces the connotation and characteristics of parametric design. Part 4 discusses the relationship between parameterization and serialization in product design and an application examples. Part 5 combines the parametric modeling design practice of the vacuum cup series, summarizes the design flow and method of implementing the product serialization with Grasshopper parameterization, and analyses the application difficulties. The last part is the conclusion.

2. Product serialization

Product serial design refers to scientific and unified planning of the size ratio, form, structure, parts, color, material, etc. of the same type of products, so that the series products are related in a certain dimension, and produce sense of series. In this paper, the range of product serialization is limited to industrial product modeling design. Therefore, the following content mainly explores the possibility of serialization from the perspective of product modeling elements.

Modeling design is an important part of industrial design. In today's highly competitive market environment, a successful product model can increase the added value of products and enhance the market competitiveness of products. Modelling design of serialization products must plan the development idea of the product series at the beginning of the design, based on market research and user needs, selectively develop variant products from basic products, set application specifications of modeling elements such as form, structure, texture, color and materials in the serialized design, and try to embody the serialized features in each product. For example, the size and performance parameters of the base product are transformed according to a certain proportional relationship; or some design elements are designed and changed by adopting design methods such as additions or deletions, replacement, reconstruction, and deformation to ensure that serialized products, no matter how to innovate or improve the design, can always maintain a unified style, achieve uniformity in change, and form a relatively stable series of spectrum and style system in the development of upgrading.

In the design process of serialized products, we must pay attention to the following three design principles: First, there should have a clear point of design creativity, it can be a theme concept, or a structural form, modeling element, collocation mode, etc. so as to make the series sense of products coherent and prominent; Second, the series sense should be distinct. In the design of serialized products, the application of design points should have priorities and contrast, so that each product in series has their own characteristics; Third, the serialized products should be unified and changed, not only related, but also innovative. Under the premise of unification, the subtle changes of the serialization specifications will be extended to each product in series to form a rich and balanced effect.

With its flexible production mode, product serialization can save the design cost and production cost, improve research and development ability and production efficiency, develop new products flexibly, maintain the long-lasting vitality of the product line and meet the diversified needs of users. At the same time, product serialization is also a differentiated design strategy that shapes corporate image and expands brand influence, which is conducive to improving product identifiability and user brand loyalty.

3. Parametric design

As early as the 1970s, parameterization has been used to automate control the production process and results in large-scale industrial production. With the continuous development of computer and numerical control technology, parameterization has become a major trend in the information age and has been widely used in various design fields, changing the way of designing and manufacturing.

The designer's traditional top-down working model has been broken, nowadays the new generation of product designers are trying to use parametric tools to control the whole design process efficiently, and bringing infinite possibilities to "creation" through digital modeling technology.

Parametric design of industrial products, that is, taking certain elements that affect product modelling design as parameters, setting a logical relationship (i.e., algorithm) through computer software to construct a digital model, that is, the prototype of product modeling. This prototype is often multi-solution and non-quantifiable, because the adjustment of parameters can generate a series of modelling, so designers need to further screen, refine and evolve, then seek the best solutions in a large number of new forms, and finally determine the exact size to complete the design.

Compared with the traditional product design methods, the fundamental difference in logic between them is that whether a design first forms a modelling in the designer's mind or is realized by a parametric system. The advantages and significance of parametric design are mainly reflected in the following two aspects: First, parametric design allows designers to liberate from traditional hand-painted conception and model simulations, especially when designing and constructing some complex surfaces and organic forms, parametric design has natural advantage, the complex rules and algorithms become the basis of modelling design. Second, the dynamic change characteristics of

IOP Conf. Series: Materials Science and Engineering **573** (2019) 012078 doi:10.1088/1757-899X/573/1/012078

parametric model greatly improves the speed and quality of product digital model generation and modification, and realizes the close connection from product design modeling to NC manufacturing. The whole design process is fully intuitive, logical, and malleable.

4. Application examples of product serialization by Parametric design

In the field of industrial design, parametric design is often used to generate product modelling, especially complex surface modelling, and to quickly realize product serialization. When designing the same series and different types of products, product designers do not need to start from scratch, but design variants based on the existing base products in this series, which greatly improves the design efficiency, shortens the design cycle and reduces the repetitive work. Parametric design thinking and modeling technology have great application potential in the serialization design of home appliances, furniture and household items.

Figure 1. "Digital Object/Triangulation Series" chair design project.

Independent digital artist and designer Zhang Zhoujie's “Digital Object/Triangulation Series” chair design project (Figure 1) is based on Rhino+ Grasshopper parametric design platform and the latest open source program Processing. Its core idea is to explore the formative form of digital objects. By setting up programs and optimizing logic, taking the computers as the subject, the computer can "freely" create art forms and make it possible to combine art design with digital manufacturing. This series of chairs is designed with stainless steel as material, triangle as feature elements, and sharp form showing sculpture sense and futuristic look. The form is derived from the points around a plane, which naturally extends downward under the action of gravity in a digital environment, forming a flowing arch structure. The edges, surfaces, scales, etc. seem to be different, but with the same material and approximate structural logic, it realizes the serialization of modelling. All chairs are automatically generated by computer programs according to the algorithm, and various forms can be generated by changing certain parameters. Computers become an art of morphological algorithms, and “digital creation” creates a series of subversive visual modeling languages.

5. Product series design based on Grasshopper parametric platform

5.1 Introduction to Grasshopper

In recent years, Grasshopper has gradually become a widely used parametric design software in the field of industrial design. It has the characteristics of visualization and dynamic. The parametric modeling technology discussed in this paper is based on this software. Grasshopper is a modeling plug-in based on Rhino environment, it has powerful parametric design function, the arithmetic unit is the core of it. It completes traditional manual modeling operations such as drawing lines or complex forms, moving, offset and arrays by means of data programming and automatic calculation. It automatically generates and modifies the model according to the algorithm through the way of input parameters and output results and connecting lines to visualize nodes.

5.2 General procedure for parametric design of series products (Figure 2)

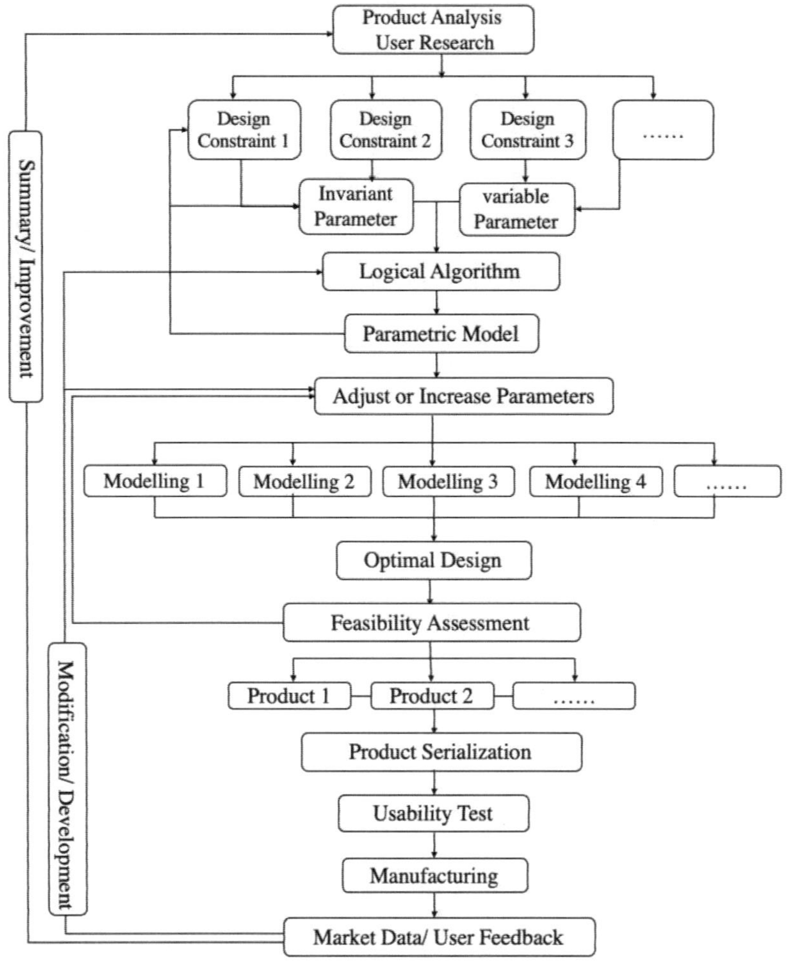

Figure 2. General procedure for parametric design of series products.

5.2.1 Plan parameters: According to the research results and design positioning, we should determine the design ideas, analyse the design requirements and constraints of the design project, digitize several key factors, set the quantitation and variable as the basis for formulating the algorithm in the parametric design.

Taking the vacuum cup design as an example, the author plans to design a personalized and modern "family" series of vacuum cup – designed for men, women and children. In the style of vacuum cup, we should not only distinguish them according to the user's characteristics and use needs, but also reflect the series sense and the integrity of "family". With "water" as the design inspiration, we can extract the water ripple as the element, analyze the constraints, set the functional parameters and morphological parameters, such as the length and width of the vacuum cup prototype model, the number of curves of the water ripple, the undulation direction and height, etc.

5.2.2 Formulate algorithm (Figure 3): The logical relationship and transfer relationship between parameters are analysed, and a set of logical algorithm for controlling parameter relationship and generating model is compiled.

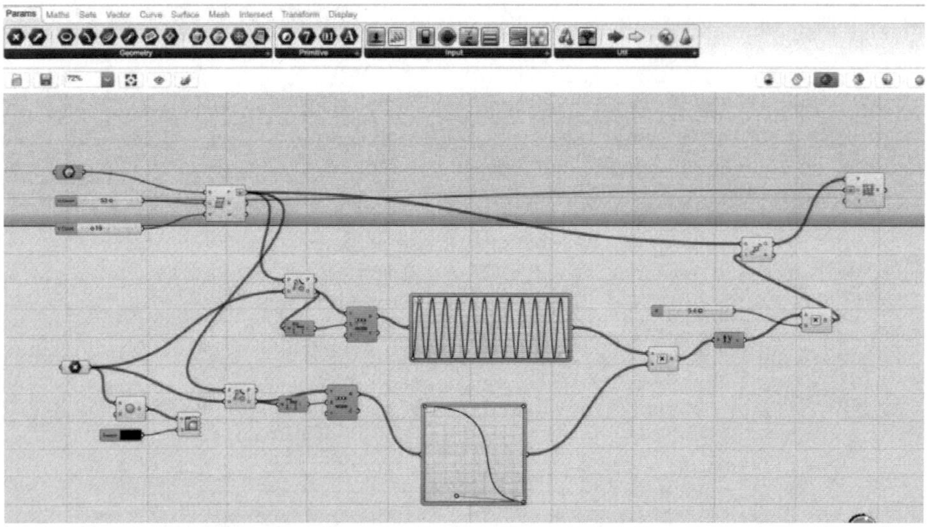

Figure 3. Writing of logic program for water ripple morphology

5.2.3 Generate model: using Rhino+ Grasshopper modeling, establishing a basic parameterized model of the vacuum cup, setting the induction points or curve, affecting the fluctuation pattern of the water ripple in the way of image interference, adjusting the parameters in the logic input or adding new parameters, various surface modelling results are dynamically generated, showing a linear trend of change (Figure 4), then getting a series of related but characteristic modeling schemes; if the model results are not ideal, return to the previous steps to modify and adjust .

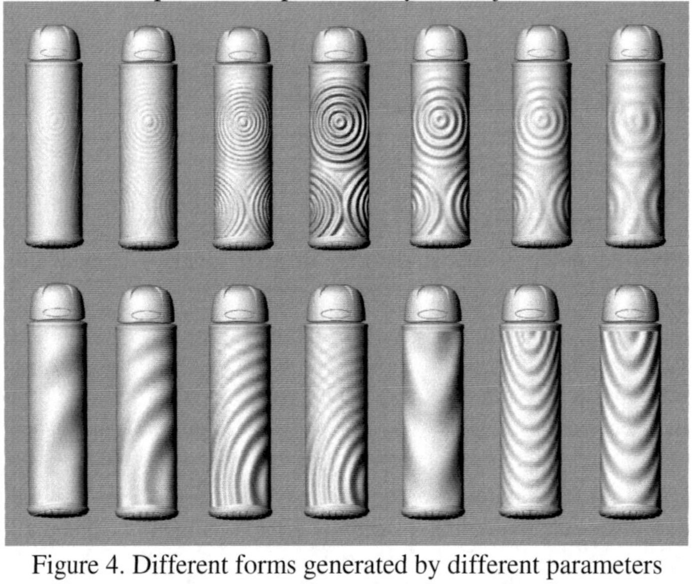

Figure 4. Different forms generated by different parameters

5.2.4 Optimizing and evaluating: According to the natural, experiential and economic attributes of the product, develop evaluation rules and select several schemes which are closest to the design objectives from a series of models.

For example, in the optimum design process of "family" series of vacuum cup (Figure 5), combined with the characteristics and functional requirements of the three targeted groups, and considering factors such as material, technology and structure, select three types of vacuum cup forms from the series modeling schemes. Then, adjusting the details, dimensions and texture accordingly. Finally, three modeling schemes are obtained. As a symbol, “water ripple” is not only a factor that reflects the series sense, but also has the function of anti-skid and heat insulation. At the same time, it also uses metaphorical method to reflect the specific connotation semantics – “water” cup hold water.

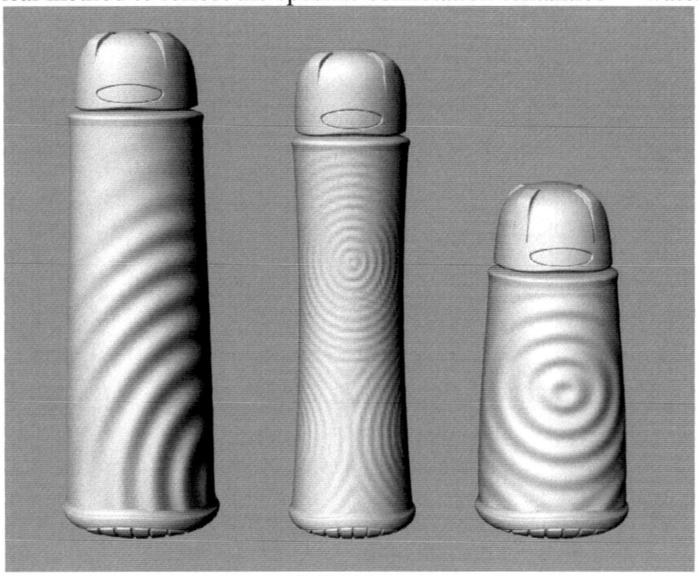

Figure 5. "family" series of vacuum cup

5.2.5 Verification and improvement: According to market data, user feedback, etc., summarize problems and improve, expand the series, and continue iteration.

5.3 Difficulties in the application of parametric design for series products

5.3.1 The logic operator in Grasshopper involves the basic mathematics of computational design, such as vector algebra, trigonometric function, solid geometry, etc., so designers need to have clear logical thinking ability and basic mathematics knowledge.

5.3.2 The generation system of the three-dimensional product model is complex. The key point of parametric product design is to establish a constraint model for product modeling and solve the constraint. Therefore, in the design process, the designer should analyze constraints among the many factors such as function, structure and technique, and coordinate local constraints and global constraints, set the invariant parameters, variable parameters, and the relationship between parameters, achieve the parameterization of product model data.

5.3.3 The first consideration in the initial modeling is not direct modeling, but how this form is generated, and what is the law governing its morphological changes. Designers need to use "macro thinking" to think about a unified solution to such problems, create a set of computational methods and spatial organization logic for generating designs.

5.3.4 A parametric model contains a large number of design variants, which contain a product family. Therefore, in determining the final product series, it is necessary to establish an effective, comprehensive and objective evaluation system to solve the multi-solution problem and retain the best product modelling to form a series.

6. Conclusion

The article discusses the concepts, design methods, application cases and their relations of serialization design and parametric design in the field of industrial design, and puts forward the viewpoint of using Grasshopper parametric design to realize serial design of product modeling, and draws the following conclusions: First, parameterization can be an efficient tool and innovation source for product serial design. The characteristics of rapid generation, dynamic adjustment and convenient modification of parametric model have obvious advantages for series product modeling, revision serialization design, multi-scheme comparison and evaluation, series product renewal iteration, etc. Second, combined with the parametric design practice of the vacuum cup series, the method and process of implementing product serialization by parameterization are preliminarily summarized, and the feasibility of the method is verified, and the product design theory is perfected, which has certain application and reference value in similar serialized product development.

Acknowledgement
Beijing Normal University, Zhuhai teachers' research capability promotion program (Item no. 201850008)

References
[1] Liu ZM, Li YX. (2008) Research on parametric design of lamps based on Grasshopper plug-in. Packaging engineering, 39(18):209-213.
[2] Zhang K, Zhang H. (2016) Application of parametric design in product form design. Design, 03:24-25.
[3] Wang N, Wang PG. (2014) Research on the application of parametric design in product modeling design. Design, 07:37-38.
[4] Chen DH, Wen L. (2011) Preliminary study on Solidworks based product serialization design system. Science and technology vision, 04:18-19.
[5] Zhang S, Yin PF. (2015) Research on parametric modeling of irregular curved surface structure based on Rhino+ Grasshopper. Civil and architectural engineering information technology, 7(05):102-106.
[6] Zhang ZJ, (2009) Digital Object / Triangulation Series, http://www.zhangzhoujie.com
[7] Dai XW. (2016) Application exploration of parametric design based on Grasshopper in product design. Design, 11:122-123.
[8] Chen DH, Wen L. (2011) Preliminary exploration of Solidworks based product serialization design system. Science and technology vision, 04:18-19.
[9] Deng QC, Liu T. Method of product serialization design based on deformation. Mechanical design and research, 03:20-22.
[10] Zhou W. (2010) Application of UG in parametric design of serialized products. Development and innovation of mechanical and electrical products, 23(04):116-117.
[11] Cao WY. (2018) Research and application of parametric design of products. North China university of technology.
[12] Zhang K, Zhang H. (2016) Application of parametric design in product form design. Design, 03:24-25.
[13] Yu SM. (2015) Significance of serialization of new product development and teaching practice of industrial design innovation. Machinery manufacturing, 53(03):1-5.

IOP Conf. Series: Materials Science and Engineering **573** (2019) 012079 doi:10.1088/1757-899X/573/1/012079

Mode of Innovative Design to Product Transformation of College Students

Xiaoxian Lu

Department of Mechanical Science and Engineering, Huazhong University of Science and Technology, Wuhan, Hubei, 430074, China

*Xiaoxian Lu's e-mail: 1769268186@qq.com

Abstract. Since Premier Li Keqiang proposed the concept of " mass entrepreneurship and innovation " in 2014, innovation and entrepreneurship have become a force that cannot be ignored in economic development. " Made in China 2025 " proposes to comprehensively upgrade and develop China's manufacturing industry, which is undoubtedly inseparable from the deep integration of innovative design. The " China Good Design Award " came into being in this entrepreneurial and innovative environment. Company, research unit and individual's independent innovation ability have received unprecedented attention. Among them, industrial design students have many innovative ideas and design creations, and much of them can be productized and put into the market. However, those high-quality innovations and designs don't have a chance to face the market. Many excellent innovations that could be used as venture capital and help the country's " double innovation " policy get lost, it is a great lost for the country, the unit and the individual. The new trend of economic development— " Internet + " has also brought new possibilities and ways for " double innovation ". Therefore, in such a vigorous environment, I hope to make full use of the existing conditions to find a way that can help college students to put their industrial innovative design products into market.

1. Research questions

1.1. Research purposes

We want to develop a model to help college students' innovative design to transform into products and help them start their own businesses. College students, especially design majors, continue to experiment with innovative designs after starting professional courses. Many colleges and universities are gradually combining teaching with practical topics. There are many innovations in student curriculum design and they have practical utility. However, in fact, also there are many innovative designs and ideas, the ratio of those designs and ideas turn into entrepreneurship is still not high. Many practical designs and ideas are lost. So I want to explore the problems and find the corresponding solution.

1.2. Research significance

We can make full use of the innovative designs and ideas of college students, and combine various resources to help students start a business. Specifically for the following seven points: 1. Students can understand the company and the market in advance and expand their knowledge. 2. Deepen their professional knowledge, train their ability to work, and have an advantage in future employment. 3. Students can also enrich their research results, increase their personal income and improve their living standards. 4. Develop professional talents for enterprises and society in advance. 5. Enterprises can also use this to develop new products, expand markets and acquire new business opportunities. 6. Respond to the " double creation " call to help economic development and increase market vitality. 7. Relieve employment pressure.

While college students and businesses gained profits, they will encourage more college students to carry out innovative design, encourage more enterprises and businesses to help students innovate and start their own business, form a virtuous circle, respond to the call for " double creation " , and help " China Manufacture 2025 " , integrate innovative designs with other industries, and play the core strength of innovative designs.

2. Overview of research at home and abroad

In China, " popular innovation and entrepreneurship " is in full swing, while foreign countries have placed innovation at the forefront of development. In 2015, the United States released the " New Strategy for American Innovation " to promote innovation of various industries and make full use of the creativity of citizens. The UK also has a strategy paper entitled " Our Growth Plan: Technology and Innovation " . Organizations and countries such as the European Union and Japan also have corresponding policies to promote innovation.

From the perspective of design, the United States had the strategy of " designing the United States " as early as 1992, and regarded design, especially industrial design, as an important means of national strength growth. In Asia, Japan and South Korea were the first countries to raise their designs to the height of national industries. Although China started late than them, the development is relatively fast. " Guidelines on Promoting Industrial Design Development " in 2010, also promoted the development of the design industry to the national level.

The new words such as innovative design and creative design are the original creations of China, and there is no corresponding words in foreign languages. It can be said that the word " design " has its own meaning of innovation and creativity. When designing education abroad, it pays special attention to practice. Whether it is a research center in the school or a joint teaching between the school and the company, the actual design work is carried out while teaching. In China, innovation and entrepreneurship are closely linked.

Generally speaking, there are four main types for students that can make their design work turn into a real product at home and abroad: 1. Course teaching combined with actual projects. 2. Independent design that echoes the surrounding environment. 3. Product design through the competitions. 4. Promotion of design results based on new environments such as Internet+ and big data. However, these models do not elaborate on the specific process of design works' transforming , and most of them are guided by company projects. Students are still only completing the design work itself, instead of turning their own innovative designs into products. Only some specific areas, such as ceramics, clothing, which are relatively simple for making and selling, can be completed by students independently. However, in general, productization needs to consider more factors: design feasibility evaluation, design requirements, materialization requirements, market evaluation, funding sources, sales channels, legal assistance and many more.

2.1. Course teaching combined with actual projects

The Art Center College in the United States directly invites famous designers or artists to teach at the school. Many well-known companies such as Nike, Sony, ADOBE, etc. will also cooperate and sponsor the school project. The University of Cincinnati pioneered the CO-OP, a paid internship program, which was emulated by more than 40 countries. A paid internship program means that students can participate in actual projects and receive salary during their school days. It not only improves the income of students, but also helps students gain professional experience in advance and increase employment opportunities after graduation. International students can earn between $40, 000 and $60, 000 in 12-18 months of paid internships. The School of Visual Arts in New York, the Bolimonta Fashion Institute in Italy, the internship base at Essen University in Germany, the studio training system in France, and the project cycle practice in Japan are all project-oriented and guide students' design. In China, Tsinghua University will invite professional designers to participate in student work review. The " Innovative Design Workshop " of the College of Art and Design of Hunan University also has many cooperations with famous companies such as Philips. Huazhong University of Science and Technology also has its own innovation base to help students innovate and start a business.

2.2. Independent design that echoes the surrounding environment

Students from the Jingdezhen Ceramic Institute in Jiangxi took their work to the creative market for sale; the Parsons School of Design in New York required students to interact with passers-by on the street during design; students from China University of Geosciences (Wuhan) collaborated with surrounding shops to conduct Jewelry design; teachers and students of Rhode Island College in the United States will sell their designs in the surrounding shops and so on. These are all using the surrounding business environment, combined with their own professional features to complete innovation and entrepreneurship.

2.3. Product design using competition form

For example, the " Huanghe Cup " , a competition jointly organized by our school and the Yellow Crane Tower Group and other universities, and various other alliance enterprises and universities and research institutes jointly held various competitions.

2.4. Promotion of results based on new environments such as Internet+ and big data

Nanjing Art Institute Shangmei College established its own website to showcase students' graduation designs. Students from the School of Textile and Apparel of Nan University will sell and promote their clothing on Taobao.

In order to strengthen the technology transfer of research results of universities and research institutions, all localities have established their own cooperation platform for industry, university and research. In addition to the physical center, the online platform for industry, university and research also plays a key role in promoting and connecting.

3. Research content

3.1. Overview of the research content

In order to help college students complete their design works' transformation and entrepreneurship of innovative achievements, build an efficient platform for communication, learning and work, and open up the communication channels between campus talents, campus teams and enterprises, so that more campus talents and talents can resonate. Let students work with each other to create teams and join teams. Let more campus talents and teams join the enterprise, let more enterprises go deep into the campus to support and create teams, establish long-term cooperation and understanding with campus talents and teams, help college students achieve innovative design and independent design, help enterprises get more business opportunity.

3.2. Innovations in research

Link students and use the power of the team to complete innovation and entrepreneurship. Industrial design students can work with other professional students to complete their innovative designs. Moreover, this can be done not only across disciplines, but also across schools and geographies, using current information and technological advantages to accomplish this collaboration.

The Department of Industrial Design of Huazhong University of Science and Technology has strong faculty, scientific research conditions. Huazhong University of Science and Technology has a good atmosphere of innovation and entrepreneurship, providing policy, economic and other aspects of help for entrepreneurial students. The business incubator of Huazhong University of Science and

Technology and the Industrial Technology Research Institute established in cooperation with the government also provided rich research materials for this study.

4. Introduction to research results - PFI platform

PFI is the abbreviation of " Platform For Innovation " . It is tentatively used to convey the core concept of this platform. It may be changed after the platform is launched.

4.1. User Analysis

4.1.1. Students

From the perspective of students, it provides students with a platform to participate in team and enterprise practice in schools, improve the overall quality of college students, cultivate project experience, and enhance the competitiveness of graduate job search: 1. Upload your own innovative design, innovative ideas and resume information to showcase your innovations and personal abilities and strengths to campus teams and companies. 2. Search for campus talents, campus teams and companies that you are interested in, and add them as friends to provide friend information push function.3. Search for university campus competition information and lecture information in real time.4. Find a part-time job and improve your ability through practice. 5. Download beautiful resume templates and a variety of contract templates.

4.1.2. Team

From a team perspective, it is convenient for the team to promote self-branding on campus. Provide a talent search platform for the school team, reduce team costs and stimulate the vitality of the school team: 1. Upload team resume information, campus talents and companies to showcase team characteristics and activities. 2. It is convenient for the campus team to search for talents in the school and recruit new personnel. 3. Facilitate cooperation and communication between campus teams and enterprises.

4.1.3. Enterprise

From the perspective of the enterprise, it is convenient for enterprises to enhance the influence of brand promotion in colleges and universities, provide a fast talent search platform for enterprises, assist enterprises to tap talents of potential colleges, and better connect talent resources in schools. Provide docking enterprise commissioning services to help enterprises develop targeted talents in the school: 1. Upload the resume information of the company, show the enterprise to the campus talents and teams, carry out brand promotion, and expand the campus market. 2. The company goes deep into the campus to support and create teams to absorb more campus talents. 3. Enterprises provide talent recruitment channels to find better and more suitable talents.

4.2. Product description

4.2.1. Platform function partition

IOP Conf. Series: Materials Science and Engineering **573** (2019) 012079 doi:10.1088/1757-899X/573/1/012079

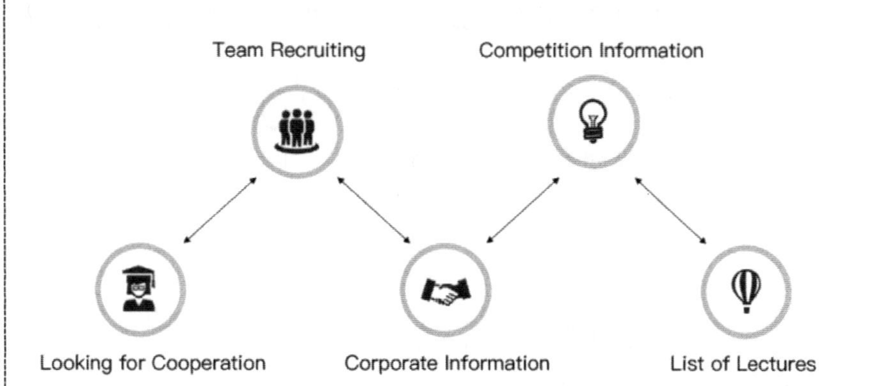

Figure 1. Platform function partition.

4.2.2. Platform function flow chart

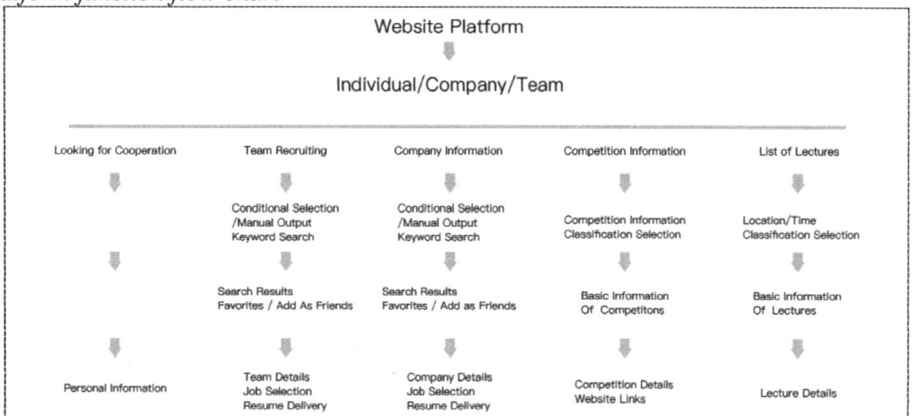

Figure 2. Platform function flow chart.

4.2.3Platform effect display

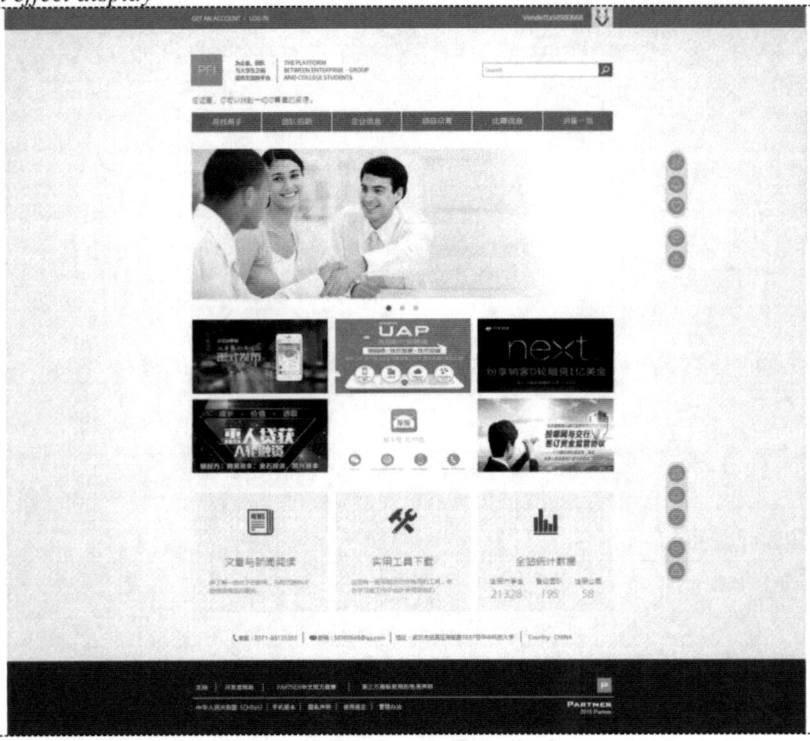

Figure 3. Platform effect display.

References

[1] Li, N. (2014) The Discussion On The Operation Mechanism Of The Transition Of Designs Of Students In University In China. D.

[2] Hanson, R. (2003) Combinatorial Information Market Design.J.Information Systems Frontiers., 5: 107–119.

[3] Anttiroiko, A., Valkama, P., Bailey, J. (2014) Smart Cities In The New Service Economy: Building Platforms For Smart Services. J. AI & Soc., 29: 323–334.

[4] Fu, J., Li, Y., Fang, Z. (2013) Construction of Innovation Platform for CEEUSRO. J. Management Science and Engineering., 7: 52–59.

[5] Shane, S., Venkataraman, S. (2014) The Promise of Entrepreneurship as a Field of Research. D.

[6] Krogn, V. (1998) Care In Knowledge Creation.J.California Management Review., 40: 133–153.

[7] Ketchen, D. J., Ireland, D., Snow, C. C. (2007) Strategic Entrepreneurship, Collaborative Innovation, And Wealth Creation.J.Strategic Entrepreneurship Journal., 1: 371–385.

[8] Hamdani, M. (2012) Advance Principles For Visual Aesthetics In Designing The Contents Of E-learning. J. Science Series Data Report., 4: 51–58.

[9] Rueda, F. G., Millot, V. (2015) Measuring Design And Its Role in Innovation. J. OECD Science, Technology and Industry Working Papers., 1-51.

A new method of product-service system design: product-based, participatory service design method

Jing Sun[1], Junjie Chu[2]

[1] Engineering Department, Ocean University of China, 266000, China

[2]Engineering Department, Ocean University of China, 266000, China

Email: sj@stu.ouc.edu.cn

Abstract: With the change of business model, market competition has gradually changed from product competition to service competition, and service design has been paid more and more attention by various industries. The ultimate goal of service design is tantamount to bring users a better experience and maximize the interests of stakeholders. Therefore, this paper will create a service design method based on service dominant logic and more emphasis on user experience. This paper sorts and summarizes the concept, content and characteristics of service design and product service system, and then find out the future development direction of service design and product service system. In this paper, various design methods are analyzed and compared, and one of the best design methods is associated with the product service system to form a new design method. After that, this new design method will be applied to the service design of children's design thinking education as a practice.

1. Introduction

Service is becoming an important commodity in economic activities. Manufacturing companies, for example, are just trying to integrate services into products to create a more complete solution [1].

Service-Dominant logic(S-D logic) describes the transformation from a product-centered market view to a service-oriented model [2]. It holds that the customer is always the co-creator of value. While the enterprise can only make a value proposition, and the customer converts value into value through use [2].

Service design provides a human-centered, holistic, co-creative methodology to develop new services [5,6,7]. In service design, designers take into account the interests and ideas of various stakeholders while optimizing the user experience. Therefore, it is extremely important to understand the suggestions of users and other stakeholders in the design.

Product service system (PSS) is a business model in which an enterprise provides sales services at the same time when selling a product [5]. Instead of providing tangible goods, it is better to provide goods plus services [4].The basic business benefit of PSS is to enhance the total value of goods by adding service elements. Product service system is a special mode in service design, which also focuses on the user experience [5].

This article will set out the three service design concepts and compare the similarities and differences. According to the final design requirements, a classic design method is selected and changed on the basis of it to form a more specific and more service-oriented product-based new service design method, which is more in line with the service economy.

IOP Conf. Series: Materials Science and Engineering **573** (2019) 012080 doi:10.1088/1757-899X/573/1/012080

2. Concept

2.1. Service Design

Service design has been rather introduced as a human-centred and creative approach to service innovation[3]. The definition of service is that service is the application of resources (primarily knowledge and skills) for the benefit of another individual or organization (the beneficiary) [6].

Service design is an emerging field, a method of cross-disciplinary, showing people a new way of thinking - service design thinking [7]. In service design thinking, to provide services, there must be user involvement. Designers need to put users at the center of the service design process, and need to have a real understanding of the user, not just statistical descriptions and empirical analysis. Everyone has creativity. Service design needs to consider different stakeholders. Each group has its own needs and expectations. Therefore, it is a good way to let them participate in the design process and develop co-creation. In addition, service design requires attention to the order and tempo of service delivery. If the tempo of the service is too slow, the customer will get bored. If it is too fast, the customer will be too stressful. However, the service is invisible and short-lived. The designer can materialize the service and extend the service experience by providing the customers with products. This way can effectively improve the customer's loyalty and encourage the customer to recommend it to others. Service design thinking supports collaboration between different disciplines. Therefore, designers need to have an overall awareness [7].

2.2. Product-Service System Design

A product-service system (PSS) is an integrated combination of products and services. This concept embraces a service-led competitive strategy, environmental sustainability, and the basis to differentiate from competitors who simply offer lower priced products [5]. Product service system solutions are a range of products and service components designed to create value for customers.

There are three different types of PSS, which are product-oriented PSS, use-oriented PSS, result-oriented PSS. The fundamental business benefit of a PSS is an improvement in total value for the customer through increasing service elements [5]. The environment also benefits from PSS since a producer becomes more responsible for its products-services through take-back, recycling, and refurbishment-reducing waste through the product’s life [9]. In this way, the company can extend the life of the product, maximize cost- and material- efficiency, and reuse product parts as far as possible after the end of the product's useful life to maximize material utilization [10].

What the customer needs is not a specific product, but a solution of the problem [11]. The main barrier to the adoption of a PSS is the cultural shift, for a consumer to place value on having a need met as opposed to owning a product [8].

2.3. Service-Dominant Logic

Service-dominant logic (S-D logic) describes the transformation from a product-centered market view to a service-oriented model. In S-D logic, it views that goods are appliances for service provision[2], marketing activity is in terms of service-for-service exchange [6,7], customers are seen as the co-creators of value and suppliers as resource integrators. S-D logic facilitates mutual value creation and argues that all businesses are service businesses. Therefore, in S-D logic, it applies to all organizations and economies.

The main content of S-D logic at present are service exchange, resource integration, value cocreation, value determination, and institutions/ecosystems [6]. In S-D logic, the fundamental purpose of the enterprise is to serve itself by serving others. The value can only be actualized in the usage process not in the production phase of the product [13] and it is always uniquely and phenomenon logically determined by the beneficiary [2]. Value is experiential. Each individual (or firm) perceives offerings and integrates them with other resources differently[12]. As a result, each user gets a personalized service. S-D logic advocates a service-ecosystems perspective, which is a relatively self-contained, self-adjusting system of resource-integrating entities that are connected by shared institutional logic and mutual value creation through service exchange [12]. It needs a sustainable, shared-based service system that can integrate connected resources to achieve a co-creation.

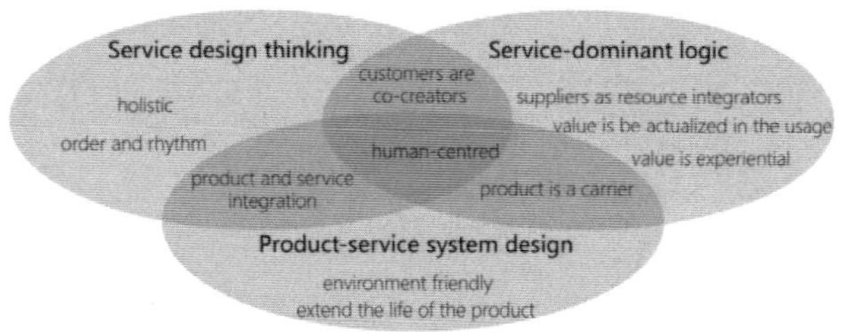

Figure 1. The comparison about SD, PSS, S-D

3. Material and methods

The design requirements of our new method are to meet the needs of users as much as possible, which requires a sufficient understanding of the user's psychology. This paper organizes some user-centric design methods and comparing these design methods in six aspects, and then finds a design method that more closely meets the design requirements. Then based on the service design thinking and S-D logic, the final method is supplemented and modified to obtain a more targeted, user-friendly, sustainable design method for the product service system.

Table 1. Some user-centric design methods.

classic design methods	method introduction	presenter	new method derived	year	1	2	3	4	5	6
The PSS approach	The PSS approach is a design method mainly applied to the product-service system.	HiCS (Highly Customized Solutions) research project	Punto X: "a system of products, services and expertise, able to offer food so-lutions that are personalized to meet the needs of specific contexts-of-use. The personalisation is obtained thanks to the flexibility in the meal composition, the organization of distribution and delivery systems, and through service/consumer interfaces." [14].	2006		√		√		
		N. Morelli	A method to design PSS on the basis of customers' active participation [15].	2006	√		√		√	
		K. Kimita et al.	A model to express the changes in customer demands from the perspective of service design [16].	2009	√	√				√
Participatory design	Participatory design is a human-centered design approach. It advocates that users and stakeholders should actively participate in all stages of the research and design process in all processes of collaborative design activities	Laura Malinverni n , Joan Mora-Guiard, Narcis Pares	The multimodal analysis constitutes an effective and coherent method to capture and analyze users' contributions across a wide range of semiotic resources, thus extending the richness of insights that can be derived from a PD workshop and communicated to the rest of the team [17].	2016	√	√			√	√
		Mariesa Nicholas,Penny Hagen,Kitty Rahilly,Nathalie Swainston	the method uses a series of co-design workshops and the role participatory design principles to engage young people in the redesign of the online service ReachOut.com [18].	2012	√	√	√		√	
		Finn Kensing, Jesper Simonsen, and Keld Bødker	The method is based on thorough participation with users and managers, and it combines the use of ethnographic techniques and intervention [19].	1998	√			√	√	

co-design	The term co-design refers to the creativity shared by two or more people, which are called collective creativity is applied across the whole span of a design process	Frank Piller , Petra Schubert , Michael Koch , Kathrin Möslein	Collaborative Customer Co-Design in Online Communities:challenges the assumption that offering customized products requires an individual (one-to-one) relationship between customer and supplier and uses interactions between customers to reduce mass confusion [20].	2005	√	√				√
		Greg Walsh, Allison Druin, Mona Leigh Guha, Beth Foss, Evan Golub, Leshell Hatley, Beth Bonsignore, Sonia Franckel	"Layered Elaboration," a new technique for co-design with children, allows design teams to generate ideas through an iterative process in which each version leaves prior ideas intact while extending concepts [21].	2010		√	√		√	
		Mona Leigh Guha, Allison Druin, and Jerry Alan Fails	Cooperative Inquiry:Cooperative Inquiry is a method of design partnering in which adults and children work together. The intense involvement of adults and children together in Cooperative Inquiry sets this method apart from informant design with children [22].	2013		√	√			

This table evaluates these design methodologies with six indicators, namely: 1: user-centered; 2: commercial; 3: users participate in the design; 4: Consider stakeholders; 5: sustainable; 6: used in the research

The PSS method mainly emphasizes people-oriented, and the designer needs to pay more attention to the user's experience; the co-design method emphasizes that the user and the designer can design together to understand the user's ideas, so that the designed product can better meet the user's needs; and the participant design The method emphasizes the user's participation and voice, and the participation of all stakeholders, which is more consistent with the requirements of SD logic, and according to the service design method, we need to consider all stakeholders, not just users. Therefore, choose a participatory design approach.

4. Methodology

The method is based on a participatory design approach, which is innovated and modified to make the new method applicable to product-oriented product service systems.The new approach needs to take into account the interests and participation of stakeholders, such as external customers, internal employees, manufacturers, and vendors.

In a service process, the service provider (such as a company) is responsible for providing a basic product, which is the carrier of the service, and services, including basic services, target services, and additional services.

There should be several characteristics in the product, which is using environmentally friendly materials, repairable, recyclable, can be mass-produced and valuable. The product cannot be disposable and its life cycle should not be too short. Therefore, the product ought to be repairable and recyclable. The service provider can collectively recycle the products at the end of the use period for reuse. Service should be personalized, but the product should be mass-produced, which not only reduces the cost of production, but also makes it easy for manufacturers to make.

Basic services include repairs, replacements and sale of individual parts of the product, and providing methods of use and value propositions. The target service is the service that the company mainly sells, and the selling price of the product is mainly determined according to the service. The additional service is timely feedback from the service provider on the design suggestions made by the stakeholders, including the reward for adopting the proposal, modification proposals, and the reason for the failure to adopt.

As resource integrator [2], service providers need to summarize the recommendations made by stakeholders on a weekly basis and show the results to stakeholders. It will increase customer loyalty and encourage customers to participate in the service process for a long time [7]. Service providers can ask stakeholders for questions on a regular basis to give some incentives to stimulate stakeholders' participation.

5. Practice

This practice uses the props in Figure 2 to develop the child's design thinking and intergenerational interactions. The child can tell a story to the old or tell the story of the old through the transformation of the shadow. The basic services provided by service providers are the recycling and replacement of parts, the target service is children's design thinking training, and the additional service is the adoption of recommendations for children's redesign. The service provider's advice is to interact with children and the elderly, reduce the generation gap, and promote intergenerational emotional communication. Value is only achieved when children use this product. The cultivation of children's design thinking lies in the children's use of props to reproduce a story, and the children's redesign of the toy itself.

Figure 2. Product parts schematic

Acknowledgments

The authors thank Mr. Chu for his guidance and error correction and thank the classmates for their help.

References

[1] Baines, T., Bigdeli, A. Z., Bustinza, O. (2017) Servitization: Revisiting the state-of-the-art and research priorities. International Journal of Operations &Production Management, 37(2): 256-278 .
[2] Lush, R. F., Vargo, S. L. (2014) Service-dominant Logic: Premises, perspectives and possibilities. Cambridge University Press, New York.
[3] Harrison, A. (2017). Design for Service. Research Technology Management, 60(4): 19–25.
[4] Manzini, E., Vezzoli, C. A. (2002) Product-service systems and sustainability: Opportunities for sustainable solutions. UNEP-United Nations Environment Programme.
[5] Baines, T. S., Lightfoot, H. W., Evans, S., Neely, A., Greenough, R., Peppard, J., et al. (2007). State-of-the-art in product-service systems. Proceedings of the Institution of Mechanical Engineers, Part B: Journal of Engineering Manufacture, 221(10): 1543-1552.
[6] Vargo, S. L., Lusch, R. F. (2017) Service-dominant logic 2025. International Journal of Research in Marketing, 34(1): 46-67.
[7] Stickdorn, M., Schneider, J., Andrews, K., Lawrence, A. (2011) This is service design thinking: Basics, tools, cases. Wiley. Hoboken.
[8] Goedkoop, M. J., Van Halen, C. J., Te Riele, H. R., Rommens, P. J. (1999) Product service systems, ecological and economic basics. Report for Dutch Ministries of environment (VROM) and economic affairs (EZ), 36(1): 1-122.
[9] Mont, O. K. (2002) Clarifying the concept of product-service system. Journal of cleaner production, 10(3): 237-245.
[10] Tukker, A. (2015) Product services for a resource-efficient and circular economy: a review. Journal of cleaner production, 97: 76-91.
[11] Boehm, M., Thomas, O. (2013) Looking beyond the rim of one's teacup: A multidisciplinary literature review of product-service systems in information systems, business management, and engineering & design. Journal of Cleaner Production, 51(0): 245-260.
[12] Greer, C. R., Lusch, R. F., Vargo, S. L. (2016) A service perspective. Organizational Dynamics, 1(45): 28-38.
[13] Kowalkowski, C. (2010) What does a service-dominant logic really mean for manufacturing firms?. CIRP Journal of Manufacturing Science and technology, 3(4): 285-292.

IOP Conf. Series: Materials Science and Engineering **573** (2019) 012080 doi:10.1088/1757-899X/573/1/012080

[14] Krucken, L., Meroni, A. (2006) Building stakeholder networks to develop and deliver product-service systems: practical experiences on elaborating pro-active ma-terials for communication. Journal of Cleaner Production, 14(17): 1502-1508.

[15] Morelli, N. (2006) Developing new product service systems (PSS): methodologies and operational tools. Journal of Cleaner Production, 14(17): 1495–1501.

[16] Kimita, K., Shimomura, Y., Arai, T. (2009) A customer value model for sustainable service design, CIRP J. Manuf. Sci. Technol. 1 (4): 254–261.

[17] Malinverni, L., Mora-Guiard, J., Pares, N. (2016). Towards methods for evaluating and communicating participatory design: A multimodal approach. International Journal of Human-Computer Studies, 94: 53–63.

[18] Swainston, N.,Rahilly, K., Hagen, P., Nicholas, M. (2012) Using participatory design methods to engage the uninterested. Proceedings of the 12th Participatory Design Conference: Exploratory Papers, Workshop Descriptions, Industry Cases - Volume 2, New York, 121-124.

[19] Kensing, F., Simonsen, J., & Bødker, K. (1998). MUST: A Method for Participatory Design. Human-Computer Interaction, 13(2): 167.

[20] Piller, F., Schubert, P., Koch, M., Möslein, K. (2005). Overcoming mass confusion: Collaborative customer co-design in online communities. Journal of Computer-Mediated Communication, 10(4):10-42.

[21] Walsh, G., Druin, A., Guha, M. L., Foss, E., Golub, E., Hatley, L., Bonsignore, E., Franckel, S. (2010). Layered elaboration: a new technique for co-design with children. In Proceedings of the SIGCHI Conference on Human Factors in Computing Systems. New York, 1237-1240.

[22] Guha, M. L., Druin, A., Fails, J. A. (2013) Cooperative Inquiry revisited: Reflections of the past and guidelines for the future of intergenerational co-design. International Journal of Child-Computer Interaction, 1(1): 14-23.

IOP Conf. Series: Materials Science and Engineering **573** (2019) 012081 doi:10.1088/1757-899X/573/1/012081

Research in the Design of Cloud Service Platform for Tea Product Design Based on Consumers' Will in the Background of Blockchain

Xin-qin Peng[1], Fei Hu[2] and Xin-qin Peng [1*]

[1] Art Design Den., Bengbu University, Bengbu City, An-hui Province, 233000, China

[2] Art Design Den., Bengbu University, Bengbu City, An-hui Province, 233000, China

[*]Corresponding author's e-mail: pxq79@126.com

Abstract. Traditional tea sets are biased on the design, manufacture and sale of handicrafts. In the context of Blockchain, the comparison of tea makers' consumer wishes and data statistics, combined with the design cloud service platform, will inevitably bring about a huge change in the design industry chain. Through two experiments, this paper reveals the relationship between consumer and product closeness, psychological distance and social distance. At the same time, based on the research and analysis of tea product design impact factor, it is uploaded to the product design cloud service platform. Let users mine design requirements, set up design teams, establish an open and shared industrial design, network interaction environment, and realize the design life cycle process through collaboration between front-end and back-end of the cloud service platform, combined with corresponding access technologies and virtualization technologies. Sharing and collaboration will also provide data and technical support for the breakthrough of the block industry chain in the later period.

1. Introduction

Blockchain is the most discussed topic in the industry at present. Its core is that each small "Block" is composed of a chain. The point-to-point transmission and encryption algorithm connects unit of transaction list-distributed data storage, Form a complete timeline industry chain. At present, the domestic industrial chain has taken shape, from platform services, security services to design services, as well as talent services and media terminals. The main areas of use in the block industry chain are smart contracts, intelligent trading, smart energy, intelligent supply chain, intelligent health record, smart passport, intelligent cross-border payment, smart home appliances, etc. According to the "2018 China Blockchain Industry White Paper" statistics, this application field will also have large-scale industrial application and expansion. As the core of various industries, industrial design and product design undoubtedly have great application space. This space will have greater development under the cloud service platform. Designing a cloud service platform as a service platform for cloud computing complication, software and platform can be used as service terminals, thereby using distributed data storage and platform transactions as a means to break through the traditional design BUG and achieve virtual reality of product design. The combination of the two, with its high reliability, versatility to achieve on-demand services and through socialized business crowd funding and other means to greatly reduce costs. Tea design is a product design category that favours traditional handicrafts. If the power of the cloud service platform is introduced, it will inevitably lead to design innovation and crowd funding, thus completing the value of its block industry chain. Based on the background of Blockchain, this paper uses consumer's willingness analysis as the argument of distributed data storage, attempts to establish networked collaborative design, and introduces customer participation in design decision-making and multi-objective quantitative evaluation system in designing cloud service platform. Through the comparison of consumers' willingness in social space and the field investigation and data statistics of tea design, the design influence factors of tea products are quantified and uploaded, so as to promote the BLOCK base point of the design cloud service platform, and draw a multi-objective creative design pair. They shared value of product design cloud platforms. Through the data exchange of WeChat, website and a mobile phone application APPS and other social platforms, the legal and brainstorming design spread, and finally improve the design of tea set products from the perspective of traditional crafts, integrate the tea set design into the cloud service platform, and improve this Block industry chain in the design field.

2. Comparison of consumer wishes for tea design

2.1. Psychological distance in the network environment affects consumers' willingness to preset
The traditional tea set products were very slow in the initial stage of the establishment of e-commerce. Many consumers have maintained a wait-and-see attitude towards the traditional tea products e-commerce, and dare not try it easily [1]. Many scholars start from different dimensions of psychological distance and explore their influence on the willingness to purchase online. Martin et al. explored how time distance affects consumers' willingness to purchase online [2]. Henderson et al. proposed that spatial distance and time distance have similar effects on human behaviour. The spatial distance is close, people tend to use specific thinking, and when the distance is far away, people tend to use abstract thinking [3]. Tangari concludes that there is an interactive influence between time distance and social distance on the purchase intention of environmentally friendly energy-saving products [4]. Lee proposes that psychological distance increases the willingness of consumers to purchase online by affecting trust [5].

Nowadays, a new model of e-commerce for tea products comes into being, namely the tea business e-commerce in the socialized business environment. It is the process of e-commerce of tea products by retail individuals through social platforms, and the tea products under the socialized business environment. E-commerce is an emotional sale. When sellers sell goods, they are also selling the feelings between friends. At present, there is relatively little research in the theoretical field: whether the acquaintance relationship can better eliminate the consumer's concern about online shopping tea products, thereby increasing consumers' willingness to purchase. Therefore, studying the influencing factors of consumers purchasing tea products on social platforms has become an urgent need for the development of e-commerce of tea products under the current socialized business environment.

2.2. Experimental design and results analysis

2.2.1. Experimental preparation. Randomized grouping: The biggest difference between the experimental method and other empirical methods are the randomization of the subjects. In order to avoid the experimental result error caused by the non-randomized groups, this time, the online experiment was used, and the subjects were randomly grouped using the questionnaire topic. Since there are several sub-experiments in this paper, it is necessary to divide the subjects into several groups.

2.2.2. Experiment: The impact of social distance on purchase intention. The purpose of the experiment was to explore the impact of the social distance between sellers and consumers on the willingness of online shopping. Verification Hypothesis 1: In the socialized business environment, the closer the seller is to the consumer society, the greater the willingness of consumers to purchase tea products.

1. Experimental design. Experiment adopts inter-group design (social distance between seller and consumer: close distance, long distance), in which social distance is manipulated as an independent

IOP Conf. Series: Materials Science and Engineering **573** (2019) 012081 doi:10.1088/1757-899X/573/1/012081

variable. Participants completed the questionnaire online, as required. The questionnaire was randomly divided into two groups by setting randomizes questions. In the first experiment, the two groups of participants will read different text materials, aiming to create different online shopping tea sets for the subjects, and let the two groups complete the questionnaire under different social distance manipulation.

(1) Variable measurement and manipulation test. First, the experiment sets the randomization problem through the questionnaire and divides the subjects into 2 groups (social distance: near, far). In this study, the experimental purpose was achieved through the manipulation of the social distance of the independent variables. The two groups of subjects were treated differently, which made them feel different from the social distance of the seller. In the first experiment, a classmate of the subject's imaginary unit sold the tea set on the QQ group, and was planning to purchase the tea set product through the Internet. The reasons for setting this scenario are as follows: The students are the main people who are in contact with each other, and the relationship between the students is relatively stable. At the same time, the QQ group is currently the largest social platform in China, with the largest number of users, and is widely used in work and life; With the rise of social commerce, buying tea products on the Internet is no longer unfamiliar to everyone, and it is entirely up to them to choose which seller to buy tea products. Therefore, the social distance between sellers and consumers can be reasonably manipulated through the setting of experimental scenarios.

Participants in the Near Social Distance group will read the following text:

“Assume that your best friend sells tea sets through the QQ group or WeChat group, the price is within your acceptable range, and you have recently planned to buy tea sets. Will you buy them?”

Participants in the Far Society Distance Group will read the following text:

“Assume that your newly-recognized classmates sell tea sets through the QQ group or WeChat group, the price is within your acceptable range, and you have recently planned to buy tea sets. Will you buy them?”

The two kinds of text materials are manipulated by different social distances to achieve the purpose of the experiment. The "best friend of the unit" corresponds to the manipulation of the social distance, and the "new classmate" corresponds to the remote social distance manipulation. The social distance manipulations commonly used by scholars before are: self, friends, members of the group, intimate roommates, etc; typical remote social distance manipulations are: others, strangers, members outside the group, same-sex students who are known in the public elective class. How to judge whether an experiment is successful depends on whether its manipulation achieves a significant difference. How to manipulate to achieve the effect is the difficulty of the experiment. The reason why this study defines the manipulation of social distance as “the best friend in the unit” is because this kind of processing can bring the social distance between consumers and sellers closer; the reason why the manipulation of far distance is defined as “new classmates” "Because the new students came to the first time, the relationship with the classmates was relatively unfamiliar, which could widen the social distance between consumers and sellers."

The manipulation of the independent variables is as follows: The near social distance group replies “How close is your friend and your intimacy in the unit”, the distance from the distance group to answer “How close is your new classmate to you?” (1 means very alienated, 4 expresses very close).

In addition, the experiment also added the item to check the validity of the questionnaire, in order to exclude the invalid questionnaire and improve the credibility of the experimental results. The item set in the questionnaire is “Which reason may be that you are not willing to buy a tea set in the best classmate/new classmate of the unit”, this question is a fill-in-the-blank question, at least two items are listed.

(2) Measurement of the dependent variable. The dependent variable of the study is the willingness to purchase online tea products. After a lot of the literature review, the willingness to purchase online is defined as: The likelihood of consumers purchasing products or services over the Internet. The measurement of this variable is mainly based on the research results of Jaeki Song et al. [6]

2. Experimental results. The total number of questionnaires retrieved in the experiment was 90. According to the question of screening invalid questionnaires, 12 invalid questionnaires were excluded. According to the answering method, the use of more than 18 minutes will be regarded as the principle of invalid questionnaires, and 5 invalid questionnaires will be excluded. Therefore, the experiment rejected a total of 25 invalid questionnaires, 73 valid questionnaires, and the efficiency of the questionnaire reached 81.1%. The male to female ratio is close to 3:5, 29 males and 44 females.

(1) Manipulation check of independent variables. First, the participants in the questionnaire responded to the best friends of the unit (new students and their own relationship.) After the data collection, the study used the independent sample t test to test the social distance manipulation of the subjects, the close group The distance group showed a significant social distance difference ($P<0.01$), and the social distance perceived by the short-distance group ($M=5.35$, $N=36$) was significantly higher than the social distance perceived by the distant group ($M=4.11$, $N=37$). From the above data, the manipulation of the social distance in Experiment caused a significant difference in the social distance between the two groups of participants and the merchants.

Table 1. Descriptive statistics of purchase intention

Social distance group	Sample size	Mean	standard deviation	standard error means
Near social distance	36	5.3494	1.56171	0.26037
Far social distance	37	4.1111	1.68959	0.26957

According to the survey data, experiment verifies the main effect of Hypothesis 1: The social distance between sellers and consumers on the willingness of online shopping tea products. Compared with the social distance, when the social distance is far away, people's willingness to purchase is weakened. At the same time, according to the survey data analysis, when the seller and the consumer are in close social distance, the seller tends to use figurative thinking, resulting in lower perceived risk, thereby further increasing the consumer's willingness to purchase; and when the seller and the consumer are far away from society Sellers tend to use abstract thinking, and the association of perceived risk is too abstract, resulting in increased perceived risk, which makes consumers' willingness to purchase decline.

Summary and Outlook: This experiment only verifies the impact of social distance on purchase intention. Social distance is only one dimension of psychological distance, and then other dimensions, such as time, distance, spatial distance and probability, will also have an impact on purchase intention.

3. Forecasting and Statistical Blockchain——Consumer Willingness of Tea Set Products under the Background of Internet of Things

The purpose of Experiment 2 is to predict and count the consumer's willingness of tea products under the background of Internet of Things, in order to facilitate the network statistics and data cloud achievement of the later consumers' willingness, paving the way for the research of tea service design cloud service platform.

In the data statistics, the survey audience will be classified, and through data comparison, the appearance, color, material, creativity, packaging and other aspects of tea set product design subdivision design points will be transmitted to the cloud platform, in order to carry out targeted product design for consumers and the market. At the same time, with the help of websites and mobile APPS applications, consumers can add and upload new design breakpoint data at any time to continuously improve the data and increase the credibility and extensiveness of data collection.

Option	subtotal	ratio
A. Online stores (such as Tmall and Jingdong)	55	75.34%
B. Online supermarket	26	35.62%
C. Mobile APPS	26	35.62%
D. Micro-business	19	26.03%
E. others	23	31.51%
This question is valid for the number of times	73	

Figure 1. Research and statistics in the form of a cloud service platform designed by tea set

Integrated research and statistical data, the current demand for the largest consumer demand for tea design is: the appearance of individuality, the color of the national and simple style of the route, the style of ceramic materials. At the same time, consumers have higher requirements for the production process and precision of tea sets designed for tea sets, accounting for 63.01%. Secondly, the appeal of the design of tea sets accounted for 56.16%; once again, the appeal for integrating local cultural characteristics accounted for 46.58%; color, material, appearance, surface pattern decoration and functional improvement are less occupied. Uploading the above-mentioned series of factors to the design cloud service platform will undoubtedly greatly improve the design of the tea set. For the formal survey of the design cloud service platform, the main choice of consumers is the service platform built by network providers, accounting for the vast majority - 75.34%; the mobile application APPS and the network super are the second choice, accounting for 35.62%, Other forms accounted for 31.51%.

4. Product design cloud platform based on consumer wishes

Based on the previous consumer willingness comparison, design impact factor research and data uploading, the user sets up the design team after the cloud service platform releases the requirements, establishes the user database, translates the user requirements into evaluation goals, and builds a scheme design based on quality function configuration model. At the same time, the designer can use the data analysis of the product cloud service platform to conduct a user-evaluation network for the next tea product design. The multi-objective particle swarm optimization algorithm based on decision preference is used to conflict the tea product design and determine the optimal innovative design.

Under the support of internet technology, the creative design of networked collaboration has begun to transform into an economical industrial design cloud service platform [7]. According to the user's release requirements, the network virtual team [8] was established, and the ideas and effectiveness of the optimal tea design innovation were verified through the modules of creative factory, cooperation space and sharing station [9].

5. Conclusion

In the construction of the tea service product design cloud service platform, users can participate in the whole process of platform demand release, design team formation, project start, project progress and project end. With the cloud service platform as the application background, it provides effective resource support by building an industrial design interaction environment with open sharing and service innovation. Through the collaboration between the front-end and back-end of the cloud service platform, combined with the corresponding access technology and virtualization technology, the sharing and collaboration of the design life cycle process is realized. In the construction of the tea equipment product design cloud platform, the process model of the system construction network evaluation product industrial design and evaluation is a core indicator. The model based on the cloud platform, through open sharing and service innovation of industrial design, interactive environment, gather information services, product design and all kinds of design knowledge base and other industrial design relevant resources, the early stage of the research and design for tea product scheme design optimization of the late group evaluation, process evaluation, such as the provision of effective resource support.

How to study the high efficiency and accuracy of tea set product design data collection and application on the cloud service platform is the next problem to be solved in this case. Future research will focus on reflecting users' personalized needs into the innovation process of designing cloud service platform, and combining cloud service platform with 3D printing personalized customization. Fundamentally improve the layout of the existing tea set design block industrial chain.

Acknowledgments

Funded projects: Humanities and Social Sciences key projects in Anhui Province (No. SK2017A0635).

References

[1] TROPE Y.LIBERMAN N. (2000) Temporal construal and time-dependent changes in preference.Journal of Personality and Social Psychology.,79:876-889

[2] Martin B A S. A Stranger's Touch (2012) Effects of Accidental Interpersonal Touch on Consumer Evaluations and shopping Time. Journal of Consumer Research. 39 (1): 174-184.

[3] FUJITA K, HENDERSON M, ENG J, TROPE Y, LIBERMAN N. (2006) Spatial distance and mental construal of social events. Psychological Science. 17:278-282.

[4] WALLER M A, WILLIAMS, B D, TANGARI A H, et al. (2010) Marking at the retail shelf: an examination of moderating effects of logistics on SKU market share. Journal of the Academy of Marketing Science., 38 (1): 105-117.

[5] LEE J, SHRUM L J. (2012) Conspicuous Consumption versus Charitable Behavior in Response to Social Exclusion: A Differential Needs Explanation. Journal of Consumer Research. 39 (3): 530-544.

[6] ZAHED F, VAN PELT W, V, SONG J. (2001) A conceptual framework for international Web design. Professional Communication, IEEE Transactions on. 44 (2): 83-103.

[7] CHU Jianjie, LI Xuerui, YU Suihuai . (2016) Key technologies of cloud service platform oriented to the whole chain of industrial design . Journal of Machine Design.,33 (11): 125-128.

[8] LIU Jing, YU Suihuai, CHU Jianjie, et al. (2017) Research in the member optimal selection of network team. Computer Integrated Manufacturing Systems., 23 (3): 1205-1215.

[9] CHU Jianjie, LI Xuerui, YU Suihuai . (2016) Key technologies of cloud service platform oriented to the whole chain of industrial design . Journal of Machine Design.,33 (11): 125-128.

IOP Conf. Series: Materials Science and Engineering **573** (2019) 012082 doi:10.1088/1757-899X/573/1/012082

The Importance of Design Management to Brand Strategy

Jia Wang[1,*], **Min Li**[1], **Tingyu Wan**[1]

[1] Art College, Jiangxi University of Finance and Economics, Nanchang City, Jiangxi Province, 330013, China

[*]Corresponding author e-mail: 784362505@qq.com

Abstract. With the rapid development of economic globalization, more and more enterprises focus their development planning on the brand strategy. An excellent brand strategy should be supported by good design management, and a successful brand strategy comes from high-level design management. In this paper, the status quo of design management and the role of brand strategy are introduced. "MUJI" is taken as an example to illustrate that design management runs through the brand strategy and the importance of design management to brand strategy.

1. Introduction

With the rapid development of economic globalization, brand strategy has become an important parameter indicator for the sustainable development of enterprises. More and more companies are focusing their development plan on the layout of brand strategy. Brand is the vitality of sustainable development of enterprise innovation. Good brand strategy should be supported by good design management. Successful brand strategy comes from high-level design management. Design is not only to establish concepts, provide technologies, and create good service qualities for corporate brands, but also to design the perfect touch of the brand to capture, reflecting its unique character. Therefore, good design ideas are gradually seen as a competitive advantage that can sustain the sustainable development of corporate brand strategy for a long time. With the rising status of design in the corporate brand strategy, design management is increasingly valued as a key point in the connection between business and design.

2. The concept, importance and status of design management

2.1. The concept and importance of design management

The concept of "design management" was first proposed by British designer Michael Farry in 1966, "Design management is to define design issues, to find the right designer, and to make the designer solve the budget problem in a timely manner within the established budget [1]". Nowadays, with the development of society, people are increasingly pursuing the value of the brand due to the excellent creative design is emerging one after another and good design brands are dazzling. Establishing a good brand image has become the consensus of many companies, and the establishment of a good brand image requires excellent design. Design is a process of integration of designer thinking innovation and behavior creation. Designers will collide with many links in the design process. One link is that the designers communicate effectively with various departments e.g., mass market, brand marketing and management. The division of labor and cooperation with each other can effectively accomplish the design goals and reflect their value.

Design management is a systematic process. It is not just about "design plus management". It is about how to integrate and coordinate a series of design activities and strategies that manage them, though communicate effectively to achieve goal consensus in final. If there is design, there will be management. In the work, designers and managers sometimes have differences, which are attributed to the great differences in the way they think about problems and the scope of their work. In general, the designer is the creative creator, and his work focus is sometimes biased towards emotional design creativity. He is the executor of research and decision making in creative performance and construction solutions. However, managers usually have a strong analytical ability. He is a project organization, deployment, and coordinator. Their rational analytical skills make them pay more attention to work steps, efficiency, comprehensive system planning, and so on. Therefore, design management plays a decisive role in the design project.

2.2. The status quo of design management in China

In the context of the digital creative industry and the creative economy's continuous upgrading and cross-border integration, the competition of enterprises has become increasingly fierce. The requirements for design have become higher and higher and for designers and managers have been further improved, therefore, design management has received extensive attention.

Design management is an emerging field in the discipline of management. Its production and development time is relatively short, and research progress at home and abroad is uneven. The research on design management is largely based on theoretical research in China. However, in South Korea, Japan, and the West, its design management has occupied an important position in enterprise development management, which requires Chinese scholars to learn from the rapid development of other countries and make up for their own shortcomings in the development of this field. The current shortcomings of design management in China are as follows: (1) the development of China's design management has not been highly valued in the status of corporate brand strategic activities and the importance of its functions. (2) design management theory rarely plays its role in the enterprise as much as possible.

In the face of these problems, we should focus on how to use the relevant theoretical knowledge of design management to make the design work play the biggest role in the corporate brand strategy and create more social benefits.

3. The role of corporate brand strategy core value

3.1. The connotation of brand strategy

The brand strategy is the enterprise's overall action plan based on the analysis and study of its own conditions and external environment with the core of brand creation, use and maintenance [2].

3.2. The role of brand core value

A brand is an intangible asset that can bring a premium to the owner and increase its value. Its carrier is a name, term, symbol, sign or design and its combination that is distinguished from other competitors' products or services. The source comes from the impression of its carrier formed in the minds of consumers [3-4]. The important role of exploring the core value of a brand can be summarized as:

(1) Enhance the cohesiveness of the enterprise. A successful brand strategy can enhance the cohesiveness of the enterprise from the inside to double the sense of identity, belonging of the employees within the company, and make them feel happy and start their own work, and willing to work together with the company to develop and progress together.

(2) Improve the influence of enterprises. Excellent brands are easy to reach people's consensus. On the one hand, they can increase their investment value. On the other hand, they can attract more talents.

(3) Improve the visibility of enterprises and strengthen the competitiveness of enterprises. Because of the intangible influence of brands, enterprises can effectively improve their manpower, material

resources and financial resources.

4. Design management runs through the brand strategy - taking Japan's "MUJI" as an example

A corporate brand is an important indicator that distinguishes its products from other products or enterprises. It is a necessary choice for an enterprise to reflect its core competitiveness [5]. For the public, it is a guarantee of reputation and quality. In today's dazzling array of products, companies create their own influence by creating brands, and the public chooses what they want through different brands and different content. Brands have a win-win situation between the sales influence of the company and the satisfaction of the public shopping. The benefits created by the brand for the company are inherent and lasting.

A successful brand can demonstrate its own brand value, it can be able to spread positive and upward core values, and have the responsibility and obligation to guide consumers to rational consumption. In this work, we are taking the Japanese "MUJI" brand as an example. It was born in a grandiose and secular society, but it adheres to its unique brand standard: "low-key, natural, environmentally friendly, and healthy". The society spreads new consumption views and values, namely: anti-fashion and anti-show concept of civilian consumption. The brand standard of MUJI is a soft standard for creating new consumption, its brand image communication influences people's consumption concept in a certain sense [6].

4.1. MUJI's era background

For the well-known brand of MUJI, it was founded in 1980 by the *Seiyou Ryohin Keikaku Co., Ltd.* of Japan. The meaning of "unprinted" in Japanese national language is "no pattern", which can be extended to no brand. In general, "good" refers to high quality and low quality products. "MUJI" is "a good product without brand". The innovative consumption concept of MUJI brand strategy finally formed a unique brand of "low-key, natural, environmentally friendly and healthy" [7].

After the Second World War, with the Japanese economy developed rapidly and luxury brands were dazzling, it was setting off a fanatical brand of hot air. In the mid-to-late 1970s, the world economy was declining and the public consumption power was low. With the development of industrial design at the time, the lack of resources and other issues, Japan has switched the views from pursuing luxury goods to the era of rational consumption. As a result, MUJI, which is cost-effective and returns to the essence of the product, officially entered the commercial stage under the establishment of *Xeiyou Co., Ltd.*, the president of *Seiji Tsusumi*.

4.2. The consumption values of MUJI brand strategy communication

The creator of MUJI has been very accurate in positioning the brand from the very beginning, and its brand strategy standards reflect its unique brand values. MUJI has emerged in the public view with its unbranded brand strategy, leading the "anti-fashion" and "anti-show" civilian consumption concept. Its products pursue minimalist design concepts from design to production, and over-packaging on packaging, remove glitz, delete complex and simple, pay attention to functional expression, which guarantees quality and advocates rational consumption. In the promotion, there is no flood of advertising for its promotion, thus, saving publicity expenses to make it simple, only the function itself and the form material are used. At the same time, MUJI advocates the concept of natural ecology, advocates green design, and recognizes nature is also the best, and conveys the sustainable development of ecological consumption concept to consumers.

4.3. Design management runs through MUJI

With a highly sensitive design management team and design management talents who are good at integrating planning, they combine with the brainy designers to create a brand with enormous influence. Design management throughout MUJI can always be reflected in the following aspects:

First of all, it has a strong team. At the beginning of the founding of MUJI, the founder of *Seiji Tsusumi* invited the artistic director Mr. *Ikko Tanaka*. The team of its powerful elite began to develop and plan and promote deep collaboration with various departments. In this regard, the brand strategic image of MUJI was initially established. During its growth and development, it also introduced industry celebrities such as product designer *Naoto Fukasawa* and art director *Kenya Hara*, and these qualified designers communicated with the senior management of MUJI to review the overall brand strategy and strictly grasp the trend of product innovation and development of MUJI. In this way, high-level personnel gathered in diverse wisdom can gather in the brand strategy of MUJI design and development companies to effectively maintain the concept that the brand wants to create and efficiently carry out management to fully shape the brand image of MUJI.

Second, MUJI has a rigorous brand and design management system. A successful brand growth and expansion will inevitably encounter ups and downs. At the beginning, MUJI, which had been in business for more than a decade, was not doing effective and comprehensive communication between the management team because of the variety of stores and products and the complacency of employees and peer competition. It has led to a decline in performance. Based on this, president *Tadamitsu Matsui* paid attention to the efficient management and maintenance of the brand image. Under his auspices, MUJI produced two books, *"MUJI GRAM"* and *"Business Standard Book"* to guide the operation of the company. Once again, it embodies the importance of design management for corporate brand strategy, clarifies the core concept, enhances team execution, and regulates a series of problems such as store operation system product collocation through design management manual. The MUJI shops can accurately convey the brand's strategic goals and attributes of MUJI to consumers.

It can be seen that in the planning and establishment of corporate brand strategy, design managers and designers should pay close attention to the forward-looking nature of service and customers and the whole brand strategy, and pay attention to the overall process of building brand strategy so that design management runs through the brand strategy.

5. Introduce and strengthen design management to build brand core values

Taking the above-mentioned MUJI in Japan as an example, it shows that design management runs through the brand strategy, and strengthening design management is very important for establishing the core value of the brand. Because the brand strategy will indirectly affect the popularity, it is a quiet thing for the growth of a company. Through design management, the company's staff recognizes the corporate brand strategy and twists into a rope for enterprise development and self-development. Over time, the launch of the corporate brand strategy creates a sense of identity and belonging through employee effort and consumer experience. In the meantime, it has become more important to establish a relatively complete design management concept system, because design management plays a coordinating and macro-control role in the whole design-related activities and project processes. In the process of brand building, design managers need to have a forward-looking vision to gain insight into the changes in the market's international market and the changes in the environment, to recognize the strengths and weaknesses of competitors, the cognitive behavior of consumers, etc. The investigation situation adjusts its own brand strategy, knows itself and responds to new situations and changes, appropriately adjusts product design and management, thus truly establishing and maintaining the core of brand value and promoting the development of the company.

6. Conclusion

In the current society, various kinds of life information appear explosively, real-time dynamic development, and diversified product models, which requires enterprises to rationally integrate and organize the use of resources. A flexible design management system is the most important strategic resource of an enterprise. The vitality of brand strategy comes from the core value connotation of the brand, which can be created and achieved through effective design managements. Design management can create brand culture to lead social life. The creation of a successful enterprise brand strategy is the inevitable path for the sustainable development of enterprises. Enterprises should grasp the relationship with design management from the perspective of brand strategy and implement effective design managements. In order to enhance the visibility of a company, strengthen brand influence and competitiveness, it is necessary to correctly understand the important role of design management in corporate brand strategy. Thereby, we can create the brand advantage of Chinese enterprises, face the challenges of social development and internationalization, and strive to accelerate from China's manufacturing to China's creation.

Acknowledgment
Fund Project: This work was supported in part by the "Twelfth Five-Year Plan" of the Social Science Research of Jiangxi Province of China (Project Approval Number: 14YS16), by the student research project of Jiangxi University of Finance and Economics (Item Number: 40).

References

[1] Zhang, W. (2006) Design and design management. Journal of Nanjing University of the Arts (Art and Design Edition), 04: 152-153+202.
[2] Chen, X.M. (2012) The significance of design management to brand strategy. Modern Economic Information, 17: 13-14.
[3] Yu, M.Y., Yang, F.P. (2005) "Brands tutorial". Fudan University Press. Shanghai.
[4] Lu, M.D., Zhi, J.F. (2006) Analysis of the role of design management in creating brand core values. Packaging Engineering, 05: 213-215+225.
[5] Qiu, M.F. (2012) Enterprise design management and brand strategy exploration. Market Modernization, 34: 115-116.
[6] Sheng, H. (2017) The Inspiration of MUJI branding setting up and design management. Design Research, 7(04): 97-103.
[7] Wang, H.Y. (2017) The brand standard of "MUJI" leads the new trend of international consumption. Foreign Trade Practice, 08: 44-47.
[8] Zhang, L.W. (2011) Design and design management in brand strategy. Literary and Artistic Contention, 12: 135-137.
[9] Dong, S.Y. (2014) Analysis of design strategy. Packaging Engineering, 35(18):86-90.

IOP Conf. Series: Materials Science and Engineering **573** (2019) 012083 doi:10.1088/1757-899X/573/1/012083

Exploration of Chinese Traditional Furniture Art Form in Practical Teaching

Xiang Wang[1*]

[1] School of Architecture and Art, North China University of Technology, Beijing, 100144, China

[*]Corresponding author's e-mail: wx@ncut.edu.cn

Abstract. This paper studies the application and implementation of Chinese traditional furniture elements in contemporary art design, particularly in practical teaching. Through exploring the related curricular experiments and the practical projects of North China University of Technology, this paper analyses and explains the effect of traditional furniture, introduces the innovative design of traditional furniture structure and contemporary innovative design of traditional furniture patterns. These teaching achievements express the understanding of the cultural connotation of furniture and realize the interpretation of traditional Chinese furniture in contemporary design. It is an effective way to inherit and develop traditional furniture culture and art from mortise and tenon structures and patterns of traditional furniture as reference for contemporary design.

1. Introduction

It's important to inherit Chinese traditional culture and art for university students whose filed is design. Traditional culture and art provide rich nutrition, including cultural connotation, art form, and design elements. It is proved that the content of practical teaching in the curriculum of art design specialty can not only promote students' interest in learning, but also help to develop design thinking. In the practical teaching, students make full use of the opportunity to complete the modelling of the design, which learn from traditional art. It is consistent with the quality teaching advocated nowadays [1].

In North China University of Technology, for example, the class of furniture design, which contains three aspects: the first is the experimental course and curriculum design of furniture design listed in the syllabus; the second is the open experimental project set up by the university; the third is the scientific and technological activities and the entrepreneurship projects for students. The open experiments projects, scientific and technological activities and entrepreneurship projects are extracurricular practical activities, and these projects supported by the certain finance resources, which help students to complete the design extremely.

In the course or relative practical teaching of furniture design, it is explored the redesign of the structure of furniture, mortise and tenon structure, furniture pattern and so on. The first step is to learn traditional furniture. Through to measure, analyse, draw, saw, mill, install, etc., the students begin to learn the basic knowledge of Chinese traditional furniture. The second step is to separate the mortise structure and redesign. The third step is to extract the pattern of traditional furniture and redesign. From the following several cases, the specific methods and achievements of practical teaching of art design are illustrated.

IOP Conf. Series: Materials Science and Engineering **573** (2019) 012083 doi:10.1088/1757-899X/573/1/012083

2. Learning Traditional Furniture

Zitan wood waist armchairs with burl wood inlayed twelve zodiac animals, (hereinafter as the "Zodiac chair"), which is the classic Beijing furniture handed down from Zhan Ning Hall in the Summer Palace. The Zodiac chair is a three-panel-screen-form back armchair, also known as Taishi chair. [2] The structure of the zodiac chair is exquisite and worth studying. According to the basic form of the traditional chair, the structure of the zodiac chair was analyzed and duplicated. The main frame is made of zitan, which is matched with the panel of back and arm carved zodiac by beautifully burl. Burl is natural hemispherical outgrowth or a knot usually at the base of a tree. It has been prized since ancient times for its beautiful looking [3]. It uses tapered locking tenon to connect the back, armrests and the seat surface. The structure should be push and lock by two components, so it also called the running horse tenon. Tapered locking tenon has the advantages of free disassemble and assemble. The connection between the chair seat and the leg frame mainly depends on a long and short tenon of embracing shoulder tenon, which is blind mortise and tenon joint with a tapered dovetail wedge. The leg frame is the joint of the chair seat, and the bottom frame, which is the combination of the horizontal and vertical components. These complex mortises and tenons make furniture components bite each other, forming a stable and solid structure (figure 1). In practical teaching, students were required to measure, analyze, draw, saw, mill, install, etc., thus a Zodiac chair is duplicated. The process of duplicated Zodiac chair is also the process of further studying.

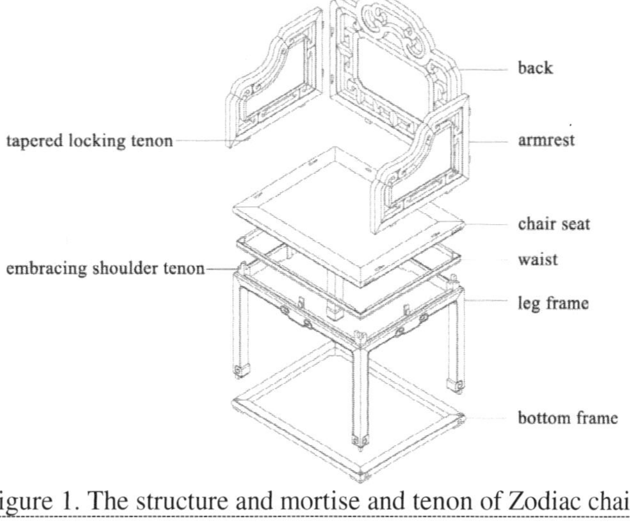

Figure 1. The structure and mortise and tenon of Zodiac chair

3. Structure

Mortise and tenon is a major feature of Chinese traditional furniture structure, which is a concave-convex connection method used on two or more wood components. The protruding part is called tenon; the concave part is called mortise (or mortise groove). When the tenon is inserted into the mortise, the mortise and tenon grip and play a connecting role. This is the main connection of ancient Chinese architecture, furniture and other wooden components. Although each component is relatively thin, it can withstand tremendous pressure as a whole. This structure does not lie in the strength of individuals, but in the combination and support of each other. Mortise and tenon is an ingenious combination wood part of more and less, high and low, long and short [4]. It can effectively restrict the twisting of wood parts in all directions.

The X-form stretcher of stool is joined by cross structure, which is kind of mortise and tenon. This structure is made of removing the half thickness of material from that part of both components where they intersect. When the components are attached, the form gains the thickness of a single component (figure 2)[3]. Referring to the cross structure of traditional furniture, apple stool in figure 3 is designed which composed of two kinds of components. In fact the outline of two parts are same, the cross structure just like figure 2. By using the occlusion of the cross structure of the eight components, smooth and seamless surface of the joints is formed. Without other auxiliary materials or hardware, it is completely supported by the connection of its own structure (figure 3). When the stool is assembled or disassembled, it is just like playing with a big toy. If it needs to be stored, the cross structure can be disassembled into eight same size components without other auxiliary connectors, which is very convenient (figure 4).

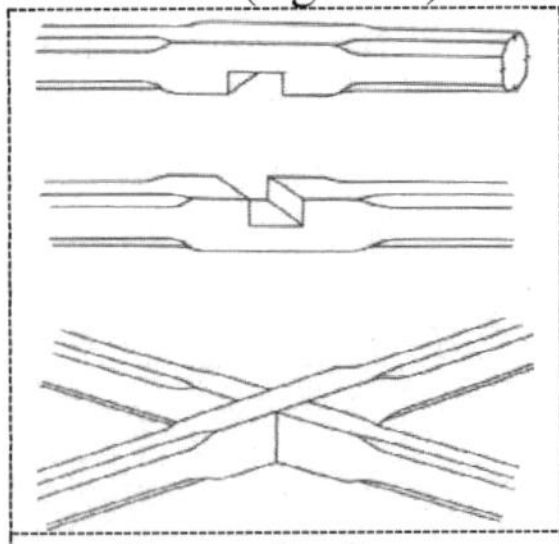

Figure 2. Cross structure

Figure 3. Apple stool

Figure 4. Assembled or disassembled apple stool

On basin stands of figure 5[5], three straight components cross at the same spot, which developed from the X-form stretcher. The centre component in this type has one-third of the thickness cut away from the top and bottom, the upper component has two-thirds of the thickness cut away only from its bottom, and the lowest component also has two-thirds of its thickness cut away from its top (figure 6). When the form is assembled, the total thickness is equal to that of a single member. [3] In the open experimental project, learning from joining of three straight members, students design the clock of snow (figure 7), which is highly portable and flexible by using structure of joining of three straight members (figure 8).

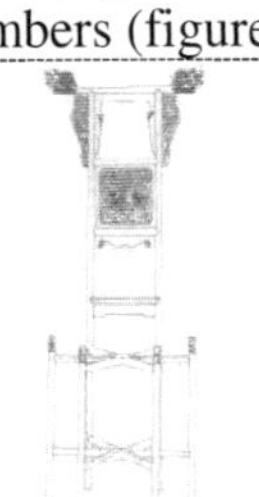

Figure 5. Basin stands

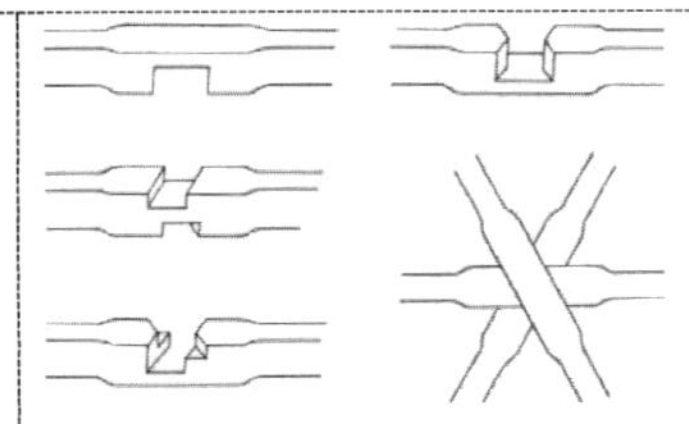

Figure 6. Joining of three straight member

Figure 7. Clock of snow

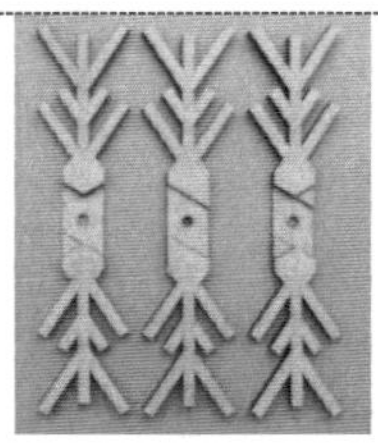

Figure 8. Components of clock

4. Motif

Traditional furniture is not only beautiful in shape and exquisite in structure, but also charming in decoration. There are three kind of patterns, the first is plant patterns such as curly grass, peony, plum blossom, chrysanthemum, lotus and so on; the second is animal patterns, just like Dragon pattern, Phoenix pattern, Kirin, Lion; the third is religious patterns and auspicious motif. All kinds of patterns are commonly used in furniture decoration, which are applied not only to hardwood furniture but also to black and gold lacquer, carved lacquer, lacquered wood inlaid with mother-of-peal. Decorative materials include wood, bamboo, various kinds of jade, pearls, metal, various colors enamel, silk, embroidery, paper and silk painting, calligraphy and so on. It can be described as dazzling, colorful [6].

Figure 9 shows Ming zitan wood (red sandalwood) fan-shaped southern official's hat armchair. It is relatively special furniture, with a large size, a fan-shaped seat surface, and tenons out. The most attractive in this special design is the peony medallion on splat (figure 10)[7]. It adds a charming and elegant to the chair. The peony is praised to be a national beauty and heavenly fragrance by Chinese, and regarded as the King of all flowers. It is the symbol of peace, happiness and prosperity. In the scientific and technological activities projects, students are asked to draw traditional furniture patterns and figure out the shape, scale and decorative characteristics of furniture through imitation. Three views drawing of furniture to help students understand the structure of furniture; perspective help students to understand the spatial and volumetric relationship of furniture; line drawing of peony flowers help students to grasp the exquisite sculpture. The exquisite peony pattern should be highlighted as a single pattern when it is used. At last, this scheme is used in leather sculpture to make small handbags (figure 11), which can be described as the ingenuity.

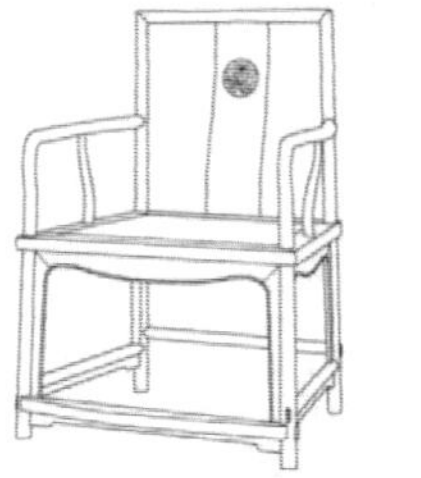

Figure 9. Southern official's hat armchair

Figure 10. Peony medallion

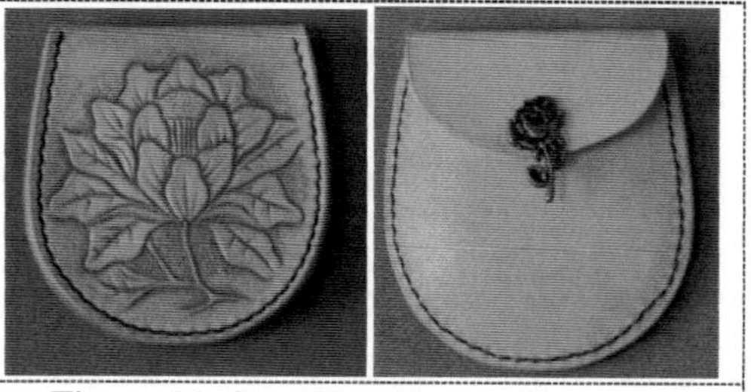

Figure 11. Small handbags with peony

The furniture in figure 12 is the midsection of carved clothes rack of Ming Huanghuali. The pattern is formed by phoenix and four-cloud unit motif. The motifs of cloud and phoenix are carved from panels of wood, which are joined with loose pins. Phoenix pattern (figure 13) is more attractive such as slender outline phoenix eyes, rolled crown, and the tail rotates around to form a circle, which is elegant [4]. In the entrepreneurship projects for students, the students were required to copy the pattern, using sketch to show the character of motif. In the period of redesign for the phoenix motif, several plans were talk about, and different materials were considered. In order to express the character of the motif, leather sculpture was decided, which can express the power of phoenix motif. As a lid of jewellery box, it endows life to the Phoenix motif (figure14).

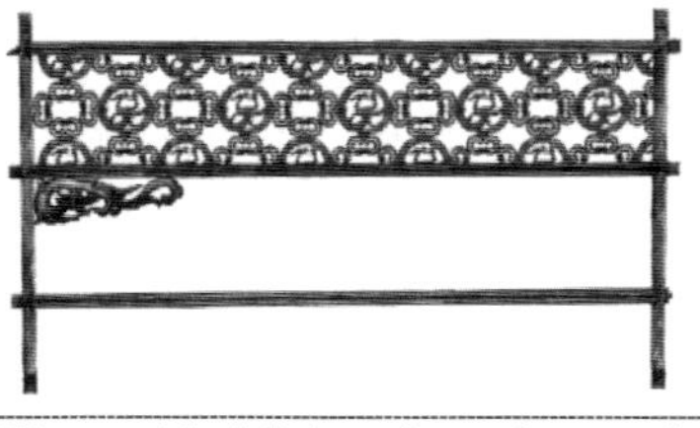

Figure 12. Midsection of carved clothes rack

Figure 13. Phoenix pattern

Figure 14. Jewellery box with Phoenix pattern

In teaching, the deeper connotation of traditional furniture decorative patterns is explored. It can not only be applied to furniture with original style, but also be extracted as decorative or design elements to interpret a wider range of design. Traditional furniture patterns show the perfect combination of traditional Chinese painting and art, and the combination of modern life and fashion, which shows its profound artistic characteristics.

5. Conclusion

These practical teaching achievements express the understanding of the cultural connotation of furniture and realize the interpretation of traditional Chinese furniture in contemporary design. Extracted elements from mortise and tenon structures and patterns of traditional furniture used, as reference for contemporary design is an effective way to inherit and develop traditional furniture culture and art.

Acknowledgments

This paper is funded by the education project of North China University of Technology in 2019, and is one of the phased achievements of *The Teaching Exploration of Art Design Course under the Background of "Internet +"*.

References

[1] Zhang G F, Gong L J. (2015) Research on Practical Teaching of Art Design Major in Colleges and Universities. Journal of The Chinese Society of Education, 274:129-130.
[2] The Beijing Summer Palace Management Office. (2013) The Ming and Qing Furniture Collected in the Summer Palace. Culture Relics Press, Beijing.
[3] Wang SX. (1990) Connoisseurship of Chinese Furniture. Joint Publishing, Hong Kong.
[4] Wang X, Yin JW, Zhang ZH (2014). Basic Form and Modelling Structure of Traditional Bench.2014 China Institute of Interior Design Journal, 163-168.
[5] Wang X. (2018) Auspicious Animals Patterns on Traditional Chinese Furniture Design.
[6] Wang X. (2019) Damei Wood Works: A Study on the Protection of Beijing Traditional Furniture. China Architecture & Building Press, Beijing.
[7] Wang SX. (2008) Appreciation of Ming Style Furniture. Cultural Relics Publishing House, Bei Jing.

Inheritance of Traditional Dragon Patterns in Modern Design

Bing bing Liu[1*], Xue ying Wang[2]

[12]School of Architecture and Art, North China University of Technology,

Beijing,100144, P.R. China

[*]Corresponding author's e-mail:gisele7811@sina.com

Abstract. Taking the traditional dragon pattern as the breakthrough point, this paper discusses its wide application in modern design fields such as logo design, poster design, packaging design, decoration design, corporate visual indentity design, product modeling design and so on. The dragon pattern is not only a symbol of ancient Chinese power, but also has more symbolic significance and symbol function. It is proposed that the use of dragon pattern is one of the means of inheriting traditional culture, and it is also the key to inheriting traditional cultural connotation.

1.Preface

In China's extensive and profound traditional art, dragon pattern is one of the most representative patterns. It is not only the product of the imagination of the Chinese working people, but also the symbol of power in ancient China. As for the image of dragon, the most common saying is "horn like deer, head like camel, eye like rabbit, neck like snake, abdomen like mirage, scale like fish, claw like eagle, palm like tiger, ear like ox". That is to say, choosing the animal body that exists in nature partially constitutes the dragon in legend and giving it powerful ability. It comes from people's worship of nature and the beautiful desire of fruitful grain. In ancient China, the image of dragon can only be used by the ruler with the highest power, which is the embodiment and symbol of power. Modern dragon pattern is no longer the monopoly of the ruler, but the significance of symbolizing dignity and power has been passed down and become the material used by modern designers.

2.Extreme simplification of dragon pattern

In ancient China, the dragon is the exclusive image of the ruler that is the symbol of power and dignity. It can only be used in the ruler's own application system, which is used relatively in small scope. This is the characteristic of the ancient Chinese imperial culture. It is important significance of the existence of the dragon pattern to deify the ruler with the image of the dragon in order to stabilize his dominant position and power.

The re-creation of dragon pattern in modern has eliminated the power connotation of dragon image. First of all, the understanding and inheritance of dragon pattern in modern design think that dragon is the symbol of the Chinese nation, and usually called the descendants of the Chinese as the descendants of the dragon. Secondly, the dragon is a symbol of dignity, and the use of dragon pattern is a subconscious enhancement of the product. The symbol of Huaxia Bank originates from the Yulong of national treasure in China （figure 1）. Yulong is a classic artifact of Hongshan culture in the Neolithic Age. As a moral image of the national spirit, the dragon pattern adopted by the symbol of Huaxia Bank also conforms to the meaning of "Huaxia". There are similarities and differences

IOP Conf. Series: Materials Science and Engineering **573** (2019) 012084 doi:10.1088/1757-899X/573/1/012084

between them. Dragon is not only the representative of the Chinese nation, but also the representative of Huaxia Bank. The simplified dragon sculpture is completed by brush, forming a graphical visual effect. The square inside the logo is the chip symbolizing modern science and technology, but also the hollow place of ancient coins. The final graphic is the fusion of traditional and modern.

Figure 1. Symbol of Huaxia Bank.

3.Technicalization of dragon pattern

There are strict restrictions on the use of traditional dragon pattern forms. Whether in the application of architectural art, etiquette and clothing, or daily necessities, it mainly serves the idea of sovereignty and divinity. At this time, the dragon pattern is mostly realistic, and can be applied in the main pattern, suitable pattern, decorative pattern, etc., regardless of the way, there is a set of strict standards.

In modern design, dragon pattern has more means of expression, especially with the popularity of computers, the use of design software to re-create traditional patterns, the addition of technical means, so that dragon pattern has more possibilities of expression. Designer Liu Nan's Qianlong in the Abyss and Flying Dragon in the Heaven are refined and processed on the basis of traditional dragon patterns （figure 2）. By using modern computer technology and incorporating scientific and technological expression techniques, the dragon patterns have a certain sense of technology. The dragon image accomplished by computer technology has abandoned the planarization and stationarization of traditional patterns. By dealing with the virtual-real relationship of points, lines and planes, it not only retains the most essential characteristics of the dragon, but also vividly expresses the dynamic sense of the dragon. It is the change brought about by science and technology, and also the new means of integrating traditional patterns with modern design.

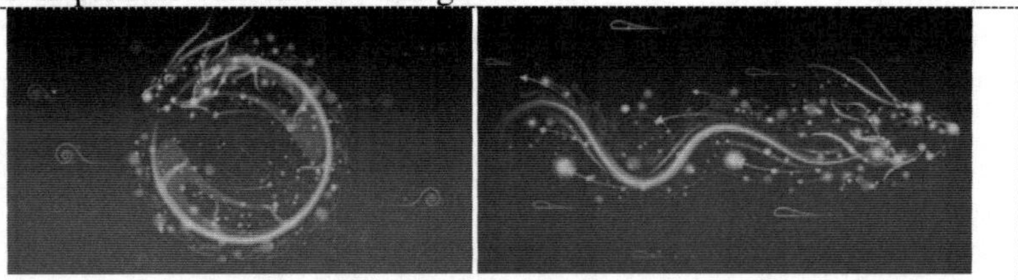

Figure 2. Designer Liu Nan's Qianlong in the Abyss and Flying Dragon in the Heaven

4.Dragon pattern in shape

The application of dragon pattern in traditional pattern is mainly around the ruler's clothing, food, shelter and transportation, for example clothing, daily necessities, furniture, architecture and so on. It is the most direct way to present the dragon pattern symbolizing the status of ruler on the carrier with specific materials and techniques without any other narration and elaboration.

For the modern application of traditional patterns, there are many application methods, such as direct application, meaning application, creative application, associative application and artistic conception application. Among them, direct application is the simplest way. When the theme coincides, the original pattern of traditional patterns is directly adopted. It is a direct way to modernize traditional patterns to save time and effort. The wine packaging design of the National Pavilion in the figure below combines realistic dragon pattern with cylindrical packaging design（figure 3）. The use of dragon pattern as a visual element is closely related to the brand's advertising slogan " Speaking for Chinese Culture ". As a symbol of the Chinese nation, dragon is one of the representative images of Chinese culture. The application of dragon pattern is a plus for the packaging of National Pavilion.

Figure 3. Packaging Design of National Pavilion

5.Branding of dragon pattern

In modern sense, the application of traditional dragon pattern is actually the design of the personal image system of the ruler. It involves all aspects of the ruler's life. By creating a personal image of the ruler as a symbol of power, its main purpose is to safeguard inviolable and supreme privileges.

In modern logo design, designers often extract valuable elements of dragon pattern to form a new visual image, and derive the new image as the core graphics into brand design or system design. Hong Kong designer Chen Youjian's new brand image for Hong Kong is based on the flying dragon as the prototype, transforming the solemn and stereotyped dragon in the traditional design into a colorful and lively ribbon dragon（figure 4）. The dragon image, which is full of vigor and the flavor of the times, shows the background of Hong Kong as a special administrative region of China, and also represents the flexibility of Hong Kong people. The blue and green color of the ribbon is the embodiment of the sustainable design concept, while the red is the Lion Mountain which represents the spirit of Hong Kong. Applying the new image to Hong Kong's public areas, such as knife flags on roads, double-decker buses and ferry vehicles in Victoria Harbour, it is an application system for brand image design.

Figure 4. Designer Chen Youjian's Hong Kong Brand Image Design

6.Product Derivation of Dragon Pattern

It is widely used in traditional dragon designs, which is applied in different fields in different ways, both in plane form and in three-dimensional form. For example, as one of the important image elements of porcelain, dragon pattern uses the shape of porcelain as the medium for plane decoration. As an important part of Chinese architectural art, the three-dimensional dragon pattern embodies the essence of dragon culture. In any case, the dragon pattern is used for the ruler's power.

In modern design, the interpretation of dragon pattern is diversified. Its shape varies with the design carrier. The transformation of dragon pattern in two-dimensional space and three-dimensional space brings a new visual experience. The work "Dragon Head" of designer Yu Chunxi adopts a highly generalized and refined method to design the abstract new product that integrates the meaning of the faucet with the practicability of the faucet（figure 5）. Designers skillfully combine the two objects that have common text element " Dragon Head" to design new products closely related to life. The faucet not only has certain practicability, but also reflects the designer's reference and Reflection on the relationship between dragon and water in traditional Chinese culture.

IOP Conf. Series: Materials Science and Engineering **573** (2019) 012084 doi:10.1088/1757-899X/573/1/012084

Figure5. Designer Yu Chunxi's Dragon Head Design

7.Conclusion

Facing the rich resources of Chinese traditional culture and art, we should deeply comprehend and appreciate the artistic spirit and aesthetic appreciation in the tradition, especially the meaning and spirit of traditional culture and art. This is the result of the long-term precipitation of the Chinese nation and the unique national soul of the Chinese nation. Only by integrating traditional art with modern design concepts, we can truly find Chinese design that belongs to our nation.

References

[1] Liu Chunxia. (2008)Application of Chinese Traditional Dragon Patterns in Modern Logo Design[D].Wuhan: Central China Normal University.

[2] Wu Guangwen.(2013) Improvement and Application of Dragon Graphic Design [D]. Hohhot:Inner Mongolia Normal University.

[3] Wang Xiaoan. (2009)Chinese traditional dragon decoration and its application in modern design [D].Hangzhou:Zhejiang Sci-Tech University.

[4] Yuan Enpei. (2002)Application of Dragon Patterns in Chinese Wine Packaging and Decoration [J]. Packaging Engineering.23(4):73-75.

[5] Zhang Laiyuan.(2008) ApplicationofChineseTraditionalCulturalElementsinGraphicDesign [J]. Packaging Engineering.29 (8):156-158.

IOP Conf. Series: Materials Science and Engineering **573** (2019) 012085 doi:10.1088/1757-899X/573/1/012085

Research on Train Exterior Design Oriented to Natural and Humanistic Environment-friendliness

Wang Zhan[1], Xu Gang[2], Yan Lei[2]

[1]School of Design Art and Media，Naning University of Science & Technology，Nanjing，210094，China;

[2]CRRC Qingdao Sifang Co., Ltd, Qingdao 266111, China

[*]Corresponding author's e-mail:wzkeeper@126.com

Abstract. Various types of rail transit trains have been integrated into the daily life of people. On the basis of meeting the technical indicators of safety, high efficiency and comfort, the exterior design of the trains should also have the natural and humanistic characteristics and theme of the environment in which it operates. This paper focuses on four aspects of the rail transit trains and their natural and humanistic environment: (1) the types, exterior characteristics and operating environment of the rail transit trains; (2) comparative study on the exterior design of the rail transit trains; (3) the identification and design application of natural and humanistic elements in the operating environment of rail transit trains; (4) the train exterior design program oriented towards natural and humanistic environment-friendliness.

1. The background

Rail transit refers to a kind of vehicle or transportation system that the operating vehicles need to run on a specific track. After hundreds of years of development, according to the differences in service requirements and scope, the rail transit has gradually developed from the original railway traffic system to three types of national railway system, inter-city rail transit and urban rail transit. From the perspective of the development of the rail transit trains, the exterior design of the trains is the concentrated embodiment of machinery manufacturing, engineering technology, social culture and the aesthetics of the times [1]. The environment-friendliness of train exterior design means that the train exterior design should show the characteristics and themes of the natural and humanistic environment in which it operates as well as good identifiable degree and formal aesthetic sense, accord with the public aesthetic and recognition of cultural identity, and have the function of spreading cultural image on the basis of meeting the technical indicators of safety, high efficiency and comfort and so on. Combined with the elements of natural and humanistic environment in which the train operates, the train exterior design is carried out according to local conditions. It is the basic goal of environment-friendliness train exterior design to highlight the characteristic theme.

2. The type, appearance characteristic and operating environment of rail transit trains

The differences in the appearance of different types of rail transit trains are mainly determined by the factors such as running speed, carrying capacity and track form and so on. Due to the influence of air resistance on the running of the train, the shape characteristics of the locomotive is also affected by the speed of the train, especially the medium-speed train and high-speed train. The aerodynamic resistance increases squarely with the increase of the train speed. Therefore, the running speed of the train has the greatest influence on the appearance characteristics. The main modeling features of the train can be divided into two parts: the locomotive and the body. The locomotive refers to the first carriage of the train (length =L), including the front end of the locomotive (length =L0) and part of the body (length =L1). Basically, all types of trains have provisions on the length (L) of the locomotive. And the length L0 of the front end of the locomotive usually has a certain range of values according to different types of trains. For example, the value for the trains on the high-speed railway is generally more than 6 meters, and the value for subway is less than 3 meters, and the value for tramcar is not more than 1.95 meters. Seen from the side view, the important contour reference points of the trains are vertex a, nose cone vertex b, diversion point c, bottom lowest point d[2], and the waist vertex e is the contour reference point of the front view. From the perspective of visual observation, the three-dimensional characteristics are more obvious in the locomotive part and the two-dimensional characteristics in the body part. The body design style is often the continuation of the locomotive design style.

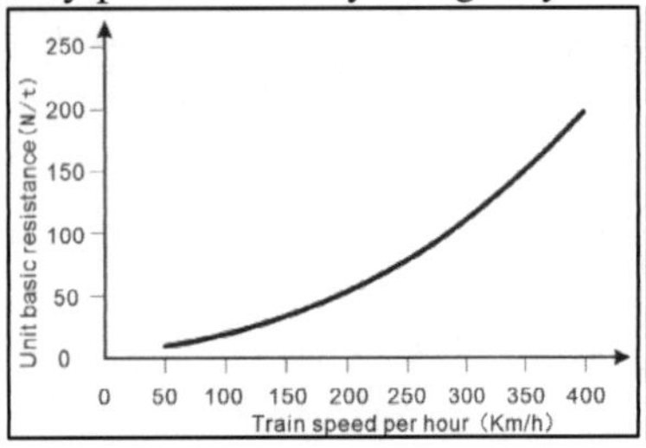

Figure 1. Relationship between train speed and air resistance

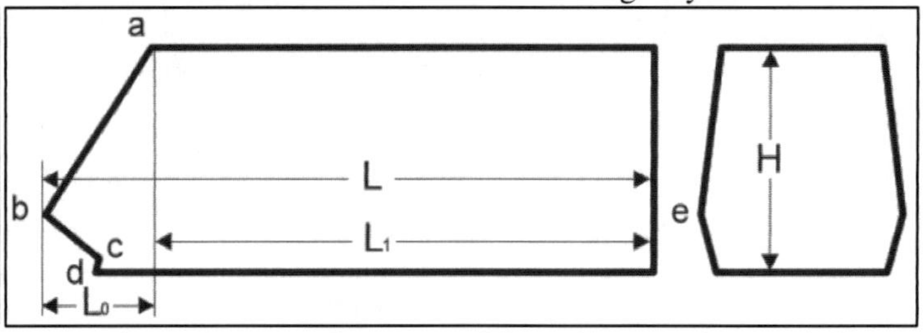

Figure 2. Key reference position of locomotive modeling

High-speed trains (excluding maglev trains) are commonly used for wide-range and long-distance transportation with a speed of more than 250 kilometers per hour, mainly including high-speed trains and inter-city trains. In order to reduce the air resistance, the front end of the locomotive the high-speed train is usually designed as a relatively flat streamlined structure, so the shape of the high-speed train must meet the aerodynamic conditions. The high-speed trains generally run for a long distance by way of different regions. The appearance design is less affected by the regional environmental characteristics, and more attention will be paid to its symbolic meaning in the design, such as whether it can represent the generalized humanistic spirit and connotation. Due to a certain region of the operation scope of the intercity trains, the operation scope is not as wide as the high-speed trains. Therefore, it can show certain style and features of regional natural and humanistic environment in the aspect of appearance design.

Urban rail trains usually include subway trains, light rail trains, monorail trains, tramcar, etc., [3] which usually run at a speed less than 80 kilometers per hour. Due to the low speed, the locomotive is less affected by wind resistance. At the same time, in order to increase the transport capacity of the locomotive, the length of the front end of the locomotive of urban rail train is shorter. As it runs within the city, the urban rail train is an excellent carrier to highlight the characteristics of the city and publicize the spirit of the city, which is of great significance in displaying the natural and humanistic environment of the city. In the exterior design of urban rail trains, the sources of modeling materials are rich and diverse. Compared with other types of trains, it has certain advantages in highlighting the characteristics of natural and humanistic environment as well as creating the train appearance with distinctive themes.

3. Comparative study on the exterior design of rail transit train

Through the research, it is found that different types of trains differ greatly in the modeling design, and the same types of trains are relatively similar in terms of the design of the main form. The reason is that the train is gradually maturing in technical requirements, function settings, manufacturing costs and other objective conditions after a long period of development and optimization. Under such circumstances, only by fully studying the types of trains and its natural and humanistic environment, and analyzing and innovating the modeling objects that make up the modeling of the train, can an

IOP Conf. Series: Materials Science and Engineering **573** (2019) 012085 doi:10.1088/1757-899X/573/1/012085

environment-friendly modeling design scheme for the train with natural and humanistic characteristics and themes be formed. At the same time, the coating design of the train body is also an important aspect of the exterior design. The design of different colors and patterns will bring significant differences to the appearance of the same type of train.

3.1. Research on the basic form of train modeling design

The two most basic characteristic lines of train modelling are respectively the front face silhouette and side-looking silhouette, and the silhouette of front face of the same type of trains is relatively close. After superposition of the front face silhouette of several typical trains of the same type, it can be found that the front face silhouette of the urban rail train (subway, tramcar) is more upright and foursquare, and the front face silhouette of the high-speed trains and inter-city trains have greater differences and richer than that of city the urban rail train. The side-looking silhouette of the train can best highlight the shape characteristics of different types of trains. Usually f1, f2 and f3 are tramcars, f2 and f3 are subway, light rail and monorail train, f4 is medium-speed and high-speed train, and f5 is high-speed train.

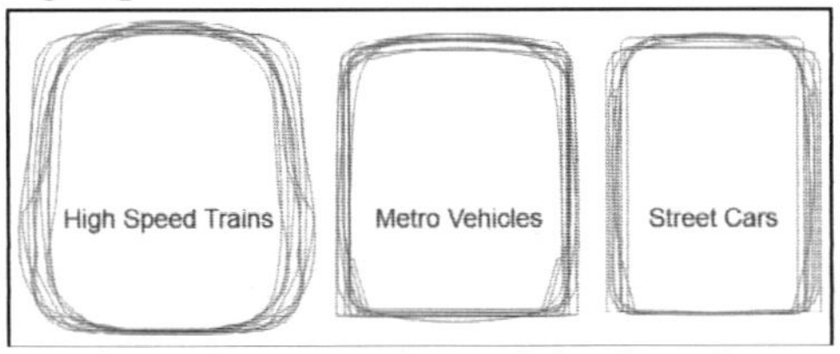

Figure 3. Comparison of the front face silhouettes of different types of trains

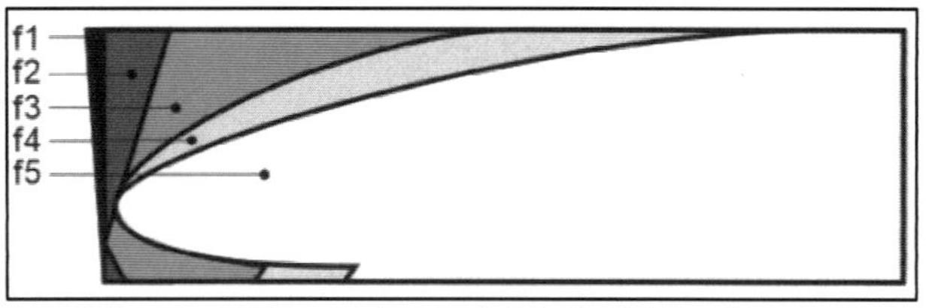

Figure 4. Comparison of side-looking silhouettes of different types of trains

3.2. Analysis of the object of train modeling design

The high-speed trains and subway trains with typical appearance characteristics are studied as the representative objects of medium-speed and high-speed trains and urban rail trains respectively. Through comparison, it can be found that, regardless of the types, the objects of train modeling design basically include roof, front window, column (both sides of the front window), headlamp sets, cab door and door and so on. The medium-speed train and high-speed train also include nose cone (cowling), diversion trench and so on. The medium-speed train and high-speed train are greatly influenced by aerodynamics. The front end of the locomotive is longer and the modeling is largely restricted by the technical requirements. But the front end of the locomotive of urban rail train is shorter and the spatial sense of modeling is not prominent. Therefore, the modeling difference expressed by the side view of the same type of train is usually not obvious. On the contrary, the frontal view of the front face of the train shows that the "expression" expressed in the frontal view of the front face of different types of trains differs from one another due to the difference in the form and position of the modeling objects although the objects of modeling design of the same type of train are the same, which has a distinct emotional representation.

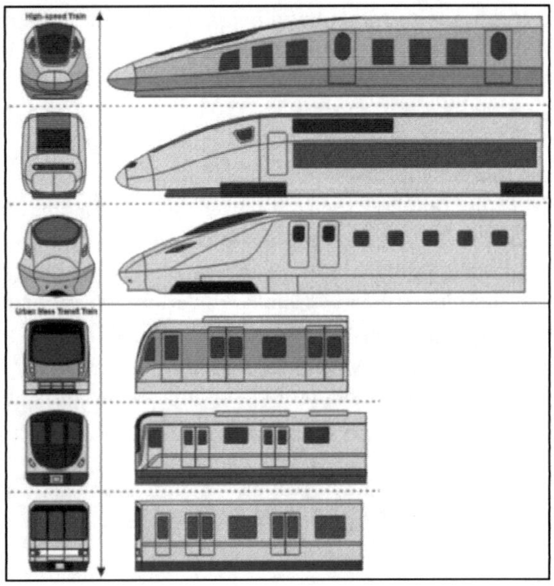

Figure 5. Comparison of front view and side view of high-speed trains and subway trains

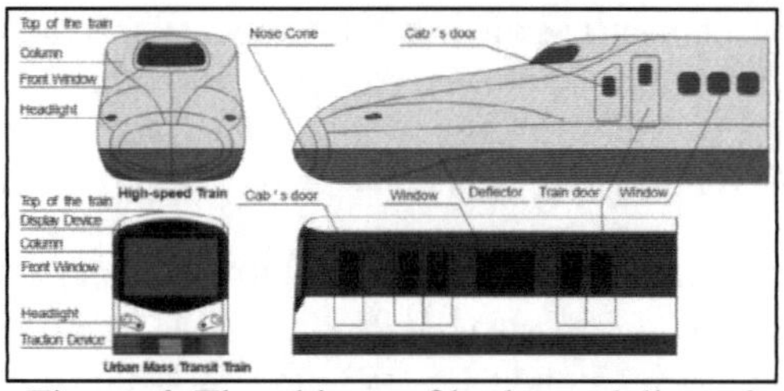

Figure 6. The objects of body modeling of high-speed trains and subway trains

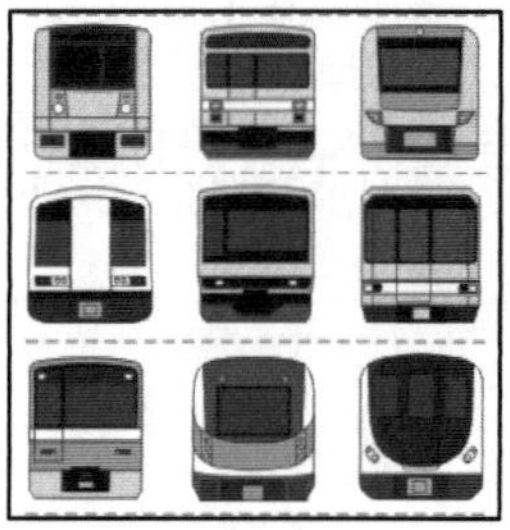

Figure 7. Comparison of headlight design of subway trains

From the frontal view of the front face, the modeling of the roof of the train is mostly arc-shaped, and the rectilinear modeling design is suitable for the train with upright and foursquare overall and detailed designs. There is a significant difference in the percentage of the area of the front window of the medium-speed train and high-speed train and urban rail trains in the area of the front view. For the medium-speed train and high-speed train, it does not exceed 25%, and that of the urban rail train is generally between 35% and 50%. The front window of the medium-speed train and high-speed train is usually designed as a whole piece of glass. Some of them add an observation window on both sides. According to the different structures of the front face of the urban rail train, the front window of the urban rail train has the design of the whole piece of glass, or the design of two pieces of glass, or three pieces of glass. The front end of the locomotive of medium-speed train and high-speed train is longer, and the gradient of the windows is large, resulting in greater bending degree of the column. And the form of column is also dominated by turning form and transition form, while the column of the urban rail train is generally flat with a small bending degree. The head lamp sets are the finishing point of the front face of the train. From the point of view of the shape, the headlamp is mainly round and rectangular; and the size of the lamp should be determined as a whole according to the requirement of illumination brightness and the proportion relation with the front face. There are three types of arrangement of the lamp sets: horizontal, vertical and oblique. Most of the headlamp sets designed in the early stage are mainly horizontally and vertically arranged. At present, the oblique arrangement of headlamps has become the mainstream. This kind of design is full of motion and change, which can make the front face look more energetic. From the perspective of position, the headlamps of high-speed trains are above 1/2 of the height of the front face, while those of urban rail trains are below the front window, which generally does not exceed 1/2 of the height of the front face. Generally speaking, the upper position of the headlamp sets seems to be more sharp and powerful than the lower position. The change of the distance between the left and right light sets of the locomotive also has a great influence on the visual perception of the front face of the train, which can make people feel the emotional feeling brought by the change of the "expression" in the front of the train. Cab doors and train doors are usually rectangular, and the trafficability characteristic, structural reliability and other factors are taken into considerations. As it is located in the front of the locomotive, the upper edge of the cab door sometimes changes with the curvature of the side profile of the front end of the locomotive. The side windows of the train are mostly rectangular with little difference in modeling, which can be regarded as the innovation point of appearance characteristics.

3.3. Analysis of train coating method

The coating design of the train is mainly divided into two types: color design and pattern design. The color design includes the whole monochrome coating and multi-color coating. The multi-color coating usually uses one color as the dominant color and other colors as the auxiliary colors, which helps to unify the colors of the train body within a range of visual colors. The use of the color belt can enhance the visual effect of the train, highlight the sense of speed and rhythm of the train. And the pattern design is usually adornment-based, but also used as a corporate advertising in some circumstances. Such as Japan's railway tourism has a cooperation with the Wildlife Adventure World in Wakayama and launches the special coating version of the Kuroshio train. With five popular giant pandas in the garden as the theme, it creates a surprise panda train, the appearance has a strong visual attraction. In the coating design of trains, attention should be paid to the types of trains. Generally, for the design of the medium-speed trains and high-speed trains, silver gray and white are used as the main color of the train, with some auxiliary colors with higher saturation. The color design of the urban rail train should focus on the color setting of the line and the characteristics of the natural and humanistic environment.

Figure 8. Standard Kuroshio train

Figure 9. Panda painted kuroshio train

4. Identification and design application of natural and humanistic elements in the operating environment of rail transit trains

According to the research, the natural and humanistic elements of the operating environment where the rail transit trains operate mainly include natural ecology, cultural tradition, historical inheritance, architectural environment, industrial economy and characteristic themes and so on. Different types of trains have different degrees of relevance and closeness with the various elements in the natural and humanistic environment. Before the design work is carried out, the natural and humanistic elements should be studied and refined in a targeted way to seek appropriate points of entry and determine the source of design elements based on the design theme.

4.1. Natural ecology

Natural ecology refers to a region's climatic conditions, geomorphic feature, species and distribution of animals and plants and so on. In the design of the train appearance, the harmony between the appearance of the train and the natural ecological environment should be considered. For example, the use of natural ecological elements as the source of design intention can often highlight the natural ecological attributes of the train appearance and obtain a wide range of affective identification. As a peninsula country, South Korea is surrounded by the sea on three sides. In the design for the new generation of "Sea Fog" high-speed train (HEMU-430X), the concept of the sea has been incorporated into its design to express strong, reliable and modern visual effects and emotional intention. It uses the image of shark for a bionic design, which implies the "Sea Fog" high-speed train is like a big land shark constantly heading towards higher speed goals.

4.2. Cultural tradition

Cultural traditions include all kinds of material, institutional and spiritual cultural entities and cultural consciousness that have existed in the past dynasties, such as traditional colors, national costumes, life

IOP Conf. Series: Materials Science and Engineering **573** (2019) 012085 doi:10.1088/1757-899X/573/1/012085

customs, concept of values, poetry, calligraphy and painting and so on. In the product design, the adoption of modern design methods with reasonable integration of the elements of traditional culture can often help the users to have a sense of cultural identity, improve the acceptance of the products. For example, in the traditional Chinese concept, the dragon and phoenix are the symbol of good luck. Since ancient times, there are also many allusions and legends about "prosperity brought by the dragon and the phoenix ". China's "Fuxing" high-speed train (models for CR400AF and CR400BF) respectively adopts the images of dragon and phoenix as the modeling elements and naturally incorporates them into the exterior design of the train. The modeling of CR400AF evolved from "dragon head", the color belt surround the locomotive came from "dragon beard", and the silver-gray body presents a strong sense of technology and industrial quality[4]. The modeling of CR400BF evolved from "phoenix head". From the side view, the front end modeling of the locomotive came from the side face of the peacock, especially the shape of the eyes, and the shape of the nose cone came from the shape of bird's beak. The overall image is as natural as though it were living! The combination of the sense of science and technology and the intention of modeling is like nature itself.

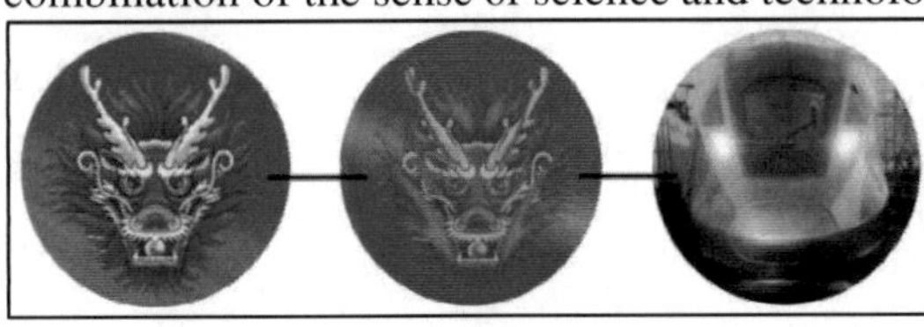

Figure 10. The evolutionary process of CR400AF modeling

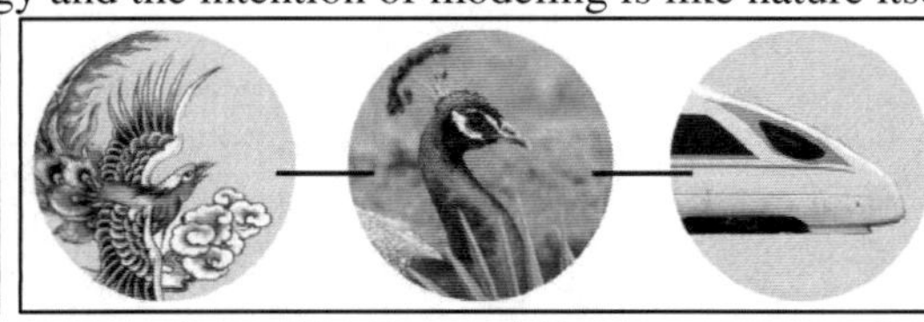

Figure 11. The evolutionary process of CR400BF modeling

4.3. Historical inheritance

Historical inheritance refers that the appearance characteristics of products formed in the early stage was gradually extended to the new generation of products as the product gene after years of development, forming the iconic features of the product appearance. Taking the Berlin subway as an example, after more than 100 years of development, the subway trains have already formed a unique style with very remarkable German characteristics. The Berlin subway train had always maintained a rational, concise and square modeling. In the early days, the front face of the train was almost a rectangular plane without any sense of hierarchy, and the design of the headlamps was also simpler. Only two round headlamps were located on both sides of the front face, and the models in the later time mostly used square headlamps. The subsequent models have made some adjustments to the front face of the train, making it more angular and adding some sense of depth, but it is still more upright and foursquare in general. In the coating design, the Berlin subway always uses yellow as the main color, and the window frames and roofs are sometimes matched with black, which is also one of the main characteristics of the Berlin subway trains.

Figure 12. Development of exterior design of Berlin subway trains

4.4. Architectural environment

The architectural environment will often affect the overall environment of the region. Therefore, the appearance of trains operating here will often adopt a relatively "harmonious" design method to correspond with the architectural style, so as to avoid the fierce conflict between the design style and the architectural environment. If the regional architecture has a heavy sense of history, then the modeling of the train and coating design are relatively conservative, like the tramcar in Goteborg of

Sweden, whose modeling and coating design are integrated with the surrounding building environment. If the operation area for the train is a modern business district or a new development area with modern buildings standing in great numbers, then the modeling and coating design of the train should reflect the sense of the times and the sense of science and technology.

4.5. Industrial economy

Industrial economy is one of the important economic characteristics of a region, which is closely related to the local industrial structure. Different cities have different industrial positioning. For example, New York, London and Hong Kong are the three major financial centers in the world, with financial and service industries as the main industries. Pittsburgh, Chicago, Houston and San Francisco are important industrial cities in the United States, with industrial manufacturing as the main industry. In the 15th century, the French city of Lyon had become one of the largest producing areas of silk fabrics in the world, and it was known as the "silk capital". The tramcars running through the downtown area of Lyon have industrial characteristics in its appearance design, and its modeling comes from the image of silkworm and white is chosen as the main color of the train. As a result of the use of bionic design techniques, the integral shape appears lovely vivid; thus it well highlights the industrial characteristics and forms a unique landscape.

Figure 13. Tramcar in Lyon

4.6. Characteristic themes

The characteristic theme refers to the design of trains into special patterns for a specific theme, which is used to strengthen the concept and connotation of the theme. In terms of train appearance, it is to carry on customized design on the appearance or coating effect of the train to make it meet the requirements of the theme. The Disney dedicated line trains in Hong Kong are more representative. In the front end of the train, the image of Donald Duck is used as the creative source of the modeling, and the image of Mickey Mouse is used for reference in the design of the shape of the window. With bright and shining coating design, it highlights a relaxed and happy theme. Generally, the train is subject to relatively large constraints in terms of modeling design, and it is also a common design method to highlight the distinctive themes through coating design.

5. The train exterior design program oriented towards natural and humanistic environment-friendliness

The work of the train exterior design oriented to natural and humanistic environment-friendliness includes the following design steps:

1.Define train type and design theme. The type of train is determined by the traffic demand, while the design theme needs to be selected according to the natural and humanistic environment where the train operates. As discussed in the third part of the article, there are many elements of the natural and humanistic environment where the train operates, and the design themes can be developed from each element. Therefore, the selection of key elements should follow the following rules: (1) distinctive and representative characteristics; (2) it has the plasticity of form; (3) the ability to display spirituality; (4) accord with the general aesthetic requirements.

2.Determine the main modeling elements. The modeling elements are mainly derived from the natural and humanistic elements that have been determined to be adopted. Through investigation, evaluation and other methods, the modeling elements suitable to be transformed into creative sources are selected from numerous formal symbols, and they will be ranked according to their close degree to the design theme and their aesthetic value. Generally, the best modeling element is chosen as the basis of modeling design, and other related elements can be coupled into the design scheme according to the design needs.

3.The modeling design is carried out around the design theme and the main modeling elements. The two-dimensional design scheme is formed by taking the front view and the side view of the train as the main part. In the modeling design, the basic requirements of train modeling aesthetics should be followed firstly. For example, the medium-speed train and high-speed train should reflect the sense of stability and speed, and the subway, tramcar and other urban rail trains should reflect the reliability and agreeableness. Secondly, modeling elements should be reasonably used to form an environmental-friendly design scheme by means of bionics, metonymy, abstraction, combination and other ways. Thirdly, the train modeling design needs to meet the general rules of modeling aesthetics, such as proportion and scale, change and unity, equilibrium and stability and so on.

4. Design and evaluation of three-dimensional modeling scheme. Based on the selected two-dimensional design scheme of train modeling, the appearance modeling is carried out with the engineering software to form a three-dimensional digital model and render the body material, color and pattern. The evaluation of the train modeling scheme is mainly the aerodynamic simulation evaluation of the three-dimensional digital model. After the scheme is passed, the methods such as wide-spectrum survey and eye movement test can be used to comprehensively evaluate the friendliness of the natural and humanistic environment of the three-dimensional digital model of the train by combining the elements of natural and humanistic environment and modeling elements.

6. Conclusion

With the construction and development of various kinds of rail transit systems, the exterior design of the train should not only consider the shaping of the form of train itself, but also consider the human-train-environment system as a whole. It is one of the ways in which the exterior design of trains can be widely recognized by the society to confirm the types and morphological characteristics of trains, reasonably incorporate natural and humanistic elements in the design of the appearance, and form an environment-friendly train design scheme with natural and humanistic characteristics and themes. The research results will provide theoretical and methodological basis for the exterior design of the train with natural and humanistic environment friendliness, provide guarantee for the quality of the train exterior design, and promote the change of the train-exterior design from “functionality” to the direction of “rationality”, “ecology”, “artistry” and “characteristic theme”.

Acknowledgments

National Key R&D Program of China, Number:2017YFB1201103-16

References

[1] Xie Xuanhui. Research on the Evolution of Urban Rail Transit Vehicle Modeling Design [J].Design, 2018，10：121-123.
[2] Xiang Zerui,Zhi Jinyi,Li Ran,etc. Research Progress and Challenge in Industrial Design of Urban Rail Train in China. [J].Journal of the China Railway Society,2019.
[3] GB/T5655-1985.[S]
[4] How were those handsome high-speed locomotives designed? China’s high-speed railway was revealed[EB/OL]. [2018-07-23]. http://www.sohu.com/a/241029871_208485

IOP Conf. Series: Materials Science and Engineering **573** (2019) 012086 doi:10.1088/1757-899X/573/1/012086

Reflection on the Education of Design Management Course

Xue ying Wang [1*], Bing bing Liu [2]

[12]School of Architecture and Art, North China University of Technology, Beijing,100144, P.R. China

[*]Corresponding author's e-mail: 357289336@qq.com

Abstract. As an important part of the education of postgraduate course in design, design management has been established and offered in some colleges and universities in China. However, as an emerging course system, it is still imperfect in the arrangement of domestic postgraduate curriculum. How to understand and identify the importance of design management course education and how to determine the knowledge structure associated with the course to ultimately facilitate the establishment of relevant systems of this course are the questions that the public concerns most now. Based on the analysis and summary of the design management course offered by North China University of Technology, this paper strives to proceed with teaching practice and further explore the significance, objectives and systematization of the design management course for postgraduate so as to promote the teaching level of design management course.

1. Necessity of design management education

Since the 20th century, the changes in modern industries and the development in global economy have been redefining the scope of design. The influence of design on enterprise innovation and economy has grown day by day, and its strategic position in enterprises and its role in the market are getting more important. This role, though not a new one, has its new significance—design becomes a valuable method for enterprises to gain competitive advantages and the most potential tool to cope with fierce competition. As the significance of design for enterprises becomes prominent around the world, design management also develops into a vital component of enterprises' business strategies.

The basic starting point of design management is to raise design efficiency. It emphatically studies the corporate design strategies and management systems worked out by enterprises under the pressure and challenges of the global market environment focusing on the demand of users and market; it deals with the means for safeguarding the path, resources and management standard for strategy planning; it centers on the integration of brand, culture, cost, technologies, and services of enterprises for innovation, as well as the management and application in its implementation. It discusses how to identify the relationship between the design development and market environment of enterprises, how to determine the design development direction and competition strategy of enterprises, and how to identify the corporate culture embodied in design. Development plan and action program shall be formulated and selected for the realization of said targets according to the overall strategic targets of enterprises.

The crucial status and role of design management accelerate its emergence as a new research field and a new discipline, which is not only the need of design, but also the need of management.

2. Objectives of design management course

2.1. To realize the specialization, centralization and marketization of education of design

In order to meet the demand by the society for the constantly increasing high quality material resources, education of design management must focuses on the specialization, detailization and centralization of design. The specialization of design is the intrinsic requirement of education of design management and the inevitable choice of design market and educational reform in design.

2.2. To develop and cultivate interdisciplinary creative talents acquainted with cultural resources of design industry

To achieve this target, it is necessary to explore regional culture and state traditional culture and establish an integrated cultivation mechanism combining cultural industry and talent resources with the education of design management, so as to train students' various qualities related with design on an overall basis.

2.3. The strategic vision is to build top discipline, foster globalized and diversified designers and design management leaders

In order to build a national top discipline in design education, integrated education combining innovative design and management shall be implemented. Reformation in English teaching programs shall be strengthened to enable students to overcome regional barriers. Long-term cooperation concerning the subject shall be promoted with the design agencies, enterprises, cultural and educational institutions at home and abroad. Students shall be guided to participate in international forums and internationally influential design competitions and more other strategies shall be adopted for the cultivation of diversified globalized talents in design management.

3. Exploration of and reflection on design management course

3.1. Crossing and integration among design management studios and colleges and enterprises

Design management studios shall be established. Colleges shall cooperate with local governments, design institutions and enterprises to develop design projects. Teachers are encouraged to guide students to enhance their practical ability and facilitate the flexible use in industries. Design management research studios can be established to foster creative talents in the cultural resources of design industry. Other measures for strengthening scientific design management include: to advance the development and production of excellent designs, to explore regional cultures and combine traditional state cultures and design education, to establish cultivation mechanism integrating cultural industry and talent resources, and to improve the branding of colleges.

3.2. Joining Design Management Institute

Design Management Institute (DMI) is an international organization of design industry in the Europe which is devoted to raising people's awareness that design is an integral essential part of business strategy. By joining DMI, students are able to acquire advanced design capability and raise international status by participating in exchange programs. DMI offers courses corresponding to its tenet which create a design education environment and foster the publishing of high quality papers.

3.3. Practical teaching based on industry-university-research cooperation

By the teaching method integrating industry, universities and research, the growth in regional innovation system (RIS) can be fostered for technical and innovative researches, and advantage can be gained for competitiveness in the market. It is required to expand students' scientific research practicality and achievements, to cultivate talents and scientific research project development required by enterprises and to establish network of contacts with industry and local enterprises.

During teaching process, the education integrating industry, universities and research is conducive to the competition among colleges, the cooperation among enterprises, the utilization of enterprises' and the government's R&D resources and the enhancement of education of students towards knowledge application in enterprises.

3.4. Establish an education composition system centering on the scientific research projects of the industry

In this respect, it is necessary to formulate relevant agreements on the industry and the exchange in industry-university education and to develop the details of agreements on industry-university-research cooperation in order to ensure the D-BASE of the industry and develop the enterprise information DB system of local enterprises. The specific processes and expected results are as follows:

Table 1. Arrangement of design management course.

Reflection on education	Contents	Expected result	
Crossing among design management centers, colleges and enterprises	Process	• Establish design management research studios • Develop DBs for enterprises and offer relevant education courses • Combine design management studios and industry-university-research cooperation	
	Expected result	Teachers	• Technical transfer in the industry and enhancement of research tasks • Increase scientific research projects and research expenditures
		Students	• Cultivate students' overall quality related to design to ensure the quality of credits obtained and increase students' employment rate
		Industry	• Enhance scientific design management and the development and production of excellent designs • Joint resource mechanism of cultural industry and talent resources
		Colleges	• Explore the combination of regional culture, traditional state culture and design education • Enhance the branding of colleges
Joining international design management institutes	Process	Participation in DMI's international activities and projects • Research exchange and collaboration • Student exchange • International seminars • Exhibitions and exchange • Students participate in international competitions • Promote scientific research projects based on exchange of teachers • Strengthen international design management projects	
	Expected result	• Teachers' research exchange and collaboration—once per year • Student exchange—more than 2 students per year • Communication and exhibitions— once or twice per year • Students take part in international competitions— more than 50 works per year • International seminars— more than once per year • Enhancement of scientific research projects by teachers exchange 1 teacher per year • Promotion of international design management projects—manage 1 international project per year	
Practical education by combining industry, universities and research	Process	• Establish network with industry • Select local enterprises as objects • Visit enterprises and provide relevant design management strategies and R&D • Formulate course components of the discipline and joint development of scientific research projects integrating industry, universities and research •Operate textbooks on industry-university-research cooperation and R&D institutions	
	Expected result	Teachers	Based on the cooperation, information sharing and internet cooperation among universities, research institutions and industry, research and development information shall be provided.
		Universities	It is conducive to the competition among colleges, the cooperation among enterprises and the ability of financial self-support for colleges, the utilization of enterprises' and the government's R&D resources.
		Students	College students can be trained in their cooperation spirit and they are able to accumulate practical experience. During this process, global market, productivity and competitiveness can be promoted. Students can get better

		opportunity of employment and can obtain more scholarship. The most important is that they may gain the opportunities to receive on-the-spot education.
Strategic vision: to build a top discipline and train global multiple designers and design managers	Process	• Formulate the highest targets for design education • Visit the universities in the US, the UK, Japan, and Korea, etc. • Visit enterprises and collect data • Research the requirements and demand of the industry at home • Adjust the development in design management teaching course • Maximize the reformation of design management discipline
	Expected result	• Enterprises to find the work flow of the exemplary enterprise they need to follow during business reformation • Enhance the cultivation of students' abilities and provide employment opportunities • Raise communication ability and English ability
Establishing a composition system of education centring on scientific research projects of the industry	Process	• Guarantee the D-BASE of the industry • Develop the DB system for enterprise information of local enterprises • Formulate relevant agreements on the industry and the exchange on industry-university education • Facilitate the details of the agreement on industry—university—research projects • The industrialization of design management and product development
	Expected result	• The practicing of industry-university-scientific research projects • Defend the copyright of design and increase colleges' earnings • Guide the reformation of local enterprises and expand employment

4. Conclusion

The core for the exploration of design management course is established based on the teaching platform integrating inter-disciplines and multiple specialties. The traditional teaching relying on specialties and skills account a lot for the fundamental training in specialties. As to the teaching of students of higher grades and postgraduates, design management teaching plays an important role for the fusion by students of the knowledge of multiple disciplines and for the cultivation of students' application of design thinking, knowledge and skills they learnt to solve enterprises' problems and problems in the market. As it is found through the exploration of this course, the education of design management will be the most important and most effective way for the further cultivation of enterprise designers and the inspiration of their creativity.

Acknowledgement

Project NO. 110052971921 / 051, Supported by "The Fudamental Research Funds for Beijing Universities"

References

[1] Griff Boyle(2003)*Design Project Management*. Ashgate Publishing Limited
[2] Edited by Prof. JR Lee（2015）*Design Management Seminar* [M] .
[3] Margaret Bruce and Rachel Cooper(1997)Marketing and Design Management. International Thomson Business Press.
[4] Xi hui, Liu(2015), *Design Managemen*t[M] .Peking University Press
[5] Martin,R(2009)., The Design of Business: Why Design Thinking Is the Next Competitive Avantage, Harvard Business Press:15-25.
[6] Exio Manzini(2016),*Design ,When Everybody Designs*,China Industry and Information Technology Publishing & Media Group[M].

IOP Conf. Series: Materials Science and Engineering **573** (2019) 012087 doi:10.1088/1757-899X/573/1/012087

Research on Interaction Design Of New Type Products Based on Users' Interest Experience

Runcong Deng

Sichuan Film and Television University, Chengdu, Sichuan, 611331, China

Abstract: User interesting experience is a method to meet users' emotional needs and an innovative idea in product interaction design. From the perspective of users' interesting interactive experience, this study will discuss and analyze the current situation of product interaction design, the aesthetic interaction concept of products, the design method and application theory of interactive products. The research method is a combination of literature research method and exploratory research method, taking the structure of "theory-practice-theory" as the research idea, and exploring new interesting product design through practical design. The practical design will design three new interactive products based on users' interesting experience, namely, a new swing bowl combining ceramic and metal technology, an integrated and convenient fruit tray, and a new interesting exhibition cabinet. This paper will combine theory and practice design to research and analysis, explore a fun interactive product design concept.

1. Introduction

1.1 The review of the present situation of the Product interaction design

The UserExperience, or UE or UX for short, began in interaction design when it was posted by Donald Norman, a cognitive psychologist in the United States. Fun design is a part of interaction design. Fun design is to add fun to the interaction design, so that users can have some fun interaction in the process of using the product and feel the fun in the user experience.

Now people talk about interaction design, mostly reflected in the human-computer interaction and other fields of science and technology, but few people pay attention to the interaction design of daily necessities, public products and other public products.However, in daily life, people are more curious about interesting products, and there is a lot of room for interesting interaction. Maslow's hierarchy of needs theory mentioned five levels of needs. Guo jipeng from tianjin university of technology proposed the hierarchy of needs of user experience based on Maslow's theory. As illustrated in Figure 1.

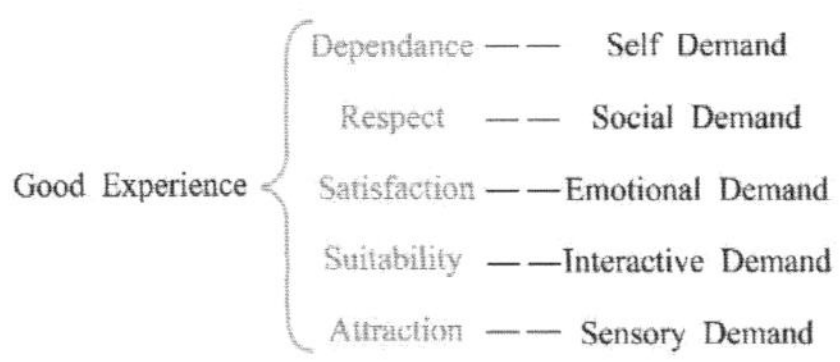

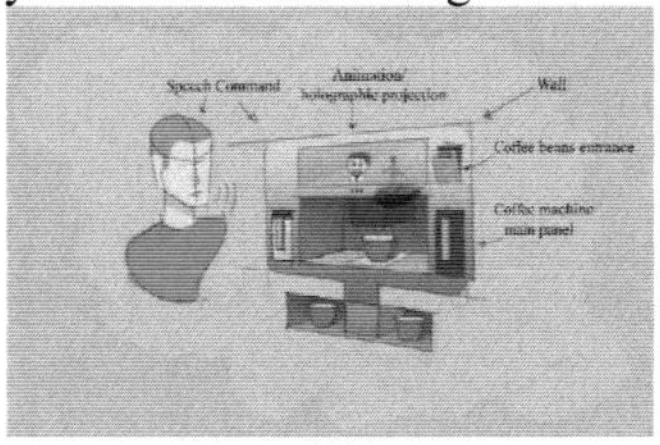

Figure 1. The hierarchy of user experience needs Figure 2. Draft of new-type coffee machine

Guo jipeng proposed that feeling is the first step for users to experience the product. After the first step is met, the interaction experience of the product can play its role, including the practicability and

IOP Conf. Series: Materials Science and Engineering **573** (2019) 012087 doi:10.1088/1757-899X/573/1/012087

functionality of the product to users, which are the basic indicators of user interaction demand experience. The third level is emotional needs. At present, most of the products can meet the needs of the first and second level. When users meet the basic material level, what they need more is a kind of psychological and emotional satisfaction brought by the product. Compared with the social demand and the self demand, the emotional demand is closer to the life of the masses. Therefore, from the perspective of innovative interaction design of products, the sense of interest, plot and interaction of products are the important perspectives of product innovation at present[1].

For example, li Yang from chongqing technology and business university proposed a new-type coffee machine with a new interactive mode. As illustrated in Figure 2. This coffee machine provides users with services by simulating human voice, and provides users with a visual service by combining animation and other technologies with art[2]. However, some scholars have pointed out that the focus of interaction design is not convenient use, but interesting use experience[3]. Although this design concept proposes a new way of coffee machine interaction, it is only grafted by using image media and digital technology. In essence, it is still human-computer interaction and electronic interaction, and it does not really carry out design innovation from the product itself. To truly carry out the innovation of product interaction design, the key is to start from interesting experience.

2. Materials and methods

2.1 Purpose and significance of the research

In the era of experience economy, consumers need not only simple functional products, but also experience brought by products. After the product design has its practicality, functionality and safety, it needs to make a breakthrough in the direction of interest. Therefore, the innovation of this study is to make new attempts on these interesting products that are rarely studied. And interesting product design breakthrough in addition to the product itself is a kind of innovation, and also for thinking, creativity, way of life is a kind of innovation, interesting products available except for the consumer market to inspire a new vitality, also inspire a new design concept.

2.2 Research contents and methods

The interaction design of new-type products based on users' interesting experience is essentially fun, interesting and interactive. It can even be exaggerated to say that this new-type product interaction design is a useful toy and a fun tool.

This study will start from the perspective of users' interesting interactive experience, take three design practice results as cases, combine literature research method and exploratory research method, and explore new-type interesting product design through specific design, so as to achieve the goal of forming interesting interactive product design concept.

3 Three research results of exploratory design practice

3.1 Design practice 1: a new type of swing bowl combining ceramic and metal technology

Ceramic bowls are indispensable in People's Daily life. However, currently in the market, people use the general ceramic bowl is static, people are not easy to hold the bowl that holding boiling hot soup. In addition, when pouring the soup, it can't concentrate on the direction of dumping, and it's not very interesting. Therefore, in order to overcome the disadvantages that the existing ceramic bowls are mostly static and too rigid, and combined with the user's fun experience, to provide an interesting new type of swing bowl, and at the same time, explored a new way of interesting diet. As illustrated in Figure 3.

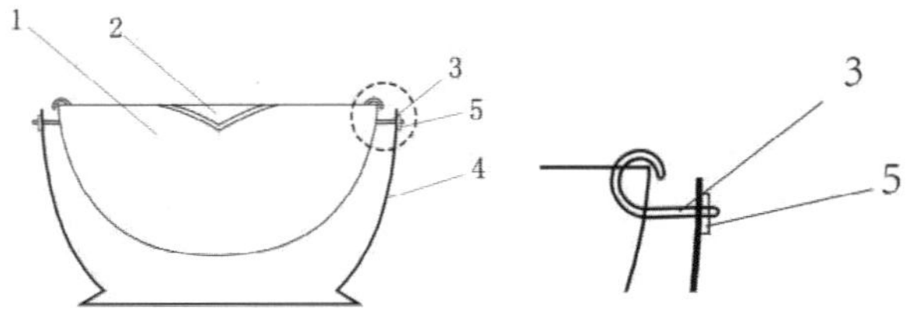

1.Ceramic bowl body 2. Ceramic bowl mouth
3. Metal fixing rod 4. Metal chassis 5. Fixed nuts
Figure 3. The illustration of the swing bowl

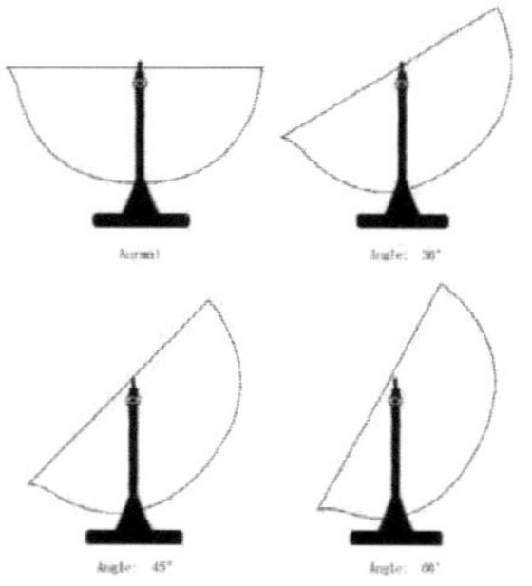

Figure 4.The illustration of the angle of swing bowl

The specific operation method of this design result is as follows: The ceramic bowl body is fixed on the metal chassis through a metal fixing rod and a fixing nut. Then pour the soup into the ceramic bowl body. When you drink it, loosen the fixing nut slightly, swing the ceramic bowl body, and drink the soup more concentrated through the ceramic bowl mouth. The ceramic bowl body can be swung 360°, and the pouring of the soup can be achieved by the different curvature of the swing. Finally, after enjoying the food, the metal fixing rod can be removed by loosening the fixing nut, and then the ceramic bowl can be removed to clean the ceramic bowl body and the ceramic bowl mouth.

In addition, this design can also be used to hold fruits, snacks, dishes, etc., the characteristics of its swaying added a lot of fun. For the swing Angle is different, the function is also different, For example, the preferred solution may be that the angle of 30 degrees is just the angle at which a person can normally drink soup. And 45 degrees can just pour the soup into other small bowls, the soup bottom can be poured out at 60 degrees. In any case, the ceramic bowl is tilted along a fixed route, effectively avoiding the phenomenon of spilling outside the soup. As illustrated in Figure 4.

These interesting designs not only solve practical problems, but also increase users' sense of experience and add interest to daily life. However, Li Jian put forward the idea of interaction design from the perspective of holism. He emphasized the interaction behavior and the interaction relationship is overall coherent [4]. Lim Y-K believes that the holistic of interaction is composed of the aesthetic of a set of "interaction gestalt", which include the aesthetic of connection, continuity, order, etc.Therefore, they propose the concept of use of aesthetic interaction integrity and instrumentality in design practice[5].

On this basis, combined with the perspective of holism, the second interactive product design practice research was carried out.

3.2 Design practice 2: An integrated and convenient fruit tray

Fruit is an indispensable healthy food in people's lives.Usually, people use fruit knives and fruit bowls every day, and the current process of enjoying fruit is generally like this: People first cut the fruit on the cutting board, then transferred the fruit to the fruit plate. After eating the fruit, they put the waste such as the core or the peel on the table or the trash can. In the case of insufficient space, such a process has obvious drawbacks, mainly as follows: On the one hand, the cutting cannot be completed without a cutting board. In the process of transferring the fruit to the fruit plate, the fruit juice will easily pollute the table or the ground. On the other hand, the waste such as core or peel has no place to put, and is not conducive to entertain a guest or keep furniture clean.

So this phenomenon brings people a bad sense of using experience. In order to solve and improve this user experience, the results of this research are intended to combine the fruit knife and the fruit plate, optimize and enhance the process, and integrate product functionality, operational safety, fun and artistic appreciation into the process, and the operation process is cyclical integration. To meet user interaction experience. As illustrated in Figure 5.

Based on the drawbacks of people cutting fruits and enjoying fruits, this design solves the problem of a series of processes for cutting fruits, placing fruits, and recycling fruit waste. At the same time, the product itself is equipped with fruit knife, serving plate and storage layer, which not only makes it convenient to use but also adds interesting point to the design. And the beautiful appearance of the product is suitable for entertaining guests, with realistic promotion value.

The specific operation method of this design result is as follows: When using the product, the user pulls the basin mouth open, places the fruit on the recycling channel, presses the basin mouth, the fruit core falls to the basin bottom directly, and the fruit falls off the basin body directly. The fruit blade has

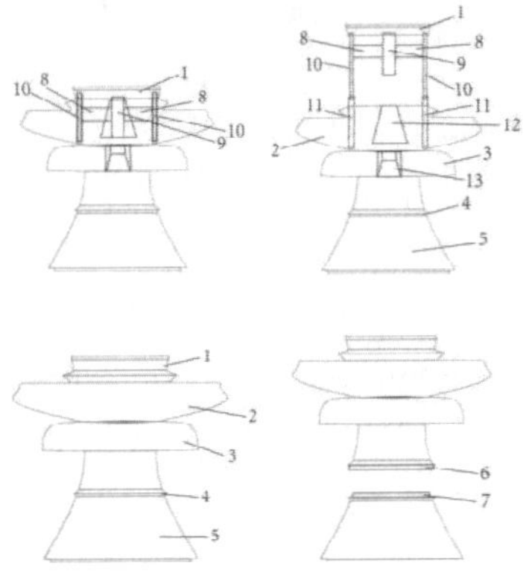

1.basin mouth; 2. basin body; 3. decorative ring; 4.Split port; 5. Basin bottom; 6. Upper half interface; 7. Lower half interface; 8. The blade; 9. Fixed link; 10. Internal lifting bracket; 11. External lifting bracket; 12. Recycling channel; 13. Rotating rod.

Figure 5. The illustration of the integrated and convenient fruit tray

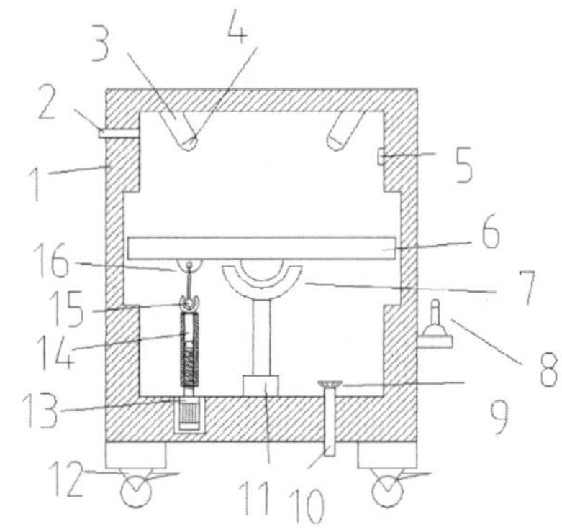

1. cabinet; 2. nitrogen outlet; 3. lamp holder; 4. lamp; 5. nitrogen concentration sensor 6. Booth 7. Rotary shaft; 8. Control button 9. Intake louver 10. nitrogen inlet; 11. fixed support; 12, wheel; 13. Motor; 14. Threaded rod and threaded tube pair; 15. Rotating shaft; 16. fixed shaft

Figure 6. The illustration of the new-type interesting exhibition cabinet

eight pieces and can cut the fruit into eight pieces. Moreover, in the process of enjoying fruits, you can manually turn the decorative ring to enjoy fruits in different directions. After the end of the enjoyment, open the split port and pour out the fruit core and other rubbish of the basin bottom.

In addition to solving practical needs, this design is also based on users' interesting experience. Starting from the circularity, the whole process of cutting, eating and pouring fruit cores is completed at one go. Compared with the previous complicated steps, this design can be described as a one-stop service. Its fun experience lies in this convenient, fashionable, exquisite use style and life attitude, reflecting the overall interaction design concept.

This result is also based on the interesting design proposed on certain problem solving. Compared with the design of the swing bowl, this design solves more practical problems, and the solution is also a fun design. Therefore, the practical and interesting design can give users a better experience. WrightP proposes that conversational interaction is an important feature of aesthetic experience. The viewpoint holds that the behavior, discourse and expression in interaction are always in an open and dynamic structure, which provides users with opportunities to obtain surprise, interest and new ideas [6]. Wallace J also believed that the value and meaning of products should be considered from the user's view, thoughts and focus of attention, and the meaning should be fed back to users [7]. Ross P also thinks from the perspective of social culture, regarding information products as the "meaningful" intermediary in the interaction between users and the world, infusing different cultural values into products, and conveying these meanings again in products[8]. These design concepts are focused on discussing the extended meaning brought by interactive products and experiencing new connotation in the interesting interaction.

IOP Conf. Series: Materials Science and Engineering **573** (2019) 012087 doi:10.1088/1757-899X/573/1/012087

Therefore, combined with this point of view, we carried out a third practical design study.

3.3 Design practice 3: A new interesting exhibition cabinet

Every family more or less will have some collections. For ordinary families, exhibition cabinets are usually placed against walls to save space. However, the other side of the collection can't be seen, and for some fine objects, a 360-degree view is a wonderful experience. And this design can not only adjust the angle, but also artificially adjust, achieve the purpose of self-adjusting the angle to watch the collection, so that the user gets a very interesting experience. At the same time, this design can adjust the height for people of different heights. Therefore, in the application to solve the actual needs, according to the requirements of the user's interest experience, it is also possible to explore and create a kind of interactive exhibition cabinet that can have the adjustment function. Especially for children and teenagers, the application of this result in the museum will be a good educational experience, so that fun teaching can be promoted. As illustrated in Figure 6.

This product is an exhibition cabinet that can manually adjust the view angle. The motor in the booth provide power for the rotation of the booth; The internal threaded rod and threaded tube pair are used for rotating the booth; The push and pull of the threaded tube causes the rotary shaft to move up and down; The fixed shaft is pushed by the connecting rod between the fixed shaft and the rotating shaft, so that the booth rotates around the rotary shaft, changing the perspective of the visitor and showing a better viewing angle for the visitors. For some of the more expensive or easily oxidized collections, the exhibition cabinet is also designed to provide nitrogen to make it easier to preserve the collection. The control button is used to control the rotation of the motor so that the booth is at an appropriate angle for visitors to see. Therefore, this new type of interesting exhibition cabinet can adjust the Angle according to the preferences of visitors, and can provide the best viewing effect for visitors of different heights in the display process, which is worthy of promotion and use.

In actual use, the user stands at the front of the booth, controls the rotation direction of the motor through the control button according to the height and angle of view, and then controls the up and down movement of the fixed shaft to make the booth Rotate around the Rotary shaft, so as to adjust the booth to the best Angle. Preferably, nitrogen concentration in the cabinet is gradually reduced due to leakage during use. In order to protect the exhibits, the cabinet can automatically replenish nitrogen through the nitrogen inlet at the bottom to protect the exhibits.

This design pay more attention to interest points of social and experiential , the user can freely adjust angle to bring the viewer into well feeling, the use of nitrogen also makes the collection are more likely to save, don't have to worry about collection of damaged, and its interactive also get a feedback, to extend interesting performance of the product itself as a more meaningful interesting development.

4. Discussion

From the perspective of users' interesting experience, the three product designs constantly improve the fun design, and then explore the meaning construction of interesting interaction. Therefore, it can be said that interesting experience is useful, and to a certain extent, it can also promote the update and iteration of products, overcome the shortcomings of functions, and increase the need for interaction. For meaning in terms of value, good fun product interaction design can feedback good results, good effects are two-way communication, people interact with products, products with interact with people, and the interaction between people and things is not just limited to digital media, but can all interact With the product itself, as small as a bowl, a cup, can be designed for product interaction, therefore, fun interactive design is nothing more than a cyclical fun experience mode. As illustrated in Figure 7.

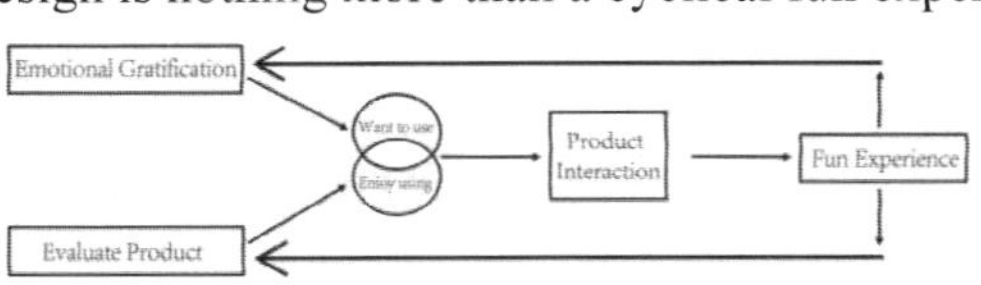

Figure 7. The mode of fun experience

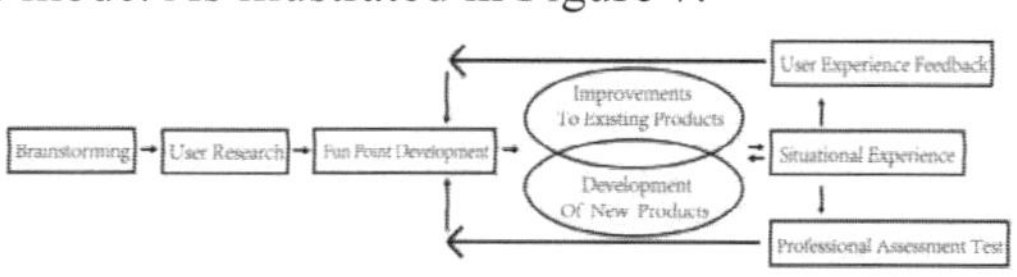

Figure 8.The mode of fun interactive design

When the user interacts with the product, the interesting points designed by the product will drive the user to further get a better interactive experience. First of all, users can get emotional satisfaction from fun experience, so that they can have the "want to use" expectation of the product, and then have the desires to interact with the product again. Second, feedback from fun experience is also very direct, Users will evaluate the product. At this level, a good fun design is bound to have the characteristics of novelty, creativity, sustainability, durability, etc， so as to get the "enjoy using" experience level for users' interesting experience, so that users can continue to use the Product.

Therefore, according to the mode of fun experience, we can apply interesting little things in daily life, interesting little ideas, or even interesting ideas in boring moments to the transformation of product interaction design in product design. When completing a series of product interaction design processes, the most important thing is to mine interesting points based on users' experience, that is, to brainstorm first and try to think of more interesting design points. Then the next step, whether it is the improvement of existing products or the development of new products, can be user fun experience as a reference, study the daily behavior of users, thinking mode, emotional vent, etc, and then the development of Interesting points of products, so that the designed products can be more "people-oriented". As illustrated in Figure 8.

After the development of fun points of the product is completed, it can not only plan for the product, but also make follow-up work based on the user experience, so that professionals can provide opinions and public users can provide suggestions and feedback, and then continue to promote the development of the product.

5. Conclusions

All the three products in this paper are designed from the perspective of users' interesting experience, and explored the fun design practice. After the discussion of the three practical designs, the fun product design gradually transitioned from the interesting small design to the integrated, systematic and completed fun product design. Aesthetic interaction design is the intrinsic spirit of interaction design ideas, in the future, interesting interaction experience will become the new trend for product design, theory of fun design aesthetics will also gradually improve, from small necessities to large industrial products, will explore the fun interaction with the user experience, make product from "availability" to "useability" evolved into "want to use" and "enjoy using", giving users an interesting emotion, when using the products meet the emotional needs of users.

Acknowledgements

The three practical designs mentioned in this paper have been successfully applied for national patents. In the process of design practice, we are committed to combining practice with theory, constantly referring to the theoretical guidance of many predecessors and the advantages of existing product design. Due to my limited level, it is unavoidable to make mistakes, welcome to correct.

References

[1] Guo,J.P.(2013). Research on product interaction design based on user experience. Tianjin university of technology.

[2] Yang,L.,li,Y.,Marcel,B.(2018) Interactive product design method for user experience and its application. Manufacturing automation,40(08):35-38.

[3] Li,Y.(2016) Theory and practice of product interaction design.Journal of chongqing technology and business university (natural science edition),33(02):106-114.

[4] Li,J.(2014) Aesthetic interaction -- the thought of interaction design from the perspective of totality.Art work,(01):13-17.

[5] Lim,Y-K.,Stolterman,E.,Jung,H., et al.(2007)Interaction Gestalt and the Design of Aesthetic Interactions.Proceedings of the 2007 conference on Designing pleasurable products and interfaces. ACM Press.239-254.

[6] Wright,P., Wallace,J., McCarthy,J.(2008)Aesthetics and Experience-Centered Design.Computer-

IOP Conf. Series: Materials Science and Engineering **573** (2019) 012087 doi:10.1088/1757-899X/573/1/012087

Human Interaction,15(11):220-241.
[7] Wallace,J.,Press,M.(2004)All This Useless Beauty: The Case For Craft Practice in Design for A Digital Age.Design Studies,7(2): 132-154.
[8] Ross,P.,Overbeeke,C.,Wensveen,S., et al.(2008)A designerly critique on enchantment.Personal and Ubiquitous Computing,12(5): 395-371.

IOP Conf. Series: Materials Science and Engineering **573** (2019) 012088 doi:10.1088/1757-899X/573/1/012088

Research on Interactive Design of Children's Public Space based on VR Technology

Yu Tang, Na Qi *, Ting Zhou

School of Art and Design, Xihua University, Chengdu, 610039, China

*Corresponding author's e-mail: happylife8183@163.com

Abstract. Children's public space is one of the important standards of the society reflecting the completeness of children's growth environment. It is the mainstream of public space development to connect VR technology with children's public space in the new era. Through the analysis of the public space which had direct interaction with the group activities of children, from the perspective of multiple sensory experiences, some interactive design principles of children's public space were summarized under the condition of VR technology and a new model of children's public space were redefined. It will beneficial to realize the emotional connection of children's new dimension, and will promote their enthusiasm for exploring unknown things and the unique creativity of self-space.

1. Introduction

In 1989, the United Nations General Assembly adopted the Convention on the Rights of the Child, (CRC), to address the lack of representation of children as subjects of society. The Convention requires that children be given the right to listen to their own views in all matters affecting them. As the creators of the future social, it is extremely important for children to transform the new spatial output mode based on the actual demand of children's growing up and provide children's public space following the trend of The Times. In the context of social informatization and intelligence, VR technology keeps growing. Integrating it into children's public space to construct an environmental base for understanding and perceiving the world will undoubtedly provide children with a new way of cognition and channels of understanding.

2. The Development trend of VR Technology in Children's Public Space

VR (Virtual reality) technology is a new era technology which integrates computer graphics technology, simulation technology, multimedia technology, sensor technology and and other interdisciplinary technologies. It has four characteristics: multi-perception, existence, autonomy and interactivity. By creating a virtual environment on the multi-dimensional digital platform that makes it difficult to distinguish the real from the fake, the user can interact in the environment and achieve the sense of immersion. The development time of virtual reality technology is relatively short in China. however, due to the vast domestic market and the great development potential of the industry, a "VR boom" has been formed in China in recent years and some achievements have been made in the fields of medicine, industry and aerospace.

VR technology has also been gradually developed in the field of children. Xina Shenkang's latest product, VRREHA, is a more interesting and efficient product for attention training, which also becomes the first domestic institution to apply VR technology to children's attention training. In addition, VR technology has produced the effect of "1+1>2" in the application of discipline training,

IOP Conf. Series: Materials Science and Engineering **573** (2019) 012088 doi:10.1088/1757-899X/573/1/012088

quality education and campus culture construction. At present, more focus is on the development of output content under the application of VR technology, and other aspects of the development is also expanding. To realize the connection between VR technology and children's public space cognition, change the old single-output state and generate a new interactive mode, so as to provide a more standardized and reasonable space for learning and entertainment.

3. Analysis on the characteristics of Children's Public Space

3.1. Importance of children's public space reconstruction

Children's public activity space refers to the special activities for children to play games, rest, study and entertainment in accordance with the children's physiological and psychological behavioral characteristics [1]. The application of VR technology has brought new opportunities to the presentation of children's public space. The creation of diversified virtual world to create different beautiful environments has aroused the enthusiasm of children and improved the efficiency of knowledge output. VR technology can build multiple different types of virtual environments in a limited space, adapt to the needs of children's various scenes and generate interesting interactions. Moreover, virtual space is not limited by space. It only needs to provide appropriate and operable physical space, coupled with the virtual environment built by VR technology, which can give children a shocking world. This immersive interactive display enriches children's horizons and creates a more imaginative future.

3.2. Analysis of multi-sensory characteristics in children

3.2.1. Audio-visual cognition. Children's vision develops with age, so the objects that children can recognize should be the smaller the age, the larger the object they can recognize [2]. Children will pay more attention to objects with obvious contour, bright colors and prominent features in the environment. Therefore, it is particularly important to form a unified style in children's public space through the reasonable collocation of colors, objects and Spaces to give children a visual impact. This also greatly improves the duration of the target object in children's brain memory layer.

After the children's auditory development is completed, the capture and archiving of music is relatively clear. However, due to children's weak receptivity to sound, complicated music is not able to timely receive and make correct response. Moreover, considering children's innocence, children's music is mainly based on relaxed and cheerful types.

Auditory content is often presented with visual content. Through background music, sound operation, role communication, sound guidance, dubbing assistance and other aspects to cooperate with the picture, so that children's audio-visual effects are more harmonious.

3.2.2. Shape characteristics of objects. Childhood is a rapid stage of physical and intellectual growth. They have a strong perception of everything in the world, are full of curiosity and questions, and have a typical active behavior and knowledge-seeking psychology at this stage [3]. Children like to feel the presence and change of known or unknown objects by touching their hands or limbs. However, children's self-management ability is insufficient, which requires that children's personal safety should be fully considered when defining the activity space. For example, the hardware facilities in the interactive space should adopt the fruity and soft curvilinear devices as much as possible. In unavoidable places, it should take protective measures to reduce potential safety hazards, such as stairs, walls, table corners and play areas.

3.2.3. Role connotation designation. Children's psychology is dominated by emotions, which usually shows variability and uncertainty, but they will be curious and explore new things [4]. Children's acceptance of unfamiliar environment needs some time to accumulate. In order to reduce children's defensiveness and quickly integrate them into the relatively unfamiliar public environment, most of them place familiar animated characters or lovely anthropomorphic images in the public space to convey friendly messages to reduce children's defensiveness and resistance.

4. Interactive design principles of children's public space under VR technology

In order to make children experience virtual environment more reasonably and efficiently in the public space under VR technology, we should start from the children's actual needs, not the actual needs just as adults think. Color, graphic language and text that are consistent with children's age characteristics are adopted for layout design. Meanwhile, safety, space use, space environment, space partition and other aspects should also be taken into consideration.Only in this way can effective combination be carried out and the overall effect of design be optimized[5]. By abstracting the characteristics of VR technology and children's sensory experience, this paper summarizes the interactive design principles that should be embodied in children's public space.

4.1. Public principle

Publicity is one of the basic characteristics of children's public space, which determines that the principle of publicity constitutes the primary basis of children's public space, aiming at enabling every child to enjoy public facilities equally and reasonably. Therefore,This requires a minimum design principle in the space design process to maximize the matching between available equipment and multiple types of children. For example, VR helmets and wearers need to meet different head sizes; Check children's cleanliness after wearing to reduce bacterial cross-infection; and spatial visual installations and virtual environments should also take into account the specific needs of children at different stages.

4.2. Children's nation principle

The principle of children's nation mainly refers to the relevant principles with children's characteristics and in line with children's use. The service role of VR technology should also be transformed into children, and the emphasis of development is on the multi-party characteristics of children. Therefore, from the basic use of color to the core of the virtual environment construction needs to meet the actual needs of children. For example, when designing a public space for children, both the appearance and content presentation of VR instruments need to adapt to the preferences of children in different stages. Color can choose to bouncing and bright colors to reflect the sense of ethereality, Also can match color and light to present the sense of technology and future content needs to be diversified. Construct different styles and levels of virtual environment to meet the multistage needs of children. Audio-visual combination is necessary. Multidirectional sound equipment can be installed to create a 360-degree surround sound environment, giving people an immersive sense of reality will increase the experience effect of children.

4.3. Interactive principle

Interactivity is one of the prominent characteristics of VR technology, which means that users interacting with objects in a virtual environment can have a real sense of entity that just like in the real environment through specific devices. In the virtual three-dimensional environment, children can communicate with objects in the environment and have different degrees of information reaction through wearing the instrument with the first perspective. Children's attention span is short, but through the two-way interaction with the content of the virtual environment, which can let children actively integrate into the environment and seriously feel the wonder of the virtual world.

4.4. Immersion principle

Immersion can make users feel "be in the world" through their own operations in the virtual environment, which is an important principle for users to immerse themselves and obtain satisfaction. Due to the limitations of objective conditions, only semi-immersion type can be realized at present. However, through 1:1 modeling between virtual environment and physical environment, steps such as space environment construction, interaction setting and multimedia equipment installation can be implemented, aiming to create a more harmonious "collision" between VR technology and children's public space.

4.5. Autonomy principle

Children, assisted by specialized instruments, can achieve relative autonomy in designated areas. That is to say, objects in a virtual environment can trigger real-time operations, such as moving, running, waving, squat and other limb movements, in accordance with the laws of physical motion of the real-world ontology. By wearing data gloves and data clothing, it is possible to achieve a high level of 1:1 reduction. The independent operation of children in the playing process is beneficial to increase the fun of children's life.

4.6. Relative mobility principle

As VR technology cannot move freely that be completely separated from the reality. it moves within a fixed-size sensing platform, and with specialized devices connected to the children to prevent them from harm. Therefore, full consideration should also be given to the actual use of child-related equipment. Keep children safe by strengthening safety arrangements and checking equipment regularly.

5. Conclusions

Virtual reality elements are integrated into the construction of children's public space, aiming to make children's growth atmosphere level in line with the overall social development. The integration of VR technology into children's public space is conducive to stimulating children's interest, enhancing children's ability to think and control themselves, and enhancing children's reaction dimension and way of thinking. Therefore, the research on the interaction design of children's public space based on VR technology has important social exploration value and development significance. Through analyzing the design characteristics of children's public space to abstracted out the concrete application principle of VR technology in children's public space. Change the previous layout definition, content display and communication mode of children's public space to realize the new form with the characteristics of The Times and promote the structural development of children's public space.

References

[1] Ye, X.Y. (2016) analysis of color visual guidance of children's public activity space. J. Art education, 03:269.
[2] Hao, J., Sun Y.Y. (2015) APP interface visual design principles for preschool children. J. Design, 03:120-121.
[3] Mo, D.P. (2018) Exploration on the functional design of children's reading space in public library under diversified needs. J. Library Work and Study, 09:124-128.
[4] Fang, H., Zhang, Y.L., Zhou, T.T., Xiang, M.P. (2016) Study on the interaction design of preschool education APP. J. Packaging engineering, 20:113-117.
[5] He, Z.Y. (2018) Guiding design and inspiration of foreign children's public space. J. Art Panorama, 06:102-103.

Study on the Sustainable Design of Wayfinding Signage System in Tourist Attraction——Taking Lushan Mountain in Jiangxi Province as an Example

Tingyu Wan[1,*], Min Li[1], Jia Wang[1]

[1] Art College, Jiangxi University of Finance and Economics, Nanchang City, Jiangxi Province, 330013, China

[*]Corresponding author's e-mail: 635177905@qq.com

Abstract. The wayfinding signage system of tourist attraction is taken as the main research object in this paper and the idea of sustainable design is used as the guiding thought. Through refining the elements of signage system construction of tourist attraction, a wayfinding signage system is constructed under the idea of sustainable design, which is combined with the design philosophy. Finally, taking Lushan Scenic Spots in Jiangxi Province as an example, the current problems are analyzed and solving strategies guided by the idea of sustainable design are proposed, so as to provide theoretical evidences for the future research on the sustainable design of signage system in tourist attraction.

1. Introduction

1.1. Background Introduction

With the rapid development of economy, people's leisure time is gradually abundant, and tourism is becoming a more and more significant part of people's life. A large number of tourists swarm into the tourist attractions, which brings about environmental problems for the scenic spots. Meanwhile, some problems have also appeared such as irrationality of many wayfinding settings, required information shortage of display panel and so on. Therefore, the construction of a scientific, reasonable and complete wayfinding signage system can not only improve the environment of the scenic spots and shape the image of the scenic spots, but also increase the cultural experience, enhance the popularity of the scenic spots, and become a window for publicizing culture to the outside world.

1.2. Research Status

From the perspective of literature search, the study on wayfinding signage design of tourist attraction started in 2007 and showed an increasing trend year by year as a whole. However, from the searching of the author, there are few studies conducted on wayfinding signage design of tourist attraction which focus on sustainable design. Through sorting out the existing academic literature, the domestic research pays more attention to the tourist attractions in the first-tier and the second-tier cities, while the research is relatively scarce for most of tourist attractions in the third-tier and the fourth-tier cities. Secondly, scholars focus more on the study of sustainable design of public environmental facilities. Finally, in terms of tourists' practical demands, it is obvious to carry out systematic research on wayfinding signage in tourist attraction. Many tourist signage contents are only closely related to the scenic spots but ignore the tourists' demands, so that the tourist information needed by tourists can not be transmitted very well. Foreign scholars have paid attention to the study on the sustainable design philosophy of wayfinding signage system in the early stage, which provides some references for this paper.

2. Overview of Wayfinding Signage System in Tourist Attraction

2.1. Definition of Wayfinding Signage

The design of wayfinding signage system refers to "using concise graphic symbols to express the accurate meaning from the angle of visual communication and carry out information transfer through manifestation pattern from the angle of environmental design"[1]. In short, it is a connecting bridge to guide tourists to conduct independent wayfinding behavior, and provides a comprehensive tourism service system for tourists in tourist attractions, which is convenient for tourists to understand each scenic spot and assist tourists to arrive at the destination safely without asking the way from others. Therefore, before constructing the whole signage system, various design philosophies should be considered comprehensively, which runs through all aspects of the whole scenic spots system.

2.2. Functional Classification of Wayfinding Signage

According to the transmission function, it can be generally divided into five types, including wayfinding signage, guidance signage, explanation signage, name signage and restriction signage(Table 1)[2]. In the planning and design of tourist attractions, it is necessary to combine the characteristics of each region to further sort out the above-mentioned five types of signage in detail, so as to keep the relevance between each other.

Table 1. Classification of Wayfinding Signage.

Location Sign	Showing the location of the facility and its relationship with the whole block
Guiding Sign	Guiding the direction to the destination
Describing Sign	Showing the intentions of managers and the content of facilities
Name Sign	Highlight different function of the facility from others
Restricting Sign	Urging people to keep safety in mind and observe public order

3. Sustainable Design Research in Wayfinding System Design

3.1. Proposing the Concept of Sustainable Design

On September 1, 1995, the concept of sustainable product design was first put forward at the international conference "Product Design for Sustainability" held in the United Kingdom[3]. The design concept has been evolved so far, with 3R principle as its core: reduce, reuse and recycle, which aims at displaying recoverability, convenience of recombination and simplification of raw material in the whole life cycle of the goods, thus achieving the purpose of environmental protection.

3.2. Sustainable Design Principle in the Wayfinding Signage System of Scenic Spot

The wayfinding signage system of tourist attraction plays a key role in guiding tourists to conduct wayfinding behaviors. Although the electronic GPS positioning system is popular at present, in the public space of scenic spots such as internal scenic spots, remote mountains, caves and so on, where the tourist experience is not good since they can not use the positioning system smoothly caused by poor signal reception.

Based on the signage system design of tourist attraction under the concept of sustainable design, this paper considers how to establish a scientific, reasonable and standardized wayfinding signage system combined with the 3R principle on the premise of reducing the pollution and damage to the resources and environment of scenic spots as far as possible, which can not only improve the environment of tourist attractions, shape the image of scenic spots and guide tourists to complete tourist activities, but also deepen tourists' appreciation of scenic spots culture, enhance the influence of tourist attractions and become a window for the publicity of scenic spots.

4. Introduction of Lushan Scenic Spots in Jiangxi Province

4.1. Geographical Location and Climate
Lushan, a national 5A scenic spot, is located in Jiujiang City, Jiangxi Province and lies in the lakeside of the Yangtze River and Poyang Lake. The total area is about $358km^2$, of which the mountain area is about $282km^2$. Lushan belongs to the northern subtropical monsoon climate, with a typical mountain climate in China, that is, humid climate, low temperature, strong wind, more fog and long frost season throughout the year. Lushan Scenic Spots have abundant rainfall capacity, and become a well-known summer tourist resort at home and abroad, where the summer is pleasantly cool, with the average temperature of 22.5 degrees centigrade[4].

4.2. Site Selection Cause
Lushan Scenic Spots mainly include three major areas: Guling Town Tourist Attraction in the mountaintop of Lushan, Southeast Foothill Tourist Attraction and Northwest Foothill Tourist Attraction. Lushan Mountain has rich ecological species, a large number of plants, birds, mammals, insects, and so on. It is worth mentioning that a significant conference, "The Eighth Plenum of Communist Party of China" was once held in Lushan Mountain, so it is also a famous political mountain. Therefore, in view of the importance of Lushan Natural Scenic Spots and its unique cultural deposits, the author decides to take Lushan Scenic Spots as a case to carry out analysis and research.

5. Existing Problems in Wayfinding Signage System of Lushan Scenic Spots in Jiangxi Province
Through field survey and interview with tourists, the problems are summarized as follows:

5.1. Lack of Uniformity of Wayfinding Signage
The wayfinding signage system in the scenic spots lacks the overall uniform coordination and normative standard, and the systematization is low. Materials, colors and styles are widely different, and the disharmony between the color used in the wayfinding signage and the environment of the scenic spots seriously affects the visual beauty of the scenic spots as a whole. For example, the wayfinding signage of Hanpoting and Yingzuishi differs greatly in form and content. The wayfinding signage explanation board of Hanpoting uses dark wooden color as the keynote. The whole style is supported by several pieces of battens, and the boards mutually combine and constitute as the main content display. The content of the explanation is only bilingual introduction in Chinese and English of the scenic spot without pictures. While the main tone of explanation board of Yingzuishi is the combination of bottle green and light green. The signage style is a modern style, consisting of a large panel surrounded by a protection framework. The layout both includes introduction text and a number of pictures presentation of scenic spots, as well as current location map.

5.2. No Updating and Maintenance in Time of Wayfinding Signage
Because of the long-time settings for many wayfinding signage in the scenic spots, the contents have been blurred. The signage boards have been faded seriously, and have broken without maintenance and updating in time. For example, the explanation signage boards in the scenic spots such as White Deer Grotto Academy, Lushan Botanical Garden and others have badly broken and the text can not be recognized clearly.

5.3. Formalization of Information Transfer in Wayfinding Signage
The font is too small for the contents of most signage boards in the scenic spots, which can not satisfy the tourists in sightseeing process of reading and play its role. For example, the Chinese characters and

IOP Conf. Series: Materials Science and Engineering **573** (2019) 012089 doi:10.1088/1757-899X/573/1/012089

English letters are too small in the signage board of tour bus passenger notice in the Lushan Scenic Spots, which is not conducive to the information transmission.

5.4. Unreasonable Layout Composing of Wayfinding Signage
The information content of wayfinding signage is too small in the scenic spots, so the tourists can not get the necessary information from the explanation signage board of the scenic spots. For example, the explanation information contents of Tiebi Peak and other scenic spots are quite limited, but the layout of guide signage board is too large. The proportion of text content and the area of signage board is unbalanced, resulting in too many blanks in the layout of explanation signage board, which leads to unnecessary waste.

5.5. Patch Placement of Wayfinding Signage System
Because the wayfinding signage is not synchronous with the development time of some scenic spots, many signage board are temporarily placed at the later stage, or several signage boards are placed in the same place, which not only destroys the overall aesthetic feeling, but also seriously wastes and damages natural resources.

6. Solving Strategies
According to the existing problems of Lushan wayfinding signage system and combined with the concept of sustainable design, the following solving strategies are put forward in this paper.

6.1. Unified Modeling
According to the system configuration of tourist attractions and considering the present situation of Lushan Scenic Spots, the style and color of wayfinding signage of tourist attractions shall have the unified design. Based on the current situation of the route in the scenic spots, the wayfinding signage of scenic spots are placed on the premise of taking into account the tourists safety and the eye-catching signage boards. In addition, the existing wayfinding signage of scenic spots shall be rectified and merged, and the unreasonable wayfinding signage of scenic spots shall be removed.

6.2. Detachable Structure Design
The structure design that can carry on repeated disassembly and recombination is used. In the assembly and connecting process of wayfinding signage, instead of using chemical glue or adhesive, it should be assembled in a clamping manner as far as possible. On the one hand, it is convenient to re-assemble and disassemble in the later stage. On the other hand, it reduces the damage to the natural environment. Combined with modular design, information content should be updated to reduce unnecessary cost waste.

6.3. Reasonable Font Size
The font size of signage board shall be adjusted reasonably. Survey data on the font size that is easy for the elderly with low vision to read shows that: "When the distance between the human and the observation is about 1 meter, the font size that is easy to recognize is more than 0.9 cm"[5]. Therefore, the font size of the wayfinding signage in the tourist attractions should be planned and designed according to the above-mentioned standards.

6.4. Applicability of Information in Signage Board
A unified design is carried out between the map orientation in the signage board and the tourist orientation. By adding auxiliary graphic symbols, the direction of sight of tourists is marked and the current position of tourists is clearly defined in order to improve the service efficiency for tourists. In view of the fact that Lushan is a national 5A scenic spot, a universal symbol graph and an explanation of English, an international language, are set up in the information of signage board of scenic spots, so as to give consideration to the use of tourists from various countries. According to the difference of the

place where the signage board is placed at present, the planning and configuration shall be carried out reasonably. In the area of introducing the whole scene,"The Wayfinding Map of the Whole Scene in the Scenic Spots"is set. "The Wayfinding Map around the Scenic Spots"is placed for the introduction to the current location of tourists around the scenic spots. "The Wayfinding Signage of Scenic Spots"is placed to guide tourists to go to the main public facilities and "The Name Signage of Scenic Spots"is placed to mark the current location of tourists. The information will be properly classified and collated so that tourists can use it accurately and clearly.

6.5. Simple Raw Material

It is suggested that more simple natural raw materials should be used in the plan and design of signage system in Lushan Scenic Spots. Because the Lushan Scenic Spots belong to typical mountain climate, the climate is humid with low temperature and more fog throughout the year. In view of the fact that the signage system is located in the scenic area of the mountain, stones with different sizes, which are not easily affected by damp and corruption, can be used as the raw materials for wayfinding signage. For example, the design of wayfinding signage system in Xilai Ancient Town, Sichuan Province, uses stones as the wayfinding signage board, and the whole style is unified and close to nature.

6.6. Combination Reduction

The public environmental facility and wayfinding signage set up in the space will be composite, which not only reduces costs but also decreases floor areas. Due to the surge of tourist volume during the peak season of tourist attractions, more space should be saved for tourists to play, so as to reduce the unnecessary waste of land in scenic spots. For example,the wayfinding signage board is combined with street lamps in the Commonwealth Games, to increase the convenience of tourists to use wayfinding signage board at night and save the installation of separate lighting signage boards. The urban wayfinding signage set up in the central district of Osaka, Japan, adopts the composite design method of public environmental facilities and road setting, thus achieving the purpose of being integrated and coordinated with the surrounding landscape.

6.7. Recoverability

In the Lushan system, recycled materials shall be used as much as possible. For example, the large stone-made wayfinding board is used in Zangyunxi Tourist Map in Wuleishan Scenic Spots. Although it is also stone, it uses relative complete natural stone whose shape has not been polished. This kind of stone is usually used to the explanation of scenic spots with relatively fixed information content. Even if the relevant information of scenic spots is updated in the later stage, it can be reused again.

7. Conclusion

Through field investigation and interview with tourists, this study obtains the existing problems of signage system in Lushan Scenic Spots, and puts forward the design strategies of unified modeling, detachable structure design, reasonability of font size, applicability of signage board information, simple raw material, combination reduction and recoverablity, so as to provide a reference for the future planning and design of signage system in the scenic spots.

References

[1] Zhang, L. (2005) Signs Design in Metropolis. Shanghai University Press, Shanghai.
[2] Tanaka, N., Iwata, M. (1999) General Design of Signage Environment--108 Viewpoints in Planning and Design. Xueyi Press, Japan.
[3] Li, H. (2012) Outdoor Public Facilities Design and Research based on Sustainable Development. Beijing Institute of Fashion Technology, 14-15.
[4] Hu, H.H. (2007) Study on Tourism Planning Method and Practice of Lushan Scenic Spot based on Sustainable Development. Northeast Forestry University, 39-40.

IOP Conf. Series: Materials Science and Engineering **573** (2019) 012089 doi:10.1088/1757-899X/573/1/012089

[5] Standard Graphic Symbol Research Association of Ministry of Land, Japan. (2001) The Wayfinding Signage that You can understand at the First Glance--A Guide of Standard Graphical Symbols. Published by Administration Department of Transportation Consumer, Integrated Policy Bureau, Ministry of Land Publishing, Japan.
[6] Li, M. (2015) Study on the Design Method of Urban Public Environment Signage System--A Case Study of Matsuyama, Japan. New Arts, 5: 120-124.
[7] Li, M. (2014) Study on the Design of Wayfinding Signage System in Urban Public Environment. New Arts, 7:109-110.
[8] Li, M., Hu, Y. (2012) Research on the Design Method of Urban Wayfinding Signage System. New Arts, 5: 68-72.
[9] Wang, X.J. (2018) Research on the Concept and Method of Green Design in the Sign System Design. Packaging Engineering, 14: 44-48

Research on Form Design of Automotive Dashboard Based on Kansei Engineering

Hong Ren, Yupeng Tan, Ningning Zhang*

Departments of industrial design, Shenyang Aerospace University, Shenyang, Liaoning, 110136, China

*Corresponding author's e-mail: znninging@aliyun.com

Abstract. Understanding the driver's emotional needs for the dashboard is crucial to the success of vehicle design. In order to reduce the psychological burden of drivers in obtaining on-board information, the image of automobile dashboard was visually described and quantitatively analyzed with applying Kansei Engineering theory and the principle of Quantitative Class I. The predictive model between the adjectives of vehicle dashboard shape and perceptual image was established by using multiple linear regression analysis method, and the relationship between perceptual image and design elements was analyzed. The results show that the main sensory factors of shape design should be gentle and comfortable in the process of automobile dashboard design. To improve the fit between automobile dashboard design and driver's perceptual needs by transforming driver's emotional perception into design elements.

1. Introduction

Comfortable and well-defined dashboard design can improve driver's attention, driving interest, driving quality and driving safety[1]. Dr. Mitsuo Nagamach believes that Kansei Engineering is a technology that converts users' emotional appeals into design elements. The purpose of this experimental study is to explore drivers' emotional demands for the form of automobile dashboard by using the Kansei Engineering method[2]. In the experimental process, Kansei Engineering and Quantitative theoretical model are used for modeling and analysis.

2. Material and Methods

2.1 Experimental method

The sensory design of automobile dashboard is user-centered. The driver's sensory images of dashboard are deeply discussed, and sensory images are transformed into design reference elements[3]. Kansei Engineering is being used to transform customers' needs and feelings into the elements of product design elements. This study applys quantitative means of Kansei Engineering to explore the relationship between human perception and dashboard[4]. It was found that most drivers pay attention to ergonomics in the design of automotive dashboard, and at the same time, automotive dashboard with morphological design made users more satisfied[5].

2.2 Materials&Constructing Sensibility Evaluation System

2.2.1 Constructing Modeling Element Space

In this experiment, 644 series of automotive dashboard patterns of more than 90 household car brands were selected through magazines and automotive home software. The researchers used Kansei Engineering to analyze the overall shape of automotive dashboards and the internal form of dashboards[6], used KJ method to classify and analyze. In this experiment, the key elements of design need to be defined to construct the intention space of modeling, and the shape of modeling should be extracted by morphological analysis[7]. In order to extract the key factors affecting the automotive dashboard, the subjects were interviewed and investigated by oral analysis and questionnaire survey. Ten representative automotive dashboard pictures were selected as the experimental materials and processed in a unified way, as shown in Figure 1.

Figure 1. Spatial sample image of styling intention.

2.2.2 Constructing Sensual Intention Space

In the experiment, perceptual intentional adjectives were collected mainly through books, magazines, reference books, website surveys and user interviews. A total of 402 perceptual image words were collected[8]. The focus group discussed the perceptual words to test the dashboard[9], excluded words with similar meanings, identified 50 words, and invited 10 graduate students to select 10 pairs of representative perceptual image vocabularies, as shown in Table 1. According to Osgood semantic difference method, 10 pairs of representative perceptual image vocabularies and 10 pictures of automobile dashboard materials are constructed into perceptual image space[10]. According to the semantics difference scale, a 7-level questionnaire was made with Likert scale.

Table 1. Sensual image adjectives for automobile dashboards.

Number	Perceptual Image Adjectives	Number	Perceptual Image Adjectives
1	Low-Luxury	6	Calm-Dynamic
2	Gentle-Strength	7	Moderate-Individual
3	Ordinary-Classical	8	Traditional-Scientific
4	Concise-Complex	9	Simple-Stylistic
5	Safe-Dangerous	10	Comfortable-Artificial

2.2.3 Selection of experimental subjects and experimental tasks

In this experiment, the researchers selected 10 people aged 20 to 35 years old, 35 to 50 years old, 50 to 65 years old on average, and a total of 30 people were surveyed. Thirty people hold driving licenses and driving experience. In the experiment, each age level represented different consumption levels and aesthetic perceptions. According to 7-level Likert Scale, The subjects were asked to observe the shape of the dashboard on the questionnaire and scored it according to their subjective feelings. Finally, the perceptual evaluation scores of the subjects on the shape of dashboard were counted[11].

IOP Conf. Series: Materials Science and Engineering **573** (2019) 012090 doi:10.1088/1757-899X/573/1/012090

3. Results

3.1 Model Construction and Factor Analysis

A total of 30 questionnaires were taken back and 28 valid questionnaires were screened out, with a valid sample rate of 93.3%. In order to ensure the reliability with the statistical data obtained from this study, the evaluation values of perceptual images were sorted out. First of all, the data reliability analysis was carried out by SPSS software[12]. When the reliability coefficient is between 0.8 and 0.9, the scale has good reliability. The reliability coefficient of this experiment was calculated to be 0.846, which was greater than 0.800, indicating the strong reliability of the questionnaire. The KMO and Bartlett sphericity tests were used to test the validity of the questionnaire. The correlation coefficients of KMO and Bartlett tests reached the significance level of 0.635 greater than 0.600. The questionnaire results in this study showed high reliability, and the experimental data were suitable for factor analysis[13]. Using SPSS software and maximum variance rotation method, factor analysis of experimental data was carried out, and the load of the two factors was obtained. After the analysis, the data were arranged into two factors load matrix according to the factor load coefficient, as shown in Table 2.

Table 2. Each style factor evaluates the load matrix of the factor after rotation.

Rotating Component Matrix		
	Factor I	Factor II
Calm	-.289	.932
Comfortable	.648	.735
Gentle	.098	-.127
Low	.956	.220
Moderate	.233	.833
Ordinary	.952	.223
Traditional	.308	.921
Simple	.851	.185
Concise	-.133	.963
Safe	.803	.580

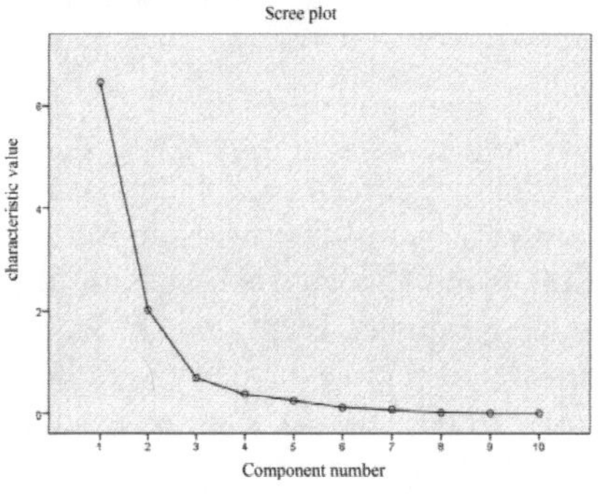

Figure 2. Characteristic scattered lithotripsy.

Through principal component analysis, the total variance of two factors was extracted to explain the cumulative proportion of total variance 83.513%. As shown in Figure 2, the former two factors eigenvalues differed greatly from the others. The former two factors had a strong generality to the automotive dashboard. 10 perceptual imagery words were divided into two categories with Combining factor load matrix and adjective meanings. And one classification was: Low-Luxury; Gentle-Strength; Ordinary-Classical; Concise-Complex; Safe-Dangerous. And the other one was Calm-Dynamic; Moderate-Individual; Traditional-Scientific; Simple-Stylistic; Comfortable-Artificial. The above results were classified and analyzed, it turned out that two kinds of factors were finally named as sensory factor and morphological factor. Sensory factor: Gentle-Strength factor; morphological factor: Comfortable-Artificial factor.

4. Discussion

4.1 Establishment of Prediction Model of Characteristic Relations of Perceptual Images

4.1.1 Parametric Characterization of Perceptual Images

In order to explore the relationship between the design elements and the overall shape of automobile dashboard, five experienced designers were invited to deconstruct and extract the shape elements from

10 sample images of automobile dashboard. The researchers extracted four design items and 16 design categories, as shown in Figure 3. Sixteen design elements were defined as independent variables, and the average perceptual evaluation of 10 pairs of perceptual image adjectives was defined as dependent variable[14]. Using SPSS 24.0 software for correlation analysis, a quantitative theory-model was used to quantify the relationship between design elements and perceptual image vocabularies in various parts of the dashboard. The deconstruction of material pictures was coded as 1, whereas the reverse was 0. In quantification theory, the sum of reactions between items and categories was 1. The load coefficients of each morphological factors were Calculated and Analyzed by Multivariate Linear Regression Analysis， as shown in Table 3.

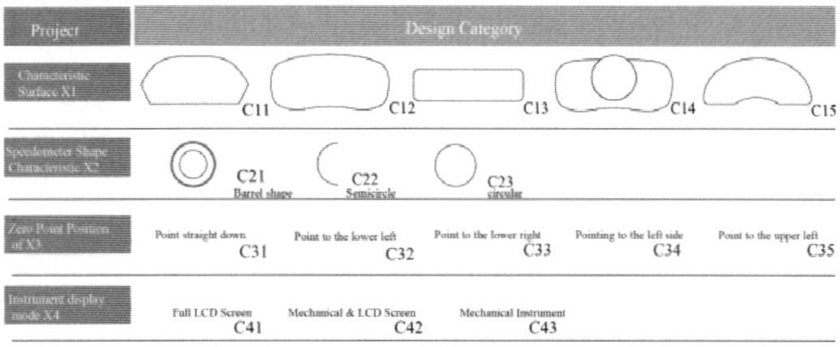

Figure 3. Classification of design elements of automotive dashboard interface.

Table 3. Load coefficients of various morphological factors

Design Project	Design Category	Comfortable-Artificial	Scoring Range	Design Project	Design Category	Gentle-Strength	Scoring Range
	C11	0.421	1.706		C11	0.316	2.321
	C12	0.350			C12	-0.325	
X1	C13	-0.415		X1	C13	-0.983	
	C14	-0.520			C14	-0.697	
	C21	-0.895	1		C21	0.288	0.596
X2	C23	-0.105		X2	C23	-0.308	
X3	C32	0.390	0.390	X3	C32	0.045	0.045
	C41	-0.415	2.275		C41	-0.808	1.091
X4	C43	-0.860		X4	C43	0.283	

4.1.2 Establish Mathematical Model

The project of automobile dashboard shape design was regarded as Item, and the design elements of automobile dashboard were regarded as Category. For the perceptual design of automobile dashboard shape, the mathematical model was established by using quantification theory I[15].

4.2 Solution of Sensory Factor Relation Model

Using SPSS multiple linear regression analysis tool and Matlab software to calculate category scores, the complex correlation coefficient R between the perceptual image of the "Gentle-Strength" interface and the evaluation vocabularies to the prediction model was 0.935, the determination coefficient R^2 was 0.875, approaching 1. The experimental results showed that the relationship model based on sensory factors could fit the real value well and meet the reliability requirements. From the prediction model, it could be seen that the highest score of morphological feature surface X4 was 2.275, the form had great influence on the design of automobile dashboard. The form of design elements C_{43} tended to be more gentle. The lowest score of X3 was 0.390, it had the least impact on the design of automotive dashboard.

4.3 Solution of Morphological Factor Relation Model

Using SPSS multiple linear regression analysis tool and Matlab software to calculate category scores, the complex correlation coefficient R between the perceptual image of the "Comfortable-Artificial" interface and the evaluation vocabulary to the prediction model was 0.968, the determination coefficient R^2 was 0.937, which was very close to 1. It showed that the overall model was more effective in predicting drivers' evaluation of the instrument panel shape factor. It could be seen from the prediction model that the highest score of morphological feature surface X1 was 2.321. And it had the greatest impact on the comfort of automobile dashboard. The design elements of shape feature surface C_{15} was more comfortable. The lowest score of X3 was 0.045. And it had the smallest impact on the comfort of the automobile dashboard interface.

5. Conclusions

In this paper, the method of Kansei Engineering is used to study the automobile dashboard. The results show that the main perceptual factors should be Gentle and Comfortable in the future design of automobile dashboard interface.

In the case of Gentle-Strength, the emotional perception of dashboard design is obvious when it has the following characteristics: the primary design element is the shape characteristic surface C_{15}. And C_{21} shape characteristic should be considered in two factors of speedometer and tachometer in the dashboard. The mechanical design should be considered as the main instrument display mode. For Comfortable-Artificial perceptual vocabulary pair, C_{13} design element makes subjects feel more comfortable and C_{23} shape is more acceptable in the speedometer and tachometer form design. When the instrument display mode is designed for full all-liquid crystal screen display, the information of the instrument can be acquired more intuitively and comfortably.

Acknowledgements

This research was financially supported by Liaoning Doctor's Initiation Fund (20170520239) and Liaoning Education Department's series of projects (L201766). I would like to express my gratitude to the teachers and co-workers for their great help in this study. I hope that the Kansei Engineering method can be used in this paper to study the form design of automobile dashboard to improve user satisfaction and usability.

References

[1] L. D. Archam, N. W. Setyanto, and A. Rahman, "Integrasi Kansei Engineering dan Struktural Equation Modeling (SEM) untuk meningkatkan kualitas produk shampo (studi kasus: Lusmas Fresh Milk Shampo), " Jurnal Rekayasa dan Manajemen Sistem Industri, vol. 1, pp. 85-96, 2013.

[2] Yang, C. C. (2011). Constructing a hybrid Kansei engineering system based on multiple affective responses: Application to product form design. Computers & Industrial Engineering, 60(4), 760–768.

[3] Mariusz Żółtowski, Michał Liss, Bogdan Żółtowski, Josef Melcer. Truss Harbor Cranes Modal Design Elements Research[J]. Polish Maritime Research, 2016, 22(4).

[4] Han, S. H., & Hong, S. W. (2003). A systematic approach for coupling user satisfaction with product design. Ergonomics, 46(13/14), 1441–1461.

[5] Han, S. H., Kim, K. J., & Yun, M. H. (2004a). Identifying mobile phone design features critical to user satisfaction. Human Factors and Ergonomics in Manufacturing, 14(1), 15–29.

[6] Mamaghani, N, K., Raihimian, E., & Mortezaei, S.-R., 2014. A succesfull statistical procedure on kansei engineering product. Iran: International Conference On Kansei Engineering And Emotion Research.

[7] A. Senthil Kumar, R. Balaji, P. Puviarasu, S. Jayakumar. Structural and morphological analysis of barium cerate electrolyte for SOFC application[J]. Materials Science-Poland, 2017, 35(1).

[8] Shen, M.S., 2004. A Study on the Combination of Multiple Kansei Words. Master thesis. National Yunlin University of Science and Technology.

[9] TANI Naho, TAKEUCHI Mariko, IKEJIRI Yoriko, MARUMO Rie, SHOJI Michiko, KUNIMOTO Chihiro, SHIRAKAWA Yuji. Factors that influence students' behavior in an

academic library: A qualitative study using focus group interviews[J]. Journal of College and University Libraries, 2016, 104(0).

[10] Chen, H.-Y., Chang, Y.-M., 2009. Extraction of product form features critical to determining consumers' perceptions of product image using a numerical definition-based systematic approach. Int. J. Ind. Ergon. 39, 133-145

[11] K.-M. Chen, Y.-P. Chou, M.-Y. Yang, Developing a system for deriving aesthetic product shapes based on shape grammar and Kansei information–the case of kettles, Int. J. Kansei Inform. 1 (1) (2010) 17–30.

[12] Fern R. Juster, Robin Camhi Baum, Christopher Zou, Don Risucci, Anhphan Ly, Harold Reiter, D. Douglas Miller, Kelly L. Dore. Addressing the Diversity-Validity Dilemma Using Situational Judgment Tests[J]. Academic Medicine, 2019.

[13] Hsiao S-W, Chen C-H. A semantic and shape grammar based ap- proach for product design[J]. Design Studies 1997, 18(3):275 - 296 .

[14] Yeh C-H, Lin Y-C. Neural network models for transforming consu- mer perception into product form design[J]. Advances in Neural Networks, 2006, 3973: 799 - 804.

[15] MISAWA Yoshitaka, TAKAHASHI Hiroshi. A Study on Peripheral Visual Stimulation Using Wide Field Angle Driving Simulator[J]. The Proceedings of the Conference on Information, Intelligence and Precision Equipment : IIP, 2017, 2017(0).

Swot analysis of sports town construction in hubei province

MaoWei

Wuhan university of science and technology

2453085739@qq.com

Abstract. This paper first shows the connotation of sports town and its significance to HuBei province,after that, the author adopts literature review and field investigation,swot analysis is carried out on the construction environment of hubei sports town,this paper points out the advantages and disadvantages, opportunities and challenges of the construction of sports town in hubei province,and according to the actual situation of hubei sports town construction,puts forward the city cluster and beautiful countryside of high planning sports characteristics of small town construction;in the sports industry, and the integration of a number of functions such as tourism;relax the investment restriction of sports town and attract the investment of social capital;with the thinking of operation to create sports town colorful sports;innovate the propaganda mode, pay special attention to the construction path that uses new media.

1. Definition of sports-themed town

At present, "characteristic town" has not been clearly defined by the academic circle, which is originated in Zhejiang and was then defined as a spatial carrier that is independent of the urban area and has definite industrial orientation, cultural connotations and tourist functions as well as certain community functions [1]. Jiangsu Administration of Sport proposed sports-themed health town and defined it as a spatial region or development platform that takes sports and health as its theme and characteristic and integrates the functions such as sports, health, tour, leisure, elderly care, culture and livability [2]. The Ministry of Housing and Urban-Rural Development, National Development and Reform Commission and Ministry of Finance jointly issued the Notice of Training of Characteristic Town in September 2016, which defined the "characteristic town" as a spatial region, national fitness and development platform or sports industry base that takes sports and leisure as the theme, has unique connotations of sports culture as well as a solid foundation of sports industry and integrates the functions such as sports and leisure, culture, health, tour, elderly care, education and training to contribute to the construction of new urbanization and healthy China as well as poverty alleviation during the process of building moderately prosperous society.

2. Significance of sports-themed town construction for Hubei

2.1 Improve town environment and construct beautiful countryside

At present, China is vigorously promoting the construction of beautiful countryside, and Hubei also pays high attention to such task, such as the Guiding Opinions on Planning and Integrating Relevant Project Funds for Pilot Construction of Beautiful and Livable Countryside issued by the General Office of Hubei Provincial Government as early as 2016 [3]. The sports-themed town industry belongs to the tertiary industry, which is favored by the public due to energy saving and environmental friendliness. Sports-themed town construction can bring about a lot of financial aids and social capital to the town infrastructure construction, and the sports-themed town has higher requirements for the environment and infrastructure, which inevitably requires that the subject of sports-themed town construction should protect the town environment to the maximum extent and improve its production and living environments as well as its public service ability [4].

2.2 Promote sports industry development and optimize provincial economic structure
The sports industry is an important part of the economic industry in Hubei. According to the 2010 statistical data, the sports and relevant industries in Hubei with prominent development achievements have basically shown the rudiment of new economic growth point, which achieve the total output value of 19.588 billion yuan, create the value added of 9.557 billion yuan and contribute 0.60% to provincial GDP. On the whole, there are a number of problems in the sports industry in Hubei. First, the sports industry due to its little output value contributes less than 1% to provincial GDP. Second, sports product manufacturing accounts for a larger proportion in the sports industry, so such irrational structure needs to be optimized. Therefore, it is of great significance for the provincial sports industry to develop sports-themed towns.

2.3 Promote Jingchu culture and province reputation
In the construction and operation of sports-themed towns, local sport and cultural resources should be rationally developed and utilized so that people can experience the charm of excellent Jingchu culture in a variety of sports activities. The Jingchu culture named after Chu State and Chu people is an important part of the Chinese national culture and occupies an important position in the development history of Chinese civilization, which is of great economic and cultural development value. Sports-themed town construction rightly provides a transmission platform and display window for the Jingchu culture, in which the historical and cultural resources that conform to the requirements of specific sports projects should be mined, further clarified and packaged for marketing and promotion, so it is a hard-won opportunity for Hubei to promote its reputation

2.4 Effectively promote supply-side structural reform and accelerate transformation of driving force
Supply-side structural reform is to adjust the economic structure so as to optimize factors allocation and improve the quality of economic growth. The demand side consists of investment, consumption and export, while the supply side is comprised of factors such as labor force, capital, institutional creation and innovation. At present, the sports industry in Hubei mainly faces the following problems in terms of supply-side structural reform, insufficient high-end and mid-to-high end supply of sports products and services, excess production capacity of low-end and mid-to-low end products and services, incomplete policies and regulations in terms of institutional supply with a view to scientific planning and guidance, and lack of top sports events and international sports products due to the insufficiency of innovation, the largest driving force of development. Hubei should seize the opportunity that China vigorously develops the sports industry and promotes sports-themed town construction to create conditions, break barriers and accelerate the flow of various production factors to the sports industry so as to actively develop the sports industry. In addition, Hubei should pay high attention to the important role of sports-themed town construction in supply-side structural reform to accelerate the transformation of driving force.

3. SWOT analysis of sports-themed town construction in Hubei

3.1 Strengths
First, sports industry development and national fitness have been raised to the level of national strategy. Second, Hubei has a strong foundation for competitive sports and mass sports, such as Wuhan International River Crossing Festival and Yichang International Dragon Boat Race, which have taken shape and produced brand effect after years of operation. Third, with relatively developed sports industry and sports education, Hubei has cultivated a number of sports enterprises with

considerable strength, such as JZX Sports with registered capital of 182,704,533 yuan and more than one thousand employees, BR Sports with registered capital of 100 million yuan and Chengchen Sports with registered capital of 13 million yuan. In terms of sports education, the sports colleges represented by Wuhan Sports University have cultivated a large number of sports professionals for the province. Finally, there are strong demands for leisure sports. Due to urban diseases such as heavily congested urban traffic, high housing prices, fast pace of life and environmental pollution, many so-called urban people have an intention to escape from the city, while sports towns closer to the urban area with beautiful and livable environment as well as leisure and community functions rightly meet such demands.

3.2 Weaknesses

Compared with developed provinces, Hubei has lower GDP and per capita GDP, which were 3,229,791,000,000 yuan and 60,198.68 yuan in 2017, when Guangdong and Jiangsu had GDP of 8,970,523,000,000 yuan and 8,590,094,000,000 yuan and per capita GDP 80,932 yuan and 107,189 yuan respectively, so the economic base is a shortcoming for Hubei to develop sports towns. The development of sports culture and the loyalty of the provincial people to sports are of great practical significance for the development and utilization of customers of sports towns in the later stage. As the soft power of sports and industrial development, sports culture is one of the important driving forces for the development of sports towns, but the development of sports culture and sports is not complete in Hubei, for there are only a few sports events every year, such as Wuhan Marathon and Wuhan Open with low popularity and incomplete services. Hubei has not formed a resultant force in the policy of sports town construction, which involves government departments of land and finance and requires their cooperation with a view to sound development, but the current situation is that Hubei Administration of Sport takes the charge, while other departments involved are of no importance, so it is difficult to achieve benign development of sports-themed towns in Hubei.

3.3 Opportunities

The central and local governments provide joint support. On the one hand, the General Administration of Sport of China has issued policy documents such as the Notice of the General Office of the General Administration of Sport of China on Promoting the Sound Development of Sports and Leisure-themed Towns, which propose definite requirements and strong support for sports-themed town construction. On the other hand, Hubei Administration of Sport has input a lot of energy and investment into sports-themed town construction. In October 2017, Hubei Administration of Sport held a training meeting in Jingmen for sports and leisure-themed town construction, when inspector Lin Xiaohua put forward specific requirements for sports-themed town construction, including promotion of start level, confirmation of subject positioning, and strengthening of organizational leadership and accountability. At present, the sports industry is deeply integrated with other industries such as tourism. Due to the consumption upgrade as per capita GDP of coastal cities and some inland developed cities exceeds 10,000 US dollars, the tourism industry has developed in the direction of deep experience and interaction. The sports industry is the most suitable target of integration and development by tourism, for sports projects provide new products and contents for tourism, which are an important industrial resource for tourism to attract and expand customers. Hubei possesses abundant tourism resources, such as world-known Wudang Mountain, which lay a solid foundation for the integration and development of sports industry and tourism.

3.4 Threats

A lot of professional talents have been cultivated under the Chinese education system, but there is a serious lack of compound talents who understand both management and sports, which seriously impacts Chinese sports-themed town construction and operation that requires not only professional talents in construction and operation but also compound talents. The research issue of sports-themed town was not started in China until 2015, while foreign sports-themed towns have developed for nearly a hundred years with suitable operation mode and mature construction experience. The existing research on sports-themed towns most deals with the construction mode and path of sports-themed towns or the implications of foreign sports-themed towns for China and has not formed a complete, systematic and scientific academic system, so in the principle that the theory guides practice, such lagging academic research has significant adverse effects on sports-themed town construction in Hubei. According to statistics, there is almost no profitable sports town in operation, but the construction requires a large amount of funds, so such predicament is a serious blow to the investment enthusiasm of social capital, which also exists in the process of sports-themed town construction in Hubei.

4. Measures and recommendations

4.1 Plan sports-themed town construction with a view to urban agglomeration and beautiful countryside

Sports-themed towns are important for beautiful countryside construction that is vigorously carried out throughout China. Sports-themed towns should integrate beautiful countryside planning into the construction process and take the idea of beautiful countryside as guidance for the development and construction of sports-themed towns. Although different from the traditional towns, sports-themed towns should also play a certain part in the urban system. On the one hand, the sports-themed town is an important base for the urbanization of rural surplus labor, and on the other hand, the sports-themed town also plays important economic and political functions in the urban system, so the planning demands of local urban agglomeration should be taken into account in the early stage of planning and construction of sports-themed towns.

4.2 Intensively cultivate the sports industry and integrate tour and other functions

As the original vitality and motivation for sports-themed town development, the sports industry should be intensively cultivated to mine its value and continuously expand its connotations in order to achieve sound and sustainable development of sports-themed towns. In addition, it has become an irresistible trend to integrate various industries for development, so sports-themed town construction should deeply integrate the sports industry with tourism, elderly care and other relevant industries to continuously expand the industrial chain and create a sound industrial ecology in which the sports industry dominates and coordinates with other industries [5].

4.3 Relax investment restrictions and attract social capital

Sports-themed town construction requires investment over ten million yuan or even hundreds of millions yuan in the early stage, which takes years to profit after in the operation stage, but due to the profit-seeking nature of capital, social capital has strict requirements for the rate of return and return period, so the competent department of sports-themed towns is required to break the barriers of investment and financing and take measures to ensure a rational rate of return of social capital to continuously attract social capital [6].

4.4 Create diverse sports projects by the operational thinking

If the sports industry is the life of the sports-themed town, then sports projects are its body. Sports projects are the most direct objects of experience and evaluation of sports-themed towns, so only with well-operated sports projects, sports-themed towns can effectively achieve their economic value as well as sustainable development, which requires the operator of sports-themed towns to take the initiative and use the operational thinking to develop sports projects [7].

4.5 Make innovations in promotion and take advantage of new media

At present, it has become an important trend in the development of the current media industry to use the Internet for promotion , so sports-themed towns that are new in China should make full use of the new media, which requires the competent department and operator of sports-themed towns to work together to make innovations in promotion and play their communication advantages in official and non-official media to promote the reputation of sports-themed towns [8].

5. Conclusion

The sports town is a good opportunity for hubei province and it has certain advantages， he article thinks hubei province has advantages in policy, talent, sports culture foundation, etc,of course, hubei province has certain challenges and disadvantages compared with coastal provinces,first of all, hubei province is economically weaker than the developed coastal provinces,Second, developed coastal cities are more open than inland ones.But hubei should seize the opportunity to rise to the challenge,we must accelerate the development of sports towns and upgrade the sports industry.

References

[1] hejiang provincial people's government. Guidance on accelerating the planning and construction of characteristic towns[Z]. 2015.
[2] angsu sports bureau. Chen gang, director of the jiangsu provincial sports bureau, describes jiangsuBlueprint of featured sports and health town [EB/OL].2016-09-30.
[3] General office of hubei provincial government's official website. The provincial people's government as a whole integration related project funds to carry out the beauty of the pilotlivableruralconstructionguidance[EB/OL].http://www.hubei.gov.cn./vfile/ezbf/201609/t20160908_1034233.shtml.
[4] Duan jin-jin, zhai ling-xin. Thoughts on the spatial production practice of characteristic small towns [J]. Journal of suzhou university (philosophy and social science edition), 2008,39(05):112-119.
[5] Journal of anhui normal university (natural science edition), 2008,41(04):405-408.
[6] Sun quansheng. Contradictions and countermeasures of urbanization in contemporary China [J]. Contemporary economic management, 2008,40(08).
[7] Su yan. Research on development strategies of characteristic towns in conghua district, guangzhou [D]. South China university of technology,2017.
[8] Zong haiyong. Value logic of spatial production and new urbanization [D]. Soochow

IOP Conf. Series: Materials Science and Engineering **573** (2019) 012092 doi:10.1088/1757-899X/573/1/012092

Design and research of somatosensory wristband based on people with language barriers

Dayong Ma, Fan Li*,

School of Design Art, Shenyang Aerospace University, Shenyang, Liaoning, 110000, China

*Corresponding author's e-mail: 381772091@qq.com

Abstract: Because of their physical defects, people with language barriers are difficult to communicate smoothly with ordinary people who are not familiar with sign language. Therefore, there is an urgent need for a tool that can realize the conversion between the sign language and normal language, so as to meet their needs of exchanging information with the outside world and freely expressing their emotions. Considering the plight faced by people with language barriers, this paper, based on the gesture recognition technology, proposes a set of solutions, which is the comprehensive application of the somatosensory wristband and mobile terminal APP.

1. Introduction

On the one hand, the needs of the audience need to be met urgently. According to the report of the World Health Organization, about 15% of the world's total population has some form of disability. Among them, 40.55% of the disabled belong to the category of language barrier. Due to physical barriers, they can only communicate in a limited and single way, such as sign language or written language supplemented by action and expression. However, physical illness indirectly increases their mental stress, making them unable to express their wishes and obtain information smoothly. To a certain extent, limiting their way to relax. The vicious circle of the past has led to the increasing vulnerability of language barriers.

On the other hand, there is a lack of emotion at the design level. In recent years, the attention of the society to the disabled has been increasing day by day. As a result, a large number of products have been created and put into use. However, most of the current designs for the disabled focus on the field of tools for the disabled and visually impaired. There are few products for people with language barriers, and they do not pay enough attention to the emotional care of the target group.

2. Material and Methods

2.1. Experimental objects

Language barrier refers to the language barrier caused by various reasons. After one year of treatment or more than two years of illness, people can not carry out normal speech communication activities, thus affecting their normal life and social life. For the congenital deaf or the unfortunate deaf before they learn the sound language, they can not imitate the sound they hear, and often can only speak sign language as the first language. The other part of the people with less language barrier, although they can use part of the voiced language to communicate, because of physiological or psychological factors, they can not communicate smoothly with others. There are also some people who have no hearing impairment, but whose vocal organs are impaired. Such as those caused by cleft palate and facial surgery. This paper will take the group of language barriers as the starting point, and put forward the design of assistant programs to meet their physiological and psychological needs.

2.2. Design position

The APP designed in this subject is mainly used in combination with the somatosensory bracelets. They together constitute a complete system of products for people with language disabilities, and they complement each other.

3. Results

3.1. Wristband appearance design

The key of product design is to give users the first impression of the emotional effect. For users, the aesthetic appearance of the somatosensory hand ring, clean color and comfortable materials directly affect their user experience.

3.1.1. Style. Cluster analysis was used to screen the words collected about the emotions of somatosensory hand rings, and determine the direction of emotional design of somatosensory hand rings for language impaired people under the guidance of several vocabulary.

Fifteen respondents were invited to classify 15 emotional words. The vocabulary that they think is the same in nature is classified into four categories. The number of words in each category is not limited, but it requires that all 15 words should be categorized without repetition or omission. The classified data are imported into SPSS software for multi-dimensional scale analysis, and the final coordinate data are obtained.

Cluster analysis was carried out with SPSS software. Fifteen words were divided into four dimensions. The results are shown in the table 1.

Choose the representative vocabulary closest to zero, and get four groups: soft, durable, unified; safe, natural, fluent; light, comfortable, friendly; modern, simple, delicate, coordinated, smooth. Among the four groups, the words with the smallest distance are: V4 soft, V7 safe, V14 light, V15 modern.

Based on the above analysis, it is concluded that the integration of soft, safe, lightweight and modern elements in the design of styling elements is more conducive to the acceptance of the product by language barriers.

Table 1. Clustering results obtained by cluster analysis.

First group	Second groups	Third groups	Fourth groups
V4（Distance 0.219）	V7（Distance 0.147）	V14（Distance 0.183）	V15（Distance 1.128）
V6（Distance 0.875）	V9（Distance 0.364）	V8（Distance 0.559）	V2（Distance 1.496）
V1（Distance 0.528）	V13（Distance 0.472）	V3（Distance 0.365）	V10（Distance 1.521）
			V11（Distance 1.587）
			V12（Distance 1.666）
			V5（Distance 1.685）

3.1.2. Color. On the one hand, color needs to meet the needs of the crowd. People's different psychological needs for products are embodied in the differences in color, weight, softness and hardness. According to the psychological needs of different characteristic groups, the emotional design of color elements of somatosensory hand rings should also be adjusted appropriately. On the other hand, color needs to consider the psychological characteristics of the crowd. Somatosensory hand rings, as an auxiliary tool for language impaired people, should fully take into account the psychological characteristics of the people who are sensitive and do not want to cause too much attention. Choose a neutral color with low profile and no publicity to give people a stable, elegant and implicit color mood and affinity.

Therefore, the design of somatosensory hand ring uses concise black and white to meet the physiological and psychological pursuit of language barriers.

Figure 1 Appearance design effect drawing of Bracelet.

3.1.3. Appearance result. The design of the somatosensory hand ring is shown in Figure 1. The front of the bracelet is a color touch screen, which facilitates user's functional interaction. It has two buttons on the side for switching machine and volume adjustment. Sensors on the back are used to detect gestures, heart rate fluctuations and so on. Then it can realize the functions of sign language translation, guitar playing, daily walking, sleep time, exercise time and so on.

3.2. APP Interface Design

3.2.1. APP Interface Flow. The design of APP is called "Sound movement", that is, "let the sound move up". By using the homonym with "vivid", I hope that the language barrier can break the psychological and physiological barriers as soon as possible and communicate vividly. Meanwhile, it also implies that voice language and gesture movements can be effectively connected through somatosensory hand rings and APP, so as to define the product target and service population. The main functions of "Sound movement" APP are divided into four modules, and its task flow chart is shown in figure2.

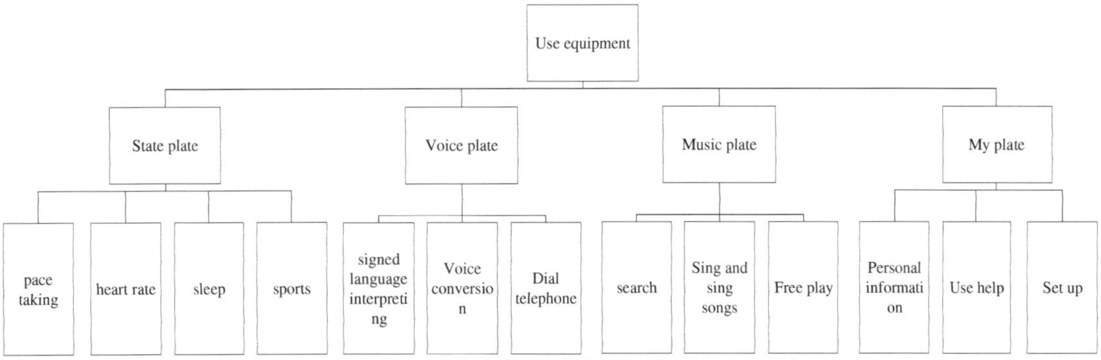

Figure 2 APP Task Flow Chart.

3.2.2. APP Interactive Design. This design uses multi-panel layout, the classification is clearer. It is convenient for people with language barriers to quickly locate the required function of the plate location, reduce the jump of redundant pages, and avoid the language barriers involving too many irrelevant pages, resulting in reduced user viscidity. The overall interaction design is shown in Figure 3.

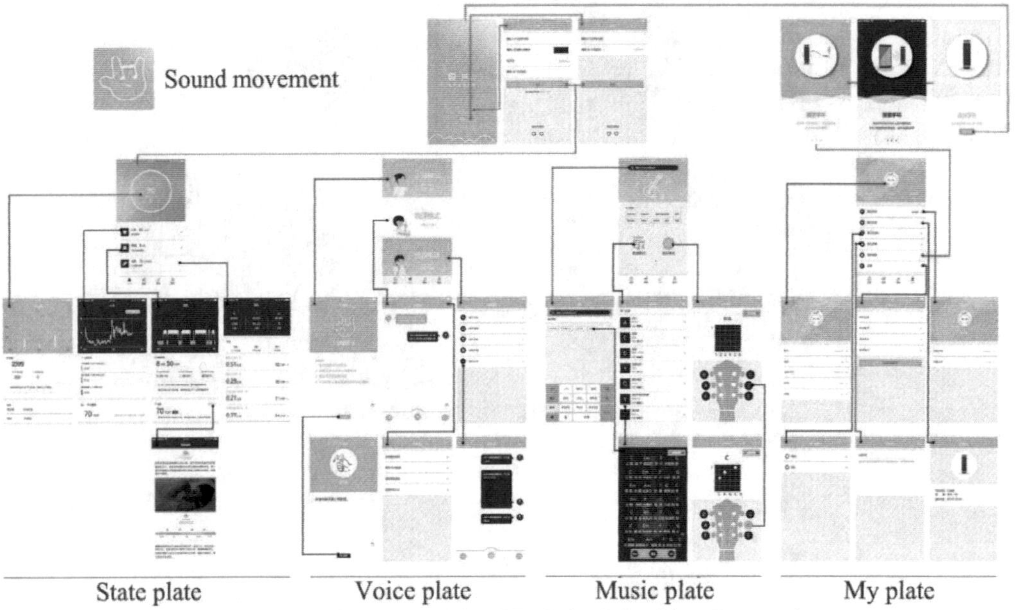

Figure 3 "Sound movement" APP Color Matching Proportion.

The first plate is the "state plate". The board monitors the users' physical health data in real time, including daily treadmill, heart rate fluctuation, sleep time, exercise time, etc. And provide corresponding health advice for users according to the data.

The voice board is the core module of APP. This section mainly provides voice assistance for users, facilitating barrier-free communication between people with language barriers and ordinary people. Among them, the handwritten translation model is suitable for the two-way communication between the language barriers familiar with sign language and ordinary people. Language handicapped people wear bracelets on both hands, make sign language gestures and recognize the gestures with somatosensory hand rings, and broadcast the translated results to ordinary people through APP to achieve the purpose of communication. Transliteration model is suitable for two-way communication between hearing and speech impaired people and ordinary people. Ordinary people use APP to translate speech into text or to present text directly by hand to hearing and speech impaired people. Hearing and speech impaired people use APP to respond by hand or translate their sign language into voice broadcasting to achieve two-way communication. Substitute telephone is suitable for people with language disabilities to seek help in emergencies. The content can be transmitted to a third-party sign language translator by means of text or sign language video through APP, in order to achieve the purpose of fast transmission of information. At the same time, APP can adjust male and female voices, language types and so on through voice settings, which is more suitable for different users' preferences.

"Music Plate" is an important module of APP. This section mainly provides users with music experience. It can realize the air performance of guitar, make the language handicapped perform rehabilitation exercises in the process of playing and singing, and obtain good entertainment emotional experience. Among them, the song mode is suitable for beginners without foundation. Users can choose songs to play and sing from the music library according to their personal preferences. Wear a bracelet to make a sound with a single wave. The accompaniment is played from APP to the rhythm of sweeping the strings with the wave. Freedom mode is suitable for users with guitar foundation or preference for freedom and free personality. This mode has many chords for users to play freely. When using the left hand to press the chord screen chord button, the right hand to wield the sweep chord can truly simulate the guitar sound. While enriching the user's entertainment needs, it can also help users to carry out oral rehabilitation training by singing loudly. At present, APP has been built in several popular inspirational songs, such as Chengdu, the brightest star in the night sky, and The Tracker. Help to alleviate psychological problems such as overpressure, loneliness and inferiority complex, anxiety and fragility.

The fourth plate is the "my plate". This section is the user's personal center, which can change the information and software settings, including personal information, my goals, my equipment, third party access, feedback, use help and settings module. To facilitate users to use according to preference settings.

4. Conclusions

This paper focuses on the production of somatosensory hand rings and APP. The function of positioning supporting products is to help language barriers and normal people communicate with each other in terms of language and emotion. Based on user research to obtain the needs of language barriers, and combined with emotional design to lay the user experience, the design positioning of somatosensory hand ring and APP is carried out. And the conclusions drawn from the study are applied to the practical development and design, and finally a set of auxiliary supplies for language barriers with hardware and software is completed. This product is of great practical significance to improve the communication barriers of language barriers. It should be a sponsor to help language barriers build self-confidence, gain respect and praise.

Acknowledgements

This research was financially supported by Educational Cooperation and Education Project of the Ministry of Education (Grant No. 201802321017). The project name is Research on the Training Model of Innovative and Entrepreneurial Talents in Animation Major.

References

[1] Chen, G. (2016) Intelligent Wearing: the Next Wind Port in the Internet of Things Era. Chemical Industry Press, Wuhan.
[2] Xu, W. (2016) Wearable Equipment: Mobile Intelligent Life. Tsinghua university press, Beijing.
[3] Xu, B.S. (2014) Zhejiang Deaf Natural Sign Language. Zhejiang University of Industry and Commerce Press, Hangzhou.
[4] Gao, J.S. (2017) UI Design Compulsory Course Game+ Software+ Website+ APP Interface Design Course. Publishing House of Electronics Industry, Beijing.
[5] Zhao, Y.C. (2012) China Disabled Persons Federation releases the latest data on the disabled population in China. Research on Disabled Persons, a01: 11.
[6] Koller, O. (2015) Deep Learning of Mouth Shapes for Sign Language. In: Workshop on Assistive Computer Vision & Robotics. Naples. pp. 165-178.

IOP Conf. Series: Materials Science and Engineering **573** (2019) 012093 doi:10.1088/1757-899X/573/1/012093

Based on Creative Thinking to Museum Lighting Design Influences to Visitors Emotional Response Levels Theory Research

Zhisheng Wang[1 2],Yukari Nagai[1*], Dan Zhu[2],Jiahui Liu[2],Nianyu Zou[2]

[1]Knowledge Science, Japan Advanced Institute of Science and Technology

Nomi, Japan

[2]Research Institute of Photonics, Dalian Polytechnic University Dalian, China

[*]Corresponding author's e-mail: ynagai@ jaist.ac.jp

Abstract. This paper expounds the synthetic design process of artificial lighting, and emphasizes that the key of museum lighting design is based on the visual factors of visitors. Combining theories of architecture, design, optics and psychology, this paper studies the emotional response level of museum visitors through the interdisciplinary integration and creative thinking. At the same time, creative thinking is applied to put forward the theoretical model of "SVOE" lighting design based on visual factors. As a function of the evolution of spatialization, visualization, optical indexing, and emotionalization, the theoretical model of lighting design innovates on the thought processes and evaluation of lighting design. Based on research related to LED light source lighting design methods,it is an urgent problem to combine functional reasonable lighting design with human emotion.This paper discusses the design and practice of emotional lighting by using the theory and research system of psychology and the design method of emotion analysis based on spatial, visual, optical engineering and psychology.

1. Introduction

Creative thinking is a way of thinking which, through the process of understanding, promotes the generation of new ideas and perspectives through analysis, synthesis, comparison, abstraction,and reasonable imagination [1].The formation of new concepts in traditional lighting design methods is the key to innovative lighting design. The environmental lighting of architectural spaces has been called the fourth dimensional space of architecture. However, one of the key reasons for this is the invention of artificial light sources. With the development of the LED light source, its excellent characteristics of volume, color temperature, brightness, color rendering and controllability make it the most ideal light source carrier. Thus, as lighting technology is rapidly developing, lighting design concepts and methods should also be improved and promoted.

Creative artificial lighting design with interdisciplinary integration takes human visual factors as the core element. The effects of rods and cones on visual factors in dark and light conditions are studied through emotional responses to the environment, including comfort assessment, visibility assessment, warmth assessment and preference assessment. Based on the conceptual optimization of emotional response level in theory, this paper proposes the definition and theme of the whole standard artificial lighting design result, as shown in Figure1.

IOP Conf. Series: Materials Science and Engineering **573** (2019) 012093 doi:10.1088/1757-899X/573/1/012093

2. Interdisciplinary integration of creative thinking

Kruith of proposed an interior lighting design method to achieve the "pleasure effect"related color temperature (CCT).The conclusion of the study on the optimization of light source in commercial psychology experiment is that the most suitable natural light source produces more color gamut with more uniform appearance and symmetry than the preference proportion [2].A visual phenomenon in which light stimuli of different wavelengths act on visual organs in two brightness ranges.When the brightness of light stimulation is above 3cd/m^2, the vision acquired by human cone cells is called light vision or cone cell sense.In the case of dark adaptation, the vision mainly acquired by the stem cells is called dark vision or stem cell vision.

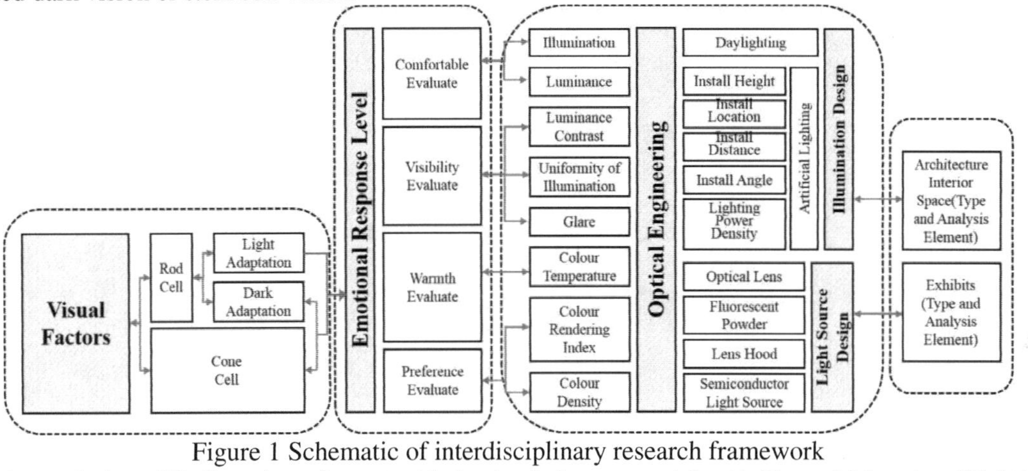

Figure 1 Schematic of interdisciplinary research framework

Appreciation of feelings is an important behavior in human social activities, which act on "lighting design" and involve the whole process of behavior. The viewer's feelings are divided into three basic stages: First, the quantity of feelings. People are looking for exhibits that they can view easily and clearly. Second, the stage of quality must be considered. People have begun to demand a certain quality of light in the environment space. In these two stages, the needs of the viewers remain at the most "basic"level, with the aim of obtaining more material-level performance in the exhibits. The third stage is that of emotional feeling.

3. Spatial research on museum lighting design

This research focuses on innovative theoretical research related to museum lighting design, which generally considers four elements:spatialization,visualization,optical engineering, and emotionalization."Spatialization"research in innovative lighting design focuses on the physical analysis and categorization of space and exhibits.

3.1 Spatialization

At present, there are different perspectives related to the relationship between Lighting Design and the Spatial Environment, and between the understanding of architecture and interior design[3].The museum building belongs is classified as a public building. The functional division and traffic flow layout in the building space are complex. The museum building space is divided into display and auxiliary spaces, based on the function of the display. The exhibition space is divided into permanent exhibition space and temporary exhibition space.The auxiliary space is the area that connects the display space, and generally includes the entrance hall, the preface hall, and the corridor space. The lighting design of these auxiliary spaces has a close correlation with the lighting environment of the exhibition space. The visual lighting design of the auxiliary space can affect the light environment of the exhibition space. Faced with such diversity within the building space, the light environment requirements of the building space are also high, and each space must be provided with a light design that meets the specific environmental requirements of that space.

The exhibition space design needs to fully consider the characteristics of the exhibits, making it more stereoscopic and comprehensive to shape and highlight the main body of the exhibition. In the museum exhibition space, we pay great attention to the spatial scale. The scale of different exhibition space will bring different visual and psychological feelings to the audience. The suitable space can increase the comfort of the exhibition, and vice versa. The specific display space structure in the exhibit area is shown in Figure 2.

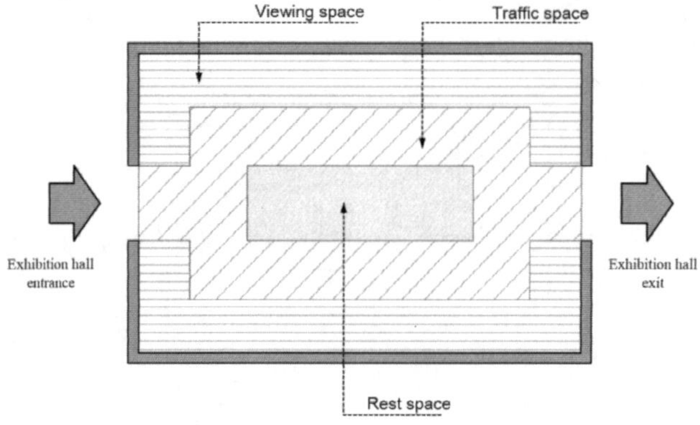

Figure 2 Typical display space structure.

3.2 Visualization and optical engineering

Consider the lighting design of visual factors from the perspective of the layout and design flow of the building space.Vision is divided into bright vision, intermediate vision and dark vision[4].The time required for the various visual transformations is different. The layout of the space structure is planned and a suitable light environment is created through the transformation of the bright, dark and middle vision, to allow the visitors to feel more comfortable. Vision is the process of observing things through the eyes,Visualization is the recognition and understanding of human physiological indicators[5]. In terms of illumination,vision is the sensation of light passing through the rod and cone cells in the retina and are the most critical components.This information then passes through the optical nerve to the visual center and the brain for final image processing[6].The color in the light is very important for evaluation of quality lighting. The evaluation of light in optical engineering primarily depends on illuminance, color temperature, color rendering index, and color tolerance.

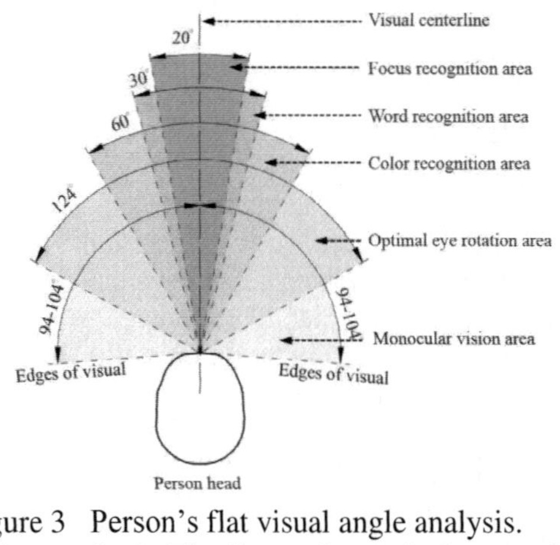

Figure 3 Person's flat visual angle analysis.

Figure 3 is a person's plane view analysis. The figure shows the human visual centreline as the axis. The distance from the centre is 10°which is the focus recognition area, 15°is the text recognition area, 30°is the colour recognition area, 62°is the eye's optimal range of rotation, and 94°~104°is the visual range of a single eye.

IOP Conf. Series: Materials Science and Engineering **573** (2019) 012093 doi:10.1088/1757-899X/573/1/012093

3.3 Emotional

The impact of the visual sense on psychological feelings has been confirmed in other spaces. For light in classrooms, hospitals, and offices, it has been proven that light environments can change people's emotions and have an important impact on learning, work, and health[7]. The museum is a public place for visitors to obtain cultural information. There are many types of exhibits in this space. Therefore, logically, this space needs to be adjusted to meet the psychological needs of human beings using the technical parameters provided by optical engineering.

The three different dimensions of "visceral, behavioral, and reflective" are intertwined in any design. It is impossible for one of the dimensions to be absent[8]. They appropriately define the three existing forms of emotion in lighting and highlight human emotions[8].After introducing the ideology of "human emotion based" into the field of emotional lighting, the "SVOE"model of innovative lighting design is perfected, and it becomes apparent that the feelings and emotions of the viewer can be influenced by a change in the lighting. The essence of lighting design is to solve the relationship between lighting, space, and people, to enable people to experience the most suitable lighting in a given space, and to appropriately address the functional needs of people. The most "Visceral" element in lighting is the human visual function; the lighting design should, therefore, also be designed to meet the most visceral needs of mankind, which is the pursuit of light.as shown in Figure 4.

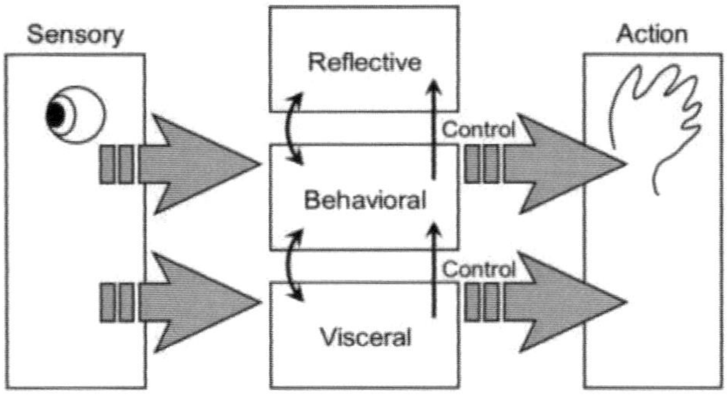

Figure 4 Visceral, behavioral, and reflective analysis

4. Museum research

In this experiment, lighting design and tourists' feelings in different functional areas of Liaoning museum were investigated by means of lighting index detection and questionnaire.

4.1 Illuminance test

The illuminance test uses the central distribution method in which the illuminance measurement area is divided into grids. The grids must be square and the illuminance must be measured at the center of each grid.

$$Eav = \frac{1}{M \bullet N} \sum Ei \qquad (1)$$

Eav is the average illuminance value, the unit is Lux (Lx), Ei is the illuminance in the Iobby,M is the longitudinal measuring point, and N is the transverse measuring point (Equation 1)

Figure 5 Illuminance distribution and real scene map of lobby illuminance test

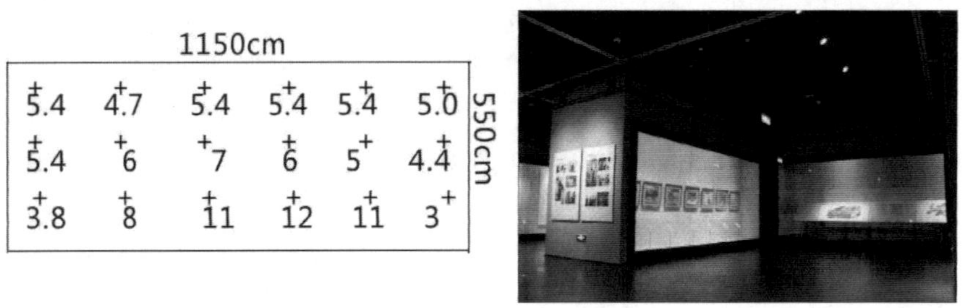

Figure 6 Local ground illuminance distribution and real scene of exhibition hall 2

By substituting the data measured in the lobby into the formula: $e = 1 / (27) \sum Ei$, the horizontal average illuminance is 139.30 Lx and the horizontal illuminance uniformity is 0.933. Hall 2 has an illuminance of 101.89 Lx, an illuminance uniformity of 0.491, as shown in Figure 5and6

4.2 Measurement of color temperature and color rendering index

From the perspective of the museum's simple and omitted function, there are three basic functions with higher requirements for the quality of light and color of lighting. One is the exhibition function, which requires lighting to achieve a higher definition of exhibits. This requirement is related to illumination, fidelity, and color gamut of lighting. Secondly, the preservation and maintenance function of cultural relics or artwork, which requires the lowest light-induced damage of illumination to the target object[9].The third is to study the educational function. Besides the fidelity index, color gamut, and relevant index mentioned above, the spectral characteristic parameters that affect the apparent color quality of LED lighting in the museum primarily include illuminance and color temperature.

A spectral radiometer was used to measure the color temperature and color rendering index of the site. The number of measuring points in each site does not fall below 9, and the arithmetic average value was used for the color temperature and color rendering index of the tested lighting site.

4.3 Sensory test questionnaire

A subjective evaluation was carried out by inviting an audience to scan the QR code on site and to complete a questionnaire, convenient and efficient for statistics. A total of 106 sets of data were collected for more than 20 people. the factors affecting the lighting of the museum's environment, such as the degree of realism of display color, the degree of light source color preference, the display detail expression, the three-dimensional expression, the display texture definition, the display outer outline definition, the display brightness acceptance, the visual adaptability, the psychological pleasure,

IOP Conf. Series: Materials Science and Engineering **573** (2019) 012093 doi:10.1088/1757-899X/573/1/012093

and the degree of light artistic preference, were selected, and through it the test questionnaire data were processed .

Through the museum lighting experiment designed by us, we found that: the artistic preference of basic display light is highly evaluated, which needs to be improved in visual adaptability and appeal.Temporary display subjective visual comfort is high, to improve the color preference of light source;The light art of the hall is highly appraised, and the appeal needs to be improved.Corridor subjective visual comfort is better, there is still room for improvement, but the appeal of the degree of preference is general, there is more room for improvement.

5. Innovative Theoretical model

The research and conclusions of the above museums provide a good foundation for us to establish theoretical models. The interdisciplinary and creative thinking merges to create a bridge between architectural space, light source and human psychology.Traditional lighting design only considers the basic lighting requirements of spaces and exhibits and has not considered the influence of spatial visualization conditions and emotional feelings. By using the lighting design theory, "SVOE" mode, to influence the viewer's mood when visually perceiving a space, a solution related to the relationship between people, space, and light will be proposed.

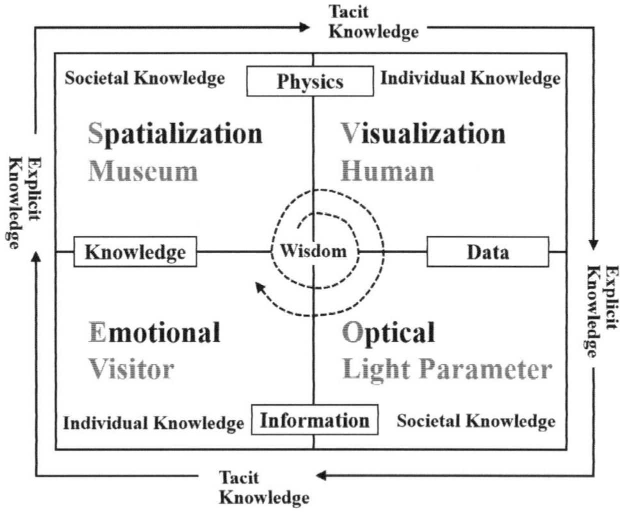

Figure 7 "SVOE" mode lighting design analysis chart

Two kinds of human knowledge, i.e. explicit knowledge and tacit knowledge, are used in this research..Explicit knowledge, also known as direct knowledge, refers to knowledge that can be clearly expressed, and is usually described as knowledge that can be expressed in written words, diagrams, and mathematical formulas[10]. In "SVOE" mode, space and optics are part of the body of explicit knowledge. Knowledge that one cannot clearly express is called tacit knowledge. Visual perception and psychological experience are part of Tacit Knowledge.

6. Conclusion

This paper summarizes a lighting design method based on creative thinking. Through field investigation of Liaoning Museum, the emotional response of museum visitors under different lighting conditions is studied, and a theoretical lighting design model which combines architectural physiology with human emotional factors is innovated.Lighting design is essentially a creative activity. Spatialization, visualization, and optical elements, which emotionally affect the characteristics of creation are the main characteristics of lighting design. Lighting design that is driven by creative thinking and innovative techniques is not limited to traditional functional design frameworks. Considering the elements of emotional lighting design, the use of the creative thinking method reflects the human-centered innovation lighting design method.This research suggests that the design of the light environment can have a significant influence on the inner feelings of human beings. The internal relationship between the elements exhibited by a comfortable light environment corresponds to human emotional patterns. The research related to innovative lighting design methods, and the theory of spatialization, visualization and optical engineering are used to establish new theories and standards of emotional lighting design; through the use of the theory of explicit and tacit knowledge, they are also used to solve the design problem of the light environment, and most importantly, to influence the emotions of visitors.

References

[1] Guo,J., Su,Q and Zhang,Q. (2011) Individual Creativity during the Ideation Phase of Product Innovation: An Interactional Perspective . John Wiley & Sons Ltd Volume ,26: 31-48.
[2] Wei,M., Houser.KW., David.A., Krames.MR .(2016) Effect of gamut shape on colorpreference . Proceedings of CIE2016 Melbourne Australia ,5:32-41.
[3] AJ., Wilkins., DPhil. (2015) A physiological basis for visual discomfort: Application in lighting design . Lighting Res. Technol ,48:22-30.
[4] W.J.M.,van Bommel., GJ van den Beld .(2004) Lighting for work: a review of visual and biological effects. Lighting Res. Technol , 36: 22-31.
[5] Liu,Z., Chen,H. (2014) Intelligent and Visualization Design of University Classroom Lighting Energy-saving Control System. Intelligent Computer and Applications , 6:64–68.
[6] Timothy Onosahwo ., Iyendo.(2014) Enhancing the Hospital Healing Environment through Art and Day-lighting for User's Therapeutic Process. International Journal of Arts and Commerce , No. 9:101–119.
[7] CIE.(2004) Control of Damage to Museum Object by Optical Radiation . CIE Pobication Vienna ,157:35-42
[8] Quang Trinh Vinh., Peter Bodrogi and Tran Quoc Khanh.(2014) Effects of dynamic lighting on office workers：First results of a field study with monthly alternating settings. Lighting Res. Technol , 42:345-360.
[9] Maher., Laura J. Mrs.(2017) Aging-In-Place Home Modification: LED Lamp Color Temperature Preference Among Adults . Electronic Theses and Dissertations, 5:107-110.
[10] Zhai,Q,Y., Luo,M., Liu,X. (2015) The impact of LED Lighting parameters on viewing fine art paintings.Lighting, Research &Technology,48: 23-31.

IOP Conf. Series: Materials Science and Engineering **573** (2019) 012094 doi:10.1088/1757-899X/573/1/012094

Modern Product Image Design from the Perspective of Corporate Culture

Guidan Yao

Huaqiao University, Quanzhou, Fujian, 362021, China

*Corresponding author's e-mail: 352628083@qq.com

Abstract. Corporate culture is playing an increasingly important role in today's society and has drawn attention from all walks of life. From the perspective of corporate culture, the article analyzes the relationship between corporate culture and modern product image design, and discusses the relevant content of corporate culture into modern product image design principles. The purpose is to reflect the value of corporate culture by studying the mutual integration of corporate culture and modern product image design, so that the product has its own characteristics, advanced nature, and the times, exploring the breakthrough point of modern product image design, promoting the design of modern product image in China, bringing new opportunities and challenges for the design of modern product image in China, and bringing different experiences to the public.

1. Introduction

With the rapid development of society, corporate culture has become more and more concerned in the process of the whole society, which has positive significance for deepening the construction of enterprises. Corporate culture needs the material carrier to spread, and modern product image design is its carrier. The two are closely related, and the connotation of the corporate culture is introduced into the modern product image design, which brings a good development space for the modern product image design. At the same time, in the context of corporate culture, optimize the image design of modern products, make products innovative, enhance the added value of products, and strengthen the recognition of social users.

2. Overview of corporate culture

The so-called corporate culture is the sum of of the values, business principles, management style, enterprise spirit, ethics and development goals of an enterprise with its own characteristics, which is gradually formed and recognized by all employees in the practice of socialist market economy [1]. Corporate culture promotes the cohesiveness of corporate members, feels the spiritual outlook of the enterprise, represents the core concept of the enterprise, and is an essential component to guide the advancement of the enterprise. Highlighting its style characteristics, and the accumulation of time can establish a good public image of the enterprise, has an important impact on the development of the enterprise, and stimulate unlimited potential. It will sever to shape the brand of the enterprise, deepen the road of enterprise construction, enable the development of enterprises to achieve a qualitative leap, and enhance the internal power of the enterprise.

IOP Conf. Series: Materials Science and Engineering **573** (2019) 012094 doi:10.1088/1757-899X/573/1/012094

3. Image communication of corporate culture

With the development of today's society and the pattern of cultural pluralism, corporate culture has played a vital role in the development process of enterprises, bringing an invisible impetus to the spiritual strength of enterprises, stimulating the vitality of enterprises, strengthening the image of enterprises, and thus promoting the sustainable development ability of enterprises. Corporate culture is the compass of corporate action, which is in an irreplaceable position in the development of enterprises and can guide the construction of corporate culture. Therefore, it requires a long-term accumulation and exploration process accumulation, so as to produce a consensus value system and build a corporate characteristics culture.

From the late 1970s to the early 1980s, the corporate world paid more and more attention to the exploration of corporate culture. In the 1970s, the rise of the Japanese economy stemmed from its commitment to the construction of corporate culture and the promotion of the exploitation of its corporate products. Corporate culture refers to the sum of values and behavioral standards with their own characteristics that have been summarized and refined by enterprises. To endow enterprises with a new cultural connotation, so that enterprises can have their own advantages in such a social environment, promote the effective integration of corporate culture and product image design, give play to the value of product design, better implement product concepts, and achieve good operation and healthy development of enterprises.

4. The relationship between corporate culture and modern product image design

4.1. Corporate culture is the spiritual core of modern product image design

The corporate culture endows the modern product image with profound connotation. Relied on the corporate culture to make the modern product image have unity and richness, and can convey certain emotions and information to the users. When users see the product, they can associate with the corporate culture and become a guide for the enterprise to design the product, so that it can be clearly reflected in the product design, which has a positive meaning and makes the product recognized by the user. Resonate with users, embody the charm of corporate culture and provide added value for product design. As a value orientation, corporate culture brings favorable support to the image of modern products.

For example, Panasonic's corporate culture is the Seven Spirits of Panasonic, which has a decisive role in promoting the image and publicity of products, so that Panasonic's products can be understood and supported by users. (Figure 1) Panasonic's product, like the eye steam massager, its product itself is integrated into Panasonic's corporate culture. The product moisturizes the eyes through steam, which not only relieves the body and mind, alleviates eye fatigue, but also has the effect of beauty. It also helps to improve the people with dark circles and can be recycled. The product has shaped the image of protecting the user's body and mind, all of which reflects the corporate culture of Panasonic. It is the artistic purpose of Panasonic's Seven Spirits that highlights the value of the product image. This is an opportunity to make users feel the intimacy and practicality of the products and reflect the happiness of Panasonic Electrical Appliances.

Figure 1. Panasonic eye steam massager

Figure 2. Amadana upright double oven

4.2. Modern product image design is an important carrier of corporate culture

Corporate culture is an invisible force that requires material carriers to achieve its tangibility, while its carrier is the image of modern product, that is, exploiting the market by virtue of the advantages of product design and enable the corporate culture to be effectively disseminated. In the process of using products, consumers are deeply impressed by the fact that products represent the beliefs and attitudes of enterprises, thus enhancing the recognition of corporate culture and making product design a booster of corporate culture. It transcends the product itself, but realizes the deep-seated significance of the corporate culture through the product as a bridge to communicate with the user. If the product design is sought after by the user, the value of the corporate culture is promoted accordingly. When the user experiences material products, they can get spiritual enjoyment through the thoughts and emotions carried by the material carrier. Excellent products can touch the user and enhance happiness, which shows the success of their corporate culture.

For example, amadana, a Japanese household appliances company, whose corporate culture is to make beautiful home appliances. Therefore, amadana presents a product image with a sense of beauty and function. Amadana's oven (figure 2) has a strong sense of design. The whole design pays attention to the design and performance, which is pleasing to the eye. The product image is plain and textured and the double-layer design shows the unique design symbols of the enterprise. The door handle of the oven is made of leather material, which shows the intimate of amadana's design. It protects the hands when it is taken, enhances the user's memory of the product, and creates a product image that is deeply rooted in the hearts of the people.

4.3. Corporate culture and modern product image design complement each other

Corporate culture and modern product image are inseparable and mutually inclusive. Modern product image performance and the inherent spiritual nature of corporate culture, both of which are you have me, I have you. Good corporate culture guides modern product image design, thus enriching modern product image, displaying high-quality products and better interpreting corporate culture. The two are the relationship between content and form. The modern product image permeates the content of corporate culture, and the corporate culture is expressed through modern product image. The user's first impression of the product depends on the image. As an effective platform for dissemination and display, the modern product image can deepen the user's understanding of the product, and combine the charm of its corporate culture to enhance the customer's goodwill and trust. It is conducive to obtaining a stable consumer group. Therefore, we should pay attention to the relationship between the two and summarize the corporate culture. Refining, summarizing, and integrating the corporate culture into modern product image design.

Japan's Craft Design Technology (CDT) stationery company, its corporate culture has been reflected in the company's name, that is, the three principles of craft, design and technical culture. Japan's traditional craftsmanship combines modern technology to serve modern design, delivering corporate culture messages to customers and pursuing superior quality assurance. (Figure 3) CDT's classic scissors are manufactured by hand, each of which is from the hands of the craftsman. The tough product image highlights the craft texture of the product. The details of the design, that is, the pointed design, is conducive to the accuracy of tailoring. At the same time, taking into account the way of holding, it has made a certain inspection. Considering the usual habits and other issues, the user is comfortable and convenient to use. The entire process of product design embodies the three principles of the corporate culture. The three principles guide the direction of this product image design. It cannot only be based on tradition, but also has a strong sense of modernity. The whole is integrated with the shape of the product image, mutual dependence and common development.

IOP Conf. Series: Materials Science and Engineering **573** (2019) 012094 doi:10.1088/1757-899X/573/1/012094

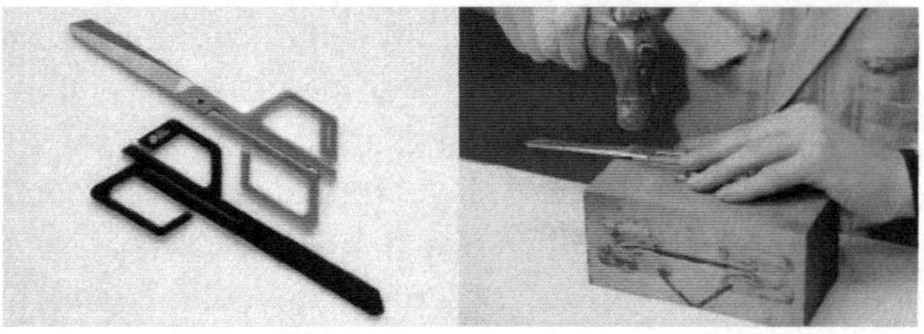

Figure 3. CDT classic scissors

5. Principles of integrating corporate culture into modern product image design

5.1. Elements of shaping modern product image

Corporate culture is a distinctive feature of an enterprise, which is distinct from other enterprises. Therefore, the design of modern product image follows the characteristics of corporate culture. From this point of departure, it has regularity and methods to follow, which is distinct from other products on the market. Avoid the tendency of product customization and sameness. With ingenuity, novelty and creativity, the image of modern products will not appear the same phenomenon. Corporate culture can promote the image of modern products to break through itself, make the whole modern product image vivid and flexible, create the product's own style, attract the attention of the users, surprise users and stimulate users' interest in the product. It is this personalized refinement, to a certain extent, enhances the taste of the product, gradually forms a unique product temperament, highlights the unique side of the product, and has an innovative spirit.

The Italian enterprise ALESSI is a typical example. It is a global household leading brand, subverting the tradition, creating the product image with its own personality according to its distinctive corporate culture, and making significant differences with other household products on the market. With the creativity, aesthetics and crafts as the criteria of corporate culture, the product image has artistic temperament and fun. From the ordinary hand workshops, after a hundred years of historical accumulation, it has gradually become an aesthetic creative workshop. (Figure 4) Alessi's mobile phone accessories are a conceptual face phone with unique style. The product is inspired by fish and fishing boats. It conveys the concept of fishing. It is different from some cold product images. It has a warmth, both practical and artistic beauty, evoking the user's emotional experience, designed to make the user unforgettable. Alessi's image of each product is closely related to the theme of corporate culture, reflecting the aesthetics of Italian life and the exploration of craftsmanship. The product image is full of new ideas, which enable users to understand its corporate culture characteristics and highlight the appeal of product image design.

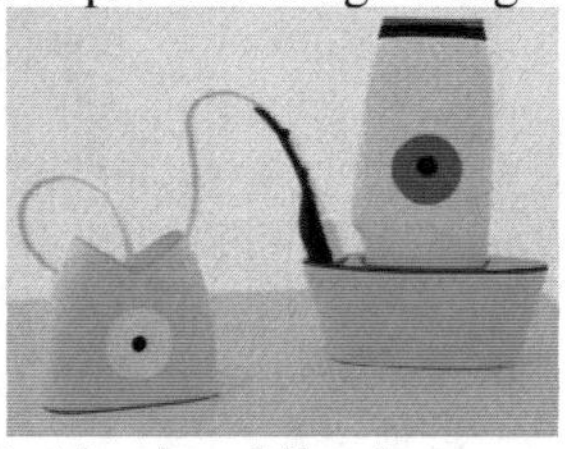

Figure 4. Alessi mobile phone accessories

5.2. Optimize the modern product image system

Corporate culture can deepen the image of modern products and help to optimize the modern product image system. When we design the product image, it will be more or less influenced by the corporate culture, which provides a support for the design of modern product image. Based on the corporate culture, we will deeply cultivate the image of modern products on the basis of purpose and direction. Innovative thinking brings design inspiration to designers, meets the psychological needs of users, and promotes the transformation of corporate culture into consumer products through the medium of products. The image of modern product is also the extension of corporate culture. The social effects produced by products are the feedback to corporate culture. In today's fierce market competition, the innovation of modern product image can effectively enhance the product brand. Users cannot only have a visual experience for the modern product image system, but also perceive the, so as to achieve interactive communication between products and people. Corporate culture adds luster to the image of modern products, which makes the image of modern products have a strong connotation, helping products to stand firm in a pluralistic society, maintaining innovative functions and cultivating sensitivity to product creation.

For instance, the ELECOM company's product note sound (Figure 5), its corporate culture is to strive for progress, challenge oneself, enthusiasm, and achieve personal growth of the enterprise and employees. This bluetooth speaker is not only cleverly designed in the form of modeling language and notes, which makes people feel pleasant, but also includes the product image system to join the near field communication pairing. It also has high-performance output power, even though it is small in size. When using, the touch is comfortable. It has passed the environmental certification, and the system of the product is perfected. Based on ELECOM's corporate culture, the product image system is optimized in many aspects. It is the cultural concept of challenging ourselves and making progress. It stimulates the potential of the enterprise in developing products, and continues to create the value of the product, continuously improving the content of each system.

Figure 5. ELECOM note sound

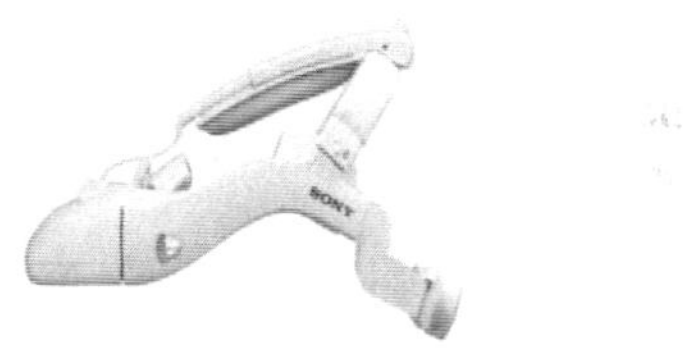

Figure 6. Sony head-mounted display system

5.3. Grasp the direction of modern product image

A good modern product image should focus on the consideration of the entire design direction. Corporate culture caters to the innovation of modern product image and is conducive to grasping the direction. Guided by corporate culture, it promotes the evolution of modern product image,develops products on this basis, makes modern product image design more planned, promotes the sharing and integration of design resources, and cultivates a perfect modern product image system, thus improving product design efficiency, and driving the development of the enterprise. Exploring the innovative ideas of modern product image, grasping the direction, making the positioning of modern product image clearer, deepening its concept, providing a broad platform for products, product design results will be more mature, so that products can be favored by the market.

(Figure 6) Sony's design and development of the product head-mounted graphics processing unit, which is the corporate culture that attaches importance to science and technology and continuous innovation. In this direction, Sony's numerous product images are full of corporate culture. This product uses a new technology to innovate, applied in medical treatment, provides doctors with endoscopic images with high definition. It has changed the traditional design, that is, the way of the external display. The internal view facilitates the doctor's surgery, and gives the doctor its fluency during the long-term surgery. At the same time, this innovative technology can also alleviate the patient's pressure. It is the corporate culture of Sony that enables an enterprise to have a clear direction in designing the image of new products, demonstrating that the products are both efficient and intelligent.

6. Conclusion

As an intangible asset, corporate culture and modern product image are interlinked and interconnected, giving the product a profound connotation. From the developed products, the semantics conveyed by them can be obtained, that is, the value orientation of corporate culture, thus enhancing the value and significance of products. It effectively promotes the innovation of modern product image, promotes the establishment of modern product image, has a positive impact, injects new vitality into it, and then creates a distinctive product. It strengthens and perfects the image of modern products, as well as enhances the user's recognition of products and exerts the potential of modern product image.

Acknowledgment
Subsidized Project for Postgraduates'Innovative Fund in Scientific Research of Huaqiao University.

References

[1] Lan, T. (2012) Corporate Brand and Culture. Huazhong University of Science and Technology Press, Wuhan.
[2] Brenneran. (2012) Vital design. China Renmin University Press, Beijing.
[3] Chen, G. (2017) Product Image Design (2nd Edition).Publishing House of Electronics Industry, Beijing.
[4] Wang, Y.Q., Liu, Z.J. (2015) Product Image Design Strategy Based on User Experience.China Packaging Industry, 6: 44-44.
[5] Lou, M., Xiao, W.Q. (2017) Research on image design strategy of mechanical products based on integration thought.Packaging Engineering, 2: 115-119.
[6] Hu, Z.G., Li N., Qiao X.L., Guo, X.Y. (2011) Research on the strategy of image shaping of industrial products.Manufacturing Automation, 21: 141-143.
[7] Zhao, H.L. (2010) Building Corporate Culture to Promote Sustainable Development of Enterprises.Economic Research Guide, 31: 27-28.
[8] Kan, C.W., Chow，C.Y.T., Chamlong, S. (2017) A Study of Brand Green Image in Relation to Eco-Branding of Fashion Products. Applied Mechanics and Materials,866:448-451.

IOP Conf. Series: Materials Science and Engineering **573** (2019) 012095 doi:10.1088/1757-899X/573/1/012095

On the Concepts of Artificial Intelligence and Innovative Design in Product Design

WANGLi, ZHANGHai-chao, WANGQiang

Dalian University of Science and Technology, China

Dalian University of Science and Technology, China

Dalian Jiaotong University, China

ABSTRACT: With the continuous development and wide application of artificial intelligence technology, new opportunities are brought for innovative product design. Artificial intelligence improves the efficiency of product design, so that besides satisfying people's daily needs, the humanization, intellectualization and experience Interaction of artificial intelligence technology inject unique sensory experience into products to meet people's spiritual needs. This paper explains the concept of AI, points out the application direction and development trend of AI in the field of design, provides more comprehensive theoretical and practical reference for product design, and promotes the construction and development of intelligent life.

1. A Survey of Artificial Intelligence

1.1. Artificial Intelligence

As a new science and technology, artificial intelligence technology mainly develops and studies the theoretical methods and technical means of simulating and expanding human intelligence. Artificial intelligence technology, nano science and technology, and genetic engineering technology are known as the three most advanced technologies in the 21st century.

The development of AI technology can be divided into four stages: gestation stage, formation stage, knowledge application stage and comprehensive integration stage. At present, AI technology in China is in the weak AI stage. That is to say, AI technology in China does not have the ability of independent reasoning, self-thinking and independent problem solving. At present, AI technology in our country does not have self-consciousness. Only when researchers develop appropriate programming language, AI will have the ability of self-thinking, self-perception and self-action.

1.2. History of Artificial Intelligence

In short, artificial intelligence is to make machine equipments have the same wisdom and ability as human beings through artificial methods. At present, because scientists still have doubts about intelligence, there is no precise definition of artificial intelligence in the scientific community. The concept of AI was first proposed by Turing, the father of computer science, and then by McCarthy, the father of AI, in 1956, with the efforts of several generations of scientists. However, due to the immaturity of theory and technology, the research of AI stagnated in the 1970s and 1980s, until the maturity of information technology and theory, AI entered a period of rapid development. Nowadays, more and more AI products come into people's lives, and at the same time, they also inject interest into life.

1.3. development status of artificial intelligence

With China's AI technology attaches great importance to and support, sound policy advantages and development environment to promote the development of China's ai technology. Especially since the 21st century, the rapid development of Internet information technology has provided solid data resource support for China's artificial intelligence technology and laid a stable foundation for the development of China's artificial intelligence technology and industrial chain.

The rapid development of artificial intelligence technology brings opportunities for the development of product design, also faces severe challenges. First, the AI talent reserve is seriously insufficient. Compared with developed countries, China's AI technology talent training system is not sound. Secondly, there is information island in the construction of data platform. Generally, the development of artificial intelligence technology cannot be separated from the support of massive data information. Finally, with the continuous development of artificial intelligence technology, social problems highlighted by artificial intelligence in some fields are increasingly aggravated, which will lead to unemployment in some industries.

2. Overview of innovative design

2.1. concept of innovative design

Innovative design is a systematic transformation process in which information is represented, perceived, concrete, abstract and cognitive in the process of ascending and descending dimensions of 11 dimensions. It is a combination of innovative ideas and design practice. In the design of science, technology, culture, art, society and economy, give full play to creative thinking, design and develop new products with novelty, creativity and practicality.

Product design follows the principle of people-oriented, user experience as the core, to meet the basic needs of users, pay attention to the mining of new functions of products, new USES, at the same time using new materials, new methods, new technology, reduce product costs, improve product quality and market competitiveness.

2.2. influence of artificial intelligence on innovative design

The information and communication technology represented by artificial intelligence not only changes people's life, but also changes people's feeling of getting the real things in the real world. In other words, the value system formed by the application of artificial intelligence technology is sympathetic to the group user experience and individual sensibility, giving birth to new aesthetic concepts and aesthetic consciousness. Under the background of artificial intelligence, it has changed the traditional business model. On the one hand, the service system of mass production in the industrial era has gradually changed to the direction of product and service customization, and from the group user experience to individual sensibility. On the other hand, individual sensibility, based on massive data information of artificial intelligence, gradually forms a collaborative symbiosis between group intelligence and sense of experience, and product design develops towards the direction of high perception, high efficiency, high value, high speed and high appearance level.

In order to make the results of artificial intelligence more consistent with human aesthetic consciousness, it is urgent to use human aesthetic consciousness to carry out innovative design, so as to meet the two basic aesthetic systems of appreciation and identification and value creation.

3. Application of artificial intelligence and innovative design in product design

3.1. change productivity and create new production tools

From the initial imitation of human intelligence to the independent formation of independent thinking mode, artificial intelligence has changed from passive supervision and compulsory to unsupervised active thinking and learning, subvert the traditional intelligent productivity, the winning relationship between artificial intelligence and production creator, and can fully express creative ideas in product

IOP Conf. Series: Materials Science and Engineering **573** (2019) 012095 doi:10.1088/1757-899X/573/1/012095

design. The way, expression, presentation and content of art are all embodied by artificial intelligence. Artificial intelligence provides natural conditions for the creation of artistic content and aesthetic forms, as well as new perspectives and approaches for innovative design and aesthetic consciousness. How to use aesthetic consciousness, choose aesthetic dimension and judge aesthetic value in artificial intelligence environment has become the key to the application of artificial intelligence in product design.

3.2. integrate massive information data to analyze public intentions

In the proposal and research stage of product design, artificial intelligence technology is applied to analyze people's intentions through big data, so as to determine the trend of future products and business development direction. For example, WeChat and weibo can provide official reference and referential data. What is the hot spot in the future, changes in user groups, and even what happens in the future will affect product design. Therefore, a sentence of big data and mathematical model can help us develop and make more appropriate decisions.

In the process of product design, the sketch stage is the most focused stage, and the charm of artificial intelligence is mainly shown here. The most creative link is given to the designer to release the designer's productivity, so as to focus on product creative design. By analyzing every sketch, accumulating image learning data, and automatically optimizing the sketch, whether it is a paper sketch or a two-dimensional rendering, artificial intelligence will greatly improve the work efficiency of product design.

With the progress of science and technology, 3d modeling software can be fully applied in product design to transform the ideas in the mind into the created and built models in real space, and then the precise digital models can be generated by the computer to further release the productivity of the creative personnel.

In the product design, the artificial intelligence will automatically check the model, save the model, model everywhere, and form a seamless connection with the renderer. With the development and application of rapid prototyping technology, the virtual product design in one second will become the real product in the next second.

3.3. real-time optimization to achieve optimal product design

At present, artificial intelligence technology in the field of product design emphasizes fast iteration. Without too much courage from the creators, it can conduct real-time review and optimization, closely combine digital model, effect drawing, packaging and production, and automatically generate variable set and animation video.

In the process of product design, production, artificial intelligence will carry out model inspection, surface analysis, assembly inspection, budget analysis and so on. At present, the field of product design in China has entered the preliminary stage of intelligent product design. We do not have to worry about the impact of creative modification on the downstream. Before production or even in the production process, artificial intelligence means and methods can be used to optimize the design scheme in real time and finally achieve the optimal product design.

Image editing and intelligent typesetting are adopted in product design and packaging stage. For graphic design, in the future, artificial intelligence will better reduce the work intensity of designers, and some skilled and repetitive work will be completed by artificial intelligence machine, while design can focus on the thinking innovation of "design" itself.

4. To summarize

Artificial intelligence is the inevitable result of scientific progress and the development of human society. It is also the inevitable trend to adapt to the fierce market competition. The extensive application of artificial intelligence changes human life, improves work efficiency and creates more value wealth. Product design with the advantages of artificial intelligence automation, not only release the designer's productivity, but also inspire creativity, improve design efficiency, make the product

IOP Conf. Series: Materials Science and Engineering **573** (2019) 012095 doi:10.1088/1757-899X/573/1/012095

more humane.

References:

[1] xu zhilei. On the concept of intelligent system and innovative design. Decoration

[2] zheng liuyang. Application and development trend of artificial intelligence in product design [J]. Furniture and interior decoration

[3] Yang peng. Development and transformation of design thinking under the background of artificial intelligence [J]. Jiangsu science and technology information

[4] Chen qingjun. Future design, walking with artificial intelligence [J]. Art observation

[5] li di, huang xuetong, xie hengyuan. Analysis of product design in the era of artificial intelligence [J]. Tomorrow fashion

[6] yu tao. On the application field of artificial intelligence and its future development prospects [J]. Science and technology innovation and application, 2017, (8) : 83

[7] shi yun. Discussion on the application and development prospect of artificial intelligence [J]. Electronics technology and software engineering, 2014, (16) : 271-272

IOP Conf. Series: Materials Science and Engineering **573** (2019) 012096 doi:10.1088/1757-899X/573/1/012096

The Teaching of Art Design Theory and the Cultivation of Students'Innovative Ability in Applied Undergraduate Colleges

ZHANGHai-chao, WANGLi, WANGQiang

Dalian University of Science and Technology, China

Dalian University of Science and Technology, China

Dalian Jiaotong University, China

Absrtact: Design is a professional course that keeps pace with the times. The teaching requires to master the market foresight and integrate the enterprise demand information and enterprise concept into the teaching. Therefore, we attach great importance to the cultivation of students'innovative ability. In the course of art design specialty, the knowledge of art design theory, as an important content, plays an important role in training students'ability of design innovation. Firstly, this paper introduces the teaching of art theory, points out its role in the process of cultivating students'innovative ability. Secondly, it puts forward corresponding countermeasures for the teaching of art design theory and the cultivation of students' innovative ability in Applied Undergraduate colleges, so as to improve students'innovative ability of design according to students' personality characteristics, so as to cultivate Art talents that meet the needs of the times.

1. Overview of Art Design Theory Teaching

1.1. Teaching Contents of Art Design Theory

An excellent design artist should not only have skilled art design skills, but also have solid knowledge of art design theory. Therefore, the teaching of art design theory can not be separated from the teaching of art design theory to continuously enrich and improve personal artistic accomplishment. The teaching content of art design theory mainly covers: design outline, design aesthetics, design thinking, design psychology research, design art history theory, design culture and so on. It plays a guiding role in cultivating students'innovative application ability in Applied Undergraduate colleges.

1.2. The Role of Art Design Theory Teaching

Firstly, the study of art design theory course helps to exercise the ability of designers to discover, analyze and solve problems; secondly, it helps to improve the ability of designers to appreciate art and criticize and correct art, to shape and improve the comprehensive design art accomplishment of designers, and ultimately to enhance the core competitiveness of art talents in the market, and to better adapt to market demand. Employment. Finally, the teaching of art design theory is the premise and foundation of cultivating and improving the creative ability of art design. It saves time for artists to explore theory in the innovation of art design, provides successful experience for artistic design innovation, and provides mature knowledge of art theory for artists'design. Therefore, Applied Undergraduate Colleges and universities cultivate students' creative ability of art. We should innovate teaching methods, constantly cultivate students'theoretical thinking ability and enhance the ability of artistic innovation and application.

2. The Current Situation of Art Theory Teaching in Applied Undergraduate Colleges and Universities

2.1. Emphasis on technique rather than theory, and undergraduate teaching tends to develop in Higher Vocational Colleges and universities.

The teaching of art theory in Applied Undergraduate Colleges pays attention to the integrity and systematicness of knowledge, and emphasizes the cultivation of students'understanding of theoretical knowledge and the way of thinking to solve problems. The mastery and promotion of theoretical knowledge follow the principle of gradual progress. The intuition and clarity of art theory knowledge teaching are different from those of art practice ability. In addition, in recent years, higher education attaches more and more importance to the training of applied talents. Art courses pay more attention to the cultivation of students'practical ability than basic theory courses.

In order to promote students'better employment, schools, parents and students themselves will focus on practical ability as a major, and even the practical ability as a standard of assessment. This talent training mode, which aims at cultivating students'practical ability and employment, leads to the setting of specialized courses in practical teaching in Application-oriented Undergraduate colleges, which attaches too much importance to practical application, neglects the importance of theoretical teaching, and tends to be higher vocational and tertiary.

2.2. The teaching method is single and the theory teaching is not unified with the professional practice teaching.

In the art teaching of Applied Undergraduate Colleges and universities, the teaching method of art theory course is single, and the combination of theoretical knowledge and professional practice content in teaching is poor, which can not arouse the students'initiative and seriously affect the teaching effect. Generally, theoretical knowledge teaching emphasizes the systematicness and theory of knowledge, but art majors generally have the characteristics of well-developed thinking in images and poor logical thinking. Therefore, abstract and systematic theoretical knowledge should be displayed in concrete images. The dull theoretical knowledge in teaching should be supported by vivid, interesting and fresh cases, which is also raised to art theory teaching. At present, the teaching of art theory in Applied Undergraduate Colleges and universities mostly adopts the methods of teaching and inculcating, mostly explaining the knowledge points of textbooks, lacking some case analysis and thematic discussion. At the same time, in the course design, the content of art theory course has less connection with other professional courses, which leads to the closeness and isolation of art theory knowledge, thus unable to ensure that students apply theoretical knowledge to the actual creation of art.

2.3. Uniformity of Assessment Method

The monotony of teaching methods directly leads to the singularity of the assessment methods of artistic theory. Examinations become the ultimate goal of teaching. Teachers focus on explaining the knowledge points of textbooks in class, and take the content of textbooks as the basic basis for assessing students, while students pass the examination by rote memorization of knowledge points. This way of assessment makes art theory teaching mere formality. Classes, examinations and credits become the contents of teachers'classes, which leads to the deviation between teaching and learning. Students and teachers do not communicate in the classroom, but regard the teaching of theory courses as a task. Students lose the motivation of learning, so they can not effectively cultivate and improve the ability of independent innovation.

3. Cultivation of Students'Innovative Ability in the Teaching of Art Theory in Applied Undergraduate Colleges and Universities

3.1. Change old teaching concepts and reform teaching contents

On the one hand, schools and teachers should attach importance to the important role of art design theory teaching in cultivating students'innovative ability, renovate the old teaching concepts of unilaterally attaching importance to skill teaching while neglecting theory teaching, optimize the curriculum structure, and strengthen art design theory teaching from the aspects of curriculum construction and teacher allocation. According to the current students'knowledge level and ability, we should adjust the content system of art theory course, integrate knowledge content, avoid repetition of teaching, strengthen the corresponding training of applied skills, give full play to the guiding significance of theoretical knowledge to practical application, understand and digest theoretical knowledge in creative practice, and form a complete system in the teaching process to help students learn art comprehensively. The theoretical knowledge of technical design will gradually improve students'logical thinking ability and innovative ability.

On the other hand, we should innovate teaching contents and cultivate students'research learning habits. Based on the weak cultural foundation of art students, the content of art design textbooks in applied undergraduate colleges can be divided into two parts: compulsory and self-study. The compulsory content is the core theoretical knowledge with high value, and it is equipped with corresponding application cases to select part of the content of design history. Teachers need not explain each knowledge point clearly, but guide students to explore and research after class. 。 The optional content can introduce more designers of different schools and works of art of different styles, guide students to study and research independently, not only lighten teachers'teaching burden, improve students' learning efficiency, but also cultivate students'research-based learning habits.

3.2. Continuous improvement of teaching methods

The reform and innovation of the teaching method of art theory course in applied undergraduate colleges should follow the principle of combining teacher's leading role with student's main body. Students are the receivers and processors of theoretical knowledge. In teaching, teachers adopt heuristic, discussion and salon teaching methods, and integrate traditional teaching methods of imparting knowledge into divergent and creative thinking. The combination of discovery method and case method will stimulate students'initiative, change passive learning into active absorption and digestion, constantly accept and learn new knowledge, and constantly improve their knowledge structure.

In teaching, teachers make comprehensive use of teaching methods such as lecture method, graphic method, discussion method and case demonstration method to increase the teaching effect, maintain interactive communication with students, and timely grasp the learning situation of each student. For example, the use of discussion method in theory teaching can help students express their views, carry out discussions and exchanges on a certain issue, and cultivate students'exploratory spirit. In the course of explaining a certain theoretical knowledge, teachers should not only explain it through picture cases, but also connect it with the reality, stimulate students'interest in learning, enrich the skills and experience of design creation, and better understand the relevant theoretical knowledge through design training, so as to facilitate the later application in design, and then enhance the creative ability of art design.

3.3. Innovating teaching methods and reforming assessment methods

Art theory courses make full use of the advantages of Internet information technology, innovate and optimize teaching methods, introduce multimedia, teaching software and network technology into the teaching process of art theory courses, stimulate students through vivid audio-visual effects such as pictures and videos, change dull theoretical knowledge into an image, create a relaxed and interesting classroom atmosphere, and then enhance students. Interest in learning and promoting interaction between teachers and students.

Reform the traditional single assessment method, pay attention to comprehensive students in class, homework, final examination and other semester performance, to scientifically and accurately examine the students'learning effect, and then improve the students' initiative in daily teaching.

4. summary

Under the background of talent cultivation mode in Applied Undergraduate Colleges and universities, the teaching of art theory course in Colleges and universities should pay attention to cultivating students'practical application ability. Through adjusting and optimizing the professional teaching content, teaching methods, teaching means and assessment methods, this paper constructs a theoretical course teaching mode suitable for the needs of market development, and Cultivates Innovative Applied talents.

Reference:

[1] Liu Chunyan. Teaching of Art Design Theory and Cultivation of Students'Innovative Ability in Applied Undergraduate Colleges. Educational Science (Citation Edition). 2017, (10): 18
[2] Li Wen, Li Jie. Discussion on the training mode of outstanding talents in art design major in Applied Undergraduate Colleges and universities. Journal of Hubei University of Economics (Humanities and Social Sciences Edition). 2015, (9): 145
[3] Sun Daosheng. Research on the Implementation Path of Excellent Talents Training in Local Applied Universities [J]. China Higher Education, 2014, (4): 25-29
[4] Chen Mo. Domestic and international research review on the cultivation of College Students'innovative ability [J]. Heilongjiang Higher Education Research. 2015, (1)
[5] Ren Hongxia, Hua Shiwang. On the Basis of Teaching Design of Applied Undergraduate Art Design Specialty [J]. Art Science and Technology. 2017, (10): 107

IOP Conf. Series: Materials Science and Engineering **573** (2019) 012097 doi:10.1088/1757-899X/573/1/012097

Countermeasure Research on the Transformation and Upgrading of the Garment Industry in Jilin Province under the National Policy of "Cultural Creativity and Design Service"

Yingying Ge[1], Tiantian Yang[2]

[1,2]School of art and design, Changchun university of technology, Changchun, Jilin, 130012, China

[2]Y_tiantian1234@163.com

Abstract: There are some outstanding problems in the development of the garment industry in Jilin Province, which are mainly reflected in the following aspects: the imperfect garment industry chain and the lacking support of pre-order garment industry; the homogenization competition and vicious competition of garment and the thin awareness of brand; the less attention to the garment industry from the government departments and the weak influence of industry associations; the rising labor costs and serious brain drain. Based on the above problems, the corresponding strategies are proposed: to play the role of the government and build a service-oriented government; to strengthen the subjective consciousness of the enterprise and accelerate the industrial agglomeration; to play the role of the garment association commerce and strengthen the service function; to attach importance to the talents cultivation and promote the integration of production, education and research.

1. Introduction

The concept of the creative industry was first published in the UK's 1998 Creative Industry Path Report. Since the 1990s, China has begun to attach importance to and develop creative industry of culture. As an emerging industry, creative industry is characterized by high knowledge, high value-added and high integration, which plays an increasingly important role in promoting industrial transformation and upgrade and driving China's economic development. As a traditional industry, the garment industry has shown a trend of serious inventory, increasing homogenization competition, and declining profits year after year in recent years. Currently, the cultural creativity and design services integrative development with the integration of related industries have become an important trend in the industrial development of China, accelerating the transformation and upgrade from the garment industry to the creative industry has become a strategic choice for the garment industry development.

2. Status quo of garment industry development in Jilin Province

Although the garment industry in Jilin Province has made great progress after the development more than 20 years, we should clearly understand that it is still relatively backward. The specific performance is that there are less than 50 medium-sized garment enterprises in Jilin Province, mostly are small and micro-garment enterprises; most enterprises are developing slowly, relying on the original equipment manufacturer(OEM) for simple processing production and tooling production mode, which means simple and extensive expansion focusing on processing and production; there are few famous brands in Jilin Province, there is 0 of Jilin Province in the list of "Top 100 Enterprises in the Garment Industry in 2017" released by the China Garment Association in July 2018, indicating that the influence of clothing brands in Jilin Province is minimal; the regional structure of the garment industry is unreasonable while the garment wholesale industry and the garment retail industry are less relevant, and the resource allocation is not sufficient.

According to relevant statistics, the total population of Jilin Province was 26.158 million, and the natural population growth rate was 0.26‰ at the end of 2017, indicating that Jilin Province has a huge consumer market and potential, which is contradictory with the current development situation of unbalanced and inadequate for the garment industry in the province.

3. Analysis of the countermeasures for the transformation and upgrade of the garment industry in Jilin Province

3.1 Strengthening the subjective consciousness of enterprises and accelerating industrial agglomeration

Enterprises are the main body of transformation and upgrade of the garment industry. The development and growth of individual enterprises is the foundation of industrial agglomeration.

3.1.1 Transforming the concept of business management and implementing differentiated competition in brands or products. Most of the garment enterprises in Jilin Province, subjected to the management level of enterprise operator and a large amount of manpower, material resources and financial resources needed by transformation, have to survive on low processing fees. Business managers should actively change the business concept, establish a sense of crisis, and make accurate position according to the actual situation of the enterprise to avoid the vicious low-price competition caused by product homogenization. Large and medium-sized garment enterprises, which have strong financial strength, certain design and research capabilities, and high-quality product quality, have gradually separated from the low-end manufacturing links of the global value chain, so it can be considered to increase their investment in research and development design and brand promotion to speed up the development of the clothing creative economy and the pace of restructuring the enterprise. As for the processing and manufacturing links with lower added value, medium and large-scale garment enterprises can implement the "slicing" out of the value chain, that is, shifting the processing and manufacturing links of garments to underdeveloped areas with relatively low labor costs. As shown in Figure 1,the small and micro garment enterprises could set up their own brand later but invest their limited capital and energy into the improvement of production environment and the updating of machinery and equipment. It improves the quality and production efficiency of the products, which can better undertake the industrial transfer of large and medium-sized garment enterprises in the future, and lay a solid foundation for the high-end link in the global value chain.

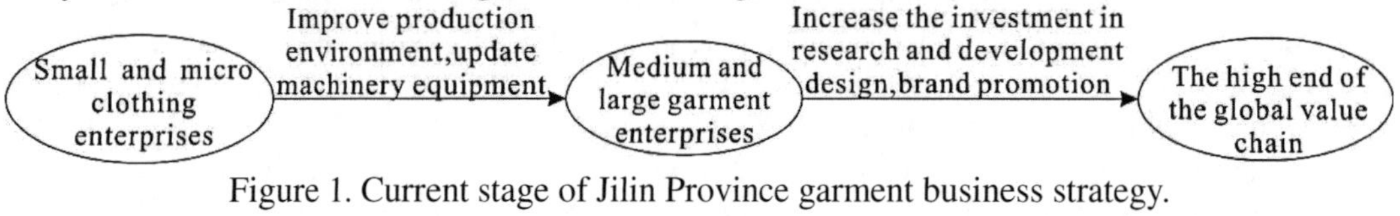

Figure 1. Current stage of Jilin Province garment business strategy.

3.1.2 Focusing on the cultivation of enterprise innovation ability. The continuous technology, development and innovation are important factors of the structural adjustment of enterprises in the process of enterprise transformation and upgrade. According to investigations, many garment enterprises in Jilin Province do not pay much attention to research and development design and technological innovation. Therefore, garment enterprises in Jilin Province should increase their capital and manpower investment in technology, development innovation and intelligent manufacturing equipment, increase excavation and application of local culture, enhance the overall competitiveness and creativity of the garment industry in Jilin Province.

As far as the research results are concerned, Jilin Province garment enterprises still have huge room for improvement in terms of management capabilities, especially in information management. Jilin Province garment enterprises should learn from the successful experience of top garment enterprises in management, such as applying information technology to procurement, design, logistics, production, sales and other aspects to comprehensively enhance the core competitiveness of enterprises: introducing OA system to achieve financial and daily administrative office management; introducing ERP system to achieve the management of funds, logistics, information; establishing a rapid response mechanism from the market to development for the characteristics of consumption diversification, differentiation, as shown in Figure 2; further considering the establishment of integrate platform coordinating design and manufacturing, that is, integrating the design platform, the networking technology with management through the Internet to realize the sharing of information which includes design, production and management.

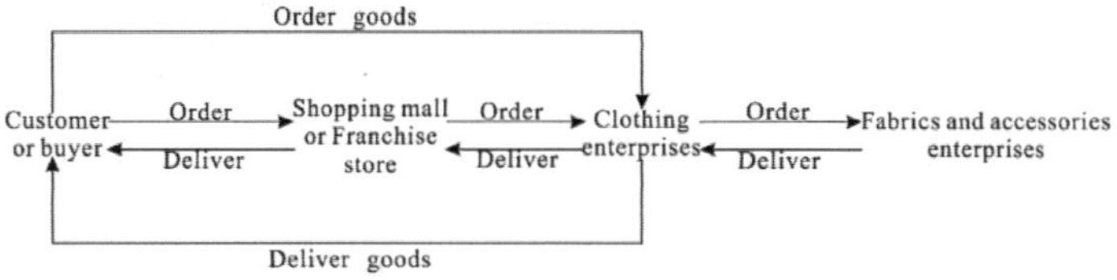

Figure 2. Fast response mechanism of garment enterprises.

3.1.3 Industry Cluster. Industrial cluster is essentially a kind of bio-aggregation phenomenon, a concrete manifestation of human population biology in the development of the garment industry. Jilin Province garment enterprises could spontaneously carry out division of labor and cooperation according to the vertical industrial chain of clothing to improve the overall competitiveness of the cluster, but this solution is difficult to achieve. Therefore, the Jilin Provincial Government and the Garment Association should actively guide the integration and reorganization of garment enterprises. The government should establish a garment industrial park, and provide enterprises with land, water and other aspects of relief and preferential treatment; formulate corresponding policies and measures to attract foreign capital and enterprises to settle in to further improve the garment industry chain; build and optimize the corresponding supporting facilities in the industrial park, such as the establishment of the Industrial and Commercial Bureau Office, the garment industry association office, etc., to provide relevant information, consulting, training and other services for enterprises. The specialized division of industry chain and the improvement of industrial specialization supporting services will promote the adjustment of regional industrial structure, technical structure and product structure, thus promoting the local industrial upgrading.

3.2 Playing the role of the government and build a service-oriented government

The main body of transformation and upgrade of the garment industry is the enterprises, while the key of the success for transformation and upgrade is the government.

3.2.1 Strengthening the guidance, leverage, and punishment of policies. The Jilin Provincial Government has formulated the Guiding Opinions of the Jilin Provincial People's Government on Accelerating the Industrial Transformation and Upgrading (Jilin provincial government issued [2014] No. 41) but has not formulated corresponding policies of transformation and upgrade for the garment industry. The government should consider incorporating the garment industry into the key industrial directory of Jilin Province to improve the efficiency of the transformation and upgrade, and have a clear objective in the process of transformation and upgrading. The implementation plan for the transformation and upgrading of the garment industry in line with the provincial conditions should be formulated, drawing on the developed provinces of garment industry such as Zhejiang and Guangdong and basing on systematic investigation of Jilin Province garment industry.

The garment creative industry is a fusion product of the garment industry and the cultural and creative industry, where the extensive and congregate knowledge is represented by garment as a carrier, which should be protected by relevant laws. Therefore, the relevant government departments should gradually improve the legislation and strictly enforce the law to ensure the development and legitimate rights and interests of the garment creative enterprises. In order to prevent the outflow of talents, the government should establish a special talent pool and introduce a series of policies to attract talents and promote the healthy flow of talents, like that in Shanghai, Jiangsu, Zhejiang and other provinces. For those zombie enterprises relying on loans to survive, the relevant government departments should increase penalties and clamp down, in order to optimize the structure of garment production capacity and stimulate the innovation of other garment enterprises.

3.2.2 Optimizing the urban humanistic environment to gather talents. In addition to the preferential measures of the government and enterprises, the urban humanistic environment is also particularly important to attract or retain talents. According to statistics, 80% of garment graduates choose to go out because of the poor humanistic environment in Jilin Province, such as the imperfect city's urban infrastructure, the underdeveloped service industry, the inconvenient information dissemination, and the outdated concept. Therefore, the Jilin Provincial Government should speed up the development of urban public transportation, information technology, medical, health care, cultural education, social security, and etc., to improve the construction of the "catering-shopping-entertainment" service system. It should develop or incubate a number of garment industrial parks facilitating the exchange of talents and the collision of inspiration, so that the comprehensive service level of the city will be improved. At the same time, the government should vigorously publicize the regional culture of Jilin Province, because the label of diverse culture will attract garment talents to explore the essence and truth of it, then integrate it into the development of the garment creative industry.

3.2.3 Innovating corporate financing forms. The transformation and upgrade of the industry requires sufficient financial support, including the structural adjustment of the garment industry. According to relevant information, the Jilin Provincial Government has introduced various forms of financing for enterprises, such as promoting the issuance of corporate bonds, middle and small-sized enterprises private placement bonds, collective bonds and collective notes. Although the government has implemented many forms of financing, the problem of corporate financing still exists. Therefore, the Jilin Provincial Government should actively explore new forms of financing based on local conditions, such as innovating the transfer of shares in unlisted companies.

3.3 Playing the bridge role of Garment Association and Chamber of Commerce and strengthening the service function

At present, the Jilin Province Garment Industry Association has shortcomings such as few member enterprises, insufficient credibility, weak industry binding force, and weak service capabilities. So it is difficult to undertake the transformation and upgrade of the garment industry. The Jilin Provincial Government should comply with the requirements of the development of the times, vigorously support the development of the garment association, and formulate the corresponding policies to ensure the work of the Garment Association and Chamber of Commerce. At the same time, as the Garment Association and Chamber of Commerce has a better understanding of the actual development of the industry, it can be tried to change the government functions, decentralize the management authority, and entrust the Garment Association and Chamber of Commerce to undertake administrative examination and approval, such as the annual inspection of garment enterprises and the approval of garment related items.

Furthermore, Jilin Province should combine with the actual situation in the province and complying with the tendency, which should integrate and reorganize the associations of business and overlapping members either, and further refine the responsibility of garment industry associations and chamber of commerce. In the meantime, formulate the development plan of the association, release and publicize relevant information, and provide related services such as training, technical exchange, event hosting and experience exchange according to the actual research and relevant policy recommendations; combine the hot issues such as the transformation and upgrading of the garment industry and the "cultural creativity and design services" policy, which can strengthen the exchanges and cooperation with other countries or regions to expand the visibility of the garment industry in Jilin Province.

3.4 Attach importance to the cultivation of talents and promote the development of integration of production, education and research

The clothing talent training in colleges and universities in Jilin Province is greatly influenced by the region, and there are problems of emphasis on theory and slight on practice. Due to the lack of practical ability, students have a large gap with the actual needs of enterprises. Therefore, when setting up and teaching the courses, the garment colleges should combine the latest developments of the industry with the local culture to enable students to understand the current situation of the industry and explore the innovative cogitation of students. In order to change the phenomenon of "no consistency of learning and using", the teachers of the garment colleges should combine the actual situation of the enterprise to carry out the teaching, and hire the garment enterprise experts to train the students in various forms, such as curriculum design or series of lectures. The garment profession should pay attention to the practice links in the teaching field, and encourage students to practice in depth and boldly innovate. At the same time, it is necessary to integrate and reorganize the repeated and useless courses, so as to reduce the burden and improve the professional level of students.

4. Conclusion

To sum up, in order to speed up the transformation and upgrading of the garment industry in Jilin province, it is necessary to optimize the services of the government and the Garment Association under the guidance of the national policy; accelerate the upgrading and innovation of technical equipment, especially intelligent manufacturing equipment; improve the investment and financing environment of Jilin province and guarantee the funds needed for the transformation and upgrading of the garment industry; accelerate the reform of the education system of the clothing colleges and universities, improve the quality and adaptability of talents; accelerate the deep understanding and research on cultural creativity, give birth to and strengthen the garment creative industry of Jilin province.

Acknowledgment

The completion of this paper is supported by the research project of Jilin Province Science and Technology Department, "Transforming and upgrading of the garment industry in Jilin province under the national policy of "Cultural Creativity and Design Services"(the project number is 20180418044FG).

Reference

[1] Q.L.Gu,T.Zhang.(2011) Internal and external evolutionary factors of the clothing industry cluster effect.J. Textile Journal, 5: 135-140.

[2] S.F.Qiu,D.L.Jiang.(2016)Multidimensional perspective on the upgrading of China's garment industry .J. Woolen Technology, 44: 58-61.

[3] P. Li,Z.B.Zhang.(2017)Research on the Countermeasures of Hebei Textile and Apparel Enterprise Cluster Competitiveness and Industrial Transformation and Upgrading. J. Woolen Technology,45: 75-78.

[4] Q.L.Ma.(2018)Research on the Reasons and Countermeasures of the Homogeneous Competition of Clothing Industry under the Vision of Supply Side Reform .J. Textile Journal, 39: 182-187.

[5] S.L.Hao.(2015)Research on low-end upgrade of clothing industry chain. J.Woolen Technology, 43: 60.

IOP Conf. Series: Materials Science and Engineering **573** (2019) 012098 doi:10.1088/1757-899X/573/1/012098

Evaluation and its Method of Green Design of Industrial Products from the Perspective of Symbiosis

Zhijin Xu[1] and Yankui Song[2*]

[1] School of Management, Capital Normal University, Beijing, 100056, China

[2] College English Department, Capital Normal University, Beijing, 100048, China

[*] Corresponding author's e-mail: sylviasong@163.com

Abstract. Green design of industrial products is an important part of the core competitiveness of enterprises. On the basis of symbiosis theory, this paper analyzes systematically the evaluation index system and the corresponding evaluation method of green design of industrial products from the research perspective of symbiotic relationship among the factors of human-machine-environment-society system. The author holds that the symbiotic relationship among products, human, technology, resources, economy, society, environment and others, which are mutual coordination, mutual integration, mutual restriction and mutual benefit, is the fundamental analysis framework and research perspective of green design evaluation of industrial products. It is the premise and foundation in the research of green design evaluation and is of important strategic significance to determine the reasonable research perspective and analysis framework. From the perspective of symbiotic relationship, the green design evaluation is a problem of multi-factor, multi-level and fuzzy comprehensive evaluation. Its evaluation index system, composed of a relatively independent set of indicators, can describe and represent the attributes and their symbiotic relationship among technology, resources, economy, society and environment. It is reasonable, feasible and has special advantages to select and apply the fuzzy comprehensive evaluation method in the green design evaluation.

1. Introduction

Green design of industrial products is the core of product development and market extension, constitutes an important part of enterprises' competitiveness and is receiving increasing attention. At present, scholars at home and abroad have achieved a lot of findings on the evaluation and its method of green design, but each has its own characteristics, showing different emphasis. Domestic scholars often focus on the interpretation of evaluation theory and the construction of evaluation model, establish the evaluation index system of green design from multi-dimension and multi-level, and emphasize rational analysis, logical proof and sufficient data, but they often have little relevance with users or pay insufficient attention to them. Expert opinions and suggestions are taken as the starting point and basis of evaluation[1-3]. While foreign scholars pay more attention to the evaluation of design process, construct the evaluation index system of green design from the perspective of the relationship among consumers, markets and products, emphasize perceptual judgment and experience summary, and take psychological experiments and users' feedback information as the starting point of evaluation and the basis of analysis so as to reflect users' personalized real feelings[4-5]. These diverse findings also embody the differences of green design in different national conditions and cultures.

This paper attempts to analyze and construct the evaluation criteria and the corresponding evaluation index system of green design of industrial products based on symbiosis theory, taking the human-machine-environment-society system and the *symbiotic relationship*[6] among products, human, technology, resources, economy, society and environment as an analysis framework and research perspective. Furthermore, Fuzzy Comprehensive Evaluation method (FCE)[7-8] is applied to the evaluation of green design and its scientific, effective and feasible evaluation method is drawn.

2. New research perspective of evaluation of green design

It is an important premise and foundation of research on the green design of industrial products to select and determine a reasonable research perspective and analysis framework. Considering that the green design of industrial products involves many factors, it is appropriate to evaluate it in the big system of human-machine-environment-society. Green design evaluation should not focus on the advantages and disadvantages of individual factors of human, machine, environment and society, but should treat products, human, environment and society as a system. It analyzes the characteristics of human, product, environment, society and etc. and their relationship among interaction, mutual coordination, mutual restriction and mutual benefit to achieve the optimization of the overall function through adjusting the system structure. In fact, a well-designed product must be compatible with human, technology, economy, society environment and others.

The relationship of coordination, ingenious integration and mutual benefit among various factors in human-machine-environment-society system constitutes a kind of symbiotic relationship. *Symbiosis* was originally a biological concept, first proposed in 1879 by Anton de Bary, a German mycologist. He defined it as "living together of different species"[6], that is to live together, being a permanent material connection among organisms to some extent. American biologist Lynn Margulis (1981) formally put forward the endosymbiosis theory in his book *Symbiosis in Cell Evolution* which gradually formed the general symbiosis theory. Since the end of the 20th century, symbiosis theory has been gradually expanded from biology to economics, management, pedagogy and other disciplines and many important findings have been achieved. So it has received more and more attention and recognition.

In this paper, it is a new choice in the research perspective to study green design evaluation of industrial products under the analysis framework of human-machine-environment-society system, starting from this symbiotic relationship of coordination, ingenious integration and mutual benefit. It is not only necessary, but also significant to take the symbiotic relationship among products, human, environment and society as a new perspective of the evaluation of green design and thus to research systematically on the evaluation criteria and methods of green design in line with the needs of sustainable development of human beings.

3. Index system construction of evaluation of green design

It is a process of abstracting, generalizing and describing the green design of industrial products, and transforming the abstract concept into a real measurable variable or carrier to establish its evaluation index system. This is a cognitive process from concrete to abstract, and then from abstract to concrete, which unifies subjectivity and objectivity.

From the new research perspective of symbiotic relationship in the human-machine-environment-society system mentioned above, an evaluation index system of green design of industrial products (see Table 1) is preliminarily constructed, which consists of 25 relatively independent indicators through extensive document research and expert consultation[9-13]. This evaluation index system can better describe, characterize and represent the multiple evaluation criteria and requirements of green design. It reflects and embodies the symbiotic relationship among products, technology, resources, economy, society, human and environment and the goal of sustainable development. Due to space limitation, there is no further explanation about the specific connotation of the indicators in the above.

IOP Conf. Series: Materials Science and Engineering **573** (2019) 012098 doi:10.1088/1757-899X/573/1/012098

Table 1. Evaluation index system of green design of industrial products.

Objects evaluated	Evaluation factors (First level indicators)	Evaluation indicators (Second level indicators)
Evaluation index system of green design of industrial products	Technical attributes (U_1)	*Quality reliability* (u_{11})
		Functional effectiveness (u_{12})
		Operational security (u_{13})
		Maintenance convenience (u_{14})
		Species diversity (u_{15})
	Resource attributes (U_2)	*Material type* (u_{21})
		Unit product resource consumption (u_{22})
		Utilization rate of resources (u_{23})
		Resource renewability (u_{24})
	Economic attributes (U_3)	*Cost* (u_{31})
		Product added value (u_{32})
		Profit benefit (u_{33})
		Industrialization degree (u_{34})
		Market share (u_{35})
	Social attributes (U_4)	*Maintenance of social security* (u_{41})
		Improvement of quality of life (u_{42})
		Permissibility of law and policy (u_{43})
		Social public satisfaction (u_{44})
	Humanistic attributes (U_5)	*Human-computer interaction* (u_{51})
		Aesthetics of Modelling (u_{52})
		Cultural integration (u_{53})
	Environmental attributes (U_6)	*Water pollution degree* (u_{61})
		Air pollution degree (u_{62})
		Noise pollution degree (u_{63})
		Waste Pollution degree (u_{64})

4. Method selection of evaluation of green design

In order to evaluate the green design of industrial products scientifically and reasonably, we must also select the evaluation method on the basis of determining reasonably its constituent factors and evaluation index system. Among all kinds of evaluation methods, we think that it is more appropriate to select FCE in green design evaluation for which is essentially a multi-object, multi-factor, multi-level and fuzzy comprehensive evaluation problem.

The evaluation indicators are various in the problem of multi-object, multi-factor and multi-level comprehensive evaluation. Some are quantifiable and others are difficult to quantify. Traditional evaluation methods are effective for quantifiable indicators, but they often encounter great difficulties for non-quantifiable ones. When both indicators coexist in an evaluation index system, the traditional methods have great limitations. We must select new more reasonable and effective evaluation methods. FCE is the suitable one among them. By abstracting, generalizing and describing the various attributes of the evaluated objects, it can well deal with the relationship between quantifiable and non-quantifiable indicators with fuzzy operation and weight distribution. Finally, it reaches scientific and reasonable comprehensive evaluation results that can both reflect the subjective value judgment of the evaluator and can be measured in practice.

In fact, FCE has specific advantages, which is accepted and applied widely in many fields. It is a correct choice to apply FCE into the evaluation of green design, which is of great strategic significance for the sustainable development of an enterprise, an industry and even a country.

5. Models of comprehensive evaluation method of green design

The theoretical basis of comprehensive evaluation method of the green design of industrial products is the FCE models provided by fuzzy mathematics.

IOP Conf. Series: Materials Science and Engineering **573** (2019) 012098 doi:10.1088/1757-899X/573/1/012098

According to FCE[7], set $U=\{u_1,u_2,...,u_n\}$, u_i ($i=1,2,...,n$) as the evaluation factor set; $V=\{v_1,v_2,...v_m\}$, v_j ($j=1,2,...,m$) as the comment set; and $A=(a_1,a_2,...a_n)$, a_i ($0\leq a_i\leq 1$) as the weight allocation; where $R=(r_{ij})=(r_{i1},r_{i2},...,r_{im})$, r_{ij} ($0\leq r_{ij}\leq 1$, $i=1,2,...,n$, $j=1,2,...,m$) as the single factor evaluation. Thus, a comprehensive evaluation model of multiple operators $M(\overset{\bullet}{*}, \overset{+}{*})$ is given:

$$AoR = B = (b_1 b_2 ... b_m) \quad (1)$$

Where,

$b_j = (a_1 r_{1j})(a_2 r_{2j})...(a_n r_{nj}) \qquad (j=1,2,...,m)$

Let o be operators of fuzzy compound operation. They are the generalized operators $\overset{\bullet}{*}$ (fuzzy *and* operations) and $\overset{+}{*}$ (fuzzy *or* operations). In theory, there are various generalized fuzzy operations, but only five specific models have been proposed up to now[8]. Operators of different fuzzy compound operations have different attributes and characteristics, play different roles and embody different comprehensive evaluation ideas and purposes. Therefore, the selection of operators of fuzzy compound operations is of important guiding significance.

To divide the factor sets into different levels, the above model can be applied to multi-level factors, and the evaluation results of each level are the input of the upper level evaluation in sequence until to the top level. Thus, the final fuzzy comprehensive evaluation results can be obtained.

Specifically, the general process of comprehensive evaluation of green design with the FCE models includes the following steps.

Firstly, establish an evaluation index system to determine the corresponding evaluation factor set U. According to Table 1, the evaluation factor set of the green design of industrial products $U=\{U_1, U_2, U_3, U_4, U_5, U_6\}$={technical attributes, resource attributes, economic attributes, social attributes, humanistic attributes, environmental attributes}.

Secondly, determine the evaluation grades and criteria so as to establish the comment set V. Generally, an evaluation group consisting of several experts (usually 9-13) of different categories will conduct peer reviews, and establish a corresponding comment set for evaluating green design of industrial products $V=\{v_k\}=\{v_1, v_2, v_3, v_4, v_5\}$={Excellent, Good, Medium, Qualified, Unqualified} ($k_{max}=5$).

Thirdly, determine the weight $A=\{a_i\}$ of evaluation indicators. It is an important link and soul in the comprehensive evaluation of green design of industrial products. There are many methods to determine the weights. Some emphasize the subjective experience, simple and easy to implement, but with poor objectivity; while others are objective, but complicated in their analysis and calculation. Therefore, they should be selected according to the actual situation.

Fourthly, evaluate the single factors and obtain the single factor fuzzy comprehensive evaluation matrix $R=\{R_i\}$. Single factor fuzzy evaluation evaluates each indicator u_{ij} separately in subset U_i, and that is, each secondary indicator u_{ij} in the factor U_i can be evaluated as the subordinate degree r_{ij} of every grade v_k.

Fifthly, determine the fuzzy comprehensive evaluation model $M(\overset{\bullet}{*}\overset{+}{*})$. When weight A and single factor fuzzy comprehensive evaluation matrix R are known, the comprehensive evaluation can be carried out by the compound operation of fuzzy matrix, and the evaluation model of fuzzy comprehensive evaluation can be obtained.

Sixthly, obtain finally the results B of fuzzy comprehensive evaluation. When matrices R_i and weights a_i of index u_{ij} are known by the single factor fuzzy comprehensive evaluation, the results $B_i= a_i \text{ o } R_i$ of single factor fuzzy evaluation can be obtained through fuzzy compound operations (1).

On the basis of single factor fuzzy evaluation, a higher level of fuzzy comprehensive evaluation can be continued. That is, take the results B_i ($i=1, 2, ..., m$) of single factor fuzzy evaluation as the matrix $R=\{B_i\}$ ($i=1, 2, ..., m$) of higher level fuzzy comprehensive evaluation, combine with weight A of single factor U_i by the fuzzy compound operation (1), and then obtain a higher level of fuzzy comprehensive evaluation results $B=A \text{ o } R$. Finally, continue the steps until the result B of the fuzzy comprehensive evaluation of green design is obtained:

$$B=A \circ R=[a_1, a_2, a_3, a_4, a_5] \circ [B_1, B_2, B_3, B_4, B_5]^T \quad (2)$$

In this way, according to the principle of maximum subordination, we can obtain and determine the comprehensive evaluation value of green design of industrial products and its results of technical, resource, economic, social, humanistic and environmental attributes which belong to different grades of excellent, good, medium, qualified and unqualified by Formula (2).

Due to space limitation, only the general process of comprehensive evaluation of green design of industrial products is shown by FCE without specific case analysis.

6. Conclusion

The evaluation of green design of industrial products is essentially a fuzzy comprehensive problem of multi-object, multi-factor and multi-level. The premise and foundation of solving this problem are to select and determine a reasonable research perspective and analysis framework. Its key is to construct the evaluation criteria and apply the evaluation method correctly. The widely applied FCE method is effective and advantageous in solving and dealing with the comprehensive problem. With this method we can better judge and evaluate whether an industrial product design is good or bad and whether it can support and meet the needs of sustainable development of human beings. It is of great important, decisive and guiding significance to allocate indicator weights and to select compound operators rationally.

In a word, the research on the evaluation and its method of green design of industrial products is important and still needs to be explored continuously for the further optimization of human-machine-environment-society system and the symbiotic development of products, human, technology, resources, economy, society and environment.

Acknowledgments

The paper is funded by the soft science project from the Talent Centre of the Ministry of Science and Technology (2017-2019).

References

[1] Ke, S.J., Liu, H.J. (2010) Research on product modelling evaluation technology based on brand image. Packaging Engineering, 18: 21–24.
[2] Liu, T.J. (2016) Application of green design in industrial product. China Packaging Industry, 02: 11–14.
[3] Pan, P., Yang, S.X. (2013) Evaluation system and evaluation method of products style design. Mechanical Design and Research, 10: 19–22.
[4] Li, Y., Chen, H. (2005) Green Design Methodology for Industrial Products. Tsinghua University Press, Beijing.
[5] Wang, L.L., et al. (2018) Application of green design in industrial product. Grass-roots Construction, 02: 17–18.
[6] Yuan, C.Q. (1998) Symbiosis Theory and Small Economy. Economic Science Publishing House, Beijing.
[7] Wang, P.Z. (1980) Introduction to fuzzy mathematics. Practice and Understanding of Mathematics, 02-03: 5–12.
[8] Liu, Y.Y., et al. (1983) Mathematical model of comprehensive evaluation. Fuzzy Mathematics, 1: 23–26.
[9] Xie, J.P., Chen, R.Q. (2003) Three-dimensional fuzzy evaluation model for green design of industrial products. Journal of Huazhong University of Science and Technology (Natural Science Edition), 06: 15–18.
[10] Wang, C.P. (2009) Constructing the evaluation index system of green design. Packaging Engineering, 12: 8–11.
[11] Liang, C.W., et al. (2013) Construction of evaluation index system for green design of industrial products in Foshan. Modern Economic Information, 11: 25–28.

IOP Conf. Series: Materials Science and Engineering **573** (2019) 012098 doi:10.1088/1757-899X/573/1/012098

[12] Wan, L. (2015) Application of green design in industrial product. Industrial Design, 08: 17–20.
[13] Xing, P.H. (2017) On the application of green design in industrial product. Design, 13: 9–12.

A New Way of Inheriting Traditional Culture: A Case Study of the Design of Cultural and Creative Products of Laolongkou Wine Industry in Shenyang City

Feng Lyu [1], Xiaoling Cheng[2*], Jinyan Yan[3]and Xiaoqing Yang[4]

Academy of Design and Art, Shenyang Aerospace University, Shenyang, Liaoning, 110136, China

*Corresponding author's e-mail: lynne0315@qq.com

Abstract. With the restructuring of commodity economy and society, the value of cultural and creative products in spreading the history and culture of the industry is increasing. By mixing culture and design, they have sufficient expressiveness in art. This paper explores the constructive significance of art language in cultural and creative products in terms of eliminating cultural dross, enhancing artistic self-confidence, and the social and market recognition by taking the national craft, expressions and artistic elements of cultural and creative products in Laolongkou wine industry.

1. Introduction

In recent years, with the rapid development of China's tourism industry, cultural and creative products have grown into a "cultural business card" of the museum, bearing the responsibility of promoting cultural characteristics. The heritage of Laolongkou wine industry is not only a recorder of many important historical nodes of Shenyang City in modernization as an important cultural and historical place in Shenyang, but also the presenter of traditional winemaking techniques and local characteristics, the witness of history and the carrier of culture. In its innovation and development, it is necessary to adhere to the principle of cultural value priority, and present the unique winemaking culture to the world through cultural and creative products, thus opening a window to visitors to see the culture in the past and learn more about winemaking in a richer, more accurate and systematic manner[1].

Properly treating the wine culture of Laolongkou is crucial to the history of Laolongkou as it shows the attitude towards the history of wine and Shenyang's industrial development. On the basis of cultural priority, scientifically using cultural and creative products to show the original industrial culture can help to exert its value in such fields as history, culture, economy, society and ecological environment and increase space for development and competitiveness of Shenyang.

2. The potential of Laolongkou wine industry in making cultural and creative products should be tackled

According to Peter Koslowski, a famous German scholar, public consumption is shifting to the aesthetic and cultural consumption. Cultural and creative products are the consumer products with profound connotations made by following the original traditional cultural elements and combining with modern design. They are characterized by practicality and aesthetic needs and market value. The advent of cultural and creative products is a novel way to inherit traditional culture by combining tradition with modernity[2]. It also explores a modern way of promotion, which helped to popularize the traditional wine culture of Laolongkou.

At present, cultural and creative products are increasingly becoming a new growth point for memorials and museums and an important channel for gathering social resources. It is understood that the annual average sales revenue of cultural and creative products of the Metropolitan Museum of the United States, the Smithsonian Museum, in recent years was more than 100 million USD. Compared with that, the development and operation of cultural and creative products in Chinese museums is still to be explored. In addition, different people have different opinions on these products while the industry is not regulated well with products lacking cultural connotation and workmanship. It is not appropriate only to pursue packaging, but the feelings and creations, and the original craft culture. The inheritance of the national craftsmanship of the Laolongkou Wine Museum is to put people in the first place with more attention paid humanity, aiming to express the traditional culture while innovating.

The economic value generated by cultural and creative products is a strong support for the development of museums. Nowadays, with the increasing demand for cultural and creative products, Laolongkou Winery should seize this opportunity to ensure that cultural and creative products can deliver the special cultural heritage of Laolongkou, thus becoming unique and collectible. In the wave of the promotion of these products, it is necessary to form a unique design concept, remain dedicated to extracting the essence of wine culture, and combine the functional, aesthetic, and craftsmanship characteristics, so as to enable the public to use these products in their daily life.

3. Artistic performance of Laolongkou cultural and creative products

With the flourishing of this industry, cultural museums, cultural tourism, cultural industry parks are all blooming. The design of Laolongkou cultural and creative products must be devoted to spread its cultural spirit, historical significance and artistic value, maintain people-oriented, and keep up with the trend of the times, so as to promote the national culture in the museums.

3.1. Cultural element selection

As time passes by and the society changes, part of the Laolongkou wine industry culture has been forgotten. For designers, inheriting regional culture is not only a responsibility but also an obligation. It is necessary to incorporate regional cultural elements into product design, pay attention to the content and significance of Laolongkou culture, and then find out the values of the culture to be preserved, and finally display it in the form of products using specific symbols. Therefore, the design of cultural and creative products based on regional cultural elements should focus on its core value and combine cultural connotation with visual beauty.

The cultural attribute requirement of cultural and creative products is to protect and inherit the culture. By combining the ethnic cultural elements of Laolongkou with the lines, symbols and colors in the art, and by drawing the comics flow chart of winemaking, the process can be displayed in a flat or three-dimensional form with high cultural, artistic and collectible values. It will also help consumers to better understand the cultural knowledge related to the museum, so as to better appreciate the collection and the cultural connotation[3].

The design of Laolongkou cultural and creative products accurately grasp the emotional tone of the products by making full use of the rich cultural elements of the wine industry. The semantic expression of cultural elements includes shape and meaning, which indicates that the design of cultural and creative products using elements of Laolongkou culture should take into consideration the matching of color and materials for the overall beauty. It is necessary to focus on the essence of Laolongkou culture, fully display the characteristics of Laolongkou wine culture, and vividly integrate into design. Selecting and characterizing the distinctive cultural characteristics of Laolongkou wine industry and make them into creative postcards is an effective means to express the cultural attributes of the Laolongkou ancient winemaking method in Shenyang City. The design adopts the five-dimensional expression method of "people, thing, object, field and condition" to summarize the basic characteristics of wine culture, including the process of mixing, koji-making, sun drying, distillation, putting into cellar, and taking out from cellar, etc. The design of the postcard (as shown in the picture1) can promote the unique winemaking process of Laolongkou more effectively .

Picture1. Laolongkou Postcard.

3.2. Shaping of cultural and creative products

The list of authors should be indented 25 mm to match the abstract. The style for the names is initials then surname, with a comma after all but the last two names, which are separated by 'and'. Initials should not have full stops—for example **A J Smith** and *not* **A. J. Smith**. First names in full may be used if desired. If an author has additional information to appear as a footnote, such as a permanent address or to indicate that they are the corresponding author, the footnote should be entered after the surname[4].

Based on the characteristics of the culture of Shenyang Laolongkou wine industry, this paper uses various forms of expression conveyed by the current creative products, so as to cognize, perform and apply its inheritance process step by step. In the design of cultural and creative products, the method of ancient winemaking is analyzed and reorganized, in this way, creative postcards, creative mobile phone shells, canvas bags(as shown in the picture2), and multi-functional creative bookmarks can help the public to have a deeper understanding of the industrial connotation of Laolongkou.

Picture2. LaologkouCreative products： U Disk and Cell Shell.

The practical design of the existing cultural and creative products takes up a relatively large proportion in the entire creative industry. This paper uses the bionic method to extract the surface texture, color and shape of the characteristics of Laolongkou wine industry, abstract the transform the cultural elements, and design the daily necessities such as mobile phone shells by combining the practical functions, canvas bag and bookmarks[5]. It not only satisfies the practical needs but also vividly and directly displays the design source so as to visually stimulates the users to understand more deeply, thus realizing the educational significance.

The design of cultural and creative products of Shenyang Laolongkou wine industry should break the boundary of simple media, materials,and the fixed mindset, and uses multiple medias and methods

IOP Conf. Series: Materials Science and Engineering **573** (2019) 012099 doi:10.1088/1757-899X/573/1/012099

to redesign, so as to highlight the characteristics of the winemaking method of Shenyang Laolongkou wine industry. These products pay attention to the style, pattern, and material in particular (such as environmentally friendly inks). For example, the creative canvas bags explore the combination of functions and cultural creativity, thus satisfying the needs of consumers in culture, innovation, function and aesthetics.

3.3. Sales of cultural and creative products

Good cultural and creative products can enhance the popularity of Laolongkou wine. At the same time, the launch of the cultural and creative product needs to form a series, covering different types of items under one cultural element. These daily cultural and creative products can promote the brand image of Shenyang, a historical and cultural city, and invite more friends to know about the history and culture of Shenyang Laolongkou wine. Cultural T-shirts, pillows, U-disks and other carriers are fashionable and popular. By combining the cultural symbol in the museum cultural collection in design, the physical commodity has both aesthetic and practical values. In addition, it is relatively cheap, making it popular among ordinary consumer and promising as a museum cultural products.

From excavating and sorting out the rich cultural resources of Laolongkou wine to developing cultural and creative products that are suitable for the public, exhibitions and information-based digital media marketing models are used to expand the popularity of these products. People can not only experience the culture of Laolongkou wine industry in the museum, but also take the aftertaste of the exhibition to home, thus extending the content of the wine culture in time and space, and expanding the cultural influence.

4. Unique cultural IP

The text of your paper should be formatted as follows:

IP is an acronym for English "Intellectual Property". The museum has rich cultural resources, and it is necessary to explore the ways to develop, protect, and make use of cultural IP. In the design of cultural IP, the author uses the elegant auspicious pattern border and the cartoon shape to create the characteristics of winemaking (as shown in the picture3), using the form, symbol and other art expressions. The image of small monster is taken as a basis to design two round images with orange and blue as the main colors.

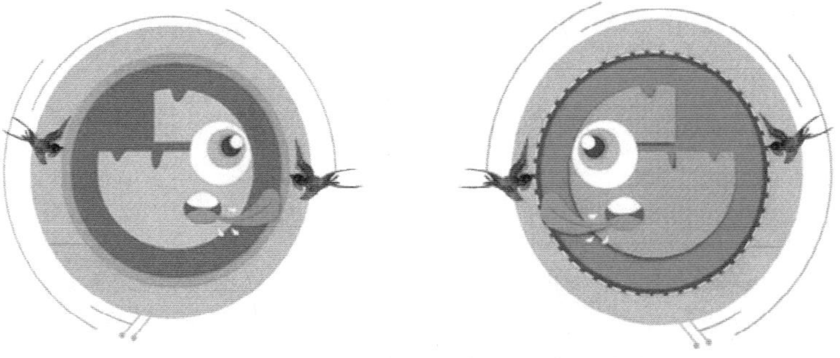

Picture3. LaologkouUnique cultural IP.

Cultural IP can improve the popularity of Laolongkou wine industry, which is conducive to later promotion. This advantage can be directly transformed into the potential for discussion and consumption in the later stage. The mascot designed for the Shenyang Industrial Relics suites the cultural texture of Laolongkou. Meanwhile, one eye and a playful tongue are representative as it integrates cultural IP into cultural and creative products so as to give these products a unique auspicious meaning, thus meeting the emotional needs of consumers, and highlighting the national cultural characteristics of Laolongkou.

It is common method in art to use “cultural metaphor” to express the meaning of cultural IP, with public cultural perception as the guidance. Laolongkou cultural and creative products featuring multiple fields, big design, beauty, and practicality, can express the concept and emotion of a national culture, and remind people of its culture through the cultural IP, thus forming an autonomous cultural brand and an active long-term mechanism. The combination of the traditional culture of Laolongkou wine industry and modern product design can turn cultural and creative products into a bridge connecting people and cultural emotions.

5. Conclusion

The heritage of Laolongkou Museum is not a historical burden for Shenyang's urban development, but a valuable cultural heritage because it is not only a historical witness of Shenyang's development over the years, but also has unique cultural values. Therefore, it is necessary to design cultural and creative products in a way that better satisfy the public's feelings and corresponds to their daily behavior. This culture-based and innovative development path can promote Laolongkou culture more effectively by creating a perfect combination of tradition and modernity, and a image of Laolongkou that is willing to keep up with the times. During design and sales of cultural and creative products, people's sense of belonging and identity to Laolongkou wine can be aroused, and they are encouraged to protect the culture. In this way, the social forces and social funds can be guided into the protection of the heritage, so that the resources of Shenyang Laolongkou wine industry can better contribute to social development.

Acknowledgments

This work was supported by Provincial Social Science and Technology Association 2019 Liaoning Province Economic and Social Development Research Project (2019lslktqn-054), 2017 Shenyang Philosophy and Social Planning Project (17001) and 2019 Shenyang Social Science Project Research Achievements (SYSK2019-11-02).

References

[1] Wang， L. (2012)Cultural and Creative Product Design: A Case Study of Tourism Souvenir Design [J]. Popular Literature and Art.pp. 59-60.

[2] Qiang, B.(2018)Innovative Design and Development of Museum Creative Products.Light industry technology. , Volume 34, Issue 3.

[3] Wu， W.(2015) Design Elements of Museum Creative Products - Taking Guangxi Museum as an Example. Research on Cultural Industry of Chinese Museum.pp. 13-18.

[4] Mo， B. (2015)Core Ideas and Design Points of Museum Literary Creation and Development. Research on Museum Cultural Industry of China.pp.392-404.

[5] Yang， M. Fan， C.G. (2015)Cultural symbols of creative products are published [J]. Chongqing Social Science.pp.81-85.

IOP Conf. Series: Materials Science and Engineering **573** (2019) 012100 doi:10.1088/1757-899X/573/1/012100

Application of Design Geometry in Minimalist Style Product Design

Xuyang Zhang[1, a], Yanmin Xue[1, b]

[1] Department of Industrial Design, Xi'an University of Technology, Xi'an, Shaanxi, 710054, China

[a]18829028651@163.com

[b]xueym@xaut.edu.cn

Abstract. From the historical overview of design geometry, this paper briefly introduces the current popular style of product design---minimalist design. The emphasis is laid on the analysis and research of design geometry in the minimalist style product design, leading to the contemporary designer's concern and thinking about design geometry. Thus, this paper establishes the importance of design geometry in minimalist style product appearance design.

1. Introduction

In the complex information age, people have to perceive and contact increasingly jumbled information, data, emotions, etc. More and more people choose a MINIMALIST lifestyle to relieve pressure, abandon flashiness, and return to simplicity and purity. Minimalism style has been favored by modern man. Minimalism was born in the 1960s. Its design concept advocates removing decoration, retaining basic functions and focusing on product details to enhance function and aesthetics so as to achieve the effect of "Simplicity but not Simplicity". It encouraged designer to pursue the most concise artistic effect in artistic creation, simplify the shape into the most basic simplified geometric form, and give priority to the single color such as black, white and grey, integrate the space into the simplest two-dimensional style, create an atmosphere of PURE NIHILITY without emotion, rendering and space, and give the audience the most essential and pure visual experience [1]. Minimalist style products are simple and delicate in appearance, focusing on the proportion and composition of product modelling. Round, square and triangle are the main contour lines and its product modelling is concise and harmonious. This coincides with the design method advocated by design geometry. By using design geometry to deal with the relationship between the proportion and scale of minimalist style product design, the product can achieve visual coordination and beauty. This paper analyses the application and influence of design geometry in product appearance modeling through the following classical minimalist style products cases.

2. Conceptual Theory of Design Geometry

2.1. Design Geometry in Nature

From the 20th to the 16th centuries B.C., a recorded proportion of gold section on the megalithic stone pillars. After that people in different industries have applied the golden section to sculpture, painting and architecture. There are also examples of golden section in nature, such as the proportion of human face and body, the proportion of salmon body, the spiral growth of sunflower and pine cone.

IOP Conf. Series: Materials Science and Engineering **573** (2019) 012100 doi:10.1088/1757-899X/573/1/012100

2.2. Golden Section Proportion
The beauty of golden section proportion can be appreciated from both nature and man-made goods. Many other golden section shapes can be obtained from the original golden section rectangle, such as golden section triangle, golden section ellipse, golden section helix and various dynamic rectangles of golden section.

2.3. Root Rectangle
The proportion of root-2 rectangle is similar to the golden section ratio. It can be infinitely divided into smaller equal-ratio rectangles. A root-2 rectangle can be divided into two equal small root-2 rectangles, and can also be divided into four equal smaller root-2 rectangle. The paper size system of German industrial standard adopts root-2 rectangle. This standard system is not only convenient, but also makes the best use of paper without any waste. Root-3 rectangle, root-4 rectangle and root-5 rectangle can also be divided into smaller similar rectangles like root-2 rectangle. These root rectangles can be divided both horizontally and vertically. There are many examples of root rectangles in life and nature, such as beehives, snowflakes, etc[2].

3. Classic product analysis of Design Geometry in Minimalist Style Product

3.1. Bauhaus
It can be said that the establishment and development of Bauhaus led to the rise of minimalist style. Just like the slogan LESS IS MORE which was raised by the designer Mies van der Rohe, minimalist style advocates the design concept of FUNCTION FIRST and SHAPE SIMPLICITY. Bauhaus advocates taking science and rationality as premise, standardization and unification as design criterion, and product form as determined by product function[3]. Its design mostly takes simple geometric figure as design element and carries on certain arrangement and combination. For example, the MT8 Bauhaus metal table lamp shown in Figure.1 has a simple shape, which is composed of a semi-circular lamp shade, a cylindrical lamp post and a disc lamp chassis. From the lamp chassis to the bottom of the lamp shade is a root-2 rectangle. A golden rectangle is formed from the lamp chassis to the middle of the semi-circular lamp shade, and the length of the lamp rope is equal to the radius distance of the lamp shade. Due to the appropriate proportion and minimalist style, it has become a representative work of Bauhaus style, which can be seen in many modern lamps.

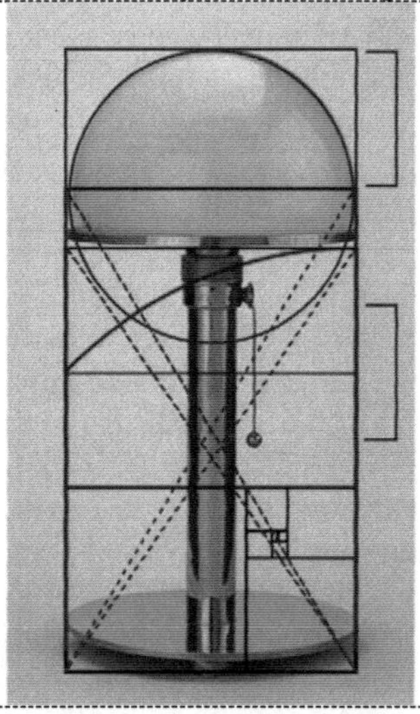

Figure.1 MT8 Bauhaus Metal Table Lamp.

IOP Conf. Series: Materials Science and Engineering **573** (2019) 012100 doi:10.1088/1757-899X/573/1/012100

3.2. Braun

German Ulm Design Institute inherits and improves the rationalistic design principles of Bauhaus Design Institute. The cooperation between German Braun Company and Ulm Design Institute has successfully expanded the minimalist design. This cooperation has produced fruitful results, so that Braun design is still regarded as one of the representatives of minimalist products. As Braun Designer Dieter Rams mentioned in the Ten Design Standards, "Good Design is Minimalist", many of Braun's excellent product designs show the characteristics of centralized function and simple shape[4].

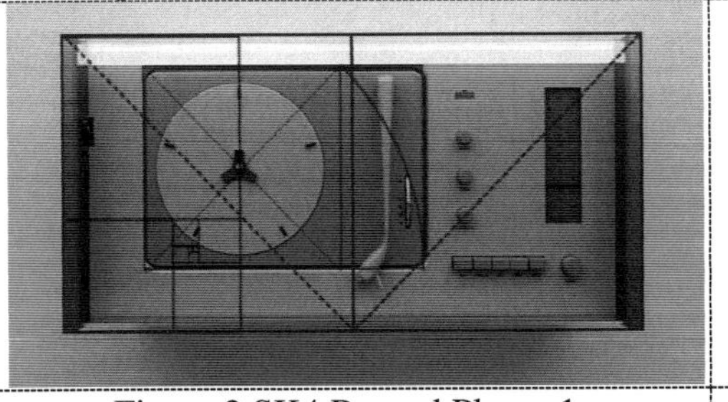

Figure.2 SK4 Record Player 1.

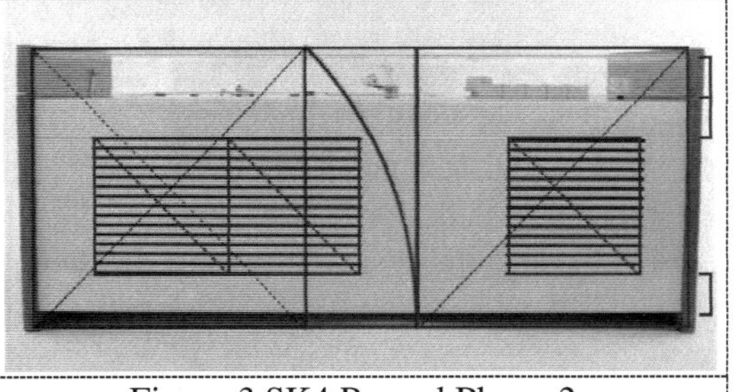

Figure.3 SK4 Record Player 2.

The most typical product is SK4 record player designed by Dieter Rams and others in 1956. As a classic work of simple design of Braun Company, its greatest feature is its unique shape. The external form of the product is a simple right-angled white hexahedron, consisting of an openable transparent plexiglass cover and a white metal box. As shown in Figure.2, the overhead area of the metal box is composed of two squares. The area of the gramophone record is a standard root-2 rectangle. The center of the circular turntable of the record is at the center of the square on the left half. A golden rectangle is formed from the center of the record to the left boundary of the box. The distribution of panel keys is regular and orderly. As shown in the side view of Figure.3, the whole area of the metal box is composed of a root-2 rectangle and a square. The window-shaped ventilation groove is designed as three squares of the same size. The height of the upper glass cover is equal to the distance from the ventilation groove to the boundary of both sides of the box. The whole product is simple and elegant in shape and delicate in structure, which breaks the aesthetic standard of people and thus gained the reputation of SNOW White's COFFIN.

3.3. Apple

Apple's product design is deeply influenced by Dieter Rams. It follows and pays tribute to Braun's minimalist design style. Minimalism is also the core force Apple adheres to and the foundation of its future development. Apple's products look exquisite and simpler, showing the aesthetic significance of design geometry.

As shown in Figure.4, Apple's first-generation iPod product is a golden rectangle with symmetrical left and right sides. The bottom part of the product screen is a square. The center of the three concentric circles that make up the button is located at the intersection of the diagonal lines of the square[5]. Its simple and unique shape and humanized operation mode were once sought after by people.

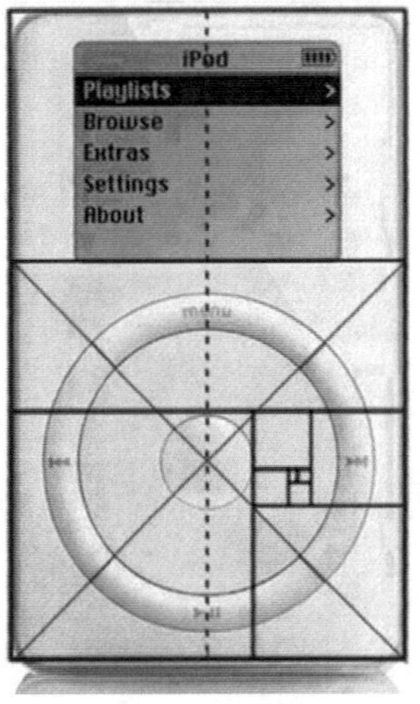

Figure.4 iPod.

Another example is Apple's iMac Intel i7 computer. As can be seen from the main view of this compact computer in Figure.5, the front frame of the whole computer is a root-2 rectangle shape. From the top to the bottom of the black display frame, the shape is a golden rectangle. The Apple logo is at the diagonal intersection of the lower white rectangular panel. As can be seen from its rear view in Figure.6, the parallel line at the connection of the computer bracket coincides with the horizontal central line of the overall shape. The bottom width of the computer bracket is one quarter of the width of the computer display screen. The Apple logo on the back is located at the diagonal intersection of the rectangle above the computer bracket. The overall shape of the computer is symmetrical, giving a sense of order and symmetrical.

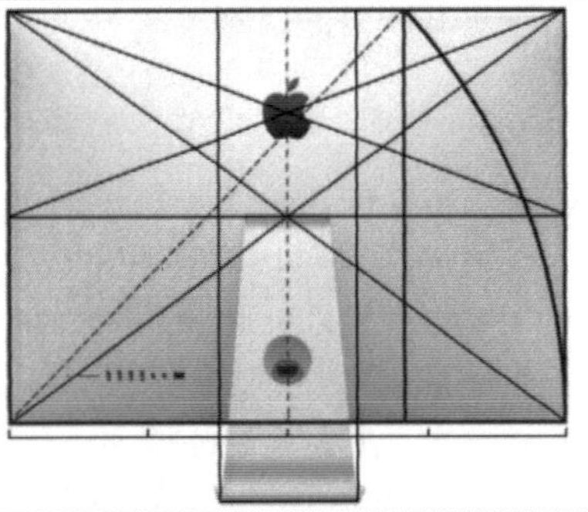

Figure.5 iMac Intel i7 1.

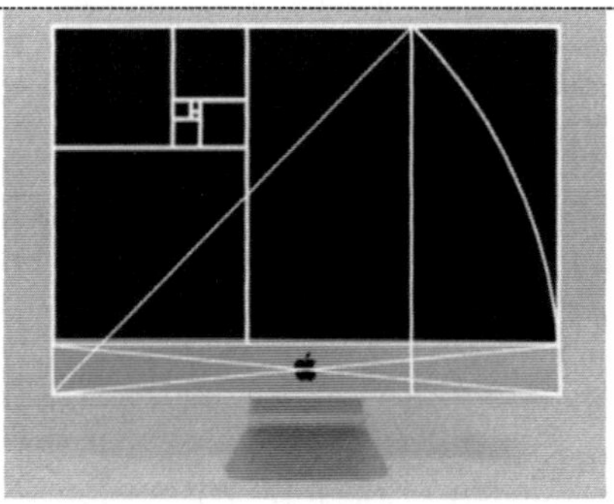

Figure.6 iMac Intel i7 2.

The whole form of the iPhone 4 in Figure.7 is a rectangle composed of two squares. The Apple logo is located on the diagonal intersection of the upper square. The main display area of the screen is a root-2 rectangle. The center of the circular key is located at the diagonal intersection of the rectangle at the bottom of the display screen. The diameter of the key is the same as the width of the auditory hole.

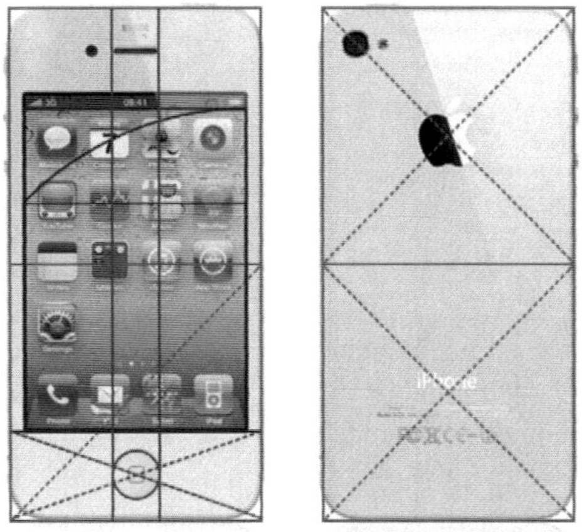

Figure.7 iPhone4.

3.4. *Other Minimalist Style Products*

In addition to the representatives of the minimalist style products mentioned above, there are also excellent NO DESIGN products like MUJI. Figure.8 is a MUJI CD player. The front of the player is modeled as a square. The CD slot is at the intersection of the square cross diagonal lines. Surface decorative holes are arranged in a radial manner with the intersection point as the center of the circle, and the four corners are round chamfers of equal size. The distance between the four circles and the intersection of the diagonal intersection of the square form a root-2 rectangle. The whole appearance of the product is simple and elegant, creating a tranquil and peaceful mood. This concise geometric design greatly improves the product identification of the MUJI.

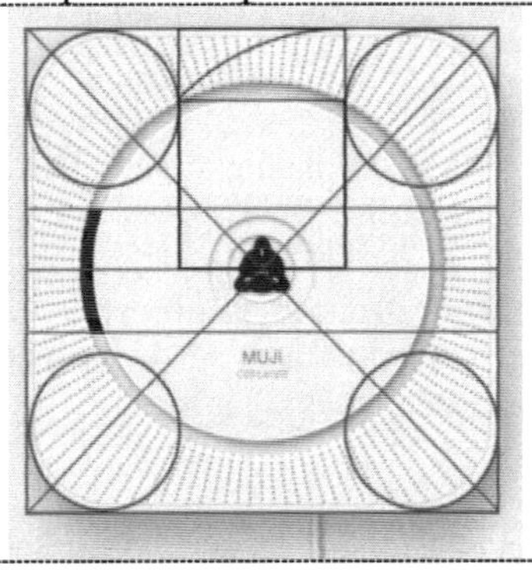

Figure.8 MUJI CD Player.

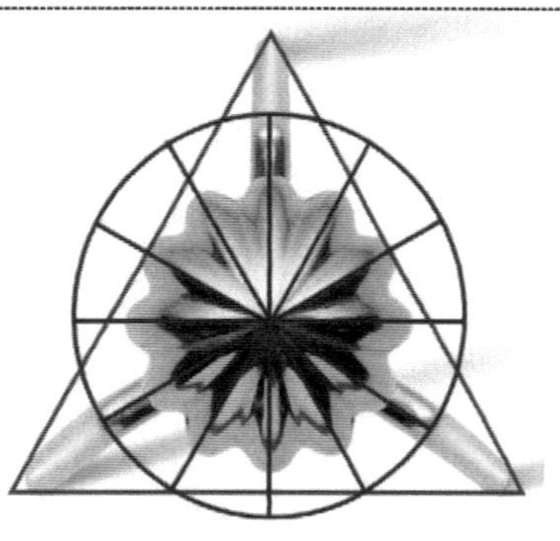

Figure.9 Juicy Salif Citrus Juicer 1.

Philippe Starck, known as the design genius, designed many products in a minimalist way. These products have a single shape, but they attach great importance to the use of geometric shapes. Their design is simple and functional. His Juicy Salif Citrus Juicer is novel and unique in shape, without any redundant decoration. In the top view of the juicer shown in Figure.9, the three supports of the juicer form an equilateral triangle, and the top part of the juice extractor is a fan of 12 sizes with the same 30 degrees. Figure.10 is the side view of the juicer. The top of the head and the support is in the golden ellipse. The drop point of the juice extracted by the juicer is in the center of the golden ellipse. The head contour of the juicer is also a golden ellipse. The contour edge line and the big golden ellipse from two golden rectangles. The distance from the bracket to the big golden ellipse is equal to the distance from the bracket to the head of the juicer, which is one fifth of the width of the whole composition. The shape of the juicer is a perfect example of applying design geometry to the minimalist style of product shape. The product has great artistic appreciation value.

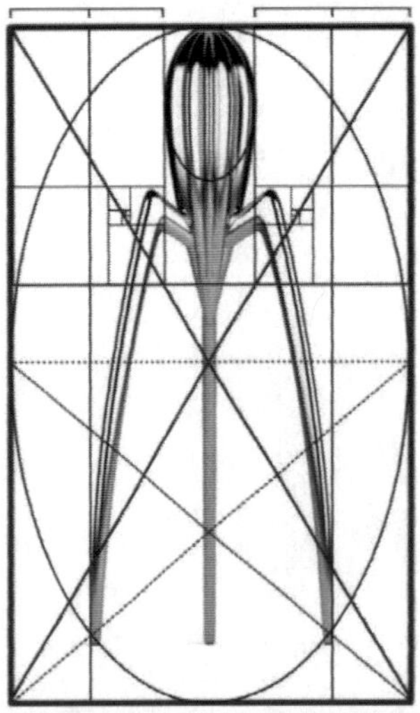

Figure.10 Juicy Salif Citrus Juicer 2.

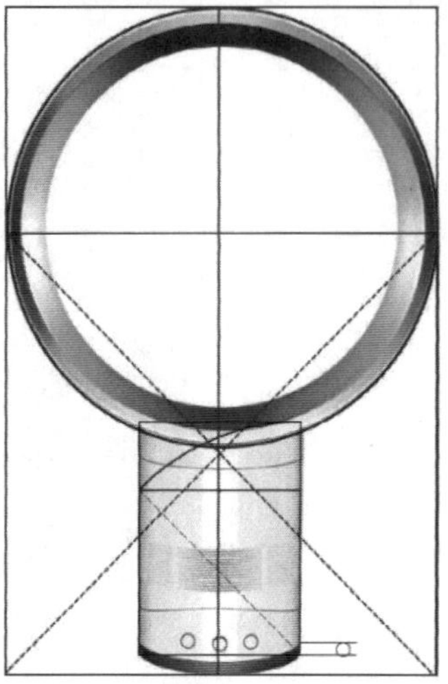

Figure.11 Dyson Air Multiplier.

Dyson Air Multiplier is also a minimalist style design product. Its unique structure principle, simple and generous shape, is the combination of science and art. Figure.11 The product is mainly composed of upper outlet component and lower motor module. The outlet is round in shape and the front view of the motor module is a root-2 rectangle. The diameter of the outlet and the base form a square, and the upper rotating point of the product is at the intersection point of the cross diagonal line of the square. The lower three keys are equal in size, circular, and equal in diameter to the height of the outer frame of the base. The overall form of the product is simple, fashionable and proportionally coordinated, which is popular indeed.

4. Conclusion

Through the analysis of the above classical minimalist style products, we can find that with the development of society and life, minimalist style products are still popular. Good design depends on many design principles[6] and designers can reasonably apply design geometry to minimalist style product modelling according to the needs of minimalist style product modelling while considering the design principles of humanization and technology of products to add visual aesthetic sense to products and enhance the aesthetic value and appreciation value of products.

Acknowledgments

This paper was supported by Chinese Ministry of education of Humanities and Social Science (project No. 17YJAZH100) and Humanities and Social Science Program of Shaanxi Province (Project No. 2018K13).

References

[1] WANG Zhao-xia. (2016) Application of Minimalism in Product Design. J. Packaging Engineering,37:167-169.

[2] Kimberly, E., LI Le-shan's Translation. (2003) Geometry of Design. China Water & Power Press, Beijing.

[3] GUO Lin-sen, YANG Ming-lang. (2015) Minimalism in the Daily Necessities Design Application Study. J. Packaging Engineering, 36:127-130.

[4] FAN Bo, LIU Jun-nan. (2016) Rational and Inheritance—Talking about the Influence of the Minimalism on German Product Design. J. Theoretical Research,112-113.

[5] HOU Nan-jing. (2018) The use of Design Geometry in Product Modeling Design. J. Theoretical Research,044-045.

[6] XUE Yan-min. (2012) Analysis of Brockmann's Posters Based on Geometry of Design. In: Advanced Materials Research.Switzerland.pp.1960-1966.

Design Creates Employment---Exploration on the Poverty Alleviation by Employment in Poverty-stricken Areas of Ethnic Minorities

Ming Xu [1], Xinyi Zeng [2] and Yijing Wang [1*]

[1] School of Architecture and Art,Central South University,Changsha, Hunan, 410083, China

[2]School of Architecture and Art,Central South University,Changsha, Hunan, 410083, China

[*]Corresponding author's e-mail: 642629097@qq.com

Abstract. It is purposed in 2019 Central Document No.1 that we should adhere to the general policy of giving priority to the development of agriculture and rural areas, and it is stressed in the chapter of "developing rural industries and expanding the channels of farmers' income "that we should "innovate and develop the rural handicraft industries with national and regional characteristics". Taking the comprehensive development and utilization of intangible cultural resource of Jianghua Yao autonomous county as the direction of targeted poverty alleviation, Central South University implemented the design, development and production of creative cultural products of Yao national group in Jianghua and transferred the right of commercial development to local enterprises, increasing the ability of producing by training the local residents' working skills. Through these methods, we can use economic confidence to drive the cultural confidence, ensure the steady employment, increase people's income and help the local ethnic groups find jobs in their hometown.

1. Introduction

Jianghua Yao autonomous county, the only autonomous county of Yao group as well as the autonomous county with the largest population of Yao group in China, is known as "the capital of Yao". It is pointed in the report of the 19th CPC National Congress that we should implement the strategy of revitalizing rural areas, promote the development of primary, secondary ad tertiary industries in rural areas, support and encourage farmers to find jobs and start their own business, and expand the channels of increasing income.However, due to the overall low cultural quality of the population in ethnic minority areas,large regional differences in traditional ethnic culture and customs, and poor transportation in remote areas, ethnic minority areas are facing greater difficulties in the process of alleviating poverty and urbanizing. As a result, poverty alleviation in ethnic minority areas is far from enough on their own, requiring more effective cooperation and support from the government, society and relevant organizations.

2. New connotation of employment and poverty alleviation in poverty-stricken ethnic minority areas

Located in the southernmost part of Hunan province and bordering on Guangdong and Guangxi, Jianghua Yao autonomous county is a typical "deficient, remote and poor"area with a impoverished population of 103,477. How to promote the large number of rural surplus labor to achieve employment, which is to help the employment of poor households, is an important topic.At present, due to the lack of employment opportunities, employability, initiative and adaptability of the labor force, the region's ability to absorb employment is not strong, and the economic structure conducive to increasing employment has not been established, which restricts the improvement of the livelihood of the minority areas along with the economic development.

In September 2018, the Strategy of Rejuvenating the Country Planning (2018-2022) issued by the Central Committee of the CPC and State Council proposed "the development of rural cultural industry", pointing out that "we should strengthen planning guide, typical demonstration, discover local talent and cultivating local culture, construct a batch of exhibit of farming culture and industry with characteristic and prominent advantage, build a batch of township village and group with characteristic culture industry".As a result, the crucial way of alleviating poverty in poverty-stricken area is to support the local workers to participate in developing and producing and to reasonably incline the working positions with effective organization through the design and production of creative cultural products of Jianghua Yao ethnic group and the transfer of the commercial rights from government to enterprises, hence to give a full play to the advantage of the local human resource. To carry out the poverty alleviation work from the CPC central, we must make effort on promoting the employment of surplus labor force in Jianghua Yao autonomous county, so that the people from ethnic group can feel the party's care and the warmth of the big family of the motherland, continue to consolidate the great situation of ethnic unity, and ensure the realization of the overall goal of social stability and long-term peace and stability.

3. "Culture": Inherent dynamic to drive the poverty alleviation in poverty-stricken areas with ethnic groups

The most direct manifestation of poverty is material poverty. People's long-term poverty and lack of material often lead to the bondage and depression in spiritual level, resulting in spiritual poverty. For the poor, lack of material is the external manifestation, and the restraint of thought is the internal source. To get rid of poverty, first of all, we should use the advantage of "culture" to change the old ideas, and then establish the sense of self-reliance and confidence of the people of ethnic minorities in poverty-stricken areas, firmly build the faith of poverty alleviation, and set up the confidence to shake off poverty from the mind-set, so as to stimulate their inherent dynamic to shake off poverty and get rich, and play the role of "assisting aspiration". Secondly, the employment platform is used to provide professional skills and employment opportunities so as to increase the income of the poor and cut the poverty root from the cultural source of poverty.

The 18th National Congress of the CPC proposed "cultural confidence", which enabled China's cultural industry to obtain an unprecedented opportunity for development. According to the survey, 70% of the creative cultural resources based on intangible heritage are concentrated in "old, deficient, remote and poor" areas. However, due to the unreasonable use of folk cultural resources and the weak local folk traditions, the Yao ethnic culture in Jianghua county has a low cognition of its own culture. Both the inheritors of intangible cultural heritage and local experts of Yao ethnic group are over 70 years old and have no successors. When our team visited Mr. Li Yibiao, the inheritor of long drum making, we found that Mr. Li lived in a shanty and had a poor life. It was difficult to make a living by making the long drum by hand. It can be seen from this that the excellent traditional national culture here cannot be transformed into material wealth, and people's main income depends on the income and expenditure of planting, forestry, animal husbandry and other first industries. Therefore, it is a practical way to protect the excellent traditional culture of Jianghua and increase people's spiritual confidence to explore and promote the national culture in the area of Jianghua Yao autonomous county. Taking Jianghua county as the research object, our team visited Jianghua for more than 10 times. We seeked and searched the Yao brocade “Babao quilt”, discussed the local culture and gathered the relevant material with the inheritors of Jianghua Yao’s long drum dance such as Zhao Minghua, Pan Shangke, Pan Xiong, Pan Ying and Li Yibiao. Therefore, relying on the cultural resources of Yao minority in Jianghua, we design and produce creative cultural products with creativity as the core. The final design products will be incubated in Jianghua, and then to achieve the goal of increasing employment in Jianghua area, to alleviate poverty through employment.

Due to the long-term poverty,the subject of poverty has negative adaptation to the status quo and forms the ideology such as dependency psychology and subservience consciousness ,which is the root of poverty culture and the internal cause of poverty. Its exterior characteristic is reflected in lacking courage, volition and confidence of shaking off poverty and becoming rich. From the perspective of poverty alleviation, this reveals the importance of endogenous poverty alleviation. Only with a strong will and enough internal motivation can external wisdom really come into play.

4. “Design” : Realistic effect to expand of poverty alleviation in poverty-stricken minority areas

Due to the low overall cultural quality of the farmers and the insufficient means of getting rich, to obtain practical skills to promote employment and solve the immediate livelihood problem is the realistic way to achieve poverty alleviation. For the poverty-stricken areas of ethnic minorities, “design” focuses on the production of designing products and vocational skills training to meet their needs of poverty alleviation, improve their economic income status, and thus realize the practical purpose of “poverty alleviation by employment”. For example, the long drum dance of the Yao minority in Jianghua is the most unique and representative dance art form among the Yao minority in China. It is one of the national intangible cultural heritage projects.

The decorative elements involved in the existing culture of ethnic minorities can be used in modern design, such as the product design in which the patterns of ethnic characteristics are introduced into the creative life. The active use of traditional elements in contemporary folk has completed the preliminary extraction and transformation from tradition to modernity, so it can be conveniently combined with modern art design. It can be said that there is a great space to develop the derivative cultural and creative design of ethnic minorities. Poverty alleviation not only means to support the material but also the spirit and culture. Many ethnic minority areas in China have outstanding cultural resources of intangible cultural heritage with characteristics and large numbers, but they are little known or even gradually disappear because no one cares about them. As the old saying goes, “It is better to teach a person to fish than to give a person a fish.” By tapping excellent cultural resources in ethnic minority areas, designing creative cultural products and training poor farmers to master relevant production and vocational skills, it plays a role of “fostering wisdom”. We will strengthen the “blood-making” function of poor farmers, provide guarantees for expanding the scope of employment and shift the focus of poverty alleviation from merely focusing on material and capital input to equally emphasizing the improvement of individual capabilities. It can be said that “poverty alleviation by design” promotes the overall quality and sustainable development ability of poor farmers, and drives them onto the road of making employment by skill and increasing income by skill.

5.“Transformation”: To form new opportunity of minority people’s local employment

Since 2016, Central South University has taken the comprehensive development and utilization of intangible cultural heritage resources in Jianghua Yao autonomous county as the direction of targeted poverty alleviation, developing tertiary industries such as culture and tourism, working hard to provide more jobs for people of all ethnic groups in poor areas, and implementing the design and production of creative and cultural products. Culture originates from life, then returns to life and decorates life. It completes this regurgitation process in the form of simple craft works.

Through the creative design, the products can be more widely introduced into modern life. For example, the long drum dance experience workshop construction organize and carry out long drum production and long drum dance on-site teaching experience. We invite the inheritors who adopt the ancient method of making long drum and folk artists to teach us how to carry out practical activities and make creative long drum by DIY (see figure 1). In modern design, ethnic patterns are more and more favored by young people and designers, and the design works applied in them are more and more popular in the market. The characteristic and ethnic style are increasingly popular among consumers. However, with the continuous development of design works with minority's pattern , there is also the situation of single pattern application, confusion and lack of cultural sources. At the present stage, the combination of ethnic cultural products and modern consumers' aesthetic vision is not enough, so that ethnic design is still only for minority consumers. Therefore, the professional members of our team designed the craft drum, and its pattern was extracted from the original Yao's pattern. While popularizing intangible cultural heritage, traditional folk culture is transformed into material wealth (see figure 2). In addition, the creative products are designed with workshop activities, and the commercial development right obtained from the design will be transferred to the enterprise in Jianghua county, so as to improve the production capacity by cultivating the labor skills of local people. We will use economic confidence to boost cultural confidence, ensure steady employment, increase people's income, and help local ethnic minorities find jobs.

Figure 1 Daily activities in long drum workshop

Figure 2 Craft long drum design

In this process of transformation, not only the government and relevant organizations need to play a leading role, but also the rural poor labor force to actively cooperate.To be specific, we need the correct guidance from the government first. The government occupies a leading position in the whole strategy of poverty alleviation by employment. In terms of policy formulation, the government should establish and improve the policy system for poverty alleviation by employment as soon as possible, and encourage more resources to be channeled into the work of poverty alleviation by employment. From the investment in the early stage of design to the tax preference in the later stage of commodity production, certain financial subsidies will be given to the poor labors to start their own businesses. At the same time, financial funds can be used to create more jobs for the poor labors. Second, strengthen both the horizontal and vertical supporting policies, give corresponding preferential loans to the enterprises and organizations that invest in the poor areas, and give discount loans or even interest-free loans to the poor labor force who actively start their own businesses. Therefore, enterprises should combine the actual situation of rural areas, choose the most suitable resources and industries for local resource endowment to invest, and create more jobs for the poor labor force. Finally, the poor households should cooperate and join in. The poverty-stricken households should abandon their dependence on relief for poverty alleviation, take an active part in employment in posts provided by the government and organizations, take the initiative to participate in various skills training and improve their own quality.

In the activities of mass entrepreneurship and innovation, the poverty-stricken labor force can actively participate in entrepreneurship activities of poverty alleviation according to their own advantages and resources, and carry out various forms of entrepreneurship by taking advantage of the Internet economy and the development of local tourism industry. Self-employment can not only help people find jobs and get rid of poverty, but also bring into full play the advantages of entrepreneurship in driving employment and lifting a number of poor households out of poverty. Income from hard work has proved to be the most effective way out of poverty.

6. Conclusion

Poverty alleviation by employment is an important way to build a moderately prosperous society in ethnic minority areas, an important measure to give full play to the rich advantages of local labor force, and a fundamental measure to achieve social stability and long-term peace and stability. Therefore, the government should let the poor people take the initiative to participate in the implementation and management of poverty alleviation projects, stimulate their initiative and enthusiasm for entrepreneurship and employment, help the poor people get rid of the narrow natural economic consciousness and the influence of poverty culture, and enable the poor people to explore the excellent traditional culture of ethnic minorities while getting economic benefits. In addition, the government should let the impoverished minority population build up "cultural confidence" and make it boost the "economic confidence". Thus to provide more employment opportunities and improve the production capacity of the job skills of the labor force, so that we can get poor labour more secure job, and guaranteed to work not only makes the poor income increased, but also extended the income stability, which help people thoroughly shake off poverty and eliminate the model of relief poverty alleviation existed in poverty-stricken ethnic minority areas for years.

Fund Project.

This paper is the project of the inheritance base of excellent Chinese traditional culture of the Ministry of Education Yao minority's long drum dance; Fund project of humanities and social science research program of the Ministry of Education Research on intangible creative cultural design of Yao minority in Hunan based on "cultural confidence"; Phased result of research project of teaching reform in colleges and universities in Hunan province: Research on the inheritance and promotion of traditional culture and art curriculum construction in colleges and universities based on "cultural confidence" --- taking the national intangible cultural heritage project in Hunan province as an example; Social science fund project of Hunan province Research on cultural creativity of Yao intangible cultural heritage in Hunan based on "cultural confidence".The Key Project of Philosophy and Social Sciences Research in Hunan Province Research on the Ways to Realize the Higher-Quality and Fuller Employment

References

[1] Zuo Mingzhang, Xiang Lei, Ma Yunpeng, Yang Dengfeng: Assisting Aspiration, Assisting Competency, Assisting Learning: Construction of "Trinity" Model of Educational Targeted Poverty Alleviation Promoted by Educational Informatization[J]. E-Education Research. 2019,40(03):13-19+33.Podani, J. (1994) Multivariate Data Analysis in Ecology and Systematics. SPB Publishing, The Hague.

[2] Zhou, Bin. Thoughts on poverty alleviation by education in southern Xinjiang[J]. New West,2018(22):63-64+67.

[3] Hu, Angang. Creating employment in rural areas and promoting rural revitalization[J]. China Fruit News,2017,34(12):2.

[4] Gong, Xiaokuan. Creative research on China's models of poverty alleviation in rural area[D]. Sichuan University, 2006.

IOP Conf. Series: Materials Science and Engineering **573** (2019) 012102 doi:10.1088/1757-899X/573/1/012102

Packaging Design of "Good Cup" Based on Visual Elements of Yao Ethnic Group

Ming Xu[1], Yijing Wang[2] and Xinyi Zeng[1*]

[1] School of Architecture and Art, Central South University, Changsha, Hunan, 410083, China

[2] School of Architecture and Art, Central South University, Changsha, Hunan, 410083, China

[*]Corresponding author's e-mail: 595975181@qq.com

Abstract. With the rapid development of globalization and technology, the inheritance and protection of intangible culture are attracting more and more attention, however, there are fault phenomena existing in the culture of Yao ethnic group. In order to further protect the excellent traditional culture of Yao minority and deeply excavate the Yao's characteristics, supported by long cultural environment of Yao minority in Jianghua, based on the theory of semiotics and chromatic, this paper aims to make extraction and design of Yao's long drum design, extracting Yao's long drum cultural elements and clothing elements then applying in the packaging design of "GOOD CUP". The development trend of product packaging design on the creative cultural products of Yao minority is analyzed rationally and discussed deeply, so as to boost the innovation of the Yao minority's traditional creative design and cultural connotation of expression, popularizing Yao's long drum culture, establishing cultural confidence, and contributing to the local economic development.

1. Introduction

The packaging of cultural and creative product belongs to a kind of special type, with the application of new materials and new invention of technology, it should not only meet the demand of the function of the general packaging, but also have the different geographical and ethnic culture idea. Only through the appropriate use of outstanding culture resources with rich connotation and a wide variety of topics, well revealing regional ethnic culture, the unique effect of product packaging can be obtained, and the chic experience and enjoy can be brought to the consumer. The author excavates the visual elements of long drum culture of the Yao minority in Jianghua, redesigns them from the perspectives of semiotics and color psychology with scientific and effective modern design methods, and applies them to the packaging design of cultural and creative product "GOOD CUP". This is not only an ingenious combination of technology and art, but also the collision between traditional cultural connotation and modern design thought. It is the best interpretation of carrying forward and reproducing the history and regional culture of the Yao minority in Jianghua.

2. Extraction of visual elements of Yao minority in Jianghua

The research on the extraction of Yao's dress patterns in Jianghua should start from different perspectives. According to the Signifier and Signified of Saussure's semiotics theory, Signifier mainly points to the modeling, composition, color and other aspects of decorative patterns, and the Signified mainly points to the cultural background, spiritual sustenance, symbolic significance and other sustenance and expression of decorative patterns. The unique cultural, regional and national characteristics of Yao minority in Jianghua are reflected in the design of cultural and creative product packaging patterns.

2.1. Historic elements of Yao minority

The first on is the long drum, which is recorded in Twelve Yao people crossing the mountain, "in front of the temple of the son of heaven, the king wore long sleeves; Yao people wore spotted clothes with red collar, carrying long wooden drum around the waist, dancing and singing with the music of Pipa[1]." Yao's long drum is used for two purposes: one is ritual vessel, offering sacrifices to heaven and earth ghosts and gods, the other is instrument, singing and dancing to liven things up. The long drum culture, no matter from the perspective of ritual procedure, generation or the origin of using, has the cultural relics of ancient totem worship, the cultural significance of entertaining people and gods, the creativity of the Yao people and the artistic expression of self-entertainment.

The second one is the twelve surnames of Yao, according to The Emperor of Ping's Document[2], Emperor of Ping used to fight with King of Gao. The emperor repeatedly failed in the war, but finally won the fight with the help of Panhu, the ancestor of Yao people. In order to reward Panhu, the emperor betroth his third daughter to him. Panhu and the princess gave birth to six sons and six daughters after their marriage. The emperor awarded 12 different surnames to the twelve children, respectively are Shen, Bao, Huang, Li, Deng, Zhou, Zhao, Hu, Tang, Lei, Feng, which is the origin of twelve surnames of Yao people. Due to the frequent historical migration, the Yao minority failed to unify their own writing system, and later they formed their own unique language under the influence of the Chinese language.

2.2. Dressing elements of Yao minority in Jianghua

Known as the "capital of Yao minority in China", Jianghua's Yao minority has formed its unique dress patterns in the process of historical changes and cultural development.

Yao's costume patterns are composed of simple basic patterns in a simple way. They choose simple, unadorned, exquisite and natural material in life as the basic pattern. The basic pattern can be roughly divided into theme of figure, plant, animal and geometric pattern. These basic patterns follow the beauty of form and the aesthetic law of inner beauty to form point or line or plane patterns through arrangement and combination. Adopting the method of symmetrical composition, the clothes are made more orderly, neat and pleasing to appreciate, which is a kind of primitive and simple aesthetic doctrine. Book of Diverse Crafts records that the sky has times, the earth has the air, the work has the skill, the material has the beauty, and then the four for good, which is a good overview of the Yao's handicraft clothing external appearance and connotation characteristics.

3. Application of visual elements in design practice

Based on the data collected in the early stage, the author extracted the historical elements, dressing elements and colour elements of the Yao minority. the author unified the traditional aesthetics with the modern aesthetics and designed the brand LOGO and series of cultural and creative packaging.

3.1. Design of brand logo

Taking the calligraphy script to highlight the theme, the logo strengthens the the characteristic of parataxis in Chinese. "GOOD CUP" means good cup in English, which is also homophonic with "cup of drum" in Chinese, producing the effect of 1+1>2 conveyed by brand cultural information. Design techniques such as modeling exaggeration and reduction of strokes are used to discretely reconstruct the glyph, and the beauty of strokes, structure and intention is reflected in the body of the text to express the meaning of the form. The left side shows the English explanation and the right side is

[1] Pipa: a traditional Chinese string instrument
[2] The Emperor of Ping's Document: The historic documents that record the history of Yao ethnic group

stamped by the Yao minority in Jianghua, which conforms to the international trend and balances the atmosphere as a whole. Figure 1 is the LOGO of GOOD CUP.

Figure 1. Logo of GOOD CUP

3.2. Application of visual elements in packaging graphic

The first one is the packaging graphics with plant theme. Yao women use flowers, fruits, roots and leaves in nature as the creative source to extract the most representative features, so that people can clearly identify different varieties of plants, which is simple and natural, and convenient for weaving and embroidery. The author extracted and redesigned the octagonal flower pattern, gardenia flower pattern, plane tree pattern and rice grain pattern. Octagonal flower is the most common pattern of Yao minority. Its symmetrical shape has a strong sense of form. And because of its pungent flavour can expel parasite and snakes, it has the implied meaning of the sun and luminosity. Gardenias grow in pairs, which means purity and beauty. As the rice is planted in pieces, Yao people abstract and simplify it to form a zigzag pattern, which symbolizes the full harvest of rice and embodies the Yao people's wish to have a good harvest and live in peace and contentment.

According to the principle of plane composition, as shown in Figure 2, a new unit shape is formed by redesigning individual patterns of plant patterns, and then combined with two consecutive and four consecutive forms of expression, giving people a sense of balance, harmony, symmetry and joy. Using simple and vivid life material, following the aesthetic rules of formal beauty and inner beauty, Yao's clothing pattern is designed through different style and basic symbols combined point, line and plane. It refine the art of the material and the induction, having a strong sense of form and aesthetic.

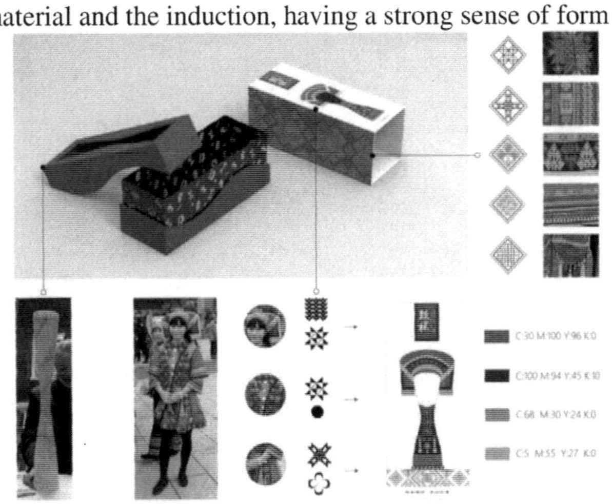

Figure 2. Packaging design of plant theme

The second one is the packaging graphics of animal themes. The animal patterns in Yao's costumes are mostly from poultry in daily life, mythical animals and animals in the mountains. Yao women are skillful and have a strong sense of national culture. With their unique aesthetic taste, they symbolize these seemingly ordinary animal elements and endow them with theme connotation, which is simple and unadorned, national but individual. The dragon dog is the ancestor of the Yao people in Yao's mythology. In Yao costumes, the spider pattern can be expressed in two forms, namely square and circle because Yao people always remember the spider's kindness to them. Bird patterns include a

IOP Conf. Series: Materials Science and Engineering **573** (2019) 012102 doi:10.1088/1757-899X/573/1/012102

single great hawk spreading wings and pairs of birds mutually looking at each other. Yao minority in Jianghua holds a "Song bird festival" every year to commemorate the Yao men and women who pursue pure love. The butterfly has a wide varieties, symbolizing happiness and love.

As shown in Figure 3, the animal pattern is redesigned to form a new unit shape, and then combined with the expression of two consecutive and four consecutive forms. Then, by extracting the Yao men's clothing elements and combining with the Yao animal patterns, the flat figure is designed by using the drum shape and applied to the packaging design.

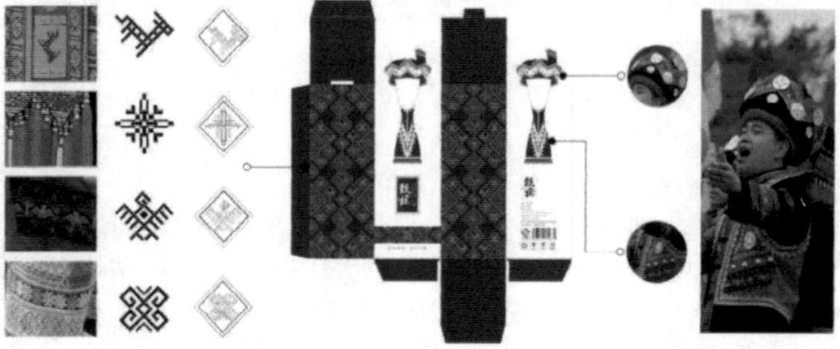

Figure 3. Packaging design of animal theme

The third is the figure theme packaging graphics. People is one of the most common forms with its action form also kaleidoscopic. Basic humanoid pattern is generally arranged and distributed in line, forming a unique form of beauty. Men's and women's patterns are simplified and refined by Yao people. Men's clothes are simple and elegant, and their heads are wrapped in headscarves. Yao women's clothes are relatively gorgeous and diverse, and Yao people only distinguish female patterns from male patterns in the head. In the process of historical migration, Yao people would work together for farming and hunting. At the same time, due to the frequent wars in ancient times, Yao people suffered a serious loss of personnel, so extending their descendants is the wish of Yao people. As a result, thread-bound or face-like humanoid pattern will be shown in clothing, trouser legs, headscarves, and so on. As shown in Figure 4, the figure patterns are applied to packaging design in two consecutive directions and four consecutive directions.

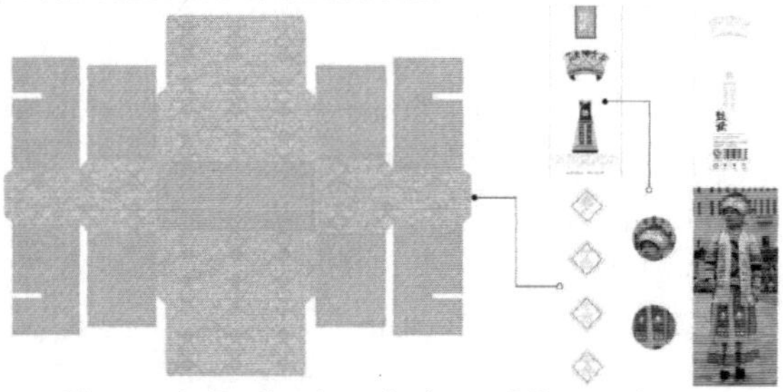

Figure 4. Packaging design of figure theme

The fourth is the packaging design of twelve Yao surnames. There are records about ancient Yao texts in the Study of Nvshu[3] Culture written by Li Qingfu. The twelve surnames are Pan, Shen, Bao, Huang, Li, Deng, Zhou, Zhao, Hu, Yang, Lei and Feng. The Yao language and the twelve surnames are combined to design a unique pattern. Figure 5 is the main graphic design of the twelve Yao surnames.

[3] Nvshu: the world's only gender-defined and now disappearing "women's language ," known as Nvshu, circulated exclusively in Jiangyong County of Hunan Province in south China

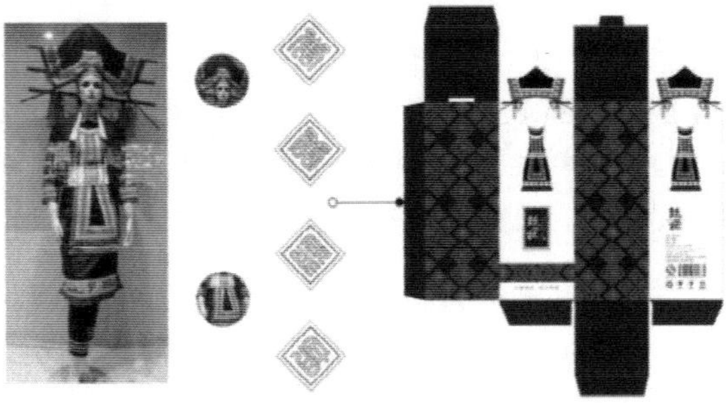

Figure 5. Packaging design of twelve Yao surnames theme

4. Conclusion

Through two field investigations in Jianghua county, the author has a preliminary exploration and understanding of the local Yao culture. After a large number of data collection and creation practice, the extraction and design of long drum culture and dressing culture, which are the most characteristic part of Yao minority, the package of GOOD CUP was made. In the design process, the theory of semiotics and colour psychology theory are applied as the support to make the package have chic modelling, unique style and functional capabilities. It puts forward a feasible plan to improve the existing visual symbols of the Yao minority, enhance the level of cultural innovation design, and achieve the goal of promoting the Yao culture to the public. In the next stage, the long drum visual elements with GOOD CUP as the carrier will no longer meet the needs of cultural promotion. The expression forms of cultural connotation should be considered in multiple dimensions and aspects, and the packaging of cultural and creative products should also be deeply considered.

Acknowledgments

Fund project. this paper is the project of the inheritance base of excellent Chinese traditional culture of the Ministry of Education Yao minority's long drum dance; Fund project of humanities and social science research program of the Ministry of Education Research on intangible creative cultural design of Yao minority in Hunan based on “cultural confidence”; Phased result of research project of teaching reform in colleges and universities in Hunan province: Research on the inheritance and promotion of traditional culture and art curriculum construction in colleges and universities based on “cultural confidence” ---taking the national intangible cultural heritage project in Hunan province as an example; Social science fund project of Hunan province Research on cultural creativity of Yao intangible cultural heritage in Hunan based on “cultural confidence”.

The Key Project of Philosophy and Social Sciences Research in Hunan Province Research on the Ways to Realize the Higher-Quality and Fuller Employment

Since the 18th National Congress of the Communist Party of China, the central government has paid great attention to the construction and enhancement of cultural confidence in contemporary China and the 19th National Congress of the Communist Party of China proposed the basic principle of Xi Jinping's new era cultural construction to promote the development of public cultural undertakings and cultural industry. With the introduction of the concept of cultural creative industry, creative cultural products are gradually entering the Chinese market. As the most representative dance art form of Yao minority, Jianghua Yao minority's long drum dance is listed as the national intangible cultural heritage. As a necessary performance prop, the drum plays an important role in Yao's traditional history and culture. However, there is a fault in the inheritance of the Yao's long drum culture, which makes how to preserve and generalize long drum culture become a significant part of exploring and developing the culture of Yao minority.

Reference

[1] Li, Xianghong. (2010) Yao Minority in Jianghua. The Ethnic Publishing House, Beijing.
[2] Zhang,Ruhua. (2004) Three main compositional design rules、legends. Jilin Fine Arts Press,

IOP Conf. Series: Materials Science and Engineering **573** (2019) 012102 doi:10.1088/1757-899X/573/1/012102

Changchun.
[3] Gao, Qinyan. (2012) Research on creative elements in the package of cultural and creative products. Packaging Engineering, 33(12):107-111.
[4] Li, Ning. (2000) Myth of Panhu and Yao Ethnic Group. Ethnic Forum, (04):38-40.

Personalized Product Design and Service System for Cloud Manufacturing

Kan Wu [1,2]**, Peiying Bian**[2]**, Yan Guo**[1]**, Xihui Yang**[3]

[1] School of Art, Xi'an University of Science and Technology, 710054, China

[2] School of Art, Xi'an University, 710065, China

[3]School of Mechano -Electronic Engineering, Xidian University

Corresponding author's e-mail: carriewuk3@hotmail.com

Abstract: Personalized product design system needs the support from cloud-manufacturing technology in order to promptly achieve. Based on the investigation on the characteristics of cloud-manufacturing system platform, a system architecture applied for the process of personalized product design service built on cloud-manufacturing platform was proposed, which was called Intelligent Cloud Manufacturing and Service System for Personalized Product Design (ICMSS-PPD), to connect both the customer needs and manufacturing resource for a better service.

1. Introduction

From a design perspective, the Product Service System (PSS) can be seen as a design-led system that integrates various services and manufacturing resources to design products that provide perfect service. In recent years, cloud manufacturing as a computing and service-oriented manufacturing model was proposed based on the purpose of resource-sharing in the manufacturing sector. Under the support of existing advanced manufacturing models, cloud computing, Internet of Things, virtualization and service-oriented technologies [1-2], cloud manufacturing aim at meeting growing demands for higher product individualization, broader global cooperation, knowledge-intensive innovation and increased market-response agility. Obviously, it is the trend that product design and service system only combine with advanced manufacturing systems—cloud manufacturing to achieve fast response to user demand for personalized design. For this purpose, Intelligent Cloud Manufacturing and Service System for Personalized Product Design (ICMSS-PPD) is proposed to explore the possibilities of an integration design service system combined with user-model and cloud manufacturing technology. ICMSS-PPD focus on providing the necessary technical support to improve design efficiency and quality during the process of personalized design service by well blend of information technology, design and manufacturing resources.

2. Theory and Methods

2.1. Product as a Service

United Nations Environment Planning Department first put forward the concept: Product Service System (PSS) in the 1990s, which also was called "functional product development". The aim of PSS is integrating type products and intangible services together as to a pre-designed system, which is concerning product PLC and sustainable development, maximize meet the specific needs of customers, to improve the competitiveness of traditional business models. This concept "product" is rather viewed as a serial of combination of functional services than a single entity. This is not only involved users, but also extends the field of the product. The earliest researches of PSS were mainly concentrated in European countries, especially in Scandinavia and Germany. Early studies were mainly focused on conceptual definitions and theoretical models such as the influencing factors of development, environmental performance evaluation.

2.2. *Service System of Industrial Products*

After 21 centuries, researchers have proposed the industrial product service system based on PSS make the PSS research from theoretical to application methods. For example, from 1998 to 2002 a lot of research about the PSS processes and methods (MEPSS) carried through under the EU Fifth Framework Programme funded; Hockerts proposed six-step innovation model in development progress of eco-efficient services in 2002, marked the beginning of PSS design practice research. Applications of PSS are mainly concentrated in the areas of service function of white goods, office furniture.

Horst Meier gives IPS2 [3] the concept of deepened understanding of the PSS in 2005. IPS2 is engaged to construct a social system suitable integration knowledge of the interaction and interdependence in process of products and services as a whole services pack to meet the user. It not only follow with interest product quality by product service providers paying more attention to it, but also the user experience involved in product life cycle services. It improves product innovation ability, develops new market space and reduces capital investment by providing users with high-quality product services. Through specialized service sharing, the "win-win" of product service providers and users can be realized.

Technically, the definition "Crowdsourcing" fist published by Jeff Howe[4] in the United States "Connection" magazine in June,2006 provides a viable approach for users' deeply involved in the way of resources share, and improve the whole process of product design and manufacturing, and thus was born a new mode of production organization. Crowdsourcing is the development and continuation of business models such as the development and continuation of outsourcing (outsourcing), open source code (open-sourcing).

By Internet, the term Crowdsourcing permits traditionally work usually supposed to be done by employees or external agencies will be openly assigned to a large group or community to complete, including the development of new technologies, design task, improving an algorithm or massive data analysis, full use of cognitive leverage collective intelligence and creativity to achieve massive information filtering, analysis and processing capabilities (such as voting mechanism), and have a certain degree of financial integration capabilities. Daren C. Brabham[5] puts forth a problem-based typology of crowdsourcing approaches: Knowledge Discovery & Management, Distributed Human Intelligence Tasking, Broadcast Search and Peer-Vetted Creative Production further explored crowdsourcing's disruptive effect to traditional product development and manufacturing.

2.3. *PSS model for cloud manufacturing*

Cloud manufacturing is service-oriented, High-quality low-cost, and smart networked manufacturing model, providing access to the network at any time for manufacturing the product life cycle process, services all types of manufacturing activities.

Cloud manufacturing has received more and more attentions. In China, a national high-tech research and development programme named 'Key Technologies in Cloud Manufacturing Platform' [1,2] was in place across academia, industry and government; In Europe, a project supported by European 7th Framework Programme called 'ManuCloud' aimed at developing distributed cloud product specification and establishing supply chain manufacturing execution infra-structure, Another on-going European Union (EU) project, CAPP-4-SMEs, is expected to realise collaborative and adaptive process planning for sustainable manufacturing environments. In the United States, Defense Advanced Research Projects Agency (DARPA) Manufacturing Experimentation and Out reach (MENTOR) programme funded a project to 'develop an integrated, distributed cloud-based design and manufacturing infrastructure that can support a progressive set of prize challenge competitions through integrated computer-aided design and manufacturing tools'. Based on the reviews, it has been found that most current literatures are focused on discussion about theoretical framework such as concepts and architecture, the key characteristics are not well understood[5,6]. There still lack research advances on cloud manufacturing, which would provide theoretical references for system implementation especially for personality product design. More importantly, from a practical perspective, more cloud manufacturing applications are needed to showcase the concept and push this new area forward. To address these issues, this paper presents an over-view of key characteristics of cloud manufacturing; then, under the analyses of infra-structure of cloud manufacturing to clarify main activities and relative resources involved in personality design process. A personalized product design-oriented cloud manufacturing and service system (ICMSS-PPD) is proposal. Finally, User model-driven personalized products and services module is illustrated.

3.Results

ICMSS-PPD is a user-centric-design service system integrated manufacturing information, digital virtual design resources, and cloud computing to provide most adequately resource utilization. Similar to the concept of cloud computing, where the service to customer of providing product function has become the main purpose, that is, the most important terms are whether customer needs and service abilities is available. Therefore, to adapt to cloud the inevitable trend of development, the user-driven product design and system development, new product concept that is transplanted to the cloud services platform manufacturing technology.

3.1. Cloud-manufactured technology platform product service system and characteristics

Tao Fei[3-7] consider the following typical characteristics of cloud manufacturing: the demand for services and manufacturing; uncertainty in the manufacturing; user participation manufacturing; transparent and integrated manufacturing; Active Manufacturing; supports multi-user manufacturing; support on-demand and pay-use manufacturing; low threshold, outsourced manufacturing; agile manufacturing; specialized manufacturing; share -based manufacturing capacity and transactions; manufacturing knowledge–based; groups of innovative manufacturing; green carbon-based manufacturing.

Meng Xiangxu[8] believed that cloud manufacturing embrace both the nature and characteristics of cloud computing and manufacturing infrastructure model. So that the basic services rramework of cloud computing as a service (IaaS), Platform as a Service (PaaS), Software as a Service (SaaS) features , such as the basic framework for the cloud model provides the service of data process. As one of manufacturing resources, cloud manufacturing reveal the essential characteristics of self-organization, collaboration, demand-driven and so on.

In addition to the above characteristics of cloud manufacturing, this paper argues that cloud manufacturing also has its own uniqueness, Such as： a) " Product as service" ； b) Multi-network collaboration services;c) " Crowdsourcing" terms ； d) User-centric design and ability available to quickly respect to individual users.

In addition, the research and application of cloud manufacturing service system is still at an initial stage, many questions have not been fully considered. For example, according to the literature [1-2][8] research point of view, the ideal cloud manufacturing system is desirable to provide a self- service resource organization optimization process, without human intervention and management. However, the responsibility for the needs of the business process may cause some problems such as safety, cost estimates, services billing and accounting , all these can not be ignored. If managed well, it could cause a system panic disorder, For concerns both customer or resource provider would trust in traditional product manufacturing way rather than cloud manufacturing system. Therefore, we consider it is necessary to establish a personalized design service cloud manufacturing systems in the cloud services platform, to be responsible for the link up between customer needs of individual design and the cloud manufacturing system.

IOP Conf. Series: Materials Science and Engineering **573** (2019) 012103 doi:10.1088/1757-899X/573/1/012103

3.2. Personalized design services for cloud manufacturing service system architecture

According to the study of Li Bohu, Zhang Lin , Tao Fei et al[9][11], system architecture of cloud manufacturing technology platform must contain three levels of resources: physical resources layer (P-Layer), cloud-manufacturing service layers of virtual resources , cloud application layer. In this hierarchy of three basic resources, personalized design is of the application layer and an intervene services ranging from a cloud service model to cloud manufacturing platforms. It needs to integrate user needs and demands and classify to determine which available service mode will be carried out, and which admissibility resources about design or manufacturing might be involved in. After the above analysis, to meet the resource-sharing characteristics of cloud manufacturing, personalized product design-oriented cloud manufacturing and service system shown in Figure 1.

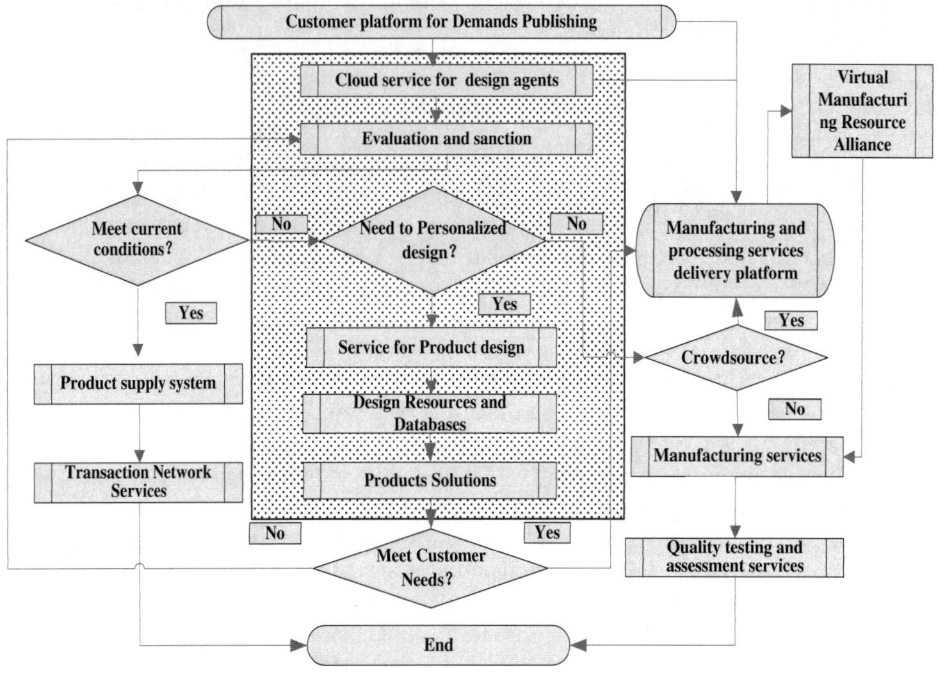

Figure 1. personalized product design-oriented cloud manufacturing and service system

4. Discussion

PSS need to establish a pre-designed program to quickly respond to the individual needs of Customer. That is design for User Model-Driven Personalized Products and Services Module (UMD-PDSM, Black in Figure 1). Cloud service for design agents (CS4DA) actual execute the function of UMD-PDSM. This part I have done in other literature [12][13], here is just a brief description of to the relation part to Personalized design services cloud platform based on the manufacture. CS4DA would be adequate for this job and consist of the design service Clouds which are part of cloud resource providers, personalized products design and services module must be considered to build a common structural model for personalized products. Its purpose is to give the design services cloud with the necessary knowledge and tools to quickly responded with the received request from Customer platform for Demands Publishing, and for further progress as organize, evaluate and allocate design resources and tasks to candidate collaborator in cloud manufacturing platform. Cloud service for design agents (CS4DA) also undertake the tasks including released crowdsourcing demand information to Manufacturing Service Platform, selection suitable partner, assessment the ability, services quality, every partner's distribution for this remands request and keep the service information with user. Detail of tasks include accepting demands—evaluation task—request for design services—request for manufacturing

Resources—process routes identified—assembling and addressing optimal—Evaluation and closing of the transaction and so on.

First step, as soon as confirm the customer request, CS4DA evaluation its' degree of innovation and rationality. If there is an available product service for ready-made, CS4D will inform Customer and transfer request task to transaction platform.

Second step, is to judge preference and creativity. Design services to solve things quickly identify and design requirements according to personas template, required design specifications and relevant Products Database. If not to find related templates to reference, generate new specific template (the highest level of customization) and follow the analyses according to customer needs. So here are three basic functions needed to support. The first is a typical example templates library, which can be invoked; The second is a rich role template library, which can provide design support functions of similar types of product template, diversified selection or partial improvement (a small amount of customization design); the third is that no design template and similar Personas template can be found to reference. CS4PD need to set up a new exception for this new user request, which tools supporting designer to analyses and rapid generate universal design specification are necessary. Figure 2 show a Subsystem of Personalized design services based on user model, more detail discussion could be review at literature.

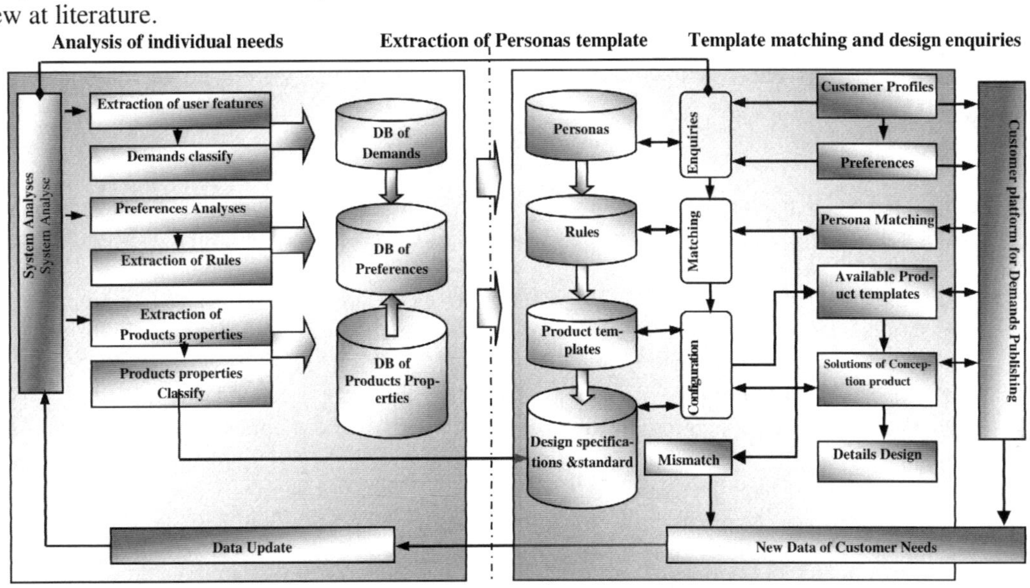

Figure 2. Subsystem of Personalized design services based on user model

After design services available, the full profile view of product design must to be determine as soon. CS4PD will be responsible for the relevant component or subsystem design of the requirements and technical details, distribute crowdsourcing information to seek collaborators, choose partners consist of Virtual Collaboration Services Unit (VCSU), formally place an order to meet the specific personality demands. CS4PD, virtual collaboration services Unit and user will sign contract or charge a certain deposit for design services charge and responsible for the user at this demands task. participating resources and Design Center for a one-to-one relationship,

Subsequent manufacture, assemble part manufactured by the design center in the cloud platform to find and form a virtual team, every member participating in this VCSU is a one-to-one relationship with CS4PD. For the sake of security, CS4PD have right to approval temporary connection between members in VCSU under its permissions if necessary helpful for further collaborative relationship.

Terms of geographic between members participating in VCSU must be considerate seriously at evaluation on service ability of VCSU in machining processes and parts were transferred and final logistics.

5. Conclusion

PSS theory regards product design and manufacturing as a more active service mode to provide users with the whole product life cycle(PLC). And it is very suitable for the design of personalized products. Building personalized design manufacturing system platform is a necessary means to ensure the rapid response and realization of personalized design. It is also a necessary means to integrate products, services, participants, networks and infrastructure. Intelligent Cloud Manufacturing and Service System for Personalized Product Design proposed in this paper can improve the productivity of personalized products design and application mode on cloud manufacturing platform. In order to better realize its functions, it is necessary to further construct a service evaluation system to evaluate the quality of service of virtual resources involved in the whole process of personalized design and cloud manufacturing services, One is to help the design service cloud agent center to effectively evaluate the selection of appropriate virtual manufacturing alliance members for users, and the other is to evaluate the value of contribution which service of every virtual resource member meet user needs after task, such as efficiency, utility and cost etc. In addition, all the evaluation of data would also provide an important credit score to every virtual resource member for next accession to cloud manufacturing platform system. Due to space limitations, this section will detail in our next paper.

Acknowledgements

This study is supported by Fund of Shaanxi Science and Technology Research and Development Plan Project（2015GY073）.

References

[1] LI Bohu, ZHANG Lin, WANG Shilong, TAO Fei,et al(2010). Cloud manufacturing: a new service-oriented manufac -turing mod-el[J].Computer Integrated Manufacturing Systems.16(1):1-7-16.

[2] L Zhang V C Venkatesh Y Luo and Y Cheng F tao(2011).Cloud manufacturing_ a computing and service-oriented manufact- uring model[J].Proc. IMechE Vol. 225 Part B: J. Engineering Manufacture, 225(2011):1969-1976.

[3] Lin Zhang Lihui Wang Fei Tao& Xudong Chai Lei ren(2014). Cloud manufacturing_ key characteristics and applications [J].International Journal of Computer Integrated Manufact- uring, pp:1-16.

[4] Yuanjun Lai, and Lin Zhang Fei tao(2013).A Parallel Method for Service Composition Optimal-Selection in Cloud Manufacturing System[C]//IEEE TRANSACTIONS ON INDUSTRIAL INFORMATICS,pp:2023-2033.

[5] Brabham, Daren C(2013). Crowdsourcing [M], MIT Press

[6] Jian Li Qingjia Huang Xiao Huang Kai Shuang Jie Wang Sen su(2013).Cost-efficient task scheduling for executing large programs in the cloud[J].Parallel Computing,2013, 39.pp: 177-188.

[7] Matthew John Greer David W. Rosen Dirk Schaefer Dazhong wu(2013). Cloud manufacturing_ Strategic vision and state-of-the-art[J].Journal of Manufacturing Systems,2013, 32.pp:564-579.

[8] Meng Xiang Xu, Liu Shijun, Wu Lei, et al(2011). Cloud manufacturing model and supporting technology [J]. Shandong University (Engineering Science), 2011 (5) :16-23.

[9] Yongliang Luo Fei Tao Bo Hu Li Lin zhang.Cloud manufacturing-a new manufacturing paradigm[J].Enterprise Information Systems,2012,8(2): 167-187.

[10] Meiera H, Uhlmann E, Krugcm,et al(2011). DynamicIPS2 networks and operations based on software Agents [J]. CIRP Journal of Manufacturing Science and Technology, 2010,3(2):165-173.

[11] Fei Tao Lin Zhang* Ying Cheng Yongliang Luo Bhaba R. Sarker Yuanjun laili(2013). A Ranking Chaos Algorithm for dual scheduling of cloud service and computing resource in priv -ate cloud[J].computer in industry,2013(64): 448-463.

[12] Wu Kan(2013), rough set role models - fuzzy clustering analysis, Computer Engineering and Applications. 2013.6
[13] Wu Kan(2013), The role of user requirements analysis model based on research, packaging, engineering .2013.5

IOP Conf. Series: Materials Science and Engineering **573** (2019) 012104 doi:10.1088/1757-899X/573/1/012104

Research on the Characteristic of Projectile Fragments Dispersion Based on Fragment Warhead Design

Yong Li[1*], Dong Chen[2], Ming Jiang[3] Zonghao Tian[4],

[1]No. 451 Huangshan Road, Hefei City, Anhui Province, China

[2]No. 451 Huangshan Road, Hefei City, Anhui Province, China

[3]No. 451 Huangshan Road, Hefei City, Anhui Province, China

[4]No. 451 Huangshan Road, Hefei City, Anhui Province, China

[*]Corresponding author's e-mail: wwwly100@foxmail.com

Abstract. Dispersion of fragments of projectiles is an important factor affecting the lethal efficiency of projectiles and one of the main objectives of optimal design of fragmented warheads. By using Taylor angle approximation, Shapiro formula and experimental measurement, the fragment dispersion characteristics are analyzed. The results show that if the mass of metal plate (fragments) detonated by explosive is constant, the characteristics of fragments scattering are basically determined; for a certain structure of shell warhead, the direction of fragments scattering is related to the initiation position and the direction of detonation wave propagation; and by marking the size of the dispersing sector, the range of fragments scattering can be determined.

1. Introduction

In the modern battlefield, the main way of warfare is to strike and destroy by fierce fire. Among them, the most important means is to use the fragments of projectiles to kill the personnel and equipment effectively, that is, the metal shell of projectiles is broken down by the explosion of internal explosive charge, resulting in the destruction of personnel, aircraft, vehicles and tank equipment. In order to make full use of the kill efficiency of fragments of projectiles, it is necessary to optimize the design of fragmented warheads in combination with the important parameters such as the number of fragments, the initial velocity of fragments, the mass distribution of fragments and the spatial distribution of fragments. As fragment dispersion is an important phenomenon after explosion of fragment warhead and various artillery shells, its spatial distribution is an important parameter to determine the fragment killing field. Therefore, the optimal spatial distribution (scattering) of the maximum number of effective fragments is the primary objective of the optimal design of fragmented warheads.

2. Fragment dispersion analysis based on Taylor angle approximation

Usually, it is necessary to assume that the moving direction of metal is perpendicular to the surface of the shell when the fragments are scattered by Gurmey equation. But in fact, this is only true when the detonation wave is incident vertically on the inner surface of the shell. When the detonation wave passes through the shell, Taylor angle approximation is needed. As shown in Figure 1, the projection of explosive detonation to metal plates.

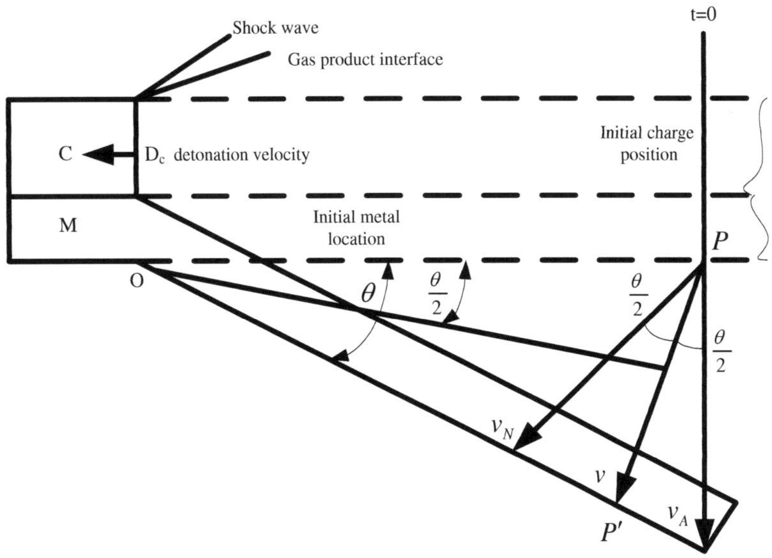

Figure 1 Projection of metal plate by explosive detonation

Analysis of the above figure shows that when the detonation wave passes over the surface of the flat plate, the flat plate deviates from the θ angle. Assuming that under steady-state conditions, the plate accelerates instantaneously from the initial position to the final velocity, and the metal plate undergoes pure rotational motion, and there is no change in length or thickness or shear flow, the original P -point plate element will reach P'-point after projection, and its length is $OP = OP'$. Vertical $\overline{PP'}$ leads from point O, because $\overline{OPP'}$ is an isosceles triangle, so this line is angular bisector, and the bisector θ. If the time of sweeping the detonation wave from point P to point O is t, then there is

$$\sin\frac{\theta}{2} = \frac{\overline{PP'}/2}{OP} = \frac{vt}{2D_e t} = \frac{v}{D_e} \tag{1}$$

The formula is Taylor's angle relation. Among them, v is the scattering velocity of metal plate, which can be obtained by Gurney equation. D_e is the detonation velocity of the explosive. From the analysis of the relationship, it can be seen that the angle $\frac{\theta}{2}$ between the scattering direction of the element and its surface normal can be determined by the formula.

Because the velocity component v_A of the initial position of the vertical plate is

$$v_A = D_e \tan\theta \tag{2}$$

Then there is

$$\frac{v}{v_A} = \frac{2D_e sin(\theta/2)}{D_e \tan\theta} = \frac{\cos\theta}{\cos(\theta/2)} \tag{3}$$

The velocity component perpendicular to the surface of the flying metal plate is

$$v_N = D_e sin\theta \tag{4}$$

It can be obtained as

$$\frac{v}{v_N} = \frac{2D_e sin(\theta/2)}{D_e \tan\theta} = \sec\frac{\theta}{2} \tag{5}$$

From the above analysis, it can be seen that, in practice, only a few percent correlation exists between v, v_A and v_N, that is, the value of $v/2D_e$ is approximately the same for most explosives in projectiles, so the value of θ is almost equal to a constant. If the mass of the driven metal plate is certain (e.g. prefabricated fragments), the velocity of the metal plate will increase with the increase of explosive detonation speed, so $v/2D_e$ is basically equal to a constant. It can be seen that if the metal plate (fragment) detonated by explosive has a certain mass, the characteristics of fragment dispersion of projectile are basically determined.

3. Fragment scattering direction calculation based on Shapiro formula

For the analysis of fragment dispersion, Taylor theory provides a basic idea for predicting the dispersion characteristics of static fragments, while Shapiro applies it to the calculation of the dispersion direction of fragments. According to Shapiro's hypothesis, the warhead is made of many rings arranged continuously, that is to say, the shell or warhead shell is superimposed by rings, and the centers of the rings are all on the symmetrical axis of the projectile body. Although this assumption is not consistent with the actual projectile shell, it still has sufficient accuracy in approximate calculation.

According to Shapiro theory, detonation wave is generated by booster or detonator and propagates outward in the form of spherical wave front. At the same time, the angle ϕ_1 between the normal of the warhead shell and the symmetrical axis of the projectile body, the angle ϕ_2 between the normal of the explosive wave front and the symmetrical axis of the projectile body, and the deviation angle of the fragment velocity vector from the normal of the projectile shell are set as ϕ_s, as shown in Fig. 2.

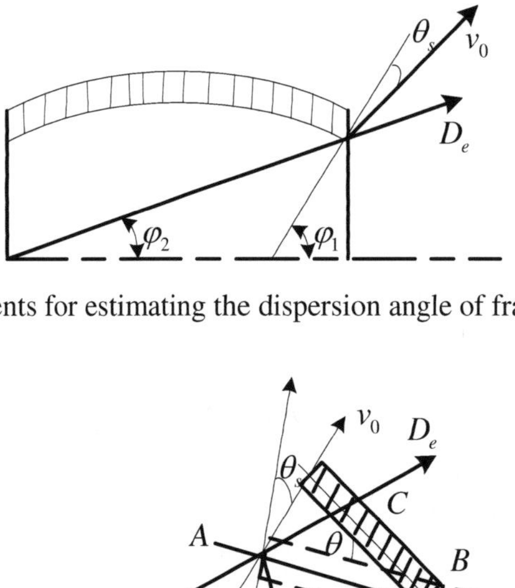

Figure 2 Elements for estimating the dispersion angle of fragment beam

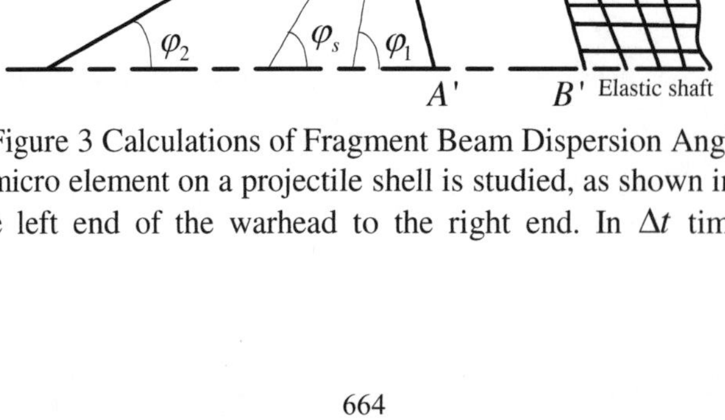

Figure 3 Calculations of Fragment Beam Dispersion Angle

The ring AB of a micro element on a projectile shell is studied, as shown in Fig. 3. The detonation wave moves from the left end of the warhead to the right end. In Δt time, the direction of the detonation wave acting on the AB shell remains unchanged. At this time, the detonation wave front moves from $A-A'$ to $B-B'$, and the distance travels is $D_e\Delta t$. The expansion velocity of A-point shell increases from 0 to v_0, and the shell AB turns θ angle. The length and thickness of the shell remain unchanged. In isosceles triangle ABC, according to sine theorem, we can get

$$\frac{\overline{AC}}{\sin\theta}=\frac{\overline{AB}}{\sin(\pi/2-\theta/2)} \quad (6)$$

Using Taylor's hypothesis, it can be seen that the acceleration of AB element to the final velocity is instantaneous. So there are

$$\overline{AC}=v_0\Delta t \quad (7)$$

$$\overline{AB}=\frac{D_e\Delta t}{\cos(\pi/2-\phi_1+\phi_2)} \quad (8)$$

By substituting the above two formulas into Formula (6), we can get the result which is

$$\sin\frac{\theta}{2}=\frac{v_0}{2D_e}\cos(\frac{\pi}{2}-\phi_1+\phi_2) \quad (9)$$

As shown in Figure 3.2, $\theta_s=\theta/2$. And when θ_s is very small, $\tan\theta_s\approx\sin\theta_s$. Therefore, the above formula can be rewritten as

$$\tan\theta_s=\frac{v_0}{2D_e}\cos(\frac{\pi}{2}-\phi_1+\phi_2) \quad (10)$$

This formula is the Shapiro formula. θ_s is the deflection angle of the fragment and v_0 is the initial velocity of the fragment. By analyzing the formula, it can be seen that the direction of fragments scattering is related to the initiation position and the propagation direction of detonation wave for a certain structure of shell warhead shell. At the same time, the scattering direction of fragments can be calculated by this method.

4. Determination of fragment scattering range based on test measurement

For the angle distribution (i.e. the range of dispersion) of fragments of projectiles, not only can the shell of projectiles be divided into a number of micro-elements calculated by theoretical method, but also can be determined by experimental results. That is to say, the target is surrounded by different distances around the warhead. After the warhead explodes, the number of fragments hit on each unit area of the target can be calculated, and the angular distribution density of fragments relative to the projectile axis can be obtained, as shown in Figure 4.

IOP Conf. Series: Materials Science and Engineering **573** (2019) 012104 doi:10.1088/1757-899X/573/1/012104

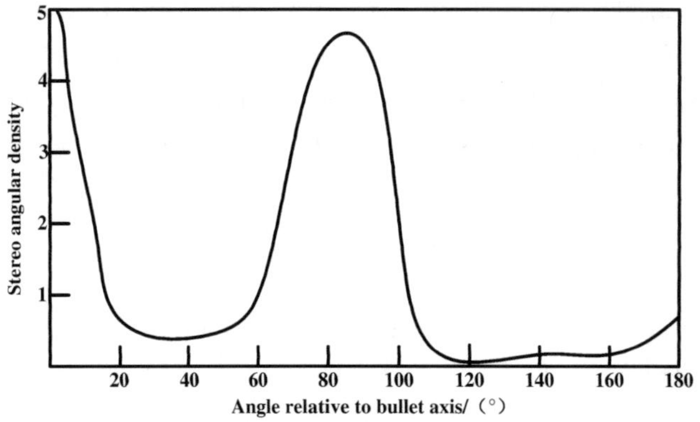

Figure 4 Typical curve of fragment dispersion range of projectile

When measuring the dispersion range of the projectile, the recovery and distribution of the fragments of the projectile are mainly analyzed. In the analysis of distribution, it is assumed that the fragments formed by warhead explosion distribute uniformly around the warhead. Therefore, fragments recovered in one sector can be considered to represent the whole distribution. The warhead is usually placed in the test field in a horizontal state, and then the recovery box, velocity target, speed recovery box and anti-jump baffle are placed in the appropriate position for determining the correlation between speed and quality. The recycling box is used to capture fragment samples intact and nondestructively, and to determine the spatial distribution of fragments accordingly. Therefore, the number of projectile fragments in any area around the warhead can be determined according to the position in the receiving box and plotted as a spatial distribution curve, as shown in Figure 5. Through the curve, the sector size of the effective fragments of the projectile can be marked, and the range of the fragments can be determined.

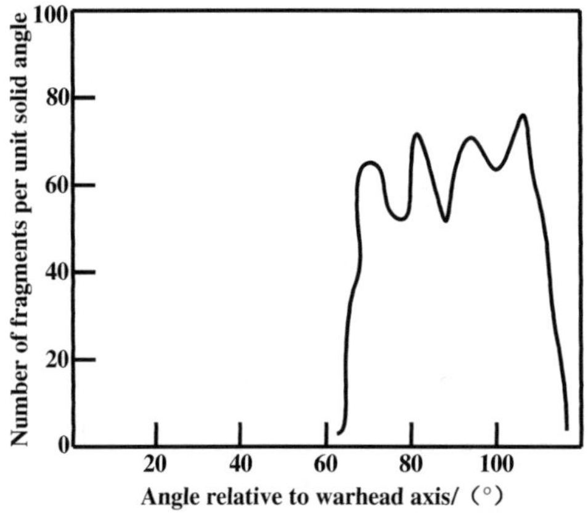

Figure 5 Spatial distribution curve of fragments

5. Conclusion

Projectile fragments are the killer elements directly acting on the target, and their effectiveness directly affects the killing effect of the target. The scattering characteristics of the projectile fragments are one of the important factors determining the killing efficiency of the projectile. When the Taylor angle approximation is used to analyze fragment dispersion, it is not necessary to assume that the moving direction of metal is perpendicular to the surface of the shell. If the mass of the metal plate (i.e. fragment) detonated by explosive is constant, the characteristics of fragment dispersion of projectile are basically determined. When Shapiro formula is used to calculate the direction of fragment dispersion, the direction of fragment dispersion is related to the detonation position and the direction of detonation wave propagation for the shell of a certain structure. Based on the experimental

IOP Conf. Series: Materials Science and Engineering **573** (2019) 012104 doi:10.1088/1757-899X/573/1/012104

measurement, the range of fragments is determined. The main purpose is to recover the fragments and draw the spatial distribution curve. The range of fragments is determined by indicating the size of the dispersed sector. Through theoretical analysis and numerical calculation of the dispersion characteristics of fragments of projectiles, it can provide theoretical reference for the optimal design of fragmented warheads, make them more in line with the needs of actual combat, and give full play to operational effectiveness.

References

[1] Kangqian Q. (2016) Study on the dispersion angle of composite damage element of prefabricated fragments and multi-explosively formed projectiles. Nanjing University of Technology, 1:52-55.

[2] Jingtian S., Jianjun S. (2018) Effect of Ammunition Flight Velocity on Direction Angle of Fragments Group. Journal of Ordnance and Equipment Engineering, 39(12): 97-100.

[3] Xiaoyue S., Hanshan L., Yun Z. (2017) Study on the density distribution and motion law of near-explosive fragments of projectiles . Mechanics and Electronics, 35(10): 36-39.

[4] Songnan L., Guowei Z., Xiaojie C., Wenbin H. (2018) Study on the influence of initiation point position on fragment dispersion direction. Journal of Ordnance and Equipment Engineering, 39(11): 49-53.

[5] Tao S., Guowei Z., Lei Q. (2017) A numerical simulation study on the dispersion law of prefabricated fragments. Journal of Ordnance and Equipment Engineering, 38(12): 45-47.

IOP Conf. Series: Materials Science and Engineering **573** (2019) 012105 doi:10.1088/1757-899X/573/1/012105

The visual expression of strength in product form design

Cong Deng[1,a], Fengyan Kan[1,b]

[1] Department of Industrial Design, Northeast Petroleum University, Daqing, Heilongjiang, 163318, China

[a]1012640374@qq.com

[b]9724844@qq.com

Abstract. The visualization of power is an important part of the semantic expression in product form design. Take visual as the pointcut, the structure, color, material and emotion are infiltrated into the design of the product appearance. By considering the design factors such as structure, color and material, and applying these factors to the design of product form reasonably, the product will have beautiful shape and more reasonable functions.

1. Introduction

The visualization of power comes from the book *Rhetoric of Design*[1]. Things can be seen everywhere in lives all show the power of nature in a certain structure and rules. Such as fierce wind, rainstorm, hail and earthquake are the concrete manifestation of the power of nature. Since the beginning of human consciousness, people have started to explore the power of nature in many ways. They apply the rules which they got from explore to production and life, it is helpful for people to improve their life experience. Therefore, when power is expressed in product form design, it should follow the rationality of mechanics, coordinate beauty and rationality, shape and function, and endow products with emotional power.

2. The visual cognition of power in product form design

Products are the mediums of communication between people, people and things, and people and the environment. They can convey different semantic information to users through appearance, which show the following characteristics: self-evident, self-instruction and semantic migrating use etc. A product, user can get the basic information, use, usage merely through its appearance. Also emotional information which product want to convey.

The external form of the product can give the user the most intuitive visual experience, and the information transmitted by different forms is slightly different. The product form of straight edge and right Angle presents simple mechanical aesthetic feeling or cold and elegant feeling; the plump product form is soft and intimate. Every function key of production, its distribution position, distribution mode, size, shape, colour, etc., will bring inspiration to the user's functional hints and operation modes. In addition, the change in the thickness of the external form of the product can be shown as the static or dynamic state of the product, the curve change shows the hardness or softness of the product, and the temperature of the colour reflects the stability or liveliness of the product [2].

The spiral structure of the pine cone makes people feel happy, and we feel positive vitality from its shape. The appearance of aircraft and automobile is mainly based on streamlined structure, which not only meets the requirements of beautiful appearance, but also solves the mechanical problems under the condition of high-speed driving. The sense of strength of product is showed by different expressional gimmick: the curve of its configuration straight, radius, coarse and smooth, the light and shade of colour collocation. When the structure and characteristics of products meet people's potential affection needs in visual balance, people will have a sense of pleasure. Therefore, the visual expression of power includes four aspects: the rationality of product structure design; the aesthetic expression of product colour; the rational choice of product material and the semantic transfer of emotion.

3. The visual expression of strength

3.1. Rationality of structural design

The products that make users satisfied and delight to use are not only the external beauty of the products, but also the overall beauty of the structure, color, function, material etc.. Whether the product is reasonable depends on the needs of users, in addition to the beautiful appearance of the form, but fully considering the internal structure of the product, ergonomic design principles and safety problem in the process of use. When the size and operation mode of the product can not conform to the user's habits, the satisfaction of user's experience will be reduced. Therefore, before designing a product, it is necessary to conduct a large number of user surveys, analyze the relevant structure knowledge system, and propose a reasonable solution to solve a certain problem. For example, the shapes of the hand tools are designed with the human hand structure as the reference and combines various functional characteristics of the hand to design, so that users can have a comfortable experience.

The product form includes the internal structure and external structure of the product [3], and the internal and external cooperation to form the whole product. The mainly structure in traditional Chinese furniture design is mortise and tenon joint structure. Each part is mutually compatible and flexible. This kind of furniture is also convenient for disassembly and combination during transportation. Like lot of furniture of Ming dynasty, woodiness is hard, the cent body that will alternate among each other makes up whole, when some among them is damaged, facilitate repair and change, although structure of this kind of furniture is flexible, but sedate, unvarnished however. IKEA has a mobile phone bracket as shown in Figure 1. The mobile phone bracket is mainly made of wood, and the traditional mortise and tenon structure is adroitly applied to the design of modern products. The reasonable structure shows the stability and strength of the product.

Figure 1. Mobile phone holder.

3.2. The color design of product

Color is one of the visual expressions of strength. Matching the color of the product clever can bring different emotional experiences to the user and arouse people's different inner emotions. Positive emotional color can stimulate people to accept the outside information, which will attract the attention of consumers, and then generate purchase behavior [4]. Nature is rich in color, dark color are heavy; and light colors reflects lightness. The power that color gives a person is reflected in the feeling of people towards temperature, weight, distance which different colors generated. The warm colors such as red and yellow make people easy to reminiscent of the sun, creating a hot and warm feeling, and that color easily brings people happy emotions. Cold colors such as green, green and blue evoke people's association with the sea and the shade of trees, leading to a sense of cold and loneliness, and easily leading to melancholy. The lightness of color affects people's judgment of weight. Warm colors with strong color have a sense of weight, while cool colors with weak color have a sense of lightness. The dark color is heavy and the light color is light [5]. For example, the lower part of the mechanical equipment mostly uses black and gray to increase the sense of stability of the product, and the upper part is mostly white which color gives people a light sense. The sense of distance that color gives people will affect the importance of different parts of the product in people's mind. Generally, the warm color with high brightness will have a sense of expansion, and there will be a sense of proximity in the distance. The cold colors with low brightness will have a sense of contraction, and the distance will give people a sense of distance.

In the color application of the product, the color gradation of the same color system or different color systems are unified, which will bring a sense of harmony to the whole product and achieve a unified visual in experience. As shown in figure 2, this series of poisonous mushroom furniture has bright and vivid colors for each individual product. Although the colors are gorgeous, the grayscale is highly unified, and the furniture also has a sense of harmony when put together. Besides, each single product has a round shape, bringing pleasure to users. Black, white, grey and silver can make people think of metal, industry and bring people hard and cold power. Blue is the color of the sky and the sea. It means deep, calm, and rational, and it can bring a serene and broad visual experience. Especially blue and white match each other, it will show a clear and refreshing feeling to people, giving them a psychological high-tech sense. According to the use environment and occasion of the product, with different colors, can add vitality to the product and market competitiveness.

Figure 2. Poisonous mushroom series furniture.

3.3. Material performance of the product

Material is one of the important elements for the perception of the external strength of a product. It combines human vision and tactile elements to bring users a rich emotional experience in the process of using the product. Artificial material such as metal, glass contains the feeling of science and technology and fashionable feeling, and the intimacy that wood, cloth and stone reflects a nature and comfortable feeling. According to the different materials, the power and emotional experience presented to the user is also different, such as: bulky and light, smooth and rough, gorgeous and elegant, hard and soft, warm and cold etc.. The combination of natural and fresh wood and industrial metal not only gives the product a sense of stability, but also makes the cold metal appear soft.

The combination of simple ceramics and natural wood enables us to enjoy a slow pace of beauty in the fast-paced modern life. The chair that make up with soft spin material and natural wood , giving people warm and steady feeling. The combination of metal and plastic, let the product produce a kind

IOP Conf. Series: Materials Science and Engineering **573** (2019) 012105 doi:10.1088/1757-899X/573/1/012105

of modern sense and high-tech, bring users a sense of refreshing visual. Therefore, the reasonable collocation of different product materials presents different visual impact.

The Japanese design starts with the essence of the product, eliminating all unnecessary processing and color, and simply leaving the material itself and the function, so that the user can get a return to the aesthetic experience. The design by Naoto Fukasawa is called "Lamp with tray", which is simple in appearance and combines different geometric shapes with unified colors. The base part of desk lamp is a tray, the purpose is to satisfy the behavior that user conveniently places key or purse when user get home. When the tray feels the gravity, it will light up by itself, giving the user a warm feeling. When the gravity disappears, the lamp will dim accordingly. This kind of interactive means, such as gesture, sound and smell, not only makes people feel interesting, but also is a natural and intuitive interactive form [6].This kind of unconscious behavior into visible design, can grasp the user's inner perception of power, will win more market.

3.4. emotional semantic transfer of product form

The user's favorable impression of the product is derived from its nifty shape, intimate copy writing and interesting cartoons, but actually it is the change in the emotional experience brought by the details to consumers. The design case of the match by Kaoru Mende , which is inspired by the scattered branches on the ground, and used the scattered branches as the basic shape of matches. The shape of the small branches that fall to the ground is beautiful, and the redesign can arouse people's impression of nature, fire and everything in the world, and they can do something before returning to the nature[7]. When people use matches, every time they open the match box, it will be a wonderful experience. This kind of original and ecological tree-like matches can convey people's feeling of getting close to nature.

Accordingly, product configuration that makes user satisfactory should be full of interest. The form of this kind of product is to apply the organic form in nature to the design of the external form of the product through the method of personification, exaggeration, combination, segmentation and other methods, so as to inject new vitality into the product. The popularization of modern new materials, new technologies and new processes enables the production of more interesting and abstract product forms, and brings more choices for users. When we see an airplane, we will think of birds flying freely in the sky. When seeing the black and white stripes, it is easy to think of the shape of zebra. These products, which can arouse users' imagination, have certain symbolic and suggestive functions. The natural sphere, for example, presents a round, soft and lovely state without any external factors. Therefore, the shape with the sphere as the basic element naturally brings the user a sense of intimacy and easy acceptance. At the same time, the ball is the product of the same volume of the most material saving shape, simple structure, reduce the number of product parts. In the design of children's products, it is often used in the form of roundness, which conveys a safe, friendly information and conforms to children's cognitive psychology.

4. Conclusion

Contemporary consumers attach importance to the satisfaction of user experience and emotional needs. Beautiful and easy-to-use products can increase users' happiness. A good product will have a reasonable structure, harmonized shape and function, beauty and rationality. Therefore, through the research of the product's structure, color, material and emotional information, it can enrich the designer's method to optimize the design of product form, and make the product design more humanized and more socially serviceable.

References

[1] Masayuki Kurokawa. ,ZHANG-Yu's Translation.(2014) Rhetoric of Design. Hebei Fine Arts Publishing House, Shijiazhuang.
[2] GAO-Liqun. (2010) Product Semantic Design. China Machine Press, Beijing.

[3] LUO-Jiali. (2009)Analysis of the Importance and Influencing Factors of Product Structure Design. J. Packaging Engineering,06:128.
[4] LI-Bo. (2011) Research on Modernity in Product Design. D. Hunan University.
[5] WU-Yongxiang. (2006)The choice of appearance color of mechanical products. J. Modern Manufacturing Engineering,10:129-130.
[6] XIE-Chuanwei. (2014)The Application of Emotional Design in User Experience. J. Design, 02:21-22.
[7] Kenya Hara. ,ZHU-E's Translation.(2006) Design in design. Shandong People's Publishing House, Jinan.

Cognitive Function of Artificial Intelligence Literature

Tao Feng[1,2*]

[1]College of Philosophy, Nankai University, Tianjin,300350, China

[2]Institute of World Literature, School of Foreign Languages, Peking University, Beijing,100871,China

[*]Corresponding author's e-mail: taofeng@mail.nankai.edu.cn

Abstract. Artificial Intelligence literature is a literary work that is generated by AI in semi-automatic or automatic method. AI literature needs the integration of symbolism and connectionism AI to finally realize general AI in literature. AI literature can also promote research in AI technology, such as emotion and creativity. The metaphor in literature provides the possibility for AI to learn and master the analogy-associative thinking of human beings.

1. History and Concept: Artificial Intelligence Literature

People have imagined the concept of automatic writing in long history. In ancient Greece, God used the mouth of the psalmist to sing poetry. People seemed to be the machines that God manipulated. In the modern art such as Dadaism and Surrealism, God gives way to the subconscious, automatic writing is to liberate the subconscious mind controlled by consciousness. After the birth of the computer, the era of programming computer programs to write literature began. In 1962, R.M. Worthy and others invented the poetry creation process, "Auto-beatnik", its poems published in magazine *Horizon*.[1] In 2018, IBM researchers created AI programs to create Sonnet poems. They input 2,600 sonnets in AI, and AI was able to generate new sonnets and pass blind tests. However, researchers acknowledge that the poems created by these machines still lack emotion.[2] In China, AI writing poetry technology has achieved significant breakthroughs. The Microsoft Xiaobing program developed by MSRA, after learning the works of 519 poets, published the poem collection in 2017.In terms of novel, in 1998, S. Bringsjord, D. Ferrucci developed the computer software "Brutus", which can write a short story in only 15 seconds. In 2013, D. Kazemi founded the NaNoGenMo (National Novel Generation Month), a conference of Writing Robot: "where people are challenged to write code that writes a novel." [3] NaNoGenMo has gradually become a composite art movement using algorithm experimental literary forms. In 2016, M. O'Neill and others created a computer program that could write a Harry Potter style novel. Script is more regular than novel and therefore automatic writing of script is smoother. In 1976, Mechan's TALESI'IN can write simple characters and straightforward narrative scripts. In 1983, M. Lebowitz developed the program "Universe" that automatically writes scripts. It focuses on the coherence and consistency factors in writing, such as role, relationship and historical background, can write a series of soap operas[4].In 2016, the first short film "Sunspring" written by AI was shot and shortlisted in the top ten of a film competition. In the writing of biographical literature, one of the authors, Feng Tao, working with researchers at Peking University to plan a project called "Dabai Write Biography".

In addition to the writing of fictional literature, the performance of AI in non-fiction writing is more eye-catching. In 2015, Tencent released a financial report written by the automatic writing program "Dreamwriter". Baidu writing programs can also produce three kinds of news: quick reports, knowledge, and information collection. The Associated Press has used the robot editor "Wordsmith" to write, but its job is mainly to produce corporate financial report. The program Saatchi can write 50 ad scripts based on location, behavioral characteristics and career data. Crystal Knows understands their personality by analyzing the email language of the user's colleagues or customers, and then it helps to optimize the user's email accordingly.

Therefore, the authors define "AI literature" as "literary work that is generated by AI in semi-automatic or automatic method". The biggest difference between AI literature and previous program writing and digital literature is that AI program has become the subjectivity of writing to some extent, and it is possible to create literary works independently. At the same time, we should also see that under weak AI technology, AI could not be completely separated from human work, pre-programming and data inputting, as well as supervised learning and output evaluation, all of these requires human assistance.

2. Art and Technology: The Technical Orientation of AI Literature

At present, the technology of AI literature mainly has two major directions, one is symbolism whose method is from top to bottom. Experts in this school believe that it is possible to analyze a certain type of knowledge of human beings, set up a framework for AI, and form an expert system and database in advance. For example, in the writing script, story can be divided into many dimensions: plot, character, setting, literary theme, writing style, imagination, and so on. The designer of Brutus said: "our approach has been to bestow the Brutus architecture with a counterpart to every substantive aspect of human literary genius."[5].The advantage of this method is that the structure is clear and the accuracy is high. The disadvantage is that the application range is narrow and the text's style is limited. The other approach is bottom-up connectionism, which is based on the latest machine learning, combined with big data technology. This school advocates to input a large amount of text data for AI, letting itself find the rules of the text and generate works. For example, Tsinghua's "Jiu Ge" program, after learning 300,000 poems, can write Chinese classical verses. AI in this way has a certain degree of independence and can generate more diverse works. However, it requires a large amount of data, and it requires a lot of manpower in the early data labeling and classification.

The authors believe that the more ideal way is the combination of symbolism and connectionism. Now some writing procedure like Baidu writing program is based on this combination. It needs to divide the core process "automatic writing" into three stages: document planning and micro-planning and surface realization, each of which requires data entry and human supervised learning. However, this low level of combination is that people need to be more involved in the creation process of AI. This is just a semi-automatic writing process. The real fusion in the future is that after people set up the initial program, AI can judge the text, collect and input data, and finally generate the work independently. AI realizes the general intelligence in writing.

At present, the main purpose of AI writing technology is not to create real literary works, but to aim to the research of AI natural language processing, affective computing and cognitive function. For example, in order to improve the interactivity of its APP, Google's AI has read 2865 romance novels for better understanding of human language. Software engineer Andrew Dai said the purpose of the work was to improve search application technology and provide feedback to users, as well as improve the "smart response" capabilities of mailboxes to make them more natural and diverse. He believes that romance novels are an excellent tool for training AI because their plots have a commonality that allows AI to focus on understanding the subtleties of human language.

AI writing uses affective computing technology, which allows AI to grasp the partial relationship between words and human emotions. In this way, it is possible to set emotional programs for AI and to understand human emotions. For example, Microsoft Xiao Bing is "based on the affection computing framework proposed by Microsoft, with EQ as the main direction, creating Conversational AI with complete sense and knowledge map"[6]. Psychologist A. Damasio found that emotions, feelings and affections are the basis of human understanding of activities and minds. Therefore, if AI wants to achieve true intelligence, truly understand human beings, and realize human-computer communication, it must understand human emotions, and even it needs emotional programs. And this kind of emotional learning can be realized through AI literature.

In addition, one purpose of AI literature is to achieve the creativity of AI. The creators of the writing program "Brutus" claim that Brutus has achieved raw origination.[5] And if AI can create literary works, then it can realize human creativity. Shen Xiangyang, the developer of Microsoft Xiaobing, put forward the "three principles of AI creation": he said, "The process of AI creation must correspond to some kind of creative behavior of human beings, not a simple substitute for human labor."[6] This kind of creative behavior is embodied in literature and art obviously. That is why the report issued by Oxford University also regards creative work such as writers as the most difficult substitute for AI. Only if AI could have real intelligence, is it truly created.

3. Metaphor and Analogy-Associative Thinking: The Cognitive Function of AI Literature Language

In addition, the authors believe that AI can also enhance its cognitive and behavioral ability through the learning and understanding of human language. The current AI requires big data and only can accomplish a single task. But humans do not need too much data, and can learn a variety of skills to complete a variety of tasks. The authors believe that the root causing of AI's big data - small tasks is that AI lacks human analogy - associative thinking. And if we want to train AI to get this ability, let it learn from the language of AI literature.

The first thing that AI literature must solve is natural language understanding. However, at the level of natural language processing, AI is far from human beings in some aspects, such as semantic problems, common sense processing, and analogical reasoning. This is because AI is very different from human beings in the way of language processing. The language of the computer is pre-defined, only in the deep learning stage, the computer realizes the induction to a certain extent, and realizes the feedback and correction of the language through the big data of network. Human language is not only innately defined, but also acquired posteriori experience, which means that human language is inseparable from human's experience. Human language contains rich semantics which are rooted in the diversity of this experience. This human's experience is the basis of human analogy-associative thinking, and this kind of thinking makes people's language have the essence of metaphor. On the one hand, humans learn the language and understand the deep meaning (metaphor) behind the language through analogy-associative thinking, on the other hand, through metaphor, to further exercise the analogy-associative thinking ability.

Both AI experts J. Hawkins and R. Kurzweil mentioned a key cognitive model of auto-association in the human cerebral cortex, which means arousal of the overall memory through a partial memory or fragment.[6] The authors believe that human analogy-associative ability is an important foundation for language with rich semantics. Language is not only the symbolic representation of objects, but also contains people's understanding, imagination and experience of the object. The speaker and the listener share through the dialogue not only the reference of a fixed object, but the understanding, experience and even value of the two parties. Therefore, the language has rich other meanings besides the object. That is what we often say "the meaning beyond words."

Metaphor has a very complex connotation. In traditional idea of metaphor, it is generally regarded as a means of rhetoric. However, the modern idea extends it to the basic characteristics of human cognition and thinking. Linguists G.Lakoff and M.Johnson believe that metaphor is a basic way of thinking throughout people's daily lives.[8] They divide metaphor into three major types: orientational metaphor, ontological metaphor, and structural metaphor. Spatial orientation is one of the foundations of human cognition. Kant believes that time and space are intuitive forms of human cognition. The establishment of the spatial sense often comes from the body's experience of the external world. In addition, entity perception transforms the abstract, opaque emotions and mental states of thinking into tangible entities, which also requires human experience. Therefore, the body has become a sign of measuring the external world and inner feelings. Structural metaphors show that human physical experience can be systematically transformed into a structure that corresponds to the conceptual system. We can also see from the character-creation of Oracle bones that the oracle bones have a lot of words related to body parts and organs. Therefore, when human experience, thinking and emotion are expressed through language, metaphor is applied everywhere. Metaphor connects the human body and the world, which is the basis of human cognition. Metaphor is the embodiment of human analogy, association, imagination and other thinking in language. At the same time, people continue to learn and strengthen these abilities in metaphor. In other words, language is not only a manifestation of human thinking ability, but it will in turn strengthen and consolidate some of human cognitive ability. This two-way feedback process is lacking in current AI.

In order to truly understand natural language, AI must understand rich semantics. Human beings can understand the implication because of human analogy, association, imagination and etc. Men can connect superficially different words through the metaphorical relationship between languages. There are two ways to bestow AI associative ability. One is the combination of human and machine as envisioned by Hawkins and Kurzweil, adding counterpart of human new cerebral cortex to the machine, because human language, memory and auto-association skills come from this part. But they also point out that human emotions, affections and feelings are not produced in the new cortex of the brain, and such mechanisms are difficult to imitate. We also know that metaphor has a lot of contents about human subjective feelings, emotions and affections. Without these experiences and participation, metaphors may be difficult to produce. Another method, which is now commonly used, is to collect all the relevant semantics of human sentences through network big data and keyword search methods, to build a corpus, and let AI grasp the meaning of these sentences through deep learning. However, this method is still only a passive imitation, unable to truly understand the language and create new words or metaphors. Keyword searching can only search from the similarity of words in a single way, and cannot consider the connection between rich images behind semantics.

The authors believe that AI natural language understanding should pay attention to the metaphorical features of human thinking and language.AI experts should study and learn the metaphorical structure of human beings in depth, set up procedures for AI agent, and put metaphorical structures in human language into it. In addition to this top-down setup, you can also let the machine learn from the bottom up. For example, we can create a human metaphorical corpus, let AI learn and master the methods of generating metaphor through deep learning methods, and then generate new metaphors. The combination of these two methods allows AI to use and generate metaphorical language like humans.AI learning and using metaphors in language can also reinforce and enhance its own analogy-association capabilities, and it is even possible to extend this capability from language to image, action and the entire representation system. And if AI really masters the analogy-associative ability, then its intelligence will move forward a big step.

In the literary text, there is a rich metaphorical corpus. Moreover, literary text is a unified organism, not a collection of word fragments. After AI learning metaphors, we also need to let AI learn in the entire literary text. In this case, AI learns not simple words-words pairing, but a more complicated connection of sentence and thinking.

4.Conclusion

With the development of AI technology, AI literature has slowly grown from simple program writing, and it is possible to create literary works relatively independently. The combination of symbolism and connectionism is the future development direction of AI literature. AI literature can promote the development of technology. Through the study of human literary works and the imitation of human writing methods, AI can enhance emotional learning, improve emotional communication skills with humans, and enhance creativity. And a large number of metaphors in literary works can provide a rich metaphorical corpus for AI. By learning human metaphor, AI can understand and learn human analogy-associative thinking to some extent. This kind of thinking enables AI to acquire more knowledge and creativity on the basis of a few experiences, thus achieving generality. AI literature reflects the combination and promotion of art and technology, and also provides a way for AI to master human cognition.

IOP Conf. Series: Materials Science and Engineering **573** (2019) 012106 doi:10.1088/1757-899X/573/1/012106

Acknowledgments

This paper was supported by the National Social Science Fund of China, the name of this project is "The Language Thought in Adorno's Philosophy". No. 16BZX118

References

[1] http://02mydafsoup-01.soup.io/post/394476511/Auto-Beatnik-Computer-Poems
[2] Jey Han Lau and others, "Deep-speare: A joint neural model of poetic language, meter and rhyme", https://arxiv.org/pdf/1807.03491.pdf
[3] https://github.com/NaNoGenMo
[4] Lebowitz,M.(1983) Creating a Story-Telling Universe. In: Bundy,A.(Eds.), Proceedings of the 8th International Joint Conference on AI. Karlsruhe, FRG, William Kaufmann, pp.63-65
[5] Bringsjord,S., Ferrucci,D.(2000)AI and Literary Creativity： Inside the Mind of Brutus, a Storytelling Machine, Lawrence Erlbaum Associates, Pxxii, PXXV
[6] MSRA. (2017)Microsoft Xiao Bing: Three Principles of AI Creation, http://www.msra.cn/zh-cn/news/features/xiaoice-20170519
[7] Hawkins, J. and others,(2005) On Intelligence, St. Martin's Griffin,p.22
[8] Lakoff, G., Johnson, M.(2003)Metaphors we live by, The university of Chicago press,p.4

Design of planting mechanism for rapid return transplanter

Zhen Jia, Lin Chen*, Yunfeng Tan, Mocheng Guo and Lihua Zhang

College of Mechanical and Electrical Engineering, Sichuan Agricultural University, Yaan, Sichuan, 625000, China

*Corresponding author's e-mail: 461769837@qq.com

Abstract Large-scale crop planting uses transplanting machines to transplant seedlings. The planting mechanism of transplanting machines is the last step of the transplanting machine seedlings. In the production practice, the planting mechanism often causes the seedlings to be turned over and loging, which affects the survival rate of the crops. In response to this problem, the mathematical model of the planting mechanism which combined with the digital design method was established. SolidWorks2017 was used to design a rod-type plant with rapid return characteristics. The rationality of the planting mechanism was tested using ANSYS 17.0. A quick-return transplanter planting mechanism with reasonable motion characteristics and reliable work performance was obtained.

1. Introduction

At present, transplanting machines are widely used in the planting process of crops, including corn in food crops, various vegetables in cash crops, etc. In order to improve the survival rate, various seedlings are transplanted into the fields using transplanting machines [1-4]. The planting mechanism of the transplanting machine is an important link in the transplanting process. The planting mechanism on the existing transplanting machine often has the situation that the seedlings are turned over and fall, which leads to the roots of the seedlings to expose and die and has directly impact on the quality of planting [5-7]. The occurrence of this kind of situation is mostly caused by the unreasonable design of the planting mechanism. The planting mechanism does not meet the requirements of "zero-speed seedling" and "low-acceleration seedling" when seedlings are planted [8-9], the seedlings flip due to its own speed and acceleration when they are falling down. The multi-rod planting mechanism is widely used in semi-automatic transplanters because of its simple structure and cost-effective.

2. The structure and working principle of the transplanting plant

2.1. Structural design

He planting mechanism is fixed on the tractor, so it has the speed of the tractor in traveling direction when throwing seedlings. The planting mechanism itself generates acceleration, and the movement process of the duckbill device from the seedling to the seedling is also accelerated with the movement of the planting mechanism. If the speed and acceleration of the duckbill device of the planting mechanism are too large, the seedlings will be turned over and fall down. Therefore, the design of the planting mechanism should meet the principle of low speed and low acceleration of the duckbill seedlings [10].

The crank rocker mechanism in the four-bar mechanism has a rapid return characteristic [11], and the working stroke speed and acceleration of the crank rocker mechanism are lower than the return stroke. Therefore, applying the crank rocker mechanism to the planting mechanism can decrease the speed and acceleration of the process of seedling throwing and reduce the probability of the seedlings turning over and lodging. moreover, the application also make the mechanism quickly leave the working soil during the return journey to avoid the fall of the seedling. On this basis, Solidworks 2017 is used to draw a three-dimensional model of the planting mechanism of the quick-return transplanter, as shown in 'Figure 1'. The quick-return transplanting planting mechanism is composed of a crank-link mechanism and a planter. The crank-link mechanism is composed of a frame, a crank, a rocker and a connecting rod, and the planter is composed of a duckbill device I, a duckbill device II, a spring, a brake line, a seedling bucket and a support frame.

2.2. Working principle

A decription of the working principle of the rapid-return transplanter planting mechanism is given with reference to 'Figure 1'. The entire planting mechanism relies on the fixed connection of the frame and the tractor. The crank rotates at a constant speed to drive the rocker to swing the rocker. The lower end of the swing bar is fixed with a planter, and the planter moves with the end of the rocker. Because the reciprocating stroke of the end of the rocker is different, it means the working stroke speed and acceleration of the rocker end are low, while the return speed acceleration is high ,and therefore the plant has a quick return characteristic during the working process, meanwhile, the planter is relatively stable from the seedling to the seedling,so The seedlings are not easy to turn over and fall down.and it also quickly leave the soil when returning, avoiding the fall of the seedlings. At the beginning of the operation, the duckbills I and II remain closed, and the seedlings can be used to pick up the fallen seedlings, and the duckbills are then sent to the soil by means of a crank rocker mechanism. When the lower end of the planter moves to the lowest point, the external force pulls the brake line, and the lower ends of the duckbills I and II are opened, and the seedlings fall into the soil. As the tractor travels ,The entire working process continues to circulate, the seedlings are orderly planted in the working soil.

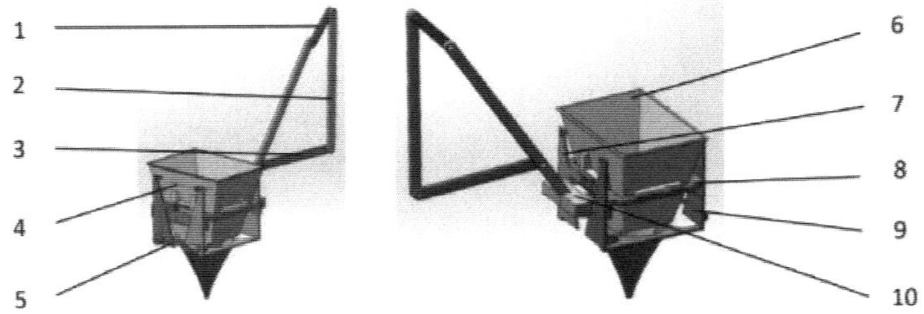

1 crank. 2 rack. 3 connecting rod. 4 brake line. 5 spring. 6 picking up the bucket. 7 duckbill I. 8 support frame. 9 duckbill II. 12 rocker.

Figure 1. Overall structure of the planting mechanism of the rapid return transplanter.

3. Theoretical analysis of the motion characteristics of the planting mechanism of the quick-return transplanter

3.1. Theoretical model

In order to facilitate the analysis of the motion characteristics of the planting mechanism, a structural diagram of the planting mechanism is established, as shown in 'Figure 2'. For racks $\overline{AD}$, the crank is indicated by $\overline{AB}$, the rocker is indicated by $\overline{BC}$, and the connecting rod is indicated by $\overline{CD}$.

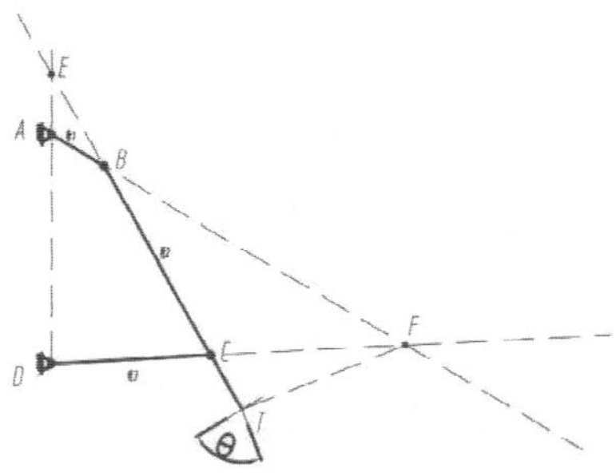

Figure 2. Quick return transplanter planting mechanism diagram.

3.2. Mathematical model

3.2.1. Speed equation. Relative to the instantaneous heart F

$$v_B = \omega_1 l_{AB} \quad (1)$$

$$v_B = \omega_2 l_{BF} \quad (2)$$

$$v_T = \frac{\omega_1 l_{AB} l_{FT}}{l_{BF}} \quad (3)$$

In the middle:

ω_1—the angular velocity of point B relative to the instantaneous center A, rad/s;
v_B—the speed of point B relative to point A, m/s;
v_T—the velocity of point T relative to point F, m/s;
ω_2—the angular velocity of the rod AB, rad/s.

3.2.2. Acceleration equation. According to the speed of the instantaneous heart A, F

$$v_{B1} = v_{B2} = \omega_1 l_{AB} = \omega_2 l_{BF} \quad (4)$$

In the middle:

v_{B1}—the speed of point B relative to the instantaneous center A, m / s;
v_{B2}—the speed of point B relative to the instantaneous center F, m/s;
Calculate the angular velocity of the rocker

$$\omega_2 = \omega_1 \frac{l_{AB}}{l_{BF}} \quad (5)$$

Differentiate time on both sides of the above formula

$$\alpha_2 = \omega_1 \frac{l_{BF}\frac{dl_{AB}}{dt} - l_{AB}\frac{dl_{BF}}{dt}}{l_{BF}^2} \quad (6)$$

Where $\omega_1$and $l_{AB}$are constants, inferred

$$\alpha_2 = -\omega_1 \frac{dl_{BF}}{dt}\frac{l_{AB}}{l_{BF}^2} \quad (7)$$

Where $\frac{dl_{BF}}{dt}$ is an unknown quantity, which is essentially the relative velocity v_{BF} of the moving point F moving on the AB extension line.

$$a_T^t = \alpha_2 l_{BT} \quad (8)$$

$$a_T^n = \frac{(v_T \cos\theta)^2}{l_{BT}} \quad (9)$$

$$\alpha_B = \frac{v_B^2}{l_{AB}} \quad (10)$$

$$a_T = a_B + a_T^n + a_T^t \quad (11)$$

In the middle:

α_2—Angular acceleration of rod 2, rad/s^2.
a_T^t—the tangential acceleration of point T, m/s^2;
a_T^n—the normal acceleration of point T, m/s^2;
a_B—the acceleration of point B, m/s^2;

a_T—the acceleration of point T, m/s^2.

4. Kinematics simulation of planting mechanism of rapid return transplanter

4.1. Motion trajectory analysis

The analysis of the trajectory of the planter take the lower end of the duckbill as analytical point, as shown in 'Figure 3'. The left half of the trajectory represents the working stroke, and the right half of the trajectory represents the return stroke. It can be directly seen that the trajectory has the characteristics of large working stroke and small return stroke, which can ensure stable work during seedling and quickly recover the planter after the seedling is completed.

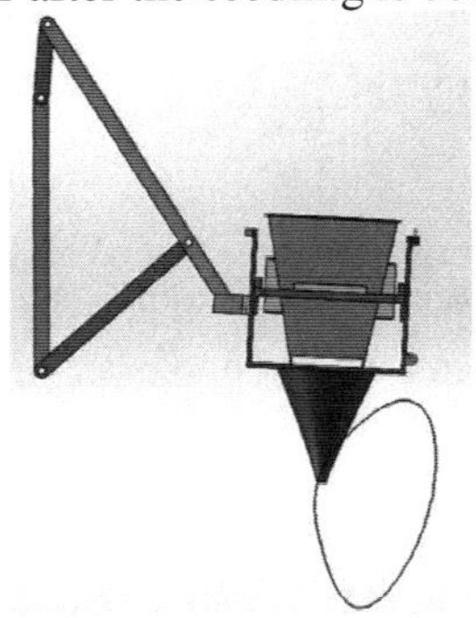

Figure 3. The working track of the lower end of the duckbill.

4.2. Speed and acceleration analysis

The lower end point of the planting duckbill can indicate the movement characteristics when the seedlings are detached from the planting mechanism. Therefore, the speed and the acceleration of the horizontal direction of the lower end of the duckbill are selected as the research object. According to the current working speed of most transplanters, the crank speed is set to 30r/min. The velocity and acceleration characteristics of the lower end of the duckbill are shown in 'Figure 4'. When the platypus opener is opened, the horizontal speed of the break point of the slap beak is about 100mm/s, and the horizontal acceleration is close to 0, so the planting mechanism can better satisfy the the requirements of "zero speed seedling" and "low acceleration". So the quality of seedling transplanting have improved .

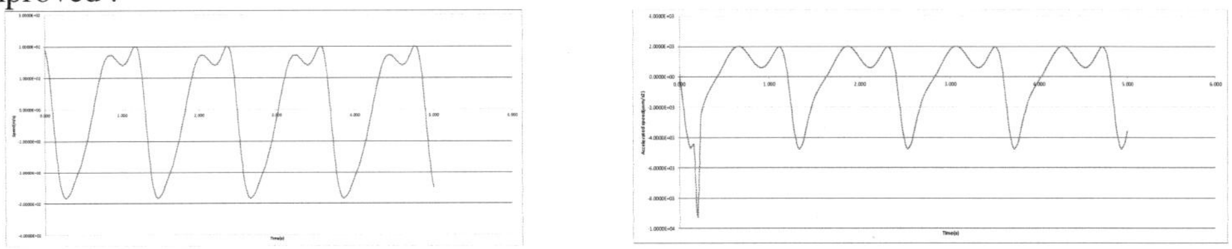

(a) speed (b) acceleration

Figure 4. Speed and acceleration diagram of the lower end of the duckbill.

5. Dynamic simulation of planting mechanism of quick-return transplanter

5.1. Crank strength check.

The crank provides power to the entire planting mechanism and is highly stressed. Static analysis of its structure is completed with ANSYS 17.0. Its stress and strain cloud diagrams are shown in 'Figure 5'. The maximum stress is 1.6955MPa, the maximum strain is 0.0030744mm. And the deformation of the material is negligible. so the structural strength of the crank is acceptable.

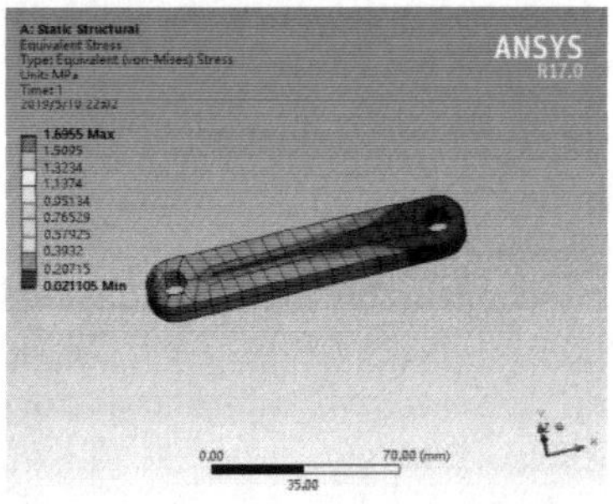

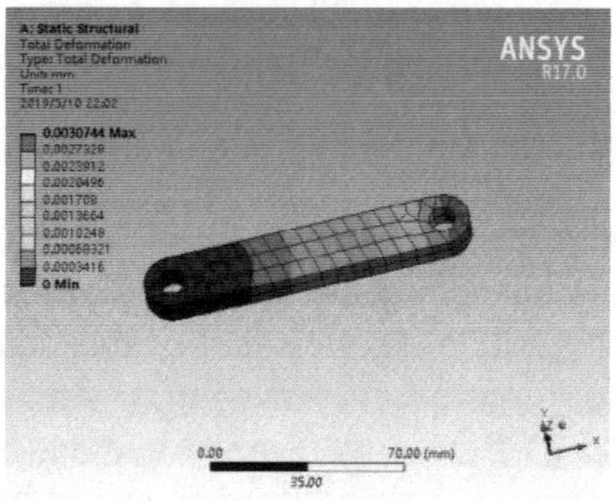

(a) stress (b) strain

Figure 5. Schematic diagram of the static structure analysis of the crank.

5.2. Modal analysis of the support frame

The 3D model of the support frame was drawn with SolidWorks2017, and the conversion format was imported into ANSYS 17.0. The modal analysis function in ANSYS Workbench was used for processing. Because considering the overall weight of the planting mechanism . Fix the hole connected to the rocker in the support frame model, apply the 5N pre-tightening force on the upper surface of the support frame to simulate the weight of the connecting part of the support frame, divide the mesh, and obtain the sixth-order mode of the support frame, as shown in Table1. When the tractor engine is working, the vibration frequency of the tractor engine is generally low. The support frame does not resonate with the engine, and the structure of the planting mechanism does not be destroyed due to resonance.

Table 1. Sixth-order modal frequency distribution table of the support frame.

Mode	Frequency[Hz]
1.	66.261
2.	143.62
3.	218.77
4.	443.96
5.	695.99
6.	724.34

6. Conclusion

According to the analysis of the mathematical model and the kinematics analysis of the model using Motion in SolidWorks2017, a rapid-return transplanter planting mechanism with reasonable motion characteristics was designed. This planting mechanism can meet the requirements of some crop seedlings transplanting. It can effectively decrease the rate of seedling lodging and increase the survival rate of seedlings. Static design analysis and modal analysis were carried out on the designed planting mechanism with ANSYS 17.0. The overall working performance of the planting parts was verified to ensure that the strength of the planting mechanism meets the requirements ,the structural characteristics are reasonable and it is not easy to fatigue and damage. This design can be reliably applied to the cultivation of crops such as corn and vegetables, improving work efficiency and further reducing the production cost of these crops.

Reference

[1] Chen Dajun, Hou Jialin, Shi Guoying, Li Tianhua, Zhang Yueshun, Chang Jinming. (2018) Analysis of the current status of domestic dryland transplanter technology[J]. Chinese Journal of Agricultural Mechanization, 39(07): 98-102.

[2] Jiang Lan, Wu Chongyou, Tang Qing, Zhang Min, Wang Gang. (2018) Kinetic model and parameter optimization of planting process of rapeseed-like seedling transplanting

machine[J]. Transactions of the Chinese Society of Agricultural Engineering, 34(21): 37-46.

[3] Shentu Liufang, Tang Lijie, Sun Xingwei, Sun Yajun. (2018) Design and Simulation of Finger-clamping Sweet Potato Transplanting Planting Mechanism[J]. Jiangsu Agricultural Sciences, 46(14): 223-226.

[4] Wang Hongbo. Design and experiment of planting mechanism of tomato automatic transplanting machine [D]. Shandong Agricultural University, 2018.

[5] Wang Pengfei. Design and analysis of planting mechanism for dryland seedling transplanting machine [D]. Shihezi University, 2017.

[6] Liu Lei, Chen Yongcheng, Zhang Wei. (2008) General Situation of Application and Development of Corps Transplanting Technology[J]. Journal of Agricultural Mechanization Research, 161(09): 240-243.

[7] Du Wei, Wang Weibing. (2016) Kinematics Analysis of Transplanting Parts under Soil Contact Conditions[J].Machinery Design & Manufacture, 299(01): 246-249.

[8] Wang Yipeng, Wang Weibing, Zhang Heng, Guo Xinfeng. (2018) Design and Simulation of Planting Mechanism of Staggered Chain Clamping Transplanter[J]. Mechanical Design and Manufacturing, 326(04): 68-71.

[9] Liu Wei, Cao Weibin, Tian Dongyang, Zhao Hongzheng. (2017) Kinematics analysis and experiment of effective zero-speed transplanting planting mechanism of seedlings[J]. Journal of Mechanical Engineering, 53(07): 76-84.

[10] Chen Jianneng, Wang Bohong, Zhang Xiang, Ren Genyong, Zhao Xiong. (2011) Kinematics model and parameter analysis of seedling mechanism of multi-rod zero speed seedling transplanting machine[J]. Transactions of the Chinese Society of Agricultural Engineering, 27(09): 7-12.

[11] Fu Qiulin, Chen Qi, Liu Jialun. (2011) Practice of Teaching Reform of Mechanical Design for Higher Vocational Education——Classroom Teaching Design of the Movement Characteristics of Crank and Rocker Mechanism[J]. Science and Technology Information, 374(18): 599-600.

IOP Conf. Series: Materials Science and Engineering **573** (2019) 012108 doi:10.1088/1757-899X/573/1/012108

Fundamentals and contributions to validation of constructive computational methodologies: a focused survey

Yongzhe Li* **and Imre Horváth**

Industrial Design Engineering, Delft University of Technology, The Netherlands

*Corresponding author's e-mail: y.li-8@tudelft.nl

Abstract. Computational mechanisms (complex structures of functionally connected algorithms) are developed very frequently by both the academia and the industry for specific applications, but the transferability to other applications is seldom addressed and systematically tested. This raises the need for application validity of constructive computational methodologies (CCMs). First, this background research paper clarifies the fundamental concepts related to application validation. Applicability validation focusses on the indicators of appropriateness with regards to a particular purpose. Then, it surveys the various approaches of application validation based on the publications available in contemporary literature. Its main finding is that applicability validation of CCMs seems to be a stepchild of academic research. The same applies to the industrial exploration of the applicability of CCMs tailored to a narrow family of applications to a wider range of applications.

1. Introduction

To support various knowledge-intensive activities during the life cycle of artifacts, constructive computational methodologies (CCMs) are developed [1]. The activities may concern requirement engineering, ideation, functional conceptualization, system architecting, detail design, simulation, optimization, implementation, operationalization, etc. The artifacts can be physical products (customer durables, electromechanical devices, working prototypes, etc.), software products (control software, middleware, application software, etc.), cyberware products (ontologies, knowledge warehouses, image repositories, etc.), and service products (mail service, transportation service, bank service, etc.). CCMs typically include five major constituents: (i) one or more underpinning theories, which provide a conceptual basis and a logical framework, (ii) a procedural scenario, which arranges the activities identified and supported by the methodology, (iii) a collection of methods, by means of which the activities can be executed in a structured and effective manner, (iv) a selection of instruments, which provide technical aids for the execution of activities, and (v) a specification of the constraints and goodness criteria for applicability and utilization of the concerned methodology (Figure 1).

The above-mentioned testing/proving of a computational methodology boils down to a sequence of justification, validation, and consolidation (J-V-C) activities. These three activities include distinct sets of confirmative actions, but complement each other from an epistemological and a methodological point of view. They focus respectively on the assessment of (i) logical properness (trustworthiness), (ii) argumentative appropriateness (adequacy), and (iii) outward relationships (agreement). The given order of consideration in the confirmative stage of research and/or development is implied by epistemological logical arguments. Though justification and validation are interconnected, they have distinct roles in demonstrating properness and appropriateness. Therefore, this paper focuses on validation only.

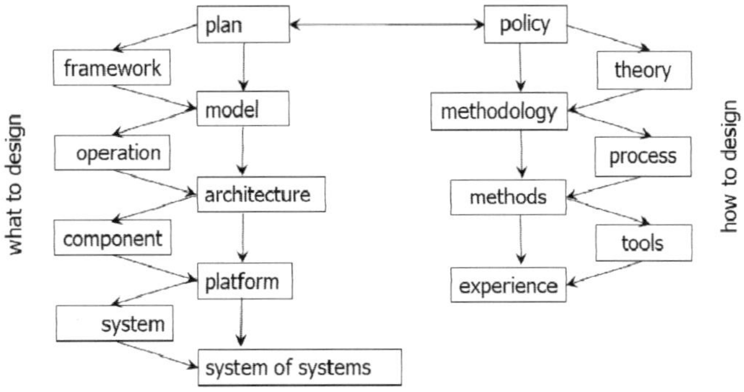

Figure 1. Constructive computational methodology in the context of system development

2. Disambiguation of the concept of validation

The word 'validation' is frequently used in publications intuitively or ambiguously. Authors often replace the term 'validation' by synonyms such as certification, attestation, authentication, or confirmation. In the tradition of scientific inquiry, validation is a multi-faceted activity focusing on confirmation of knowledge. As Barlas, Y. and Carpenter, S. stated, while verification refers to internal consistency, validation refers to the appropriateness of knowledge claims [2]. In the process of establishing scientific theories, validation is done to test and prove their appropriateness and utility for a purpose in some (application) context. It seems that theoretical and pragmatic interpretations of these two notions exist concurrently. Pragmatic interpretation is predominant in the engineering and design practice, where verification simply means testing against requirements, while validation is checking the fulfillment of the expected functionality [3].

In simple words, the main question of validation is whether a new body of knowledge (theory, framework, methodology, etc.) does what it is supposed to do? This may not be achievable because of biases or lack of adequacy. Thus, validation should focus on the critical factors (possible sources and forms of biases) that influence the conduct as well as the status of the outcome (findings) of research in contexts (various real-life situations). The major concepts and the related representative approaches are graphically visualized in Figure 2. Normally, internal validation and external validation are distinguished. Internal validation aims at exploring and evaluating biases, which cause a theory not doing what it is supposed to do. The conduct of internal validation can be characterized by its temporal dimension and periodicity, respectively. External validation checks issues related to generalizability and reusability and provides information for consolidation of theory.

In general, consolidation evaluates how strongly the conducted research and its outcome relate to other investigations of a phenomenon and the state of knowing. Consolidation also has internal and external aspects. The former is related to the investigation of the theory in the subsequent phases of the conducted research, whereas the latter concerns the relationship of the new theory with the current epistemology. Consolidation also informs about what may be expected to occur in other research contexts. Thus, consolidation can be done from the perspective of a conducted research project, but it may also be directed towards more general disciplinary issues and contexts (i.e., how the theory fits into a broader picture?). Due to this dual objective, it may involve both specialization and generalization of the validated knowledge.

Many aspects of justification and validation of CCMs have been addressed and covered with knowledge in the related literature. Applicability validation is among them, but not with specific attention to constructive computational methodologies. This is an important fact since the input information for methodology development is typically obtained by considering only a limited number of cases, while the broadest possible range of applications is expected when the methodology has been developed. This is known as the methodology incongruity phenomenon in the literature. Though important from the perspective of industrial use of methodologies, application validation is challenging due to the shortage of efficient testing approaches. Therefore, an extensive literature study and web search have been conducted to see the available approaches and to get insights into the opportunities, constraints and major issues.

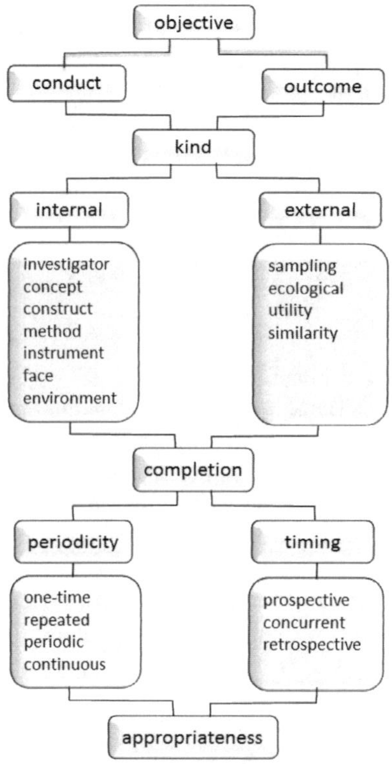

Figure 2. Taxonomic concepts and representative approaches of validation

3. Developments concerning systematic and rigorous validation

Five subject areas for validation have been identified, which are validation of: (i) data, information, and knowledge, (ii) concepts, theories, and models, (iii) objects, structures, and systems, (iv) actions, processes, and services, and (v) methods, methodologies, and tools. This served as the reasoning model for systematic exploration.

3.1. Validation of data, information, and knowledge

Input data and output data validations are routinely done in computer science. The former ensures that data inserted into an application satisfy the defined formats and other input criteria, whereas the latter focuses on the prognostic output from numerical models or computational algorithms. The applied methods are jointly determined by the type of data and the purpose of validation. For real-time validation of high-frequency data, Horn, W. et al. considered: (i) time-point-based methods, (ii) time-interval-based methods, (iii) trend-based methods, and (iv) time-independent methods [4]. Sensor data/information validation is a typical task in the context of CPSs, which includes two operations: (i) basic validation (i.e. detecting the presence of a fault), and (ii) separation (isolation of faulty elements). Ibarguengoytia, P.H. et al. developed an anytime algorithm for this purpose [5]. Information validation is a critical semantic evaluation of data with respect to the degree of agreement among sources. Mengshoel, O.J. made an inventory of the principles and practice of knowledge validation [6]. Donald, J.G. discussed disciplinary differences in knowledge validation [7]. On the computational side, Garcia, A.C. and Vivacqua, A.S. argued that knowledge validation is still a challenge at constructing knowledge-based systems [8]. As a specific issue of knowledge validation, Centobelli, P. et al. proposed a 3D fuzzy logic methodology to evaluate the level of misalignment between an enterprise's knowledge and the knowledge management systems it adopts [9].

3.2. Validation of concepts, theories, models

Model validation is about how closely a model used for exploration, prediction and/or prescription mirrors the represented reality (fidelity), and consequently, how well the information obtained by processing the model complies with reality (reliability). As McCarl, B.A., and Nelson, A.G. argued, fidelity and reliability imply the need for a priori tests and post priori tests [10]. This resonates with the opinion of Pace, D.K., who argued that validation has (i) conceptual validation aspect (when the anticipated fidelity of a conceptual simulation model is assessed), and results validation aspect (when the results of simulation by the implemented model are compared with an appropriate reference) [11]. If both end with positive confirmation, then the model supports the intended use. Balci, O. emphasized that verification, validation, testing, accreditation, certification and credibility assessment activities of models and simulations primarily deal with the measurement and assessment of their accuracy [12]. Thus, model validation is substantiating that a model, within its domain of applicability, performs with satisfactory accuracy consistent with the objectives and that it eventually boils down to the issue of building the right model. Malak R.J., and Paredis, C.J. identified three complementary validation problems, namely: (i) validity characterization, (ii) compatibility assessment, and (iii) adequacy assessment [13]. They claimed that individually these provide insight into the properties of predictive behavioral models, and together solve the validation problem and give sufficient information for validity descriptions.

3.3. Validation of objects, structures, and systems

Validation of physical objects typically involves empirical testing, whereas validation of (computer generated) virtual objects is based on computational analyses and simulations. Software validation extends to both the physical and the virtual domains. Many issues concerning system validation are in the focus of recent studies. To identify defects in the system specification, Brings, J. et al. dealt with the issue of supporting early validation of cyber-physical system (CPS) specifications based on model-based prototype development [14]. Gonzalez, A.J. and Barr, V. exposed the differences to be taken into account in the verification and validation of intelligent systems [15]. Feth, P. et al. focused on the validation of open and heterogeneous systems, such as CPSs in the automotive domain, and proposed a simulation-based framework, which integrates AUTOSAR applications [16]. The virtual validation concerned the functional behavior and the performance of the software. Bradley, T.H. claimed that validation of system models within a multi-disciplinary design framework has three primary components (i) independent validation of the contributing analyses, (ii) validation of the overall performance and behavior on system-level, and (iii) validation for utility with respect to design decision making [17]. Goubault, E. et al. dealt with the precision problem of validation in the case of engineered systems and found that they can be precise enough only if the "semantics" of the physical environment is modeled accurately [18].

3.4. Validation of actions, processes, and services

Heikkinen, H.L. et al. proposed five principles for validation of actions in research: (i) principle of historical continuity, (ii) principle of reflexivity, (iii) principle of dialectics, (iv) principle of workability, and (v) principle of evocativeness [19]. White, J.D. offered criteria for validating interpretations of administrative actions and showed how action theory can be rescued from misplaced claims of subjectivism by locating the meaning of an act with the intention of the actor [20]. Validation of mental actions differs from validation of motor actions. For instance, in the case of inferring or reasoning, the prevailing reasoning model has a strong influence on the outcome of mental action [21]. Validation of a process checks if its plan and manifestation meet the predetermined specifications and quality attributes. The approach differs if it has done on the basis of the process itself or using a process-oriented model. Barth, A. et al. argued that validation is essential to design science and is best used by researchers to guide the development and evaluation of new methods [22]. As sketched up by Kelly, B. et al., the early service validation practice focused on the front end of service provisioning [23]. It checked if the service specification meets its requirements and interacts sensibly with other services at the specification level. Combes, P., and Renard, B. reviewed the early approaches for validation of network services and exposed their dependence on the service specification, modeling, and implementation. Since service validation has rapidly gained importance, many specialized validation methods for digital and utility servicing have been proposed [24].

3.5. Validation of methods, methodologies, and tools

Method validation is one of the universally recognized elements of any rigorous research conduct [25]. For this reason, the number of techniques, protocols, and guidelines for research method validation is large. Their objective is to demonstrate whether the method is fit for a particular constructive or analytical purpose. Design methods are intangible artifacts. Consequently, their validation and testing require different approaches than those used in the case of tangible artifacts. In engineering contexts, methods have been proposed for both empirical and virtual validation. For instance, Mejía-Gutiérrez, R. and Carvajal-Arango, R. discussed four types of prototypes that are used in design validation during product development, namely: (i) abstract prototypes, (ii) virtual prototypes, (iii) functional prototypes, and (iv) physical prototype [26]. Validation of computational methods is largely influenced by the purpose of their application and the purpose. There are both experimental and analytical (logical) approaches to the validation of computational methods. The approach should be logically rigorous, internally consistent, and mathematically correct. Olewnik, A.T., and Lewis, K.E. elaborated on validation of design decision methodologies [27]. According to them, the complexity of prescriptive models makes their validation a difficult task. Validation of a software engineering methodology was seen as a challenging issue by Lee, S.W., and Rine, D.C., since it involves testing of data, algorithms, usability, and performance [28]. Ahmad, W. et al. analyzed validation techniques for safety-critical software such as (i) functional failure analysis, (ii) HAZard and OPerability Studies (HAZOP), (iii) failure modes and effects analysis), and (iv) fault trees analysis, and found that HAZOP was the most prominent one for safety validation by the industrial stakeholders [29].

4. Concluding remarks

In the tradition of scientific inquiry, internal validation and external validation goes hand-in-hand. Because, a body of knowledge may suffer from internal biases due to lack of procedural rigor, and/or may not be appropriate for or has serious limitations with regards to its intended purpose. In the context of design methods, validation is driven by the question of whether the method provides design solutions ‘correctly’ (effectiveness) and whether it provides ‘correct’ design solutions (efficiency). Correct in this context are design solutions with acceptable operational performance, which are designed and realized with less cost and/or in less time.

The above demonstrative survey provided evidence that validation works well as an alternative of mathematical modeling-based or real life experiments-based validation, which are difficult to realize in the case of design methods proposed by engineering research. The validity of a body of knowledge, that conveys the know-how of a design method, is determined by many more measures than just its usefulness. Validation has to do also with exploration and reduction of biases and errors (credibility provided by internal validation) and testing potentials and implications (transferability delivered by external validation). Nevertheless, usefulness is difficult to disprove as a pragmatic objective of validation of design methods.

Furthermore, with regards to the validation of constructive methodologies, the literature offers only limited insights. A robust approach to rationally investigate the appropriateness of a design method for a specific purpose is needed. The rational robustness can be explained by a principle of logic, which hides in the logical background and provides a ‘virtual’ verification in an application independent fashion. This is the principle of syllogism that makes validation of design methods possible based on their outcome (i.e. reasoning with consequences). Consequently, a design method should be logically correct to result in logically correct results. If the outcome experienced as correct (i.e. not conflicting with the empirically observable reality), then the method was correct. Notwithstanding, it is also an issue that no biases are supposed to be created by the way of applying the method and its results.

IOP Conf. Series: Materials Science and Engineering **573** (2019) 012108 doi:10.1088/1757-899X/573/1/012108

References

[1] Li, H., Van Oystaeyen, F., Taft, E., Nashed, Z. (2000) A Primer of Algebraic Geometry. Boca Raton: CRC Press.
[2] Barlas, Y., Carpenter, S. (1990) Philosophical roots of model validation: Two paradigms. System Dynamics Review. 6(2): 148-166.
[3] Du Bois, E., Horváth, I. (2013) Operationalization of the quadrant-based validation in case of a designerly software development methodology. In Proceedings of 19th International Conference on Engineering Design. Seoul, Korea. pp. 365-376.
[4] Horn, W., Miksch, S., Egghart, G., Popow, C., Paky, F. (1997) Effective data validation of high-frequency data: Time-point-, time-interval-, and trend-based methods. Computer in Biology and Medicine, 27(5): 389-409.
[5] Ibarguengoytia, P.H., Sucar, L.E., Vadera, S. (2001) Real time intelligent sensor validation. IEEE Transactions on Power Systems, 16(4): 770-775.
[6] Mengshoel, O.J. (1993) Knowledge validation: Principles and practice. IEEE Expert, 8(3): 62-68.
[7] Donald, J.G. (1995) Disciplinary differences in knowledge validation. New Directions for Teaching and Learning, 64: 6-17.
[8] Garcia, A.C. Vivacqua, A.S. (2018) Grounding knowledge acquisition with ontology explanation: A case study. Journal of Web Semantics. (In Press)
[9] Centobelli, P., Cerchione, R., Esposito, E. (2018) How to deal with knowledge management misalignment: A taxonomy based on a 3D fuzzy methodology. J. Knowl. Manag., 22(3): 538-566.
[10] McCarl, B.A., Nelson, A.G. (1984) Model validation - An overview with some emphasis on risk models. Review of Marketing and Agricultural Economics, 52(3): 153-173.
[11] Pace, D.K. (2004) Modeling and simulation verification and validation challenges. Johns Hopkins APL Technical Digest, 25(2): 163-172.
[12] Balci, O. (1998) Verification, validation, and accreditation. In Proceedings of the Winter Simulation Conference. pp. 41-48.
[13] Malak, R.J., Paredis, C.J. (2004) On characterizing and assessing the validity of behavioral models and their predictions. In Proceedings of ASME Design Engineering Technical Conferences, Salt Lake City, Utah, pp 1-13.
[14] Brings, J., Bohn, P., Bandyszak, T., Föcker, F., Daun, M. (2016) Model-based prototype development to support early validation of cyber-physical system specifications. In Proceedings of the REFSQ Workshops. Göteborg, Sweden. pp. 1-15
[15] Gonzalez, A.J., Barr, V. (2000) Validation and verification of intelligent systems - What are they and how are they different? J. Exp. Theor. Artif. In., 12(4): 407-420.
[16] Feth, P., Bauer, T., Kuhn, T. (2015) Virtual validation of cyber physical systems. In Software-Engineering and Management, pp. 201-206.
[17] Bradley, T.H. (2008) Modeling, design and energy management of fuel cell systems for aircraft. Doctoral dissertation, Georgia Institute of Technology.
[18] Goubault,E., Martel, M., Putot, S. (2006) Some future challenges in the validation of control systems. In Proceedings of European Congress on Embedded Real Time Software. pp. 1-11.
[19] Heikkinen, H.L., Huttunen, R., Syrjälä, L., Pesonen, J. (2012) Action research and narrative inquiry: Five principles for validation revisited. Edu. Act. Res., 20(1): 5-21.
[20] White, J.D. (1987). Action theory and literary interpretation. Admin. Soc., 19(3): 346-366.
[21] Ouyang, T.Y., Forbus, K.D. (2006) Strategy variations in analogical problem solving. In Proceedings of 21st National Conference on Artificial Intelligence. pp. 446-451.
[22] Barth, A., Caillaud, E., Rose, B. (2011) How to validate research in engineering design? In Proceedings of International Conference on Engineering Design. Copenhagen. pp. 1-10.
[23] Kelly, B., Crowther, M., King, J., Masson, R., DeLapeyre, J. (1995). Service validation and testing. FIW, 173-184.
[24] Combes, P., & Renard, B. (1999) Service validation. Comput. Netw., 31(17): 1817-1834.

IOP Conf. Series: Materials Science and Engineering **573** (2019) 012108 doi:10.1088/1757-899X/573/1/012108

[25] Frey, D.D., Dym, C.L. (2006). Validation of design methods: Lessons from medicine, Res. Eng. Des., 17: 45–57.
[26] Mejía-Gutiérrez, R., Carvajal-Arango, R. (2017) Design verification through virtual prototyping techniques based on Systems Engineering. Res. Eng. Des., 28(4): 477-494.
[27] Olewnik, A.T., Lewis, K. (2005). On validating engineering design decision support tools. Concurrent Eng., 13(2): 111-122.
[28] Lee, S.W., Rine, D.C. (2004) Case study methodology designed research in software engineering methodology validation. In Proceedings of SEKE Conference, pp. 117-122.
[29] Ahmad, W., Qamar, U., Hassan, S. (2015) Analyzing different validation and verification techniques for safety critical software systems. In Proceedings of 6th International Conference on Software Engineering and Service Science. Beijing, China. pp. 367-370.

IOP Conf. Series: Materials Science and Engineering **573** (2019) 012109 doi:10.1088/1757-899X/573/1/012109

Intelligent Savings Equipment Design Based on Urban and Rual Empty Nest Elderly

Lin Su, Na Qi and Xiaoying Wei

School of Art and Design, Xihua university, Chengdu, 610039, China

*Corresponding author's e-mail: haapylife8183@163.com

Abstract. China's aging population and large base, which brings a lot of social problems. They have unique psychological and physical characteristics, which make them lonely in a social corner. They are the predecessors who have contributed their youth and wisdom to the construction of a better country. Therefore, we should extend our most sincere thanks and do our best to create a comfortable environment for them. What needs to be studied in this paper is the urban and rural empty nester who need more attention. They are a group separated from the intelligent age. In theIntelligent era, they have to use smart Banks. Therefore, it is particularly important to design a savings device that is more suitable for urban and rural empty nester. This is a qualitative research to design an intelligent savings device for urban and rural empty nester.

1.Empty nests in urban and rural China

The empty nest theory was first proposed by Hill and Hansen in the 1930s. In 1949, in his paper, Gurlik divided each family population into six stages: formation、expansion、 stabilization、contraction、empty nest and disintegration according to the beginning and end of each family population. The concept of an empty nester refers to an elderly person who has no children to take care of, lives alone or lives with a couple. It can be divided into three situations: the first is the childless, widowed elderly without a wife. The second is the elderly who have children but live apart from them. The last one is that the children are away from home and the elderly have to be separated from them. The empty nesters in this case are couples with children over 60 years old who have been living alone outside for a long time or single elderly people.

According to the population data released by the national bureau of statistics in past years, the proportion of people over 65 years old in China's total population was 4.9 percent in 1982, 5.6 percent in 1990, 7.1 percent in 2000, 8.9 percent in 2010 and 10.1 percent in 2014. As shown in figure 1-2, the proportion of the elderly population in China is increasing rapidly, and the aging of the population is accompanied by the problem of empty nesters, which has become a major social problem.

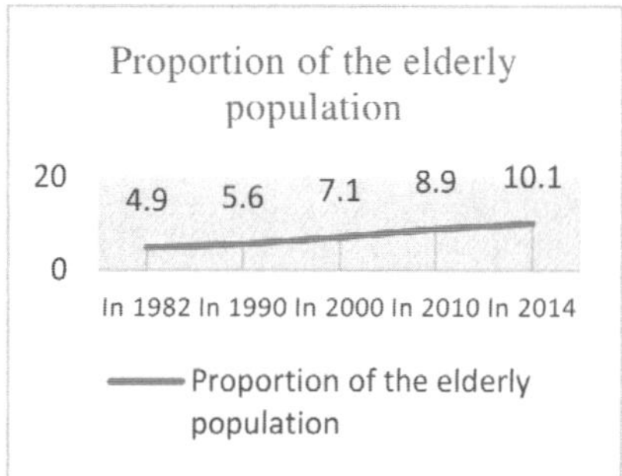

Pigure 1: population proportion of elderly population in China

2. Empty nesterrs old man analysis the actuality and problems

In China, mobile payment is active in all aspects of life, bringing great convenience to the nation. However, this group does not include empty nesters, who are a specific group of people in an intelligent society. They live in an intelligent environment, but are excluded from this group.

In this paper, the application of empty nesters in saving facilities was investigated. Found them in handling savings business, the main business required is withdrawal.The withdrawal of capital source roughly divided into four types, the first category is retirement pay, the second class of social security, the third class for children of remittance, the fourth class for national subsidies, such as veterans、low income、returning farmland to forest、the old man over the age of 70 and other governmental subsidies.

In the process of investigation, eight questions were analyzed and summarized. (1) Most young people go out for work, leaving elderly people to raise young children, and regularly remit money to their families every month to maintain their basic living and the money a child needs to grow up, which requires elderly groups to enter the savings activities. (2) Most empty nesters use the red passbook instead of the bank *card. (3) Most of the elderly in the unconscious of the default intelligent era of* convenience, but in the actual operation process is standing in the second perspective, the need to operate the intelligent savings equipment with the help of staff.(4)In the view of talking about the intelligent savings device is all the elderly are proposed interface font small.(5) The old man also said that the interface is complicated, there are some unnecessary options and voice reminder is not clear enough.(6)In the survey there are male elderly withdrawals more than the number of female elderly withdrawals.(7)Most empty nesters don't use counters or ATM, so the money they need to withdraw is remitted by their children to relatives, who then transfer it to the elderly.

The elderly and vulnerable groups without digital capability are part of the social group and also the group that needs to be benefited. Because the existence of this group is impossible to disappear even in the case of the strong third-party mobile payment, it is a trend to rely on the empty nesters and apply them to the savings devices.

3.User characteristics

Design is born to serve people, not users to adapt to design. Therefore, the design should start from the physiology and psychology of the user group, explore the characteristics of this group, and start from their needs to design products that meet the needs of users.

3.1. Psychological characteristics of empty nesters

The common psychological characteristics of empty nesters are inability to do anything， gradually rising sense of incompetence and loss、psychological distrust of the world，and the degeneration of nervous system, behavior and perception.

3.2. Physiological characteristics of empty nesters

Psychologically, the status of the elderly has changed in their eyes this is the decline of family status, so the sense of loss, inferiority is obvious, have random ideas. The degradation of some physical functions leads to the slow behavior, and thus produces a sense of decline in the intangible psychological, resulting in greater psychological pressure.

3.3. Personal characteristics of users

Empty nest old man of the family will be by gender、age、social background and affected by the degree of education, formed by the different to the "science and technology acceptance" the old man under the condition of the physical characteristics degradation will produce "technology phobia", for intelligent product is strange, it makes them unable to use life type of basic science and technology service facilities, and more alienation intelligent products. Many of the existing savings devices are smart products that empty nesters dread.

4.Unreasonable part of existing savings equipment

4.1.Unreasonable part of labor in existing saving equipment

Restricted by various social conditions at that time, in the villages and towns of empty nesters old man by the education level is generally low,many can't read alone, so these groups store exchange activities need to rely on the counter staff guidance to complete.

(1)The counter can solve the problem of difficulty for old people to withdraw money, but it has low timeliness. Simple operation activities will affect the customers who queue up for other complicated businesses.(2)It takes a long time to deal with business at the counter, but the rest area is limited, which is beyond the physical capacity of the elderly. (3) Counter staff are trained and enter this industry, but there is no lack of distrust and preciseness towards empty nesters, which makes the staff irritable, and the staff have a bad attitude when handling business for the elderly, which even hurts the self-esteem of the elderly.(4) The long-term cost of the staff is high, and they have holidays and rest time, so they cannot provide services at any time.

4.2.The existing savings are the irrational part of the intelligent equipment machine

With the progress of science and technology, the intelligent era has begun to creep in, and the emergence of intelligent savings devices in major Banks has greatly enriched the daily lives of ordinary people. Under the condition that the timeliness of the counter is insufficient, the intelligent saving device is favored by the user group, but new problems appear in the use process.

(1)There is an age gap in the setting of the existing savings equipment. The setting group does not take into account the elderly group, and the setting process is complex. (2)The display interface of the saving device is complex, with unnecessary Settings. (3) Whether it is the number keys or the interface font display is too small to meet the needs of the elderly.(4) The prompt is not obvious and clear. (5)The color of the machine is often cold, which makes the empty nesters' cold and lonely mood even weaker and paler.(7)The basic design of small towns is not perfect and the social resources are insufficient. There are basically such Banks as post offices and rural commercial Banks.The machine setting is less, there will often be part of equipment damage and too late to repair, in the "migratory birds return" when the flow of large and affect efficiency.

5.New intelligent saving equipment

5.1.Direction of intelligent savings devices based on empty nesters

Savings devices should be designed to lower standards and be limited to people who have difficulty in using them, so that the design is suitable for the general population. The direction of this design is based on the above problems and in combination with characteristics of the old man's psychological physiology design can satisfy the urban empty nesters old man use smart saving equipment, design a new urban and rural old man smart saving equipment.

IOP Conf. Series: Materials Science and Engineering **573** (2019) 012109 doi:10.1088/1757-899X/573/1/012109

5.2.Product presentation

As shown in figure 5-1, this product is divided into two parts, the display screen and the money box. The operation bayonet ports are arranged in the upper part of the machine according to the custom. The operation steps are arranged in the position according to the orientation, and the bankbook length ranks them at the top separately for this purpose. According to the azimuth habit from bottom to top next, it is bank card socket、ID card socket and voucher small bill exit respectively. Replacing complex text forms with numbers and simple patterns.Because traditional buttons for plastic material, in with the touch of the people and corrosion, which is formed by sweat and paint, so the breakthrough tradition button in the form of a password, password now popular touch screen interface, and heavy work lead to deformation of the fingers and bulky, button in the form of a password will be next to accidentally touch the Numbers, so take the touch screen instead of buttons. Slow movement of the elderly in the use of slow reaction and long time to lower the head of the operation will lead to physical discomfort, so the password keyboard arranged in the product with its eyes on the part, to alleviate fatigue.

The display interface is divided into operation steps, digital keys and display interface by region. The serial number marked with operation is displayed in the form of a pattern on each socket. Setting up the middle of the smooth plane is convenient for the elderly to let go and relieve part of the weight, put carry-on items. The light prompt with arrow display is designed beside each required operation jack, and the green light arrow indicates that the user will operate or is operating, and the light prompt and voice prompt are combined to achieve better effect.

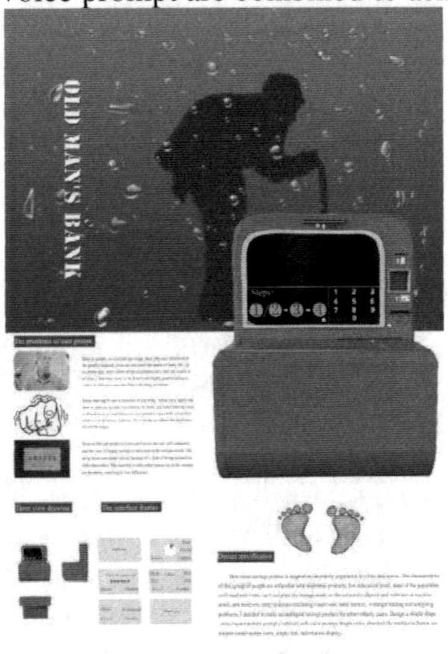

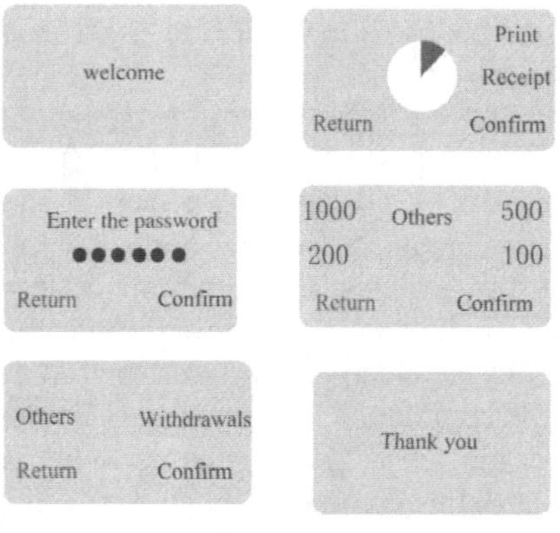

Pigure 2: shows the figure Pigure 3 :new smart savings device interface

Difficulties in the operation of intelligent savings machine are mainly concentrated on the display of the interface. The complexity of the interface will lead to difficulties in the process of product promotion, which will greatly blow the self-esteem of the elderly and make the interface as clear as possible.

As shown in figure 5-2, the interface display of this product opens the eye-protection mode. Interface simplification is one of the aims of this design.Numbers are presented in combination with simple fonts, complex fonts are distinguished by red and green in social language, and red characters are used to indicate return and green characters to indicate confirmation to simplify more complex fonts.Interface items are mainly simple withdrawal items, other items in the other inside to make the interface simple and clear. It takes time to wait in the card taking link, so there is a ring counting second device in the card taking link to express the remaining processing time and feed back the information that the customer is running.

6.Conclusions

The problem of empty nesters in Chinese empty nester families has become a major social problem. Empty nesters have become a major social problem in China. We should pay attention to this group, integrate them into the public, and turn technology into a tool for application. Therefore, we should start from small things, always pay attention to the inconvenience in their lives, design intelligent products that they can use, and feel the beauty brought by science when using the convenience brought by intelligent technology products.

References

[1] Wang Zhe. Difference analysis of influence of personal characteristics on technology acceptance -- a case study of mainland China and Taiwan [J]. Science and technology management research,2014,34(03):141-145.
[2] Yan Wei. Usability and design planning analysis of ATM for the elderly [J]. Inner Mongolia science & technology & economy,2013(11):98-100.
[3] Huang Lucheng, Zhang Ertao, Miao Hong. An empirical study on the needs of technology for the well-being of the elderly [J]. Science and technology management research,2016,36(13):102-105.
[4] Chen lili. Research on common psychological problems and coping strategies of rural empty-nesters [J].Journal of chongqing vocational college of electronic engineering,2017,26(05):23-26.
[5] Li Yanfang. Research progress on the living conditions of rural empty-nesters [J]. Chinese geriatric healthcare medicine,2015,13(04):84-86.
[6] Nie Miao, Wang Quanhai, Yao Yingshui. Research progress on psychological problems of left-behind and empty-nesters in China [J]. Chinese journal of gerontology,2011,31(12):2364-2366.
[7] Liu Yanjie. ATM cash demand forecast based on ga-svr [D]. Jinan university,2016.
[8] Sun Dongsheng, Zhao Ruotong, Xu Ka. Research on customer perceived value dimensions based on ATM human-machine interface context [J]. Rural finance research,2016(04):43-48.
[9] Yu Honghu. Semantic research on household refrigerator products for the elderly in China [D]. Beijing forestry university,2008.
[10] Qiu Zhijuan. A study on the family structure of China's urban and rural elderly population and the differences in its old-age care problems [J]. Hubei social science,2018(07):46-52.
[11] LuoJunbo. Research on intelligent product design suitable for the elderly [D]. East China university of science and technology,2016.
[12] Jin Wei, Li Jinlong, Wang Rong. "marginal people" in the age of intelligence: how middle-aged and elderly people fit into the age of intelligence [J]. Finance age,2016(29):223-228.

IOP Conf. Series: Materials Science and Engineering **573** (2019) 012110 doi:10.1088/1757-899X/573/1/012110

Interface Design of Children's Education APP Home Page Based on Emotion Measurement

TingtingWu[1], LiPeng[1]

[1]School of Mechanica Engineering, Guizhou university, GuiYang, GuiZhou, 550025, China

TingtingWu: 1442713864@qq.com LiPeng: 41091119@qq.com

Abstract. Take the home page of children's education APP as the research object,the experimental samples were morphologically deconstructed by the method of human-computer interface component deconstruction, and the interactive interface feature space was constructed.In addition, PrEmo experiment was conducted on children users to obtain the emotional image index of users and construct the emotional cognitive space.By using the method of multiple linear regression, the relationship model between the characteristic elements of interactive interface design and the index of user perceptual image is constructed, so as to guide the design of the home page of educational interactive products in line with children's emotional image.This method can obtain the deeper emotional image of children users, guide the design of products more in line with children's emotional needs, so as to improve children's user experience.

1. Introduction

With the in-depth contact of mobile Internet to all walks of life, APP also gradually enters the field of education. According to the 2016 survey data released by ASO big data research center, educational APP ranks the 9th in user demand [1]. Children's education market has always been a rigid demand, now the young parents are open-minded, they are in the era of Internet influence, network education has a high degree of recognition.

At present, most of the domestic and foreign children's APP design focuses on product functionality [2]. However, few people pay attention to the emotional experience of children users, and the implicit emotional appeals of children users are not included in the process of product development and design [3]. In addition, the homogenization of children's APP in function is serious, and it is difficult to highlight the feature points compared with similar products. Therefore, it will become a new trend to improve the competitiveness of products by improving the emotional experience of children users [4]. In recent years, many researchers at home and abroad have taken users' emotional appeals as the research focus. Lin li [5] et al. proposed to take human-machine interface components as the starting point of product morphological deconstruction method to construct KE model based on multi-dimensional modeling features. Liu zhenghong et al. [6] proposed a KE model for constructing multidimensional variables by constructing MMFs space and emotional cognitive space. Xu qiaoqiao [7] et al. analyzed children's emotional needs based on Norman's three-level theory of emotion, and guided children's icon design with the semantic difference method in kansei engineering.

2. Analysis of Children's Emotional Characteristics

Debra Levin Gelman[8], a famous American designer of children's products, divided children into different age groups in his book children's product design in the digital age. Children aged 6-8 years are called "big children". Children at this stage begin to learn to reflect, their thinking starts to spread, but they are still curious about new things. They begin to have their own thoughts and emotional images, and are no longer influenced by others. Young children like exploratory learning, but children aged 6-8 prefer intuitive information to help them complete the task.

Therefore, the research object of this paper is "big children" aged 6-8 years old. Children in this period have the most vigorous emotional brain development and are willing to express their real thoughts. However, the acceptability of emotional expression is also low. In view of this, in order to explore methods for measuring children's emotions, the research framework process of this paper is summarized as shown in figure 1.

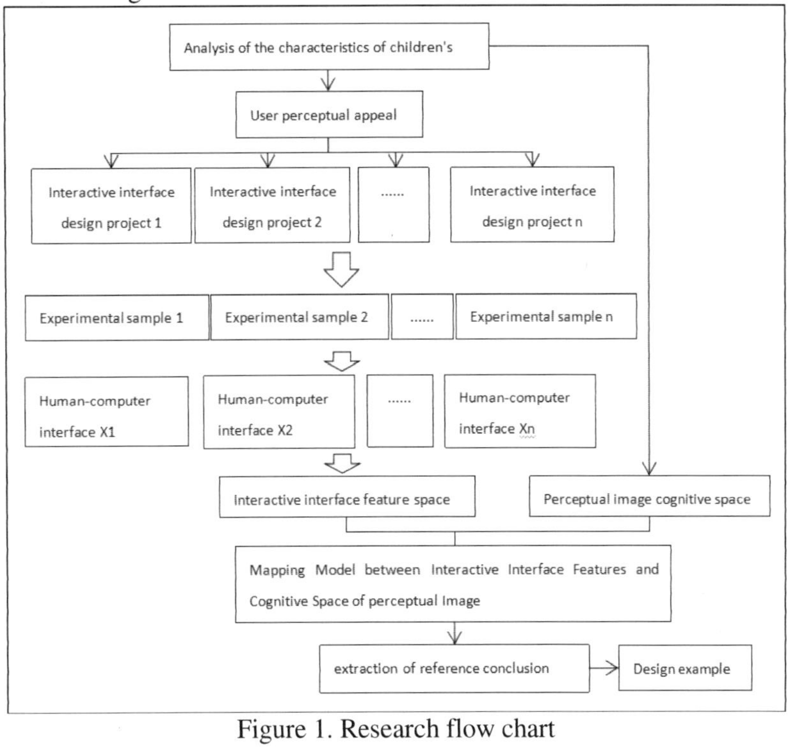

Figure 1. Research flow chart

3. Construction of Children's Emotional Cognitive Space

3.1 Select representative samples

First of all, 20 children's apps were collected in the APP Store as samples. The sample images were of high resolution and covered the main themes of children's education apps.In order to elaborate the research process of this topic, the homepage of children's education APP was selected as the research example, and representative samples were selected.Choice of 21 (4 children product designers with more than 3 years working experience, 17 parents aged 26 to 30) as the research object, screen the homepage interfaces of 20 children's apps, Group them into groups that are considered to have the same or similar layout and interface tone, according to their own image to determine the number of groups and within the number of groups.The investigation results were sorted out and analyzed, The Times of any two samples being divided into the same group were counted, the similarity degree matrix of the lower triangle was sorted out, and then it was transformed into the distance matrix in the multi-dimensional scale. After the multi-dimensional scale analysis with SPSS software, the six-dimensional spatial coordinates of 20 samples were obtained. After the coordinate was analyzed by clustering analysis method, 4 representative samples were finally determined, as shown in figure 2.

Figure 2. Representative samples

3.2 Build an interactive interface feature space

The nature of products on the home page of children's educational apps is deconstructed by using the method of man-machine interface element (partially) deconstruction proposed by Lin li [5] et al., and the interface flow is not included in this study because the research objective of this paper is the interface layout design of the home page.The man-machine interface component is an important carrier to convey perceptual information between users and products. Based on the human-machine interface components, the objective is to deconstruct the smallest perceptual information unit.The nature is the design feature points that users can see and feel, and it is the smallest unit where the emotional interaction between users and the product can occur [5]. The global nature (human-machine interface elements) method is used to deconstruct the five dimensions of the subject, background, icon and text of the children's APP home page in terms of form, color, quality, texture and style.First, the main body and background of the interface are non-functional parts, while icons and text are functional parts.Then list the input features (such as number of icons, icon size) and output type features (such as background color), the second one by one, the non-functional and functional components from the Angle of the senses and rational Angle HIEs segmentation,deconstructed 7 non-functional HIEs, as shown in table 1, deconstructed 10 functional HIEs, as shown in table 2.

Table.1 Non-functional

Category	Number	HIEs
Subject	X1	Number of subjects
	X2	The scale of the subject
	X3	The number of main colors
	X4	The hue of the main color
	X5	The material gloss of subjects
Background	X6	The light and shade of the background color
	X7	Background style texture

Table.2 Functional HIEs

Category	Number	HIEs
Icon	X8	Number of ICONS
	X9	Icon style
	X10	Size of the icon
	X11	Icon color tone

IOP Conf. Series: Materials Science and Engineering **573** (2019) 012110 doi:10.1088/1757-899X/573/1/012110

	X12	Number of icon colors
	X13	Background texture of icon
Characters	X14	Writing style
	X15	Number of words
	X16	Text size
	X17	Text color

3.3 Construct Emotional Cognitive Space

Based on the above analysis of children's emotional cognitive characteristics, children's cognitive ability to things is still in the development stage, they cannot accurately understand and distinguish the meaning of words, and their verbal expression ability is relatively weak. Therefore, in order to obtain more accurate emotional imagery of children, the PrEmo measurement method proposed by foreign scholar Desmet PMA et al. [9] was used to extract emotional imagery indicators of children users.

PrEmo measurement method is a measurement tool for measuring users' emotional imagery proposed by Desmet PMA and other scholars from delft university in the Netherlands. It is a psychological reaction testing technology based on psychology. PrEmo is a non-verbal self-reporting tool that measures 14 emotions often triggered by a product. This instrument has been developed because existing (verbal and non-verbal) tools cannot measure the subtle emotions usually triggered by product design [10].Seven of the 14 emotions represented pleasure (Desired, satisfied, interesting, fascinated, inspired, appreciative and surprised) and seven represented unhappiness (Disgust, dissatisfied, unpleasant, scared, disappointed, bored and contemptuous). Instead of relying on words, interviewees can use animated expressions and sounds to express their emotions. Each of the 14 emotions measured was represented by an animated mannequin, as shown in figure 3. Each animation was rated according to 5 rating scales. The rating criteria were as follows: 1 represents that I did not feel the emotion expressed in the animation at all; 5 represents that I did feel this emotion. The emotional imagery of the subjects can be measured by the scale method.

PrEmo is a cross-cultural and cross-age emotional measurement method. In terms of application, Desmet et al. also applied PrEmo emotional measurement method to the front face design of cars and the design of child seats [10-11]. Therefore, PrEmo has been widely used abroad, but has not been popularized and studied in China.

Figure 3. PrEmo emotional sketchpad

Using PrEmo tool to construct the cognitive space of emotional image is based on the investigation of the target user experiment. First, according to the 4 selected experimental samples, and the 17 interactive interface design characteristics (characteristics) obtained by the deconstruction of the experimental samples through the method of global nature (elements of the man-machine interface), are shown in table 1 and table 2. Purposefully, each design feature corresponds to the 5-level emotional animation scale in the PrEmo tool. The diagram of the scale is shown in figure 3 to form the image cognition experiment questionnaire. Three senior children's product designers, three children's educators and 24 children aged 6-8 years were then invited to participate in the experiment. According to the experimental principles of PrEmo, participants were allowed to evaluate one by one on the 5 subscales of each nature of each sample, in which 24 child subjects were guided by 3 child educators.

Through the partially cognitive experiments of the nature's emotional imagery, the evaluation value of the nature of each sample can be obtained, thus constructing a multi-dimensional cognitive space of emotional imagery.

3.4 Parametric treatment of children's emotional needs

The importance coefficients of adjacent components are compared sequentially by using sequential scoring method, so as to judge the main features. Taking no. 3 emotion (" interesting ") as an example, in sample 1, it is concluded through DARE method that components 1, 2, 5, 7, 8, 9, 12, 15 have the largest influence coefficient on no. 3 emotion.

The above eight main influence elements were parameterized. In sample 1, the main influence elements obtained by the sequential method included the elements 1, 2, 5, 7 of the non-functional elements and the elements 8, 9, 12, 15 of the functional elements. Of the nature data, 7 are "interesting", =1; If the stereotype data with only the 12th main feature is not "interesting", then =0; Therefore, the main characteristic parameterized array of "interesting" is:

$$([Main(8)])_{1,1} = \{1,1,1,1,1,1,0,1\}$$

The mean value of the user's emotional evaluation of the no.3 emotional animation (" interesting ") in sample 1 obtained by PrEmo measurement is: $\frac{\sum_{m=1}^{30}\sum_{j=1}^{17} A_{1,j}^{m,1}}{30\times 17}$

3.5 Build a mapping model from multi-dimensional interactive design features to design perceptual space

Using multiple linear regression technique, the correlation model between user emotion image and interaction design elements is constructed. According to this calculation method and the analogy, in the process of data processing, in order to avoid the occurrence of multicollinearity, the multicollinearity was tested by the variable inflation factor, and it was stipulated that the item variable inflation factor included in the linear regression model should not exceed 10.The values of the main characteristic parameters under each emotion animation are calculated: $([Main(L)])_{i,n}$ and the emotional image evaluation mean of each sample under each emotional animation: $\frac{\sum_{m=1}^{q}\sum_{j=1}^{r} A_{i,j}^{m,n}}{qr}$ By inputting SPSS software for data processing, the mapping relationship model from the main features of the interactive interface to the perceptual design space can be completed, thus building a multi-dimensional emotional engineering model.Among them,this formula: $([Main(L)])_{i,n}$ represents the parameterized array corresponding to all the main influence characteristics of the i-th evaluation sample of the emotional animation in the NTH PrEmo experiment.Q is the total number of subjects.R is the total number of nature (nature of human-machine interface elements). $A_{i,j}^{m,n}$ is the emotional evaluation value of the MTH interviewee on the JTH nature of the ith sample under the NTH emotional animation.

According to the correlation theory of linear regression analysis, the linear relationship between emotional animation and interactive interface design elements is established, and then the quantitative analysis is carried out. It is assumed that the dependent variable Y1 is linearly related to the independent variable Xi, the dependent variable is the perceptual evaluation value of the sample, and the independent variable is the main characteristic design element, and the functional relationship is established. Taking no.3 emotional animation (" fun ") as an example, the multi-dimensional kansei engineering model is established by calculation as follows:

$$Y_1 = -0.891x_2 + 0.691x_5 + 0.493x_7 - 0.970x_8 + 0.327x_9 - 0.751x_{12} - 0.900x_{15} + 4.351$$

Y_1 in the above formula is the kansei engineering model of no.3 emotional animation (" interesting "). Xi is the ith nature; The coefficient of X_i represents the contribution of the human-machine interface element to the influence of no.3 emotional animation (" interesting "), where positive value represents positive influence and negative value represents negative influence. Positive impact components play a promoting role in guiding the design process to the corresponding emotions, and negative impact components play a hindering role in the design process. Therefore, the design should be based on the emotional needs of the positioning to avoid or reduce such components can also play a promoting role in the corresponding emotions. On the type, element 5, 7, 9 to 3 sentiment animation (" interesting ") have a positive impact, element 2, 8, 12, 15 to 3 sentiment animation (" interesting ") have a negative impact, reduce or reduce these elements can make children education class interactive product page design more "interesting", can effectively improve the user experience of children.

3.6 Interface home page design

According to the experiment, the user's emotional image was obtained and a multi-dimensional kansei engineering model was constructed. The key interactive interface design elements were summarized and analyzed to find the interactive interface design elements and their main features suitable for the homepage of children's education APP. The reference conclusions for the interface design of the homepage of children's education APP were obtained as shown in table 3:

Table3 Reference conclusions of interactive interface design

Category	Number	Main feature element	Reference conclusion
Subject	X1	The number of subjects	The number of subjects is kept at the best of two, and excessive number of subjects will cause visual error.
	X2	The picture proportion of subjects	The proportion of the main picture is 1:18 is the most suitable
	X3	The number of subject's colors	The number of main colors is three best
	X5	The material gloss of subjects	The subject should have a certain sense of material quality
Background	X6	The light and dark of the background color	Background color selection warm tone
	X7	Background style texture	Interface background suitable for texture filling
Icon	X8	Numbers of icons	The number of icons is 2-3 best
	X9	Styles of icon	Icon style selection line icon
	X10	Size of icon	Icon size is moderate
	X12	Number of icon colors	The number of icon colors is the best in the same color
Characters	X15	Number of words	Control the number of words at the best of six
	X16	Text size	The best text size is 13

The interactive interface design reference conclusion in table 3 as important theoretical basis for children's education class APP homepage design, combining with the building of various samples of multi-dimensional correlation coefficient in perceptual engineering model of comprehensive consideration, determine the children's education class reference design for the homepage of interactive products: the body of the interface color warm color number and not too much; The number of subjects in the interface is moderate, not too much; The proportion of the main page should not be too large; The main material should have a certain luster; Interface background should have texture sense; ICONS should not be too many, easy to cause visual interference; Icon style is linear icon; Too many and miscellaneous icon colors are likely to cause negative emotions of children users. Therefore, the number of ICONS should be uniform and consistent. Text size moderate;

According to the above reference content, the theme function of the APP is defined as an insect knowledge popular science education APP, which is designed in combination with the development size of the Android system and the corresponding design specifications of the interactive interface

IOP Conf. Series: Materials Science and Engineering **573** (2019) 012110 doi:10.1088/1757-899X/573/1/012110

design. The final design scheme of the APP homepage for popular science education of children's insect knowledge was obtained, as shown in figure 4.

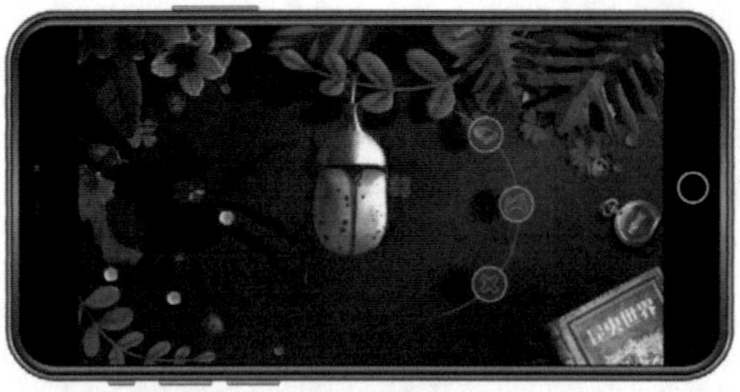

Fig.4 Children nsect knowledge education APP home page

4. Conclusion

This paper takes the home page of children's education APP as an example to deconstruct the interface elements of the home page through the method of human-machine interface component deconstruction. Secondly, emotional image indicators of children users are collected by using PrEmo emotion measurement experiment. The mapping relationship model between emotional image and interactive interface design elements is constructed by using multiple linear regression analysis. Finally, based on the relationship model, the reference conclusions for the interface design of the home page of children's education apps that conform to users' emotional images are derived, and innovative design is carried out. Research shows that the design of children's education APP in the digital age should pay attention to children's emotional appeal while considering product content. Through PrEmo experiment, emotion animation was used to measure the emotion of children users instead of semantic difference method.In this experiment, the emotional needs of children users and the design elements of the interface are considered comprehensively, and the semantic difference method has more reliable emotional prediction ability.Therefore, it can play an effective guiding role in the perceptual design stage of children's products, improve the emotional experience of children users, and provide reference value for the interactive interface design of other children's digital products.

References

[1] Sadi research institute of the ministry of industry and information technology.White paper on the development of mobile Internet(2016)[N].China computer news,2017-09-11(005).

[2] NIELSEN J. Designing web usability [J]. New Riders Indianapolis, 2000(1): 86-88.

[3] HADIANA A. Kansei analysis of interface's elements for mobile commerce application [C]//International Conference on Information and Communication Technology. New York: IEEE Prees, 2016: 1-4.

[4] FUG, WEI L L, CAO Y. Optimization design of a webpage based on kansei engineering [J]. Human Factors & Ergonomics in Manufacturing & Service Industries, 2016, 26(1): 110-126.

[5] LIN L, XUE C Q, WANG H Y, et al.To optimize the product form deconstruction method of KE model[J].Journal of southeast university (natural science editio n), 2010, 40 (06):1353-1357.

[6] LIU Z H, XIE Q S, HUANG H S, et al.Multidimensional variable kansei engineering model construction and its performance evaluation[J].Journal of sichuan university (engineering science edition),2016,48(02):198-206.

[7] XU Q Q, ZHANG B C, XU Hao Y.Research on the design of APP ICONS for children based on emotion analysis[J].Packaging engineering,2018, 39 (16) :212-216.

[8] Debra Levin Gelman. 《Design of children's products in the digital age》 [M].WuHan.Huazhong university of science and technology press,2017.

[9] Desmet P M A. To love and not to love: why do product elicit mixided emotions? [C].The First International Conference on Design and Emotion, Delft, 1999: 67-73.

[10] Desmet P M A, Hekkert P, Jacobs J J. When a car makes you smile: development and application of an instrument to measure product emotions [J]. In: S.J. Hoch and R.J. Meyer (Eds.), Advances in Consumer Research, 2000, 27: 111-117.
[11] Desmet P M A.A wheelchair can be fun: a case of emotion-driven design.[J].2003,1:704-711.

IOP Conf. Series: Materials Science and Engineering **573** (2019) 012111 doi:10.1088/1757-899X/573/1/012111

New paradigm of packaging design in Internet scenario

Dingwei Zhang [1], Huijie Peng [2]

[1]College of Art and Design, Shanghai University of Engineering Science， Shanghai 201620, China

[2] School of Business, East China University of Science and Technology，Shanghai 200237, China

E-mail: zdw8good@163.com

Abstract. Based on the new design context, this paper studies the development trend of packaging design with Internet thinking from the perspective of strategic marketing and new design paradigm, and considers the influence of Internet design context on packaging design. Iterative innovative design can not only satisfy consumption, but also create and guide demand. The creative packaging design paradigm of the new era pays more attention to the application of humanistic elements in the process of design-driven innovation. Culture, brand building, visual entertainment, Internet of everything and other elements are integrated and innovated in packaging design. Change is the way to innovate. Therefore, packaging designers should grasp the new trend of packaging design era and strengthen the ability of data mining and technology implementation, and maintain the design sensitivity and responsibility to inject new power into the packaging design of the new era, so that our life will be better.

1. Introduction

Internet thinking is a way of thinking about the complex social practice innovation design in the context of Internet technology. Packaging, as a form of marketing, has the function of promoting product sales. Modern packaging design should not only play the role of protection and promotion, but also pay more attention to the application of humanistic elements in the context of The Times. In modern packaging design system, all kinds of design elements combine and create new value to seek innovative development in the change. In order to stand out from other similar products, the packaging design of products needs to be very outstanding [1].

2.New paradigm of package marketing in Internet scenario

The Internet scenario of artificial intelligence and information sharing makes modern packaging design have prominent new changes and development compared with traditional packaging design.

2.1Diverse communication methods

In the context of Internet, some new modern marketing information communication platforms have appeared in the modern product sales mode. Compared with the traditional sales communication channels, these platforms are more diversified and precise.

With the prosperous development of information network technology, designers can use Internet tools to collect a large number of consumers' personal preference information data, which is the advantage of precise marketing of packaging design in the context of the Internet. By mining users' online browsing habits and relevant personal information data, the website will match and place the optimal advertisements according to different users' behaviour habits, interests and other information [2]. For example, when a young mother and a retired teacher visit the page of the same website, the website will show the best advertisement information based on their browsing history and professional hobbies.

In the era of "Self media", WeChat, QQ, Facebook, Instagram, Line, microblog and other social media have changed the way we live, study and work in our daily life. Such interactive social platforms have also become effective communication and publicity platforms for product packaging.

Case: “Le Chun” -- the first yogurt in the world to meet another foodie with a story!

Le Chun yoghurt is a popular product packaging design case with social attributes. In terms of content marketing, “Le Chun” has a high standard for copywriting design and brand construction. The back of the Le Chun yoghurt box becomes a matchmaking medium. Businesses voted from thousands of submissions to produce a new packaging theme. Through the collaborative participation of consumers in packaging design, the company realizes the continuous update of product packaging design. You can get to know another food lover with a story through the QR code on the lid of the scan music box. You all love to travel and be unconventional, also have your same hobbies and get in touch with him/her immediately. The Le Chun team also specially filmed relevant packaging promotion videos to show the social entertainment features of Le Chun. As is shown in Figure 1, “Pure food – Le Chun” is a typical example of the mutual achievement of Internet thinking in social platforms and packaging design [3].

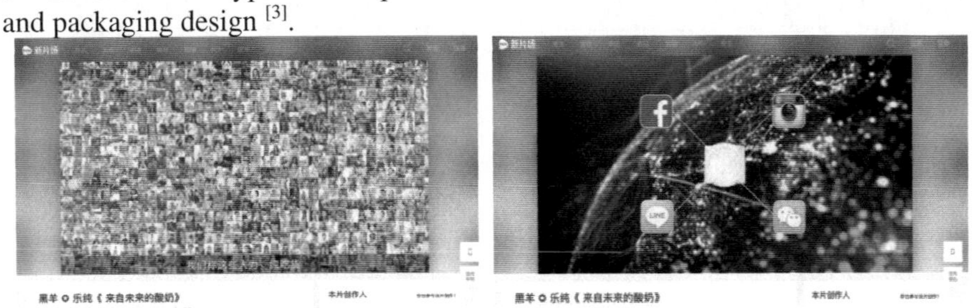

Fig.1 Video "Yogurt from the future"

2.2 Humanized marketing methods

People's lifestyle and shopping style have changed significantly because of Internet technology. Nowadays online shopping becomes a fashion, so the brand design of online visual is much more important. Now we have entered the era of Product Identity System (PIS) from the past Company Identity System (CIS) era. Young people will buy this product because of friends' recommendation or web celebrity hot style. For example, when choosing snack foods, we will choose "You Bai Jia" in the candy category, "Three squirrels" in the nut category. Why is that? That's the brand. The essence of creating a product brand is to provide customers with different consumption experience: user experience, visual experience, communication experience and service experience. Taking "Jiang Xiao Bai" brand as an example, its copywriting planning and operation strategy innovation have grasped the characteristics of young people's preference of simple life, extremely rebellious spirit and customized demand, and provided more consumer services [3].

2.3 Intelligent transportation and packaging

The optimization of Internet technology and product marketing eliminates the limitation of region and time, enabling overseas online shopping to be realized. Products can be traded freely on the Internet. Taking Jing Dong logistics as an example, Jing Dong logistics management pays more attention to the timeliness and the design of information privacy protection for consumers. Jing Dong shopping can realize the service of same-day arrival of shopping in the same city. Transportation packaging adopts mobile phone ordering and door-to-door blocking business, which ensures the effectiveness of sending and receiving express mail and the real-time tracking design and management of logistics information. Thus this prevents the leakage customer information and the express mail loss event occurrence.

3. The influence of Internet thinking on product packaging design

3.1Strengthen visual innovation

The application of design thinking makes the packaging design much visually impactful and able to highlight the artistic characteristics of packaging to achieve outstanding reasonable and unexpected visual effects. Consumers will generate visual impulse in the first three seconds of watching video, and the emotional impact of product information determines whether the product can achieve a transaction. Therefore, it is of great significance to enhance the visual impact of modern packaging design for the successful completion of sales. For example, designer Ban Hu's newly designed high-grade gift series packaging and visual marketing plan for Liang Pin store expresses the geometric aesthetics of contemporary design and profound Chinese traditional culture through visual design, which is presented in front of us across media, as shown in figure 2.

The visual image design of packaging marketing design focuses on the explosive power of visual impact, which is a new understanding of contemporary product packaging under the Internet domain. The goal of packaging design is to promote sales, not just the beauty of the product. In the context of contemporary Internet, packaging marketing plan upgrading and visual image design strengthening become the key components of design.

Fig. 2 high-end zero food packaging design of BESTORE

3.2 Optimization of marketing strategy

Internet marketing is a subversive marketing mode based on the application of Internet technology, which integrates brand marketing, network marketing, micro marketing and other marketing modes. In the era of Internet DT, the success of a brand requires two aspects: 1. Online, brand operators use Internet technology to precisely meet the needs of customers. 2. Offline, brands operators realize the implementation of offline marketing by combining brand consensus, creative marketing and copywriting design [4]. In the context of Internet, the construction of packaging design brand is coordinated by online and offline.

Since modern young people pay more attention to the consumption experience, innovative planning document is more conducive to brand promotion in the Internet scenario. Social media platforms and online word-of-mouth effect are all effective drivers of brand building. Under the Internet thinking, the typical cases of successful online marketing brands include Mushroom Street, Vip.com, Tmall.com and so on.

3.3 Enhancement of technology realization

At present, the improvement of packaging materials and various packaging printing and molding technologies has promoted the innovation of packaging design. The application of anticorrosive technology in packaging solves the problem of environmental protection in food packaging storage and transportation. The use of edible glutinous rice paper in candy packaging and pastry packaging and the innovative application of aluminum beer bottle packaging are the results of the development of modern molding technology. In the era of Internet of everything, in addition to the basic functions of storage, protection, transportation and sales, the design of modern transport packaging should take whether it is environmentally friendly and recyclable as an important indicator of the quality of

IOP Conf. Series: Materials Science and Engineering **573** (2019) 012111 doi:10.1088/1757-899X/573/1/012111

packaging design. The relationship among ecology, society, economy and culture needs to be balanced in the future packaging design.

4.New paradigm of creative packaging design in the future

(1) Green packaging design and packaging follow-up function design. Green packaging design is the designer's reflection and design practice on the environment and ecological destruction of modern civilization activities. The influence of sustainable development concept and designers' application of environmental protection and sustainable design materials will contribute to the healthy and sustainable development of our living environment. The recycling of packaging design materials and the derivation of the function of subsequent use of packaging will promote the development of green and sustainable packaging design in the future [5].

(2) Interesting packaging design that meets the psychological needs of consumers. Interesting packaging has nothing to do with age or gender. For example, packaging designed for children is more interesting. However, the premise of interesting design should first ensure the practicality of packaging design, and then satisfy the consumer's entertainment consumption psychology [6]. As shown in Figure 3: "creative packaging design of potato chips", creative potato chip packaging design captures the appetite and entertainment psychological characteristics of children and female consumer groups. This kind of packaging changes the traditional bagging way of potato chips in the past, realizes the deformation structure of barrel and the function of jigsaw puzzle, and adds interest while enjoying delicious food.

Fig.3 Creative packaging design for potato chips

(3) Packaging Bionic design is the designer's re-creation of natural things. It is the design practice of the concept of "Green, ecological and creative". Bionic packaging design combines the contemporary design context, service object and aesthetic taste, and makes the products stand out from other similar products with the concrete bionic design form. Bionic design can deepen the intimacy between consumers and products. As is shown in fig.4, "Welcome to Jian De" lotus seed wine packaging design ", the designer makes a bottle of lotus seed according to its shape, together with the ceramic bottle body and wooden bottle cap, which magnifies the attributes and characteristics of lotus seed wine.

Fig.4 "Welcome to Jian De" lotus seed wine packaging design

(4) Advances in technology have made it possible to design customized packaging services. Taking the liquor brand "Jiang Xiao Bai" as an example, its main consumer groups are young people full of vigor and vitality and the pursuit of individuality. With the maturity of digital printing technology and the popularity of small data programs, "Jiang Xiao Bai" brand realizes customized services with the help of WeChat public account platform. In addition, consumers can also share their consumption experience through WeChat circles of friends based on the customized service. The visual effect of the cartoon image of the post-80s young people combined with the short perceptual, humorous and uninhibited advertising quotations makes the packaging design of “Jiang Xiao Bai” liquor distinctive [7].

(5) Innovative packaging design emphasizes the use of humanistic elements. The essence of design is Human-oriented. Designers display a culture through packaging design in order to find the cultural identity and resonance between product design and consumers. Figure 5，Taking the rice packaging design of "Five grains are plentiful" and "Fish every year" as an example, this kind of packaging design utilizes strong traditional Chinese culture and regional cultural elements, showing a strong humanistic concern. The application of cultural elements and graphic design of words reflects the recall of history, which is a kind of caring and increases the emotional communication between products and consumers, thus promoting the sales of products.

Fig.5 "Five grains are plentiful" and "Fish every year" rice packaging design

5.Characteristics of excellent product packaging design

The sustainable development of packaging design brand requires the ability of iterative innovation to promote the sustainable sales and design innovation for packaging design. The genetic characteristics of packaging brands are cultural genes that iteratively innovate designs. For example, For example, PepsiCo uses blue and red elements. The blue and red elements are also used in bottles and cans of Pepsi or Pepsi related products. Therefore blue and red are the cultural genes of its Pepsi brand. The better the packaging brand, the easier it is to identify. For example, almost everyone can recognize the "M" of McDonald's and the "KFC" logo of KFC. Excellent packaging design for the promotion of products and the derivative of peripheral products cannot be ignored. Take the packaging design of "Lai Yi Fen" and "Bai Cao Wei" as examples. Especially in the era of modern intelligent Internet, the prosperity and development of e-commerce make online packaging design also have the same characteristics [8].

6. Conclusion

Packaging designers should be good at adjusting marketing strategies and design methods to seek innovation in the Internet information age and strive to adapt to new design principles. Packaging design not only needs breakthrough innovation of external structure, shape, graphic design, but also fits the culture and consumer psychology. In the context of a new design paradigm, design is innovated through iteration to create and guide requirements. Designers should be familiar with the creative packaging design paradigm of the new era, pay more attention to the driving innovation and application of modern packaging design, and integrate and seek change in the development process of packaging design. Designers should not only have design feelings, but also maintain the sensitivity to design and responsibility to explore the collaborative development of packaging design in the future.

References

[1] D. Wu. Packaging design in the era of perceptual consumption [J]. China packaging industry, 2013(02): 13-14.
[2] D. W. Zhang, L. S. Chen. Innovative design of product packaging in the context of contemporary Internet [J]. Packaging engineering, 2008,39(24): 110-114.
[3] Y. Shu. When Packaging Has More Powerful Social Functions. (2016-09-06) https://www.digitaling.com/articles/30202.html.
[4] SOHU. From 0 to 1 Billion Sales in 5 Years the Four Secrets. (2017-08-06) http://www.SOHU.com/a/162648960_806626.
[5] C. G. Xie. Research on Sustainable Development Strategy of Modern Packaging Design [J]. Packaging Engineering, 2015, 36(10): 101—103.
[6] Q. Chen. Research on brand marketing with visual consistency in the era of mobile Internet [D]. Southwest university, 2017.
[7] J. Y. Wang. Jiang Xiao Bai marketing model SWOT analysis. （2013-05-01） https://m.baidu.com/from=0/bd_page_type=1/ssid=0/u.
[8] X. S. Huang. A Brief Discussion on the Development Trend of Packaging Design in E-commerce Era [J]. China Packaging Industry, 2016(6): 152.

The Mechanism of Intra-Industry Technology Spillover Effects of Up-Gradient ODI and Down-Gradient ODI in China

Xiaofei Luo[1], Yonghui Han[2*] and Jinchuan Huang[3]

[1] Faculty of Logistics, Guangdong Mechanical and Electrical Technical College, Guangzhou, Guangdong, 510515, China

[2] Guangdong Institute for International Strategies, Guangdong University of Foreign Studies, Guangzhou, Guangdong, 510420, China

[3] School for English of International Business, Guangdong University of Foreign Studies, Guangzhou, Guangdong,510420, China

[*] Corresponding Author： hanyonghui2006@foxmail.com

Abstract. With the fast development of the global division of labour, intra-industry technology spill over has become a significant manifestation of both Foreign Direct Investment (FDI) and its impact on the domestic industrial structure, as well as the key to industrial value chain upgrading. Meanwhile, China's outbound foreign direct investment (ODI) has developed rapidly since the reform and opening up, with a large number of both up-gradient ODI in developing countries and down-gradient ODI in developed countries. Therefore, it is essential to figure out the relation between ODI and industrial technology change. This paper intends to theoretically clarify the adjustment function of ODI to intra-industry technology change and to construct the mechanism of this process. It provides a theoretical reference for the empirical analysis of the role of direct investment in intra-industry upgrading.

1. Introduction

The Chinese economy has achieved rapid growth due to the comprehensive implementation of reform and opening-up. Back in 1978, China's gross domestic product (GDP) was about 364.5 billion Yuan, and the per capita GDP was 381 Yuan. They then soared significantly during the years, and in 2015, GDP had reached 68.55 trillion Yuan with per capita GDP of 50,000 Yuan[1]. However, the rapid growth is accompanied by high investment and massiv econsumption. Take the manufacturing industry as an example; the contribution rate to its growth is 63.04% for capital, 26.23% for the labor force, but only 10.72% for total factor productivity (Zheng Ruogu, 2009). Therefore, its growth is mainly driven by factor input. This has a serious negative externality, indicating an unreasonable industrial structure, a low-position in the global value chain, and a lack of mutual support among industries, which have hindered the industrial development in China. Thus, transforming and upgrading industrial structure is urgent and important. One solution is to put forward the Belt and Road Initiative (B&R), which helps achieve new and sustainable economic growth by investing in the countries along the Road, as well as making use of their complementary advantages.

Many scholars have investigated into the competitiveness and complementarity between China and the counties along the Road, yet few went deeper into the interaction of competition and cooperation,

[1] Data resources: *China Statistical Yearbook*

which can achieve capacity cooperation, coordinated upgrading, and mutual benefit. For example, when examining the economic cooperation between China and B&R countries, scholars focus mainly on the cooperation of energy and resources (Liu Jiajun, 2013), but fail to develop it at a multi-pattern and multi-aspect level. Thus, further study is of great importance. If China's ODI in both developing countries and developed countries are of equal significance (Chen Juncong and Huang Fanhua, 2014), it should be divided into two types and studied separately: the up-gradient ODI to developing countries and the down-gradient ODI to developed countries. Due to the limitations, this paper develops solely from the perspective of intra-industry technology spillover effects to set up a theoretical framework for the impact of ODI on intra-industry technological change in China. The specific empirical analysis will be introduced later.

For up-gradient ODI, the existing studies mainly include the Flying Geese Pattern theory from Kaname Akamatsu (1962), the Product Life Cycle theory from Raymond Vernon (1966) and the Marginal Industrial Expansion theory from Kiyoshi Kojima (1978). When applying those theories, it can be deduced that with the on-going international division of labour and the worsening environmental problems, the competitive advantage of traditional factor endowments of developing countries has changed, so has their positioning of core competitiveness. Therefore, the developing countries transfer the low value-added intermediate industries or production departments to the less-developing countries through up-gradient ODI, so that they can shift the focus onto research and development, design, marketing, and other high value-added intermediate industries or departments. This can better promote the domestic industrial upgrading. As for down-gradient ODI, developing countries mainly invest with the purpose of seeking advanced technology, because the investment object is more developed than the home country enterprises. With reference to those theories, this paper embeds the consideration of foreign direct investment cost from Helpman et al (2004), as well as consideration of technology factors from Kohler (2004), in the Feenstra and Hanson (1996) model, to construct the theoretical mechanism of the effect of up-gradient ODI and down-gradient ODI on industry technology change.

2. Model Construction

2.1. Theoretical Framework of Up-gradient ODI and the Determination of the Scope of ODI

Assume that there are only two countries in the world: the more developed country D and the less developed country U; the two factors of production: capital K and labour L, in which labour is divided into high-skilled labour H and low-skilled labour L; an industry Y: an integration of a series of continuous intermediate inputs z. Each unit of intermediate input z needs $b_L(z)$ unit of low-skilled labour, $b_H(z)$ unit of high-skilled labour and $b_K(z)$ unit of capital, so the total factor demand of intermediate input is L(z), H(z) and K(z), respectively. Besides, assume that z is ordered as technological intensity increases, that is $\partial\left(b_H(z) / b_L(z)\right) / \partial z > 0$, to index all commodities z to a continuous set of commodities [0,1]. What's more, a supplementary assumption is made to the order of z that it is in accordance with the order in which the capital intensity increases.

Here we assumed the production function of intermediate input z is compound by Leontief and C-D, that is:

$$x(z) = A_i[\min\left\{\frac{L(z)}{b_L(z)}, \frac{H(z)}{b_H(z)}\right\}]^{\theta}[\frac{K(z)}{b_K(z)}]^{1-\theta}, i=D, U \quad (1)$$

where $0 < \theta < 1$ and A is a constant.

The two countries have Hicks-neutral technical difference, that is $A_D > A_U$, indicating that the technical level of the more developed country D is higher than that in the less developed country U. The final product is assembled at no cost according to the production technology:

$$\ln Y = \int_0^1 \alpha(z)\ln x(z)dz, \quad (2)$$

where $\int_0^1 \alpha(z)dz = 1$ and $\alpha(z)$ is the proportion of intermediate inputs in total inputs (2)

A country's low-skilled labour endowment, high-tech labour endowment, and capital endowment are defined as L_i, H_i and K_i respectively as the factor price corresponding to w_i, q_i and r_i. Assume that the technological gap and factor endowment differences between countries D and U are large enough so that factor prices will not be equalized.

Since the more developed country D has more sufficient capital endowment and high-tech labour, while the less developed country U has more low-skilled labour, country D has a lower capital return rate($r_D < r_U$), and a higher wage return rate ratio of high-tech labor to low-skilled labour($q_U/w_U > q_D/w_D$)compared with country U, according to the law of supply and demand. At the same time, suppose that the supply of high-skilled labour and low-skilled labour is in comparison to the relative wage, then$L_i'(q_i/w_i) \leq 0$and $H_i'(q_i/w_i) \geq 0$can be proved, since the increase of relative wage will lead to the decrease of the low-skilled labour supply, as more of them become high-skilled labour by means of learning and technological training.

According to the principle of cost minimization, the cost of producing a unit of commodity z by countryi can be expressed as:

$$c_i(w_i,q_i,r_i;z) = B_i[w_i b_L(z) + q_i b_H(z)]^\theta [b_K r_i]^{1-\theta} \qquad (3)$$

where $B_i = \theta^{-\theta}(1-\theta)^{-(1-\theta)}A_i^{-1}$, while $c_D(w_D, q_D, r_D; z)$ and $c_U(w_U, q_U, r_U; z)$ are the cost of producing a unit of commodity z by country D and U respectively.

About the principle of cost minimization, we can deduce the production cost line C_DC_D for country D and C_UC_U for country U, with all commoditiesz $\in [0.1]$. Since country D has competitive advantage over high-skilled labor-intensive product, while country U has that over low-skilled labor-intensive product, C_UC_U must be higher than C_DC_D when producing the high-skilledlabor-intensive product. It could be applied in the production of a low-skilledlabor-intensive product that C_UC_U must be lower than C_DC_D, as shown in Figure 1. Since all the intermediate inputz $\in [0.1]$, the proportion of capital in cost for the two countries is both$(1-\theta)$. Therefore, the two lines will have only one point that equates the two costs, as z^* in Figure 1, which satisfies:

$$c_D(w_D,q_D,r_D;z^*) = c_U(w_U,q_U,r_U;z^*) \qquad (4)$$

Here we add a method to determine whether the more developed country D has a competitive advantage of saving cost in producing commodity z:

$$r(w,q,r;z) = \frac{c_D(w_D,q_D,r_D;z)}{c_U(w_U,q_U,r_U;z)} = \frac{A_U}{A_D}\left(\frac{w_D}{w_U} \times \frac{1+\frac{q_D b_H(z)}{w_D b_L(z)}}{1+\frac{q_U b_H(z)}{w_U b_L(z)}}\right)^\theta \left(\frac{r_U}{r_D}\right)^{1-\theta} \qquad (5)$$

If $r(w,q,r;z)>1$, that is,$c_D(w_D,q_D,r_D;z) > c_U(w_U,q_U,r_U;z)$ and $z<z^*$, it means that the less developed country U has the advantage of saving cost when producing commodityz $\in [0,z^*)$. On the contrary, if$r(w,q,r;z)<1$, that is, $c_D(w_D,q_D,r_D;z) < c_U(w_U,q_U,r_U;z)$ and $z>z^*$, it means that the more developed country D has the advantage of saving cost. Yet if $r(w,q,r;z)=1$, that is, $c_D(w_D,q_D,r_D;z) = c_U(w_U,q_U,r_U;z)$ and $z=z^*$, there would be no difference in cost for the production of z in the two countries. Meanwhile, since$q_D/w_D < q_U/w_U$ and$\partial\left(b_H(z) / b_L(z)\right) / \partial z > 0$, when the price of the two production factors and technology remain constant,$r(w,q,r;z)$ decreases as z increases.

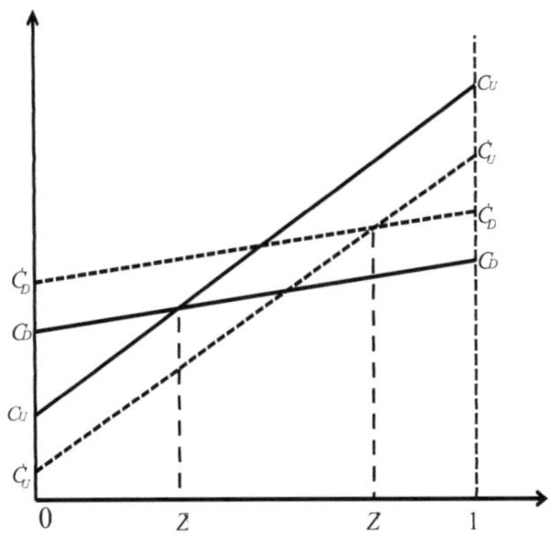

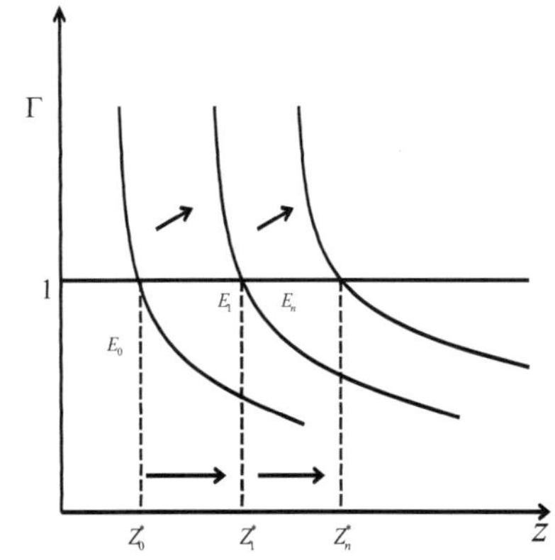

Figure 1 Determination A for the Scope of ODI Figure 2 Determination B for the Scope of ODI

Then we consider the cost of ODI referring to the idea of Helpman et al. (2004). Lots of input costs are incurred in the process of ODI, such as contract signing, information exchange, plant construction and so on. When the cost of ODI is lower than that of trade, a country will choose the former. Suppose this input cost is$\tau(z) > 1$, which means that when the country U produces one unit of commodity z domestically, the foreign direct investment in country U should produce$\tau(z)$unit of z. Thus, the investment cost will lead to a certain loss for cross-broader production. At the same time, given that ODI will have reverse technological spillover effect$\rho(z) > 1$, which is generated by the Industrial Combined effect, Forward and Backward Connection Effect, as well as Synergy Effect, the home country can achieve economy of scale and technology advancement. Meanwhile, the country receiving foreign direct investment would benefit from technological spillover effect$\lambda(z)$, since the FDI inflow is accompanied by the transfer of integration of technology, management experience, and human resources, which ultimately generate technological spillover through Connection Effect. To simplify the analysis, we refer to the methodology of Zheng Ruogu (2011) to assume a Hicks-neutral technological spillover, whose impact on capital and labour is proportional.

Taking the cost of ODI and the reverse technological spillover into consideration, the competitive advantage of cost in producing commodity z is determined by:

$$\Gamma(w,q,r;z) = r(w,q,r;z) = \frac{A_U(\lambda(z))}{A_D(\tau(z),\rho(z))}(\frac{w_D}{w_U} \times \frac{1+\frac{q_D b_H(z)}{w_D b_L(z)}}{1+\frac{q_U b_H(z)}{w_U b_L(z)}})^{\theta}(\frac{r_U}{r_D})^{1-\theta} \tag{6}$$

In this equation, A_D is the function of$\tau(z)$ and$\rho(z)$. Since$\partial A_D / \partial_\tau < 0$ and$\partial A_D / \partial_\rho > 0$, the increase of ODI cost has a negative impact on the production technology of country D's multinationals, while the reverse technological spillover effect can positively promote technology growth for them. On the other hand, A_U is the function of$\lambda(z)$, and$\partial A_U / \partial_\lambda > 0$ means that positive technological spillover can boost technology development in the host state being invested.

Given that the cost of ODI and the reverse technological spillover effect exist, we assume the more developed country D has a higher level of "technology" than the less developed country U, $A_D(\tau(z), p(z)) > A_U(\lambda(z))^2$. Therefore,$A_D(\tau(z), p(z))$ would have a lower limit, which represents the overall "technology" level of country D. In this way, we can assure that $\Gamma(w, q, r; z)$ has the same z monotonicity as that of $r(w, q, r; z)$.

Thus, we can determine the scope of industries for the ODI of a more developed country based on their competitive advantage, as shown in Figure 4.2. To be specific, if $\Gamma(w, q, r; z) > 1$, the less developed country U has a competitive advantage in producing commodity$z \in [0, z^*)$; On the contrary, if$\Gamma(w, q, r; z) < 1$, the more developed country D gains a competitive advantage in producing commodity$z \in [z^*, 1)$. If$\Gamma(w, q, r; z)$ =1, there would be no difference between the two countries to produce commodity z.

It can also be proved that, when the more developed country encounters a rapid growth of technology, a reducing ODI cost, and rising trade barriers, the scope of ODI would be enlarged, as shown in Figure 2, a movement from z^* to $z^{'}$.

2.2. *Theoretical Framework of Down-gradient ODI and the Determination of the Scope of ODI*

Similar to the examination of up-gradient ODI, we still assume the only two countries in the world: the more developed country C and the less developed country D, and other assumption being the same as that of up-gradient ODI. Also, suppose the production function of intermediate input z is compounded byLeontief and C-D, that is:

$$x(z) = A_i[\min\left\{\frac{L(z)}{b_L(z)}, \frac{H(z)}{b_H(z)}\right\}]^{\theta}[\frac{K(z)}{b_K(z)}]^{1-\theta}, i=D, C \tag{7}$$

where $0 < \theta < 1$ and A is a constant.

The two countries have Hicks-neutral technological differences that $A_C > A_D$, which means the technology level of the more developed country C is higher than that of the less developed country D. The final product is combined with no cost according to production technology:

$$\ln Y = \int_0^1 \alpha(z)\ln x(z)dz, \text{ where } \int_0^1 \alpha(z)dz = 1 \tag{8}$$

According to the cost minimization principle and the analysis of up-gradient ODI, it can be deduced that the less developed country U have a competitive advantage when producing commodity $z \in [0, z^*)$, while the more developed country C gains a competitive advantage when producing $z \in [z^*, 1)$. Furthermore, when the more developed country encounters a rapid growth of technology, a reducing ODI cost, and rising trade barriers, the scope of ODI would be enlarged.

2.3. *The Impact of ODI on Industrial Technological Change*

2.3.1. The Analysis of Technological Spillover Effect of Up-gradient ODI. Here we loosen the assumption that the technical level of all intermediate inputs for the more developed country D has a lower limit of AD. Since there is not only reverse technology spillover in ODI but also technology spillover effect on host country being invested. That means$\rho(z) > 1$ will lead to the increase of AD and AU of country D and U, which ultimately lower the cost of producing commodity z for both countries. Thus, as shown in Figure 4-1, the cost line of CDCD and CUCU moves downward, and the intersection of the two cost lines is a marginal scope of ODI, which is also the intersection of 1 and every curve of competitive advantage of cost. This means that ODI is the function of both positive technology spillover and reverse technology spillover. Furthermore, taking into account the expansion of ODI (increasing z*), which is to increase an amount of dz on the originalz_0^*basis, so that the previous equilibrium of E0is broken and a new equalization of E1 is reached, the equal-cost line can be expressed as:

$$\Gamma(w, q, r; z_0^* + dz) = r(w, q, r; z_1^*) = \frac{A_U(\lambda(z_1^*))}{A_D(\tau(z_1^*), \rho(z_1^*))}(\frac{w_D}{w_U} \times \frac{1 + \frac{q_D b_H(z_1^*)}{w_D b_L(z_1^*)}}{1 + \frac{q_U b_H(z_1^*)}{w_U b_L(z_1^*)}})^{\theta}(\frac{r_U}{r_D})^{1-\theta} = 1 \tag{9}$$

Due to the technology spillover that$\partial A_D(\tau, \rho)/\partial\rho > 0$ and$\partial A_U(\lambda)/\partial\lambda > 0$, the cost lines of both two countries continue to move downward and reach a new equilibrium of E2. However, whether it is bigger or smaller depends on the degree of movement. In this way, after many rounds of adjustment, it will finally reach the equilibrium of En, with an outsourcing critical point ofz_n^*. Thus, it can be deduced that if the cost reduction from positive technology spillover generated by investing abroad is greater than the cost reduction from reverse technology spillover to the home country, the scope of ODI will be enlarged. In contrast, if the cost reduction from positive technology spillover generated by investing abroad is less than the cost reduction from reverse technology spillover to the home country, the scope of ODI will shrink.

2.3.2. The Analysis of Technological Spillover Effect of Down-gradient ODI. Similar to the examination of up-gradient ODI, we loosen the assumption that the technical level of all intermediate inputs for the more developed country D has a higher limit of$\overline{A_D}$. Since$\rho(z) > 1$, the A_D and A_C for country D and C move upward, so that a new equilibrium is reached. Due to the technology spillover effect that$\partial A_D(\tau,\rho)/\partial\rho > 0$, as well as$\partial A_U(\lambda)/\partial\lambda > 0$, the cost lines of both two countries continue to move downward and form a new equilibrium. Again, whether the new level is bigger or smaller depends on the level of movement of the two cost lines. In this way, after many rounds of adjustment, it will finally reach the equilibrium of E_n, with an outsourcing critical point ofz_n^*. However, what is different from up-gradient ODI is that down-gradient ODI seeks more for reverse technology spillover and less for the spillover to the host country. What's worse, the host country may even adopt various barriers to hinder down-gradient ODI. Therefore, we deduce that the reverse technology spillover effect is much larger than the spillover to host country in down-gradient ODI.

3. Conclusion

Based on the above models, the internal mechanism of intra-industry technological change for up-gradient ODI and down-gradient ODI is deduced as follows:

Firstly, the technology spillover effect generated from the expansion of ODI scope is determined by the relative degree of positive technology spillover effect and reverse technology spillover effect.

Secondly, since there is a positive technology spillover effect on the host country and a reverse technology spillover effect on the home country, the technology change generated from the expansion of the ODI scope is determined by the relative degree of the positive technology spillover effect and the reverse technology spillover effect.

Acknowledgements

This work was supported by Guangdong Province Humanities and Social Sciences Key Research Project & Annual Project of Institute of Studies for the Great Bay Area of Guangdong University of Foreign Studies (No.2018YGA002),Institute of City Strategy Studies Research Base Project (No.JD201801),Key Grant Projects of Philosophy and Social Sciences Research, Ministry of Education of China (No.16JZD018),National Natural Science Foundation of China (No.71873041 ; No.71603060), Ministry of education of Humanities and Social Science Project (No.16YJC790023), Natural Science Foundation of Guangdong, China (No. 2017A030313422).

References

[1] Zheng R G 2009 Industrial Agglomeration, Growth Factor and Regional Disparity—An Empirical Analysis of China's Manufacturing Industry Since Joining the WTO. Econ. Manage. J., 12: 14-22.

[2] Liu, J.J., Li, X.H., Shi D. (2014) Study on the Shift of CO2 Emissions Gravity Center and Driving Factors. Fin.& Trade Econ., 12:112-123.

[3] Chen, J.C., Huang, F.H. (2014)Foreign Direct Investment and Upgrading of Export Structure. J. Int. Trade, 3:113-122.

[4] Akamatsu, Kaname (1962) A Historical Pattern of Economic Growth in Developing Countries. The Devel. Econ, Preliminary Issue No.1, pp.3-25.

[5] Vernon, R. (1966) International Investment and International Trade in the Product Cycle. The Quar. J. Econ., 80: 190-207.

[6] Kiyoshi Kojima (1978) Direct Foreign Investment: A Japanese Model of Multinational Business

Operations. CroomHelm, London.
[7] Helpman, E., Melitz, M. andYeaple, S.(2004) Export versus FDI with Heterogeneous Firms. Ame. Econ.Rev.,94:300—316.
[8] Kohler, Wilhelm (2004) International Outsourcing and Factor Prices with Multistage Production. Econ. J., 114: 166-185.
[9] Feenstra, Robert, C., Gordon H, Hanson. (1996) Globalization, Outsourcing, and Wage Inequality.Ame.Econ. Rev., 86: 240-245.

IOP Conf. Series: Materials Science and Engineering **573** (2019) 012113 doi:10.1088/1757-899X/573/1/012113

Functional Structure Modeling and Assembly Practice of Ditching Fertilizer Based on Standardized Module Design

Zhang Yuan[1], Deng Yiguo[1*], Wang Yeqin[1], Chen Peimin[1], Yan Bo[1], Zou Xiangjun[2*] , Zheng Yong[3],Wu Sihao[3]and Zhu Huixian[2]

[1] Tropical Agricultural Machinery Research Institute, Chinese Academy of Tropical Agricultural Sciences, Zhanjiang, Guangdong, 524091, China

[2] South China Agricultural University, Guangzhou, Guangdong, 510642, China

[3] Fundamental Research Funds for Rubber Research Institute, Chinese Academy of Tropical Agricultural Sciences, Danzhou, Hainan, 571737, China

*Corresponding author's e-mail: zhang198414@126.com

Abstract. In the existing agricultural machinery field, the agricultural machinery is generally cumbersome, the transportation, disassembly and assembly, maintenance procedures are more complicated, and the functions are single, and the versatility is not good. This article combines the industrial design standardization and modularization ideas for the trenching fertilizer application machine. Model construction and assembly practice were carried out. The whole machine was built according to independent functions, and divided into racks, fertilisers, trenching assemblies, suspension systems, transmission systems, etc. Modeling of the rack, modeling of the fertilizer turntable, and plowing the modeling process is elaborated. By assembling the functional components and standard parts, the modular design idea can be reused and exchanged in the design of the trenching fertilizer machine. The relevant modules can form the final product through the arrangement and combination. The final assembly entity can be a variety of products, which can better meet the customer's customization needs; the repetitive use of similarity can make the whole product more convenient in manufacturing and maintenance, and the overall machine coordination and appearance are more beautiful.

1. Introduction

With the development of agriculture's demand for modernization, China has developed a variety of digging and fertilizing machinery, which is mainly used in soybean, rice, sugar cane and other crops, and generally adopts methods such as outer groove wheel type, scraper type and throwing type. The machine has its own supporting frame, trenching plough, fertilizer discharging mechanism, etc. Although their respective structures are relatively simple, they basically adopt functional structural design methods, lack of modular and standardized design ideas, and their fertilization process, function replacement, and parts maintenance. All need to be done manually, and fine control cannot be done [1].

Most of the ditching and fertilizing machines in foreign countries are large-scale machinery, which tends to be dedicated to special planes, but generally reflects the elements of standardized design. The CASE860 chain trencher and the T-1255 chain trencher developed by the American company have been equipped with more advanced devices, but can only achieve a single function of trenching [2]. The centrifugal spreader developed in Australia is relatively simple in structure and high in production efficiency. However, it is not suitable for domestic use due to cost and environmental compatibility [3-6]. The ultrasonic sensor system developed by Saeys et al. is used in the ditching and fertilizing machine to automatically adjust the ditch depth [7-9].

With the increasing demands of the society for modern agriculture, under the limited land resources, it is impossible to achieve the cost-saving increase only by satisfying the existing fertilization technology, and it is necessary to carry out the development of agricultural machinery from the standardization and modularization. Designed to maximize cost and maintenance, functional interchange and other benefits.

The development of precision fertilization technology is inseparable from the development of standardized fertilization machinery. The new fertilization machinery should have the characteristics of simple structure, simple operation, saving input cost, improving use efficiency, etc., which is convenient for farmers to maintain and repair. The core components are rack and trenching plow. Body, fertilizer mechanism, etc. The fertilizer-discharging mechanism can replace the artificial fertilization and automatically apply a certain amount of fertilizer to the gum tree. With the continuous development of intelligent agricultural machinery, more and more intelligent agricultural machinery has entered the field to replace human-machine collaboration. The new fertilizing machinery will continue to develop along the direction of intelligence, continuously improve the level of precision fertilization technology, and promote the development of precision fertilization technology [10-14].

2. Functional Structure Modeling of Modular Ditch Fertilizer

Here, SolidWorks2016 is used to model the three-dimensional solids of the trench fertilizer applicator, focusing on the modeling process of the rack, fertilizer turntable, deep trench plow and the control mechanism, and the structural design of the control mechanism to make it have multiple Angle adjustment function, which provides reference for researching standardized and modular fertilization technology.

2.1. Rack modeling

The rack is the basic part of the entire machine and serves to communicate with other systems. A fertilizing system, a hydraulic system, and a three-point suspension system are installed above the rack, and a trenching system is installed below. The modeling of the entire rack can be divided into the creation of the frame beam, the creation of a three-point suspension system, the creation of a fertilizer assembly, and the creation of a rear suspension.

The frame beam can be simply constructed as a plurality of square steel models, and the modeling process can be done with a stretched rectangular parallelepiped. Select the "Extruding Boss" command, select the upper reference plane as the reference plane, draw a rectangular sketch and define the required size. After exiting the sketch, enter the desired stretch length value to complete the creation of the square steel model. Use the same steps to draw other square steel models on the same datum plane. It is important to note that each time you draw an entity, you need to uncheck the "Merge Results", as shown in Figure 1.

Figure 1. Supporting bracket model.

Further improve the model, add stiffeners, and draw two support frame holders for the model, as shown in Figure 4. The modeling idea of the fixture still uses the "stretching boss" command to create a drum, and then the side of the adjacent square steel model is used as a reference plane, and the intermediate small round hole is drawn by stretching and cutting, and the small round hole is supplied to the bolt.

Then use the "mirror direction" command to select the symmetry plane "right-view datum plane" from the mirror face. Select the model to be the mirrored entity to create the support frame holder on the other side. The frame beam assembly is shown in Figure 2.

IOP Conf. Series: Materials Science and Engineering **573** (2019) 012113 doi:10.1088/1757-899X/573/1/012113

Figure 2. Frame beam assembly.

For the same reason, create another datum, as shown in Figure 3. Using the datum plane as the sketching plane, complete the sketching of the left lower dangling point of the three-point suspension system, and stretch the model of the left lower dangling point by the "stretching boss" command, and take the "right-view datum" as the mirror direction. The bottom right hanging point is generated by mirroring.

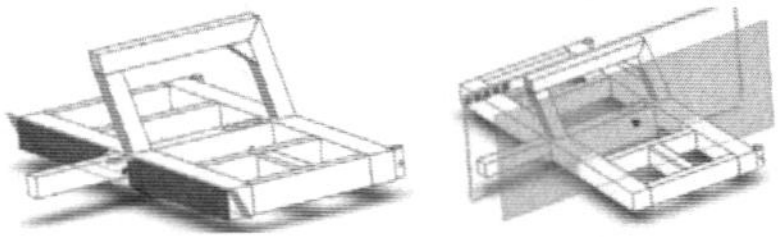

Figure 3. Datum creation

After completing the creation of each suspension point, the three-point suspension system is supplemented with the drawing of the reinforcing ribs. The final rendering is shown in Figure 4.

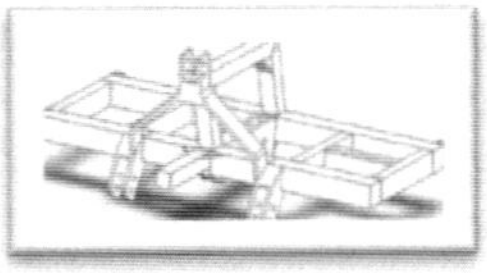

Figure 4. Hanging point creation renderings

The fertilizer assembly mainly includes a hopper, a barrel fixing ring and a worm and a worm reducer base and a fixed ring. The modeling idea can be to first draw the base and the fixed ring of the worm gear reducer, and then draw the barrel fixed ring and the hopper.

The bottom plate of the reducer is a simple plate-shaped model, which can be drawn by a command of "stretching boss". The creation of the reference plane of the fixed ring of the reducer can be created by referring to the reference plane of the three-point suspension system, and the same stretching command can be drawn. Since the two-part model is relatively simple, its creation process is similar to the previous model creation process, and will not be described here.

Call the "stretch boss" command to draw a sketch of the closed part of the hopper. After exiting the sketch, enter the value of the stretch length. Use the "shell" command to enter the thickness of the shell and select the shell surface to perform the shelling process on the model just drawn, as shown in Figure 5.

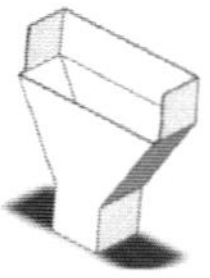

Figure 5. Hopper model

Referring to the above process, after the entire rack is modeled, the effect diagram is shown in Figure 6.

Figure 6. Rack assembly model

2.2. Fertilizer turntable modeling

The fertilizer turntable is an important part of the fertilization system. The bottom is connected with the drive plate by bolts, and the drive wheel is rotated by the worm gear reducer to drive the fertilizer to rotate.

Call the Stretched Boss command, use the Top View Datum as the sketching plane, draw a circular outline and define the dimensions, and enter the stretch length value after exiting the sketch. Use the Shell command to shell the model you just created. Use the "Right-view datum" as the sketching plane to draw a right-angled triangle sketch, and use the "Rotate Boss" command to complete the rotation of the above sketch. Finally, create four threaded holes at the bottom of the model, call the Shaped Hole Wizard command, select the specifications of the desired hole, edit the hole position sketch, and complete the creation of the threaded hole. The effect diagram of the fertilizer turntable model is shown in Figure 7.

Figure 7. Fertilizer model

The drive plate is a key component for connecting the worm gear reducer and the fertilizer chassis. The structure is relatively simple. It can be simply viewed as consisting of a drum and a disc. The modeling idea can draw the drum model first, and draw on this basis. Out of the disc structure. The modeling process of the drive disc is mainly done by two commands: the tension boss and the stretch cut. As shown in Figure 8.

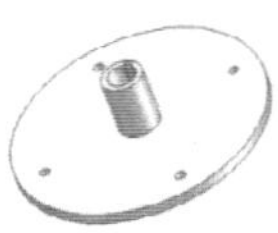

Figure 8. Drive plate model

2.3. Plow modeling

The deep groove plough is the core component of the trenching system. The upper two holes of the main shaft can be adjusted by the bolts and the frame beam holes to adjust the different positions, thereby adjusting the depth of the deep groove plough. The difficult part of modeling the deep trench plough model is the creation of two wings. The modeling idea of the model can use the two commands of "scan" + "stretch cut".

Use the "Right-view datum" as the sketching plane to draw a sketch of the main shaft outline, and use the "Stretching Boss" command to stretch the boss symmetrically on both sides. Take the front end surface of the main shaft as the sketching plane, and combine the two commands of stretching and cutting to draw the small boss connected with the plow head. The effect diagram is shown in Fig. 9.

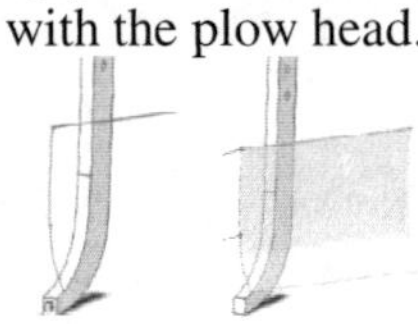

Figure 9. Plow column creation renderings

After the wing is created, a similar extension is created for the front end of the wing in a similar manner. Use the "mirror direction" command to mirror the wing and add stiffeners to the middle of the two wings to meet the stiffness design. The effect diagram is shown in Figure 10.

IOP Conf. Series: Materials Science and Engineering **573** (2019) 012113 doi:10.1088/1757-899X/573/1/012113

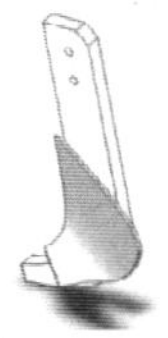

Figure 10. Wing creation rendering

3. Functionalized module assembly practice

The composition of the trenching fertilizer applicator can be divided into four parts: the frame, the ditching system, the hydraulic system and the fertilizing system. The other standard parts are the parts in the standard parts library of the 3DSource parts library.

Create a new "Assembly", browse the components you need to insert, and insert the component into the assembly interface, as shown in Figure 11.

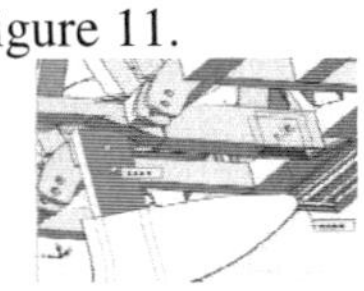

Figure 11. Example of an assembly interface

First assemble the rack and trench system. Insert the frame model and the deep groove plough model into the assembly interface, adding a "width" fit to the two models, limiting the three degrees of freedom of the X-axis movement, the Y-axis rotation, and the Z-axis rotation. Then, the rear side of the deep groove plough main shaft and the rear side of the inner beam of the frame are added with a "coincidence" fit to limit the two degrees of freedom of the Z-axis movement and the X-axis rotation. Finally, a "coaxial" fit is added to the upper hole of the slotted plough spindle and the middle beam hole of the frame to limit the movement of the Y-axis by one degree of freedom, and the cooperation between the frame and the deep groove plough is completed.

In the same way, the assembly of the small grooved plow, the grooved disc and the frame on the left and right sides is completed in turn, and finally the bolts and nuts of the target size are inserted into the frame and the trenching system from the 3DSource parts library, and the corresponding position is added. The six-degree-of-freedom constraint is completed, and the effect is shown in Figure 12.

Figure 12. Assembly renderings

4. Discussion

In the structural design process of the trenching fertiliser machine, based on the idea of standardized module design, we have independent functions for each part, with a uniform geometric connection interface and a unit of consistent input and output interfaces, and modular design, the results found the same kind of modules can be reused and exchanged in the product family. The arrangement of the relevant modules can form the final product. The final assembly entity is shown in Figure 13. Through the combined configuration of the modules, it is possible to create products with various requirements to meet the customization needs of customers; the reuse of similarities can make the entire product manufacturing and maintenance more convenient.

Figure 13. Assembly machine

In order to replace a certain functional module to achieve quality or efficiency improvement, without changing the entire structure, only need to change the corresponding module, the workload will be significantly reduced, so modular design and application, is the industrial design ideas Introducing an effective attempt to design the structure of the trenching fertiliser, the design and assembly practices show the following characteristics:

1. The structure is more reasonable. Due to the modular design, the internal structure of the trenching fertiliser machine is more reasonable and more scientific, so that the function of each component can be well played and the service life of the whole machine can be extended;

2. Maintenance is simpler. Each part of the trenching fertilizer application machine is a module. If there is a problem with the machine, it can be diagnosed quickly. It only needs to replace the corresponding module, and the operation is simple. It can be replaced without the need of professionals, which saves farmers time and effort!

Acknowledgments
This work was financially supported by the Natural Science Foundation of Hainan Provincial(518QN279);Natural Science Foundation of Guangdong Provincial(2018A030307067); Chinese Academy of Tropical Agricultural Sciences(1630132017003);Public Welfare Research and Capacity Building of Guangdong Province (2018A03023)and Fundamental Research Funds for Rubber Research Institute, CATAS(1630022019006).
I would like to thank the members of the research group for their selfless help, especially the communication author Deng Yiguo&Zou Xiangjun,and the guangba farms in Hainan Province for the project test of strong support.

References

[1] Liu Wei, Xiao Hongru, Song Zhiwei, et al. Current status and development trend of orchard fertilization machinery [J]. Agricultural Mechanization Research, 2017 (11): 263-268.
[2] ZHANG Qi,WANG Wei,LIAO Jie-an.Research Status of Fertilizer Ditching Machine in Orchard at Home and Abroad[J].Journal of Agricultural Mechanization Research,2016(5):264–268.
[3] Dawn M. Geske. Same size, more power from newest Vermeer rockwheel trencher [J]. Diesel Progress North American Edition. 2004, 70(1): 70.
[4] Zhang Yuan, Wei Lijiao, Deng Yiguo et al. Test and effect evaluation of 3XSP-1 rubber fertilizer applicator [J]. Modern Agricultural Equipment, 2017 (1): 37–41.
[5] Zhang Yuan, Deng Yiguo, Wei Lijiao et al. Pilot production and experiment of rubber plant ditching machine[J]. Anhui Agricultural Sciences, 2015, 43(26): 364–365, 372.
[6] [6] Zhang Yuan, Deng Yiguo, Wei Lijiao et al. Development and experimental analysis of rubber planting trench fertilizer applicator [J]. Modern agricultural equipment, 2016 (5): 50–54.
[7] Saeys W, Engelen K, Ramon H, et al. An auto-matic depth control system for shallow manure system [J]. Biosystems Engineering, 2007, 98(10): 146–154.
[8] Saeys W, Wallays C, Engelen K, et al. An automatic depth control system for shallow slurry injection, part 2: Control design and field validation [J]. Biosystems Engineering, 2008, 99(2): 161–170 .
[9] Gong Yuebo. Current status and development direction of China's rubber machinery industry [J]. China Machinery, 2013 (7): 284.
[10] Chaohai C, Jieli D, Guoqi Y, et al. Research status and development countermeasure of ditching machine in orchard[J]. Modern Agricultural Equipment, 2013, 2: 64-68.
[11] Qi Z, Wei W, Jiean L. Design of Fertilizing Machine's Furrow Opener in Jujube Orchard and Finite Element Analysis[J]. Journal of Agricultural Mechanization Research, 2016, 4: 006.

[12] Chen X, Luo X, Wang Z, et al. Experiment of synchronous side deep fertilizing technique with rice hill-drop drilling[J]. Transactions of the Chinese Society of Agricultural Engineering, 2014, 30(16) : 1-7.
[13] Chen M, Hewei M, Za K, et al. The Research Current Situation and Development Countermeasure of the Orchard Organic Fertilizer Deep Application of Disc Ditching Machine [J]. Journal of Agricultural Mechanization Research, 2017, 10: 004.
[14] Qi Z, Wei W, Jie'an L. Study status of fertilizing and ditching machine in orchard at home and abroad [J]. Journal of Agricultural Mechanization Research, 2016, 5: 055.

IOP Conf. Series: Materials Science and Engineering **573** (2019) 012114 doi:10.1088/1757-899X/573/1/012114

Practical Exploration of Association Rules Mining Algorithm in Open Education

Shaozhen Huang[1]

[1] Hainan Open University, Hainan, Haikou, 570100

Abstract: Open education is a new type of education, which is produced along with the development of modern information technology. It is a teaching model based on modern information technology. With the continuous reform of China's education system, the autonomy of students in major colleges and universities is increasing. However, because students do not understand the course, it is very difficult to select courses. Therefore, it is especially important to excavate the university teaching platform database. This paper briefly introduces the concept of association rules mining algorithm, mainly studies the application of association rules mining algorithm in the open education learning system, in order to provide meaningful guidance for students' elective courses, thus improving students' academic performance.

1. Introduction

The characteristic of open education is that students learn independently, make full use of the various resources provided by modern distance education, change traditional educational concepts, and gradually transform into modern open and distance education. And from the teacher-centered passive learning to the student-centered autonomous learning. Students can learn at any time and any place without delaying their work, while learning knowledge and improving their quality. Organize teaching with "courses as a unit" to enable students to conduct courses more independently and flexibly. However, students' self-selected courses will inevitably lead to the problem of blind selection of students. We must improve them in order to provide meaningful guidance for students' electives and facilitate their selection. Data mining, as a new technology that can automatically and efficiently extract valuable information and knowledge from massive data to effectively support decision-making. Applying it to open education can provide useful guidance for student learning[1].

Data mining is the use of pattern recognition, statistics, and mathematical techniques to filter out new and meaningful relationships, patterns, and trends from a large amount of stored data. It is the process of extracting hidden and potentially useful information and knowledge from a large number of incomplete, noisy, fuzzy, and random data. It relies mainly on artificial intelligence, machine learning and statistical techniques to inductively infer data and uncover potential patterns. Predict future trends and support decision making.

2. Association rule mining algorithm

The concept of association rules is first of all by R. Agrowal et al. propose a rule that describes the (potential) relationship between data items (attributes, variables) in a database. It has become a very important research direction in data mining.

There are four steps to discover the association rules: 1) Pre-processing the data related to the mining task. According to the requirements of the specific problem, the database is operated accordingly to form a standardized database D. 2) For D, find all the itemsets that satisfy the minimum support, that is, the large itemsets. Since the database is generally large, this step is the core of the algorithm. 3) Generate a rule that satisfies the minimum confidence to form a rule set R. 4) Explain and output R.

Apriori, a classical association rule mining algorithm, is a basic algorithm for finding frequent itemsets. The core of the algorithm is mainly to find frequent itemsets, mainly based on the Apriori nature: all non-empty subsets of frequent itemsets must also be frequent. This property can be used to effectively compress the search space. The main idea of the algorithm: To find L_k, connect L_{k-1} with itself to generate a set of candidate k itemsets. The set of candidates is recorded as C_k, and then C_{k+1} is empty. When C_k, k = 1, 2, ···, k are generated, C_k is compressed using a pruning strategy. The use of any infrequent (k-1)-item set is unlikely to be the Apriori property of frequent k-item sets, deleting those k-candidate itemsets whose (k-1)-subsets are not in L_{k-1}[1].

The Apriori core algorithm is described as follows:

Input: database D, minimum support threshold min_sup
Output: frequent item set L in D
method:
L1={find_frequent_1-itemsets(D)}
For(X=2; L_{x-1}<>C/;X++)do begin
C_x=Apriori_gen(L_{x+1},min_sup)
For all tranctions∈D do begin
C_t=subset(C_x,t)
For all candidates c∈C_t do begin
c. count++
end

3. Application of association rules technology in student score analysis

There will be more or less links between the course itself, between the course and the course, and between the course setting and the course grade. Through data mining technology, it is very intuitive to find valuable information hidden behind student achievement, so as to better guide teachers to complete teaching work. The implementation process of student achievement data mining is shown in Table 1:

Table 1: Mining implementation table

Serial number	Step
Steps 1	Determine the mining object
Steps 2	Selected model
Steps 3	Data collection
Steps 4	Data preprocessing
Steps 4	Data mining
Steps 6	Analysis result
Steps 7	Knowledge application

Specifically speaking, firstly, taking the scores of computer majors in a university as an example, we can find out the influence of the excellence of a certain course on other courses. Then, using the association rule mining algorithm Apriori analysis to explore the degree of influence between the students' grades. A similar conclusion is drawn: the excellent performance of the "C Language Programming" course directly affects the excellent performance of the "Data Structure" course. Then collect the usual grades and test scores of the student's course, and the relevant teachers will provide relevant data.

Preliminary processing of the obtained data: 1) Integrated data: the data of the collected multiple databases are combined. Randomly select some course scores from students. 2) Clean up the data: When counting the data of the student's score source table, there will be some cases where the attribute lacks the attribute value. For these attributes lacking the attribute value, the cleanup technique is used to fill the vacant data value. 3) Convert data: use a unified format to represent grade data for data mining. This paper explores the excellent relationship between the various courses of computer majors. The use of association rules to mine performance data requires logical data, so the attribute value greater than 90 points is defined as "1", otherwise it is "0"[3].

Data mining using Appriori algorithm: 1) Establish transaction data table D. According to the mining requirements, only excellent grades are retained, and excellent student records are not obtained for deletion. 2) Frequent 1-item set data table frequent-1-itemsets. The table is used to store the project name X1, X2 and the number of records containing this project. 3) Delete the record with the support count less than the minimum support min-sup in the data table frequent-1-itemset, and get the final frequent 1-item set. 4) Seeking succession of frequent itemsets. 5) Delete the records whose final frequent items are less than the minimum confidence threshold, thereby generating rule information[4].

Finally, the analysis results, the transaction number of this paper is 11, set the minimum support degree of 30%, the confidence level is 50%. 1) The set L1 of the frequent -1 item set is {X1}, {X2}, {X3}, and all the frequent itemsets are obtained based on this. 2) The candidate set C2 is {X1, X2}, {X1. X3}, {X2, X3}. 3) The candidate set C3 is {X1, X2, X3}. 4) The final result: C3 is the frequent set L3, which is the maximum frequent set, deposited in the data sheet 2. X1, X2, X3 represent different computer courses

Table 2： Association rule data table

Xuehao	X1	X2	X3
0903020006	0	1	0
0903020027	0	1	1
0903020020	1	0	0
0903020016	1	1	1
……	……	……	……

Calculate the final confidence, delete the record less than the minimum confidence threshold, and get the final association rule, which is stored in the data table 3.

Table 3： Result data sheet

Serial number	specific contents
X1 is excellent and X2 is excellent	X3 has more than 71% excellent possibilities
X1 is excellent and X3 is excellent	X2 has more than 100% excellent possibilities
X2 is excellent and X3 is excellent	X1 has more than 100% excellent possibilities
Excellent X1	X2, X3 have more than 71% and excellent possibilities
Excellent X2	X1, X3 have more than 50% and excellent possibilities
Excellent X3	X1, X2 have more than 83% and excellent possibilities

Implementation effect: Using the mining algorithm Apriori analysis to explore the degree of influence between the students' grades. In turn, as a reference, it is possible to mine valuable information between the course itself, the curriculum and the course, and the curriculum setting and course achievement, so as to guide the teachers to achieve greater achievements in teaching reform, professional talent training program development, and teaching quality improvement[5].

4. Conclusions

In general, the analysis of student achievement in open education is a complete data mining process. The user can automatically analyze the selected database as long as the user provides the necessary data. And return the information that the user needs, help the user to make decisions, thus helping the students to choose courses and further study, which plays a very good role in promoting and has certain practical value. The association rule mining algorithm is applied in open education, and the function of the distance learning system is improved through continuous data mining, so that learners have a clear understanding of their own learning. Complete the study in constant reflection and persistence, and truly achieve independent learning. At the same time, teachers have a global understanding of teaching to improve teaching.

Acknowledgements

Fund Project: Hainan Higher Education Research Funding Project
Project No.: Hnky2018-62

References

[1] Bing L, Hsu W, Ke W, et al. Visually Aided Exploration of Interesting Association Rules[C]// Pacific-asia Conference on Methodologies for Knowledge Discovery & Data Mining. 1999.
[2] Chen Q, Zhou K, Dan W. Exploration and Practice of the Computer Innovation Education in Mining Engineering[C]// International Workshop on Education Technology & Computer Science. 2009.
[3] Zhong R, Wang H. Research of Commonly Used Association Rules Mining Algorithm in Data Mining[C]// International Conference on Computational Intelligence & Industrial Application. 2010.
[4] Zhi L, Guoming S, Mingyu L. A Vector Operation Based Fast Association Rules Mining Algorithm.[C]// International Joint Conference on Bioinformatics. 2009.
[5] Yan Y, Wang D, Ge Y U. Association Rules Mining Algorithm of Web Pageviews for Personalized Recommendation[J]. Computer Engineering, 2005, 31(1):79-81.

IOP Conf. Series: Materials Science and Engineering **573** (2019) 012115 doi:10.1088/1757-899X/573/1/012115

Extraction and Analysis of Rock Painting Color Data Based on Data Association of Color Psychology

Shang guannan

(Jilin Teachers Institute of Engineering and Technology 130052)

Abstract: We should continue to dig and develop the breakthrough point of color composition in rock color painting based on color psychology. This paper objectively evaluates the present situation of rock color painting in contemporary China. We should conform to the development of contemporary multiculturalism. We should promote and inherit the development of rock paintings, so as to broaden the manifestations of rock paintings and truly meet the aesthetic demands and value pursuits of contemporary people in the new context.

1. Introduction

The correspondence between color in color psychology and color in physics is not only a simple proportional relationship, but also has different characteristics. And how to apply this perception difference to practice, so as to promote communication with the audience, this point has also begun to be explored and studied by scholars.

2.THE CURRENT DEVELOPMENT OF ROCK PAINTING

Rock color painting has a long history, extensive and profound. It is not only an important part of traditional Chinese painting, but also one of the contemporary art forms of painting. Color composition is a kind of expression language in painting, which is the result of artists' aesthetic creation and the key to creating ever-changing color effects. Nowadays more and more painters begin to pay attention to the role of color composition factors in creating picture atmosphere and expressing personal feelings[1]. And as the theoretical basis of guiding painting, combining with the characteristics of rock color painting, it makes various creative attempts.

Rock color painting, as a unique form of expression, traces its development history and has a deep origin with our country. China is an ancient civilized country, which carries thousands of years of civilization, and painting is an important means and media to reflect the development of its spiritual civilization. Rock paintings are painted with early mineral pigments. Huang murals, a treasure of our country, flourished in the Han and Tang Dynasties. As an important part of Huang culture, rock paintings are an important bridge and hub connecting early heavy-color paintings with later rock paintings.

3.EXTRACTION AND ANALYSIS OF ROCK COLOR DATA BASED ON DATA ASSOCIATION OF COLOR PSYCHOLOGY

The design of color matching scheme recommendation model links emotional data collected and analyzed with the recommendation of color matching scheme in the process of system operation, which requires the selection of color classification, emotional classification and how color system and emotion correspond in the system. Regarding the classification of color systems, the relevant color scientists have developed quite mature theories. The classification and selection of emotions and the corresponding problems of color systems and emotions in the recommendation model need to be combined with the development level of current data emotional analysis technology and the related research of color psychology[2]. As shown in Table 1.

Table 1. Basic emotional categories

intensity of feeling	Basic emotional categories							
	1	2	3	4	5	6	7	8
Grade: weak	calm	Care	annoyance	boredom	Sentimental	To be puzzled	Worry	accept
Grade: medium	happy	expect	get angry	Hate	Grief	surprised	Fear	trust
Grade: strong	An ecstasy	Be vigilant	Rage	hate	Grieved	Amazed	Thriller	Reverence

3.1 CHOICE OF COLOR SYSTEM FOR RECOMMENDED MODEL OF DESIGN MATCHING SCHEME

Because all the color systems in the visible spectrum belong to the color system, the basic color systems can be combined into other color systems through different mixing relations, plus the black, white and Frey colors in the non-color system. Based on the relevant research results of color system, the recommended model of color matching scheme is designed. Finally, the black, white and Frey color systems in the colorless system and the seven color systems of red, orange, yellow, green, cyan, blue and purple in the colorless system are determined. Ten basic color systems are selected as the color system selection of the recommended model of color matching scheme[3].

3.2 SELECTION OF BASIC EMOTIONS IN THE RECOMMENDATION MODEL FOR DESIGNING COLOUR MATCHING SCHEME

The famous psychologist Robert Protozoic, who put forward the emotional wheel, gives the basic relationship between emotional types and emotions. He puts forward that there are two kinds of emotions: one is the basic psychological state in daily life, the other is the feedback psychological state based on the changes of life. Chopstick believed that emotional change was born in accordance with the theory of biological evolution and had an important impact on the survival of the fittest among species. Based on this understanding, in the 1980s, Hitchhiker produced a "emotional wheel" that could reflect emotional changes[4]. In the "emotional roulette", he first put forward eight basic emotions, their emotional intensity is the strongest, namely ecstasy, vigilance, rage, hatred, grief, surprise, thriller, reverence. "Emotional Wheel" can be divided into two types: plane display and cube display, which are used to understand the changes and differences between emotions. In the social field, emotional roulette often helps people understand the law of emotional change.

4. APPLICATION OF ROCK PAINTING COLOR DATA BASED ON DATA ASSOCIATION OF COLOR PSYCHOLOGY

4.1 IMPROVING COLOR MATCHING SCHEME BY MEANS OF BIG DATA TECHNOLOGY

In the modern design industry, learning the theory of color and skilled application is a very necessary basic skill. In the traditional design industry, the use of color generally depends on the experience of designers and limited research to complete, but the actual situation is far more complex, relying on experience and small-scale research and selection of color schemes can not always accurately reflect the public's expectations of design works. If the reference information related to the design work can be obtained by technical means before the design, and the designer can improve the color matching scheme based on the reference information, the deviation of design caused by experience can be effectively avoided[5]. As shown in Figure 1.

IOP Conf. Series: Materials Science and Engineering **573** (2019) 012115 doi:10.1088/1757-899X/573/1/012115

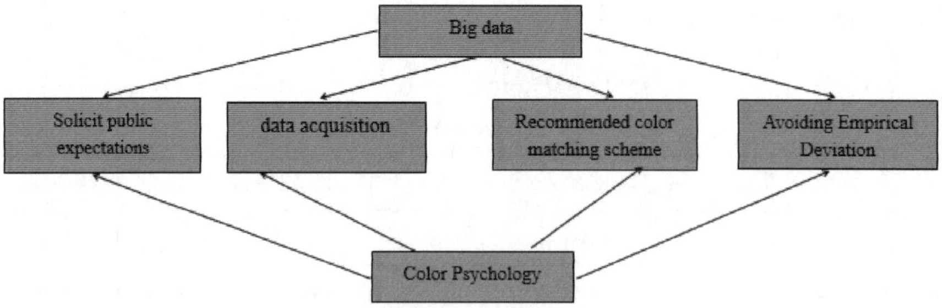

Figure 1. Examples of Colour Matching Scheme Selection in Big Data Technology Intervention Design

4.2 USING INTERDISCIPLINARY THINKING MODE TO OPTIMIZE THE DESIGN PROCESS

Big data technology has brought new ideas to traditional problem-solving methods, and has achieved many results in search optimization, machine translation, advertising and other fields. Traditional design methods tend to rely on the experience of designers to make strategies, while big data technology can pay attention to details that designers do not notice through data analysis, relying on its powerful logic. Analytical ability can often make designers produce new design ideas. For the field of art design, the innovation of thinking and working mode brought by big data technology is more revolutionary than the breakthrough in technology.

Originally, color is the most important expression language in Chinese painting. It is very important in artistic characteristics and artistic value. Nowadays, the age has given rock color and color discourse power, and recasting color innovation is the inevitable development of contemporary Chinese rock color painting. It is not only the need of China's diversified development, but also helpful to find new growth points of Chinese traditional art in the modern process, activate the present perception in the profound and powerful Chinese tradition, and develop toward a broader performance space.

5.CONCLUSION

To sum up, we should innovate and improve on the basis of understanding and inheriting the color style of traditional Chinese paintings. We should neither stick to stereotypes and old traditional ideas, nor blindly pursue visual impact in order to express individuality, leading to the loss of the color tradition of Chinese characteristics in Chinese rock paintings. In the development of rock color painting, find out the style of rock color painting with Chinese characteristics, truly create more excellent works with contemporary characteristics and social spirit, truly lead people to improve artistic aesthetics, and thus convey their true feelings, and free to gallop in the new field and space of Chinese painting.

References

[1] Che E, Olsen M J. Mufti-scan segmentation of terrestrial laser scanning data based on normal variation analysis[J]. Sprigs Journal of Photographer & Remote Sensing, 2018.

[2] Chi X, B A, Li T, ET AL. Effects of Mindfulness-Based Stress Reduction on Depression in Adolescents and Young Adults: A Systematic Review and Meta-Analysis.[J]. Frontiers in Psychology, 2018, 9:1034.

[3] Bale B, Mature M, Withholding A. Contrasting Gist-Based and Template-Based Guidance During Real-World Visual Search.[J]. Journal of Experimental Psychology Human Perception & Performance, 2018, 44(3):367.

[4] Jr D T L, Garcia-Reid P, Reid R J. The interacting effects of psychological empowerment and ethnic identity on indicators of well-being among youth of color[J]. Journal of Community Psychology, 2018(2).

[5] Bhangra S, Merrick C, Ch H, ET AL. Involuntary Entry Into Consciousness From the Activation of Sets: Object Counting and Color Naming.[J]. Frontiers in Psychology, 2018, 9:1017.